Lecture Notes in Statistics **169**

Edited by P. Bickel, P. Diggle, S. Fienberg, K. Krickeberg, I. Olkin, N. Wermuth, and S. Zeger

Springer
New York
Berlin
Heidelberg
Hong Kong
London
Milan
Paris
Tokyo

Yu. I. Ingster
I.A. Suslina

Nonparametric Goodness-of-Fit Testing Under Gaussian Models

Springer

Yu. I. Ingster
Department of Applied Mathematics
St. Petersburg State Transport University
9 Moskowskii Avenue
190031 Saint Petersburg
Russia
ingster@pdmi.ras.ru

Irina A. Suslina
Saint Petersburg Institute of Fine Mechanics
and Optics (Technical University)
14 Sablinskaya Street
197101 Saint Petersburg
Russia
suslina_i@mail.ru

Library of Congress Cataloging-in-Publication Data
Ingster, Yu. I.
Nonparametric goodness-of-fit testing under Gaussian models / Yu. I. Ingster, I.A. Suslina.
p. cm. — (Lecture notes in statistics ; 169)
Includes bibliographical references and index.
ISBN 0-387-95531-3 (softcover : acid-free paper)
1. Goodness-of-fit tests. 2. Nonparametric statistics. I. Suslina, I.A. II. Title.
III.Lecture notes in statistics (Springer-Verlag) ; v. 169
QA277 .I54 2002
519.5'4—dc21 2002070731

ISBN 0-387-95531-3 Printed on acid-free paper.

Printed in the United States of America.

9 8 7 6 5 4 3 2 1 SPIN 10883426

Typesetting: Pages created by the authors using a Springer $\TeX$ macro package.

www.springer-ny.com

Springer-Verlag New York Berlin Heidelberg
A member of BertelsmannSpringer Science+Business Media GmbH

Preface

In this book we would like to present some aspects of the modern theory of goodness-of-fit testing developed since the 1980s. Certainly we do not try to consider the subject in full. The selection of material corresponds to the author's interests. In particular, we consider the mathematical side of the problem; more exactly, we study asymptotically minimax setting in the problem (however, a lot of results are not asymptotic) and we do not discuss any related applied problems.

To explain the main direction of our study we would like to make some general remarks. There are two main problems in Statistics: estimation theory and hypothesis testing. For the classical finite-parametric case these problems were studied in parallel. The studies are based on the common grounds: mainly they are Fisher's Maximum Likelihood principle and Le Cam's Local Asymptotical Normality property.

On the other hand, a lot of statistical problems are not parametric in the classical sense: the objects of estimation or testing are functions, images, and so on. These can be treated as unknown infinite-dimensional parameters which belong to specific functional sets. This approach to nonparametric estimation under asymptotically minimax setting was started in the 1960s–1970s by Chentsov, Ibragimov and Khasminskii, Pinsker, among others. It was developed very intensively for wide classes of functional sets and loss functions by Donoho, Johnstone, Kerkyacharian, Picard, Lepski, Mammen, Spokoiny, Tsybakov, among others.

We present an analogous approach to the nonparametric goodness-of-fit testing which was started in the 1980s by one of the authors.

We study nonparametric hypothesis testing problems analogous to estimation problems. There are many connections between nonparametric estimation and hypothesis testing. However, the nonparametric hypothesis testing theory is essentially different from the estimation theory. There is a lot of new effects in nonparametric hypothesis testing theory, in comparison to parametric hypothesis testing as well as to nonparametric estimation.

In the main, this book is addressed to mathematical statisticians who are interested in the theory of nonparametric statistical inference.

The setting is based on the asymptotic variant of the minimax approach. Here we follow the modern nonparametric estimation theory. It is well-known from Wald's Statistical Decision Theory that minimax problems correspond to the Bayesian problems for the *least favorable priors*. However, there are no sufficiently general methods for finding these priors.

The key element of the presented theory is the method of constructing *asymptotically least favorable priors* for a wide enough class of nonparametric hypothesis testing problems. These methods are based on the asymptotic reduction of the hypotheses testing problem to the convex extreme problem in specific Hilbert space, and on the study of this extreme problem. We would like to demonstrate that this is a good tool for the study of a wide class of such problems. Throughout the book we try to explain these methods starting with the simplest cases.

As one of the main results, we apply these methods to construct asymptotically minimax tests and to obtain sharp asymptotics of minimax error probabilities for alternatives which are determined by the power and Besov norms in the sequence space. Jointly with wavelet transform, they give the structure of minimax consistent tests and the rate asymptotics for alternatives which are determined by Sobolev and Besov norms in the functional space.

These methods are well applicable for the study of hypothesis testing problems of growing dimensions and for "adaptive" versions of these problems.

Note that we are interested in principal phenomena in the problem. For this reason the study is based on the Gaussian models. This case is of importance and is the simplest from a technical point of view.

On the other hand, using Le Cam's theory of asymptotic equivalence, one can translate the main results to other statistical models, such as the regression function model and the sample model with an unknown probability density. Mainly we deal with testing the simple null-hypothesis in this book. The main ideas and results are extended for the testing of composite null-hypotheses as well. We refer to the papers which contain these extensions for other models and for parametric and various nonparametric null-hypotheses.

We hope that the book will be of interest to specialists who are dealing with the applied nonparametric statistical problems that are actual in signal detection and transmission, in technical and medical diagnostic, etc.

These readers could find the test procedures, which provide asymptotically best or near-to-best rates of testing, with regards to a wide class of nonparametric alternatives. We present adaptive versions of the tests which provide almost the same rates of testing. We give new lower bounds for possible minimax error probabilities and rates of testing which would be of importance for the statement of technical requirements for testing.

Usually the test procedures that we propose provide essentially better rates of testing than those based on the optimal estimators of unknown parameters. This means we can distinguish between the null-hypothesis and the alternative which are close, such that the distance between them is much smaller than the possible accuracy of estimation of an unknown infinite-dimensional parameter. This is one of the principal differences, among others, between nonparametric estimation and hypothesis testing. It will be of importance in signal detection, technical, and medical diagnostic problems because this could provide the detection of a signal (a defect or an illness) *before* it is possible to estimate what the signal is.

The authors are very grateful to Professor I. Ibragimov who was interested in and concerned with this research which started in the 1980s. We would like to thank all the participants of the Seminar on Probability Theory and Mathematical Statistics in Saint Petersburg (Leningrad) a branch of Steklov's Mathematical Institute where the authors had extensive talks and very helpful discussions of the subject. We are greatly appreciative to Professor O. Lepski and Professor V. Spokoiny for many suggestions and discussions on the subject in the 1990s which provided essential progress in this research. We are grateful to Professor M. Nussbaum who supported these studies in WIAS (Berlin, Germany) and who suggested we write this book. We thank Professor A. Korostelev, Professor Ya. Nikitin, Doctor C. Pouet, and reviewers for many helpful remarks and suggestions to the text of the book.

Research was partially supported by INTAS Grants NN 93-1585, 93-1585-EXT, by RFBI Grants NN 94-01-00301, 96-01-00684, 96-15-96199, 99-01-00111, 00-15-96019, 02-01-00262, and by RFBI–DFG Grants NN 98-01-04108, 02-01-04001.

Saint Petersburg, Russia,
August 15, 2002

Yuri I. Ingster
Irina A. Suslina

Contents

Preface v

1 Introduction 1
- 1.1 Tests 4
- 1.2 One-Dimensional Parameter 5
- 1.3 Multidimensional Parameter 9
- 1.4 Infinite-Dimensional Parameter 15
  - 1.4.1 Classical Goodness-of-Fit Tests 15
  - 1.4.2 Nondistinguishability in the Problem 22
  - 1.4.3 Two-Sided Constraint Problem 23
  - 1.4.4 Comparing with the Minimax Estimation Problem 25
- 1.5 Problems of the Study and Main Results 28
- 1.6 Methods of the Study 32
- 1.7 Structure of the Book 35

2 An Overview 38
- 2.1 Models 38
  - 2.1.1 Gaussian Models in Sequence Space 38
  - 2.1.2 Functional Gaussian Model 39
- 2.2 Hypothesis Testing Problem 40
  - 2.2.1 Hypothesis, Alternative, Tests 40
  - 2.2.2 How to Compare the Tests? 41
  - 2.2.3 Simple Hypothesis and an Alternative 42
  - 2.2.4 L_1-Distance on the Space of Measures 44

2.3 Bayesian Approach in Hypothesis Testing 47
2.3.1 General Setting 47
2.3.2 Bayesian Approach in the Gaussian Model 48
2.4 Minimax Approach in Hypothesis Testing 53
2.4.1 General Setting 53
2.4.2 Connection with the Bayesian Approach 55
2.4.3 Convex Hypothesis and an Alternative under the Gaussian Model 58
2.4.4 Convex Symmetric Admissible Set 61
2.5 Asymptotics in Hypothesis Testing 62
2.5.1 Asymptotic Approach 62
2.5.2 Asymptotic Minimaxity 63
2.5.3 Asymptotic Minimaxity in the Parametric Case . 65
2.6 Minimax Distinguishability in Goodness-of-Fit Problems 67
2.6.1 The Parametric Case 68
2.6.2 Classical Distinguishability in the Nonparametric Case 68
2.6.3 Nondistinguishability in the Nonparametric Case 69
2.7 Norms and Wavelet Transform 71
2.7.1 Norms in the Sequence Space 72
2.7.2 Norms in the Functional Space 73
2.7.3 The Wavelet Transform 76
2.8 Short Overview of Minimax Estimation 77
2.9 Constraints of Interest 79
2.10 Rates in Estimation and in Hypothesis Testing 79

3 Minimax Distinguishability **81**
3.1 Minimax Properties of Test Families 81
3.1.1 Consistency Conditions and Gaussian Bounds . . 81
3.1.2 Tests of χ^2-Type 83
3.1.3 Tests of χ^p-Type 85
3.1.4 Supreme Tests 89
3.2 Asymptotic Minimaxity for Square Norms 92
3.2.1 Lower Bounds and Inner Radii 93
3.2.2 Upper Bounds and Kolmogorov Diameters 95
3.3 Bayesian Approach under a Gaussian Model 97
3.3.1 L_1- and L_2-Distances for Mixtures 97
3.3.2 Product Priors 98
3.3.3 Hilbert Structure 99
3.3.4 Two-Point Factors 102
3.3.5 Symmetric Three-point Factors 103
3.3.6 Minimax Distinguishability for l_p^n-Balls, $p < \infty$. 104
3.3.7 Semiorthogonal Priors 106
3.3.8 Minimax Distinguishability for l_p^n-Balls, $p = \infty$. 108
3.4 Triviality and Classical Asymptotics 109

3.4.1 Minimax Triviality 109
3.4.2 Triviality for l_p-Balls 109
3.4.3 Triviality and Classical Asymptotics for Power Norms 110
3.4.4 Triviality and Classical Asymptotics for Besov Norms 115
3.4.5 Triviality and Classical Asymptotics for a Positive Alternative 116
3.4.6 Nonempty and Triviality for Power Norms: Two-Sided Constraints 117
3.4.7 Triviality for Besov Norms: Two-Sided Constraints 123
3.4.8 Triviality under the Functional Gaussian Model . 127
3.5 Distinguishability Conditions for Smooth Signals 128
3.5.1 Rates and Tests for Smooth Signals 128
3.5.2 Proof of the Lower Bounds 130
3.5.3 Proof of the Upper Bounds 134
3.5.4 Rates for Positive Smooth Signals 135

4 Sharp Asymptotics. I **136**
4.1 Tests Based on Linear Statistics and Convex Alternatives 137
4.2 Two-Sided Constraints for the Positive Alternatives, $p \leq 1, q \geq p$ 141
4.2.1 System of Equations for an Extreme Problem . . 142
4.2.2 Solution of an Extreme Problem: Case $p < q < \infty$ 144
4.2.3 Solution of an Extreme Problem: Case $p = q$ . . . 146
4.2.4 Solution of an Extreme Problem: Case $p < q = \infty$ 147
4.2.5 Constraints of Exponential Type with $p \leq 1, q \geq p$ 150
4.3 Sharp Asymptotics of Gaussian Type: Product Priors . . 154
4.3.1 Asymptotic Normality for Product Priors 155
4.3.2 Symmetric Two-Point Factors and Tests of χ^2-Type 159
4.3.3 The Power Norms: $p \leq 2,\ q \geq p$ 162
4.3.4 Besov Norms: $p \leq 2,\ q \geq p$ 168
4.3.5 Smooth Signals: Case $p \leq 2,\ q \geq p$ 171
4.4 Sharp Asymptotics: Asymptotic Degeneracy 173
4.4.1 Asymptotic Degeneracy in Bayesian Problems . . 173
4.4.2 Degenerate Asymptotics for Power Norms 176
4.4.3 Degenerate Asymptotics for Besov Norms 180
4.4.4 Degenerate Asymptotics for Smooth Signals . . . 184

5 Sharp Asymptotics. II **185**
5.1 Tests Based on Log-Likelihood Statistics and Thresholding 185
5.1.1 Test Procedures 185
5.1.2 Reduction to the Nontruncated Case 191
5.2 Extreme Problem in the Space of Sequences of Measures 194
5.2.1 Statement of the Problem 194

5.2.2 Moment Constraints 196
5.2.3 Convex Properties of the Extreme Problem . . . 200
5.3 Separation of the Problem 203
5.3.1 Case $p = q$ 204
5.3.2 Case $p \neq q$ 204
5.4 Solution of One-Dimensional Problems 208
5.4.1 Lemma on Convexity 209
5.4.2 Case $p = q$ 212
5.4.3 Case $p \neq q$ 214
5.4.4 Reduction of the Extreme Problems (5.76), (5.85) 218
5.5 Sharp Asymptotics for l^n-Balls 224
5.5.1 One-Sided Constraint 224
5.5.2 Two-Sided Constraints 226

6 Gaussian Asymptotics for Power and Besov Norms 231
6.1 Extreme Problems 232
6.2 Principal Types of Gaussian Asymptotics 234
6.2.1 Principal Types of Gaussian Asymptotics for Power Norms 235
6.2.2 Principal Types of Gaussian Asymptotics for Besov Norms 237
6.3 Frontier Log-Types of Gaussian Asymptotics 240
6.3.1 Frontier Types of Asymptotics for Power Norms . 240
6.3.2 Frontier Types of Asymptotics for Besov Norms . 241
6.4 Graphical Presentation 243
6.5 Remarks on the Proofs of Theorems 6.1–6.4 246
6.6 Proof of Theorems 6.1 and 6.3 for $p \leq 2,\ q \geq p$, and $p = q$ 249
6.6.1 Case $p \leq 2,\ q \geq p$ 249
6.6.2 Case $p = q > 2$ 250
6.7 Extreme Problem for Power Norms: $p \neq q$ 253
6.7.1 Equations System for $q < \infty$ 254
6.7.2 Equations System for $q = \infty$ 255
6.8 Properties of the Extreme Sequences for Power Norms . 256
6.8.1 Continuity Properties 257
6.8.2 Correlation Properties 258
6.9 Extreme Problem for Besov Norms 258

7 Adaptation for Power and Besov Norms 262
7.1 Adaptive Setting 263
7.1.1 Idea of the Study 263
7.1.2 Statement of the Problem 265
7.1.3 Main Results 266
7.2 Lower Bounds 268
7.2.1 Asymptotically Orthogonal Priors 269
7.2.2 Asymptotic Sharp Lower Bounds 270

7.2.3 Lower Bounds for Power Norms 276
7.2.4 Nondistinguishability Conditions for Besov Norms 278
7.3 Upper Bounds for Power Norms 280
7.3.1 Methods of Constructions 280
7.3.2 The Construction of Collections of Tests 282
7.4 Upper Bounds for Besov Norms 285
7.4.1 Test Procedure 285
7.4.2 Type I Errors 287
7.4.3 Type II Errors for Particular Tests 288

8 High-Dimensional Signal Detection **291**
8.1 The Bayesian Signal Detection Problem 292
8.1.1 Limit Theorems and Infinitely Divisible Distributions 293
8.1.2 Some Remarks on the Study of Log-Likelihood Ratios 294
8.1.3 Limit Distributions for $\tau \geq 2$ 297
8.1.4 Limit Distributions for $\tau \in (1,2)$ 299
8.1.5 Limit Distributions for $\tau \leq 1$ 304
8.1.6 Graphic Representation 306
8.2 Multichannel Signal Detection Problems 307
8.2.1 Reduction to the Bayesian Problem 308
8.2.2 Proof of Theorem 8.1 309
8.3 Minimax Signal Detection for l^n-Balls 314
8.3.1 Main Results 314
8.3.2 Examples 316
8.3.3 Proof of Lower Bounds 318
8.4 Proof of Upper Bounds 319
8.4.1 Upper Bounds for $\tau \geq 2$ 321
8.4.2 Upper Bounds for $\tau \in (1,2)$ 325
8.4.3 Study of the Tests (8.41) 334
8.5 Testing a Hypothesis which Is Close to a Simple Hypothesis 337
8.5.1 Statement and Main Results 337
8.5.2 Proof of Lower Bounds 338
8.5.3 Upper Bounds 344
8.5.4 Extension to the Functional Gaussian Model . . . 350

A Appendix **354**
A.1 Proof of Proposition 2.16 354
A.2 Proof of Proposition 5.3 356
A.2.1 Properties of Statistics under Alternatives. 356
A.2.2 Evaluations of Type II Errors 359
A.3 Study of the Extreme Problem for Power Norms 364
A.3.1 Solution of the System (6.86), (6.87) 364
A.3.2 Solutions of the Extreme Problem for $\kappa \in \Xi_\Delta^-$ . . 369

A.3.3 Proofs of Theorems 6.1, 6.3 for $\kappa \in \Xi_{\Delta}^{-}$ 372
A.3.4 Solution of the Extreme Problem (6.88) 376
A.3.5 Solution of the Extreme Problem for $\kappa \in \Xi_{\Delta}^{+}$ . . . 379
A.3.6 Proofs of Theorems 6.1, 6.3 for $\kappa \in \Xi_{\Delta}^{+}$ 385
A.3.7 Extreme Problem for $q = \infty$ 389
A.3.8 Proofs of Propositions 6.1, 6.2 391
A.4 Study of the Extreme Problem for Besov Norms 396
A.4.1 Solution of the System (6.110), (6.111) 396
A.4.2 Solution of the Extreme Problem (6.112) 402
A.4.3 Solution of the Extreme Problem for $\tau \in \Gamma_{\Delta_1}^{-}$ . . 407
A.4.4 Solution of the Extreme Problem for $\tau \in \Gamma_{\Delta_1}^{+}$ . . 409
A.4.5 Upper Bounds 419
A.4.6 Lower Bounds 425
A.4.7 Proof of Proposition 6.3 432
A.5 Proof of Lemma 7.4 433
A.5.1 Case $\kappa \in \Xi_{G_1}$ 436
A.5.2 Case $\kappa \in \Xi_{G_2}$ 438
A.6 Proofs of Lemmas 8.2, 8.3, 8.4, 8.6 439

References **444**

Index **450**
Subject Index 450
Parameter and Function Index 453

1 Introduction

This book is concerned with nonparametric goodness-of-fit testing problems. However, we would like to start with a general review of the situation.

Let a statistician have a realization $x \in \mathcal{X}$ of a random observation X. The observation is generated by a probability measure P_θ, $\theta \in \Theta$, which depends on θ and is defined on the space $\mathcal{X}$ with σ algebra $\mathcal{A}$. Let a subset $\Theta_0 \subset \Theta$ be given; Θ_0 is nonempty and $\Theta \neq \Theta_0$.

A goodness-of-fit testing problem is to test the null-hypothesis

$$H_0 : \theta \in \Theta_0.$$

Usually the set Θ_0 corresponds to some desirable properties of the object of study. If the set Θ_0 consists of one point, i.e., $\Theta_0 = \{\theta_0\}$, then the hypothesis is called *simple*; in the opposite case it is called *composite*.

An unknown parameter θ could be a real value belonging to an interval Θ (finite or infinite) in the real line R^1; it could be an n-dimensional vector from a region Θ in R^n; the problem is called *(finite)-parametric* in these cases. The set Θ_0 usually corresponds to a subspace or a submanifold of dimension $0 \leq m < n$. In this case, at least locally, the parameter θ could be presented in the form $\theta = (\theta^{(u)}, \theta^{(n)})$ and the null-hypothesis is defined by the equality $\theta^{(u)} = \theta_0^{(u)}$, that is, the tested hypothesis corresponds to a given value of the parameter $\theta^{(u)}$. The parameter $\theta^{(n)}$ is called *nuisance* in this case.

On the other hand, a parameter θ could be of infinite dimension, for example, it could be a function belonging to some functional set Θ. In this case, the set Θ_0 could correspond to a subspace of finite or infinite codimension. In particular, $\theta = (\theta^{(u)}, \theta^{(n)})$ where the nuisance parameter $\theta^{(n)}$ is of finite or infinite dimension. In the first case the problem is called *semiparametric*, in the second case it is called *nonparametric*.

Classical goodness-of-fit testing problems correspond to observation of an identical and independently distributed (i.i.d.) sample $(X_1, \dots, X_N)$ of size N with an unknown distribution function $F(t)$. Let a distribution function $F_0(t)$ be given and we test the null-hypothesis $H_0 : F = F_0$. Let possible alternatives be assumed belonging to a finite-parametric class $F(t) = F(t;\theta)$, $\theta \in \Theta \subset R^n$ which contains $F_0(t) = F(t;\theta_0)$. Then we have the finite-parametric problem. A typical example is $F_0(t) = \Phi(t)$ where $\Phi(t)$ is a distribution function of the standard Gaussian law $N(0,1)$; we can consider a two-parametric alternative set

$$F(t;a,\sigma) = \Phi((t-a)/\sigma), \quad a \in R^1, \quad \sigma > 0,$$

i.e., $F(t;a,\sigma)$ is a distribution function of the Gaussian law $N(a,\sigma^2)$. If there is not a natural parametric class, then one can consider a function F or its density f as unknown infinite-dimensional parameters which belong to some sets $\Im$ of $\mathcal{F}$ including F_0 or its density f_0.

A typical nonparametric null-hypothesis is the hypothesis of symmetry of a distribution function $F(t)$ or of a density $f(t)$; here Θ_0 corresponds to a set of distribution functions $\Im_0$ or of densities $\mathcal{F}_0$ which consist of functions such that $F(t) = 1 - F(-t)$ or $f(t) = f(-t)$. One can consider various classes of alternatives $\Im$ or $\mathcal{F}$ in the problem. Often the functions $F(t) + F(-t)$ or $f(t) - f(-t)$ could be considered as a nuisance parameter. The next example is hypothesis of homogeneity $H_0 : F_1 = F_2$ of two i.i.d. samples $(X_1, \dots, X_{N_1})$ and $(Y_1, \dots, Y_{N_2})$ which correspond to distribution functions $F_1(t)$ and $F_2(t)$; here the nuisance parameter is $F_1 - F_2$. Another example is the hypothesis of independence of coordinates of a vector-valued sample.

Goodness-of-fit testing arises in the problem of the detection of a signal which is observed in a noise of level $\varepsilon > 0$: the null-hypothesis corresponds to the case where a signal is absent and the alternative corresponds to various possible signals which could be presented. Similar problems arise in technical or medical diagnostics: the null-hypothesis corresponds to the case where a defect or an illness is absent.

Analogous problems correspond to various other statistical models, such as a regression model, a spectral density model, etc.

Goodness-of-fit problems are studied very intensively in Statistics. Certainly, these problems are very different and, usually, very difficult. Essential progress in the problems is possible under an asymptotic approach, i.e., one studies the problems assuming the sample size is large or the noise level is small; more exactly, one studies the limit properties of the problems as $N \to \infty$ or $\varepsilon \to 0$. Under the asymptotic approach one can find many common properties in the problems: generally, these are similar to the problems under various versions of the Gaussian model. In particular, under the general regularity conditions which provide Le Cam's Local Asymptotic Normality properties, one can reduce a wide class of finite-parametric problems to analogous problems under an n-dimensional Gaussian model.

This model corresponds to the observation $X = v + \xi$ where ξ is a standard n-dimensional Gaussian vector and $v \in R^n$ is an unknown mean vector; the nuisance parameter corresponds to a linear subspace and it could be excluded by projection into orthogonal subspace without loss.

This is the foundation for the asymptotic finite-dimensional hypothesis testing theory which was developed in the 1940s–1970s simultaneously with the finite-parametric estimation theory. These theories give estimators and test procedures which have various asymptotically optimal properties and, often, these are very useful in practical problems where parametric assumptions are sufficient.

The next natural step is to develop similar theories for nonparametric problems. This has been done for estimation problems in the 1980s-2000s. The scheme was analogous: first, one studies the problem for an infinite-dimensional Gaussian model and, next, the results are translated into other statistical models of interest.

It has been found that, even for a Gaussian model of infinite-dimension, there are various principal difficulties, and to overcome these one needs to use new (for Statistics) mathematical tools, such as Function Theory, Approximation Theory, Information Theory, etc. To translate the results from Gaussian models into other statistical models of interest, one has to use probability tools such as the Theory of Random Processes and the Martingale Theory. On the other hand, a general enough Theory of Statistical Equivalency has been developed and is still developing at present. This theory provides the needed translation; however, it does not give constructible statistical procedures at present.

There are many practical applications of procedures based on the modern nonparametric estimation theory.

The main aim of this book is to present a similar theory for goodness-of-fit testing problems. More exactly, we are dealing with infinite-dimensional or with increased-dimensional Gaussian models. The main reasons follow. First, analogously to estimation theory, the main phenomena can be found under the Gaussian models and these could be translated into other models by using the Theory of Statistical Equivalency formally. Next, the study is rather technical for other models of interest. In the following text we give references to papers where analogous results have been obtained. We also study the simple null-hypothesis mainly for the same reasons.

Note that a full spectrum of results is not translated into all models and null-hypotheses of interest at present. The translation is the problem of interest and of importance for further development of the theory.

In this chapter we consider some examples to illustrate the problems in goodness-of-fit testing and the main results we would like to present. We often omit technical assumptions and proofs here; these are given in the next chapters or in the References.

1.1 Tests

We need to construct a decision rule with the values 0 (this means the null-hypothesis is accepted) or 1 (the null-hypothesis is rejected). The decision rule $\psi = \psi(x)$ is a measurable function of an observation and it is called a *nonrandomized test.* For a nonrandomized test the set

$$\mathcal{X}^{\text{cr}} = \{x \in \mathcal{X} : \psi(x) = 1\}$$

is called *critical* and the set

$$\mathcal{X}^{\text{ad}} = \{x \in \mathcal{X} : \psi(x) = 0\}$$

is called *admissible.* Clearly, this defines the partition of the observation space $\mathcal{X}$ into two measurable subsets

$$\mathcal{X}^{\text{cr}} \cup \mathcal{X}^{\text{ad}} = \mathcal{X}, \quad \mathcal{X}^{\text{cr}} \cap \mathcal{X}^{\text{ad}} = \emptyset,$$

and any measurable partition defines a test. Sometimes it is convenient to consider a randomized test: this is a measurable function $\psi(x)$, $x \in \mathcal{X}$, with values in the interval $[0, 1]$: the null-hypothesis is rejected with the probability $\psi(x)$ and is accepted with the probability $1 - \psi(x)$.

A solution based on random data could be incorrect. Two types of errors are considered in hypothesis testing. The type I error is to reject the hypothesis H_0 whenever it is true. The type II error is to accept the hypothesis H_0 whenever the null-hypothesis does not hold. This corresponds to the alternative $H_1 : \theta \in \Theta_1$ where the set $\Theta_1 = \Theta \setminus \Theta_0$ is the "natural" alternative set in goodness-of-fit testing. The quality of a test ψ is characterized by type I and type II error probabilities. The type I error probability is the function defined on the set Θ_0 by the equality

$$\alpha(\psi, \theta) = E_\theta(\psi), \quad \theta \in \Theta_0,$$

for a nonrandomized test this is of the form $\alpha(\psi, \theta) = P_\theta(\mathcal{X}^{\text{cr}})$, $\theta \in \Theta_0$. The type II error probability is the function defined on the set Θ_1,

$$\beta(\psi, \theta) = E_\theta(1 - \psi), \quad \theta \in \Theta_1,$$

for a nonrandomized test one can write $\beta(\psi, \theta) = P_\theta(\mathcal{X}^{\text{ad}})$, $\theta \in \Theta_1$. Often, instead of a type II error probability function, one considers the power function

$$1 - \beta(\psi, \theta) = E_\theta(\psi), \quad \theta \in \Theta_1.$$

Certainly, one wants to construct a test which provides possibly minimal error probabilities or maximal power. To explain the difficulties in the problem let us consider some particular cases.

1.2 One-Dimensional Parameter

Let us start with a simple Gaussian model. Observe the random variable

$$X = v + \xi,$$

where $\xi \sim N(0,1)$ is the standard Gaussian random variable, v in an unknown real-valued parameter. This corresponds to the Gaussian measure $P_v = N(v,1)$ with unknown mean v (we denote the unknown parameter by v under the Gaussian model). Let us test the null-hypothesis $H_0 : v = 0$.

First, let us consider a simple alternative $H_1 : v = v_1$ where $v_1 > 0$ is given. The quality of a test ψ is characterized by two values,

$$\alpha(\psi) = \alpha(\psi, 0), \quad \beta(\psi) = \beta(\psi, v_1),$$

and one would like these to be small. Usually, however, if one tries to decrease (increase) the type I error probability $\alpha(\psi)$, then the type II error probability $\beta(\psi)$ increases (decreases).

There are two ways out of this situation. The first one is to consider some function in $\alpha(\psi)$ and $\beta(\psi)$, for example, the sum

$$\gamma(\psi) = \alpha(\psi) + \beta(\psi)$$

and try to minimize it. This leads to the optimal test

$$\psi^+ = \mathbf{1}_{\{X > v_1/2\}}.$$

Here and below we denote as $\mathbf{1}_A$ the indicator function of the event A. For this test one has

$$\alpha(\psi^+) = \beta(\psi^+) = \gamma(\psi^+)/2 = \Phi(-v_1/2).$$

Recall that $\Phi(t)$ stands for the distribution function of the standard Gaussian law. This test depends on the alternative v_1.

The second way corresponds to the Neyman–Pearson approach. Fixing the significance level $\alpha \in (0,1)$, introduce the set Ψ_α which consists of all tests of the level α, i.e., such that $\alpha(\psi) \le \alpha$, and let us try to minimize $\beta(\psi)$ over $\psi \in \Psi_\alpha$. This leads to the test

$$\psi_\alpha^+ = \mathbf{1}_{\{X > T_\alpha\}},$$

i.e., one accepts the null-hypothesis when $x > T_\alpha$ and one rejects it in the opposite case. Here and below T_α is a $(1-\alpha)$-quantile of the standard Gaussian law,

$$\Phi(T_\alpha) = 1 - \alpha.$$

For this test one has

$$\alpha(\psi_\alpha^+) = \alpha, \quad \beta(\psi_\alpha^+, v_1) = \Phi(T_\alpha - v_1), \tag{1.1}$$

Note that this test does not depend on the alternative $v_1 > 0$. Therefore ψ_α^+ is a *uniformly most powerful test* of the level α for the right-hand side alternatives $V_1^+ = \{v > 0\}$: this means it provides the maximal power (or minimal type II error probability) among all tests of level α.

For these reasons the Neyman-Pearson approach and its various modifications are very popular in hypothesis testing.

However one has a different test for the negative alternative $v_1 < 0$:

$$\psi_\alpha^- = \mathbf{1}_{\{X<-T_\alpha\}}$$

and

$$\alpha(\psi_\alpha^-) = \alpha, \quad \beta(\psi_\alpha^-, v_1) = \Phi(T_\alpha + v_1). \tag{1.2}$$

The test ψ_α^- is uniformly most powerful for the left-hand side alternative $V_1 = \{v < 0\}$.

Note that it is senseless to use the test ψ_α^+ for negative alternatives or to use the test ψ_α^- for positive alternatives: the power of the tests is smaller than α.

However, the natural set Θ in the problem (we denote this by V under the Gaussian model) is $V = R^1$, i.e., it corresponds to the two-sided alternative $V_1 = R^1 \setminus \{0\}$. For this case there is no existing uniformly most powerful test and one needs to add some additional constraints to select the best test for the problem.

The test ψ is called *unbiased* if

$$\alpha(\psi, \theta_0) + \beta(\psi, \theta_1) \le 1 \quad \forall\, \theta_0 \in \Theta_0, \quad \theta_1 \in \Theta_1.$$

For any $\alpha \in (0,1)$ there exists a uniformly most powerful unbiased test of the level α in the problem of the form

$$\psi_\alpha^\pm = \mathbf{1}_{\{|X|>T_{\alpha/2}\}}.$$

For this test we have

$$\alpha(\psi_\alpha^\pm) = \alpha, \quad \beta(\psi_\alpha^\pm, v_1) = \Phi(T_{\alpha/2} - |v_1|) - \Phi(-T_{\alpha/2} - |v_1|) \quad \forall\, v_1 \in R^1. \tag{1.3}$$

Also the results are extended to the models $Y = \theta + \sigma\xi$ and to the sample models $Y_i = \theta + \sigma\xi_i$, $i = 1, \dots, N \ge 1$, where ξ_i are i.i.d. standard Gaussian and $\sigma > 0$ is known. In fact, by changing variables $X = Y/\sigma, v = \theta/\sigma$ or by passing to the sufficient statistics $X = N^{-1/2}\sigma^{-1}\sum_{i=1}^N Y_i$ and to the parameter $v = N^{1/2}\sigma^{-1}\theta$, we go to the model above. In particular, one can obtain small error probabilities if and only if the quantity $N^{1/2}\sigma^{-1}\theta$ is large.

For a composite null-hypothesis one considers the quantity

$$\alpha(\psi) = \alpha(\psi, \Theta_0) = \sup_{\theta\in\Theta_0} \alpha(\psi, \theta)$$

which is called the *level of a test*. Define the set Ψ_a by the constraint $\alpha(\psi) \le \alpha$. The test is called *similar*, if the type I error probability function is constant for all $\theta \in \Theta_0$.

The results are extended to an n-dimensional Gaussian model

$$X = v + \xi, \quad v \in R^n, \quad \xi = (\xi_1, \dots, \xi_n); \quad \xi_i \sim N(0,1) \text{ i.i.d.} \tag{1.4}$$

This corresponds to the n-dimensional Gaussian measure $P_v = N(v, I_n)$ with mean vector v and unit covariation matrix I_n.

Let a vector $r \in R^n$, $|r| = 1$, of unit length be given and test the simple null-hypothesis $H_0 : v = 0$ against the one- or two-sided alternatives

$$H_1 : (v,r) > 0; \quad H_1 : (v,r) < 0; \quad H_1 : (v,r) \neq 0,$$

where $(v,r) = \sum_{i=1}^n v_i r_i$ is the inner product and $|v|^2 = (v,v)$ is the norm in R^n. The uniformly most powerful tests ψ_α^+, ψ_α^- in the first and second problems and the uniformly most powerful unbiased test $\psi_\alpha^\pm$ in the third problem are of analogous form and are based on the projection (X,r) of an observation X to the direction r; the type II error probability functions are determined by the relations (1.1)–(1.3) where v_1 is changed by the projection (v,r) of a vector v to the direction r.

These results are extended to a composite null-hypothesis $H_0 : (v,r) = 0$, where the nuisance parameter corresponds to the orthogonal to r subspace R^{n-1} consisting of vector $u : (u,r) = 0$. The same tests ψ_α^+, ψ_α^- are similar uniformly most powerful in the first and second problems and $\psi_\alpha^\pm$ is the similar uniformly most powerful unbiased test in the third problem.

The next extension is possible for the exponential families of distributions. We have an i.i.d. sample $X_1, \dots, X_N$, $N \ge 1$, from a distribution having density (with respect to some probability measure on some probability space) of the form

$$f(x,\theta) = C(\theta) \exp\left(\sum_{j=1}^n \theta_j \phi_j(x)\right), \quad \theta = (\theta_1, \dots, \theta_n),$$

where $C(\theta)$ is a normalizing factor and $\phi_j(x)$, $j = 1, \dots, n$, are linear independent statistics. Let us test the null-hypothesis $H_0 : \theta_1 = \theta_1^{(0)}$ on the first coordinate of an n-dimensional parameter θ (i.e., if $n > 1$, then $(\theta_2, \dots, \theta_n)$ is the nuisance parameter). Under the general assumptions for any $\alpha \in (0,1)$ there exists a similar uniformly most powerful test for the one-sided alternatives $\theta_1 > \theta_1^{(0)}$ or $\theta_1 < \theta_1^{(0)}$ and a similar uniformly most powerful unbiased test for the two-sided alternative $\theta_1 \neq \theta_1^{(0)}$ (see Lehmann [72]). In particular, under the sample model $Y_i = \theta + \sigma\xi_i$, $i = 1, \dots, N$; $\xi_i \sim N(0,1)$, with nuisance parameter $\sigma > 0$, we obtain the well-known Student test based on the statistic

$$S_N = \sqrt{N}\hat{Y}_N / \sigma_N; \quad \hat{Y}_N = N^{-1} \sum_{i=1}^N Y_i; \quad \sigma_N^2 = (N-1)^{-1} \sum_{i=1}^N (Y_i - \hat{Y}_N)^2.$$

For the general case, the test statistics and thresholds are determined in terms of conditional distribution. However, one does not have a simple formula for the power or type II error probability such as (1.1)–(1.3).

Unfortunately, it is impossible to construct the most powerful tests for the nonexponential case. However, one can do it under the *asymptotic approach.*

This means that one has not got an observation X but a family of observations $X_\varepsilon \in \mathcal{X}_\varepsilon$, which are generated by probability measures from a family of measures $P_{\varepsilon,\theta}, \theta \in \Theta$, defined on the probability spaces $(\mathcal{X}_\varepsilon, \mathcal{A}_\varepsilon)$, where $\varepsilon \to \varepsilon_0$ is an asymptotic parameter. One considers not one test $\psi = \psi(x)$ but a family of tests $\psi_\varepsilon = \psi_\varepsilon(x_\varepsilon)$, and one is interested in its limit properties as $\varepsilon \to \varepsilon_0$.

Under the Neyman–Pearson approach, one changes the constraint $\alpha(\psi) \le \alpha$ to its asymptotic version $\alpha(\psi_\varepsilon) \le \alpha + o(1)$. Under very general assumptions for any $\alpha \in (0,1)$, one can construct *asymptotically uniformly most powerful tests of the asymptotic level* α in the problem of testing the simple null-hypothesis $H_0 \ : \ \theta = \theta_0$ on a one-dimensional parameter: these tests provide asymptotically maximal power or asymptotically minimal type II error probability (up to a vanishing item) either for all $\theta \in \Theta_0^+ = \Theta \cap (\theta_0, \infty)$ or for "local" alternatives θ_ε (which tend to θ_0 with a suitable rate) and uniformly over any compact $K \subset \Theta_0^+$; the same holds for the left-hand side and for the two-sided alternatives.

In particular, for an i.i.d. sample model of sample size $N \to \infty$ with a density $f(x,\theta)$, $\theta \in \Theta$, where Θ is an open interval in the real line, under the general regularity conditions one can use the maximum likelihood tests or tests based on the maximum likelihood estimators θ_N^*: these are of the form

$$\psi_{N,\alpha}^+ = \mathbf{1}_{\{t_N > T_\alpha\}}, \quad \psi_{N,\alpha}^- = \mathbf{1}_{\{t_N < -T_\alpha\}}, \quad \psi_{N,\alpha}^\pm = \mathbf{1}_{\{|t_N| > T_{\alpha/2}\}}, \tag{1.5}$$

where $t_N = \sqrt{NI(\theta_0)}(\theta_N^* - \theta_0)$ and

$$I(\theta) = E_\theta\left(\frac{\partial \log f(X,\theta)}{\partial\theta}\right)^2$$

is the Fisher information at point θ. For "local" alternatives of the form $\theta_N = N^{-1/2}\delta$ one can use analogous tests based on the statistics

$$t_N = (NI(\theta_0))^{-1/2} \sum_{i=1}^N \left.\frac{\partial \log f(X_i,\theta)}{\partial\theta}\right|_{\theta=\theta_0}.$$

Setting $u_N(\theta) = \sqrt{NI(\theta_0)}(\theta - \theta_0)$, one has an asymptotic extension of the relations (1.1)–(1.3) for the tests (1.5),

$$\alpha(\psi_{N,\alpha}^+) = \alpha + o(1), \quad \beta(\psi_{N,\alpha}^+, \theta) = \Phi(T_\alpha - u_N(\theta)) + o(1),$$
$$\alpha(\psi_{N,\alpha}^-) = \alpha + o(1), \quad \beta(\psi_{\alpha,N}^+, \theta) = \Phi(T_\alpha + u_N(\theta)) + o(1),$$

$$\alpha(\psi_{N,\alpha}^{\pm}) = \alpha + o(1), \quad \beta(\psi_{N,\alpha}^{\pm}, \theta) = \Phi(T_{\alpha/2} - |u_N(\theta)|) + o(1).$$

Therefore one can obtain type II error probabilities which tend to 0, if and only if $u_N(\theta) \to \infty$. Analogous results hold true for the case when the nuisance parameters are presented.

Thus, under general assumptions, the results for the problem of the testing of hypothesis on a one-dimensional parameter are similar asymptotically to those in the simplest Gaussian model.

1.3 Multidimensional Parameter

The situation is different for testing the hypothesis on an n-dimensional parameter, $n \geq 2$. Let us consider an n-dimensional Gaussian model (1.4) and let us test the null-hypothesis $H_0 : v = 0$ against the alternative $H_1 : v \in \breve{R}^n$ where we set $\breve{R}^n = R^n \setminus \{0\}$.

It is difficult to define what the best tests are. There do not exist uniformly most powerful or uniformly most powerful unbiased tests in the problem. It is simpler to say what tests should be excluded from the study. Namely, the test ψ is called *inadmissible* if there exists a test $\tilde{\psi}$ such that

$$\alpha(\tilde{\psi}, \theta_0) \leq \alpha(\psi, \theta_0), \quad \beta(\tilde{\psi}, \theta_1) \leq \beta(\psi, \theta_1) \quad \forall\, \theta_0 \in \Theta_0, \quad \theta_1 \in \Theta_1,$$

and there exist either $\theta_0 \in \Theta_0$ or $\theta_1 \in \Theta_1$ such that the correspondent inequality is strict. If the test ψ is not inadmissible, then it is called *admissible*. Clearly, admissible tests only are of mathematical interest.

It is possible to describe the complete class of admissible tests. Namely, an admissible test has a convex admissible set $C \subset R^n$ in the problem, i.e., it is of the form $\psi = \mathbf{1}_{\overline{C}}$ almost surely (a.s.); here $\overline{C}$ is the complement of the set C in R^n. The test of level α corresponds to the convex set C such that $P_0(C) = 1 - \alpha$. An unbiased test corresponds to a symmetric convex set C (with respect to the transform $x \to -x$).

Analogous results hold true for the case of exponential families of distributions. We cannot prefer one admissible test to any other admissible test. Therefore we need some additional criteria to select the "best" test in the problem.

This situation is typical in the statistical decision theory developed by Wald [99]. Two main criteria have been proposed in statistical decision theory for this situation: the Bayesian criterion and the minimax criterion. For the hypothesis testing problem with the simple null-hypothesis one can present these criteria in the following forms.

Bayesian Criterion

Let a probability measure (prior) π concentrated on the set Θ_1 be given. For a test ψ let us consider the mean of type II error probability

$$E_\pi \beta(\psi) = \int \beta(\psi, \theta)\pi(d\theta).$$

Set

$$\beta(\pi, \alpha) = \inf_{\psi \in \Psi_\alpha} E_\pi \beta(\psi).$$

We call this test $\psi_{\pi,\alpha} \in \Psi_\alpha$ *Bayesian*, if it minimizes $E_\pi \beta(\psi)$ over all tests of the level α, i.e.,

$$\beta(\pi, \alpha) = E_\pi \beta(\psi_{\pi,\alpha}) \le E_\pi \beta(\psi) \quad \forall\, \psi \in \Psi_\alpha.$$

Certainly, the Bayesian test depends essentially on the prior π and there are many Bayesian tests. Moreover, the set of Bayesian tests and their limits is the same as the set of admissible tests.

Fortunately, for the Gaussian model we can use invariance arguments for the case when the null-hypothesis and the alternative under consideration are invariant with respect to rotations $v \to Gv$ where G is an orthogonal matrix. In this case one needs a prior π on $\breve{R}^n$ which is rotation-invariant (this means $\pi(A) = \pi(GA)$ for a region $A \subset \breve{R}^n$ and an orthogonal matrix G). It is easily seen that this test depends on the norm $|X|$ or on $|X|^2$ only, i.e., on χ^2-statistics

$$t_n(X) = \sum_{i=1}^n X_i^2 = |X|^2.$$

Therefore this constraint selects the χ^2-test of the level α which is of the form

$$\chi^2_{\alpha,n} = \mathbf{1}_{\{t_n > T_{n,\alpha}\}}, \tag{1.6}$$

where $T_{n,\alpha}$ is a $(1-\alpha)$-quantile of χ^2_n-distribution with n degrees of freedom. We have the relations

$$\alpha(\chi^2_{\alpha,n}) = \alpha; \quad \beta(\chi^2_{\alpha,n}, v) = G_n(T_{n,\alpha}, |v|^2) \quad \forall\, v \in R^n. \tag{1.7}$$

Here $|v|$ is the Euclidean norm in R^n and $G_n(x, t^2)$ is the distribution function of a noncentral χ^2_n-distribution with the parameter of noncentrality t^2.

Minimax Criterion

For an alternative $H_1 : \theta \in \Theta_1$ and for a test ψ let us consider the maximal type II error probability

$$\beta(\psi, \Theta_1) = \sup_{\theta \in \Theta_1} \beta(\psi, \theta).$$

Set

$$\beta(\alpha, \Theta_1) = \inf_{\psi \in \Psi_\alpha} \beta(\psi, \Theta_1). \tag{1.8}$$

We call the test $\psi_\alpha \in \Psi_\alpha$ *minimax* if it minimizes $\beta(\psi, \Theta_1)$ over all tests of the level α, i.e.,

$$\beta(\alpha, \Theta_1) = \beta(\psi_\alpha, \Theta_1) \leq \beta(\psi, \Theta_1) \quad \forall\, \psi \in \Psi_\alpha.$$

The quantities $\beta(\alpha, \Theta_1)$, $\alpha \in (0,1)$, characterize a guarannted quality of testing. Note the inequality

$$0 \leq \beta(\alpha, \Theta_1) \leq 1 - \alpha \quad \forall\, \alpha \in (0,1).$$

In fact, the left-hand side inequality is evident and to obtain the right-hand side it suffices to consider the trivial randomized test $\psi \equiv \alpha$ which does not depend on an observation. We say that the minimax hypothesis problem is *trivial*, if $\beta(\alpha, \Theta_1) \equiv 1 - \alpha \ \forall\, \alpha \in (0,1)$.

Let $\Theta_2 \subset \Theta_1$. Then for any test ψ we have the inequality $\beta(\psi, \Theta_1) \geq \beta(\psi, \Theta_2)$ which yields monotonicity:

$$\beta(\alpha, \Theta_1) \geq \beta(\alpha, \Theta_2) \quad \text{for any } \Theta_2 \subset \Theta_1. \tag{1.9}$$

There are often some questions to the statement of the minimax problem. Sometimes the minimax criterion seems too pessimistic because it is determined by the worst alternatives, whenever, for other alternatives, one could obtain an essentially better quality of testing. However, often the worst alternatives are of the main interest. If this is not so, then one needs to remove these from consideration in the problem. A statistician also needs to study properties of the minimax test, in particular, to study its type II error function or power function.

However, the main difficulty is that it is not a simple problem to construct the minimax test. It is known that, under general assumptions, the minimax test is the Bayesian test for the *least favorable prior*, that is, for the prior π which is concentrated on the set Θ and provides a minimum mean power or maximum mean of type II error probability. Unfortunately, there are no general methods to construct the least favorable priors.

The following example is of importance for our study below. Under the n-dimensional Gaussian model (1.4) let us test the simple null-hypothesis $H_0 : v = 0$ against the alternative $H_1 : v \in V$ where $V \subset R^n$ is a closed convex set. Let v^* be the point of minimum length:

$$|v^*| = \inf_{v \in V} |v| > 0. \tag{1.10}$$

Then the least favorable prior is a Dirac mass δ_{v^*} at the point v^*, and the minimax test is of the form $\psi_\alpha = \mathbf{1}_{\{(X, r^*) > T_\alpha\}}$ and is based on the projection of an observation to the direction $r^* = v^*/|v^*|$; the quantity $|v^*|$ is the minimum of the projection of the set V to the direction r^*,

$$|v^*| = \inf_{v \in V} (r^*, v), \tag{1.11}$$

see Example 2.3 below for details.

Certainly this problem is not of the "goodness-of-fit" type. However, the main results we would like to present in this book, are based, roughly, on the reduction of a wider class of goodness-of-fit problems to the problem similar to the above and to an extreme problem similar to (1.10).

There is a peculiarity for using the minimax approach in goodness-of-fit testing. Typically, the minimax problem is trivial when we test the simple null-hypothesis $H_0 : \theta = \theta_0$, where θ_0 is the interior point of the set Θ, against the alternative $H_1 : \theta \neq \theta_0$. The reason is that the alternative set $\Theta_1 = \Theta \backslash \{\theta_0\}$ contains points too close to the point θ_0.

Often the way out of a situation is to remove some neighborhood U of the point θ_0. The neighborhood could be a ball in some norm $|\cdot|_{(1)}$ on the set Θ, i.e., we consider an alternative of the form

$$\Theta_{1,\rho} = \{\theta \in \Theta \ : \ |\theta - \theta_0|_{(1)} \geq \rho\} \tag{1.12}$$

and we would like to study *how small the radius* $\rho > 0$ *could be to obtain a good minimax quality of testing* (i.e., for the quantities $\beta(\alpha, \Theta_{1,\rho})$ to be small).

Analogously, to test the composite hypothesis $H_0 : \theta \in \Theta_0$ one needs to remove some neighborhood of the set Θ_0.

Sometimes under the Gaussian model (1.4), we can overcome the difficulties noted above by using invariance arguments. For $\rho > 0$ let us set

$$\breve{D}_\rho^n = \{v \in R^n : \ |v| \geq \rho\},$$

i.e., we remove a spherical neighborhood of the null-hypothesis. Then the least favorable prior is the uniform probability measure on the frontier sphere $S_\rho^{n-1} = \{v \in R^n : \ |v| = \rho\}$ of the radius $\rho > 0$ and χ^2*-test* $\chi^2_{\alpha,n}$ *is the minimax for alternatives of the form* $H_1 : v \in \breve{D}_\rho^n$ *with any* $\rho > 0$ and

$$\beta(\breve{D}_\rho^n, \alpha) = \beta(S_\rho^{n-1}, \alpha) = G_n(T_{n,\alpha}, \rho^2). \tag{1.13}$$

Moreover this test provides optimality under the following criterion. Given $\eta \in \Theta \backslash \Theta_0$ denote by $\hat{\beta}(\eta, \alpha)$ the minimum type II error probability for testing the level α in the problem of testing the null hypothesis against the simple alternative $H_1 \ : \ \theta \ = \ \eta$. The function $1 - \hat{\beta}(\theta, \alpha)$ is called an *envelope power function.* For a test $\psi \in \Psi_\alpha$ the difference $\delta(\psi, \theta; \alpha) = \beta(\psi, \theta) - \hat{\beta}(\theta, \alpha)$ is called a *shortcoming function.* Then *the test* $\chi^2_{\alpha,n}$ *minimizes the maximal and the mean shortcoming over any sphere* S_ρ^{n-1}.

It is of importance for what follows to consider the case $n \to \infty$. Then one can use the well-known Gaussian approximations of the central and noncentral χ^2-distribution function

$$G_n(x, 0) = \Phi((x - n)/(2n)^{1/2}) + o(1),$$
$$G_n(x, t^2) = \Phi((x - n - t^2)/(2n + 4t^2)^{1/2}) + o(1), \quad n \to \infty,$$

which are uniform over t, x. Using these relations, for any $\alpha \in (0,1)$, one gets

$$G_n(T_{n,\alpha}, \rho^2) = \Phi(T_\alpha - \rho^2/\sqrt{2n}) + o(1) \text{ uniformly over } \rho, \quad n \to \infty, \tag{1.14}$$

(see Section 3.1.2 for details). Comparing (1.13) and (1.14) we see that, if n is large, then a good minimax quality of testing is possible, if the radii $\rho = \rho_n$ increase, at least, at the rate $\rho_n^* = n^{1/4}$ only. Namely, for any $\alpha \in (0,1)$, one has

$$\beta(\alpha, \breve{D}_\rho^n) \to 0 \qquad \text{iff} \quad \rho_n/\rho_n^* \to \infty; \tag{1.15}$$

$$\beta(\alpha, \breve{D}_\rho^n) \to 1-\alpha \quad \text{iff} \quad \rho_n/\rho_n^* \to 0. \tag{1.16}$$

This rate characterizes a "*payment for dimension.*"

The results above are extended to the sample model of the form

$$X_i = v + \sigma\xi_i, \quad v \in R^n, \quad i = 1, \dots, N,$$

where $\xi_i \sim N(0, I_n)$ are i.i.d. and $\sigma > 0$ is known. Analogous results hold true for the case when a nuisance parameter is presented, i.e., $v = (v^{(1)}, v^{(2)})$, $v^{(1)} \in R^m$, $v^{(2)} \in R^{n-m}$, $0 < m \leq n$, and we test the null-hypothesis $H_0 : v^{(1)} = 0$. In this case we consider a χ^2-test based on the first m coordinates of observations and the degree of freedom is m in the problem.

One can obtain asymptotic extensions of these results for the case when the dimension n of an unknown parameter is fixed. In particular, consider an i.i.d. sample model with a density $f(x,\theta)$, $\theta \in \Theta$, where Θ is a region in R^n. Let θ_0 be an interior point in Θ and we test the simple null-hypothesis $H_0 : \theta = \theta_0$. Then the maximum likelihood test provided an asymptotic version of the optimal properties analogous to those noted above. The same properties are also provided by the test based on the maximum likelihood estimator θ_N^*:

$$\psi_{N,\alpha} = \mathbf{1}_{\{t_N^2 > T_{n,\alpha}\}}, \quad t_N^2 = N(\theta_N^* - \theta_0)^T I(\theta_0)(\theta_N^* - \theta_0), \tag{1.17}$$

where $I(\theta)$ is the Fisher information matrix at the point θ (i.e., the test statistics t_N is determined by the distance between the null-hypothesis and the estimator θ_N^* and the distance is measured in Fisher's metric).

In particular, these tests are asymptotically Bayesian for any prior π having density $p(\theta)$ which is positive and continuous at the point θ_0. Set

$$u_N(\theta)^2 = N(\theta - \theta_0)^T I(\theta_0)(\theta - \theta_0)$$

(this is Fisher's distance at the point θ_0). These tests are asymptotically minimax for alternatives of the form $H_1 : \theta \in K_{N,r}$ where $K_{N,r} = \{\theta \in K,\ u_n(\theta) > r^2\}$ and $K \subset \Theta$ is a compact set which contains some neighborhood of the point θ_0.

We have an asymptotic extension of the relations (1.7) for the tests (1.17),

$$\alpha(\psi_{N,\alpha}) = \alpha + o(1), \quad \beta(\psi_{N,\alpha}, \theta) = G_n(T_{n,\alpha}, u_N^2(\theta)) + o(1). \tag{1.18}$$

The relations (1.18) are uniform over $\theta \in K$ for any compact $K \subset \Theta$.

For example, let $f(x, \theta)$, $\theta \in \Theta$, correspond to a polynomial distribution: an observation X takes the values $1, \ldots, n$ with probabilities $p_1, \ldots, p_n$; here $\theta = (p_1, \ldots, p_n)$ is a probability vector and $\Theta = \mathcal{P}_{n-1}$ is an $(n-1)$-dimensional simplex which consists of probability vectors such that

$$p_j > 0, \quad j = 1, \ldots, n; \quad \sum_{j=1}^{n} p_j = 1.$$

Let us test the null-hypothesis which corresponds to $\theta_0 = (p_{0,1}, \ldots, p_{0,n})$. Then the relation (1.17) determines Pearson's χ^2-statistics

$$t_N^2 = \sum_{j=1}^{n} \frac{(k_{N,j} - Np_{0,j})^2}{Np_{0,j}}, \quad k_{N,j} = \#\{i : X_i = j\}; \tag{1.19}$$

the Fisher distance at the point θ_0 is of the form

$$u_n^2(\theta) = N \sum_{j=1}^{n} \frac{(p_j - p_{0,j})^2}{p_{0,j}}. \tag{1.20}$$

The difference is that the degree of freedom is $n-1$ because it is the dimension of the probability simplex.

For "local" alternatives of the form $\theta_N = \theta_0 + N^{-1/2}\delta$, $\delta \in \breve{R}^n$, one can use analogous tests based on the quadratic forms

$$t_N^2 = N^{-1/2} \sum_{i=1}^{N} l_i^T I^{-1}(\theta_0) l_i, \quad l_i = \text{grad}_\theta(\log f(X_i, \theta))|_{\theta=\theta_0},$$

where grad_θ is the vector which consists of partial derivatives with respect to the variables θ_j, $j = 1, \ldots, n$. For these tests we can rewrite the relations (1.18) in the form

$$\alpha(\psi_{N,\alpha}) = \alpha + o(1), \quad \beta(\psi_{N,\alpha}, \delta) = G_n(T_{n,\alpha}, u^2(\delta)) + o(1). \tag{1.21}$$

where we set $u^2(\delta) = \delta^T I(\theta_0)\delta$. The relations (1.21) are uniform over $\theta \in K$ for any compact $K \subset R^n$.

Analogous results also hold true for the case when the nuisance parameters are presented. Moreover, analogous results hold true for some semiparametric hypotheses problems as well (see Hájek and Sidak [30]). These also hold for other parametric families. Note the parametric white Gaussian noise model which has been studied by Ibragimov and Khasminskii [38] and corresponds to the observation of a signal $s(t, \theta)$, $t \in [0, 1]$, in the white Gaussian noise of level $\varepsilon > 0$; $\varepsilon \to 0$. Formally, this is of the form

$$dX_\varepsilon(t) = f(t, \theta)\, dt + \varepsilon\, dW(t), \quad t \in [0, 1]; \quad s(\cdot, \theta) \in L_2(0, 1), \tag{1.22}$$

where $W(t)$ is the standard Wiener process. Many parametric models of interest are presented in the book by Ibragimov and Khasminskii on estimation theory [38].

The results noted above have been developed since the 1930s. We do not give a historical report here. Note only that the modern version of these results are based on the Local Asymptotic Normality properties formulated by Le Cam, and on some regularity conditions that have been proposed by Ibragimov and Khasminskii [38]. See also [4], [30], [33], [71], [72], and [42], [45].

1.4 Infinite-Dimensional Parameter

The unknown probability density or distribution function is one of the main examples of the infinite-dimensional parameters studied in Statistics. One does not assume that these densities or functions belong to a given finite-parametric class.

1.4.1 Classical Goodness-of-Fit Tests

One of the classical goodness-of-fit testing problems corresponds to the testing of the simple null-hypothesis $H_0 : F = F_0$ on unknown distribution function $F(t)$, $t \in R^1$, of i.i.d. real-valued random observations $X_1, \dots, X_N$, where $F_0 = F_0(t)$ is a given known continuous distribution function. By making the transform $y = F_0(x)$ we can assume without loss of generality that $X_i \in [0,1]$ and $F_0(t) = t$ for $t \in [0,1]$, i.e., the observations are uniformly distributed on the interval $[0,1]$. Let us consider the hypothesis testing problem

$$H_0 : F = F_0 \quad \text{against} \quad H_1 : F \neq F_0; \quad F_0(t) = t, \quad t \in [0,1]. \tag{1.23}$$

Let an unknown distribution function be assumed to be absolutely continuous. Then we can rewrite the problem in terms of an unknown probability density

$$H_0 : f = f_0 \quad \text{against} \quad H_1 : f \neq f_0; \quad f_0(t) \equiv 1, \quad t \in [0,1]. \tag{1.24}$$

where $f(t)$, $t \in [0,1]$, is a probability density with respect to Lebesgue measure on the interval $[0,1]$. The most popular tests in the problem are Kolmogorov tests, Cramér–von Mises–Smirnov tests, and Pearson χ^2-tests.

Kolmogorov and Cramér–von Mises–Smirnov Tests

These tests are based on a distance between the empirical distribution function

$$F_N^*(t) = F_N^*(t; X_1, \dots, X_N) = N^{-1} \sum_{i=1}^{N} \mathbf{1}_{\{X_i < t\}}$$

and the known distribution function $F_0(t)$. Kolmogorov tests are based on the supreme distance

$$k_N = \|F_N^* - F_0\|_\infty = \sup_{t \in [0,1]} |F_N^*(t) - t|$$

and Cramér–von Mises–Smirnov tests are based on the squared distance

$$s_N^2 = \|F_N^* - F_0\|_2^2 = \int_0^1 (F_N^*(t) - t)^2 \, dt.$$

Here and below we denote by $\|\cdot\|_p$ the L_p-norm in the functional space $L_p(0,1)$, $1 \le p \le \infty$; often $\|\cdot\|$ stands for the L_2-norm.

Taking a nonrandom threshold T_N, we accept the null-hypothesis if $t_N < T_N$ and we reject it if $t_N \ge T_N$; we set $t_N = k_N$ or $t_N = s_N$ here and below in this subsection.

Under the Neyman–Pearson approach, to obtain the test of level α, the threshold $T_N = T_{N,\alpha}$ should be a $(1-\alpha)$-quantile of the statistics t_N in $P_{F_0}^N$-probability. It is not a simple problem to calculate the quantities $T_{N,\alpha}$ for fixed N. Fortunately, one can do it asymptotically. By the Kolmogorov and Smirnov theorems, the statistics $\sqrt{N} t_N$ have nondegenerate continuous limit distribution functions under the null-hypothesis, as $N \to \infty$,

$$P_{F_0}^N(\sqrt{N}\|F_N^* - F_0\|_\infty < t) \to K(t), \quad P_{F_0}^N(N\|F_N^* - F_0\|_2^2 < t) \to S(t);$$

where $K(t)$, $S(t)$ are special continuous distribution functions that are strictly increasing on the half-line $(0, \infty)$. Therefore, to obtain the tests $\psi_{N,\alpha}$ of the asymptotic level $\alpha_N \to \alpha$, one can take $T_{N,\alpha} = N^{-1/2} k_{1-\alpha}$ or $T_{N,\alpha}^2 = N^{-1} s_{1-\alpha}^2$ where $k_{1-\alpha}$ and $s_{1-\alpha}^2$ are $(1-\alpha)$-quantiles of the distribution functions $K(t)$ and $S(t)$, i.e., $K(k_{1-\alpha}) = S(s_{1-\alpha}^2) = 1 - \alpha$.

One is often interested in a large sample size N and one can use various limit theorems connected with the empirical distribution function to evaluate the probabilities of errors. By the Glivenko–Cantelli theorem, under the null-hypothesis $F = F_0$, the statistics k_N tend to 0 a.s. and in probability as $N \to \infty$. Moreover, by the Kolmogorov and Smirnov theorems the rates of t_N are $N^{-1/2}$ in probability under the null-hypothesis. On the other hand, for any fixed alternative $F \ne F_0$ the statistics k_N or s_N tend to the positive distances $\|F - F_0\|_\infty$ or $\|F - F_0\|_2$. By $T_{N,\alpha} \to 0$, for any $\alpha > 0$, the type II error probabilities vanish:

$$\beta_N(\psi_{N,\alpha}, F) = P_F^N(t_N < T_{N,\alpha}) \to 0 \quad \text{as } N \to \infty.$$

Here P_F^N is the probability that corresponds to an i.i.d. sample of size N from the distribution function F. This means that the Kolmogorov tests and the Cramér–von Mises–Smirnov tests are *consistent* in problem (1.23) for any $F \neq F_0$.

Certainly, the consistency properties are not uniform over the alternatives $F \neq F_0$ in problem (1.23): if the distance between F and F_0 is of the rate $r_N = o(N^{-1/2})$, then one cannot obtain a small type II error probability $\beta_N(\psi_{N,\alpha}, F)$. On the other hand, using these tests, one can distinguish between the null-hypothesis and the alternatives F_N which are bounded away from F_0 in the supreme distance or in the squared distance with $r_N \gg N^{-1/2}$.

Moreover, taking a sequence $r_N > 0$ such that $N^{1/2} r_N \to \infty$, set (compare with (1.12))

$$\Im_N = \{F : \|F - F_0\|_\infty \geq r_N\} \quad \text{or} \quad \Im_N = \{F : \|F - F_0\|_2 \geq r_N\} \tag{1.25}$$

and let us consider the alternatives $H_1 : F \in \Im_N$. Then we get

$$\sup_{F \in \Im_N} \beta_N(\psi_{N,\alpha}, F) \to 0. \tag{1.26}$$

This means that, for alternatives of the form (1.25), the Kolmogorov tests and the Cramér–von Mises–Smirnov tests provide good minimax distinguishability, i.e., the rate of testing is $N^{-1/2}$.

One can show that this is the best rate of testing for the alternatives (1.25) because of the same rate one had in the finite-parametric problems, see Section 1.3. We call this rate *classical.*

Let us consider the hypothesis testing problem (1.24) on a probability density of an i.i.d. sample. Recall that this problem is equivalent to problem (1.23) with absolutely continuous distribution functions and we can rewrite the properties of the Kolmogorov and Cramér–von Mises–Smirnov tests for this problem. In particular, consistence property noted above holds true in the problem (1.24) for any fixed alternative $f \neq f_0$. Analogous properties also hold for any sequence of alternatives f_N of the form

$$f_N(t) = f_0(t) + r_N \delta(t), \quad f_0(t) \equiv 1; \quad r_N > 0, \tag{1.27}$$

with any fixed function $\delta(t)$, $t \in [0, 1]$, such that

$$\int_0^1 \delta^2(t)\, dt > 0, \quad \int_0^1 \delta(t)\, dt = 0, \quad \inf_{t \in [0,1]} \delta(t) > -\infty.$$

Taking the tests of an asymptotically fixed level $\alpha \in (0, 1)$, one cannot distinguish between the null-hypothesis and the alternatives for $r_N = o(N^{-1/2})$ but one can do it for $r_N \gg N^{-1/2}$. The reason is that the supreme and squared distances between F_0 and the distribution functions F_N for the densities of the form (1.27) are of the same rate r_N.

However, the alternatives of the form (1.25) are not natural in problem (1.24). To remove the alternatives which are "too close" to the null-hypothesis, it is natural to use the distances not between distribution functions but between densities. This means in problem (1.24) that it is natural to consider the alternatives $H_1 : f \in \mathcal{F}_N$ for the sets of densities of the form

$$\mathcal{F}_N = \{f : \|f - f_0\|_\infty \geq r_N\} \quad \text{or} \quad \mathcal{F}_N = \{f : \|f - f_0\|_2 \geq r_N\}. \quad (1.28)$$

The problem for alternatives (1.28) is essentially different from the problem for alternatives (1.25). In particular, the relation analogous to (1.26) does not hold for any small enough r_N for alternatives (1.28). The reason is that the supreme and squared distances between distribution functions can be much smaller than the distances between probability densities. It suffices to consider the collection of probability densities

$$f_k(t) = f_0(t) + C\, r_N \sin(2\pi kt) \in \mathcal{F}_N, \quad k = 1, 2, \dots; \quad Cr_N < 1, \quad (1.29)$$

where $C = 1$ for the supreme norm and $C = \sqrt{2}$ for the squared norm. For the corresponding distribution functions one has

$$\|F_k - F_0\|_\infty = \|F_k - F_0\|_2^2 = (2\pi k)^{-1} r_N \to 0, \quad k \to \infty.$$

It is easily seen that for the Kolmogorov tests $\psi_{N,\alpha}$ and for alternatives (1.28) with a supreme norm one has

$$\sup_{f \in \mathcal{N}_N} \beta_N(f) = 1 - \alpha_N \quad \forall\, N, \quad r_N \in (0, 1). \quad (1.30)$$

One can obtain an analogous relation for the Cramér– von Mises-Smirnov tests and for alternatives (1.28) with the squared norm. This means that the Kolmogorov and Cramér–von Mises-Smirnov tests provide bad minimax distinguishability for alternatives of the form (1.28).

One can ask: *Is it possible to construct tests which provide better minimax properties for alternatives* (1.28) *in this problem?*

The answer is: *no, it is not.* One can show that, if an alternative contains the collection of densities (1.29), then relation (1.30) holds true for any test (see [52]). This means that it is impossible to distinguish between the null-hypothesis and alternatives of form (1.28) in the minimax sense.

Note that the collection (1.29) contains functions with frequencies of the sine curves tending to infinity and derivatives of the functions are unbounded over the collection.

For this reason one can ask: *Would the situation be better under some regularity constraints on the class of densities under consideration?*

The answer is: *yes, it would be.*

This is the main object of study in the book. Typically, regularity constraints are formulated in terms of constraints on some derivatives of the underlying function and the rates of testing are slower, than the classical rates. Moreover, the Kolmogorov tests or the Cramér–von Mises-Smirnov tests do not give the best rates of testing under typical regularity constraints. Sometimes we can use the Pearson χ^2-tests to provide the best rates.

The Pearson χ^2-Tests

These are of the following form. Fixing an integer $n \geq 1$, let us take a partition $\Delta_n = \{\delta_1, \ldots, \delta_n\}$ of the interval $(0,1]$ to n subintervals $\delta_j = (t_{j-1}, t_j]$, $j = 1, \ldots, n$; $t_0 = 0$, $t_n = 1$, of the length $p_{0,j} = P_0(\delta_j) = t_j - t_{j-1} > 0$ and let $k_{N,j}$ be the number of observations X_i contained in δ_j, $j = 1, \ldots, n$. The random vector $\bar{k}_N = (k_{N,1}, \ldots, k_{N,n})$ corresponds to the polynomial distribution and, by passing to $\bar{k}_N$, we change the nonparametric hypothesis testing problems (1.23) or (1.24) to a parametric hypothesis testing problem on a probability vector $\theta = p$ belonging to the simplex $\Theta = \mathcal{P}_{n-1}$,

$$H_0 : p = p_0 \quad \text{against} \quad H_1 : p \neq p_0, \quad p \in \mathcal{P}_{n-1}. \tag{1.31}$$

We can use the Pearson $\chi^2_{n,\alpha}$-tests in problem (1.31) which are based on statistics (1.19). It was noted in Section 1.3 that these tests provide various asymptotic optimal properties. An asymptotic relation of type II error probability is given by (1.18) with the quantity $u_N(\theta)$ defined by (1.20). This relation is uniform over $\theta \in \mathcal{P}_n$.

However, there is an essential difference between problems (1.31) and the original problems (1.23), (1.24). To simplify, we consider the case of equispaced partition into n cells $\delta_{n,j} = ((j-1)/n, j/n]$ with $p_{j,0} = 1/n$. For any $\varphi \in L_1(0,1)$, let $\varphi_{n,j} = n \int_{\delta_{n,j}} \varphi(t)\,dt$ be the mean value of the function φ over the interval $\delta_{n,j}$ and let

$$Pr_n\varphi = \sum_{j=1}^{n} \varphi_{n,j} \mathbf{1}_{\delta_{n,j}}$$

be the $L_2(0,1)$-projection to the n-dimensional space which consists of a step function on $[0,1]$ for the partition Δ_n. The quantity $u_N(\theta)$ can be presented in the form

$$u_N(\theta) = Nn \sum_{j=1}^{n} (p_j(f) - 1/n)^2 = N\|Pr_n(f - f_0)\|_2^2, \quad p_j(f) = \int_{\delta_{n,j}} f(t)\,dt; \tag{1.32}$$

i.e., this is determined by an L_2-norm of the projection of $(f - f_0)$, and it is possible that $f - f_0$ is larger but that $\|Pr_n(f - f_0)\|$ is small or vanishing.

To overcome this difficulty one can try to increase n jointly with N. Combining (1.18), (1.20), (1.32), and (1.14) we see that the Pearson $\chi^2_{n,\alpha}$-tests of level α with growing $N \to \infty$ and $n \to \infty$ provide vanishing type II error probabilities if and only if

$$n^{-1/2} N \|Pr_n(f - f_0)\|_2^2 \to \infty.$$

It is easily seen that

$$\|Pr_n(f - f_0)\|_2 \to \|f - f_0\|_2, \quad \text{as } n \to \infty \quad \forall\, f \in L_2(0,1).$$

This yields that *the Pearson $\chi^2_{n,\alpha}$-tests are consistent for any fixed $f \in L_2(0,1)$, $f \neq f_0$, and for any sequence $n \to \infty$ such that $n^2 = o(N)$.* However, the consistency property is not uniform over alternatives f and depends on the regularity of a function f. In particular, for any n let us consider the density $f_n(t) = 1 + \sqrt{2}r(\sin(2\pi ln)t)$ of form (1.29) with $k = ln$, l is an integer. Then $\|f_n - f_0\|_2 = r$, but $p_j(f_n) = p_{0,j}\ \forall j = 1, \ldots, n$, and $\|Pr_n(f_n - f_0)\|_2 = 0$. So, using the $\chi^2_{n,\alpha}$-tests, it is impossible to distinguish between f_0 and f_n.

The first study of a minimax nonparametric problem of hypotheses testing related to the present subject appears to be that of Mann and Wald [82], concerning testing for uniformity on $[0,1]$ against alternatives with distribution functions $F(t) \in \Im_N$ under constraints (1.25) with the supreme norm.

They studied the minimax properties of the sequences of Pearson χ^2-tests $\chi^2_{n,\alpha}$ with equispaced partitions of $[0,1]$ into $n = n_N \to \infty$ intervals in the problem. The results in [82] imply, in particular, that a necessary and sufficient condition for $\beta_N(\chi^2_{n,\alpha}, \Im_N) \to 0$ is $r_N N^{2/5} \to \infty$, and an optimal number of intervals is of order $n_N \asymp N^{2/5}$. These results could be extended to constraints (1.25) with the squared norm as well. These lead χ^2-tests are worse than Kolmogorov tests or Cramér–von Mises–Smirnov tests in the problem with alternative (1.25).

On the other hand, Pearson χ^2-tests are better than Kolmogorov tests or Cramér–von Mises–Smirnov tests in the problem with alternative (1.28) under some additional regularity constraint on a density (see [43], [48], [52]).

Namely, suppose an unknown density has bounded derivative of order η, i.e., we consider the alternative of the form

$$\mathcal{F}_{N,\eta} = \left\{ f \in \mathcal{F} : \ \|f - f_0\|_2 \geq r_N, \ \sup_{t \in [0,1]} |f^{\eta}(t)| \leq C \right\} \tag{1.33}$$

for some constant $C > 0$ and an integer η. Then the Pearson χ^2-tests $\chi^2_{n,\alpha}$ with equispaced partitions of $[0,1]$ into $n \asymp N^{2/(4\eta+1)}$ intervals provides minimax distinguishability, (i.e., $\beta_N(\chi^2_{n,\alpha}, \mathcal{F}_{N,\eta}) \to 0$), if

$$r_N \gg r_N^* = N^{-2\eta/(4\eta+1)}$$

and r_N^* is the best rate in the problem (i.e., for any $\alpha \in (0,1)$ there does not exist any sequence of tests of asymptotic level α which provides a vanishing maximal type II error probability).

On the other hand, for Kolmogorov tests or Cramér–von Mises–Smirnov tests one has minimax distinguishability for alternative (1.33) only if $r_N \gg N^{-\eta/(2\eta+2)}$ (see [43]).

The above noted optimality properties of Pearson χ^2-tests hold true for the noninteger $\eta > 0$ as well. For example, for $\eta \in (0,1]$, this means that we replace the constraint on the derivative in (1.33) by a Lipschitz condition of order η:

$$|f(t+h) - f(t)| \le Ch^\eta \quad \forall\, h \in (0,1), \quad t \in [0, 1-h].$$

It is worth mentioning that for $\eta < \frac{1}{4}$ (the case of fairly irregular functions) the number n_N of cell intervals turns out to be much greater than the sample size, $n_N/N \to \infty$, i.e., the majority of cells are empty. This contradicts the traditional recommendations on the use of Pearson χ^2-tests, nevertheless, χ^2-tests of this type are optimal in this case with regard to a minimal order of r_N required for minimax distinguishability . Note, however, that the condition $n_N/N^2 \to 0$, whose role for the Pearson χ^2-tests was pointed out in [3], is met here.

Suitable versions of the Pearson χ^2-tests provide optimal rates of minimax distinguishability for classical composite nonparametric null-hypotheses: the hypothesis of symmetry of a density, the hypothesis of homogeneity of two samples, the hypothesis of independence; see [43], [47], [49].

Some different minimax properties of classical goodness-of-fit tests were studied by Ermakov [24], [25], [27], [28].

Remark. Note that various other criteria of asymptotic optimality and asymptotic efficiency are studied in the statistical literature as well, such as first-, second- and higher-order optimality, Pitman's, Hodges–Lehmann's, Bahadur's efficiency, Kallenberg's intermediate efficiency, asymptotics of shortcoming, etc. In particular, returning to the problem of testing uniformity of an unknown density f of an i.i.d. sample of size $N \to \infty$. Taking a significance level $\alpha \in (0,1)$ and an alternative $f = f_N$, one can compare the power (or type II error probabilities $\beta = \beta_N$) of studied tests $\psi_{N,\alpha}$ of level α with the power (or type II error probability) of the optimal Neyman–Pearson tests $\psi_{N,\alpha;f}$ for alternative f. Also, taking type I and type II error probabilities α, β, a sequence of tests $\psi_{N,\alpha}$ of level α, and an alternative f, let $N(\alpha, \beta, f; \psi_{N,\alpha})$ be the sample size which provides the type II error probability β for alternative f. Then, by fixing two among three parameters α, β, f, one can compare the sample sizes $N(\alpha, \beta, f; \psi_{N,\alpha})$ of the studied tests $\psi_{N,\alpha}$ with the sample sizes $N(\alpha, \beta, f; \psi_{N,\alpha;f})$ of the Neyman–Pearson tests $\psi_{N,\alpha;f}$. Pitman's efficiency deals mainly with a local alternative of the form $f(x) = 1 + N^{-1/2}p(x)$, fixed $\alpha \in (0,1)$, and a function $p(x)$. Bahadur's efficiency deals with a fixed alternative $f \ne 1$, and $\beta \in (0,1)$ which corresponds to small $\alpha = \alpha_N$ such that $\log \alpha_N^{-1} \asymp N$. The Hodges–Lehmann efficiency deals with fixed alternative $f \ne 1$, and α which corresponds to small $\beta = \beta_N$ such that $\log \beta_N^{-1} \asymp N$. The Kallenberg intermediate efficiency deals with a small levels $\alpha = \alpha_N \to 0$ such that $\log \alpha_N^{-1} \asymp N^\tau$ for some $0 < \tau \le 1$, and $\beta = \beta_N$ bounded away from 0 and 1.

We do not present here exact statements and results for these criteria and refer the reader to Nikitin [84], and to [2], [66], [39], [40] where the reader can find many results and references.

Recall that we study the minimax criterion $\beta(\alpha, \Theta_1)$, of form (1.8) for nonparametric sets Θ_1 and related criteria, in this book. We assume a significance level $\alpha \in (0,1)$ to be fixed or bounded away from 0 and 1; however, the results could be extended to the case $\alpha = \alpha_N \to 0$, but not too quickly; under an i.i.d. sample model this corresponds to $\log \alpha_N^{-1} = o(N^\tau)$ for all $\tau > 0$. Note that different results and phenomena are possible in the opposite case.

In this book we study the problems under the Gaussian models. Recall that the main principal phenomena could be extended to other statistical models of interest, in particular, to an i.i.d. sample model of size $N \to \infty$.

1.4.2 *Nondistinguishability in the Problem*

Let us consider an infinite-dimensional Gaussian model

$$X = v + \xi; \quad v = (v_1, \dots, v_n, \dots) \in l^2; \quad \xi = (\xi_1, \dots, \xi_n, \dots); \quad \xi_i \sim N(0,1) \text{ i.i.d.}, \tag{1.34}$$

where l^2 is a Hilbert space which consists of real-valued sequences with the inner product $(u,v) = \sum_i u_i v_i$ and the norm $|v|^2 = \sum_i v_i^2$. Sometimes this model is considered in some different form which corresponds to the Gaussian noise of the known level ε: we observe a sequence

$$Y = \theta + \varepsilon\eta, \quad \theta \in l^2, \tag{1.35}$$

where θ is an unknown parameter. By changing variables $X_i = Y_i/\varepsilon$, $v_i = \theta_i/\varepsilon$ we see that the models (1.34) and (1.35) are equivalent.

The model (1.35) is equivalent to the functional white Gaussian noise model (compare with (1.22)):

$$dX_\varepsilon(t) = s(t)\,dt + \varepsilon\, dW(t), \quad t \in [0,1], \quad s \in L_2(0,1). \tag{1.36}$$

It suffices to fix any orthonormal basis $\{\varphi_i\}$ in $L_2(0,1)$ and consider Fourier coefficients:

$$\theta_i = (s, \psi_i) = \int_0^1 \varphi_i(t) s(t)\,dt, \quad Y_i = \int_0^1 \varphi_i(t)\,dX_\varepsilon(t). \tag{1.37}$$

Let us test the null-hypothesis $H_0 : v = 0$ and consider the alternative $H_1 : v \neq 0, v \in l^2$. There do not exist uniformly most powerful tests in the problem. There are no natural priors and we cannot use orthogonal invariance arguments because there do not exist rotation-invariant priors concentrated on the Hilbert space l^2.

We can try to use a minimax approach. As was noted above, one needs to remove some neighborhood of the null-hypothesis. However it is often not sufficient to obtain a nontrivial problem.

In fact, analogously to the finite-dimensional case, let us consider the alternative $H_1 : v \in V$ where $V = \breve{D}(\rho) = \{v \in l^2 : |v| \geq \rho\}$. We can consider all n-dimensional models as submodels of (1.34) and the sets $\breve{D}_n(\rho)$

as subsets $\breve{D}(\rho)$. By the inequality (1.9) we get $\beta(\alpha, \breve{D}(\rho)) \geq \beta(\alpha, \breve{D}_n(\rho))$ and for all $\alpha \in (0,1)$, $\rho > 0$, using the relations (1.13) and (1.14) we have

$$\begin{aligned}\beta(\alpha, \breve{D}(\rho)) &\geq \sup_n \beta(\alpha, \breve{D}_n(\rho)) = \sup_n G_n(T_{n,\alpha}, \rho^2) \\ &\geq \Phi(T_\alpha - \rho^2/\sqrt{2n}) + o(1) \to 1 - \alpha, \quad \text{as } n \to \infty.\end{aligned}$$

Thus $\beta(\alpha, \breve{D}(\rho)) \equiv 1 - \alpha$ for all $\alpha \in (0,1)$, $\rho > 0$.

This fact was first noted by Ibragimov and Khasminskii [35]. It follows from Burnashev [8] that the analogous fact holds for alternatives of the form $\breve{D} + r(\rho) = \{v \in l^2 : |v|_p \geq \rho\}$ for all $p > 0$ where $|\cdot|_p$ is an l^p-norm (quasi-norm for $p < 1$)

$$|v|_p = \left(\sum_i |v_i|^p\right)^{1/p}; \quad |v|_\infty = \sup_i |v_i|.$$

Under the functional Gaussian model (1.36) the same holds for testing the null-hypothesis $H_0 : s = 0$ against the alternative $H_1 : \|s\|_p \geq r$, $s \in L_2(0,1)$, $1 \leq p \leq \infty$ where $\|\cdot\|_p$ is an L_p-norm in the space of functions on $[0,1]$.

The same holds for an i.i.d. sample model on the $[0,1]$. Let $\mathcal{F}$ be the set of probability densities on the interval $[0,1]$ with respect to Lebesgue measure and let $f_0(t) \equiv 1$, $t \in [0,1]$. Using i.i.d. observations $X_1, \ldots, X_N$ one tests the null-hypothesis $H_0 : f = f_0$ against the alternative of the form $H_1 : f \in \mathcal{F}_{N,p}$ for the sets of densities of the form

$$\mathcal{F}_{N,p}(r_n) = \{f : \|f - f_0\|_p \geq r\}. \tag{1.38}$$

Then $\beta(\alpha, \mathcal{F}_{N,p}) \equiv 1 - \alpha$ for all $\alpha \in (0,1)$, $N \geq 1$, and $r \in (0,1)$.

One can find the proofs of these results in [52].

1.4.3 Two-Sided Constraint Problem

The facts noted above are not surprising. Analogous effects arise in nonparametric minimax estimation problems (see [36]). A way out of this situation is to add some additional constraints. Typically, these are of the form of some regularity conditions and could be formulated as constraints on derivatives of an unknown functional parameter θ (compare with (1.33)). Often these constraints could be presented in terms of some norm $|\theta|_{(2)} \leq R$, i.e., jointly with the constraint (1.12), we consider the alternative sets under two-sided constraints

$$|\theta - \theta_0|_{(1)} \leq r, \quad |\theta - \theta_0|_{(2)} \geq R; \tag{1.39}$$

under an asymptotic approach the radii $r = r_\varepsilon$, $R = R_\varepsilon$ could depend on an asymptotic parameter.

Let us consider an example of a two-sided constraint problem. Given $\eta > 0$, set

$$V_\eta(\rho_\varepsilon, R_\varepsilon) = \left\{ v \in l^2 : \sum_{i=1}^{\infty} v_i^2 \geq \rho_\varepsilon^2, \sum_{i=1}^{\infty} i^{2\eta} v_i^2 \leq R_\varepsilon^2 \right\}, \tag{1.40}$$

where

$$R_\varepsilon = R/\varepsilon, \quad \rho_\varepsilon = r_\varepsilon/\varepsilon. \tag{1.41}$$

This corresponds to the observation of the vector $\theta \in l^2$ in the Gaussian noise of level ε under model (1.35) and constraints $\theta \in \Theta_\eta(r_\varepsilon, R)$,

$$\Theta_\eta(r_\varepsilon, R) = \left\{ \theta \in l^2 : \sum_{i=1}^{\infty} \theta_i^2 \geq r_\varepsilon^2, \sum_{i=1}^{\infty} i^{2\eta} \theta_i^2 \leq R^2 \right\}. \tag{1.42}$$

The second constraint in (1.42) determines the ellipsoid $\Theta_\eta(R)$ of the decreasing semiaxes $d_i = Ri^{-\eta}$:

$$\Theta_\eta(R) = \left\{ \theta \in l^2 : \sum_i (\theta_i/d_i)^2 \leq 1 \right\}. \tag{1.43}$$

Assume R is fixed. Let us consider the alternative $H_1 : v \in V_\eta(\rho_\varepsilon, R_\varepsilon)$, i.e., $H_1 : \theta \in \Theta_\eta(r_\varepsilon, R)$. We are interested in small ε.

The question is: *How small could r be to distinguish between the null-hypothesis and the alternative?*

Let us try to give an asymptotic answer assuming $r = r_\varepsilon \to 0$, $\varepsilon \to 0$, i.e., $R_\varepsilon = R/\varepsilon \to \infty$.

Let us obtain the lower bound for a minimax quality of testing in this problem. Taking into account relations (1.13) and (1.14), let us try to embed spheres $S^{n-1}(\rho_\varepsilon)$ into the set $V_\eta(\rho_\varepsilon, R_\varepsilon)$ in order to minimize the function $u_\varepsilon(\rho_\varepsilon, n) = \rho_\varepsilon^2/\sqrt{2n}$. Set

$$n(r_\varepsilon) = \max\{n : d_n \geq r_\varepsilon\} \sim (r_\varepsilon/R)^{-1/\eta}, \quad \text{as } r_\varepsilon \to 0.$$

Since the set $V_\eta(\rho_\varepsilon, R_\varepsilon)$ contains the n-dimensional spheres of the radii d_n/ε, we get

$$u_\varepsilon = \min u_\varepsilon(\varepsilon^{-1} r_\varepsilon, n) = \varepsilon^{-2} r_\varepsilon^2 / \sqrt{2n(r_\varepsilon)} \sim 2^{-1/2} \varepsilon^{-2} r_\varepsilon^{2+1/2\eta} R^{-1/2\eta}.$$

This leads to the lower bounds

$$\beta_\varepsilon(\alpha) = \beta(\alpha, V_\eta(\rho_\varepsilon, R_\varepsilon)) \geq \Phi(T_\alpha - u_\varepsilon) + o(1). \tag{1.44}$$

On the other hand, let us try to use $\chi^2_{n,\alpha}$-tests based on the projection $X_n = Pr_n(X) = (X_1, \ldots, X_n)$ of the observed sequence X to n-dimensional subspaces. Set $\rho_n^2(v) = \sum_{i=1}^n v_i^2$. Taking into account (1.7), (1.14) we have

$$\beta(\chi^2_{\alpha,n}, v) = \Phi(T_\alpha - \rho_n^2(v)/\sqrt{2n}) + o(1).$$

For $v \in V_\eta(\rho_\varepsilon, R_\varepsilon)$ we have the relations

$$\rho_n^2(v) = |v|^2 - \sum_{i=n+1}^{\infty} v_i^2, \quad |v|^2 \ge \rho_\varepsilon^2,$$

$$\sum_{i=n+1}^{\infty} v_i^2 \le (d_{n+1}/\varepsilon)^2 \sum_{i=1}^{\infty} (\varepsilon v_i/d_i)^2 \le (d_{n+1}/\varepsilon)^2. \tag{1.45}$$

Thus we have $\rho_n^2(v) \ge \rho_\varepsilon^2 - (d_{n+1}/\varepsilon)^2$.

To select the best dimensions $n = n_\varepsilon$ for the projection, let us try to maximize in n the function $h_\varepsilon(r_\varepsilon, n) = \varepsilon^{-2}(r_\varepsilon^2 - d_{n+1}^2)/\sqrt{2n}$. Assuming $n \to \infty$, we get

$$h_\varepsilon = \max_n h_\varepsilon(r_\varepsilon, n) \sim 2^{-1/2}\varepsilon^{-2} r_\varepsilon^{2+1/2\eta} R^{-1/2\eta} c(\eta),$$

where $c(\eta) = 4\eta(1+4\eta)^{-(1+1/4\eta)} < 1$, and optimal $n = n_\varepsilon$ are of the form

$$n_\varepsilon \sim c_1(\eta)(R/r_\varepsilon)^{1/\eta}, \quad c_1(\eta) = (1+4\eta)^{1/2\eta}.$$

Therefore we get the upper bounds

$$\beta(\alpha, V_\eta(\rho_\varepsilon, R_\varepsilon)) \le \beta(\chi^2_{\alpha,n_\varepsilon}, V_\eta(\rho_\varepsilon, R_\varepsilon)) \le \Phi(T_\alpha - c(\eta)u_\varepsilon) + o(1). \tag{1.46}$$

Certainly we have a difference between the lower bounds (1.44) and the upper bounds (1.46). To obtain sharp asymptotics and to construct asymptotically minimax tests we need more powerful tools. We do this in Section 4.3.3.

However, using (1.44) and (1.46), we can obtain the rates of testing in the problem. Set

$$r_\varepsilon^* = \varepsilon^{4\eta/(4\eta+1)}. \tag{1.47}$$

Then, analogously to (1.15) and (1.16), we get

$$\beta(\alpha, V_\eta(\rho_\varepsilon, R_\varepsilon)) \to 0 \quad \text{iff} \quad r_\varepsilon/r_\varepsilon^* \to \infty; \tag{1.48}$$

$$\beta(\alpha, V_\eta(\rho_\varepsilon, R_\varepsilon)) \to 1-\alpha \quad \text{iff} \quad r_\varepsilon/r_\varepsilon^* \to 0. \tag{1.49}$$

The rates r_ε^* in (1.47) are essentially different from the classical rates $r_\varepsilon^* = \varepsilon$ which correspond to the simple or finite-dimensional alternative for the noise level $\varepsilon \to 0$ and they depend on the parameter $\eta > 0$. This is not surprising because analogous phenomena arise in minimax estimation problems.

1.4.4 Comparing with the Minimax Estimation Problem

Recall the minimax estimation problem. Let us keep in mind model (1.35). Using an observation Y one would like to construct an estimator $\theta_\varepsilon = \theta_\varepsilon(Y)$ of an unknown $\theta \in \Theta$. To characterize the quality of an estimator one considers some norm $|\cdot|_{(1)}$ and loss function $l(\theta, \theta_\varepsilon) = F(|\theta - \theta_\varepsilon|_{(1)})$. Here

$L(t)$ is a nondecreasing function in $t \geq 0$, $L(0) = 0$. One would like to minimize the maximal risk:

$$R_\varepsilon(\theta_\varepsilon, \Theta) = \sup_{\theta \in \Theta} E_{\varepsilon,\theta} l(\theta, \theta_\varepsilon), \quad R_\varepsilon(\Theta) = \inf_{\theta_\varepsilon} R_\varepsilon(\theta_\varepsilon, \Theta),$$

where the infimum is taken over all possible estimators. Often it is very difficult to construct a minimax estimator that provides the infimum $R_\varepsilon(\Theta)$. For this reason, assuming ε is small, one studies the asymptotics of the minimax risk $R_\varepsilon(\Theta)$ as $\varepsilon \to 0$. The estimators θ_ε are called *asymptotically minimax* if $R_\varepsilon(\theta_\varepsilon, \Theta) \sim R_\varepsilon(\Theta)$. The sharp asymptotic problem in minimax estimation is to construct asymptotically minimax estimators and to study the sharp asymptotics of the minimax risks. The rate asymptotics problem is often studied: to construct estimators θ_ε such that $R_\varepsilon(\theta_\varepsilon, \Theta) \asymp R_\varepsilon(\Theta)$ and to study the rates of the minimax risks. Typically $R_\varepsilon(\Theta) \asymp L(r_\varepsilon^{\text{est}})$ and the quantities $r_\varepsilon^{\text{est}}$ are called *rates of estimation.*

Usually it is impossible to construct minimax consistent estimators without some constraints on the parameter set. The set Θ often is determined by a constraint in some other norm $|\cdot|_{(2)}$, i.e., (compare with (1.43))

$$\Theta = \{\theta : |\theta|_{(2)} \leq R\}.$$

Let us consider the square losses $l(\theta, \theta_\varepsilon) = |\theta - \theta_\varepsilon|^2$. For the sets $\Theta_\eta(R)$ determined by the constraint (1.43), the rates in this problem have been calculated by Ibragimov and Khasminskii [36], [38]. It was shown that

$$r_\varepsilon^{\text{est}} \asymp \varepsilon^{2\eta/(2\eta+1)}. \tag{1.50}$$

These rates are provided by the projection estimators $\theta_\varepsilon = \theta_{n_\varepsilon}(Y) = Pr_{n_\varepsilon} Y$. The optimal dimensions n_ε are of the rates $n_\varepsilon \asymp \varepsilon^{-2/(2\eta+1)}$.

The construction of lower bounds for the rates (1.50) are based on methods of information theory (see [69], for example). However, the upper bounds are based on simple calculations. Let us consider the squared errors of the estimators θ_n,

$$E_\theta |\theta_n - \theta|^2 = \varepsilon^2 n + b_n^2(\theta), \quad b_n^2(\theta) = \sum_{i=n+1}^{\infty} \theta_i^2;$$

where the first item corresponds to the random error and the second item is the squared bias. Analogously to (1.45) the squared bias $b_n^2(\theta)$ is controlled by $d_{n+1}^2 \asymp n^{-2\eta}$ uniformly over $\theta \in \Theta_\eta(R)$, and the optimal n_ε is determined by the standard balance relation

$$\varepsilon^2 n_\varepsilon \asymp n_\varepsilon^{-2\eta}. \tag{1.51}$$

It seems surprising that the rates in minimax estimation and in hypothesis testing are essentially different:

$$r_\varepsilon^* = o(r_\varepsilon^{\text{est}}). \tag{1.52}$$

This means that *we can detect a vector θ of a length that is essentially smaller than the possible accuracy of estimation of θ*. Moreover, *the "plug-in" test based on the optimal estimators θ_n provide essential losses in rates of testing*. This is not typical for parametric problems. Below we show that it is typical for nonparametric cases.

The analytical reason is that we have different relations for the selection of the optimal dimensions for the projection in the problems. Namely, let us return to the calculations above. We can see that the optimal dimensions and the rates of testing satisfy the asymptotic relations

$$r_\varepsilon^* \asymp d_{n_\varepsilon}, \quad \varepsilon^2 n^{1/2} \asymp d_{n_\varepsilon}^2,$$

while the balance equation for the optimal dimensions in the estimation leads to the relations

$$r_\varepsilon^{\text{est}} \asymp d_{n_\varepsilon}, \quad \varepsilon^2 n \asymp d_{n_\varepsilon}^2.$$

The difference between the exponents of n_ε in the second relations determines the distinction in the results related to the estimation and to the hypothesis testing.

On the other hand, it is not surprising from a general point of view. Note that an estimator θ_ε contains essentially larger information on an unknown parameter θ than a test ψ_ε. Therefore we need to pay for this in loss of accuracy.

The minimax estimation problem has been studied intensively during the last two or three decades. The main model for the study was the functional Gaussian noise model (1.36). First, the results are obtained for the sequence Gaussian model and, then, these are translated to the functional Gaussian noise model.

In particular, let $\{\varphi_i\}$ be the standard Fourier basis in $L_2(0,1)$. Then the constraints (1.42) determine the set which is closely related to the set $S_\eta(r,R) \subset L_2(0,1)$ which consists of 1-periodical functions on R such that

$$\|s\|_2 \geq r, \quad \|s\|_{\eta,2} \leq R,$$

where $\|\cdot\|_2$ is the norm in $L_2(0,1)$ and $\|s\|_{\eta,2}$ is the L_2-norm of an η-derivative of the function s.

There are many other norms of interest in the functional space. Often the losses are measured in the L_p-norm, i.e., $|s|_{(1)} = \|s\|_p$, and the constraints are determined by the Sobolev norm $|s|_{(2)} = \|s\|_{\eta,q}$ which corresponds to the L_q-norm of η-derivative of the function s. Sometimes it is of interest to estimate not a function s but its derivatives up to order σ in the L_p-norm. This corresponds to the losses $|s|_{(1)} = \|s\|_{p,\sigma}$.

It is more convenient to use Besov norms:

$$|\cdot|_{(1)} = \|\cdot\|_{\sigma,p,h}, \quad |\cdot|_{(2)} = \|\cdot\|_{\eta,q,t}; \quad p,q,h,t \geq 1, \quad \sigma \geq 0, \quad \eta > 0.$$

Besov norms are analogous to Sobolev norms. They have the additional parameter h to measure the smoothness. Exact definitions and embedding relations between Sobolev and Besov norms will be given in Section 2.7.2. The main reason for using Besov norms is that the norms $\|\cdot\|_{\sigma,p,h}$ in functional space are closely related to Besov norms $\|\cdot\|_{r,p,h}$ in the sequence space. These norms are considered in the sequence space l^2 which consists of sequences of pyramidal structure $\theta = \{\theta_{ij}\}$, $j \geq 1$, $1 \leq i \leq 2^j$. A slightly simplified form of this norm is

$$|v|_{r,p,h} = \left(\sum_{j=1}^{\infty} 2^{jrh} \left(\sum_{i=1}^{2^j} |v_{ij}|^p \right)^{h/p} \right)^{1/h}. \tag{1.53}$$

The relationship between Besov norms in the functional space and in the sequence space is provided by a wavelet transform (see Section 2.7.3).

It is also convenient to consider the power norms

$$|v|_{r,p} = \left(\sum_{i=1}^{\infty} |v_i i^r|^p \right)^{1/p},$$

which are a natural extension of the norms in the constraint (1.42).

1.5 Problems of the Study and Main Results

Bearing in mind the analogy with the estimation problem and the reasons noted above, the main problem we study in this book is to test the null-hypothesis $H_0 : v = 0$ in the sequence space or H_0, $s = 0$ in the functional space against alternatives $H_1 : v \in V_\varepsilon$ or $H_1 : s \in S_\varepsilon$ for the sets which are determined by the two-sided constraints

$$\begin{aligned} V_\varepsilon &= V(\kappa, R_\varepsilon, \rho_\varepsilon) = \left\{ v \in l^2 : |v|_{(1)} \geq \rho_\varepsilon, \ |v|_{(2)} \leq R_\varepsilon \right\}, & (1.54) \\ S_\varepsilon &= S(\kappa, R, r_\varepsilon) = \left\{ s \in L_2(0,1) : \|s\|_{(1)} \geq r_\varepsilon, \ \|s\|_{(2)} \leq R \right\}. & (1.55) \end{aligned}$$

Here we denote by κ the parameters which determined the norms in the constraints. Recall that these constraints are usually needed for the minimax hypothesis testing problem to be nontrivial. The first one is required to remove "too small" alternatives. The second one is needed so that an alternative is not "too wide". The main norms under consideration are the power and Besov norms in the sequence space and Besov and Sobolev norms in the functional space. Our main aims are to study:

- how small can $\rho_\varepsilon = \rho(R_\varepsilon)$ be as $R_\varepsilon \to \infty$ or r_ε as $\varepsilon \to 0$ to obtain a good quality of minimax testing, i.e., $\gamma_\varepsilon \to 0$;
- how this depends on the norms which determine the problem; and
- what are the test procedures that provide minimax consistent, if that is possible, or asymptotically minimax testing.

Mainly, we are dealing with the problems in the sequence space (1.54) under model (1.34) and then we translate the result to the functional space (1.55). The translation is provided by the renormalization (1.41) and the relation

$$r = \sigma + 1/2 - 1/p, \quad s = \eta + 1/2 - 1/q, \tag{1.56}$$

at least for $1 \le p, q, h, t \le \infty$, $\sigma \ge 0$, $\eta > 0$.

Remark. Note that somewhat different criteria of minimax efficiency are considered in [55]. For the case of alternative in the sequence space, these criteria correspond to the replacing of constraints (1.54) by the loss functions $l_\varepsilon(v)$ which characterize the losses for the case when a statistician accepts the null-hypothesis whenever the alternative $v \in l^2$ is true. We want to minimize the maximal mean losses,

$$\inf_{\psi \in \Psi_\alpha} \sup_{v \in l^2} E_v(l_\varepsilon(v)(1 - \psi(x))) = \inf_{\psi \in \Psi_\alpha} \sup_{v \in l^2} \beta(\psi, v) l_\varepsilon(v).$$

In particular, for the power norms we can put

$$l_\varepsilon(v) = r(|v|_{r,p}^p / \rho_\varepsilon^p, |v|_{s,q}^q / R_\varepsilon^p),$$

where $r(x, y)$ is a nonnegative function on the set $R_+^2 = \{x \ge 0, y \ge 0\}$. The case (1.54) corresponds to the indicator $r(x, y) = \mathbf{1}_{\{(x,y) \in R_+^2 \,:\, x \ge 1, y \le 1\}}$.

It was shown in [55] that the main results presented in this book are extended to the case when $\log r(x, y)$ is a concave function.

There are two main reasons that we consider a wide class of norms $|v|_{(1)}$ or $|s|_{(1)}$ which determine the first constraint. The first one is the analogy with the estimation problem: as was noted above, the norm $|s|_{(1)} = \|s\|_{p,\sigma}$ corresponds to the estimation of a σ-derivative of an unknown function.

The second reason is the following. Return to problems (1.25) and (1.28) under the sample model. The difference between these problems is the type of norms $|f - f_0|_{(1)}$ which are used to define the removing neighborhood of the null-hypothesis, however, this leads to essentially different results in the problems: classical distinguishability in problem (1.25) and nondistinguishability in problem (1.28). Note that one can consider the difference between distribution functions as "derivative of order -1" of the difference between densities. Interestingly: *What is "the frontier" between these type of results; in particular, what is the "maximal order of derivative" which provides classical distinguishability?* The problems in question are connected with negative σ for the functional space or negative r for the sequence space. We will study the problems in the sequence space.

Our methods are based on the construction of asymptotically least favorable priors for the problems in the sequence space under model (1.34). We obtain sharp asymptotics of minimax error probabilities $\beta(\alpha, V_\varepsilon)$ and construct asymptotically minimax tests. In the functional space we obtain the rates of type (1.48) and construct minimax consistent tests $\psi_{\varepsilon,\alpha}$ such that $\beta_\varepsilon(\psi_{\varepsilon,\alpha}, S_\varepsilon) \to 0$, if that is possible.

There are various types of asymptotics in the problem. These depend mainly on the parameters $\kappa = (r, s, p, q)$ which determine the norms of interest. We present graphically in Section 6.4 the partition of the planes of parameters $\{r, s\}$ to the regions of the asymptotics of different types for fixed values p, q. In particular, if r is negative and small enough, i.e.,

$$r < r_p^* = \begin{cases} 1/4 - 1/p, & p \leq 2, \\ -1/2p, & p > 2, \end{cases}$$

then the second constraint is not required and we have the *classical rates* $r_\varepsilon^* = \varepsilon$. Under the functional Gaussian noise model this rate is analogous to what we have for testing of the hypothesis on a distribution function under an i.i.d. sample model: if we have a sample size $N \to \infty$ and consider alternatives, such that its distribution functions are bounded away from the null-hypothesis at the rate r_N in any L_p-norm, then the rate of testing is $N^{-1/2}$ (see Section 1.4.1 above and Section 2.6.2 below).

Sometimes the second constraint is not sufficient to obtain the nontrivial problem. We call this type *trivial*. In particular, these hold for $s \leq r$, $r \geq 0$, $p \geq q$.

The most interesting types seem to be *Gaussian and degenerate types of asymptotics*. For Gaussian asymptotics, asymptotically least favorable priors are of the product type

$$\pi^\varepsilon = \prod_i \pi_{\varepsilon,i} \tag{1.57}$$

and the Bayesian log-likelihood ratios are asymptotically Gaussian. The factors $\pi_{\varepsilon,i} = \pi(h_{\varepsilon,i}, z_{\varepsilon,i})$ are symmetric three-point measures on the real line supported on the points 0 and $\pm z_{\varepsilon,i}$:

$$\pi(h, z) = (1 - h)\delta_0 + h(\delta_z + \delta_{-z})/2. \tag{1.58}$$

The sequences $\{h_{\varepsilon,i} \in [0, 1],\ z_{\varepsilon,i} \geq 0\}$ are the solutions of specific extreme problems and depend essentially on the parameters κ, ρ_ε, R_ε. For the Gaussian asymptotics we have the relations analogous to (1.1),

$$\beta(\alpha, V_\varepsilon) = \Phi(T_\alpha - u_\varepsilon) + o(1),$$

where the quantities u_ε are determined by the sequences $\{h_{\varepsilon,i},\ z_{\varepsilon,i}\}$

$$u_\varepsilon^2 = 2\sum_i h_{\varepsilon,i}^2 \sinh^2(z_{\varepsilon,i}^2/2).$$

Asymptotically minimax tests are based on the statistics

$$t_\varepsilon = t_\varepsilon(X) = u_\varepsilon^{-1} \sum_i h_{\varepsilon,i}\xi(X_i, z_{\varepsilon,i}), \quad \xi(t, z) = e^{-z^2/2}\cosh(tz) - 1. \tag{1.59}$$

Analogously to (1.52), the rates of testing in the functional space r_ε^* are smaller than the rates of estimation $r_\varepsilon^{\text{est}}$ for parameters κ which correspond to the Gaussian asymptotics.

Degenerate asymptotics These asymptotics arise for $p = \infty,\ s > r \geq 0$ and for $\infty > p > q,\ s > r > 0,\ sq \leq rp$. Asymptotically least favorable priors supported on specific orthogonal collections $\{v_{\varepsilon,1}, \dots, v_{\varepsilon,n}\} \subset V_\varepsilon,\ n = n_\varepsilon \to \infty$, and the Bayesian likelihood ratios are asymptotically constant. On the other hand, for this case we have the same rates of testing as in the minimax estimation problem, i.e., $r_\varepsilon^* \asymp r_\varepsilon^{\text{est}}$. Moreover, asymptotically minimax tests are based on simple thresholding and do not depend on parameters κ which correspond to degenerate asymptotics. Recall that for the Gaussian asymptotics the families $h_{\varepsilon,i} \in (0,1],\ z_{\varepsilon,i} \geq 0$, depend essentially on parameters $\kappa = (p,q,r,s)$ which determine norms in constraints and the dependence is not simple. Exact knowledge of these parameters is not usually available a statistician.

This situation is analogous to what it is in the estimation theory where the structure of rate-optimal estimators depends essentially on parameters which determine losses and constraints. This has stimulated the study of adaptive problems in estimation: to construct adaptive estimators which provide the best or near best rates for wide enough sets of parameters $\kappa = (\sigma, \eta, p, q)$.

It was shown by Lepski [73], [74], [75] that adaptation is often impossible without some losses in minimax rates (the losses are of logarithmic type). Later, adaptive nonparametric estimation problems were studied by Donoho, Johnstone Kerkyacharian, and Picard [18], Lepski, Mammen, and Spokoiny [77], Lepski and Spokoiny [81], Juditsky [64], Tsybakov [97], among others.

The first, adaptive setting in hypothesis testing was considered by Spokoiny [90], [91] for Besov or Sobolev norms with the L_p-balls removed, $p = q$. It was shown that it is not possible to distinguish between hypothesis and alternative without losses in efficiency (type of $\log\log$-factor). The lower bounds for $p = 2$ and upper bounds for fixed $p \geq 1$ had been obtained in these papers.

We study the adaptive problems in hypothesis testing as well. We construct adaptive tests which provide good minimax properties for wide enough regions of unknown parameters. For power norms, formal setting corresponds to alternatives of the form (we assume a family $R_\varepsilon \to \infty$ is given)

$$V_\varepsilon(K) = \bigcup_{\kappa \in K} V(\kappa, \rho_\varepsilon(\kappa), R_\varepsilon),$$

where $K \subset R^4$ is a compact which consists of parameters $\kappa = (r,s,p,q)$ corresponding to the Gaussian asymptotics for the power norms and $\rho_\varepsilon(\kappa)$ is a family of radii of removing balls. This means we would like to obtain good distinguishability *for all* $\kappa \in K$. Certainly we need to assume $\rho_\varepsilon(\kappa)$

to be such that the quantities $u_\varepsilon(K) = \inf_{\kappa\in K} u_\varepsilon(\kappa, \rho_\varepsilon(\kappa), R_\varepsilon)$ are large enough.

Under small assumptions we obtain the relation analogous to what we have for degenerate asymptotics:

$$\beta(V_\varepsilon(K)) = (1-\alpha)\Phi(\sqrt{2\log\log R_\varepsilon} - u_\varepsilon(K)) + o(1)$$

which leads to small losses in the rates (the losses are of log log-type).

We construct asymptotically minimax adaptive tests families. The construction is based on the "Bonferroni method": we combine collections of tests based on the statistics $l_{\varepsilon,k} = l_{\bar{h}_{\varepsilon,k},\bar{z}_{\varepsilon,k}}$, $k = 1,\dots,M = M_\varepsilon \to \infty$, of form (1.59) for specially selected sequences $\bar{h}_{\varepsilon,k}, \bar{z}_{\varepsilon,k}$ and increase thresholds up to $T_\varepsilon = \sqrt{2\log M_\varepsilon}$.

We obtain rate adaptive asymptotics for Besov norms in the sequence space and for Sobolev and Besov norms in the functional space as well.

Note that the adaptive problem for the sample model and alternatives of form (1.33) with unknown η have been studied in [57]. Analogously to results for the Gaussian model, it was shown in [57] that one can construct rate optimal adaptive tests in the problem by a combination of Pearson's χ^2_n-tests with numbers of cells $n = 2^j$, $j = j_0, j_0+1,\dots,j_n$ where

$$j_0 \sim B_1 \log\log N, \quad j_n \sim 2\log_2 N - B_2 \log\log\log N,$$

where B_1, B_2 are positive constants.

1.6 Methods of the Study

Let us try to explain heuristically the key ideas for the study of Gaussian asymptotics.

For the product priors (1.57) the log-likelihood ratio is the sum

$$l_{\pi^\varepsilon}(X) = \log\frac{dP_{\pi^\varepsilon}}{dP_0}(X) = \sum_i \log(1 + g(X_i, \pi_{\varepsilon,i})), \tag{1.60}$$

where, for probability measure on R_1, we set

$$g(x,\pi) = \int\left(\frac{dP_u}{dP_0}(x) - 1\right)\pi(du) = \int\left(e^{-u^2/2+xu} - 1\right)\pi(du).$$

It seems natural that the log-likelihood ratio is asymptotically Gaussian.

To calculate the parameters of a limit Gaussian distribution under the null-hypothesis let us formally write the relation

$$l_{\pi^\varepsilon}(X) \approx \sum_i g(X_i, \pi_{\varepsilon,i}) - \frac{1}{2}\sum_i g^2(X_i, \pi_{\varepsilon,i}). \tag{1.61}$$

For the standard Gaussian variable ξ and any $u, v \in R^1$ we have

$$E(e^{-u^2/2+\xi u} - 1) = 0, \quad E(e^{-u^2/2+\xi u} - 1)(e^{-v^2/2+\xi v} - 1) = e^{uv} - 1.$$

Therefore under the null-hypothesis the items $g(X_i, \pi_{\varepsilon,i})$ have zero means and these variances are of the form

$$E_{P_0} g^2(X_i, \pi_{\varepsilon,i}) = \int_{R^1}\int_{R^1} (e^{uv} - 1)\pi_{\varepsilon,i}(du)\pi_{\varepsilon,i}(dv).$$

Let us suppose that the first term in (1.61) is asymptotically Gaussian and change the second term to its expectation. This yields that the statistics l_{π^ε} are asymptotically Gaussian $N(-\|\bar{\pi}_\varepsilon\|^2/2, \|\bar{\pi}_\varepsilon\|^2)$. Here we denote by $\bar{\pi}_\varepsilon$ the sequence which consists of the factors in (1.57) and

$$\|\bar{\pi}_\varepsilon\|^2 = \sum_i \|\pi_{\varepsilon,i}\|^2 = \sum_i \int_{R^1}\int_{R^1} (e^{uv} - 1)\pi_{\varepsilon,i}(du)\pi_{\varepsilon,i}(dv). \qquad (1.62)$$

Thus the likelihood ratio is asymptotically the same as for the testing of the simple null-hypothesis $H_0 : v = 0$ against the alternative $H_1 : v = v_1 = \|\bar{\pi}_\varepsilon\|$ under a one-dimensional Gaussian model. These lead to the asymptotics (compare with Section 1.2)

$$\beta(\alpha, \pi^\varepsilon) = \Phi(T_\alpha - \|\bar{\pi}_\varepsilon\|) + o(1). \qquad (1.63)$$

Therefore to obtain the best lower bounds in the minimax problem we need to minimize the functional $\|\bar{\pi}_\varepsilon\|$ over the sequences $\bar{\pi}_\varepsilon$ such that $\pi^\varepsilon(V_\varepsilon) = 1$.

We consider the functional $\|\pi\|$ as a Hilbert norm on the space which consists of sequences of signed measures on the real line. This corresponds to the inner product

$$(r, \pi) = \int_{R^1}\int_{R^1} (e^{uv} - 1) r(du)\pi(dv).$$

Certainly we need to identify the measures π and $\tilde{\pi}$ such that $\pi - \tilde{\pi} = a\delta_0$.

To obtain the upper bounds let us consider tests $\psi_{\varepsilon,\alpha} = \mathbf{1}_{\{t_\varepsilon > T_\alpha\}}$ based on the statistics

$$t_\varepsilon = \|\bar{\pi}_\varepsilon\|^{-1} \sum_i g(X_i, \pi_{\varepsilon,i})$$

(jointly with a thresholding). Since $E_{P_0}(t_\varepsilon) = 0$, $E_{P_0}(t_\varepsilon^2) = 1$, assuming asymptotic normality of the statistics t_ε, we get $\alpha(\psi_{\varepsilon,\alpha}) = \alpha + o(1)$. On the other hand, let us calculate the expectation of the statistics t_ε under an alternative $v \in l^2$. For the standard Gaussian random variable ξ and any u, $v \in R^1$ we have the equality

$$E(e^{-u^2/2 + (\xi + v)u} - 1) = e^{uv} - 1.$$

This leads to the equality which we could present in terms of the inner product above:

$$E_v t_\varepsilon = \|\bar{\pi}_\varepsilon\|^{-1} \sum_i \int_{R^1} (e^{uv_i} - 1)\pi_{\varepsilon,i}(du) = \|\bar{\pi}_\varepsilon\|^{-1}(\bar{\pi}_\varepsilon, \bar{\delta}_v).$$

Here $\bar{\delta}_v$ is the sequence which consists of the masses δ_{v_i} in the coordinates of the point v.

Let us forget that the variance $\mathrm{Var}_v(t_\varepsilon)$ depends on v and suppose asymptotic normality of the statistics t_ε under a family of alternatives $v = v_\varepsilon \in l^2$. This leads to the relations

$$\beta(\psi_{\varepsilon,\alpha}, v) = P_v(t_\varepsilon \le T_\alpha) = \Phi(T_\alpha - h(\bar{r}_\varepsilon, v)) + o(1),$$

where $h(\bar{r}_\varepsilon, v) = E_v t_\varepsilon$ could be presented in the form

$$h(\bar{r}_\varepsilon, v) = (\bar{r}_\varepsilon, \bar{\delta}_v), \quad \bar{r}_\varepsilon = \bar{\pi}_\varepsilon / \|\bar{\pi}_\varepsilon\|.$$

Therefore

$$\beta(\psi_{\varepsilon,\alpha}, V_\varepsilon) = \Phi(T_\alpha - h(\bar{r}_\varepsilon, V_\varepsilon)) + o(1),$$

where

$$h(\bar{r}, V_\varepsilon) = \inf_{v \in V_\varepsilon} h(\bar{r}, \bar{\delta}_v).$$

Let us change the set $\Delta_\varepsilon = \{\bar{\delta}_v,\ v \in V_\varepsilon\}$ to a convex set $\bar{\Pi}_\varepsilon$ such that $\Delta_\varepsilon \subset \bar{\Pi}_\varepsilon$. If we do this, then we obtain the inequality

$$h_\varepsilon(\bar{r}_\varepsilon, V_\varepsilon) \ge h_\varepsilon(\bar{r}_\varepsilon, \bar{\Pi}_\varepsilon) = \inf_{\bar{\pi} \in \bar{\Pi}_\varepsilon} (\bar{r}_\varepsilon, \bar{\pi}). \tag{1.64}$$

Let us consider an extreme problem (compare with (1.10))

$$u_\varepsilon = \inf_{\bar{\pi} \in \bar{\Pi}_\varepsilon} \|\bar{\pi}\|. \tag{1.65}$$

Suppose there exists a minimum point $\bar{\pi}_\varepsilon^*$ in (1.65) such that $u_\varepsilon = \|\bar{\pi}_\varepsilon^*\| > 0$. Let us take $\bar{r}_\varepsilon^* = \bar{\pi}_\varepsilon^* / \|\bar{\pi}_\varepsilon^*\|$. Then we have the equality (compare with (1.11))

$$h_\varepsilon(\bar{r}_\varepsilon^*, \Pi_\varepsilon) = \inf_{\bar{\pi} \in \bar{\Pi}_\varepsilon} (\bar{r}_\varepsilon^*, \bar{\pi}) = u_\varepsilon. \tag{1.66}$$

So, we have the upper bounds

$$\beta(\alpha, V_\varepsilon) \le \beta(\psi_{\varepsilon,\alpha}, V_\varepsilon) \le \Phi(T_\alpha - u_\varepsilon) + o(1).$$

On the other hand, suppose that the product prior π_*^ε for the extreme sequence $\bar{\pi}_\varepsilon^*$ is concentrated on V_ε, i.e., $\pi_*^\varepsilon(V_\varepsilon) = 1$ (it suffices that $\pi_*^\varepsilon(V_\varepsilon) = 1 - o(1)$). Then using (1.63) we obtain the same lower bounds. So, we see that the tests $\psi_{\varepsilon,\alpha}$ are asymptotically minimax.

Let us turn to the problem: *How to choose convex sets* $\bar{\Pi}_\varepsilon$*?*

Certainly we can take convex hull of the set Δ_ε. However, it is not convenient for the study of the extreme problem (1.65). We use the method analogous to the one by Donoho and Johnstone [17] in minimax estimation. For example, under the power norms, the set V_ε is determined by the constraints

$$V_\varepsilon = \left\{ v \in l^2 : \sum_i i^{rp} |v_i|^p \ge \rho_\varepsilon^p,\ \sum_i i^{sq} |v_i|^q \le R_\varepsilon^q \right\}.$$

Let us change here $|v_i|^p$ with $E_{\pi_{\varepsilon,i}}|v_i|^p$ and $|v_i|^q$ with $E_{\pi_{\varepsilon,i}}|v_i|^q$, i.e., we set

$$\bar{\Pi}_\varepsilon = \left\{ \bar{\pi} \in \bar{\Pi} : \sum_i i^{rp} E_{\pi_{\varepsilon,i}}|v|^p \geq \rho_\varepsilon^p,\ \sum_i i^{sq} E_{\pi_{\varepsilon,i}}|v|^q \leq R_\varepsilon^q \right\},$$

where $\bar{\Pi}$ is the set of sequences which consist of probability measures. Since the functionals under constraint are linear in π_i, we obtain a convex set. Certainly, it contains all the sequences $\bar{\delta}_v$, $v \in V_\varepsilon$. We use an analogous method for Besov norms, and if $h \leq p$, $t \geq q$, then we obtain the convex set $\bar{\Pi}_\varepsilon$ as well.

Thus we asymptotically translate the minimax hypothesis testing problem for alternatives of interest to the signal detection problem. However "a signal" is not a real-valued sequence v from the convex set V in l^2, but a measure-valued sequence $\bar{\pi}$ from the convex set $\bar{\Pi}_\varepsilon$ which is contained in a very wide Hilbert space. The extreme sequences in (1.65) determine asymptotically the least favorable priors in the problem.

Certainly the extreme problem (1.65) is difficult enough and the study is not simple. First, we separate the extreme problem to *one-dimensional problems*

$$\phi(\eta, \zeta) = \inf\{\|\pi\|^2 : E_\pi|x|^p \geq \eta^p,\ E_\pi|x|^q \leq \zeta^q\}. \tag{1.67}$$

We show that the minimum in (1.67) is attained in symmetric three-point measures (1.58) which are characterized by two parameters $z \geq 0$ and $h \in [0,1]$. Therefore we can rewrite the extreme problem in terms of sequences $\bar{z} = \{z_i\}$, $\bar{h} = \{h_i\}$. We call these extreme problems *two-sequence*. These problems are studied by using the classical Lagrange method.

1.7 Structure of the Book

The structure of this book is as follows.

In Chapter 2 we discuss general statements of the problem and related topics. Recall that we are dealing with the Gaussian models and with the simple null-hypothesis; we refer the reader to [21], [23], [32], [43], [44], [46], [47], [48], [50], [52], [53], [54], [57] where analogous hypotheses testing problems are studied for other models and for various parametric and nonparametric null-hypotheses.

In Chapter 3 we present first results. We consider minimax properties for various classes of tests which provide upper bounds for the minimax quality of testing. Joined with simple lower bounds, based on asymptotically the minimax properties of χ^2-tests, they give the rate asymptotics for ellipsoids in square norms in terms of the geometrical characteristics of alternatives (such as inner radii and Kolmogorov diameters). However, this is not sufficient to study wider classes of norms. We specialize the Hilbert structure in the space of sequences of measures and study two main types of priors which are used for lower bounds: priors of product type and priors of orthogonal type. Based on these results, we study the frontier between classical asymptotics and triviality for alternatives determined by one-sided constraints for the power and Besov norms in the sequence space. We also describe triviality conditions for alternatives determined by two-sided constraints.

In Chapter 4 we study sharp asymptotics for simple cases. We start with the problem of the detection of a positive signal. Under constraints determined by the power norms with $p \leq 1,\ q \geq p$ or by Besov norms with $h \leq p \leq 1,\ t \geq q \geq p$, these correspond to the convex sets V_ε and lead to convex extreme problems in the space of nonnegative sequences. Analogous extreme problems are obtained for the power norms with $p \leq 2,\ q \geq p$ or Besov norms with $h \leq p \leq 2,\ t \geq q \geq p$ without the constraint that a signal is positive. We study the asymptotics of degenerate type as well.

General Gaussian asymptotics are studied in Chapter 5. Here we realize the scheme noted in Section 1.6. We asymptotically reduce hypothesis testing problems of interest to convex extreme problems in the space of sequences of measures (1.65). We separate the extreme problems to one-dimensional problems and to extreme problems in a two-sequence space. We solve one-dimensional problems and show that the solutions are symmetric three-point measures of the form (1.58) (or symmetric two-point measures with $h = 1$ for $p \leq 2,\ q \geq p$).

In Chapter 6 we study extreme problems in a two-sequence space and obtain asymptotically sharp solutions of the extreme problem for the power and Besov norms. We obtain the structure of asymptotically minimax or minimax consistent tests and the asymptotics of minimax error probability. Based on the study of an extreme problem for a Besov norm, using a wavelet transform and known embedding theorems, we obtain the rate asymptotics and the structure of minimax consistent tests under the Gaussian model in the functional space for alternatives determined by Sobolev or Besov norms. We present a partition of the space of parameters which determine norms of interest to the regions of asymptotics of various types.

In Chapter 7 we study adaptive hypothesis testing for the power and Besov norms. The methods, which have been developed for degenerate asymptotics, are very applicable to adaptive problems.

In Chapter 8 we study some additional phenomena in the problem. This is based on n-dimensional Gaussian model with $n \to \infty$. Of course, it is the special case of the Gaussian model (1.34) in the sequence space. However, it is possible to study the "transient effects" between various types of asymptotics under this model. In particular, we obtain the asymptotics of special infinite-divisible types. We study the alternatives which correspond to an l_q^n-ball with an l_p^n-ball removed (l^n-ball problem) and the alternatives which correspond to a special variant of a multichannel signal detection problem. The reason is that there exist common asymptotically least favorable priors in these problems.

Finally, we consider composite null-hypotheses which correspond to small l_p- or L_p-balls. We show that this leads to new phenomena in the problem. In particular, if p is not even, then we obtain the rates which are different from those for the simple null-hypothesis.

In the Appendix we carried over some technical elements of the proofs.

2
An Overview

In this chapter we consider the statements of hypotheses testing problems and the main known results which are basic for the study below.

2.1 Models

Let us have an observation $X \in \mathcal{X}$ which is generated by a probability measure $P_\theta, \theta \in \Theta$. The parameter θ is unknown, however, the set Θ is given. Certainly we assume that σ-algebra $\mathcal{A}$ on the observation set $\mathcal{X}$ is given and all probability measures P_θ are defined on a measurable space $(\mathcal{X}, \mathcal{A})$. The triple $(\mathcal{X}, \mathcal{A};\ P_\theta, \theta \in \Theta)$ is called the *statistical experiment* and characterizes the *statistical model.*

2.1.1 Gaussian Models in Sequence Space

This is the main object of the study below. We observe a realization of the Gaussian random vector with an unknown mean vector v:

$$X = v + \eta, \quad v \in R^n \quad \text{or} \quad v \in l^2. \tag{2.1}$$

Here η is either the standard Gaussian n-dimensional vector or sequences of the standard Gaussian variables $\{\eta_i,\ i \in I\}$ with an infinite denumerable set I; typically, we assume that $I = \mathbf{N}$ is the set of natural integers, R^n is the Euclidean space, and l^2 is the Hilbert space which consists of sequences $v = \{v_i,\ i \in I\}$ with the norm $|v|^2 = \sum_i v_i^2 < \infty$. This corresponds to the sets of Gaussian measures P_v with mean v and unit covariance matrix (or operator)

$$\mathcal{P}_n^G = \{P_v,\ v \in R^n\} \quad \text{or} \quad \mathcal{P}_\infty^G = \{P_v,\ v \in l^2\}.$$

The measures P_v are defined on an n-dimensional Euclidean space R^n with Borel σ-algebra $\mathcal{B}^n$ or on the space which consists of real-valued sequences $x = \{x_i,\ i \in I\}$, and $\mathcal{B}^\infty$ is a minimal σ-algebra containing inverse images of all σ-algebras $\mathcal{B}^n$ under the natural projections corresponding to the first n coordinates.

A vector v is an unknown parameter θ; we denote the parametric set Θ by V; this could be either R^n or l^2 or some subset in these spaces. Sometimes we call an unknown vector v *a signal* to be detected, because this model is often used in signal detection problems.

All measures $P_{v_0},\ P_{v_1}$ are absolutely continuous and

$$\frac{dP_{v_1}}{dP_{v_0}}(x) = \exp((|v_0|^2 - |v_1|^2)/2 + (x, (v_1 - v_0)));$$

in particular,

$$\frac{dP_v}{dP_0}(x) = \exp(-|v|^2/2 + (x, v)). \tag{2.2}$$

Here and below we set

$$(x, a) = \sum_i x_i a_i, \quad a \in R^n \quad \text{or } a \in l^2.$$

If n is finite, then this is the scalar product in R^n and if $n = \infty$, then the series converges almost surely and in mean square. In any case, (X, a) is the Gaussian random variable with the mean (a, v) and the variance $|a|^2$ with respect to measure P_v.

Model (2.1) can be presented in some different form (see (1.35)). We observe an n-dimensional or infinite-dimensional random vector Y:

$$Y = a + \varepsilon\eta, \quad a \in R^n \quad \text{or} \quad a \in l^2, \tag{2.3}$$

where $\varepsilon > 0$ is the (known) *noise level.* The relationship between models (2.1) and (2.3) is determined by the linear transform

$$X = \varepsilon^{-1} Y, \quad v = \varepsilon^{-1} a. \tag{2.4}$$

2.1.2 Functional Gaussian Model

This mathematical model was introduced and studied by Ibragimov and Khasminskii (see [38]). This model corresponds to an observation of an unknown signal in the white Gaussian noise (see (1.36)): we observe a realization of the random process $X_\varepsilon = X_\varepsilon(t),\ t \in [0, 1]$, satisfying

$$dX_\varepsilon(t) = s(t)\,dt + \varepsilon\,dW(t), \quad s \in L_2(0, 1), \tag{2.5}$$

where $W(t)$ is a realization of the standard Wiener process, $s = s(t)$ is an unknown signal, and $\varepsilon > 0$ is the noise level. We denote by $P_{\varepsilon,s}$ the probability measure which corresponds to observation (2.5). A signal s is parameter θ, and the set Θ is a subset in the functional space $L_2(0,1)$. We denote the L_2-norm in $L_2(0,1)$ by $\|\cdot\|$.

This model is close to the sequence Gaussian model. In fact, fix an orthonormal basis $\{\varphi_i,\ i \in I\}$ in $L_2(0,1)$ and consider the sequences $Y = F_\varphi(X_\varepsilon),\ a = F_\varphi(s)$ of Fourier coefficients

$$Y_i = \int_0^1 \varphi_i(t)\, dX_\varepsilon(t), \quad a_i = \int_0^1 \varphi_i(t)s(t)\, dt. \tag{2.6}$$

These correspond to model (2.3), and the normalized Fourier coefficients $X = F_{\varphi,\varepsilon}(X_\varepsilon),\ v = F_{\varphi,\varepsilon}(s)$,

$$X_i = \varepsilon^{-1}\int_0^1 \varphi_i(t)\, dX_\varepsilon(t), \quad v_i = \varepsilon^{-1}\int_0^1 \varphi_i(t)s(t)\, dt \tag{2.7}$$

correspond to model (2.1) in the following sense: for any measurable function $h(x)$ on the sequence space $(R^\infty, \mathcal{B}^\infty)$ and any $\varepsilon > 0,\ v \in l^2$, one has

$$P_{\varepsilon,s}(h(F_{\varphi,\varepsilon}(X_\varepsilon)) < H) = P_v(h(X) < H), \quad \frac{dP_{\varepsilon,s}}{dP_{\varepsilon,0}}(X_\varepsilon) = \frac{dP_v}{dP_0}(X), \tag{2.8}$$

with $s(t) = \varepsilon\sum_{i\in I} v_i\varphi_i(t)$. Analogous relations hold for any orthonormal system $\{\varphi_i,\ i \in I\}$. In particular, if $\{\varphi_i,\ i = 1,\ldots,n\}$ is a finite orthonormal collection in $L_2(0,1)$, then n-dimensional subspaces generated by finite collections of coefficients (2.7) correspond to the n-dimensional Gaussian models (2.1): we have (2.8) with $s(t) = \varepsilon\sum_{i=1}^n v_i\varphi_i(t)$.

2.2 Hypothesis Testing Problem

2.2.1 Hypothesis, Alternative, Tests

There are two main problems in statistics: estimation and hypothesis testing. In the estimation problem, a statistician would like to construct an estimator $\hat\theta = \hat\theta(X)$ which is close to θ in a certain sense. In the hypothesis testing problem, for a given subset $\Theta_0 \subset \Theta$, a statistician would like to test a hypothesis on an unknown parameter which could be presented in the form $H_0 : \theta \in \Theta_0$. Of course, one should bear in mind an alternative which could be presented in the form $H_1 : \theta \in \Theta_1$, where Θ_1 is an other subset in Θ which is disjoint with Θ_0. In the goodness-of-fit problem it is assumed that $H_1 : \theta \notin \Theta_0$. Formally, by choosing a set Θ, one can assume $\Theta_1 = \Theta\backslash\Theta_0$.

We give the main definitions without any parametrization. Denote by $\mathcal{P}$ the set of probability measures on a given measurable space $(\mathcal{X}, \mathcal{A})$; here $\mathcal{A}$ is a σ-algebra of subsets of $\mathcal{X}$.

Let two nonempty disjoint subsets $\mathcal{P}_0$ and $\mathcal{P}_1$ in $\mathcal{P}$ be given. We call $\mathcal{P}_0$ *hypothesis* (or null-hypothesis) and $\mathcal{P}_1$ *alternative*. The hypothesis (or alternative) is called *simple*, if $\mathcal{P}_0$ (or $\mathcal{P}_1$) consists of one point P_0 (or P_1) in $\mathcal{P}$, in the opposite case this is called *composite*.

Of course, if we have a parametrization (i.e., an injective map $\Theta \to \mathcal{P} : \theta \mapsto P_\theta$), then we identify the sets

$$\mathcal{P}_0 = \{P_\theta : \theta \in \Theta_0\}, \quad \mathcal{P}_1 = \{P_\theta : \theta \in \Theta_1\},$$

with parametric sets Θ_0 and Θ_1.

A statistician would like to construct the decision rule $\psi = \psi(X)$ with the values 0 or 1: one accepts the null-hypothesis if $\psi(X) = 0$, and one rejects it if $\psi(X) = 1$. We call this rule a (*nonrandomized*) *test* .

From a mathematical point of view it is convenient to consider a wider class of tests. We call a measurable function, $\psi : (\mathcal{X}, \mathcal{A}) \to ([0,1], \mathcal{B})$ with values $\psi(X)$ in the interval $[0,1]$, a (randomized) *test*. A randomized test determines the random decision rule: using an observation x one accepts the null-hypothesis (or rejects an alternative) with the probability $\psi(x)$ and accepts an alternative (or rejects a null-hypothesis) with the probability $1 - \psi(x)$. We denote by Ψ the set of all tests ψ. This is a convex subset in the linear space $\mathcal{D}$ which consists of all bounded functions on $(\mathcal{X}, \mathcal{A})$.

There are two types of errors in hypothesis testing. The type I error is the rejection a null-hypothesis whenever it holds. The type II error is the acceptance a null-hypothesis whenever it does not hold. Of course, tests of interest should be such that errors are small in a certain sense.

2.2.2 How to Compare the Tests?

The statistical properties of a test ψ are characterized by probabilities of type I and of type II errors. They are the functions of unknown parameters:

$$\alpha(\psi, P) = E_P\psi, \quad P \in \mathcal{P}_0; \quad \beta(\psi, P) = E_P(1-\psi), \quad P \in \mathcal{P}_1.$$

Often the power function is considered

$$\delta(\psi, P) = E_P\psi = 1 - \beta(\psi, P), \quad P \in \mathcal{P}_1.$$

This is the probability to reject the null-hypothesis, if it does not hold.

The errors functions $\alpha(\psi, P)$, $P \in \mathcal{P}_0$, and $\beta(\psi, P)$, $P \in \mathcal{P}_1$, define *partial ordering* on the set Ψ: we say $\psi_1 \succeq \psi_2$ if $\alpha(\psi_1, P) \leq \alpha(\psi_2, P) \ \forall P \in \mathcal{P}_0$ and $\beta(\psi_1, P) \leq \beta(\psi_2, P) \ \forall P \in \mathcal{P}_1$; we say $\psi_1 \succ \psi_2$ if one of these inequalities is strict at some point P. Of course, we would like to find "the best" test ψ^* such that $\psi^* \succ \psi \ \forall \psi \in \Psi$. However, usually, there does not exist such a test, except the case when all measures $P_0 \in \mathcal{P}_0$ are orthogonal to all measures $P_1 \in \mathcal{P}_1$, i.e., $P_0(A)P_1(A) = 0 \ \forall P_0 \in \mathcal{P}_0, \ \forall P_1 \in \mathcal{P}_1, \ A \in$

$\mathcal{A}$. On the other hand, we would like to consider tests ψ such that there do not exist "better" tests $\tilde{\psi} \succ \psi$. Such tests are called *admissible.* We can consider the set Ψ_{ad} which consists of all admissible tests. However, usually, the set Ψ_{ad} is very wide.

2.2.3 Simple Hypothesis and an Alternative

Given measures P_0, P_1, the functions $\alpha(\psi, P_0)$, $\beta(\psi, P_1)$ are the quantities $\alpha(\psi)$, $\beta(\psi)$ and we can omit the measures in notations. One can consider two closely related approaches to define what is "the best test."

The first one is the following. Fix a positive value t. One characterizes a test by the quantities

$$\gamma_t(\psi) = \gamma_t(\psi; P_0, P_1) = t\alpha(\psi, P_0) + \beta(\psi, P_1).$$

If $t = 1$, then $\gamma(\psi) = \gamma_1(\psi)$ is the sum of the type I and type II error probabilities. Denote

$$\gamma(t) = \gamma(t; P_0, P_1) = \inf \gamma_t(\psi), \tag{2.9}$$

where the infimum is taken over all tests $\psi \in \Psi$. We call the test $\psi^{(t)}$ *optimal* (under the criterion $\gamma(t)$), if this provides the infimum in (2.9).

The second approach is due to Neyman and Pearson. Fix a value $\alpha \in (0, 1)$ which is called the *significance level.* Consider the set of tests

$$\Psi_\alpha = \Psi_\alpha(P_0) = \{\psi \in \Psi : \alpha(\psi) \le \alpha\}$$

and put

$$\beta(\alpha) = \beta(\alpha, P_0, P_1) = \inf_{\psi \in \Psi_\alpha(P_0)} \beta(\psi, P_1). \tag{2.10}$$

We call the test $\psi_\alpha \in \Psi_\alpha$ *optimal* (under the criterion $\beta(\alpha)$), if this provides the infimum in (2.10).

The function $\gamma(t)$ is nondecreasing in t and $\beta(\alpha)$ is nonincreasing in α. Observe the inequalities

$$0 \le \gamma(t) \le \min(1, t), \quad 0 \le \beta(\alpha) \le 1 - \alpha. \tag{2.11}$$

The lower inequalities in (2.11) are evident. One obtains the upper inequalities by taking trivial tests $\psi(x) \equiv c$ with $c = 0,\ 1,\ \alpha$.

The structure of the optimal test is determined by the Neyman–Pearson lemma. Let the measures P_0, P_1 be dominated by a measure P (one can take $P = (P_0 + P_1)/2$) and let

$$f_0(x) = \frac{dP_0}{dP}(x), \qquad f_1(x) = \frac{dP_1}{dP}(x),$$

be the densities. Introduce the sets

$$\begin{aligned} \mathcal{X}_t^- &= \{X \in \mathcal{X} : f_1(X) < t f_0(X)\}, \\ \mathcal{X}_t^+ &= \{X \in \mathcal{X} : f_1(X) > t f_0(X)\}, \end{aligned}$$

$$\mathcal{X}_t^0 = \{X \in \mathcal{X} : f_1(X) = t f_0(X)\}.$$

Then the following results hold:

Lemma 2.1 (Neyman and Pearson)
(1) *The optimal tests* $\psi^{(t)}$ *are of the form*

$$\psi^{(t)}(X) = \begin{cases} 0, & \text{if } X \in \mathcal{X}_t^-, \\ 1, & \text{if } X \in \mathcal{X}_t^+, \end{cases} \tag{2.12}$$

(the values of the test $\psi^{(t)}$ *on the set* $\mathcal{X}_t^0$ *are not essential).*
(2) *The optimal tests* ψ_α *are of the form* (2.12) *where the quantity* $t = t_\alpha$ *and the values of the test* ψ_α *on the set* $\mathcal{X}_t^0$ *satisfy*

$$\alpha(\psi_\alpha) = P_0(\mathcal{X}_{t_\alpha}^+) + \int_{\mathcal{X}_{t_\alpha}^0} \psi_\alpha(x) f_0(x) = \alpha; \tag{2.13}$$

here, and below in the proof, integrals are taken over the measure P.

In the main examples the sets $\mathcal{X}_t^0$ are of probability 0 under hypothesis and alternative, and one can present the optimal tests in nonrandomized form $\psi^{(t)}(X) = \mathbf{1}_{\{L(X)>t\}}$, where $L(x) = f_1(x)/f_0(x)$ is the *likelihood ratio* statistic.

Proof of Lemma 2.1. At first consider tests of the form (2.12) and note that for any $\alpha \in (0,1)$ there exists $t = t_\alpha \geq 0$ such that (2.13) holds: one can put $t_\alpha = \sup\{t : P_0(\mathcal{X}_t^-) < 1-\alpha\}$. Note that the function $F^-(t) = P_0(\mathcal{X}_t^-)$ is continuous on the left and $F^+(t) = P_0(\mathcal{X}_t^+)$ is continuous on the right. These imply

$$P_0(\mathcal{X}_{t_\alpha}^-) = 1 - P_0(\mathcal{X}_{t_\alpha}^+) - P_0(\mathcal{X}_{t_\alpha}^0) \leq 1 - \alpha;$$
$$P_0(\mathcal{X}_t^-) \geq 1 - \alpha; \quad P_0(\mathcal{X}_t^+) \leq 1 - P_0(\mathcal{X}_t^-) \leq \alpha \quad \forall t > t_\alpha.$$

Therefore

$$P_0(\mathcal{X}_{t_\alpha}^+) \leq \alpha, \quad P_0(\mathcal{X}_{t_\alpha}^+) + P_0(\mathcal{X}_{t_\alpha}^0) \geq \alpha.$$

Observe the relation for the tests $\psi^{(t)}, \psi_\alpha$ determined by Lemma 2.1,

$$\gamma_t(\psi^{(t)}) = 1 - \int_{\mathcal{X}_t^+} (f_1(x) - t f_0(x)), \quad t > 0, \tag{2.14}$$

$$\beta(\psi_\alpha) = 1 - \int_{\mathcal{X}_t^+} (f_1(x) - t f_0(x)) - t\alpha, \quad t = t_\alpha. \tag{2.15}$$

Denote $z_+ = \max(z, 0)$, $z \in R^1$. For any $\psi \in \Psi$ using (2.14) one has

$$\begin{aligned} 1 - \gamma_t(\psi) &= \int_{\mathcal{X}} (f_1(x) - t f_0(x))\psi(x) \leq \int_{\mathcal{X}} (f_1(x) - t f_0(x))_+ \\ &= \int_{\mathcal{X}_t^+} (f_1(x) - t f_0(x)) = 1 - \gamma_t(\psi^{(t)}). \end{aligned}$$

Analogously, for any $\psi \in \Psi_\alpha$ and $t = t_\alpha$ using (2.15), one has

$$1 - \beta(\psi) \leq \int_{\mathcal{X}} \psi(x) f_1(x) + t(\alpha - \alpha(\psi)) = \int_{\mathcal{X}} (f_1(x) - t f_0(x))\psi(x) + t\alpha$$
$$\leq \int_{\mathcal{X}} (f_1(x) - t f_0(x))_+ + t\alpha = \int_{\mathcal{X}_t^+} (f_1(x) - t f_0(x)) + t\alpha = 1 - \beta(\psi_\alpha).$$

□

Example 2.1 Under the Gaussian model (2.1) let us test the simple null-hypothesis $H_0 : v = 0$ against the simple alternative $H_1 : v = v^*$ which corresponds to a given "signal" $v^* \in l^2$. The likelihood ratio is of the form (see (2.2)):

$$L(x) = \exp(-|v^*|^2/2 + (x, v^*)).$$

The optimal tests could be based on any increasing function in the likelihood ratio. It is convenient to take the statistic

$$t(X, r) = (X, r), \quad r = v/|v|$$

which is the projection of the vector X to the direction $r = v^*/|v^*|$ of the signal v. This statistic is standard Gaussian $N(0,1)$ under the null-hypothesis and is Gaussian $N(|v|, 1)$ under the alternative v. The statistic $t(X, r)$ and the length of the signal $u = |v^*|$ determine the quality of optimal tests:

$$\psi^{(1)} = \mathbf{1}_{\{(X,r)>u/2\}}, \quad \psi_\alpha = \mathbf{1}_{\{(X,r)>T_\alpha\}}, \tag{2.16}$$
$$\gamma = 2\Phi(-u/2), \quad \beta(\alpha) = \Phi(T_\alpha - u). \tag{2.17}$$

In particular, $\gamma \to 0$ and $\beta(\alpha) \to 0$ for any $\alpha \in (0,1)$, if the length of the signal $|v^*| \to \infty$.

The tests (2.16) depend on the direction of the signal v. If we use the test ψ_α for an alternative $\tilde{v}$, then we obtain type II errors of the form

$$\beta(\psi_{\alpha,r}, \tilde{v}) = \Phi(T_\alpha - (r, \tilde{v})) = \Phi(T_\alpha - \tau|\tilde{v}|), \quad \tau = (v, \tilde{v})/|v||\tilde{v}|; \tag{2.18}$$

here $\tau \leq 1$ characterizes losses with respect to an optimal test for the alternative $\tilde{v}$.

2.2.4 L_1-Distance on the Space of Measures

Let $\mathcal{M}$ be the linear space which consists of signed measure $R_1 - R_2$ defined on $(\mathcal{X}, \mathcal{A})$; here R_1 and R_2 are finite nonnegative measures on $(\mathcal{X}, \mathcal{A})$. For nonnegative measure R we set $|R|_1 = R(\mathcal{X})$. Any signed measure can be presented in the form $R_1 - R_2$ where the nonnegative measures R_1 and R_2 are orthogonal, i.e., there exists a partition $\mathcal{X} = \mathcal{X}_1 \cup \mathcal{X}_2$, $\mathcal{X}_1 \cap \mathcal{X}_2 = \emptyset$ such that

$$R_1(\mathcal{X}_1) = |R_1|_1, \quad R_2(\mathcal{X}_2) = |R_2|_1, \quad R_1(\mathcal{X}_2) = R_2(\mathcal{X}_1) = 0.$$

Using this presentation we set

$$|R_1 - R_2|_1 = |R_1|_1 + R_2|_1.$$

This relation determines the total variation norm on the linear space $\mathcal{M}$.

Let measures P_0, P_1 be probability measures dominated by a measure P. Then

$$|P_1 - P_0|_1 = \int_{\mathcal{X}} |f_1(x) - f_0(x)|P(dx)$$

is the L_1-distance (or total variation distance) between the measures P_1 and P_0. This does not depend on P and could be presented in the form

$$|P_1 - P_0|_1 = \sup_{\tilde{\psi}\in\tilde{\Psi}} \int_{\mathcal{X}} \tilde{\psi}(x)(f_1(x) - f_0(x))P(dx); \tag{2.19}$$

the set $\tilde{\Psi}$ consists of measurable functions $\tilde{\psi}(x)$ such that $\sup_x |\tilde{\psi}(x)| \le 1$. The relation $\tilde{\psi}(x) = 2\psi(x) - 1$ gives one-to-one correspondence between $\tilde{\Psi}$ and Ψ. The supreme in (2.19) is attained in $\tilde{\psi}(x) = \mathrm{sign}(f_1(x) - f_0(x))$ which corresponds to the optimal test $\psi(X) = \psi^{(1)}(X) = \mathbf{1}_{\{f_1(X)>f_0(X)\}}$. This implies the relation for the quantity $\gamma = \gamma(1)$ which is of interest from a geometrical point of view:

$$\gamma = \gamma(P_0, P_1) = 1 - \frac{1}{2}|P_1 - P_0|_1 \tag{2.20}$$

(this follows from (2.14) as well). Analogously, one can write

$$\gamma(t) = \gamma(t, P_0, P_1) = \frac{t+1}{2} - \frac{1}{2}|P_1 - tP_0|_1. \tag{2.21}$$

If P_0 dominates P_1 (this means $P_1(A) = 0$ for all $A \in \mathcal{A}$ such that $P_0(A) = 0$), then

$$|P_1 - P_0|_1 = E_{P_0}|L - 1| = 2E_{P_0}(L - 1)\psi^{(1)}, \tag{2.22}$$

where $L(x)$ is the likelihood ratio. Recall other definitions for the total variation distance:

$$|P_1 - P_0|_1 = 2 \sup_{A\in\mathcal{A}} |P_1(A) - P_0(A)| \le 2 \tag{2.23}$$

and it follows from (2.19) that for any function ψ with values in $[0, 1]$ one has

$$|E_{P_1}\psi - E_{P_0}\psi| \le \frac{1}{2}|P_1 - P_0|_1. \tag{2.24}$$

Using the equalities (2.20) and (2.17) we can write the relation for the total variation distance between Gaussian measures P_{v_1}, P_{v_2}; $v_1, v_2 \in l^2$:

$$|P_{v_1} - P_{v_2}|_1 = 2(1 - 2\Phi(-|v_1 - v_2|/2)). \tag{2.25}$$

Let ξ be the standard Gaussian variable. For any $u \geq 0$, one has the relations

$$2(1-2\Phi(-u/2)) = 2P(|\xi| < u/2) = \sqrt{\frac{2}{\pi}} \int_{-u/2}^{u/2} e^{-x^2/2}\, dx \leq u\sqrt{\frac{2}{\pi}}.$$

This yields the inequality

$$|P_{v_1} - P_{v_2}|_1 \leq |v_1 - v_2|\sqrt{2/\pi}. \tag{2.26}$$

In view of (2.8), analogous relations hold for the functional Gaussian model: if we test $H_0 : s = 0$ against $H_1 : s = s^*$, then (2.16), (2.17) hold with

$$r = s^*/\|s^*\|; \quad (X, r) = \varepsilon^{-1} \int_0^1 r(t)\, dX_\varepsilon(t), \quad u = \|s^*\|/\varepsilon. \tag{2.27}$$

The quality of testing is determined by a *signal–to–noise ratio* $u = \|s^*\|/\varepsilon$.

It follows from relations (2.20), (2.21) that the total variation distance is of importance in hypothesis testing theory. In particular, the functions $\beta(\alpha, P_0, P_1)$ and $\gamma(t, P_0, P_1)$ are continuous in (P_0, P_1) under the total variation distance. This follows from the following statement:

Proposition 2.1 *Let P_0, P_1; Q_0, Q_1 be probability measures on $(\mathcal{X}, \mathcal{A})$. Then, for all $\alpha \in (0,1)$, $t > 0$, one has*

$$|\gamma(t, P_0, P_1) - \gamma(t, Q_0, Q_1)| \leq \frac{1}{2}(t|P_0 - Q_0|_1 + |P_1 - Q_1|_1), \tag{2.28}$$

$$|\beta(\alpha, P_0, P_1) - \beta(\alpha, Q_0, Q_1))| \leq \frac{1}{2\alpha}|P_0 - Q_0|_1 + \frac{1}{2}|P_1 - Q_1|_1. \tag{2.29}$$

Proof. The inequality (2.28) follows directly from (2.21). Let $\psi_{\alpha,1}$ be optimal tests of P_0-level α for alternative P_1 and let $\psi_{\alpha,2}$ be optimal tests of Q_0-level α for alternative Q_1. Denote $\tilde{\alpha} = E_{P_0}\psi_{\alpha,2}$. It follows from (2.24) that

$$|\alpha - \tilde{\alpha}| = |E_{Q_0}\psi_{\alpha,2} - E_{P_0}\psi_{\alpha,2}| \leq \frac{1}{2}|P_0 - Q_0|_1.$$

First, let $\tilde{\alpha} \leq \alpha$, i.e., $\psi_{\alpha,2} \in \Psi_\alpha(P_0)$. Using the optimality of the tests and (2.24) we get

$$\begin{aligned} \beta(\alpha, P_0, P_1) &= E_{P_1}(1 - \psi_{\alpha,1}) \leq E_{P_1}(1 - \psi_{\alpha,2}) = E_{Q_1}(1 - \psi_{\alpha,2}) \\ &\quad + \left(E_{Q_1}\psi_{\alpha,2} - E_{P_1}\psi_{\alpha,2}\right) \leq \beta(\alpha, Q_0, Q_1) + |P_1 - Q_1|_1/2. \end{aligned}$$

Let $\tilde{\alpha} > \alpha$. Setting $\tilde{\psi}_{\alpha,2} = (\alpha/\tilde{\alpha})\psi_{\alpha,2}$ we get $E_{P_0}\tilde{\psi}_{\alpha,2} = \alpha$ and, analogously,

$$\begin{aligned} &\beta(\alpha, P_0, P_1) = E_{P_1}(1 - \psi_{\alpha,1}) \leq E_{P_1}(1 - \tilde{\psi}_{\alpha,2}) \\ &= \frac{\tilde{\alpha} - \alpha}{\tilde{\alpha}} + \frac{\alpha}{\tilde{\alpha}} E_{P_1}(1 - \psi_{\alpha,2}) \leq \frac{\tilde{\alpha} - \alpha}{\alpha} + \beta(\alpha, Q_0, Q_1) + \frac{1}{2}|P_1 - Q_1|_1 \\ &\leq \beta(\alpha, Q_0, Q_1) + \frac{1}{2}|P_1 - Q_1|_1 + \frac{1}{2\alpha}|P_0 - Q_0|_1. \end{aligned}$$

Analogously, taking $\tilde{\alpha} = E_{Q_0}\psi_{\alpha,1}$, we get

$$\beta(\alpha, Q_0, Q_1) \le \beta(\alpha, P_0, P_1) + \frac{1}{2}|P_1 - Q_1|_1 + \frac{1}{2\alpha}|P_0 - Q_0|_1. \qquad \square$$

2.3 Bayesian Approach in Hypothesis Testing

It was noted in Section 1.3 that, even if the null-hypothesis is simple, there are no general satisfactory recommendations as to how to select optimal tests for the case when the alternative is composite and of dimension more than one. Of course, the main reason is that the collection of admissible tests is too wide. These difficulties hold in estimation theory as well.

The natural way out of this situation is to characterize the test, not by the error functions $\alpha(\psi, P)$, $P \in \mathcal{P}_0$, $\beta(\psi, P)$, $P \in \mathcal{P}_1$, but by some monotone functionals $F_1(\psi) = F_1(\alpha(\psi))$, $F_2(\psi) = F_2(\beta(\psi))$ defined on these functions. There are two main types of functionals which correspond to the *Bayesian* and to the *minimax* approach, according to Wald [99], in general statistical decisions theory.

2.3.1 General Setting

Fix a Borel σ-algebra on the space $\mathcal{P}$ of probability measures with respect to the total variation distance (2.19) and assume $\mathcal{P}_0$ and $\mathcal{P}_1$ to be Borel subsets. Fix two probability measures π_0 and π_1 with supports in $\mathcal{P}_0$ and $\mathcal{P}_1$ which are called *priors*. Let us consider *average errors* with respect to given priors:

$$F_1(\psi) = \alpha(\psi, \pi_0) = E_{\pi_0}\alpha(\psi, P), \quad F_2(\psi) = \beta(\psi, \pi_1) = E_{\pi_1}\beta(\psi, P).$$

Then, analogously to Section 2.2.3, we can define

$$\begin{aligned} \gamma_t(\psi; \pi_0, \pi_1) &= t\alpha(\psi, \pi_0) + \beta(\psi, \pi_1), \\ \gamma(t; \pi_0, \pi_1) &= \inf_{\Psi} \gamma_t(\psi; \pi_0, \pi_1), & (2.30) \\ \beta(\alpha; \pi_0, \pi_1) &= \inf_{\Psi_{\alpha,\pi_0}} \beta(\psi, \pi_1), & (2.31) \end{aligned}$$

where, for given $\alpha \in (0, 1)$,

$$\Psi_{\alpha,\pi_0} = \{\psi \in \Psi : \alpha(\psi, \pi_0) \le \alpha\}.$$

The tests $\psi^{(t)}$ or ψ_α are called *Bayesian*, if they provide an infimum in (2.30) or in (2.31).

Let us consider the *mixtures* P_{π_0} and P_{π_1} which are defined by

$$P_\pi(A) = E_\pi P(A), \quad A \in \mathcal{A}.$$

Then the Bayesian hypothesis testing problem is reduced to testing of the simple hypothesis $P_0 = P_{\pi_0}$ against the simple alternative $P_1 = P_{\pi_1}$ because $\alpha(\psi, \pi_0) = \alpha(\psi, P_{\pi_0})$, $\beta(\psi, \pi_1) = \beta(\psi, P_{\pi_1})$, and so on. Therefore by using the Neyman–Pearson lemma 2.1 we obtain the structure of the Bayesian tests. Particularly, if P_{π_0} dominates P_{π_1}, then the Bayesian tests are based on the likelihood ratio statistic $L = dP_{\pi_1}/dP_{\pi_0}$ and

$$\gamma = \gamma(1) = 1 - \frac{1}{2}E_{P_{\pi_0}}|L-1|.$$

Note the inequality for L_1-distances between mixtures, which follows from (2.23), (2.24):

$$|P_{\pi_0} - P_{\pi_1}|_1 = 2\sup_{A \in \mathcal{A}} |E_{\pi_0}P(A) - E_{\pi_1}P(A)| \le |\pi_0 - \pi_1|_1. \qquad (2.32)$$

Note the case when we have parametrization $P = P_\theta$ which corresponds to an injective map $\Theta \to \mathcal{P}$. We assume that a σ-algebra $\mathcal{T}$ of subset Θ is given such that the map $(\Theta, \mathcal{T}) \to (\mathcal{P}, \mathcal{B})$ is measurable. Usually, a topology on Θ is given. Then we assume $\mathcal{T}$ to be a Borel σ-algebra and the parametrization is continuous. The priors π_0, π_1 we assume to be defined on $(\Theta, \mathcal{T})$.

The Bayesian approach is of importance from a theoretical point of view. Particularly, it was shown by Wald [99] that under general assumptions the closure of the collection of the Bayesian tests (for various priors) in the coarse topology is the set of admissible tests.

Unfortunately, there are two main difficulties in applying the Bayesian approach. The first one is that for most models of interest it is a very difficult task to study the optimal tests for hypotheses related to mixtures of measures. An exception arises when both π_0 and π_1 are Dirac masses (i.e., each concentrated at a single point). This is the case for contamination and some closely related models [33], [34] which allow their successful analysis. Unlike that, in the problems under consideration, π_0 is a delta-measure, but π_1 has extended support.

The second difficulty is that usually it is not clear how to choose the priors. In principle, they should present "an information which is not based on the experience." However, often, either there is no such information or it is difficult to present it in the form of a prior distribution.

2.3.2 Bayesian Approach in the Gaussian Model

Under the Gaussian model (2.1) we consider the priors π_0, π_1 on $(R^n, \mathcal{B}^n)$ for a finite-dimensional case or on $(l^2, \mathcal{B})$; here $\mathcal{B}$ is a Borel σ-algebra. We assume without loss of generality that the index set $I = \mathbf{N}$ is the set of natural numbers. Continuous properties hold in this case by the inequality (2.26) and the mixtures are dominated by the measure P_0.

We are interested in testing the simple null-hypothesis $H_0 : v = 0$. For this case, $\pi_0 = \delta_0$ is a Dirac mass at the point 0. Let $\pi = \pi(dv)$ be the prior and let P_π be the mixture (we omit the index 1 in the prior in the case of the simple null-hypothesis H_0). The likelihood ratio is of the form

$$L(x) = L_\pi(x) = \frac{dP_\pi}{dP_0}(x) = E_\pi \frac{dP_v}{dP_0}(x) = \int \exp(-|v|^2/2 + (x,v))\pi(dv). \tag{2.33}$$

Consider a finite-dimensional case. We can change the n-dimensional variable $x \in R^n$ in (2.33) by the n-dimensional complex variable $z = x + iy \in C^n$ where $x \in R^n$, $y \in R^n$, and $i = \sqrt{-1}$. Then relation (2.33) defines the function $L_\pi(z)$, $z \in C^n$.

Proposition 2.2 *Let π be a prior on $(R^n, \mathcal{B}^n)$. Then the function $L_\pi(z)$ is an analytical function of the n-dimensional complex variable $z \in C^n$.*

Proof. It suffices to verify that the power series for $L_\pi(z)$ converge for any $z = x + iy \in C^n$. To control this, we introduce a probability measure r on $(R^n, \mathcal{B}^n)$ by the relations

$$r(dv) = C^{-1}\pi(dv)e^{-|v|^2/2}, \quad C = \int e^{-|v|^2/2}\pi(dv) = L_\pi(0). \tag{2.34}$$

Using a power expansion for the exponent we get

$$L_\pi(z) = C\sum_{k=0}^{\infty} \mu_r^k(z)/k!,$$

where

$$\mu_r^k(z) = \int (z,v)^k r(dv) = C^{-1}\int (z,v)^k e^{-|v|^2/2}\pi(dv).$$

Since $|(z,v)| \le |z||v|$ (here $|x+iy|$ is the length of the $2n$-dimensional vector in R^{2n}), using the inequality

$$u^k e^{-u^2/2} < k^{k/2}e^{-k/2}, \quad u \ge 0$$

and the Simpson formula, we have, for large enough k,

$$|\mu_r^k(z)| < |z|^k k^{k/2}e^{-k/2}, \quad |\mu_r^k(z)|/k! < B|z|^k e^{k/2}/k^{(k+1)/2}.$$

This yields the required convergence. □

Proposition 2.3 *Let π_0, π_1 be two priors. Then*

$$P_{\pi_0} = P_{\pi_1} \quad \textit{iff} \quad \pi_0 = \pi_1.$$

Proof. Clearly, if $\pi_0 = \pi_1$, then $P_{\pi_0} = P_{\pi_1}$. Let $P_{\pi_0} = P_{\pi_1}$. Then $L_{\pi_0}(x) = L_{\pi_1}(x)$ a.s. First, consider the finite-dimensional case $x \in R^n$. Consider the probability measures r_0, r_1 determined by (2.34) for the priors π_0, π_1. The functions $L_{\pi_l}(x)$, $l = 0, 1$, are analytical by Proposition 2.2. Thus we get the equality for the complex variable: $L_{\pi_0}(z) = L_{\pi_1}(z)$, $z \in C^n$. For $z = ix$, $x \in R^n$, this corresponds to the equality of characteristic functions of probability measures r_0, r_1: $f_{r_0}(x) = f_{r_1}(x)$, $x \in R^n$. Therefore $r_0 = r_1$, which yields $\pi_0 = \pi_1$.

For an infinite-dimensional case these considerations yield the equality for all finite-dimensional marginal measures $\pi_{0,n}, \pi_{1,n}$ which yields $\pi_0 = \pi_1$ as well. □

Let us return to the finite-dimensional case.

Proposition 2.4 *Let $\pi \neq \delta_0$ be a prior on $(R^n, \mathcal{B}^n)$. Then the Bayesian tests $\psi^{(t)}$, ψ_α under the criteria (2.30), (2.31) are nonrandomized almost surely, i.e., they are of the form $\mathbf{1}_\mathcal{X}$ a.s. for some measurable set $\mathcal{X} \subset R^n$.*

Proof. Let $\psi(x)$ be a Bayesian test. It follows from the Neyman–Pearson Lemma 2.1 that $\psi(x) \in (0,1)$ is possible for $x \in \mathcal{X}_T$ with some $T \geq 0$ only; here $\mathcal{X}_T = \{x \in R^n : L(x) = T\}$ and $L(x)$ is the likelihood ratio. Recall that $L(x)$ is analytical by Proposition 2.2. Note the following statement:

Lemma 2.2 *Let $L(x) \neq \text{const}$ be an analytical function on R^n and let $\mathcal{X}_T = \{x \in R^n : L(x) = T\}$. Then $\text{mes}_n(\mathcal{X}_T) = 0$ for all $T \in R^1$; here mes_n is the Lebesgue measure on R^n.*

Proof of Lemma 2.2. We prove this lemma using induction on n. Let $n = 1$. If $\text{mes}_1(\mathcal{X}_T) > 0$, then one can find a sequence $x_i \in \mathcal{X}_T$, $x_i \to x_0$, $x_i \neq x_0$, $L(x_i) = L(x_0) = T$ which yield $L(x) = T \; \forall\, x$ (see [9], Section 1.4.4).

Now we have proved the lemma for dimension $n-1$. To prove this for dimension n, set $x = (y,t)$, $y \in R^{n-1}$, $t \in R^1$, and let $\mathcal{X}_T^t$ be an $(n-1)$-dimensional cross-section of the set $\mathcal{X}_T$ for fixed t. Suppose $\text{mes}_n(\mathcal{X}_T) > 0$. We have

$$\text{mes}_n(\mathcal{X}_T) = \int_{R^1} \text{mes}_{n-1}(\mathcal{X}_T^t)\, dt > 0.$$

Therefore there exists $t_0 \in R^1$ and a sequence $t_i \to t_0$, $t_i \neq t_0$ such that

$$\text{mes}_{k-1}(\mathcal{X}_T^{t_i}) > 0$$

which yields $L(y,t_i) = T$ for all $y \in R^{n-1}$, $i = 0,1,\ldots$ Suppose there exists $x^* = (y^*,t^*)$ such that $L(x^*) \neq T$. Consider the function $L_{y^*}(t) = L(y^*,t)$ which is analytical in $t \in R^1$ as well. Since $L_{y^*}(t_i) = T \; \forall\, i$, this yields $L_{y^*}(t) = T \; \forall\, t$ which contradicts the assumption $L(x^*) \neq T$. Thus $L(x) = T$ for all $x \in R^n$. □

Return to the proof of Proposition 2.4. By Lemma 2.2, if $\text{mes}_n(\mathcal{X}_T) > 0$, then $L(x) \equiv T$. Since $E_{P_0} L(x) = 1$, this yields $L(x) \equiv 1$, i.e., $P_\pi = P_{\delta_0} = P_0$ which contradicts to Proposition 2.3. □

The likelihood ratio in (2.33) is of a difficult structure, and it is not a simple problem to study the Bayesian tests. The following example is one of the most important examples where we can overcome the difficulties noted above.

Example 2.2 Let the set V be an $(n-1)$-dimensional sphere

$$S^{n-1}(\rho) = \{v \in R^n : |v| = \rho\}$$

of radius $\rho > 0$, $n \geq 2$, i.e., we want to detect n-dimensional signals of length ρ. Let $\pi = \pi^n$ be the uniform probability measure on the sphere $S^{n-1}(\rho)$.

Let us calculate the Bayesian likelihood ratio. Denote $e_1, \dots, e_n$ as the standard basis in R^n. By making a rotation which translates the vector X to the vector $e_1|X|$, we can present the required integral in the form

$$\begin{aligned} L_\pi(X) &= e^{-\rho^2/2} \int_{S^{n-1}(\rho)} e^{(v,X)} \pi(dv) \\ &= C(n,\rho) \int_{-\rho}^{\rho} e^{t|X|}(\rho^2 - t^2)^{(n-3)/2}\, dt \\ &= 2C(n,\rho) \int_0^{\rho} \cosh(|X|t)(\rho^2 - t^2)^{(n-3)/2}\, dt, \end{aligned}$$

where $C(n,\rho)$ is a constant which does not depend on X. This yields that the Bayesian likelihood ratio is an increasing function in the norm $|X|$ or in $|X|^2$, i.e., or in χ^2-statistics

$$t_n(X) = \sum_{i=1}^{n} X_i^2 = |X|^2.$$

Under the null-hypothesis the statistics t_n have χ_n^2-distribution with n degrees of freedom. Therefore, for any level $\alpha \in (0,1)$, the Bayesian test is a χ^2-test:

$$\chi^2_{\alpha,n} = 1_{\{t_n > T_{n,\alpha}\}}, \tag{2.35}$$

where $T_{n,\alpha}$ is a $(1-\alpha)$-quantile of the χ_n^2-distribution with n degrees of freedom. For any $v \in R^n$, we have

$$\beta(\chi^2_{\alpha,n}, v) = G_n(T_{n,\alpha}, |v|^2). \tag{2.36}$$

Here $G_n(h, t^2)$ is a distribution function of the noncentral χ_n^2-distribution with noncentrality parameter t^2. Therefore the type II error probability is constant on any sphere $S^{n-1}(\rho)$ and we have

$$\beta(\chi^2_{\alpha,n}, \pi) = G_n(T_{n,\alpha}, \rho^2). \tag{2.37}$$

In view of (2.8), Example 2.2 could be extended to the functional Gaussian model (2.5). Let $\bar\varphi = \{\varphi_1, \dots, \varphi_n\}$ be an orthonormal collection in $L_2(0,1)$ and let $S_{\bar\varphi}^{n-1}(r)$ be the sphere of radius r in the n-dimensional subspace $L_n \subset L_2(0,1)$ generated by $\bar\varphi$. Let π be the uniform probability measure on $S_{\bar\varphi}^{n-1}(r)$. Then the Bayesian tests for the alternative

$S_\varepsilon = S_{\bar\varphi}^{n-1}(r)$ are based on χ^2-statistics of the form

$$t_\varepsilon = \sum_{i=1}^n X_i^2, \quad X_i = \varepsilon^{-1}\int_0^1 \varphi_i(t)dX_\varepsilon(t), \quad i = 1, \dots, n;$$

the quality of testing is determined by n and by the signal-to-noise ratio $\rho = r/\varepsilon$.

In Example 2.2 the admissible set is a ball in R^n. This set is convex. This property is extended to all Bayesian tests for the simple hypothesis.

Proposition 2.5 *Let $L = L_\pi$ be defined by* (2.33). *Then the sets $\mathcal{X}_T = \{x : L(x) < T\}$ are convex for all T.*

Proof. By (2.33), the likelihood ratio $L(x)$ is an integral of convex functions $e^{(x,v)}$ over the positive measure $\exp(-|v|^2/2)\pi(dv)$. Therefore $L(x)$ is the convex function in x. This yields the statement of proposition 2.5. □

Let $G = \{g\}$ be a subgroup of the group of linear orthogonal transforms of R^n, i.e., such that $g^{-1} = g^*$. We call a prior π, *G-invariant,* if $\pi(A) = \pi(gA)$ for all $A \in \mathcal{B}^n$, $g \in G$. We call a set $\mathcal{X} \in \mathcal{B}^n$, *G-invariant,* if $g\mathcal{X} = \mathcal{X}$ for all $g \in G$.

Proposition 2.6 *Let π be G-invariant. Then the $\mathcal{X}_T = \{x : L(x) < T\}$ are G-invariant for all T. Moreover, $P_v(\mathcal{X}_T) = P_{gv}(\mathcal{X}_T)$ for all $v \in R^n$.*

Proof. The first statement follows from the equalities

$$\begin{aligned} L(gx) &= \int_{R^n} \exp(-|v|^2/2 + (gx, v))\pi(dv) \\ &= \int_{R^n} \exp(-|v|^2/2 + (x, g^*v))\pi(dv) \\ &= \int_{R^n} \exp(-|g^*v|^2/2 + (x, g^*v))\pi(dg^*v) \\ &= \int_{R^n} \exp(-|u|^2/2 + (x, u))\pi(du) = L(x). \end{aligned}$$

The second relation follows from the g-invariance of the measure P_0 and from relations

$$\begin{aligned} P_{gv}(\mathcal{X}_T) &= \int_{\mathcal{X}_T} \exp(-|gv|^2/2 + (x, gv))P_0(dx) \\ &= \int_{\mathcal{X}_T} \exp(-|v|^2/2 + (g^*x, v))P_0(dg^*x) \\ &= \int_{g^*(\mathcal{X}_T)} \exp(-|v|^2/2 + (y, v))P_0(dy) \\ &= P_v(g^*(\mathcal{X}_T)) = P_v(\mathcal{X}_T). \end{aligned}$$

□

2.4 Minimax Approach in Hypothesis Testing

2.4.1 General Setting

Let us consider not average, but maximum errors,

$$F_1(\psi) = \alpha(\psi) = \alpha(\psi, \mathcal{P}_0) = \sup_{P \in \mathcal{P}_0} \alpha(\psi, P),$$
$$F_2(\psi) = \beta(\psi) = \beta(\psi, \mathcal{P}_1) = \sup_{P \in \mathcal{P}_1} \beta(\psi, P).$$

Sometimes it is convenient formally to set $\alpha(\psi, \mathcal{P}_0) = 0$ if $\mathcal{P}_0 = \emptyset$ and $\beta(\psi, \mathcal{P}_1) = 0$ if $\mathcal{P}_1 = \emptyset$. Given $\alpha \in (0,1)$, we consider

$$\Psi_\alpha = \Psi_{\alpha, \mathcal{P}_0} = \{\psi \in \Psi : \alpha(\psi, \mathcal{P}_0) \leq \alpha\}.$$

We can define the quantities

$$\gamma_t(\psi) = \gamma_t(\psi, \mathcal{P}_0, \mathcal{P}_1) = t\alpha(\psi) + \beta(\psi),$$
$$\gamma(t) = \gamma(t, \mathcal{P}_0, \mathcal{P}_1) = \inf_{\Psi} \gamma_t(\psi), \tag{2.38}$$
$$\beta(\alpha) = \beta(\alpha, \mathcal{P}_0, \mathcal{P}_1) = \inf_{\Psi_\alpha} \beta(\psi), \tag{2.39}$$

The tests $\psi^{(t)}$ or ψ_α are called *minimax*, if they provide an infimum in (2.38) or in (2.39).

The quantities $\gamma(t)$ and $\beta(\alpha)$ characterize the guaranteed total errors or type II errors which are possible in the problem. Of course, it is possible that these values are large enough. Maybe there exist such elements $P_0 \in \mathcal{P}_0$, $P_1 \in \mathcal{P}_1$ and such a test ψ, that the errors are essentially smaller for these elements. However this is the best that is possible *for all elements* $P_0 \in \mathcal{P}_0$ and $P_1 \in \mathcal{P}_1$.

Let a parametrization $\Theta \to \mathcal{P}$ be given. We use the notations $\beta(\alpha) = \beta(\alpha, \Theta_0, \Theta_1)$, $\gamma(t) = \gamma(t, \Theta_0, \Theta_1)$, etc.; here the parametric sets Θ_0, Θ_1 correspond to the null-hypothesis and to the alternative. For $t = 1$ we omit t and use the notation $\gamma = \gamma(\Theta_0, \Theta_1)$. Under the sequence Gaussian model, if we test the simple hypothesis $H_0 : v = 0$, then we use the notations $\beta(\alpha) = \beta(\alpha, V), \gamma(t) = \gamma(t, V)$, etc, where the set V corresponds to the alternative. Observe some properties of the quantities $\beta(\alpha), \gamma(t)$ determined by (2.38), (2.39). Set

$$\gamma(0) = 0, \quad \beta(1) = 0, \quad \beta(0) = \sup_{\alpha \in (0,1)} \beta(\alpha).$$

Analogously to the case of the simple null-hypothesis and alternative, the function $\gamma(t)$ nondecreases in t and the function $\beta(\alpha)$ nonincreases in α and the inequalities (2.11) hold. Note the inequalities (compare with (1.9)): if $\mathcal{P}_0 \subset \tilde{\mathcal{P}}_0$, $\mathcal{P}_1 \subset \tilde{\mathcal{P}}_1$, then

$$\gamma(t, \mathcal{P}_0, \mathcal{P}_1) \leq \gamma(t, \tilde{\mathcal{P}}_0, \tilde{\mathcal{P}}_1), \quad \beta(\alpha, \mathcal{P}_0, \mathcal{P}_1) \leq \beta(\alpha, \tilde{\mathcal{P}}_0, \tilde{\mathcal{P}}_1). \tag{2.40}$$

Observe the following topologic properties. We consider the total variation distance in the linear space $\mathcal{M}$ which consists of signed measures on $(\mathcal{X}, \mathcal{A})$. Denote by $\mathcal{M}_{\mathcal{P}}$ the linear subspace in $\mathcal{M}$ generated by measures $P \in \mathcal{P}$. We consider the weak topology with respect to $\mathcal{P}$ on the linear space $\mathcal{D}$ which consists of a bounded measurable function on $(\mathcal{X}, \mathcal{A})$: this means

$$f_n \to f \text{ iff } \int f_n(x)R(dx) \to \int f(x)R(dx) \quad \forall R \in \mathcal{M}_{\mathcal{P}}.$$

Let $\mathcal{P}$ be dominated by the measure P_0. Then the weak topology is equivalent to

$$f_n \to f \text{ iff } \int f_n(x)g(x)P_0(dx) \to \int f(x)g(x)P_0(dx) \quad \forall g \in L_\infty(\mathcal{X}, \mathcal{A}, P_0).$$

Clearly, we have

Proposition 2.7 *The functions $\alpha(\psi, P_0)$, $\beta(\psi, P_1) - 1$, $\gamma_t(\psi, P_0, P_1) - 1$ are linear continuous functions in $\psi \in \Psi \subset \mathcal{D}$ and in P_0, $P_1 \in \mathcal{M}_{\mathcal{P}}$.*

Observe the following convex properties:

Lemma 2.3

(1) *The functions $\alpha(\psi, \mathcal{P}_0)$, $\beta(\psi, \mathcal{P}_1)$, $\gamma_t(\psi, \mathcal{P}_0, \mathcal{P}_1)$ are convex in $\psi \in \Psi$.*

(2) *The function $\beta(\alpha)$ is convex nonincreasing continuous in $\alpha \in [0, 1]$, the function $\gamma(t)$ is concave nondecreasing continuous in $t \in [0, \infty)$. They are connected by the relations*

$$\gamma(t) = \inf_{\alpha \in [0,1]} (t\alpha + \beta(\alpha)), \quad \beta(\alpha) = \sup_{t \in [0,\infty)} (\gamma(t) - t\alpha). \tag{2.41}$$

Proof of Lemma 2.3. The first statement follows from convexity of a supremum of convex functions. Analogously, the function $\gamma(t)$ is concave in $t > 0$ as an infimum of concave (linear in t) functions $\gamma_t(\psi)$. To continue the proof, let us observe the following proposition which we give in more general form than we need at present.

Proposition 2.8

(1) *Let $f(x)$, $g_l(x)$, $l = 1, \ldots, k$, be convex functions defined on convex set X. Then $h(\bar{a}) = \inf\{f(x) : x \in X,\ g_l(x) \le a_l,\ l = 1, \ldots, k\}$ is a convex function in $\bar{a} = (a_1, \ldots, a_k) \in R^k$.*

(2) *Let $f(x)$ be strictly convex, and let $A \subset R^k$ be a convex set contained in the image $G(X)$ where $G(x) = (g_1(x), \ldots, g_k(x))$. Assume, for any $\bar{a} \in A$, that the infimum $h(\bar{a})$ is attained in some $x \in X$ and that it corresponds to equalities $g_l(x) = a_l$, $l = 1, \ldots, k$. Then $h(\bar{a})$ is strictly convex on A.*

Here and below we assume that an infimum over the empty set is ∞ and we consider convex functions with possible values ∞, see [63], [94]. We call the function $f(x)$ *strictly convex*, if $f(px_1 + (1-p)x_2) < pf(x_1) + (1-p)f(x_2)$ for any $p \in (0, 1)$, $x_1 \neq x_2$ when $f(x_1)$, $f(x_2)$ are finite.

Proof of Proposition 2.8. To prove the statement (1) for any $\bar{a}_1 = \{a_{1,l}\}$, $\bar{a}_2 = \{a_{2,l}\}$, $p \in (0,1)$, put $\bar{a} = p\bar{a}_1 + (1-p)\bar{a}_2$; $\bar{a} = \{a_l\}$. For any $x_1 \in X$, $x_2 \in X$, such that $g_l(x_1) \leq a_{1,l}$, $g_l(x_2) \leq a_{2,l}$, $l = 1, \ldots, k$, it suffices to find $x \in X$, such that $g_l(x) \leq a_l, l = 1, \ldots, k$, and $f(x) \leq pf(x_1) + (1-p)f(x_2)$. Put $x = px_1 + (1-p)x_2$. Then the required inequalities follow from convex assumptions. Statement (2) is proved by analogous arguments. □

Return to the proof of Lemma 2.3. The convexity of the function $\beta(\alpha) = \inf\{\beta(\psi, \mathcal{P}_1) : \psi \in \Psi, \alpha(\psi, \mathcal{P}_0) \leq \alpha\}$ in $\alpha \in (0,1)$ follows from Proposition 2.8. In fact, the set Ψ is convex as well as the functions $\alpha(\psi, \mathcal{P}_0)$, $\beta(\psi, \mathcal{P}_1)$ are convex on ψ as a supremum of convex (linear) functions $\alpha(\psi, P_0)$, $\beta(\psi, P_1)$. The continuity and convexity in $\alpha \in [0,1]$ and in $t \geq 0$ follow from convexity in $\alpha \in (0,1)$ and in $t > 0$, from monotonicity and inequalities (2.11).

We leave to the reader the proof of the first relation in (2.41) which is based on standard arguments. To prove the second relation introduce the functions on R^1:

$$\hat{\beta}(\alpha) = \begin{cases} \beta(\alpha), & a \in [0,1], \\ \infty, & \alpha \notin [0,1], \end{cases} \qquad \hat{\gamma}(t) = \begin{cases} -\gamma(-t), & t \leq 0, \\ t, & t > 0, \end{cases}$$

by the above $\hat{\beta}(\alpha)$, $\hat{\gamma}(t)$ are closed convex functions on R^1 (see [63]). In view of the first relation in (2.41) one has

$$\hat{\gamma}(t) = \sup_{\alpha \in R^1} (t\alpha - \hat{\beta}(\alpha)), \quad t \in R^1,$$

which means that $\hat{\gamma}(t)$ is a Legendre transform of the function $\hat{\beta}(\alpha)$) (see [63]). By the properties of the Legendre transform for a closed convex function ([94]) one has

$$\hat{\beta}(\alpha) = \sup_{t \in R^1} (t\alpha - \hat{\gamma}(t)), \quad \alpha \in R^1,$$

which implies the second relation in (2.41). □

2.4.2 Connection with the Bayesian Approach

Return to the general case. There exists a close connection between the Bayesian and minimax approach. The following assertions go back essentially to Wald [99]. Let

$$|A, B|_1 = \inf\{|P - Q|_1 : P \in A, Q \in B\}$$

stand for the L_1-distance between families of measures A and B, and let $[A]$ be the convex hull of A.

Theorem 2.1 (i) *One has*

$$\gamma(t) \geq \sup\{\gamma(t; P_0, P_1) : P_0 \in [\mathcal{P}_0], P_1 \in [\mathcal{P}_1]\}, \tag{2.42}$$

$$\beta(\alpha) \geq \sup\{\beta(\alpha; P_0, P_1) : P_0 \in [\mathcal{P}_0], \ P_1 \in [\mathcal{P}_1]\}, \tag{2.43}$$

particularly,

$$\gamma = \gamma(1) \geq 1 - \frac{1}{2}\|[\mathcal{P}_0], [\mathcal{P}_1]\|_1. \tag{2.44}$$

(ii) *Let $\mathcal{P}_0$ and $\mathcal{P}_1$ be dominated families of measures on $(\mathcal{X}, \mathcal{A})$ and let σ-algebra $\mathcal{A}$ be countably generated. Then, for any $t > 0$, $\alpha \in (0,1)$, the infima in* (2.42), (2.43) *are attained at some tests, and the equalities hold in relations* (2.42)–(2.44).

Moreover the following statement holds:

(iii) *Let families $\mathcal{P}_0$ and $\mathcal{P}_1$ be compact in the L_1-distance. Then there exist pairs of priors $\pi_{0,t}$, $\pi_{1,t}$ and $\pi_{0,\alpha}$, $\pi_{1,\alpha}$ such that the minimax tests $\psi^{(t)}$ and ψ_α are optimal tests for testing the simple hypotheses $P_{\pi_{0,t}}$ or $P_{\pi_{0,\alpha}}$ against the simple alternatives $P_{\pi_{1,t}}$ or $P_{\pi_{1,\alpha}}$.*

The priors $\pi_{0,t}$, $\pi_{1,t}$ and $\pi_{0,\alpha}$, $\pi_{1,\alpha}$ are called *least favorable.*
We will not give the proof of (iii) because we will not use this statement directly.

Proof of Theorem 2.1. The inequality (2.44) follows from the relations

$$\begin{aligned}
\gamma &= \inf_{\psi} \sup \left[\gamma(\psi, P_0, P_1) : P_0 \in \mathcal{P}_0, \ P_1 \in \mathcal{P}_1\right] \\
&= \inf_{\psi} \sup \left\{\gamma(\psi, P_0, P_1) : P_0 \in [\mathcal{P}_0], \ P_1 \in [\mathcal{P}_1]\right\} \\
&\geq \sup \left\{\gamma(P_0, P_1) : P_0 \in [\mathcal{P}_0], \ P_1 \in [\mathcal{P}_1]\right\} \\
&= \sup \left\{1 - \frac{1}{2}|P_0 - P_1|_1 : P_0 \in [\mathcal{P}_0], \ P_1 \in [\mathcal{P}_1]\right\}.
\end{aligned}$$

The first equality is a consequence of linearity in P_0 and P_1 of the functional $\gamma(\psi, P_0, P_1)$; the next inequality is a general fact and the last equality follows from the identity (2.20). The proof of inequalities (2.42), (2.43) is analogous.

If $\mathcal{P}_0$ and $\mathcal{P}_1$ are dominated and σ-algebra $\mathcal{A}$ is countably generated, then the set Ψ is compact under weak topology (see [1], Chapter 19, §3, for example). Therefore $\Psi_\alpha = \Psi_\alpha(\mathcal{P}_0)$, $\alpha \in (0,1)$ are compact as well, and the infimum is attained. The equalities in (2.42)–(2.44) follow from

Theorem 2.2 On minimax [29], [89].

Let X, Y be convex subsets in linear spaces, and let $f(x,y)$ be convex in $x \in X$ and concave in $y \in Y$. Let X be compact, and let f be semicontinuous in x for any $y \in Y$. Then

$$\min_{x \in X} \sup_{y \in Y} f(x,y) = \sup_{y \in Y} \min_{x \in X} f(x,y).$$

Analogously, let Y be compact, and let f be semicontinuous in y for any $x \in X$. Then

$$\inf_{x\in X} \max_{y\in Y} f(x,y) = \max_{y\in Y} \inf_{x\in X} f(x,y).$$

In fact, to apply Theorem 2.2, we put $X = \Psi$, $X = \Psi_\alpha$, $Y = [\mathcal{P}_0] \times [\mathcal{P}_1]$, $f(x,y) = \gamma_t(\psi, P_0, P_1)$, $f(x,y) = \beta(\psi, P_1)$. This yields (ii). □

We will use inequalities (2.42)–(2.44) for the construction of lower bounds for the quantities $\gamma(t)$, $\beta(\alpha)$, γ. We can rewrite them in the form

$$\gamma(t) \geq \limsup \gamma(t; P_{\pi_{0,n}}, P_{\pi_{1,n}}), \quad \beta(\alpha) \geq \limsup \beta(\alpha; P_{\pi_{0,n}}, P_{\pi_{1,n}}), \tag{2.45}$$

as $n \to \infty$,

$$\gamma \geq 1 - \frac{1}{2} \liminf |P_{\pi_{0,n}} - P_{\pi_{1,n}}|_1, \tag{2.46}$$

where $\pi_{0,n}$, $\pi_{1,n}$ are any sequences of priors

$$\pi_{0,n}(\mathcal{P}_0) = 1, \qquad \pi_{1,n}(\mathcal{P}_1) = 1. \tag{2.47}$$

It is convenient to replace the constraints (2.47) by a wider constraints

$$\pi_{0,n}(\mathcal{P}_0) \to 1, \qquad \pi_{1,n}(\mathcal{P}_1) \to 1; \tag{2.48}$$

this means that $\pi_{0,n}$, $\pi_{1,n}$ are asymptotically concentrated on the null-hypothesis and on the alternative.

Proposition 2.9 *Assume* (2.48). *Then* (2.45), (2.46) *hold.*

Proof. Let us consider conditional priors

$$\tilde{\pi}_{0,n} = \pi_{0,n;\mathcal{P}_0}, \quad \tilde{\pi}_{1,n} = \pi_{1,n;\mathcal{P}_1},$$

defined by

$$\pi_B(A) = \pi(A \cap B)/\pi(B); \quad \pi_B(B) = 1,$$

and

$$\begin{aligned} |\pi_B - \pi|_1 &= \int_B |\pi_B(dP) - \pi(dP)| + \pi(\bar{B}) = \int_B \frac{1-\pi(B)}{\pi(B)} \pi(dP) + \pi(\bar{B}) \\ &= \frac{1-\pi(B)}{\pi(B)} \pi(B) + \pi(\bar{B}) = 2(1-\pi(B)), \end{aligned} \tag{2.49}$$

where $\bar{B}$ is the complement of B. By continuity $\gamma(t; P_0, P_1)$, $\beta(\alpha; P_0, P_1)$ on P_0, P_1 (see Proposition 2.1), it suffices to check that

$$|P_{\tilde{\pi}_{l,n}} - P_{\pi_{l,n}}|_1 \to 0, \quad l = 0, 1.$$

The last relation follows from (2.32):

$$|P_{\tilde{\pi}_{l,n}} - P_{\pi_{l,n}}|_1 \leq |\tilde{\pi}_{l,n} - \pi_{l,n}|_1 = 2(1 - \pi_{l,n}(\mathcal{P}_l)) \to 0, \quad l = 0, 1. \quad \square$$

Statement (ii) of Theorem 2.1 reduces the problem of studying the minimax sum error γ to the calculation of the L_1-distance between convex hulls of measures in the null-hypothesis and in the alternative. Usually this is difficult.

Statement (iii) reduces the minimax hypothesis testing problem to the case of the simple Bayesian hypothesis and alternative. However, in general, it is very difficult to use these results because there are no general methods to construct the least favorable priors $\pi_{1,t}$, $\pi_{0,t}$ or $\pi_{1,\alpha}$, $\pi_{0,\alpha}$ and to calculate the L_1-distance between convex hulls. Sometimes one can verify that a given pair of measures is least favorable. The following condition is helpful:

Theorem 2.3 *Let π_0 and π_1 be priors and let $S_0 \subset \mathcal{P}_0$, $S_1 \subset \mathcal{P}_1$ be measurable sets such that $\pi_0(S_0) = \pi_1(S_1) = 1$. Let, moreover, the Bayesian test $\psi^{(t)}$ or ψ_α for testing P_{π_0} against P_{π_1} satisfy*

$$\alpha(\psi_\alpha, \mathcal{P}_0) = \alpha = \alpha(\psi_\alpha, P_0), \quad \beta(\psi_\alpha, \mathcal{P}_1) = \beta(\psi_\alpha, P_1),$$

or $\gamma_t(\psi^{(t)}; \mathcal{P}_0, \mathcal{P}_1) = \gamma(\psi^{(t)}; P_0, P_1)$ for any $P_0 \in S_0$, $P_1 \in S_1$. Then (π_0, π_1) is a pair of the least favorable priors under the criteria $\gamma(t)$ or $\beta(\alpha)$ and the tests $\psi^{(t)}$ or ψ_α are minimax.

Proof. Consider the criterion $\beta(\alpha)$ (considerations for the criterion $\gamma(t)$ are analogous). It is clear that $\psi_\alpha \in \Psi_\alpha$, $\beta(\psi_\alpha, \pi_1) = \beta(\psi_\alpha, \mathcal{P}_1)$. Let $\psi' \in \Psi_\alpha$. Then, since the average value is no larger than the maximum value, one has $\alpha(\psi', \pi_0) \le \alpha(\psi', \mathcal{P}_0) \le \alpha$ and, since the average value is no smaller than the minimum value, one has

$$\beta(\psi', \mathcal{P}_1)) \ge \beta(\psi', \pi_1) \ge \beta(\alpha, \pi_1) = \beta(\psi_\alpha, \pi_1) = \beta(\psi_\alpha, \mathcal{P}_1). \qquad \square$$

In particular, using Theorem 2.3, we obtain the minimaxity of χ^2-tests in Example 2.2. However, it is impossible to extend this example to the case $n = \infty$ (see Section 1.4.2).

2.4.3 Convex Hypothesis and an Alternative under the Gaussian Model

Sometimes the least favorable priors are Dirac masses. In such cases the problem is simple enough.

Example 2.3 Let $V \subset R^n$ be a closed convex set. Let v^* be the point of minimum length:

$$|v^*| = \inf_{v \in V} |v| > 0. \tag{2.50}$$

Then the least favorable prior is Dirac mass δ_{v^*} at the point v^* and the minimax test is an optimal test for the alternative v^*.

In fact, using the inequality (1.9) and the equality (2.17) we get the lower bound:

$$\beta(\alpha, V) \geq \beta(\alpha, v^*) = \Phi(T_\alpha - u^*), \quad u^* = |v^*|.$$

On the other hand, let us consider tests ψ_{α,r^*} of the form (2.16) based on the linear statistic $t^* = (X, r^*)$ where $r^* = v^*/|v^*|$ is the direction of the point v^*. Using relation (2.18) we get that the quality of testing for the alternative v is determined by the projection of v to the direction r^*,

$$\beta(\psi_{\alpha,r^*}, v) = \Phi(T_\alpha - (v, r^*))$$

and the minimax risk of testing is determined by the minimum of projections

$$\beta(\psi_{\alpha,r^*}, V) = \Phi(T_\alpha - h_{r^*}(V)), \quad h_{r^*}(V) = \inf_{v \in V} (r^*, v).$$

Now note the equality

$$\inf_{v \in V} (r^*, v) = |v^*|. \tag{2.51}$$

In fact, by $(r^*, v^*) = |v^*|$, we need to prove that $(r^*, v) \geq |v^*|$ for any $v \in V$. Suppose there exists $v_0 \in V$ such that $(r^*, v_0) < |v^*|$, i.e., $\lambda = (v^* - v_0, v^*) > 0$. Set $v(t) = tv_0 + (1-t)v^* = v^* + t(v_0 - v^*)$ and consider the interval $I = \{v(t),\ t \in [0,1]\}$, which is contained in V because of convexity. Then

$$|v(t)|^2 = |v^*|^2 - 2t\lambda + t^2|v_0 - v^*|^2 < |v^*|^2$$

for small $t > 0$ because of $\lambda > 0$. This contradicts the fact that v^* provides the minimum in (2.50).

The equality (2.51) yields that the test ψ_{α,r^*} is minimax and

$$\beta(\psi_{\alpha,r^*}, V) = \Phi(T_\alpha - |v^*|).$$

Clearly, this corresponds to the prior $\pi = \delta_{v^*}$.

This example is extended on composite null-hypothesis with the convex set V_0. Under the sequence Gaussian model (2.1) let us test the hypothesis $H_0 : v \in V_0$ against the alternative $H_1 : v \in V_1$ where V_0 and V_1 are convex disjoint sets. [1] Suppose there exist such $v_0^* \in V_0^c$ and $v_1^* \in V_1^c$ which minimize the distance between V_0 and V_1:

$$u = |v_1^* - v_0^*| = \inf\{|v_1 - v_0| : v_0 \in V_0,\ v_1 \in V_1\} > 0;$$

where V^c is the closure of V. Then the least favorable priors are Dirac masses at the points v_0^* and v_1^*.

In fact, let us consider the optimal test $\psi^{(1)}$ under the criterion $\gamma = \gamma(1)$ for testing the simple hypothesis $H_0 : v = v_0^*$ against the simple alternative

[1] This case is studied in detail by Burnashev in [8].

$H_1 : v = v_1^*$. Analogously, with Example 2.1, this is of the form

$$\psi^{(1)}(X) = \mathbf{1}_{\{(X,r)>h\}}, \quad r = d/u, \quad d = v_1^* - v_0^*, \quad h = (|v_1^*|^2 - |v_0^*|^2)/2u.$$

It easily follows from convexity (compare with the consideration in Example 2.3), that

$$\sup_{v \in V_0} (r, v) = (r, v_0^*), \quad \inf_{v \in V_1} (r, v) = (r, v_1^*),$$

which imply the inequalities

$$\alpha(\psi^{(1)}, v) = \Phi((v,r) - h) \leq \Phi((v_0^*, r) - h) = \Phi(-u/2),$$
$$\beta(\psi^{(1)}, v) = \Phi(h - (v,r)) \leq \Phi(h - (v_0^*, r)) = \Phi(-u/2).$$

The criterion $\beta(\alpha)$ is considered in a similar way. The minimax quality is defined by

$$\gamma(V_0, V_1) = 2\Phi(-u/2), \quad \beta(\alpha, V_0, V_1) = \Phi(T_\alpha - u). \tag{2.52}$$

If we test the simple hypothesis $H_0 : v = 0$ against $H_1 : v \in V$ and V is convex, then we get (2.52) with

$$u = |v^*| = \inf_{v \in V} |v|. \tag{2.53}$$

Example 2.4 *Positive alternative.* Let $V = V_n = \breve{D}^+_{n,p}(\rho_n)$ be a "positive part" of the complement of an l^n_p-ball: this is the set

$$v \in R^n_+, \quad |v|_p = \left(\sum_{i=1}^n v_i^p\right)^{1/p} \geq \rho_n, \quad \rho_n > 0, \quad p > 0;$$

here and below we denote $R^n_+ = \{v \in R^n : \ v_i \geq 0 \ \forall i\}$. Assume $p \in (0,1]$. Then $\breve{D}^+_{n,p}(\rho_n)$ is convex and the extreme point v_n^* is

$$v_n^* = (a, \dots, a), \quad a = \rho_n n^{-1/p}; \quad |v_n^*| = \rho_n n^{1/2-1/p}.$$

This implies

Proposition 2.10 *Let $V_n = \breve{D}^+_{n,p}(\rho_n)$, $p \in (0,1]$. Then*

$$\beta(\alpha, V_n) = \Phi(T_\alpha - u_n); \quad \gamma(V_n) = 2\Phi(-u_n/2); \quad u_n = \rho_n n^{1/2-1/p},$$

and the minimax tests esare of the form

$$\psi_{n,T} = \mathbf{1}_{\{t_n > T\}}, \quad t_n = n^{-1/2} \sum_{i=1}^n X_i,$$

and either $T = T_\alpha$ for the criterion $\beta(a)$ or $T = u_n/2$ for the criterion γ.

In particular, $\gamma(V_n) \to 0$ iff $\rho_n/\rho_n^* \to \infty$ and $\gamma(V_n) \to 1$ iff $\rho_n/\rho_n^* \to 0$ where the rates ρ_n^* are of the form $\rho_n^* = n^{1/p-1/2}$.

One can extend this example to the infinite-dimensional case. Denote by l^2_+ the set of nonnegative sequences in l^2. Let $\bar{a} = (a_1, \dots, a_n, \dots)$ be a positive sequence and let the set $V = \breve{D}_{\bar{a},p}(\rho) \cap l^2_+$, $p \in (0,1]$, be defined by the inequalities

$$v_i \geq 0, \quad |v|_{\bar{a},p} = \left(\sum_i |v_i a_i|^p\right)^{1/p} \geq \rho.$$

Assume $A = \sum_i a_i^{2p/(2-p)} < \infty$. In particular, this holds for $a_i = i^r$, $i \in \mathbf{N}$, and $r < 1/2 - 1/p$. Using the Lagrange multiplier rule [94], [63] one gets equations for coordinates of the extreme point v^* in (2.53):

$$v_i^* = \lambda a_i^{p/(2-p)}, \quad \sum_i (v_i^* a_i)^p = \rho^p,$$

which imply

$$v_i^* = \rho A^{-1/p} a_i^{p/(2-p)}; \quad |v^*| = \rho A^{-(2-p)/2p}.$$

If the series A diverge, then $|v^*| = 0$ and it is impossible to distinguish the null-hypothesis and alternative. In particular, let $a_i = i^r$, $i \in \mathbf{N}$, and $r \geq r^*_{p,+} = 1/2 - 1/p$. Then $A = \infty$.

The reason is that the closure of the set V contains the point 0. To obtain the nontrivial minimax problem we need to consider "smaller" alternatives.

Remark. The study of nonparametric hypotheses testing problems of analogous type under the general models was initiated by Huber (see [33], Huber and Strassen [34]). These are problems of testing a composite hypothesis against a composite alternative which correspond to some neighborhoods $U_{\varepsilon,0}$ and $U_{\varepsilon,1}$ of simple hypotheses θ_0 and θ_1 (settings of ε-contamination type). The main result here is basically that for neighborhoods of a certain form (analogous to disjoint convex sets in the signal detection problem), one can find a pair of points on their boundaries such that the optimality in testing the corresponding simple hypotheses entails the minimaxity in testing the composite hypotheses.

However, this approach in general fails in problems treated here where Θ "surrounds" θ_0.

2.4.4 Convex Symmetric Admissible Set

Let $x = \{x_i, \ i \in I\} \in \mathcal{X}$ be a real-valued sequence or let $x \in R^n$ and $g_k x$ be the change of the sign of the kth coordinates of x. Let G be the group generated by g_k, $k \in I$ or $k = 1, \dots, n$. We call the set $A \subset \mathcal{X}$ or $A \subset R^n$ *symmetric*, if it is G-invariant, i.e., $gx \in A$ for all $x \in A$, $g \in G$.

Often the tests we study below have a convex symmetric admissible set A, i.e., they are of the form $\psi = \mathbf{1}_{\overline{A}}$ where $\overline{A}$ is the complement of A. The following statement is convenient to study their minimax properties.

Lemma 2.4 *Let $A \subset \mathcal{X}$ be convex and symmetric. Let $v,\ v' \in l^2$ or $v,\ v' \in R^n$ be such that $|v_i| \le |v'_i|$ for all i. Then $P_v(A) \ge P_{v'}(A)$. If $|v_{i_0}| < |v'_{i_0}|$ for some i_0 and the sets $\{x_{i_0} : x \in A\}$ are bounded, then $P_v(A) > P_{v'}(A)$.*

This statement is analogous to Anderson's lemma (see [38]).

Proof. Consider the case $I = \mathbf{N}$. Let $v_n, v'_n \in R^n$ be n-dimensional vectors corresponding to the first n coordinates of $v, v' \in l^2$. By inequality (2.26) we have $P_{v_n}(A) \to P_v(A)$, $P_{v'_n}(A) \to P_{v'}(A)$ as $n \to \infty$. Therefore it suffices to consider the finite-dimensional case and to assume

$$v = (t, u), \quad v' = (t', u); \quad t,\ t' \in R^1, \quad |t| \le |t'|; \quad u \in R^{n-1}.$$

For $x = (x_1, y)$, $x_1 \in R^1$, $y \in R^{n-1}$, let

$$A_y = \{x_1 : (x_1, y) \in A\}, \quad X_{1,A} = \bigcup_y A_y,$$

i.e., $X_{1,A}$ is the projection of the set A to the first coordinate and let $Y_A = \{y : (x_1, y) \in A\}$ be the projection of A to R^{n-1}. Clearly, the sets A_y are symmetric intervals in R^1 which are nonempty for $y \in Y_A$; $P_v(dx) = P_t(dx_1)P_u(dy)$ and all A_y are bounded if $X_{1,A}$ is bounded.

Note the following property of the one-dimensional Gaussian model. Let $\Delta = (-a, a)$ be a symmetric interval on the real line and set

$$\phi(t) = P_{1,t}(\Delta) = \Phi(a - t) - \Phi(-a - t).$$

Then the function $\phi(t)$ is symmetric $\phi(t) = \phi(-t)$; it is strictly decreasing in $|t|$ and it has the maximum at $t = 0$. By $|t| \le |t'|$, this yields the inequality $P_t(A_y) \ge P_{t'}(A_y)$ and, using Fubini theorem, we get

$$P_v(A) = \int_{Y_A} P_t(A_y)P_u(dy) \ge \int_{Y_A} P_{t'}(A_y)P_u(dy) = P_{v'}(A).$$

If the intervals A_y are bounded and $t < t'$, then the inequality is strict. □

2.5 Asymptotics in Hypothesis Testing

2.5.1 Asymptotic Approach

A general enough theory of statistical inference is developed under the asymptotic approach: a concrete statistical problem is considered as an element of the family of problems which depend on the asymptotic parameter $\varepsilon \to \varepsilon_0$. The asymptotic parameter may be of various types: $\varepsilon \to 0$ is a noise level in the Gaussian models (2.3), (2.6); it is a dimension $\varepsilon = n \to \infty$ in the Gaussian models (2.1), (2.3) for finite n; it is a sample size $\varepsilon = N \to \infty$ in the probability density model (usually one takes an asymptotic parameter of the form $\varepsilon = N^{-1/2} \to 0$), and so on. Usually $\varepsilon \to \varepsilon_0$ corresponds to an increase of information in the problem. Under asymptotic setting the asymptotic properties of statistical procedures are studied, as $\varepsilon \to \varepsilon_0$. In the hypothesis testing problem the setting is as follows. We assume an observation $X_\varepsilon \in \mathcal{X}_\varepsilon$ is generated by the probability measure $P_{\varepsilon,\theta}$ on the measurable space $(\mathcal{X}_\varepsilon, \mathcal{A}_\varepsilon)$ and the parameter $\theta \in \Theta$ is unknown. Given families of subsets $\Theta_{\varepsilon,0}$ and $\Theta_{\varepsilon,1}$ in Θ_ε, we would like to test a family of null-hypotheses $H_0 : \theta \in \Theta_{\varepsilon,0}$ against a family of alternatives $H_1 : \theta \in \Theta_{\varepsilon,1}$. We consider the families $\{\psi_\varepsilon\}$ which consist of (randomized) tests $\psi_\varepsilon : (\mathcal{X}_\varepsilon, \mathcal{A}_\varepsilon) \to ([0,1], \mathcal{B})$, $\psi_\varepsilon \in \Psi_\varepsilon$; here Ψ_ε is the set which consists of all tests on the space $(\mathcal{X}_\varepsilon, \mathcal{A}_\varepsilon)$ for fixed ε. To simplify notations we denote a family $\{\psi_\varepsilon\} \in \Psi$ as ψ_ε without braces.

The statistical properties of tests ψ_ε are characterized by families of functions of type I and type II errors

$$\alpha_\varepsilon(\psi_\varepsilon, \theta) = E_{\varepsilon,\theta}\psi_\varepsilon, \quad \theta \in \Theta_{\varepsilon,0}, \quad \beta_\varepsilon(\psi_\varepsilon, \theta) = E_{\varepsilon,\theta}(1 - \psi_\varepsilon), \quad \theta \in \Theta_{\varepsilon,1},$$

or

$$\gamma_{\varepsilon,t}(\psi_\varepsilon, \theta_0, \theta_1) = t\alpha_\varepsilon(\psi_\varepsilon, \theta_0) + \beta_\varepsilon(\psi_\varepsilon, \theta_1).$$

We are interested in the asymptotics of $\alpha_\varepsilon, \beta_\varepsilon$ or $\gamma_{\varepsilon,t}$ as $\varepsilon \to \varepsilon_0$. Of course, there are the same difficulties in comparing families of tests as in a nonasymptotic setting. We deal with the asymptotic variant of the minimax approach.

2.5.2 *Asymptotic Minimaxity*

The minimax properties of the family of tests ψ_ε are characterized by the asymptotics of the maximum of type I and type II errors

$$\alpha_\varepsilon(\psi_\varepsilon) = \sup_{\theta \in \Theta_{\varepsilon,0}} \alpha_\varepsilon(\psi_\varepsilon, \theta), \quad \beta_\varepsilon(\psi_\varepsilon) = \sup_{\theta \in \Theta_{\varepsilon,1}} \beta_\varepsilon(\psi_\varepsilon, \theta),$$

or by their sum $\gamma_{\varepsilon,t}(\psi_\varepsilon) = t\alpha_\varepsilon(\psi_\varepsilon) + \beta_\varepsilon(\psi_\varepsilon)$. Denote, as above,

$$\gamma_\varepsilon(t) = \inf \gamma_{\varepsilon,t}(\psi_\varepsilon), \quad 0 \le \gamma_\varepsilon(t) \le \min(1, t),$$

where the infimum is taken over all tests $\psi_\varepsilon \in \Psi_\varepsilon$. The family of tests $\psi_\varepsilon^{(t)}$ is called *asymptotically minimax*, if $\gamma_{\varepsilon,t}(\psi_\varepsilon^{(t)}) = \gamma_\varepsilon(t) + o(1)$, as $\varepsilon \to \varepsilon_0$. Under the Neyman and Pearson approach we consider families of test ψ_ε satisfying $\alpha_\varepsilon(\psi_\varepsilon) \le \alpha + o(1)$. The family of test $\psi_{\varepsilon,\alpha}$ is called *asymptotically minimax*, if

$$\alpha_\varepsilon(\psi_{\varepsilon,\alpha}) \le \alpha + o(1), \quad \beta_\varepsilon(\psi_{\varepsilon,\alpha}) = \beta_\varepsilon(\alpha) + o(1); \quad \varepsilon \to \varepsilon_0.$$

We call the *sharp asymptotic problem* the problem of studying the asymptotics of $\gamma_\varepsilon(t)$ or $\beta_\varepsilon(\alpha)$ and of constructing the asymptotically minimax families of tests $\psi_\varepsilon^{(t)}$ and $\psi_{\varepsilon,\alpha}$. In view of the nonasymptotic relations between $\gamma(t)$ and $\beta(\alpha)$ given in Lemma 2.3, we are interested in $\gamma_\varepsilon = \gamma_\varepsilon(1)$ and in $\beta_\varepsilon(\alpha)$ and in the corresponding asymptotic minimax families of tests only.

The sharp asymptotic problem is difficult for nonparametric case. Therefore we are often interested in the simpler *rate asymptotic problem* (or *distinguishability problem*):

- To obtain conditions for *minimax distinguishability*, i.e., for $\gamma_\varepsilon \to 0$, and to construct *minimax consistent* families of test ψ_ε such that $\gamma_\varepsilon(\psi_\varepsilon) \to 0$. In view of Lemma 2.3 this implies $\gamma_\varepsilon(t) \to 0,\ \beta_\varepsilon(\alpha) \to 0$ for any $t > 0,\ \alpha \in (0,1)$.
- To obtain conditions for *minimax nondistinguishability*, i.e., for $\gamma_\varepsilon \to 1$. In view of Lemma 2.3 this implies $\gamma_\varepsilon(t) \to \min(1,t),\ \beta_\varepsilon(\alpha) \to 1-\alpha$ for any $t > 0,\ \alpha \in (0,1)$.

The study of these problems contains two steps. The first one is to obtain *the lower bounds*:

$$\gamma_\varepsilon \geq \gamma_\varepsilon^* + o(1), \quad \beta_\varepsilon(\alpha) \geq \beta_\varepsilon^*(\alpha) + o(1).$$

The second one is to construct families of tests $\psi_{\varepsilon,\alpha} : \alpha_\varepsilon(\psi_{\varepsilon,\alpha}) \leq \alpha + o(1)$ or ψ_ε which provide *upper bounds*, i.e.,

$$\gamma_\varepsilon \leq \gamma_\varepsilon(\psi_\varepsilon) + o(1), \quad \beta_\varepsilon(\alpha) \leq \beta_\varepsilon(\psi_{\varepsilon,\alpha}) + o(1).$$

The methods are based on an asymptotic variant of the Bayesian approach in the minimax problem. To construct the lower bounds we can reformulate Proposition 2.9.

Proposition 2.11 *Let $\pi_{\varepsilon,0},\ \pi_{\varepsilon,1}$ be families of priors such that*

$$\pi_{\varepsilon,0}(\Theta_{\varepsilon,0}) \to 1, \quad \pi_{\varepsilon,1}(\Theta_{\varepsilon,1}) \to 1. \tag{2.54}$$

Then

$$\beta_\varepsilon(\alpha) \geq \beta_\varepsilon(\alpha; P_{\pi_{\varepsilon,0}}\ P_{\pi_{\varepsilon,1}}) + o(1), \tag{2.55}$$

$$\gamma_\varepsilon \geq \gamma_\varepsilon(P_{\pi_{\varepsilon,0}}\ P_{\pi_{\varepsilon,1}}) + o(1) = 1 - \tfrac{1}{2}|P_{\pi_{\varepsilon,0}} - P_{\pi_{\varepsilon,1}}|_1. \tag{2.56}$$

These relations provide the lower bounds

$$\gamma_\varepsilon^* = 1 - \frac{1}{2}|P_{\pi_{\varepsilon,0}} - P_{\pi_{\varepsilon,1}}|_1 + o(1), \quad \beta_\varepsilon^*(\alpha) = \beta_\varepsilon(\alpha; P_{\pi_{\varepsilon,0}}\ P_{\pi_{\varepsilon,1}}).$$

It is often difficult to evaluate the total variation distance (L_1-distance). Let us consider a (nonsymmetric) L_2-distance $|P_1 - P_0|_2^2$ which is defined by the equality $|P_1 - P_0|_2^2 = \infty$, if P_0 does not dominate P_1 and, if P_0 dominates P_1, then

$$|P_1 - P_0|_2^2 = E_{P_0}(L-1)^2, \quad L = dP_1/dP_0. \tag{2.57}$$

Below we assume the measures $P_{\varepsilon,0}$ dominate all measures $P_{\varepsilon,\theta},\ \theta \in \Theta$. This is fulfilled under the Gaussian models.

One has the following minimax nondistinguishability condition in terms of L_2-distance.

Proposition 2.12
(1) *Let* $|P_{\varepsilon,1}-P_{\varepsilon,0}|_2 \to 0$. *Then* $|P_{\varepsilon,1}-P_{\varepsilon,0}|_1 \to 0$ *and* $\gamma_\varepsilon(P_{\varepsilon,0}, P_{\varepsilon,1}) \to 1$.
(2) *Moreover, let* $|P_{\varepsilon,1}-P_{\varepsilon,0}|_2 = O(1)$. *Then* $\limsup |P_{\varepsilon,1}-P_{\varepsilon,0}|_1 < 2$ *and* $\liminf \gamma_\varepsilon(P_{\varepsilon,0}, P_{\varepsilon,1}) > 0$.

Proof. The first statement follows from the inequality

$$|P_1 - P_0|_1 = E_{P_0}|L-1| \le \left(E_{P_0}|L-1|^2\right)^{1/2} = |P_1 - P_0|_2.$$

Let us establish the inequality

$$|P_1 - P_0|_2 \ge |P_1 - P_0|_1/2(1 - \frac{1}{2}|P_1 - P_0|_1)^{1/2}, \tag{2.58}$$

which yields the second statement. Let us consider the optimal test $\psi = \psi^{(1)} = \mathbf{1}_{\{L>1\}}$ for testing $H_0 : P = P_0$ against $H_1 : P = P_1$. By (2.22) we have

$$\gamma(\psi) = \gamma(P_0, P_1) = 1 - \frac{1}{2}|P_1 - P_0|_1 = 1 - E_{P_0}(L-1)\psi.$$

Then (2.58) follows from relations

$$\begin{aligned} \frac{1}{2}|P_1 - P_0|_1 &= E_{P_0}(L-1)\psi \le \left(E_{P_0}|L-1|^2\right)^{1/2}\left(E_{P_0}\psi^2\right)^{1/2} \\ &= |P_1 - P_0|_2\alpha^{1/2}(\psi) \le |P_1 - P_0|_2\gamma^{1/2}(\psi) \\ &= |P_1 - P_0|_2(1 - \frac{1}{2}|P_1 - P_0|_1)^{1/2}. \qquad \square \end{aligned}$$

Thus, in view of Proposition 2.11, to obtain the lower bounds we need to study the asymptotics in the Bayesian hypothesis testing problem for a wide enough class of families of priors that are asymptotically supported on the parametric sets $\Theta_{\varepsilon,0}$, $\Theta_{\varepsilon,1}$, and to choose families of priors which give asymptotically maximum lower bounds.

In particular, to obtain the nondistinguishability conditions we need to minimize the L_1-distance between mixtures $P_{\pi_{\varepsilon,0}}$, $P_{\pi_{\varepsilon,1}}$ under the asymptotic constraints $\pi_{\varepsilon,0}(\Theta_{\varepsilon,0}) \to 1$, $\pi_{\varepsilon,1}(\Theta_{\varepsilon,1}) \to 1$. In view of Proposition 2.12 we can try to minimize not the L_1- but the L_2-distance.

To construct the upper bounds, the Bayesian approach motivates us to study families of tests based on the Bayesian likelihood ratio for the priors which provide asymptotically maximum lower bounds. We will try to realize this scheme in the following chapters.

2.5.3 Asymptotic Minimaxity in the Parametric Case

One can extend the main results from a finite-dimensional Gaussian model to the general parametric case. Let Θ be a compact set in R^n and let $\theta_0 \in \operatorname{Int}\Theta$, subject to certain regularity conditions [38], [42], [45], including Le Cam's local asymptotic normality (LAN) at $\theta = \theta_0$ [30], [70]. The latter assumes that

$$\frac{dP_{\varepsilon,\theta_0+\phi_\varepsilon u}}{dP_{\varepsilon,\theta_0}} = \exp\left\{-|u|^2/2 + (\Delta_\varepsilon, u) + \delta_\varepsilon(u)\right\}$$

for $u \in R^n$, with some nondegenerate $n \times n$ matrices $\phi_\varepsilon = \phi_\varepsilon(\theta_0)$, converging to zero as $\varepsilon \to \varepsilon_0$, random n-vectors Δ_ε which are asymptotically standard Gaussian n-dimensional random vectors under P_{ε,θ_0}-probability, and random variables $\delta_\varepsilon(u) \to 0$ in P_{ε,θ_0}-probability as $\varepsilon \to \varepsilon_0$.

Usually the matrix $\phi_\varepsilon = \phi_\varepsilon(\theta_0)$ satisfies

$$\phi_\varepsilon^\top(\theta_0)\phi_\varepsilon(\theta_0) = I_\varepsilon^{-1}(\theta_0), \tag{2.59}$$

where $I_\varepsilon(\theta_0)$ is the Fisher information matrix at point θ_0,

$$I_\varepsilon(\theta_0) = E_{\varepsilon,\theta_0} l_{\varepsilon,\theta_0}^\top l_{\varepsilon,\theta_0}, \quad l_{\varepsilon,\theta_0} = \frac{\partial}{\partial\theta}\log\frac{dP_{\varepsilon,\theta}}{dP_{\varepsilon,\theta_0}};$$

one can assume $\phi_\varepsilon(\theta_0)$ to be a square root of $I_\varepsilon(\theta_0)$ and $\Delta_\varepsilon = \phi_\varepsilon(\theta_0) l_{\varepsilon,\theta_0}$ to be the normalized gradient of the log-likelihood ratio.

Let us test the simple hypothesis $H_0 : \theta = \theta_0$ against the simple alternative $H_1 : \theta = \theta_0 + \phi_\varepsilon u$, $u = u_\varepsilon \in R^n$, $|u_\varepsilon| = O(1)$. Then, analogously with (2.17), one has

$$\gamma_\varepsilon = 2\Phi(-|u|/2) + o(1), \quad \beta_\varepsilon(\alpha) = \Phi(T_\alpha - |u|) + o(1),$$

and the asymptotically optimal tests are based on statistics $l_\varepsilon = (\Delta_\varepsilon, u)/|u|$.

Moreover, let us test the simple hypothesis $H_0 : \theta = \theta_0$ against the alternative $H_1 : \theta \in \Theta_\varepsilon$ where Θ_ε be obtained from Θ by removing a spherical neighborhood of θ_0 of radius ρ_ε in the Fisher distance, i.e., in the norm $|x|_\varepsilon = |\phi_\varepsilon x|$ with $|\cdot|$ denoting the Euclidean norm in R^n; this means

$$\Theta_\varepsilon = \{\theta \in \Theta : \theta = \theta_0 + \phi_\varepsilon u, \ |u| \geq \rho_\varepsilon\};$$

under (2.59) this is equivalent to

$$\Theta_\varepsilon = \{\theta \in \Theta : |\theta - \theta_0|_\varepsilon^2 = (\theta - \theta_0)^T I_\varepsilon(\theta_0)(\theta - \theta_0) \geq \rho_\varepsilon^2\}.$$

Then, analogously to Example 2.2 (see [98]; [88], Chapter 6, §§2, 4, 6; [5], Chapter 3, §9)

$$\beta_\varepsilon(\alpha) = G_n(T_{n,\alpha}, \rho_\varepsilon^2) + o(1),$$

and the asymptotically minimax tests can be obtained as likelihood ratio tests or as tests of the form

$$\psi_{\varepsilon,\alpha} = \mathbf{1}_{\{|\theta_\varepsilon^* - \theta_0|_\varepsilon^2 \geq T_{n,\alpha}\}}, \tag{2.60}$$

where θ_ε^* is the maximum likelihood estimator for $\theta \in \Theta$ or an asymptotically equivalent estimator and $|\theta_\varepsilon^* - \theta_0|_\varepsilon^2 = (\theta_\varepsilon^* - \theta_0)^T I_\varepsilon(\theta_0)(\theta_\varepsilon^* - \theta_0)$ is the Fisher distance between θ_ε^* and θ_0.

In particular, under the parametric functional Gaussian model (2.6) with regular parametrization $s = s(t,\theta)$, $t \in [0,1]$, $\theta \in \Theta \subset R^n$, the Fisher information matrix is of the form

$$I_\varepsilon(\theta) = \varepsilon^{-2} I(\theta), \quad I(\theta) = \int_0^1 \frac{\partial s(t,\theta)}{\partial \theta}^\top \frac{\partial s(t,\theta)}{\partial \theta} dt.$$

Under a parametric i.i.d. sample model with a density $f = f(x,\theta)$, $x \in \mathcal{X}$, $\theta \in \Theta \subset R^n$, one has $I_N(\theta) = NI(\theta)$, where $I(\theta)$ is the Fisher information matrix for $N = 1$,

$$I(\theta) = E_\theta \left(\frac{\partial \log f(x,\theta)}{\partial \theta}^\top \frac{\partial \log f(x,\theta)}{\partial \theta} \right).$$

These results are extended to the composite null-hypothesis $H_0 : \theta \in \Theta_0$ for smooth submanifolds $\Theta_0 \subset \Theta$, see [42], [45].

2.6 Minimax Distinguishability in Goodness-of-Fit Problems

Let us test the null-hypothesis $H_0 : \theta \in \Theta_{\varepsilon,0}$, $\Theta_{\varepsilon,0} \subset \Theta$ and Θ is a subset of the linear space L. In goodness-of-fit problems alternatives of interest are $H_1 : \theta \notin \Theta_{\varepsilon,0}$. However, usually, it is impossible to use the minimax approach in these problems directly, because (nonasymptotically)

$$\gamma_\varepsilon(\Theta_{\varepsilon,0},\ \Theta_{\varepsilon,1}) = 1 \quad \forall\, \varepsilon.$$

The reason is that the alternative sets $\Theta_{\varepsilon,1} = \Theta \backslash \Theta_{\varepsilon,0}$ contain points $\theta_{\varepsilon,1}$ too close to $\Theta_{\varepsilon,0}$, and if the parametrization $\theta \to P_{\varepsilon,\theta}$ is continuous in the L_1-norm, then one can find sequences $\theta^n_{\varepsilon,0} \in \Theta_{\varepsilon,0}$, $\theta^n_{\varepsilon,1} \notin \Theta_{\varepsilon,0}$ such that

$$|P_{\varepsilon,\theta^n_{\varepsilon,1}} - P_{\varepsilon,\theta^n_{\varepsilon,0}}|_1 \to 0, \quad \text{as } n \to \infty.$$

Therefore we need to consider some smaller sets of alternatives

$$\Theta_{\varepsilon,1} = \{\theta \in \Theta : h_1(\theta, \Theta_{\varepsilon,0}) \geq r_\varepsilon\}$$

where

$$h_1(\theta, \Theta_{\varepsilon,0}) = \inf_{\theta_0 \in \Theta_{\varepsilon,0}} h_1(\theta, \theta_0)$$

is "the distance" between the point θ and the set $\Theta_{\varepsilon,0}$ determined by the "distance function" $h_1(\theta,\theta_0)$; usually the distance function is a norm (quasi-norm) $|\theta - \theta_0|_{(1)}$ in L. This means we "remove" neighborhoods of $\Theta_{\varepsilon,0}$ in the norm h_1.

The problem is: *What are the radii r_ε in order to obtain minimax distinguishability:* $\gamma_\varepsilon \to 0$?

Let H_0 be the simple null-hypothesis $\Theta_{\varepsilon,0} = \{\theta_0\}$ and

$$\Theta_{\varepsilon,1} = \{\theta \in \Theta : \ |\theta - \theta_0|_{(1)} \geq r_\varepsilon\}. \tag{2.61}$$

2.6.1 *The Parametric Case*

Usually this is not a difficult problem in the parametric case. In fact, let $\Theta \subset R^n$, let the regularity assumptions hold (including LAN), and let all the eigenvalues $\eta_{\varepsilon,i}$ of matrices $\phi_{\varepsilon,\theta_\varepsilon,0}$ satisfy

$$br_\varepsilon^* \le \eta_{\varepsilon,i} \le Br_\varepsilon^*, \tag{2.62}$$

where b, B are positive constants and r_ε^* is a positive family. It follows from the minimax properties of the maximum likelihood tests that distinguishability conditions are of the form

$$\gamma_\varepsilon \to 0, \quad \text{iff} \quad r_\varepsilon/r_\varepsilon^* \to \infty; \tag{2.63}$$

$$\gamma_\varepsilon \to 1, \quad \text{iff} \quad r_\varepsilon/r_\varepsilon^* \to 0. \tag{2.64}$$

We call the family r_ε^* in (2.63), (2.64) the *rates* (or the *critical radii)*).

In particular, under regular parametric white Gaussian model the rates family is $r_\varepsilon^* = \varepsilon$; under regular parametric density model the rates sequence is $r_N^* = N^{-1/2}$. We call the distinguishability conditions (2.63), (2.64) with these rates *classical.*

2.6.2 *Classical Distinguishability in the Nonparametric Case*

There exists a wide class of nonparametric problems where classical distinguishability conditions hold.

In particular, under the the i.i.d. sample model with a density on the interval $[0, 1]$ let us test the null-hypotheses $H_0 : f = f_0$ against the alternative $H_1 : f \ne f_0$ where $f_0(t) \equiv 1$. Consider the L_p-norm for the distribution function

$$|f_1 - f_0|_{(p)} = \|F_{f_1} - F_{f_0}\|_p, \quad 1 \le p \le \infty,$$

where F_f is the distribution function for the probability density f (we call these *norms of Kolmogorov type*). Given a positive sequence r_N, set

$$\mathcal{F}_{r_N,p} = \{f \in \mathcal{F} : \|F_f - F_{f_0}\|_p \ge r_N\}. \tag{2.65}$$

If $N^{1/2}r_N = u_N = O(1)$, then it is impossible to distinguish between the null-hypothesis and the alternative, i.e., $\liminf_{N\to\infty} \gamma_N > 0$; moreover, if $u_N = o(1)$, then $\gamma_N \to 1$. In fact, take a function $\delta(t)$, $t \in [0, 1]$, such that

$$\tau = \|\delta\|_2 > 0, \quad \|\delta\|_\infty < \infty, \quad h = |\delta|_{(p)} = \left\| \int_0^t \delta(u)du \right\|_p > 0$$

and set

$$f(t, \theta) = f_0(t) + \theta\delta(t), \quad f_N(t) = f(t, N^{-1/2}u_N/h).$$

Then $f_N \in \mathcal{F}_{r_N,p}$ which implies $\gamma_N \ge \gamma(f_0, f_N)$. The one-parametric family $f(t, \theta)$, $|\theta| \le \theta_0$, is regular for small enough $\theta_0 > 0$ and, by making simple calculations, we get $I(0) = \tau^2$. This implies that

$$\gamma(f_0, f_N) = 2\Phi(-u_N\tau/2h) + o(1),$$

and these quantities is bounded away from 0 for any $u_N = O(1)$ and tends to 1 for $u_N = o(1)$. On the other hand, consider the tests

$$\psi_N = \mathbf{1}_{\{\|F_N^* - F_0\|_p > r_N/2\}}. \tag{2.66}$$

where F_N^* is the empirical distribution function; if $p = \infty$, then (2.66) determines the Kolmogorov tests, if $p = 2$, then we obtain the Cramér–von Mises–Smirnov tests (see Section 1.4.1).

Proposition 2.13 *Let $N^{1/2}r_N \to \infty$. Then the tests* (2.66) *are minimax consistent.*

Proof. Recall that the random variables $k_n = N^{1/2}\|F_N^* - F_f\|_\infty$ are distribution free and bounded in P_f^N-probability in view of the Kolmogorov theorem which yields to

$$P_f^N(N^{1/2}\|F_N^* - F_f\|_p \le t) \ge P_f^N(k_n \le t) \to 1 \text{ as } N \to \infty,\ t \to \infty. \tag{2.67}$$

Let $f = f_0$. Then $\|F_N^* - F_f\|_\infty = O(N^{-1/2})$ in $P_{f_0}^N$-probability; by (2.67) this yields $\alpha_N(\psi_N) \to 0$. Let $f \in \mathcal{F}_{r_N}$. Then, uniformly over $f \in \mathcal{F}_N$,

$$\|F_N^* - F_0\|_p \ge \|F_f - F_0\|_p - \|F_N^* - F_f\|_\infty = r_N(1 + o(1))$$

in P_f^N-probability. These yield $\beta_N(\psi_N, \mathcal{F}_{r_N}) \to 0$. □

Therefore we get

Proposition 2.14 *Under the i.i.d. sample model let us test $H_0 : f = f_0$ against $H_1 : f \in \mathcal{F}_{r_N,p}$. Then distinguishability conditions are of classical form.*

2.6.3 Nondistinguishability in the Nonparametric Case

Usually, the situation is different in the nonparametric case.

Let $S^{n-1}(\rho)$ and $\breve{D}^n(\rho)$ be the sphere of radius ρ and the complement of the ball of radius ρ in R^n. Then we have (see Example 2.2 and the relations (1.14); a detailed proof for more general setting follows from Lemma 3.1 and Corollary 3.4 below):

Proposition 2.15 *Let us test the simple null-hypothesis $H_0 : v = 0$ against the alternative $H_1 : v \in \breve{D}^n(\rho_n)$ or $H_1 : v \in S^{n-1}(\rho_n)$, $n \to \infty$. Then the distinguishability conditions are of the form* (2.63) *with $\rho_n^* = n^{1/4}$, and if $\rho_n \asymp \rho_n^*$, then*

$$\gamma_n = 2\Phi(-u_n/2) + o(1), \quad \beta_n(\alpha) = \Phi(T_\alpha - u_n) + o(1); \quad u_n^2 = \rho_n^4/2n. \tag{2.68}$$

If $n \to \infty$, then $u_n \to 0$ for any $\rho > 0$. These yield the following:

Corollary 2.1 (Compare with Section 1.4.2)

(1) *Under the sequence Gaussian model* (2.1) *let us test the simple null-hypothesis* $H_0 : v = 0$ *against the alternative* $H_1 : v \in \breve{D}_{(l)}(\rho)$ *where* $\breve{D}_{(l)}(\rho) = \{v \in l^2 : |v| \geq \rho\}$ *is the complement of the ball of radius* ρ *in the sequence space* l^2, *or* $H_1 : v \in S_{(l)}(\rho)$ *where* $S_{(l)}(\rho) = \{v \in l^2 : |v| = \rho\}$ *is the sphere of radius* ρ *in the sequence space* l^2. *Then* $\gamma = 1$ *for any* $\rho > 0$.

(2) *Analogously, under the functional Gaussian model* (2.6), *let us test the simple null-hypothesis* $H_0 : s = 0$ *against the alternative* $H_1 : s \in \breve{D}_{(L)}(r)$ *where* $\breve{D}_{(L)}(r) = \{s \in L_2(0,1) : \|s\| \geq r\}$ *is the complement of the ball of radius* r *in the functional space* $L_2(0,1)$, *or* $H_1 : s \in S_{(L)}(r)$ *where* $S_{(L)}(r) = \{s \in L_2(0,1) : \|s\| = r\}$ *is the sphere of radius* r *in the functional space* $L_2(0,1)$. *Then* $\gamma = 1$ *for any* $r_\varepsilon > 0$, $\varepsilon > 0$.

The statements of Corollary 2.1 mean that the problems under consideration are *trivial* in the minimax sense.

One can extend these statements to other norms and models. In particular, under the sequence Gaussian model (2.1), introduce the norm (or quasi-norm)

$$|v|_p = \begin{cases} \left(\sum_i |v_i|^p\right)^{1/p}, & \text{if } 0 < p < \infty, \\ \sup_i |v_i|, & \text{if } p = \infty. \end{cases}$$

We call the *l_p-ball of radius ρ* the set $D_{p,l}(\rho) \subset l^2$ defined by the inequality $|v|_p < \rho$. Of course, it is possible that $|v|_p = \infty$ for $p < 2$. If $p < 1$, then $|\cdot|_p$ is a quasi-norm because the triangle inequality does not hold; the balls $D_{p,l}(\rho)$ are not convex in this case. Introduce the sets $S_{p,l}(\rho)$, $\breve{D}_{p,l}(\rho)$. These are "the boundary" (this corresponds to equality $|v|_p = \rho$) and the complement of $D_{p,l}(\rho)$ (this corresponds to inequality $|v|_p \geq \rho$). Below we will establish the following statements:

Let us test $H_0 : v = 0$ *against* $H_1 : v \in V$ *with* $V = \breve{D}_{p,l}(\rho)$ *or* $V = S_{p,l}(\rho)$. *Then* $\gamma(V) = 1$ *for any* $\rho > 0$, $0 < p \leq \infty$.

Under the functional Gaussian model consider the L_p-norm $\|\cdot\|_p$, $1 \leq p \leq \infty$, and let the *L_p-ball of radius r* be the set $D_{p,L}(r) \subset L_2(0,1)$ defined by inequality $\|s\|_p < r$. Let the sets $S_{p,L}(r)$, $\breve{D}_{p,L}(r)$ be the boundary and complement of $D_{p,L}(r)$.

Let us test $H_0 : s = 0$ *against* $H_1 : s \in S$ *with* $S = \breve{D}_{p,L}(r)$ *or* $S = S_{p,L}(r)$. *Then* $\gamma_\varepsilon(S) = 1$ *for any* $r > 0$, $\varepsilon > 0$, $0 < p \leq \infty$.

Proofs of these statements will be given in the following Chapters where we give more powerful methods to study minimax distinguishability.

First the triviality of the problem was shown by Ibragimov and Khasminkii [35] for $p = 2$; for $p \neq 2$ this follows from Burnashev [8].

Analogous statements hold for the i.i.d. sample model with an unknown probability density as well. Consider densities on $[0,1]$ and set $f_0(t) \equiv 1$,

$$\breve{D}_{p,f}(r, f_0) = \{f \in \mathcal{F} : \|f - f_0\|_p \geq r\}, \; S_{p,f}(r) = \{f \in \mathcal{F} : \|f - f_0\|_p = r\}.$$

Let us test $H_0 : f = f_0$ against $H_1 : f \in \mathcal{F}_N$ with $\mathcal{F}_N = \breve{D}_{p,f}(r, f_0)$ or $\mathcal{F}_N = S_{p,f}(r, f_0)$. Then $\gamma_N(\mathcal{F}_N) = 1$ for any $r \in (0,1)$, $N \geq 1$, $0 < p \leq \infty$.

The proof of this statement is presented in [52].

Proposition 2.14, Corollary 2.1, and the statements above show that minimax distinguishability depends essentially on which norm $|\cdot|_{(1)}$ is used "to remove" neighborhoods of the null-hypothesis in (2.61). If we use norms of Kolmogorov type, then we go to the classical distinguishability conditions.

On the other hand, if we use norms of L_p- or l_p-type to bound away the null-hypotheses and alternatives, then minimax distinguishability is not possible; the main reason is that an alternative (2.61) is "too wide" and we need to introduce additional constraints of type $h_2(\theta) \leq R$. In the cases of interest, usually, this is some other norm or quasi-norm: $h_2(\theta) = |\theta - \theta_0|_{(2)}$, and we obtain alternatives of the type

$$\Theta_{\varepsilon,1} = \{\theta \in \Theta : |\theta - \theta_0|_{(1)} \geq r_\varepsilon, \; |\theta - \theta_0|_{(2)} \leq R\}. \tag{2.69}$$

The hypothesis testing problems of this type are the main topic of this book.

2.7 Norms and Wavelet Transform

Let us consider the norms or quasi-norms of importance in the sequence space and in the functional space which can be used in (2.69). Recall that the *norm* in the (real) linear space A is a function $|a|$, $a \in A$ (or $\|a\|$) such that $|a| \geq 0$, $|\lambda a| = |\lambda||a|$, $\lambda \in R^1$, and (i): $|a| = 0$ iff $a = 0$; (ii): $|a_1 + a_2| \leq |a_1| + |a_2|$. If (i) does not hold, we call $|\cdot|$ a *semi-norm*; this determines a norm on the set of cosets with respect to the equivalence relation $a_1 \sim a_2$, if $|a_1 - a_2| = 0$. If (ii) is replaced by $|a_1 + a_2| \leq D(|a_1| + |a_2|)$, $D > 1$, then we call $|\cdot|$ a *quasi-norm* (with constant D); a semi-quasi-norm is defined a similar way. Below usually we omit the prefix "quasi-," because it is not essential to our aims. Often it is convenient to admit $|a| = \infty$ and to consider the linear space $A^* = \{a \in A : |a| < \infty\}$. We call two norms (or semi- and quasi-norms) $|\cdot|_{(1)}$, and $|\cdot|_{(2)}$, *equivalent*, if there exist positive constants c_1, c_2, that do not depend on $a \in A$, such that $c_1|a|_{(1)} \leq |a|_{(2)} \leq c_2|a|_{(1)}$ for any $a \in A$.

2.7.1 Norms in the Sequence Space

Let us consider the sequence space l which consists of sequences $v = \{v_i,\ i \in I\}$ where I is the finite or denumerable infinite index set. Let a positive sequence $\bar{a} = \{a_i, i \in I\}$ be given. Define the norm or quasi-norm, $|\cdot|_{\bar{a},p}$,

$$|v|_{\bar{a},p} = \begin{cases} (\sum_{i\in I} |v_i a_i|^p)^{1/p}, & \text{if } 0 < p < \infty, \\ \sup_{i\in I} |v_i a_i|, & \text{if } p = \infty. \end{cases} \tag{2.70}$$

If $p < 1$, then this is the quasi-norm; if $p \geq 1$, then $|\cdot|_{\bar{a},p}$ is the norm. We call the *l^p-ellipsoid* the ball $D_{\bar{a},p}(\rho) = \{v \in l^2 : \ |v|_{\bar{a},p} < \rho\}$. Denote by $\breve{D}_{\bar{a},p}(\rho)$, $S_{\bar{a},p}(\rho)$ the complement of the $(\bar{a},p)$-ellipsoid and the sphere in the norm $|\cdot|_{\bar{a},p}$. If $p = 2$, then this is the traditional ellipsoid in l^2 of the semi-axes $\{\rho/a_i,\ i \in I\}$. If $p < 1$, then $|\cdot|_{\bar{a},p}$ is the quasi-norm and the l^p-ellipsoids are not convex.

Note the case $a_i = i^r$, $i \in I = \mathbf{N}$, $-\infty < r < \infty$. In this case, we use the notation

$$|v|_{\bar{a},p} = |v|_{r,p} = \begin{cases} (\sum_{i=1}^{\infty} |v_i|^p i^{rp})^{1/p}, & \text{if } 0 < p < \infty, \\ \sup_{1\leq i<\infty} |v_i| i^r, & \text{if } p = \infty. \end{cases} \tag{2.71}$$

We denote by $D_{r,p}(\rho)$ the balls of radius ρ in the norm $|\cdot|_{r,p}$, analogously, $\breve{D}_{r,p}(\rho)$, $S_{r,p}(\rho)$. We call this the *power norm* with parameters $p > 0, r \in R^1$. If $r = 0$, then the power norm $|\cdot|_{0,p}$ is the l_p-norm $|\cdot|_p$; $D_{0,p}(\rho) = D_p(\rho)$ is the l_p-ball of radius ρ.

The following scale of norms is of importance below. Fix integers $K \geq 1, J_0 \geq 0$ and consider the sequence space l of pyramidal structure which consists of sequences

$$v = \{v_k,\ 1 \leq k \leq K,\ v_{ij},\ j > J_0, 1 \leq i \leq 2^j\}.$$

We call the *jth resolution* the set I_j of indexes $1 \leq k \leq K$ for $j = J_0$ and $1 \leq i \leq 2^j$ for $j > J_0$, and let L_j be the linear subspace generated by the coordinates (i,j), $i \in I_j$ (i.e., L_j is an n-dimensional linear space with $n = K$ for $j = J_0$ and $n = 2^j$ for $j > J_0$). Denote by Pr_j the projector to the jth resolution space L_j: $Pr_j(v) = \{v_k,\ k \in I_{J_0}\}$ if $j = J_0$ and $Pr_j(v) = \{v_{ij},\ i \in I_j\}$ if $j > J_0$. Consider the sequence

$$v^{(r,p)} = \{|Pr_{J_0}(v)|_p;\ 2^{rj}|Pr_j(v)|_p,\ j > J_0\}.$$

The *Besov norms* $|\cdot|_{r,p,h}$, $-\infty < r < \infty$, $0 < p, h \leq \infty$, in the sequence space are defined by $|v|_{r,p,h} = |v^{(r,p)}|_h$. Therefore, for $0 < p, h < \infty$,

$$|v|_{r,p,h} = \left(\left(\sum_{k=1}^{K} |v_k|^p \right)^{h/p} + \sum_{j=J_0+1}^{\infty} 2^{jrh} \left(\sum_{i=1}^{2^j} |v_{ij}|^p \right)^{h/p} \right)^{1/h}, \tag{2.72}$$

with standard modifications for $p = \infty$ or $h = \infty$. We denote $B_{r,p,h}(\rho)$ and call the *Besov body* the ball of radius ρ in the norm $|\cdot|_{r,p,h}$; we denote

$\breve{B}_{r,p,h}(\rho)$, $S_{r,p,h}(\rho)$ the complement of the Besov body and the sphere in the norm $|\cdot|_{r,p,h}$.

Note that if $p = h < \infty$, then the Besov norm $|\cdot|_{r,p,p}$ is equivalent to the power norm $|\cdot|_{r,p}$.

2.7.2 Norms in the Functional Space

We consider the functional parameters $f = \{f(t), t \in T\}$ (such as signals, probability density) defined on the interval $T = [0,1]$ only.

Recall that the L_p-norm is defined by the relation

$$\|f^{(m)}\|_p = \left(\int_0^1 |f(x)|^p\,dx\right) \text{ for } 0 < p < \infty, \quad \|f^{(m)}\|_\infty = \sup_{x\in[0,1]} |f(x)|.$$

This is the usual norm for $p \geq 1$, and this is the quasi-norm for $0 < p < 1$.

The main norms of interest in the functional space characterize a smoothness of functions. We consider different norms up to equivalence.

Let there exist the derivative $f^{(m)}(t), t \in [0,1]$, for an integer $m \geq 1$. The *Sobolev (quasi) semi-norm and norm* are defined by the relation

$$\|f\|^0_{m,p} = \|f^{(m)}\|_p, \quad \|f\|_{m,p} = \|f\|^0_{m,p} + \|f\|_p, \quad 0 \leq p \leq \infty. \tag{2.73}$$

If $m = 0$, we put $\|f\|^0_{0,p} = \|f\|_p$.

If $p \geq 1$, then $\|f\|^0_{m,p}$ is the semi-norms and $\|f\|_{m,p}$ is the norms. If $p = \infty$, then the semi-norm $\|f\|^0_{m,p}$ characterizes the maximum of the mth derivative of f, if $p = 2$, then this corresponds to the mean-square value of the mth derivative. The cases with $p < 2$ are used to characterize inhomogeneous smoothness. If $p \leq 1$, then these are the quasi-semi-norm and quasi-norm.

It is convenient to define the semi-norms $\|f\|^0_{\sigma,p}$ and norms $\|f\|_{\sigma,p}$ for a noninteger $\sigma = m+\delta$, $\delta \in (0,1)$. Let us consider the functions $f \in \tilde{\mathcal{F}}$ under periodical constraints: $f^{(l)}(0) = f^{(l)}(1)$, $l = 0,\ldots,m$, and the case $p = 2$. Then the definition is given in terms of Fourier coefficients $\theta_l = (f, \varphi_l)$ for the a standard Fourier basis in $L_2(0,1)$:

$$\varphi_0(t) = 1; \quad \varphi_{2l-1}(t) = \sqrt{2}\sin(2\pi l t), \quad \varphi_{2l}(t) = \sqrt{2}\cos(2\pi l t), \quad l \geq 1. \tag{2.74}$$

We put

$$\|f\|^0_{\sigma,2} = \left(\sum_{l=1}^{\infty}(2\pi l)^{2\sigma}(\theta^2_{2l-1} + \theta^2_{2l})\right)^{1/2}, \quad \|f\|_{\sigma,2} = \left(\theta_0^2 + (\|f\|^0_{\sigma,2})^2\right)^{1/2}. \tag{2.75}$$

Note that definitions (2.75) are extended to the case $\sigma < 0$. These correspond to functions with possible jumps and discontinuities. It is convenient to define the semi-norms $\|f\|^0_{\sigma,p}$ and norms $\|f\|_{\sigma,p}$ for noninteger and negative σ and $p \in (0,\infty]$ and without periodical constraints. In particular, a probability density $f(x)$ is the derivative of distribution function $F(x)$ and this means that the possible "smoothness" of $f(x)$ may be $\sigma \in (-1,0)$.

It is convenient to use Besov (semi- and quasi-) norm to measure a smoothness. The scale of Besov norm $\|\cdot\|_{\sigma,p,h}$ and semi-norm $\|\cdot\|^0_{\sigma,p,h}$ is defined for all $-\infty < \sigma < \infty$, $0 < p,h \leq \infty$. If $p,h \geq 1$, then these are norms and semi-norms. First, the definition is formulated for functions defined on R^1 (more generally, on R^n) in terms of the Fourier transform; the case of functions defined on $[0,1]$ corresponds to minimum norms of possible extensions on R^1.

There are many definitions which give equivalent norms. In particular, let $\sigma > 0$, $\sigma = m + \delta$, $\delta \in (0,1]$, $m \geq 0$ is an integer; $p \geq 1$. Denote by $\Delta_x(f,t)$ the first difference

$$\Delta_x(f,t) = f(t+x) - f(t), \quad x \in (0,1), \quad 0 \leq t \leq 1-x,$$

and by $\Delta_x^{(2)}(f,t)$ the second difference of the function f,

$$\Delta_x^{(2)}(f,t) = \Delta_x\Delta_x(f,t) = f(t+2x) - 2f(t+x) - f(t), \quad 0 \leq t \leq 1-2x.$$

Define the modulo of continuity of f in the L_p-norm by

$$\omega_p^{(1)}(f,y) = \sup_{0\leq x\leq y} \|\Delta_x(f,\cdot)\|_{L_p(0,1-x)}, \quad y \in (0,1),$$

$$\omega_p^{(2)}(f,y) = \sup_{0\leq x\leq y} \|\Delta_x^{(2)}(f,\cdot)\|_{L_p(0,1-2x)}, \quad y \in (0,1/2).$$

The Besov semi-norm is defined by the relations

$$\|f\|^0_{\sigma,p,h} = \left(\int_0^{1/2}\left(\frac{\omega_p^{(2)}(f^{(m)},y)}{y^\delta}\right)^h \frac{dy}{y}\right)^{1/h}, \quad h < \infty; \tag{2.76}$$

$$\|f\|^0_{\sigma,p,\infty} = \sup_{0<y<1/2} \frac{\omega_p^{(2)}(f^{(m)},y)}{y^\delta}. \tag{2.77}$$

The Besov norm is defined by the equality

$$\|f\|_{\sigma,p,h} = \|f\|_p + \|f\|^0_{\sigma,p,h}. \tag{2.78}$$

Remark. Other definitions of the Besov norm correspond either to the substitution of the modulo of continuity $\omega_p^2(f^m,y)$ of the mth derivative by L_p-norm of the difference

$$\|\Delta_y^{(2)}(f^m,\cdot)\|_{L_p(0,1-2y)}$$

(without the supremum of the L_p-norm) or by the L_p-norm of the r-th difference of the function $\|\Delta_y^{(r)}(f,\cdot)\|_{L_p(0,1-2y)}$, $r > \sigma+1$; or by the L_p-norm of local oscillation $\operatorname{osc} f(\cdot,y)$ defined by

$$\operatorname{osc} f(x,y) = \inf \frac{1}{2h}\int_{|x-t|\leq y} |f(t) - P_r(t)|\,dt,$$

where the infimum is taken over all polynomials P_r of the order $r \le m+1$. If $\delta \in (0,1)$ (i.e., σ is not an integer), then one can substitute $\omega_p^{(2)}$ by $\omega_p^{(1)}$, or the norm of the second-order difference $\|\Delta_y^{(2)}(f^m,\cdot)\|_{L_p(0,1-2y)}$ by the norm of the first-order difference $\|\Delta_y(f^m,\cdot)\|_{L_p(0,1-y)}$ in the definition of the Besov semi-norm (2.76). If $p=h=\infty$, $\sigma>0$, $\sigma=m+\delta\notin \mathbf{N}$, then Besov norm is equivalent to Hölder norm:

$$\|f\|_{\sigma,\infty,\infty} = \|f\|_p + \sup_{x\in[0,1],\ y\in(0,1-x)} y^{-\delta}|f^{(m)}(x+y)-f^{(m)}(x)|.$$

The Besov norm with $p=h=2$ is equivalent to the Sobolev norm. The case $h=\infty$ corresponds to the Nikol'skii norm.

We denote by $B_{p,h}^{\sigma}(R)$ or $S_p^{\sigma}(R)$ the balls of radius R in the Besov norm $\|\cdot\|_{\sigma,p,h}$ or in the Sobolev norm $\|\cdot\|_{\sigma,p}$.

Note the inequalities between Besov norms with different parameters and between Besov norms and Sobolev norms with $p,h\ge 1$, $\sigma\ge 0$:

$$\|f\|_{\sigma_1,p_1,h_1} \le c\|f\|_{\sigma_2,p_2,h_2} \quad \text{for } \sigma_1\le\sigma_2,\ \ p_1\le p_2,\ \ h_1\ge h_2; \tag{2.79}$$

$$c_1\|f\|_{\sigma,p,\infty} \le \|f\|_{\sigma,p} \le c_2\|f\|_{\sigma,p,1}; \tag{2.80}$$

$$\|f\|_{\sigma,p,h} \le c_3\|f\|_{\sigma_1,p_1,h} \quad \text{for } \sigma_1=\sigma-1/p+1/p_1,\ \ p>p_1\ge 1; \tag{2.81}$$

the positive constants c, c_1, c_2, c_3 depend on parameters σ, p, h, but not on f. Analogous to (2.79), (2.81) inequalities hold for the Sobolev norms as well.

In particular, if $\eta=\sigma-1/p>0$, then by (2.79), (2.81) we have the inequality for the Hölder norm:

$$\|f\|_{\eta,\infty,\infty} \le c_4\|f\|_{\sigma,p,h},\ \ \|f\|_{\eta,\infty,\infty} \le c_5\|f\|_{\sigma,p}. \tag{2.82}$$

The following property is helpful. Consider an equispaced partition of the interval $[0,1]$ to intervals $\delta_{1,n},\dots,\delta_{n,n}$ of length $l=1/n$. For $f\in L_1(0,1)$ denote by $Pr_n f$ the projection of f to a linear space which consists of the step functions which are constant in intervals $\delta_{i,n}$, $i=1,\dots,n$,

$$Pr_n f(t) = \sum_{i=1}^{n} f_i \mathbf{1}_{\delta_{i,n}}(t);\quad f_i = n\int_{\delta_{i,n}} f(u)\,du. \tag{2.83}$$

Proposition 2.16 *Let $p\ge 1, \sigma>0$. There exist positive constants C_1, C_2 which depend on p,σ only, such that for any $f\in B_{p,h}^{\sigma}(R)$ or $f\in S_p^{\sigma}(R)$ one has*

$$\|Pr_n f\|_p \ge C_1\|f\|_p - C_2 R n^{-\sigma}.$$

This inequality shows that the step functions, though having poor approximation properties for $\sigma>1$, possess fairly good "projection" properties which facilitate the construction of consistent tests.

Proof of Proposition 2.16 Given in the Appendix, Section A.1.

2.7.3 The Wavelet Transform

It was noted in Section 2.1 that there is one-to-one correspondence between the Gaussian model in sequence space (2.3) and the Gaussian model in functional space (2.5), which is determined by the transform $f \to a = \{a_i = (f, \varphi_i)\}$ for any fixed orthonormal basis $\{\varphi_i, i \in I\}$. However, the problem is to describe the given set $V \subset L_2(0,1)$ in terms of Fourier coefficients $\{a_i\}$. If the set V is determined in terms of norms $\|\cdot\|_{(1),(2)}$, then we need to find a basis such that there is a simple enough description of the norms $\|\cdot\|_{(1),(2)}$ in terms of Fourier coefficients.

In particular, the Sobolev norms with $p = 2$ under the periodical constraints (2.75) are determined by the discrete Fourier transform under the standard Fourier basis (2.74). The Sobolev semi-norms $\|\cdot\|^0_{\sigma,2}$ are equivalent to the norm $|\cdot|_{\sigma,2}$ in the space of Fourier coefficients for all $\sigma \in R^1$.

In the cases of general Besov norms the *wavelet transform* has good properties.

First the wavelet transform was used in nonparametric estimation problems by Donoho and Johnstone (see [17]). This method is very helpful in studing inhomogeneous estimation problems.

The wavelet transform corresponds to the specific orthonormal basis of a pyramidal structure in $L_2(0,1)$,

$$\{\varphi_{ij} \text{ where } 1 \le i \le K, \text{ if } j = J_0; \quad 1 \le i \le 2^j, \text{ if } j > J_0\},$$

generated by the specific functions $\psi(t)$, $\varphi(t)$ (called "father" and "mother"):

$$\varphi_{iJ_0}(t) = 2^{J_0/2}\psi(Kt - i); \quad \varphi_{ij}(t) = 2^{j/2}\varphi(2^j t - i), \ j > J_0; \quad t \in [0,1]$$

with some correction near points $t = 0$ and $t = 1$. Under the wavelet transform the Besov norm $\|\cdot\|_{\sigma,p,h}$ is equivalent to the norm $|\cdot|_{r,p,h}$ in the space of wavelet coefficients. More exactly, under general constraints on the functions $\psi(t), \varphi(t)$ (these have finite support and are orthogonal to polynomials of degree m for large enough m — we call such a wavelet basis *m-regular*) there exists positive constants $b_l = b_l(\eta^*)$, $l = 1, 2$, such that for any $1 \le p, h \le \infty$, $1/p - 1 < \sigma \le \sigma^*$ one has:[2]

$$b_1|a|_{r,p,h} \le \|f\|_{\sigma,p,h} \le b_2|a|_{r,p,h}, \quad r = \sigma + 1/2 - 1/p, \tag{2.84}$$

where $a = \{a_{ij} = (f, \varphi_{ij})\} \in l^2$ is the sequence of wavelet coefficients of a function $f \in L_2(0,1)$ (see [31], [11], [12], [17], [18]).

In view of relations (2.84) the main part of this book deals with the Gaussian model, not in functional space, but in sequence space.

---

[2]We are not sure that the constraint $\eta \ge 0$ is necessary to satisfy (2.84). The essential part of results below could be extended, if it were possible to loosen this constraint. The authors would greatly appreciate any comments on this topic.

2.8 Short Overview of Minimax Estimation

It seems useful to compare asymptotic minimax estimation and hypothesis testing problems. In the minimax estimation problem, given a family $\Theta_\varepsilon \subset \Theta$, we would like to construct an estimator $\theta_\varepsilon = \theta_\varepsilon(x_\varepsilon)$ of an unknown parameter $\theta \in \Theta_\varepsilon$. Given the loss function $l(\theta_\varepsilon, \theta)$, let

$$R_\varepsilon(\theta_\varepsilon, \theta; l) = E_{\varepsilon,\theta} l(\theta_\varepsilon, \theta)$$

be the average risk and let

$$R_\varepsilon(\theta_\varepsilon; l) = \sup_{\theta \in \Theta_\varepsilon} R_\varepsilon(\theta_\varepsilon, \theta; l)$$

be its maximum value. The problem is to study the asymptotics of *minimax risk*

$$R_{\varepsilon,l} = \inf R_\varepsilon(\theta_\varepsilon; l)$$

where the infimum is taken over all estimators θ_ε, and to construct *asymptotically minimax estimators* $\hat{\theta}_\varepsilon$, such that $R_\varepsilon(\hat{\theta}_\varepsilon; l) \sim R_{\varepsilon,l}$, or *rate minimax estimators* $\hat{\theta}_\varepsilon$, such that $R_\varepsilon(\hat{\theta}_\varepsilon; l) \asymp R_{\varepsilon,l}$.

Typically, the losses are of the form $l(\theta_\varepsilon, \theta) = L(|\theta_\varepsilon - \theta|_{(1)})$, where $L(t)$ is a nondecreasing function in $t \geq 0$, $L(0) = 0$ and $|\cdot|_{(1)}$ is a norm (quasi-norm). Often $F(t) = t^c$, $c > 0$, is a power function which implies $R_\varepsilon(\theta_\varepsilon; l_t) = t^c R_\varepsilon(\theta_\varepsilon; l)$ where $l_t(\theta_\varepsilon, \theta) = F(t\,|\theta_\varepsilon - \theta|_{(1)})$. Given the positive family $r_\varepsilon \to 0$, let us consider the *normalized losses*

$$l_\varepsilon(\theta_\varepsilon, \theta) = L(r_\varepsilon^{-1}|\theta_\varepsilon - \theta|_{(1)}). \tag{2.85}$$

For the power function L one has $R_\varepsilon(\theta_\varepsilon; l) = r_\varepsilon^c R_\varepsilon(\theta_\varepsilon; l_\varepsilon)$. This motivates the definition. We call *minimax rates of estimation* the family $r_\varepsilon^{\text{est}}$ such that $R_{\varepsilon,l_\varepsilon} \asymp 1$ for normalized losses (2.85).

The minimax rates of estimation do not often depend on function L and they characterize the possible accuracy of estimation in the norms $|\cdot|_{(1)}$ for the sets Θ_ε. The norm $|\cdot|_{(1)}$ is analogous to the norm in the first constraint in (2.69). Moreover, typically the set family Θ_ε in the estimation problem is determined by the constraint $h_2(\theta) \leq R_\varepsilon$ where the distance $h_2(\theta) = |\theta - \theta_0|_{(2)}$ is determined by some other norm for some $\theta_0 \in \Theta$. This is analogous to the second constraint in (2.69). These yield to analogy between minimax estimation and hypothesis testing problems.

Minimax estimation problems have been studied very intensively. We refer the reader to by Ibragimov and Khasminskii [38] where the parametric case has been studied in detail. In particular, under certain regularity assumptions (including LAN and (2.62)) the minimax rates $r_\varepsilon^{\text{est}} \asymp r_\varepsilon^*$ correspond to classical rates in the functional Gaussian and density models. The maximum likelihood and Bayesian estimators are asymptotically minimax.

Under the functional Gaussian model it was shown by Ibragimov and Khasminskii [35] that if $|\cdot|_{(1)} = \|\cdot\|_2$, then it is impossible to construct minimax consistent estimators (such that $r_\varepsilon^{\text{est}} \to 0$) without constraints on sets Θ_ε (type of $h_2(\theta_\varepsilon) \le R_\varepsilon$). Some important nonparametric estimation problems were considered in [38] under the functional Gaussian model and i.i.d. sample models with unknown density. In particular, it was shown that if $|\cdot|_{(1)} = \|\cdot\|_2$ and $|\cdot|_{(2)} = \|\cdot\|_{2,\eta}$ are norms of Sobolev type (with $\eta > 0$ being the smoothness in the L_2-norm and $R_\varepsilon = R$ does not depend on ε), then the rates are of the nonclassical form:

$$r_\varepsilon^{\text{est}} \asymp \varepsilon^{2\eta/(2\eta+1)}, \quad r_N^{\text{est}} \asymp N^{-\eta/(2\eta+1)}. \tag{2.86}$$

The same rates have been obtained, for the norms $|\cdot|_{(1)} = \|\cdot\|_p$ and $|\cdot|_{(2)} = \|\cdot\|_{q,\eta}$ with $2 \le p < \infty$, by Bretagnolle and Huber [7] and by Ibragimov and Khasminskii [36]. However, if $p = \infty$, then

$$r_\varepsilon^{\text{est}} \asymp (\varepsilon^2 \log(\varepsilon^{-1}))^{\eta/(2\eta+1)}, \quad r_N^{\text{est}} \asymp (N^{-1}\log N)^{\eta/(2\eta+1)}. \tag{2.87}$$

The general inhomogeneous cases for Besov norms $|\cdot|_1 = \|\cdot\|_{\sigma,p,h}$ and $|\cdot|_2 = \|\cdot\|_{\eta,q,t}$ and for Sobolev norms have been studied by Donoho and Johnstone [17], Donoho, Johnstone, Kerkyacharian, and Picard [18], Lepski, Mammen, and Spokoiny [77], and Delyone and Juditsky [14]. The methods are based either on the wavelet transform and on the estimation in the sequence space or on the pointwise adaptive procedure. These results give the minimax rates of estimation under the functional Gaussian model, at least for

$$1 \le p, q, h, t \le \infty, \quad \sigma \ge 0, \quad \eta - \sigma > \min(1/p - 1/q, 0).$$

These have the form

$$r_\varepsilon^{\text{est}} \asymp \begin{cases} \varepsilon^\delta, & \text{if } \eta/p - \sigma/q > 1/2q - 1/2p, \\ (\varepsilon\sqrt{\log(\varepsilon^{-1})}\,)^{\delta'}, & \text{if } \eta/p - \sigma/q < 1/2q - 1/2p, \end{cases} \tag{2.88}$$

where

$$\delta = (\eta - \sigma)/(\eta + 1/2), \quad \delta' = (\eta - \sigma - 1/q + 1/p)/(\eta - 1/q + 1/2).$$

For the frontier case $\eta/p - \sigma/q = 1/2p - 1/2q$ the upper and lower bounds for rates have been obtained: these give the exact power of order ε^δ but different quantities for the power of logarithmic orders. In density or regression models analogous results have been obtained: one should substitute ε by $N^{-1/2}$ or $n^{-1/2}$.

The principal rates do not depend on the parameters h, t of Besov norms. Therefore, using relations (2.80) between Sobolev and Besov norms, one can translate the rates to the Sobolev norms $|\cdot|_{(1)} = \|\cdot\|_{\sigma,p}$ and $|\cdot|_{(2)} = \|\cdot\|_{\eta,q}$.

Sharp asymptotics have been studied for two main cases only. The first case is periodic functions with $L(|\cdot|_{(1)}) = \|\cdot\|_2^2$, $h_2(\cdot) = \|\cdot\|_{\eta,2}$ which correspond to ellipsoids in $L_2(0,1)$ and square losses, i.e., $p = q = 2$ (see Pinsker [87] and Efroimovich and Pinsker [20]). Recently these results were extended to the case $L(|\cdot|_{(1)}) = \|\cdot\|_p^p$, $p > 0$ (see Tsybakov [96]). The second case is Hölder balls and the losses in the supreme-norm $|\cdot|_{(1)} = \|\cdot\|_\infty$ i.e., $p = q = \infty$ (see Korostelev [68] and Donoho [16]).

2.9 Constraints of Interest

In view of the relations between minimax estimation and hypothesis testing problems it seems natural to study the norm $|\cdot|_{(2)} = \|\cdot\|_{\eta,q}$ of Sobolev or Besov type, jointly with the norm $|\cdot|_{(1)} = \|\cdot\|_p$ which defines the constraints in (2.69). However, it is of interest to extend the setting to the case of the norm $|\cdot|_{(1)} = \|\cdot\|_{\sigma,p}$ and of the analogous Besov norm which corresponds to estimation of the σ-derivative of a signal under L_p-losses.

In view of these arguments, one of the main problem we study in this book corresponds to the sequence Gaussian model (2.1) and the alternatives determined by power norms

$$V_\varepsilon = \{v \in l^2 : \ |v|_{r,p} \geq \rho_\varepsilon; \ |v|_{s,q} \leq R_\varepsilon\}, \tag{2.89}$$

and by Besov norms

$$V_\varepsilon = \{v \in l^2 : \ |v|_{r,p,h} \geq \rho_\varepsilon; \ |v|_{s,q,t} \leq R_\varepsilon\}. \tag{2.90}$$

The results are extended from alternatives (2.89), (2.90) to the functional Gaussian model (2.5) and alternatives in $L_2(0,1)$ defined by Sobolev and Besov norms in functional space

$$S_\varepsilon = \{s \in L_2(0,1) : \ \|s\|_{(\sigma,p)} \geq r_\varepsilon; \ \|s\|_{(\eta,q)} \leq R\} \tag{2.91}$$

with substitutions

$$r = \sigma + 1/2 - 1/p, \quad s = \eta + 1/2 - 1/q; \quad \rho_\varepsilon = r_\varepsilon/\varepsilon, \quad R_\varepsilon = R/\varepsilon;$$

here $\|\cdot\|_{(\sigma,p)}$, $\|\cdot\|_{(\eta,q)}$ are either Sobolev norms $\|\cdot\|_{\sigma,p}$, $\|\cdot\|_{\eta,q}$ or Besov norms $\|\cdot\|_{\sigma,p,h}$, $\|\cdot\|_{\eta,q,t}$ with any $h,t \geq 1$; usually (except some frontier cases) the rates do not depend on the parameters h,t.

2.10 Rates in Estimation and in Hypothesis Testing

Let the sets Θ_ε be determined by the second constraint in (2.69): let $\Theta_\varepsilon = \{\theta \in \Theta : \ |\theta - \theta_0|_{(2)} \leq R\}$ and $\hat{\theta}_\varepsilon$ be estimators such that $R_\varepsilon(\hat{\theta}_\varepsilon; l_\varepsilon) \asymp 1$ for the sets Θ_ε where l_ε are normalized losses (2.85). One can construct the "plug-in" tests

$$\psi_\varepsilon = \mathbf{1}_{\{|\hat{\theta}_\varepsilon - \theta_0|_{(1)} > b r_\varepsilon/2\}}. \tag{2.92}$$

Proposition 2.17 *Let $r_\varepsilon^{\text{est}}$ be the minimax rate of estimation,*

$$l(\hat{\theta}_\varepsilon, \theta) = L(|\hat{\theta}_\varepsilon - \theta|_{(1)}),$$

where L is a nondecreasing nonnegative function, $L(t) \to \infty$ as $t \to \infty$ and r_ε is a family in (2.69) such that $r_\varepsilon / r_\varepsilon^{\text{est}} \to \infty$. Then "plug-in" tests (2.92) provide minimax distinguishability for testing $H_0 : \theta = \theta_0$ against alternatives (2.69). Here $b = D^{-1}$ is constant for a quasi-norm $|\cdot|_{(1)}$.

Proof. Using the Markov inequality $P(X > T) \leq E(X)/T$ for a nonnegative random variable X, one has

$$\begin{aligned}\alpha(\psi_\varepsilon) &= P_{\varepsilon,\theta_0}(|\hat{\theta}_\varepsilon - \theta_0|_{(1)} > br_\varepsilon/2)\\ &\leq P_{\varepsilon,\theta_0}(L((r_\varepsilon^{\text{est}})^{-1}|\hat{\theta}_\varepsilon - \theta_0|_{(1)}) > L(br_\varepsilon/2r_\varepsilon^{\text{est}}))\\ &\leq E_{\varepsilon,\theta_0}(L(\hat{r}_\varepsilon^{-1}|\hat{\theta}_\varepsilon - \theta_0|_{(1)}))/L(br_\varepsilon/2r_\varepsilon^{\text{est}}) = O(1/L(br_\varepsilon/2r_\varepsilon^{\text{est}})) \to 0,\end{aligned}$$

and by a quasi-norm pseudo-triangle inequality with $b = D^{-1}$ we have: if $\theta \in \Theta_{\varepsilon,1}$, then

$$|\hat{\theta}_\varepsilon - \theta_0|_{(1)} \geq b|\theta - \theta_0|_{(1)} - |\hat{\theta}_\varepsilon - \theta|_{(1)} \geq br_\varepsilon - |\hat{\theta}_\varepsilon - \theta|_{(1)},$$

and uniformly over $\theta \in \Theta_{\varepsilon,1}$ one has

$$\begin{aligned}\beta_\varepsilon(\psi_\varepsilon,\theta) &= P_{\varepsilon,\theta}(|\hat{\theta}_\varepsilon - \theta_0|_{(1)} \leq br_\varepsilon/2) \leq P_{\varepsilon,\theta}(|\hat{\theta}_\varepsilon - \theta|_{(1)} \geq br_\varepsilon/2)\\ &\leq E_{\varepsilon,\theta}(L((r_\varepsilon^{\text{est}})^{-1}|\hat{\theta}_\varepsilon - \theta|_{(1)}))/L(br_\varepsilon/2r_\varepsilon^{\text{est}})\\ &= O(1/L(r_\varepsilon/r_\varepsilon^{\text{est}})) \to 0. \quad \square\end{aligned}$$

Note that tests (2.60) and Kolmogorov tests (2.66) are just of plug-in type (2.92) and provide the sharp rate of testing: $\gamma_\varepsilon \to 0$ iff $r_\varepsilon/r_\varepsilon^{\text{est}} \to \infty$. One can imagine that the analogous properties are fulfilled for a general enough case, i.e., one can provide the best rates of testing by using good enough estimators. This holds in regular parametric or nonparametric hypothesis testing problems with distinguishability conditions of classical type. In this case the nonparametric hypothesis testing problem seems to be unimportant, because good estimators give essentially more information on the unknown parameter θ_ε that good tests would give.

We show below that, usually, this does not hold in nonparametric hypothesis testing problems. The important exclusion is the case with distinguishability conditions of "degenerate type" (this holds, in particular, for the supreme-norm $|\cdot|_{(1)} = \|\cdot\|_\infty$). In other cases the minimax rates of testing r_ε^* are much smaller than the minimax rates of estimation

$$r_\varepsilon^* \ll r_\varepsilon^{\text{est}}. \tag{2.93}$$

Moreover, test procedures (2.92) do not provide minimax distinguishability, if $r_\varepsilon^* \ll r_\varepsilon \ll r_\varepsilon^{\text{est}}$.

Relation (2.93) means that it is possible to test the null-hypothesis $H_0 : \theta = \theta_0$ against such alternatives $\theta_{\varepsilon,1}$ that difference the $|\theta_{\varepsilon,1} - \theta_0|_{(1)}$ is much smaller than the possible accuracy of estimation $r_\varepsilon^{\text{est}}$!

3

Minimax Distinguishability

In this chapter we begin the realization of the scheme noted in Section 2.5.2. For some problems we obtain sharp or rate asymptotics.

To obtain the upper bounds we introduce and study minimax properties of various tests, in particular, of χ^2-tests, of their extension χ^p-tests, and of supreme-tests. To obtain the lower bounds we introduce and study some classes of prior families, namely, product and semiorthogonal priors. For the main problems under consideration we describe the frontier between the triviality and classical asymptotics under one-sided constraints. We also describe the conditions of triviality under two-sided constraints.

3.1 Minimax Properties of Test Families

We would like to study some classes of test families under the Gaussian models. We start with some general remarks.

3.1.1 Consistency Conditions and Gaussian Bounds

First, let us test $H_0 : v = 0$ against $H_1 : v \in V$ under the sequence Gaussian model (2.1). Let us consider linear tests $\psi_{r,T} = \mathbf{1}_{\{t_r > T\}}$ based on the linear statistics t_r and on the nonrandom threshold T,

$$t_r = (x, r) = \sum_i x_i r_i, \quad r \in l^2; \quad |r|^2 = |r|_2^2 = (r, r) = 1. \tag{3.1}$$

Recall that $t_r \sim N(0,1)$ in P_0-probability. This yields

$$\alpha(\psi_{r,T}) = P_0((x,r) \geq T) = 1 - \Phi(T) = \Phi(-T).$$

One has $t_r \sim N((r,v),1)$ in P_v-probability. This implies

$$\beta(\psi_{r,T},v) = P_v((x,r) < T) = \Phi(T-(r,v)),$$
$$\beta(\psi_{r,T},V) = \sup_{v \in V} \Phi(T-(r,v)) = \Phi(T-h(V)),$$

where

$$h(V) = \inf_{v \in V}(r,v). \tag{3.2}$$

Analogously, under the functional Gaussian model (2.5), let us test the null-hypothesis $H_0 : s = 0$ against the alternative $H_1 : s \in S \subset L_2(0,1)$. Let us consider the tests based on the linear statistics

$$\psi_{r,T} = \mathbf{1}_{\{t_r > T\}}, \quad t_r = \varepsilon^{-1} \int_0^1 r(t)\, dX_\varepsilon(t); \quad r \in L_2(0,1), \quad \|r\| = 1.$$

We get the same relations with the change $h(V)$ by

$$h(S) = \varepsilon \inf_{s \in S}(r,s). \tag{3.3}$$

We would like to extend these results to a general case under an asymptotic setting. Let us test $H_0 : \theta \in \Theta_{\varepsilon,0}$ against $H_1 : \theta \in \Theta_{\varepsilon,1}$. Consider a family of tests $\psi_\varepsilon = \psi_{\varepsilon,T_\varepsilon} = \mathbf{1}_{\{t_\varepsilon \geq T_\varepsilon\}}$ based on statistics t_ε and on nonrandom thresholds T_ε.

Proposition 3.1 *Assume that $E_{\varepsilon,\theta} t_\varepsilon^2 < \infty \ \forall\, \theta \in \Theta_{\varepsilon,0} \cup \Theta_{\varepsilon,1}$ and that T_ε be such that the following relations hold:*

$$\inf_{\theta \in \Theta_{\varepsilon,0}} \frac{T_\varepsilon - E_{\varepsilon,\theta} t_\varepsilon}{\sqrt{\mathrm{Var}_{\varepsilon,\theta}(t_\varepsilon)}} \to \infty, \qquad \inf_{\theta \in \Theta_{\varepsilon,1}} \frac{E_{\varepsilon,\theta} t_\varepsilon - T_\varepsilon}{\sqrt{\mathrm{Var}_{\varepsilon,\theta}(t_\varepsilon)}} \to \infty. \tag{3.4}$$

Then $\gamma_\varepsilon(\psi_\varepsilon; \Theta_{\varepsilon,0}, \Theta_{\varepsilon,1}) \to 0$.

Proof. This follows from the Chebyshev inequality. First, uniformly over $\theta \in \Theta_{\varepsilon,0}$, one has

$$\begin{aligned} \alpha_\varepsilon(\psi_\varepsilon,\theta) &= P_{\varepsilon,\theta}(t_\varepsilon > T_\varepsilon) = P_{\varepsilon,\theta}(t_\varepsilon - E_{\varepsilon,\theta} t_\varepsilon > T_\varepsilon - E_{\varepsilon,\theta} t_\varepsilon) \\ &\leq \mathrm{Var}_{\varepsilon,\theta}(t_\varepsilon)/(T_\varepsilon - E_{\varepsilon,\theta} t_\varepsilon)^2 \to 0. \end{aligned}$$

Next, uniformly over $\theta \in \Theta_{\varepsilon,1}$,

$$\begin{aligned} \beta_\varepsilon(\psi_\varepsilon,\theta) &= P_{\varepsilon,\theta}(t_\varepsilon \leq T_\varepsilon) = P_{\varepsilon,\theta}(E_{\varepsilon,\theta} t_\varepsilon - t_\varepsilon \geq E_{\varepsilon,\theta} t_\varepsilon - T_\varepsilon) \\ &\leq \mathrm{Var}_{\varepsilon,\theta}(t_\varepsilon)/(E_{\varepsilon,\theta} t_\varepsilon - T_\varepsilon)^2 \to 0. \end{aligned}$$

This yields the statement of Proposition 3.1. □

Let $H_0 : \theta = \theta_0$ be a simple null-hypothesis and an alternative of the form $H_1 : \theta \in \Theta_\varepsilon$. Without loss of generality we can assume statistics t_ε

to be centered and normalized with respect to the null-hypothesis

$$E_{\varepsilon,\theta_0} t_\varepsilon = 0, \quad \mathrm{Var}_{\varepsilon,\theta_0}(t_\varepsilon) = 1. \tag{3.5}$$

Set

$$h_\varepsilon(\theta) = E_{\varepsilon,\theta}(t_\varepsilon), \quad h_\varepsilon = \inf_{\theta\in\Theta_\varepsilon} E_{\varepsilon,\theta}(t_\varepsilon). \tag{3.6}$$

Corollary 3.1 *Assume $h_\varepsilon \to \infty$ and the following inequality holds:*

$$\mathrm{Var}_{\varepsilon,\theta}(t_\varepsilon) \le B_\varepsilon + \delta_\varepsilon E^2_{\varepsilon,\theta}(t_\varepsilon) \ \ \forall\, \theta\in\Theta_\varepsilon; \quad B_\varepsilon = O(1),\ \delta_\varepsilon = o(1) \tag{3.7}$$

(it suffices $\mathrm{Var}_{\varepsilon,\theta}(t_\varepsilon) \le B + \left(E_{\varepsilon,\theta}(t_\varepsilon)\right)^\tau$ for some $\tau \in (0,2)$). Then $\beta_\varepsilon(\psi_{\varepsilon,T_\varepsilon},\Theta_\varepsilon) \to 0$ for any $T_\varepsilon = O(1)$. Moreover, let

$$T_\varepsilon \to \infty, \quad T_\varepsilon \le (1-c)h_\varepsilon, \quad c\in(0,1)$$

(one can choose $T_\varepsilon = h_\varepsilon/2$). Then $\gamma_\varepsilon(\psi_{\varepsilon,T_\varepsilon}) \to 0$.

Proof. It suffices to consider the case $T_\varepsilon \to \infty$. By (3.5) this yields the first relation (3.4). The second relation follows from assumptions (3.7) and $T_\varepsilon \le (1-c)h_\varepsilon$. □

Corollary 3.2 *Assume that (3.7) holds. Moreover, setting $t_{\varepsilon,\theta} = t_\varepsilon - h_\varepsilon(\theta)$, assume $t_{\varepsilon,\theta}$ to be asymptotically $N(0,1)$-Gaussian in $P_{\varepsilon,\theta}$-probability for $\theta=\theta_0$, uniformly over $\theta\in\Theta_\varepsilon$ such that $h_\varepsilon(\theta) = O(1)$. Then*

$$\alpha_\varepsilon(\psi_{\varepsilon,T_\alpha}) = \alpha + o(1), \quad \beta_\varepsilon(\psi_{\varepsilon,T_\alpha}) \le \Phi(T_\alpha - h_\varepsilon) + o(1),$$
$$\gamma_\varepsilon(\psi_{\varepsilon,T_\varepsilon}) \le 2\Phi(-h_\varepsilon/2) + o(1), \quad T_\varepsilon = h_\varepsilon/2.$$

Proof. The relation for type I errors follows directly from the asymptotic normality of t_ε under the null-hypothesis.

In view of Corollary 3.1, to study type II errors one can assume $h_\varepsilon = O(1)$. Consider a family $\theta = \theta_\varepsilon \in \Theta_\varepsilon$. If $h_\varepsilon(\theta) \to \infty$, then, analogously to Corollary 3.1, we have $\beta(\psi_{\varepsilon,T},\theta) \to 0$ for $T = T_\alpha$ and $T = h_\varepsilon/2$. Let $h_\varepsilon(\theta) = O(1)$. Using the asymptotic normality of $t_{\varepsilon,\theta}$, we have, for $T_\varepsilon = h_\varepsilon/2$,

$$\begin{aligned}
\beta(\psi_{\varepsilon,T_\alpha},\theta) &= P_{\varepsilon,\theta}(t_\varepsilon \le T_\alpha) = P_{\varepsilon,\theta}(t_\varepsilon - h_\varepsilon(\theta) \le T_\alpha - h_\varepsilon(\theta)) \\
&= \Phi(T_\alpha - h_\varepsilon(\theta)) + o(1) \le \Phi(T_\alpha - h_\varepsilon) + o(1), \\
\alpha(\psi_{\varepsilon,T_\varepsilon}) &= \Phi(-h_\varepsilon/2) + o(1), \quad \beta(\psi_{\varepsilon,T_\varepsilon},\theta) = P_{\varepsilon,\theta}(t_\varepsilon \le T_\varepsilon) \\
&= P_{\varepsilon,\theta}(t_\varepsilon - h_\varepsilon(\theta) \le h_\varepsilon/2 - h_\varepsilon(\theta)) \le \Phi(-h_\varepsilon/2) + o(1).
\end{aligned}$$ □

3.1.2 Tests of χ^2-Type

Return to the Gaussian model in a sequence space (2.1). Let us test $H_0 : v = 0$ against $H_1 : v \in V_\varepsilon;\ V_\varepsilon \subset l^2$. Let $w_\varepsilon = \{w_{\varepsilon,i}\}$ be a family of nonnegative real-valued sequences. We call *tests of χ^2-type* the tests $\psi_{\varepsilon,T_\varepsilon}$

based on statistics of χ^2-type:

$$\psi_{\varepsilon,T_\varepsilon} = \psi_{\varepsilon,w_\varepsilon;T_\varepsilon} = \mathbf{1}_{\{t_\varepsilon > T_\varepsilon\}}, \quad t_\varepsilon = \sum_i w_{\varepsilon,i}(x_i^2 - 1). \tag{3.8}$$

Set

$$h_\varepsilon(v) = h_\varepsilon(v, w_\varepsilon) = \sum_i w_{\varepsilon,i} v_i^2; \quad t_\varepsilon(v) = t_\varepsilon - h_\varepsilon(v), \quad |w_\varepsilon| = \sup_i w_{\varepsilon,i}. \tag{3.9}$$

Lemma 3.1 *Let the family w_ε satisfy*

$$w_{\varepsilon,i} \geq 0, \quad \sum_i w_{\varepsilon,i}^2 = 1/2. \tag{3.10}$$

Then

$$\mathrm{Var}_v(t_\varepsilon) \leq 1 + 4|w_\varepsilon| h_\varepsilon(v).$$

Assume, additionally, $|w_\varepsilon| = o(1)$. Then the statistics $t_\varepsilon(v)$ are asymptotically Gaussian $N(0,1)$ uniformly over $v \in l^2$ such that $h_\varepsilon(v) = o(|w_\varepsilon|^{-1})$.

Proof of Lemma 3.1. Let $x = \eta + t$ where $\eta \sim N(0,1)$, $t \in R^1$. Then one has

$$E(x) = t, \ E(x^2) = 1 + t^2, \ \mathrm{Var}(x^2) = 2 + 4t^2, \ E(x^2 - E(x^2))^4 \leq c(1 + t^4), \tag{3.11}$$

where c is an absolute constant. These yield to the relations

$$E_v t_\varepsilon = h_\varepsilon(v), \ \mathrm{Var}_v(t_\varepsilon) = 1 + 4\sum_i w_{\varepsilon,i}^2 v_i^2, \ \mathrm{Var}_v(t_\varepsilon) - 1 \leq 4|w_\varepsilon| h_\varepsilon(v) = o(1), \tag{3.12}$$

which are uniform over $v \in l^2$. Moreover, if $\xi_{\varepsilon,i} = w_{\varepsilon,i}(x_i^2 - 1)$, then uniformly over $v \in l^2$, we have

$$\sum_i E_v(\xi_{\varepsilon,i} - E_v(\xi_{\varepsilon,i}))^4 \leq c \sum_i w_{\varepsilon,i}^4 (1 + v_i^4) \leq c|w_\varepsilon|^2 (1 + h_\varepsilon^2(v)) = o(1).$$

Thus asymptotic normality follows from the Central Limit Theorem and Lyapunov conditions. □

Using Lemma 3.1 and Corollary 3.2 we get

Corollary 3.3 *For tests* (3.8) *assume* (3.10) *and denote*

$$h_\varepsilon = h_\varepsilon(w_\varepsilon) = \inf_{v \in V_\varepsilon} \sum_i w_{\varepsilon,i} v_i^2. \tag{3.13}$$

Then for $T_\varepsilon = h_\varepsilon/2$ one has $\gamma(\psi_{w_\varepsilon,T_\varepsilon}, V_\varepsilon) \to 0$, as $h_\varepsilon \to \infty$.
Assume, additionally, $|w_\varepsilon| = o(1)$. Then

$$\beta(\psi_{\varepsilon,T_\alpha}, V_\varepsilon) \leq \Phi(T_\alpha - h_\varepsilon) + o(1), \quad \gamma(\psi_{\varepsilon,T_\varepsilon}, V_\varepsilon) \leq 2\Phi(-h_\varepsilon/2) + o(1).$$

The following corollary describes the asymptotic minimax properties of χ^2_n-tests of increased dimensions. Let an integer-valued family $n_\varepsilon = n \to \infty$ be given. Consider the tests

$$\chi^2_{n,T_\varepsilon} = \mathbf{1}_{\{t_n > T_\varepsilon\}}, \quad t_n = (2n)^{-1/2} \sum_{i=1}^{n} (x_i^2 - 1). \tag{3.14}$$

Let P_n be a canonical projection of l^2 to R^n, i.e., $P_n(v) = (v_1, \ldots, v_n)$.

Corollary 3.4 *Under model* (2.1) *let* $H_0 : v = 0,\ H_1 : v \in V_\varepsilon \subset l^2$. *Set*

$$u_\varepsilon = (2n)^{-1/2} \inf_{v \in V_\varepsilon} |P_n(v)|^2.$$

Then for the tests (3.14) *with* $T_\varepsilon = T_\alpha$ *one has*

$$\alpha(\chi^2_{n,T_\alpha}) = \alpha + o(1), \quad \beta(\chi^2_{n,T_\alpha}, V_\varepsilon) \le \Phi(T_\alpha - u_\varepsilon) + o(1), \tag{3.15}$$

and for $T_\varepsilon = u_\varepsilon/2$ *one has*

$$\alpha(\chi^2_{n,T_\varepsilon}) = \Phi(-u_\varepsilon/2) + o(1), \quad \beta(\chi^2_{n,T_\varepsilon}, V_\varepsilon) \le \Phi(-u_\varepsilon/2) + o(1). \tag{3.16}$$

Proof. This follows from Corollary 3.2 and from Corollary 3.3 with

$$w_{n,i} = \begin{cases} (2n)^{-1/2}, & \text{if } \ i \le n, \\ 0, & \text{if } \ i > n, \end{cases}$$

because $h_n(v) = (2n)^{-1/2} |P_n(v)|^2$. □

Let us translate Corollary 3.4 to the functional Gaussian model.

Corollary 3.5 *Under model* (2.5) *let* $H_0 : s = 0,\ H_1 : s \in S_\varepsilon \subset L_2(0,1)$, *and a family of orthonormal collections*

$$\bar{\varphi}_{\varepsilon,n} = \{\varphi_{\varepsilon,i}\}, \quad 1 \le i \le n = n_\varepsilon \to \infty,$$

be given. Let us consider tests (3.14) *where*

$$x_i = \varepsilon^{-1} \int_0^1 \varphi_{\varepsilon,i}(t)\, dX_\varepsilon(t).$$

Let Pr_{L_n} *be the projection of* $L_2(0,1)$ *to the linear hull* L_n *of* $\bar{\varphi}_{\varepsilon,n}$ *and let*

$$u_\varepsilon = \varepsilon^{-2} (2n)^{-1/2} \inf_{s \in S_\varepsilon} \|Pr_{L_n}(s)\|_2^2.$$

Then for the tests $\chi^2_{n,T_\varepsilon} = \mathbf{1}_{\{t_\varepsilon > T_\varepsilon\}}$ *with* $T_\varepsilon = T_\alpha$ *one has* (3.15) *and for* $T_\varepsilon = u_\varepsilon/2$ *one has* (3.16).

3.1.3 Tests of χ^p-Type

Here we give some extensions of the χ^2-tests. These tests do not permit us to obtain asymptotically exact minimax errors or risks, but they are suitable for studies of minimax distinguishability.

We call *tests of* χ^p*-type* the tests $\psi^p_{\varepsilon,T_\varepsilon} = \mathbf{1}_{\{t_\varepsilon > T_\varepsilon\}}$ based on the statistics

$$t_\varepsilon = \sum_i w_{\varepsilon,i}(|x_i|^p - c_p), \quad p \geq 2. \tag{3.17}$$

The family $w_\varepsilon = \{w_{\varepsilon,i}\}$, $i \in I$, is such that

$$w_{\varepsilon,i} \geq 0, \quad d_p \sum_i w^2_{\varepsilon,i} = 1, \tag{3.18}$$

where $c_p = E(|\eta|^p)$, $d_p = \mathrm{Var}(|\eta|^p)$, and η is the standard Gaussian random variable.

It is clear that under null-hypothesis $E_0(t_\varepsilon) = 0$, $\mathrm{Var}_0(t_\varepsilon) = 1$. To evaluate the means and variations of statistics t_ε under the alternatives we use the following:

Lemma 3.2 *Let* $\eta \sim N(0,1)$. *Then for any* $p \in [2, \infty)$ *there exist constants* $B = B_p$, $C = C_p$, *depending on* p *only, such that for all real* m *one has*

$$E(|\eta + m|^p) - c_p \geq B(m^2 + |m|^p), \ \mathrm{Var}(|\eta + m|^p) \leq C(1 + |m|^{2p-2}), \ p \geq 2.$$

Proof. Let us verify the first inequality. It suffices to verify that for some $B_0 = B_0(p) > 0$,

$$f_p(m) = E(|\eta + m|^p) - c_p \geq B_0 m^2, \quad E(|\eta + m|^p) \geq |m|^p/2. \tag{3.19}$$

In fact, take $M = (8c_p)^{1/p}$. Then under (3.19) one has

$$f_p(m) \geq \begin{cases} B_0 m^2/2 + |m|^p/8, & \text{if } |m| \geq M \\ B_0 m^2/2 + B_0|m|^p/2M^{p-2}, & \text{if } |m| < M. \end{cases}$$

One can easily see that $f_p(m)$ is an even infinitely differentiable function in $m \in R^1$ and $f_p^{(1)}(0) = 0$. Expanding $f_p(m)$ by Taylor's formula up to the second term at point 0 we get

$$f_p(m) = B_p(\tilde{m})m^2/2, \quad |\tilde{m}| \leq |m|, \quad B_p(m) = p(p-1)E(|\eta + m|^{p-2}).$$

It is clear that $B_p(m)$ is positive and continuous in m. Therefore, $B_p(m)$ is bounded away from 0 over $|m| \leq 2$. If $|m| > 2$, then

$$B_p(m) \geq p(p-1)E(|\eta + m|^{p-2}\mathbf{1}_{\{|\eta| \leq 1\}}) \geq p(p-1)(\Phi(1) - \Phi(-1)).$$

These yield the first inequality in (3.19). The second inequality in (3.19) follows from the evaluation

$$E(|\eta + m|^p) \geq |m|^p P(\mathrm{sign}\,(m\eta) > 0) = |m|^p/2.$$

Let us verify the second inequality in Lemma 3.2. Expanding $|\eta + m|^p$ by Taylor's formula at the point $\eta = 0$ we get

$$||\eta + m|^p - |m|^p| = p(|\eta|(|\tilde{\eta} + m|^{p-1}) \leq p2^{p-2}|\eta|(|\eta|^{p-1} + |m|^{p-1}), \tag{3.20}$$

where $|\tilde{\eta}| \leq |\eta|$ and we have used the inequality

$$|a + b|^q \leq 2^{p-1}(|a|^q + |b|^q), \quad q \geq 1,$$

which follows from convexity of the function $|x|^q$, $q \geq 1$. By squaring and averaging inequality (3.20) we obtain the needed inequality. □

Using Lemma 3.2, the Hölder inequality, and the inequality $\sum_i a_i^2 \leq (\sum_i a_i)^2$ for $a_i \geq 0$ we get

$$\begin{aligned} E_v(t_\varepsilon) &\geq B\left(\sum_i w_{\varepsilon,i}(v_i^2 + |v_i|^p\right); \\ \mathrm{Var}_v(t_\varepsilon) &\leq C\left(\sum_i w_{\varepsilon,i}^2(1 + |v_i|^{2p-2}\right) \\ &\leq C\left(d_p^{-1} + \left(\sum_i w_{\varepsilon,i}^2\right)^{1/p}\left(\sum_i w_{\varepsilon,i}^2|v_i|^{2p}\right)^{1-1/p}\right) \\ &\leq C_1\left(d_p^{-1} + d_p^{-1/p}(E_v(t_\varepsilon))^{2-2/p}\right). \end{aligned}$$

Using Corollary 3.1 and the evaluations above we get the condition for the minimax consistency of tests based on statistics (3.17).

Proposition 3.2 *Assume* (3.18) *and*

$$u_\varepsilon^p = \inf_{v \in V_\varepsilon} \sum_i w_{\varepsilon,i}|v_i|^p \to \infty. \tag{3.21}$$

Then the tests $\psi_{\varepsilon,T_\varepsilon}^p$ *with* $T_\varepsilon = B_p u_\varepsilon^p/2$ *are minimax consistent, i.e.,*

$$\gamma_\varepsilon(\psi_{\varepsilon,T_\varepsilon}^p, V_\varepsilon) \to 0.$$

Given an integer-valued family $n_\varepsilon = n \to \infty$, let us consider the tests

$$\chi_{n,T_\varepsilon}^p = \mathbb{1}_{\{t_n > T_\varepsilon\}}, \quad t_n = (d_p n)^{-1/2}\sum_{i=1}^n (|x_i|^p - c_p). \tag{3.22}$$

Analogously to Corollary 3.4 we get

Corollary 3.6 *Let* P_n *be a canonical projection* l^2 *to* R^n *and let*

$$u_\varepsilon = n^{-1/2p} \inf_{v \in V_\varepsilon} |P_n(v)|_p \to \infty, \ p \geq 2.$$

Then the tests $\chi_{\varepsilon,T_\varepsilon}^p$ *with* $T_\varepsilon = B_p u_\varepsilon^p/2$ *are minimax consistent.*

We extend Corollary 3.6 to the functional Gaussian model for orthonormal collections $\bar\varphi_{\varepsilon,n}$ generated by indicator functions of the equispaced partition of $[0,1]$ to intervals $\delta_{i,n} = [(i-1)/n, i/n]$ of length $1/n$:

$$\varphi_{\varepsilon,i} = n^{1/2}\mathbb{1}_{\delta_{i,n}}, \quad i = 1,\ldots,n = n_\varepsilon \to \infty.$$

Let $Pr_n f$ be the projection of f to the linear hull L_n of $\bar\varphi_{\varepsilon,n}$ defined by (2.83). Note the relation between the l_n^p-norms in the space of normalized coordinates $v_i(f) = \varepsilon^{-1}(f, \varphi_{\varepsilon,i})$, $i = 1, \dots, n$, and the L_p-norm in L_n:

$$\varepsilon^p |v(f)|_p^p = \sum_{i=1}^n |(f, \varphi_{\varepsilon,i})|^p = n^{-p/2} \sum_{i=1}^n |f_i|^p = n^{1-p/2} \|Pr_n(f)\|_p^p. \quad (3.23)$$

Let us consider tests $\chi^p_{\varepsilon,T_\varepsilon}$ with

$$x_i = \varepsilon^{-1} \int_0^1 \varphi_{\varepsilon,i}(t)\, dX_\varepsilon(t) = \varepsilon^{-1} n^{1/2}(X_\varepsilon(i/n) - X_\varepsilon((i-1)/n). \quad (3.24)$$

In view of Corollary 3.6 these imply

Corollary 3.7 *Under model* (2.5) *let* $H_0 : s = 0,\ H_1 : s \in S_\varepsilon \subset L_2(0,1)$. *Assume*

$$u_\varepsilon = \varepsilon^{-1} n^{1/2p-1/2} \inf_{s \in S_\varepsilon} \|Pr_n(s)\|_p \to \infty; \quad 2 \le p < \infty.$$

Then the tests χ^p_{n,T_ε} *with* $T_\varepsilon = B_p u_\varepsilon^p/2$ *are minimax consistent.*

Under the sequence Gaussian model consider differently the case of the positive alternative and $p \in [1,2]$. We call by χ^+-*tests* the tests $\psi^+_{\varepsilon,T_\varepsilon} = \mathbf{1}_{\{t_\varepsilon > T_\varepsilon\}}$ based on a combination of linear and square statistics:

$$t_\varepsilon = \sum_i w_{\varepsilon,i} \zeta(x_i), \quad \zeta(x) = x^2 + x - 1, \quad \text{where } w_{\varepsilon,i} \ge 0,\ \sum_i w^2_{\varepsilon,i} = 1/3. \quad (3.25)$$

Let $\eta \sim N(0,1)$. Then $E(\zeta(\eta)) = 0,\ \mathrm{Var}(\zeta(\eta)) = 3$ which imply

$$E_0(t_\varepsilon) = 0, \quad \mathrm{Var}_0(t_\varepsilon) = 1.$$

Moreover, for all $m \in R^1$, one has

$$E(\zeta(\eta+m)) = m^2 + m, \ \mathrm{Var}(\zeta(\eta+m)) = 4m^2 + 4m + 3 = 3 + 4E(\zeta(\eta+m))$$

which imply, for any $v \in l^2_+$,

$$\begin{aligned} E_v(t_\varepsilon) &= \sum_i w_{\varepsilon,i}(v_i^2 + v_i); \\ \mathrm{Var}_v(t_\varepsilon) &= \sum_i w^2_{\varepsilon,i}(3 + 4(v_i^2 + v_i)) \le 1 + 3^{1/2} 4 E_v(t_\varepsilon). \end{aligned}$$

Using Corollary 3.1, the evaluations above, and the inequality

$$t + t^2 \ge t^p, \quad \text{for } t \ge 0, \quad p \in [1,2],$$

we get sufficient conditions for the minimax consistency of χ^+-tests based on statistics (3.25).

Proposition 3.3 *Let* $V_\varepsilon \subset l^2_+$. *Assume*

$$u_\varepsilon^p = u^p_{w,\varepsilon} = \inf_{v \in V_\varepsilon} \sum_i w_{\varepsilon,i} v_i^p \to \infty; \quad p \in [1,2]; \quad T_\varepsilon = u_\varepsilon^p/2.$$

Then tests $\psi^+_{\varepsilon,T_\varepsilon}$ *are minimax consistent:* $\gamma_\varepsilon(\psi^+_{\varepsilon,T_\varepsilon}, V_\varepsilon) \to 0$.

For a given integer-valued family $n_\varepsilon = n \to \infty$ let us consider the tests

$$\chi^+_{n,T_\varepsilon} = \mathbf{1}_{\{t_n > T_\varepsilon\}}, \quad t_n = (3n)^{-1/2} \sum_{i=1}^n (x_i^2 + x_i - 1). \tag{3.26}$$

Analogously to Corollary 3.6 we get

Corollary 3.8 *Under model* (2.1) *let* $H_0 : v = 0,\ H_1 : v \in V_\varepsilon \subset l^2$. *Let* P_n *be a canonical projection* l^2 *to* R^n *and assume*

$$P_n(V_\varepsilon) \subset l^2_+, \quad u_\varepsilon = n^{-1/2p} \inf_{v \in V_\varepsilon} |P_n(v)|_p \to \infty, \quad T_\varepsilon = u^p_\varepsilon/2, \quad p \in [1,2].$$

Then the tests (3.26) *are minimax consistent.*

Under the functional Gaussian model let us consider the orthonormal collections $\bar\varphi_{\varepsilon,n}$ generated by indicator functions of an equispaced partition of $[0,1]$ to intervals $\delta_{i,n} = [(i-1)/n, i/n]$ of length $1/n$; $n = n_\varepsilon \to \infty$. Let $Pr_n f$ be the projection of f to the linear hull L_n of $\bar\varphi_{\varepsilon,n}$ defined by (2.83). Denote by $L_2^+[0,1]$ the set of almost surely nonnegative $f \in L_2(0,1)$ and observe that $f_i \geq 0$ for all $f \in L_2^+[0,1]$. Using relation (3.23) we get a distinguishability condition for nonnegative signals.

Corollary 3.9 *Under model* (2.5) *let* $H_0 : s = 0,\ H_1 : s \in S_\varepsilon \subset L_2^+[0,1]$, *and consider tests 3.22 with* $T_\varepsilon = u^p_\varepsilon/2$ *and* x_i *defined by* (3.24). *Assume*

$$u_\varepsilon = \varepsilon^{-1} n^{1/2p-1/2} \inf_{s \in S_\varepsilon} \|Pr_n(s)\|_p \to \infty; \quad p \in [1,2].$$

Then the tests χ^+_{n,T_ε} *are minimax consistent.*

3.1.4 Supreme Tests

Under model (2.1) let $H_0 : v = 0,\ H_1 : v \in V_\varepsilon \subset l^2$. Let the nonrandom positive family of sequences $\bar T_\varepsilon = \{T_{\varepsilon,i}\}$, $T_{\varepsilon,i} \leq \infty$ is given. We call *supreme tests*, the tests based on the thresholding

$$\psi_\varepsilon = \psi_{\varepsilon,\bar T_\varepsilon} = \mathbf{1}_{\mathcal{X}_\varepsilon}, \quad \mathcal{X}_\varepsilon = \left\{x : \sup_i |x_i|/T_{\varepsilon,i} > 1\right\}. \tag{3.27}$$

Here and below, without any ambiguity, we admit $T_{\varepsilon,i} = \infty$. Formally, one can exclude all items with $T_{\varepsilon,i} = \infty$.

Calculate type I error probabilities:

$$\alpha(\psi_\varepsilon) = P_0(\mathcal{X}_\varepsilon) = 1 - \prod_i (1 - 2\Phi(-T_{\varepsilon,i})). \tag{3.28}$$

In particular,

$$\alpha(\psi_\varepsilon) \to 0 \text{ if and only if } \sum_i \Phi(-T_{\varepsilon,i}) \to 0; \tag{3.29}$$

and it is necessary that $\inf_i T_{\varepsilon,i} \to \infty$.

Calculate type II error probabilities. Let $\overline{\mathcal{X}}_\varepsilon$ be the complement of the set $\mathcal{X}_\varepsilon$. We get

$$\begin{aligned}\beta(\psi_\varepsilon, v) &= P_v(\overline{\mathcal{X}}_\varepsilon) = (1-\alpha(\psi_\varepsilon))\prod_i \frac{\Phi(T_{\varepsilon,i}-|v_i|)-\Phi(-T_{\varepsilon,i}-|v_i|)}{1-2\Phi(-T_{\varepsilon,i})}\\ &\le \begin{cases}(1-\alpha(\psi_\varepsilon))\inf_i\left(\Phi(T_{\varepsilon,i}-|v_i|)/(1-2\Phi(-T_{\varepsilon,i}))\right),\\ \inf_i(\Phi(T_{\varepsilon,i}-|v_i|).\end{cases}\end{aligned} \tag{3.30}$$

Denote

$$H_\varepsilon = H_\varepsilon(\bar{T}_\varepsilon, V_\varepsilon) = \inf_{v\in V_\varepsilon}\sup_i(|v_i| - T_{\varepsilon,i}). \tag{3.31}$$

Using these evaluations we get

Proposition 3.4
(1) *Assume* $\sum_i \Phi(-T_{\varepsilon,i}) \to 0$. *Then*

$$\gamma(\psi_\varepsilon, V_\varepsilon) \le \Phi(-H_\varepsilon) + o(1).$$

In particular, suppose $H_\varepsilon \to \infty$. *Then the tests* (3.27) *are minimax consistent, i.e.,*

$$\gamma(\psi_\varepsilon, V_\varepsilon) \to 0.$$

(2) *Assume* $\bar{T}_\varepsilon = \bar{T}_{\varepsilon,\alpha}$ *is such that*

$$\prod_i(1-2\Phi(-T_{\varepsilon,i})) = 1-\alpha+o(1), \quad \inf_i T_{\varepsilon,i}\to\infty.$$

Then

$$\alpha(\psi_\varepsilon) = \alpha + o(1), \quad \beta(\psi_\varepsilon, V_\varepsilon) \le (1-\alpha)\Phi(-H_\varepsilon) + o(1).$$

We can extend this proposition. Let $\bar{r}_\varepsilon = \{r_{\varepsilon,l}\}$ be a family of finite or denumerable collections of the vectors $r_{\varepsilon,l} \in l^2$, $|r_{\varepsilon,l}| = 1$, and let $\bar{t}_\varepsilon = \{t_{\varepsilon,l}\}$ be collections of linear statistics $t_{\varepsilon,l} = (x, r_{\varepsilon,l}) = \sum_i x_i r_{\varepsilon,l,i}$ which are Gaussian and

$$E_0(t_{\varepsilon,l}) = 0, \quad E_v(t_{\varepsilon,l}) = (v, r_{\varepsilon,l}), \quad \mathrm{Var}_v(t_{\varepsilon,l}) = 1.$$

Given a family $\bar{T}_\varepsilon = \{T_{\varepsilon,l}\}$ such that $\sum_l \Phi(-T_{\varepsilon,l}) \to 0$, consider the following variant of supreme tests:

$$\psi_{\varepsilon,r} = \psi_{\varepsilon,\bar{r}_\varepsilon,\bar{T}_\varepsilon} = \mathbf{1}_{\mathcal{X}_{\varepsilon,r}}, \quad \mathcal{X}_{\varepsilon,r} = \left\{x : \sup_l |t_{\varepsilon,l}|/T_{\varepsilon,l} > 1\right\}, \tag{3.32}$$

which correspond to (3.27) for $r_{\varepsilon,l} = e_l$, $l \in I$, where $\{e_i\}$ is the canonical basis in l^2. Analogously to the above we get

$$\begin{aligned}\alpha(\psi_{\varepsilon,r}) &\le \sum_l P_0(|t_{\varepsilon,l}|/T_{\varepsilon,l} > 1) = 2\sum_l \Phi(-T_{\varepsilon,l}) \to 0,\\ \beta(\psi_{\varepsilon,r}, v) &\le \inf_l P_v(|t_{\varepsilon,l}|/T_{\varepsilon,l} \le 1) \le \inf_l \Phi(T_{\varepsilon,l} - |(v, r_{\varepsilon,l})|),\\ \beta(\psi_{\varepsilon,r}, V_\varepsilon) &\le \sup_{v\in V_\varepsilon}\inf_l \Phi(T_{\varepsilon,l} - |(v, r_{\varepsilon,l})|) = \Phi(-H_{\varepsilon,r}),\end{aligned}$$

where

$$H_\varepsilon = H_{\varepsilon,r} = H_\varepsilon(\bar{r}_\varepsilon, \bar{T}_\varepsilon, V_\varepsilon) = \inf_{v \in V_\varepsilon} \sup_l (|(v, r_{\varepsilon,l})| - T_{\varepsilon,l}). \tag{3.33}$$

For given $\alpha \in (0,1)$ let us consider the randomized tests

$$\psi_{\varepsilon,r,\alpha} = (1-\alpha)\psi_{\varepsilon,r} + \alpha.$$

Using the evaluations above we get

$$\alpha(\psi_{\varepsilon,r,\alpha}) = \alpha + (1-\alpha)\alpha(\psi_{\varepsilon,r}) = \alpha + o(1),$$
$$\beta(\psi_{\varepsilon,r,\alpha}, v) = (1-\alpha)\beta(\psi_{\varepsilon,r}, v); \quad \beta(\psi_{\varepsilon,r,\alpha}, V_\varepsilon) \le (1-\alpha)\Phi(-H_{\varepsilon,r}).$$

Using these evaluations we get

Proposition 3.5 *Assume* $\sum_i \Phi(-T_{\varepsilon,l}) \to 0$. *Then, for any* $\alpha \in (0,1)$, *we have* $\alpha(\psi_{\varepsilon,r,\alpha}) \le \alpha + o(1)$,

$$\beta(\psi_{\varepsilon,r,\alpha}, V_\varepsilon) \le (1-\alpha)\Phi(-H_{\varepsilon,r}), \quad \gamma(\psi_{\varepsilon,r}, V_\varepsilon) \le \Phi(-H_{\varepsilon,r}) + o(1),$$

In particular, let $H_{\varepsilon,r} \to \infty$. *Then the tests* (3.32) *are minimax consistent:* $\gamma(\psi_{\varepsilon,r}, V_\varepsilon) \to 0$.

Corollary 3.10 *Let* $\bar{T}_\varepsilon$ *be such that*

$$\inf_l T_{\varepsilon,l} \to \infty, \quad \sum_l \exp(-T^2_{\varepsilon,l}/2) \asymp 1. \tag{3.34}$$

Then

$$\beta(\alpha, V_\varepsilon) \le (1-\alpha)(\Phi(-H_\varepsilon) + o(1)), \quad \gamma(V_\varepsilon) \le \Phi(-H_\varepsilon) + o(1),$$

where H_ε *are defined by* (3.31) *or* (3.33) .

Proof of Corollary 3.10. Recall the well-known formula

$$\Phi(-x) \sim e^{-x^2/2}/x\sqrt{2\pi}, \quad x \to \infty. \tag{3.35}$$

By (3.35) the assumptions (3.34) yield to $\sum_i \Phi(-\tilde{T}_{\varepsilon,l}) \to 0$. Then we use Propositions 3.5. □

Let a family $n = n_\varepsilon \to \infty$ be given, $I = \mathbf{N}$. Consider the families $\bar{T}_\varepsilon = \bar{T}_n$:

$$T_{\varepsilon,i} = \{\sqrt{2\log n}, \text{ if } i \le n; \ \infty, \text{ if } i > n\},$$

and $\bar{T}_\varepsilon = \bar{T}_{n,\alpha}$, $\alpha \in (0,1)$,

$$T_{\varepsilon,i} = \{T_{n,\alpha}, \text{ if } i \le n; \ \infty, \text{ if } i > n\},$$

where $T_{n,\alpha}$ is defined by $2\Phi(-T_{n,\alpha}) = 1 - (1-\alpha)^{1/n}$; it is clear that $T_{n,\alpha} = \sqrt{2\log n} + o(1)$. In this case, we get tests

$$\psi_n = \mathbf{1}_{\mathcal{X}_n}, \quad \mathcal{X}_n = \left\{x : \max_{1 \le i \le n} |x_i| > \sqrt{2\log n}\right\}, \tag{3.36}$$

$$\psi_{n,\alpha} = \mathbf{1}_{\mathcal{X}_{n,\alpha}}, \quad \mathcal{X}_{n,\alpha} = \left\{x : \max_{1 \le i \le n} |x_i| > T_{n,\alpha}\right\}. \tag{3.37}$$

Denote

$$R_n = R_n(V_\varepsilon) = \inf_{v \in V_\varepsilon} |P_n(v)|_\infty,$$

where $|v|_\infty = \max_{1 \le i \le n} |v_i|$, $v \in R^n$, and P_n is a canonical projection l^2 to R^n. Then, analogously to (3.31), we have

$$H_n(\bar{T}_n, V_\varepsilon) = R_n - \sqrt{2 \log n}, \quad H_{n,\alpha}(\bar{T}_n, V_\varepsilon) = R_n - T_{n,\alpha}.$$

Clearly, $H_{n,\alpha}(\bar{T}_n, V_\varepsilon) = H_n(\bar{T}_n, V_\varepsilon) + o(1)$ and we get

Corollary 3.11

$$\alpha(\psi_{n,\alpha}) = \alpha, \quad \beta(\psi_{n,\alpha}, V_\varepsilon) \le (1-\alpha)\Phi(\sqrt{2\log n} - R_n)(1 + o(1)),$$
$$\gamma(\psi_{n,\alpha}, V_\varepsilon) \le \Phi(\sqrt{2\log n} - R_n).$$

In particular, $\gamma(\psi_{n,\alpha}, V_\varepsilon) \to 0$, *as* $R_n - \sqrt{2\log n} \to \infty$.

One can translate this corollary to the functional Gaussian model (2.5). Consider the orthonormal collections $\bar{\varphi}_{\varepsilon,n}$ generated by the indicator functions of the equispaced partition of $[0,1]$ to intervals $\delta_{i,n}$ of length $1/n$; $n = n_\varepsilon \to \infty$. Let $Pr_n f$ be the projection of f to the linear hull L_n of $\bar{\varphi}_{\varepsilon,n}$ defined by (2.83). Analogously to relation (3.23) we have

$$\varepsilon |P_n(v(f))|_\infty = \max_{1 \le i \le n} |(f, \varphi_{\varepsilon,i})| = n^{-1/2} \max_{1 \le i \le n} |f_i| = n^{-1/2} \|Pr_n(f)\|_\infty.$$

Consider the tests ψ_n with x_i defined by (3.24). We get

Corollary 3.12 *Under model* (2.5) *let* $H_0 : s = 0$, $H_1 : s \in S_\varepsilon \subset L_2[0,1]$. *Denote*

$$R_\varepsilon = \inf_{s \in S_\varepsilon} \|Pr_n(s)\|_\infty$$

and assume

$$\varepsilon^{-1} n^{-1/2} R_\varepsilon - \sqrt{2 \log n} \to \infty;$$

in particular, this holds if $\liminf R_\varepsilon / \varepsilon\sqrt{n \log n} > \sqrt{2}$. *Then the tests* ψ_n *are minimax consistent.*

3.2 Asymptotic Minimaxity for Square Norms

For the problems of interest, some results on asymptotic minimaxity could be obtained by using the above results for the spherical alternatives and the results on the asymptotic properties of χ^2-tests (see Proposition 2.15 and Corollaries 3.4, 3.5).

Under the functional Gaussian model (2.5) let $S \subset L_2(0,1)$. Given a positive family $\rho_\varepsilon \to 0$ let us consider the null-hypothesis $H_0 : s = 0$ and the alternative

$$H_1 : s \in S(r_\varepsilon) = \{s \in S : \|s\|_2 \ge r_\varepsilon\}.$$

This means we remove neighborhoods of the null-hypothesis in the square norms $\|\cdot\|_2$ of radii r_ε.

Of course, for any orthonormal basis $\{\varphi_i,\ i \in I\}$ in $L_2(0,1)$, this problem is equivalent to testing $H_0 : v = 0$ against the alternative

$$H_1 : v \in V_\varepsilon(\rho_\varepsilon) = \{v \in V_\varepsilon \ :\ |v|_2 \geq \rho_\varepsilon\}, \quad \rho_\varepsilon = r_\varepsilon/\varepsilon,$$

under the sequence Gaussian model (2.1); here $V_\varepsilon \subset l^2$ and $V_\varepsilon(\rho_\varepsilon) \subset l^2$ are the sets which consist of ε^{-1}-normalized sequences of Fourier coefficients of signals $s \in S$ or $s \in S(r_\varepsilon)$ with $\rho_\varepsilon = r_\varepsilon/\varepsilon$.

We give the conditions for distinguishability and nondistinguishability in terms of geometrical properties of the set S. These properties are determined by the sequences of *inner radii* and *Kolmogorov diameters* of the set S.[1] If S is an ellipsoid corresponding to a ball in the $|\cdot|_{\bar{a},2}$-norm with the power increasing sequence $\bar{a}$, then we obtain the rate asymptotics; if sequence $\bar{a}$ is exponential increasing, then we get the sharp asymptotics.[2]

3.2.1 Lower Bounds and Inner Radii

For the set $S \subset L_2(0,1)$ let us call *inner radii* $r_n = r_n(S)$ *of dimension* n *at point* $s = 0$ the supremum of radii r of n-dimensional balls $D^n(r)$ which are contained in S (compare with [94]).

Assume $r_n > 0\ \forall n,\ r_n \to 0$ as $n \to \infty$. Put

$$n(r_\varepsilon) = \max\{n \ :\ r_n > r_\varepsilon\}.$$

It is clear that $n(r_\varepsilon) \to \infty$ as $\varepsilon \to 0$. One can find an n-dimensional subspace $L_2(0,1)$, and a ball $D^n(r) \subset L_n$, here L_n, such that $D^n(r) \subset S$ and $r_n \geq r > r_\varepsilon$. Thus the sphere $S^{n-1}(r_\varepsilon)$ is contained in $S(r_\varepsilon)$. Let us take the orthonormal basis $\{\varphi_i\}$ in $L_2(0,1)$ such that $\varphi_1, \ldots, \varphi_n$ is an orthonormal basis in L_n. By passing to normalized Fourier coefficients $v_i = \varepsilon^{-1}(s, \varphi_i)$ we get

$$S^{n-1}(\rho_\varepsilon) \subset V_\varepsilon(\rho_\varepsilon),\ \rho_\varepsilon = r_\varepsilon/\varepsilon,$$

where $n = n(r_\varepsilon)$ and $S^{n-1}(\rho_\varepsilon)$ is an $(n-1)$-dimensional sphere which corresponds to the first n coordinates in l^2. By Proposition 2.15 one has

$$\beta_\varepsilon(\alpha, S(r_\varepsilon)) = \beta(\alpha, V_\varepsilon(\rho_\varepsilon)) \geq \Phi(T_\alpha - u_\varepsilon) + o(1), \tag{3.38}$$

$$\gamma_\varepsilon(S(r_\varepsilon)) = \gamma(V_\varepsilon(\rho_\varepsilon)) \geq 2\Phi(-u_\varepsilon/2) + o(1), \tag{3.39}$$

$$u_\varepsilon = r_\varepsilon^2/\varepsilon^2\sqrt{2n(r_\varepsilon)}. \tag{3.40}$$

---

[1] Distinguishability conditions in terms of inner radii and Kolmogorov diameters were obtained in [41].

[2] These seem to be the first results on the sharp asymptotics of nonclassical type which were obtained for the sample model in [50].

Lower Bounds for Ellipsoids

Let a positive sequence $\bar{d} = \{d_i\}$ (it is possible $d_i = \infty$) and an orthonormal basis $\{\varphi_i\}$ in $L_2(0,1)$ be given, here $i \in \mathbf{N}$, $i \geq i_0$. Let $S = D_{\bar{d}}$ be *the ellipsoid of semi-axes* d_i determined by the inequality

$$D_{\bar{d}} = \left\{ s \in L_2(0,1) : \sum_i (\theta_i/d_i)^2 \leq 1 \right\}, \quad \theta_i = (s, \varphi_i); \tag{3.41}$$

clearly, the corresponding set $V_\varepsilon \subset l^2$ is an ellipsoid of the semi-axes $\bar{d}_\varepsilon = \bar{d}/\varepsilon$.

Assume d_i be nonincrease, and $d_i \to 0$, as $i \to \infty$. Then $r_n = d_{n+i_0-1}$.

Let the semi-axes be of power type $d_i \asymp i^{-\eta}$, $\eta > 0$. Then relations (3.38), (3.39) hold with

$$n(r_\varepsilon) \asymp r_\varepsilon^{-1/\eta}, \quad u_\varepsilon \asymp r_\varepsilon^{(4\eta+1)/2\eta} \varepsilon^{-2}. \tag{3.42}$$

These imply the nondistinguishability conditions

$$\gamma_\varepsilon \to 1, \quad \text{if} \quad r_\varepsilon = o(r_\varepsilon^*); \quad \liminf \gamma_\varepsilon > 0, \quad \text{if } r_\varepsilon = O(r_\varepsilon^*), \tag{3.43}$$

where

$$r_\varepsilon^* = \varepsilon^{4\eta/(4\eta+1)}. \tag{3.44}$$

Example 3.1 *Smooth signals.* Let $S = \tilde{S}_{\eta,2}(R)$ be the ball of radius R in the norm $\|\cdot\|_{\eta,2}$ (see (2.75)) which consists of an 1-periodical η-smooth functions in the L_2-norm on the interval $[0,1]$. This is determined in terms of Fourier coefficients $\theta_l = (s, \varphi_l)$:

$$\|s\|_{\eta,2}^2 = \sum_{l=0}^{\infty} \theta_l^2 + \sum_{l=1}^{\infty} (2\pi l)^{2\eta} (\theta_{2l-1}^2 + \theta_{2l}^2) \leq R^2. \tag{3.45}$$

Here $\{\varphi_l\}$ is the standard Fourier basis (2.74) in $L_2(0,1)$. Thus $\tilde{S}_{\eta,2}(R)$ is the ellipsoid of the semi-axes

$$d_0 = R, \quad d_{2l-1} = d_{2l} = R/((2\pi l)^{2\eta} + 1)^{1/2}, \quad l \geq 1; \quad d_l \asymp l^{-\eta}, \quad \text{as } l \to \infty.$$

These yield (3.43), (3.44).

Let the semi-axes be of exponential type $d_i \sim He^{-i\mu}$, $\mu > 0$. Then relations (3.38), (3.39) hold with

$$n(r_\varepsilon) = \mu^{-1} \log(H/r_\varepsilon) + O(1), \quad u_\varepsilon \sim (r_\varepsilon/\varepsilon)^2((2/\mu)\log r_\varepsilon^{-1})^{-1/2}. \tag{3.46}$$

These imply the nondistinguishability conditions

$$\gamma_\varepsilon \to 1, \quad \text{if } r_\varepsilon = o(r_\varepsilon^*); \quad \liminf \gamma_\varepsilon > 0, \quad \text{if } r_\varepsilon = O(r_\varepsilon^*), \tag{3.47}$$

where

$$r_\varepsilon^* = \varepsilon(\log \varepsilon^{-1})^{1/4}. \tag{3.48}$$

Example 3.2 *Analytical signals.* Consider the set $S = \tilde{S}(\mu, H)$ of 1-periodical real-valued functions $s(t)$ which have an analytical extension $s(z), z = x + iy$, to the region in the complex plane $|y| \leq \mu/\pi$ such that

$$\|s(\cdot + i\mu/\pi)\|^2_{L_2(0,1)} + \|s(\cdot - i\mu/\pi)\|^2_{L_2(0,1)} \leq 2H^2. \tag{3.49}$$

Consider the complex Fourier expansion

$$s(z) = \sum_{j=-\infty}^{\infty} c_j \exp(ij2\pi z), \quad \bar{c}_{-j} = c_j.$$

Then inequality (3.49) is equivalent to

$$\sum_j (\exp(4j\mu) + (\exp(-4j\mu))|c_j|^2 \leq 2H^2. \tag{3.50}$$

Passing to a real Fourier expansion, we see that $S(\mu, H)$ is the ellipsoid of semi-axes

$$d_0^2 = 2H^2, \quad d_{2j-1}^2 = d_{2j}^2 = H^2/(e^{4j\mu} + e^{-4j\mu}), \quad j \geq 1, \quad d_j \sim He^{-j\mu}.$$

Thus we get (3.38), (3.39) with (3.46) as well as (3.47), (3.48).

3.2.2 Upper Bounds and Kolmogorov Diameters

Denote by

$$\delta(S, L) = \sup_{s\in S} \inf_{u\in L} \|s - u\|_2 = \sup_{s\in S} \|s - Pr_L(s)\|_2, \tag{3.51}$$

the accuracy of approximation of the set $S \subset L_2(0, 1)$ by the linear subspace $L \subset L_2(0, 1)$; here Pr_L is the projection to L which gives best approximation of the point s in L. One has

$$\|Pr_L(s)\|_2^2 \geq \|s\|_2^2 - \delta^2(S, L). \tag{3.52}$$

We call the *Kolmogorov diameter* $R_n = R_n(S)$ *of dimension* n the best accuracy of approximation of the set S by linear subspaces L_n of dimension n:

$$R_n(S) = \inf_{L_n \subset L_2(0,1)} \delta(S, L_n),$$

Put

$$\eta_\varepsilon = \sup_n (r_\varepsilon^2 - R_n^2(S))/\varepsilon^2\sqrt{2n}. \tag{3.53}$$

Consider a family $L_{n_\varepsilon} \subset L_2(0, 1)$ of "approximately best" n_ε-dimensional subspaces such that for $s \in S(r_\varepsilon) = \{s \in S : \|s\|_2 \geq r_\varepsilon\}$ one has

$$\|Pr_{L_{n_\varepsilon}}(s)\|_2^2 \geq \|s\|_2^2 - R_{n_\varepsilon}^2(S) - o(\varepsilon^2\sqrt{2n_\varepsilon}) \geq \eta_\varepsilon\varepsilon^2\sqrt{2n}(1 - o(1)). \tag{3.54}$$

Let us take an orthonormal basis $\varphi_{1,n},\dots,\varphi_{n,n}$ in L_n, $n=n_\varepsilon$, and consider $\chi^2_{n,T}$-tests (3.14) based on the statistics

$$X_i=\varepsilon^{-1}\int_0^1 \varphi_{i,n}(t)\,dX_\varepsilon(t),\quad i=1,\dots,n.$$

In view of Corollary 3.5 we get $\alpha(\chi^2_{n,T_\alpha})=\alpha+o(1)$,

$$\beta(\chi^2_{n,T_\alpha},V_\varepsilon(\rho_\varepsilon)) = \beta_\varepsilon(\chi^2_{n,T_\alpha},S(r_\varepsilon))\le\Phi(T_\alpha-\eta_\varepsilon)+o(1), \quad (3.55)$$

$$\gamma(\chi^2_{n,T_n},V_\varepsilon(\rho_\varepsilon)) = \gamma_\varepsilon(\chi^2_{n,T_n},S(r_\varepsilon))\le 2\Phi(-\eta_\varepsilon/2)+o(1), \quad (3.56)$$

where $T_n=\eta_\varepsilon/2$, $\rho_\varepsilon=r_\varepsilon/\varepsilon$. In particular,

$$\beta(\alpha,V_\varepsilon(\rho_\varepsilon)) = \beta_\varepsilon(\alpha,S(r_\varepsilon))\le\Phi(T_\alpha-\eta_\varepsilon)+o(1), \quad (3.57)$$

$$\gamma(V_\varepsilon(\rho_\varepsilon)) = \gamma_\varepsilon(S(r_\varepsilon))\le 2\Phi(-\eta_\varepsilon/2)+o(1). \quad (3.58)$$

Upper Bounds for Ellipsoids

Let $S=D_{\bar d}$ be an ellipsoid of the semi-axes d_i, $i\ge i_0$, defined by (3.41), let $d_i>0$ be the nonincrease, $d_i\to 0$ as $i\to\infty$. Then

$$R_n^2(S)=\sup_{s\in S}\sum_{i=n+i_0-1}^{\infty}\theta_i^2\le d^2_{n+i_0-1}\sum_{i=n+i_0-1}^{\infty}(\theta_i/d_i)^2\le d^2_{n+i_0-1}.$$

First, let the semi-axes be of power type $d_i\asymp i^{-\eta}$, $\eta>0$. Then (compare with Section 1.4.3)

$$\eta_\varepsilon=\sup_n(r_\varepsilon^2-R_n^2(S))/\varepsilon^2\sqrt{2n}\ge\sup_n(r_\varepsilon^2-d^2_{n+i_0-1})/\varepsilon^2\sqrt{2n}\asymp\varepsilon^{-2}r_\varepsilon^{2+1/2\eta} \quad (3.59)$$

and the second supremum is attained in $n_\varepsilon\asymp\rho_\varepsilon^{-1/\eta}$ (compare with (3.42)). Thus (3.55)–(3.58) hold with $\eta_\varepsilon\asymp\varepsilon^{-2}r_\varepsilon^{2+1/2\eta}$. These imply distinguishability conditions

$$\gamma_\varepsilon\to 0,\quad\text{if } r_\varepsilon/r_\varepsilon^*\to\infty;\quad \limsup\gamma_\varepsilon<1,\quad\text{if}\quad \liminf r_\varepsilon/r_\varepsilon^*>0; \quad (3.60)$$

where r_ε^* are determined by (3.44). Joined with (3.43) these mean that r_ε^* are the rates in the problem.

Example 3.3 *Smooth signals.* Analogously to Example 3.1, let $S=\tilde S_{\eta,2}(R)$ be the set which consists of an 1-periodical η-smooth functions. Recall that S is the ellipsoid of semi-axes $d_i\asymp i^{-\eta}$, and we get the distinguishability conditions (3.43), (3.60) with rates (3.44).

Next, let the semi-axes be of exponential type $d_i\sim He^{-i\mu}$, $\mu>0$. Then

$$\eta_\varepsilon=\sup_n\frac{r_\varepsilon^2-R_n^2(S)}{\varepsilon^2\sqrt{2n}}\ge\sup_n\frac{r_\varepsilon^2-H^2e^{-2n\mu}}{\varepsilon^2\sqrt{2n}}\sim(r_\varepsilon/\varepsilon)^2((2/\mu)\log r_\varepsilon^{-1})^{-1/2}, \quad (3.61)$$

and the second supremum is attained in $n_\varepsilon\sim\mu^{-1}\log(H/r_\varepsilon)$ (compare with (3.46)). Thus (3.55)–(3.58) hold and we can substitute η_ε in the right-hand side of (3.61). However, the right-hand side of (3.61) is the same as u_ε in (3.46). In view of (3.38), (3.39) these give sharp asymptotics. Certainly these imply distinguishability conditions (3.60) with rates r_ε^* defined by (3.48) as well.

Example 3.4 *Analytical signals.* Let $S = \tilde{S}(\mu, H)$ be the set defined in Example 3.2; recall that the set $S(\mu, H)$ is the ellipsoid of semi-axes $d_j \sim H\exp(-j\mu)$. Thus we get

$$\beta(\alpha, V_\varepsilon(\rho_\varepsilon)) = \beta_\varepsilon(\alpha, S(r_\varepsilon)) = \Phi(T_\alpha - u_\varepsilon) + o(1); \tag{3.62}$$

$$\gamma(V_\varepsilon(\rho_\varepsilon)) = \gamma_\varepsilon(S(r_\varepsilon)) = 2\Phi(-u_\varepsilon/2) + o(1), \tag{3.63}$$

with u_ε defined by (3.46).

3.3 Bayesian Approach under a Gaussian Model

Unfortunately, the methods of Section 3.2 are not applicable to obtain the lower bounds for a wider class of norms which should determine the alternatives of interest. The main reason is that spherically supported priors are "too large." Here we introduce and study some classes of priors which are applicable to an essentially wider class of problems.

3.3.1 L_1- and L_2-Distances for Mixtures

Return to the sequence Gaussian model (2.1). Let us test the simple hypothesis $H_0 : v = 0$ against the alternative $H_1 : v \neq 0$. In this case, $\pi_0 = \delta_0$ is a Dirac mass at point 0. Let the prior $\pi_1 = \pi(dv)$ be the probability measure on Borel sets in R^n or l^2. We will omit the index 1 in the priors in the case of a simple hypothesis H_0. Recall that the likelihood ratio is of the form

$$L(x) = dP_\pi/dP_0 = E_\pi(dP_v/dP_0) = \int \exp(-|v|^2/2 + (x, v))\pi(dv). \tag{3.64}$$

It is difficult to study directly the L_1-distance between P_π and P_0 which determines the minimax criterion $\gamma = \gamma(1)$. In view of Proposition 2.12 let us evaluate the nonsymmetric L_2-distance (2.57) between P_π and P_0. Note the equality: for a sequence $\eta = \{\eta_i\}$, which consists of independent standard Gaussian variables, one has

$$E(\exp(\eta, v)) = \exp(|v|^2/2) \quad \forall\, v \in l^2. \tag{3.65}$$

Using (3.65) one can write the formula

$$\begin{aligned} |P_\pi - P_0|_2^2 &= E_{P_0}(E_\pi(dP_v/dP_0))^2 - 1 \\ &= E_{P_0} \iint (\exp(-(|u|^2 + |v|^2)/2 + (x, u+v)))\pi(du)\pi(dv) - 1 \end{aligned}$$

$$= \iint \exp(-(|u|^2 + |v|^2)/2 + |u+v|^2/2)\pi(du)\pi(dv) - 1$$

$$= \iint e^{(u,v)}\pi(du)\pi(dv) - 1. \tag{3.66}$$

3.3.2 Product Priors

The following priors are used extensively below. Let the sequence $\bar{\pi} = \{\pi_i, i \in I\}$ of probability measures π_i on Borel sets in the real line R^1 be given. Set

$$\pi(dv) = \prod_{i \in I} \pi_i(dv_i) \tag{3.67}$$

(if $n = \sharp(I)$ is finite, then we consider finite products).

Using (2.57) and the inequality $e^t - 1 \geq t$ we get

$$\begin{aligned} |P_\pi - P_0|_2^2 &= E_{P_0}\left(\frac{dP_\pi}{dP_0} - 1\right)^2 = \prod_i \int_{R^1}\int_{R^1} e^{uv}\pi_i(du)\pi_i(dv) - 1 \\ &= \prod_i \left(\int_{R^1}\int_{R^1} (e^{uv} - 1)\,\pi_i(du)\pi_i(dv) + 1\right) - 1 \\ &\leq \exp\left(\sum_i \int_{R^1}\int_{R^1} (e^{uv} - 1)\,\pi_i(du)\pi_i(dv)\right) - 1 \\ &= \exp\left(\|\bar{\pi}\|^2\right) - 1; \end{aligned} \tag{3.68}$$

here we set

$$\|\bar{\pi}\|^2 = \sum_i \|\pi_i\|^2 = \sum_i \int_{R^1}\int_{R^1} (e^{uv} - 1)\,\pi_i(du)\pi_i(dv). \tag{3.69}$$

Remark. Note that if $n = \infty$, then the product prior π may be not concentrated on l^2. However, if $\|\bar{\pi}\| < \infty$, then the mixture P_π is correctly defined as a limit of mixtures P_{π^n} which corresponds to a product prior for the factors from a finite index sets I_n, where I_n are increasing index subsets such that $\bigcup_n I_n = I$. In fact, let

$$I = \mathbf{N}, \quad \pi^n(dv) = \prod_{i=1}^{n} \pi_i(dv_i) \times \prod_{i=n+1}^{\infty} \delta_0(dv_i).$$

Using analogous evaluations one gets for any $m > n$

$$\begin{aligned} |P_{\pi^m} - P_{\pi^n}|_1^2 &= \left(E_{P_{\pi^n}}|dP_{\pi^m}/dP_{\pi^n} - 1|\right)^2 \leq E_{P_{\pi^n}}\left(dP_{\pi^m}/dP_{\pi^n} - 1\right)^2 \\ &= E_{P_{\pi^n}}(dP_{\pi^m}/dP_{\pi^n})^2 - 1 = E_{P_0}\prod_{i=n+1}^{m}(dP_{\pi_i}/dP_0)^2 - 1 \\ &\leq \exp(\sum_{i=n+1}^{m}\|\pi_i\|^2) - 1 \to 0, \quad \text{as} \quad m, n \to \infty. \end{aligned}$$

Therefore the sequences P_{π^n} converge in the complete space $\mathcal{P}^*$ which consists of probability measures P on $(R^\infty, \mathcal{B}^\infty)$ dominated by the Gaussian measure P_0 with an L_1-norm (this space is isomorphic to subset of probability density in the Banach space $L_1(R^\infty, \mathcal{B}^\infty, P_0))$.

On the other hand, for our purposes it suffises to consider a finite product, i.e., there are a finite number of items $\pi_i \neq \delta_0$ in (3.67).

These considerations, together with Propositions 2.9, 2.12, give

Proposition 3.6 *Let* $H_0 : v = 0$, $H_1 : v \in V_\varepsilon$ *and let a family of the sequences* $\bar{\pi}_\varepsilon$ *be such that* $\pi^\varepsilon(V_\varepsilon) \to 1$, *where* π^ε *are the product priors corresponding to* $\bar{\pi}_\varepsilon$. *If* $\|\bar{\pi}_\varepsilon\| = o(1)$, *then* $\gamma(V_\varepsilon) \to 1$, *and if* $\|\bar{\pi}_\varepsilon\| = O(1)$, *then* $\liminf \gamma(V_\varepsilon) > 0$.

In view of relationship (2.8), these considerations are extended into the functional Gaussian model (2.5). Let $\bar{\varphi} = \{\varphi_i,\, i \in I\}$ be an orthonormal system in $L_2(0,1)$ (finite or denumerable, maybe, depending on ε) and let $\bar{\pi}_\varepsilon = \{\pi_{\varepsilon,i},\, i \in I\}$ be given. Consider an inclusion map

$$v \to s_{\varepsilon,v} = \varepsilon \sum_{i \in I} v_i \varphi_i. \tag{3.70}$$

Under the embedding (3.70) the product prior $\pi^\varepsilon(dv)$ generates the prior $\pi^\varepsilon(ds)$ which is concentrated on the closure L_I of linear hull $\bar{\varphi}$ and corresponds to the independently π_i-distributed coordinates v_i, $i \in I$, in (3.70).

Analogously to Proposition 3.6:

Proposition 3.7 *Let* $H_0 : s = 0$, $H_1 : s \in S_\varepsilon$ *and* $\pi^\varepsilon(S_\varepsilon) \to 1$, *where* π^ε *are product priors for* $\bar{\pi}_\varepsilon$. *If* $\|\bar{\pi}_\varepsilon\| = o(1)$, *then* $\gamma(S_\varepsilon) \to 1$, *and if* $\|\bar{\pi}_\varepsilon\| = O(1)$, *then* $\liminf \gamma(S_\varepsilon) > 0$.

3.3.3 Hilbert Structure

Return to quantities (3.69). We would like to consider $\|\bar{\pi}\|$ and $\|\pi\|$ as norms of Hilbert type.

To construct the needed spaces, first, let us consider a linear space which consists of a finite signed measure $r = r_1 - r_2$ defined on Borelian sets of the real line R^1; here r_1, r_2 are finite nonnegative orthogonal measures, $0 \le r_k(R^1) < \infty$, $k = 1, 2$. Consider the bilinear form

$$\langle r_1, r_2 \rangle = \int_{R^1} \int_{R^1} e^{uv} r_1(du) r_2(dv). \tag{3.71}$$

Let $\mathcal{L}$ be the set which consists of signed measures $r = r_1 - r_2$ such that

$$\langle r_k, r_k \rangle < \infty, \quad k = 1, 2.$$

For $r \in \mathcal{L}$ we denote $\|r\|^2_{\mathcal{L}} = \langle r, r \rangle$. Let $\mathcal{P} \subset \mathcal{L}$ be the subset which consists of probability measures.

Proposition 3.8 *The bilinear form* (3.71) *is positive definite on $\mathcal{L}$, i.e., $\|r\|^2_{\mathcal{L}} \geq 0$, and if $\|r\|^2_{\mathcal{L}} = 0$, then $r = 0$.*

Proof. Observe the relation

$$\|r\|^2_{\mathcal{L}} = E\left(\int e^{-u^2/2+\eta u} r(du)\right)^2 \geq 0,$$

where η is the standard Gaussian variable. Thus the bilinear form $\langle r_1, r_2 \rangle$ is positive semidefinite. Consider a presentation $r = a_1\pi_1 - a_2\pi_2$ where π_k are the orthogonal probability measures, $a_k \geq 0$, $k = 1, 2$. Let $\|r\|^2_{\mathcal{L}} = 0$. Setting

$$L(\eta) = a_1 L_1(\eta) - a_2 L_2(\eta), \quad L_k(x) = dP_{\pi_k}/dP_0, \quad k = 1, 2,$$

we get $EL_k(\eta) = 1$, $EL^2(\eta) = 0$. These relations yield $a_1 = a_2$, and if $a_1 > 0$, then $P_{\pi_1} = P_{\pi_2}$; using Proposition 2.3 we get $\pi_1 = \pi_2$. This means $\pi_1 = \pi_2 = 0$ by orthogonality. □

Using the power expansion for exponent

$$e^x = \sum_{k=0}^{\infty} x^k/k!\ ,$$

we obtain the equality

$$\|r\|^2_{\mathcal{L}} = \sum_{k=0}^{\infty} \frac{1}{k!} \int_{R^1} \int_{R^1} (uv)^k r(du) r(dv) = \sum_{k=0}^{\infty} \frac{\mu_k^2(r)}{k!}, \tag{3.72}$$

where

$$\mu_0(r) = r(R^1), \quad \mu_k(r) = \int_{R^1} u^k r(du), \quad k \geq 1.$$

If $r \in \mathcal{P}$, then $\mu_0(r) = 1$, and $\mu_k(r)$ is the kth moment of probability measure r.

The bilinear form (3.71) generates a Hilbert structure in the space $\mathcal{L}$. It follows from (3.72) that the map

$$r \to \bar{\mu}(r) = \{\mu_k(r),\ k = 0, 1, \ldots\}$$

is an isomorphism of the space $\mathcal{L} = \{r\}$ to a linear subspace in sequence space $\ell = \{\bar{\mu}\}$ with the norm

$$|\bar{\mu}|^2_\ell = \sum_{k=0}^{\infty} \mu_k^2/k!.$$

It is convenient to consider a somewhat different space. Let

$$\mathcal{L}_0 = \{a\delta_0,\ a \in R^1\}$$

be a linear subspace generated by the Dirac mass at point 0; here and below δ_t is the Dirac mass at point $t \in R^1$. Introduce the relation of equivalence: $r_1 \sim r_2$, if $r_1 - r_2 \in \mathcal{L}_0$, and consider the factor-space $\tilde{\mathcal{L}} = \mathcal{L}/\mathcal{L}_0$ of cosets.

Let the linear functional $\phi(r) = \int f(u)r(du)$, defined on some linear subspace $\mathcal{L}_f \subset \mathcal{L}$, be given and let $f(u)$ be a Borel function such that $f(0) = 0$. Then we can define the functional $\phi(r)$ on the subspace $\tilde{\mathcal{L}}_f = \mathcal{L}_f/\mathcal{L}_0$. If $f(u,v)$ is a function such that $f(0,v) = f(u,0) = 0$ for all $u, v \in R^1$, then the same holds for the bilinear functional

$$\phi(r_1, r_2) = \int_{R^1}\int_{R^1} f(u,v) r_1(du) r_2(du).$$

Set $\breve{r} = r - r(R^1)\delta_0$. It follows from Proposition 3.8 that the relation

$$(r_1, r_2) = \langle \breve{r}_1, \breve{r}_2 \rangle = \int_{R^1}\int_{R^1} (e^{uv} - 1) r_1(du) r_2(du); \quad \|r\|^2 = (r, r), \tag{3.73}$$

defines a positive definite bilinear form in factor-space $\tilde{\mathcal{L}}$.

Let $\tilde{\mathcal{P}}$ be the set of cosets which contain a probability measure. Note that any coset $\tilde{\pi} \in \tilde{\mathcal{P}}$ contains one and only one $\pi \in \mathcal{P}$; therefore we can identify the sets $\tilde{\mathcal{P}}$ and $\mathcal{P}$. Namely, $\tilde{\pi} \in \tilde{\mathcal{P}}$ if and only if

$$a = \tilde{\pi}(\breve{R}) \in [0,1], \quad \breve{R} = R^1 \backslash \{0\}, \quad \int f(u)\tilde{\pi}(du) \geq 0,$$

for any bounded Borel function f such that $f(u) \geq 0$, $f(0) = 0$. Therefore, if $\tilde{\pi} \in \tilde{\mathcal{P}}$, then $\pi = \tilde{\pi} + (1 - \tilde{\pi}(R^1))\delta_0 \in \mathcal{P}$.

Note that $\tilde{\mathcal{P}}$ is a *convex star set* (the last means that if $\tilde{\pi} \in \tilde{\mathcal{P}}$, then $a\tilde{\pi} \in \tilde{\mathcal{P}}$ for any $a \in [0,1]$).

Thus relation (3.73) defines a norm $\|\pi\|$ for $\pi \in \mathcal{P}$.

Let us extend the definitions to sequences of measures. Denote by L and $\tilde{L}$ the product spaces which consist of sets of sequences $\bar{r} = \{r_i, i \in I\}$ of signed measures $r_i \in \mathcal{L}$ on the real line or of $\{r_i \in \tilde{\mathcal{L}}\}$. These generate the positive definite bilinear forms

$$\begin{aligned} \langle \bar{r}_1, \bar{r}_2 \rangle &= \sum_i \langle r_{i,1}, r_{i,2} \rangle = \sum_i \int_{R^1}\int_{R^1} e^{uv} r_{i,1}(du) r_{i,2}(dv), \\ (\bar{r}_1, \bar{r}_2) &= \sum_i (r_{i,1}, r_{i,2}) = \sum_i \int_{R^1}\int_{R^1} (e^{uv} - 1) r_{i,1}(du) r_{i,2}(dv), \end{aligned}$$

which determine the Hilbert structure in the linear spaces L and $\tilde{L}$.[3]

Denote by $\tilde{\Pi} \subset \tilde{L}$ the subset which consists of sequences with elements $\pi_i \in \tilde{\mathcal{P}}$. Then the set $\tilde{\Pi}$ can be identified with the set $\bar{\Pi} = \{\bar{\pi}\}$ which consists of the sequences of probability measures (priors) on the real line with a finite norm $\|\pi_i\|$. This corresponds to definition (3.69) and to $\|\pi\| = \|\pi - \delta_0\|_{\mathcal{L}}$.

Note the cases of importance below.

[3]This Hilbert structure been introduced in [55].

3.3.4 Two-Point Factors

Consider product priors defined by the two-point measures $\pi(dt;\, z, h)$ on the real line (or one-point measures for $h = 1$):

$$\pi(z,h) = (1-h)\delta_0 + h\delta_z, \quad h \in [0,1], \quad z \in R^1, \tag{3.74}$$

which are concentrated on points z and (if $h < 1$) on point 0. Introduce the function

$$\nu(x,z) = e^{-z^2/2+xz} - 1. \tag{3.75}$$

Note the properties of random variables $\nu(\eta, z)$ for the case when $\eta \sim N(0,1)$ is the standard Gaussian. Using the equality

$$\nu(x,z_1)\nu(x,z_2) = e^{z_1 z_2}\nu(x, z_1+z_2) - \nu(x,z_1) - \nu(x,z_2) + e^{z_1 z_2} - 1, \tag{3.76}$$

and since $E(e^{z\eta}) = e^{z^2/2}$, one has

$$\begin{aligned} E(\nu(\eta,z)) &= 0, \quad E(\nu(\eta,z_1)\nu(\eta,z_2)) = e^{z_1 z_2} - 1, &(3.77)\\ E(\nu^3(\eta,z)) &= (e^{z^2}-1)^2(e^{z^2}+2); &(3.78)\\ E(\nu^4(\eta,z)) &= (e^{z^2}-1)^2(e^{4z^2}+2e^{3z^2}+3e^{2z^2}-3)\\ &\le 4(e^{z^2}-1)^2 e^{4z^2}. &(3.79) \end{aligned}$$

Therefore, one has

$$\|\pi(z,h)\|^2 = h^2 E(\nu^2(\eta,z)) = h^2(e^{z^2}-1). \tag{3.80}$$

The product prior with factors of this type is of the form

$$\pi(dv;\, \bar{z}, \bar{h}) = \prod_i \pi(dv_i;\, z_i, h_i), \quad \bar{z} = \{z_i\}, \quad \bar{h} = \{h_i\}, \quad i \in I,$$

and

$$\|\bar{\pi}\|^2 = \sum_i h_i^2(\exp(z_i^2) - 1).$$

These correspond to the random vector $v = \sum_i \tau_i z_i e_i$. Here $\{e_i, i \in I\}$ is a canonical orthonormal basis in l^2 and $\{\tau_i, i \in I\}$ is a sequence of independent Bernoulli $\{0,1\}$-valued random variables,

$$P(\tau_i = 1) = h, \quad P(\tau_i = 0) = 1 - h.$$

If $\sup_i |z_i| \le B$, then there exists a constant $C = C(B) > 1$ such that

$$\sum_i h_i^2 z_i^2 \le \|\bar{\pi}\|^2 \le C \sum_i h_i^2 z_i^2; \quad C(B) \to 1 \quad \text{as } B \to 0, \tag{3.81}$$

the last relation follows from $\exp(z^2) - 1 \sim z^2$ as $z \to 0$.

3.3.5 Symmetric Three-point Factors

Consider product priors defined by the three-point measures $\pi(dt;\, z, h)$ on the real line (or two-point measures for $h = 1$):

$$\pi(z,h) = (1-h)\delta_0 + \frac{h}{2}(\delta_{-z} + \delta_z), \quad h \in [0,1], \quad z \geq 0, \tag{3.82}$$

which are concentrated on points $\pm z$ and (if $h < 1$) at point 0.[4] Introduce the function

$$\xi(x,z) = (\nu(x,z) + \nu(x,-z))/2 = e^{-z^2/2}\cosh(xz) - 1. \tag{3.83}$$

Note the properties of random variables $\xi(\eta, z)$ for the case when $\eta \sim N(0,1)$ is the standard Gaussian random variable. Analogously to (3.77), and using the equality

$$\begin{aligned}\xi(x,z_1)\xi(x,z_2) &= \frac{1}{2}(e^{z_1z_2}\xi(x,z_1+z_2) + e^{-z_1z_2}\xi(x,z_1-z_2)) \\ &- \xi(x,z_1) - \xi(x,z_2) + 2\sinh^2(z_1z_2/2)\end{aligned}$$

one has

$$E(\xi(\eta,z)) = 0, \quad E(\xi(\eta,z_1)\xi(\eta,z_2)) = 2\sinh^2(z_1z_2/2), \tag{3.84}$$

$$\begin{aligned}E(\xi^4(\eta,z)) &= 2\sinh^4(z^2/2)(e^{4z^2} + 4e^{3z^2} + 10e^{2z^2} + 12e^{z^2} + 3) \\ &\leq 15(2\sinh^2(z^2/2))^2 e^{4z^2}.\end{aligned} \tag{3.85}$$

Therefore, one has

$$\|\pi(z,h)\|^2 = h^2 E\xi^2(\eta,z) = 2h^2\sinh^2(z^2/2). \tag{3.86}$$

The product prior with factors of this type is of the form

$$\pi(dv;\, \bar{z}, \bar{h}) = \prod_i \pi(dv_i;\, z_i, h_i), \quad \bar{z} = \{z_i\}, \quad \bar{h} = \{h_i\}, \quad i \in I,$$

and

$$\|\bar{\pi}\|^2 = 2\sum_i h_i^2 \sinh^2(z_i^2/2).$$

These correspond to the random vector $v = \sum_i \tau_i z_i e_i$, where $\{e_i, i \in I\}$ is a canonical orthonormal basis in l^2 and $\{\tau_i, i \in I\}$ is a sequence of independent random variables with values 0, ± 1,

$$P(\tau_i = 1) = P(\tau_i = -1) = h/2, \quad P(\tau_i = 0) = 1 - h.$$

If $\sup_i z_i \leq B$, then there exists a constant $C = C(B) > 1$ such that

$$\sum_i h_i^2 z_i^4 \leq 2\|\bar{\pi}\|^2 \leq C\sum_i h_i^2 z_i^4; \quad C(B) \to 1, \quad \text{as } B \to 0, \tag{3.87}$$

---

[4]Product priors with symmetric two- and three-point factors were used in hypothesis testing problems for Gaussian models in [49] and for the sample model in [46].

the last relation follows from $2\sinh^2(z^2/2) \sim z^4/2$ as $z \to 0$.

It should be noted that for finite n the product prior π with $h_i < 1$, $i = 1, \dots, n$, is senseless in a nonasymptotic setting, because π is not concentrated on the alternative in this case: $\pi(\{0\}) = (1-h_1)\dots(1-h_n) > 0$. However, product priors of this type will be very helpful in the asymptotic setting below.

3.3.6 *Minimax Distinguishability for l_p^n-Balls, $p < \infty$*

As a first corollary of the results above consider the following problem. Under the sequence Gaussian model (2.1) let us test $H_0 : v = 0$ against $H_1 : v \in V_n^+$ or $H_1 : v \in V_n$ where

$$V_n = \breve{D}_{n,p}(\rho_n) \quad = \{v \in R^n : \ |v|_p \geq \rho_n\}, \tag{3.88}$$

$$V_n^+ = \breve{D}_{n,p}^+(\rho_n) \quad = \{v \in R^n : \ |v|_p \geq \rho_n, \ v_i \geq 0 \quad \forall i\}. \tag{3.89}$$

Let $n \to \infty$ and let all limits in this section be assumed as $n \to \infty$.

We would like to obtain to conditions for minimax distinguishability and nondistinguishability in the problem.

For $0 < p < \infty$ introduce the quantities

$$\rho_{n,p}^* = \begin{cases} n^{(4-p)/4p}, & \text{if } \ 0 < p \leq 2, \\ n^{1/2p}, & \text{if } \ 2 < p < \infty, \end{cases} \tag{3.90}$$

and

$$\rho_{n,p}^{*+} = \begin{cases} n^{(2-p)/2p}, & \text{if } \ 0 < p \leq 1, \\ n^{1/2p}, & \text{if } \ 1 < p < \infty. \end{cases} \tag{3.91}$$

Proposition 3.9

(1) If $\rho_n = o(\rho_{n,p}^*)$, then $\gamma(V_n) \to 1$, and if $\rho_n = O(\rho_{n,p}^*)$, then $\liminf \gamma(V_n) > 0$. *If $\rho_n = o(\rho_{n,p}^{*+})$, then $\gamma(V_n^+) \to 1$, and if $\rho_n = O(\rho_{n,p}^{*+})$, then* $\liminf \gamma(V_n^+) > 0$.

(2) If $\rho_n/\rho_{n,p}^* \to \infty$, then $\gamma(V_n) \to 0$. If $\rho_n/\rho_{n,p}^{*+} \to \infty$, then $\gamma(V_n^+) \to 0$. These mean that the families $\rho_{n,p}^*$ and $\rho_{n,p}^{*+}$ are rates in the problems.

Proof. The case of positive alternatives V_n^+ with $p \in (0,1]$ follows from Example 2.4. In fact, we have $|v_n^*| = \rho_n n^{1/2-1/p} = \rho_n/\rho_{n,p}^{*+}$. These yield to either $|v_n^*| \to 0$ as $\rho_n = o(\rho_{n,p}^{*+})$, or $|v_n^*| = O(1)$ as $\rho_n = O(\rho_{n,p}^{*+})$, or $|v_n^*| \to \infty$ as $\rho_n/\rho_{n,p}^{*+} \to \infty$.

Let us establish the lower bounds, i.e., let us prove statement (1) for other cases. For alternatives V_n^+ with $p \in (1,\infty)$ it suffices to assume $\rho_n \to \infty$ because in an other case one can increase ρ_n such that $\rho_n \to \infty$, $\rho_n = o(\rho_n^*)$ and $\gamma(V_n^+)$ does not increase. Consider sequences of two-point factors and the product priors $\pi^n = (\pi(z_n, h_n))^n$ with $z_n = z > 1$, $h_n = n^{-1}\rho_n^p = o(1)$, which corresponds to $\bar{\pi}_n = (\pi(z_n, h_n), \dots, \pi(z_n, h_n))$ (n items), $\pi(z,h) = (1-h)\delta_0 + h\delta_z$.

First, observe that $\pi^n(V_n^+) \to 1$. In fact, π^n is concentrated in the positive octant R_+^n in R^n, where $R_+^n = \{x \in R^n,\ x_i \geq 0\ \forall i\}$. The function $F_n = \sum_{i=1}^n v_i^p$ is a random variable with respect to the probability measure π^n,

$$E_{\pi^n}(F_n) = nE_{\pi(z_n,h_n)}(v^p) = nh_n z_n^p = z^p \rho_n^p,$$
$$\operatorname{Var}_{\pi^n}(F_n) = n \operatorname{Var}_{\pi(z_n,h_n)}(v^p) = nh_n(1-h_n)z_n^{2p} \leq z^{2p}\rho_n^p,$$

and by the Chebyshev inequality

$$\begin{aligned} 1-\pi^n(V_n^+) &\leq \pi^n(F_n < \rho_n^p) = \pi^n(E_{\pi^n}(F_n) - F_n > E_{\pi^n}(F_n) - \rho_n^p) \\ &\leq \operatorname{Var}_{\pi^n}(F_n)/(E_{\pi^n}(F_n) - \rho_n^p)^2 \leq z^{2p}/(\rho_n^{3p}(z^p-1)^2) \to 0. \end{aligned}$$

Next,

$$\|\bar{\pi}_n\|^2 = n\|\pi(z_n,h_n)\|^2 \asymp nh_n^2 = (\rho_n/\rho_n^*)^{2p}.$$

Thus we can use Proposition 3.6.

Consider the alternative V_n. If $p \geq 2$, then, by $\rho_{n,p}^* = \rho_{n,p}^{*+}$ and $V_n^+ \subset V_n$, statement (1) of Proposition 3.9 follows from the result for V_n^+. Let $p \in (0,2]$ and $\rho_n = O(\rho_{n,p}^*)$. Consider the sequence of the symmetric two-point factors and product priors $\pi^n = (\pi(z_n,h_n))^n$ with $z_n = n^{-1/p}\rho_n = o(1)$, $h_n = 1$, which correspond to $\bar{\pi}_n = (\pi(z_n,1),\ldots,\pi(z_n,1))$ (n items), $\pi(z,1) = (\delta_z + \delta_{-z})/2$. Clearly, $\pi^n((V_n) = 1$. Next,

$$\|\bar{\pi}_n\|^2 = n\|\pi(z_n,h_n)\|^2 \sim nz_n^4/2 = (\rho_n/\rho_n^*)^4/2.$$

Therefore statement (1) follows from Proposition 3.6.

The upper bounds follow from results of Section 3.1. For alternative V_n with $p \leq 2$ we apply χ^2-tests (3.14) and use Corollary 3.4. Using the inequalities

$$\left(n^{-1}\sum_{i=1}^n v_i^2\right)^{1/2} \geq \left(n^{-1}\sum_{i=1}^n |v_i|^p\right)^{1/p} \geq n^{-1/p}\rho_n \gg n^{-1/p}\rho_n^* = n^{-1/4}$$

(here and below the notation $a \gg b$ is equivalent to $b = o(a)$) and since $P_n(v) = v$ we get

$$u_n = (2n)^{-1/2} \inf_{v\in V_n} \sum_{i=1}^n v_i^2 \to \infty.$$

For alternative V_n^+ with $1 < p < 2$ we apply χ^+-tests (3.26) and use Corollary 3.8:

$$u_n = n^{-1/2p} \inf_{v\in V_n^+} |P_n(v)|_p \geq n^{-1/2p}\rho_n \gg n^{-1/2p}\rho_n^* = 1.$$

If $p > 2$, then we consider a wide alternative V_n and apply χ^p-tests (3.22) and, analogously, we use Corollary 3.6. □

Remark. Note that if $p = 2$, then we obtain the same rates $\rho_n^* = n^{1/4}$, as in Proposition 2.15. These were based on spherical priors which provided lower bounds, and on χ^2-tests, which gave upper bounds. If $p \le 2$, then we use χ^2-tests as well, however, spherical priors are "too massive" to provide good lower bounds here. We use product priors with symmetric two-point factors; these priors are concentrated on 2^n points

$$v^* = (\pm z_n, \dots, \pm z_n) \in R^n; \quad z_n = n^{-1/p}\rho_n,$$

which implies $|v^*|_p = \rho_n$. Therefore for $p = 2$ the supports are contained in $S^{n-1}(\rho_n)$ but this set is essentially smaller than the sphere $S^{n-1}(\rho_n)$; these provide essentially better lower bounds for $p < 2$.

In particular, consider two-sided constraints

$$V_n = \{v \in R^n : |v|_p \ge \rho, \ |v|_q \le R\}; \quad p \le 2, \quad q \ge p, \tag{3.92}$$

with $R \ge \rho n^{1/q-1/p} = |v^*|_q$ such that V_n is nonempty. Since $v^* \in V_n \subset \breve{D}_{n,p}(\rho)$, we get the same rates in the problem as in Proposition 3.9 for $p \le 2$. Therefore, we get

Corollary 3.13 *Let us test $H_0 : v = 0$ against $H_1 : v \in V_n$, where V_n is defined by the two-sided constraints (3.92), $\rho = \rho_n$, $R = R_n$, $n \to \infty$. Then the rates in the problem are $\rho_n^* = n^{(4-p)/4p}$.*

3.3.7 Semiorthogonal Priors

Return to the sequence Gaussian model (2.1). Let a family of collections of vectors $\bar{v}_\varepsilon = \{v_{\varepsilon,j}, \ j \in J_\varepsilon\}$, $v_{\varepsilon,j} \in l^2$, be given; here J_ε is finite or a denumerable index set.

We call the collections $\bar{v}_\varepsilon$ *orthogonal, semiorthogonal, and asymptotically semiorthogonal* if, respectively, $(v_{\varepsilon,k}, v_{\varepsilon,j}) = 0$, $(v_{\varepsilon,k}, v_{\varepsilon,j}) \le 0$ for any $k \ne j$, $k, j \in J_\varepsilon$, or $\sup_{k \ne j}(v_{\varepsilon,k}, v_{\varepsilon,j}) \le o(1)$.

Let $\bar{p}_\varepsilon = \{p_{\varepsilon,j}, \ j \in J_\varepsilon\}$ be a family of probability vectors

$$p_j \ge 0, \quad \sum_{j \in J_\varepsilon} p_{\varepsilon,j} = 1.$$

Consider the priors and the mixtures

$$\pi^\varepsilon = \pi(\bar{v}_\varepsilon, \bar{p}_\varepsilon) = \sum_{j \in J_\varepsilon} p_{\varepsilon,j}\delta_{v_{\varepsilon,j}}, \quad P_{\pi^\varepsilon} = \sum_{j \in J_\varepsilon} p_{\varepsilon,j}P_{v_{\varepsilon,j}}.$$

We call the priors $\pi(\bar{v}_\varepsilon, \bar{p}_\varepsilon)$ *orthogonal, semiorthogonal, or asymptotically semiorthogonal*, if the collections $\bar{v}_\varepsilon$ are of these types.[5]

[5] Semiorthogonal priors were introduced by Burnashev [8]. These were used to prove the triviality for complements of L_2-balls under the functional Gaussian model.

The likelihood ratio is of the form

$$L_\varepsilon(x) = \frac{dP_{\pi^\varepsilon}}{dP_0}(x) = \sum_{j\in J_\varepsilon} p_{\varepsilon,j}\exp(-|v_{\varepsilon,j}|^2/2 + (x, v_{\varepsilon,j})).$$

Let $\pi^\varepsilon = \pi(\bar v_\varepsilon, \bar p_\varepsilon)$ be asymptotically semiorthogonal. It follows from equality (3.65) that

$$\begin{aligned} |P_{\pi^\varepsilon} - P_0|_1^2 \le |P_{\pi^\varepsilon} - P_0|_2^2 &\le E_{P_0} L_\varepsilon^2(x) - 1 = \sum_{k,j} p_{\varepsilon,k}p_{\varepsilon,j}(e^{(v_{\varepsilon,k},v_{\varepsilon,j})} - 1) \\ &\le o(1)\sum_{k\ne j} p_{\varepsilon,k}p_{\varepsilon,j} + \sum_j p_{\varepsilon,j}^2(e^{|v_{\varepsilon,j}|^2} - 1) \\ &\le \sum_j p_{\varepsilon,j}^2(e^{|v_{\varepsilon,j}|^2} - 1) + o(1). \end{aligned} \tag{3.93}$$

Let us consider finite asymptotically semiorthogonal collections

$$\bar v_n = (v_{\varepsilon,1}, \dots, v_{\varepsilon,n}), \quad v_{\varepsilon,j} \in l^2,$$

such that $|v_{\varepsilon,j}| = \rho_n > 0$ and $p_{\varepsilon,j} = 1/n$, $j = 1, \dots, n = n_\varepsilon \to \infty$. Then

$$|P_\pi^n - P_0|_1^2 \le n^{-1}(e^{\rho_n^2} - 1) + o(1) = n^{-1}e^{\rho_n^2} + o(1). \tag{3.94}$$

Analogously to the Remark in Section 3.3.2, if $\#(J_\varepsilon) = \infty$ (say, $J_\varepsilon = \mathbf{N}$), then the asymptotically semiorthogonal priors π^ε may not be concentrated on l^2, and we consider the mixture P_{π^ε} as an L_1-limit of mixtures P_{π_n} which correspond to the finite priors

$$\pi_n = \sum_{j=1}^n \tilde p_{\varepsilon,j}\delta_{v_{\varepsilon,j}}, \quad \tilde p_{\varepsilon,j} = p_{\varepsilon,j}\left(\sum_{j=1}^n p_{\varepsilon,j}\right)^{-1}.$$

We leave the calculations to the reader.

These considerations give

Proposition 3.10 *Let $H_0 : v = 0$, $H_1 : v \in V_\varepsilon$ and let $\pi_\varepsilon = \pi_\varepsilon(\bar v_\varepsilon, \bar p_\varepsilon)$ be a family of asymptotically semiorthogonal priors such that*

$$v_{\varepsilon,j} \in V_\varepsilon \quad \forall\, j \in J_\varepsilon, \quad \sum_j p_{\varepsilon,j}^2(\exp(|v_{\varepsilon,j}|^2) - 1) = o(1).$$

Then $\gamma(V_\varepsilon) \to 1$, $\beta(\alpha, V_\varepsilon) \to 1 - \alpha \quad \forall\, \alpha \in (0,1)$.

These considerations could be extended into the functional Gaussian model (2.5). Let $\bar s_\varepsilon = \{s_{\varepsilon,j},\ j \in J_\varepsilon\}$ be a family of collections in $L_2(0,1)$. We call these orthogonal, semiorthogonal, and asymptotically semiorthogonal if, respectively, $(s_{\varepsilon,k}, s_{\varepsilon,j}) = 0$, $(s_{\varepsilon,k}, s_{\varepsilon,j}) \le 0$ for any $k \ne j$, $k, j \in J_\varepsilon$ or $\sup_{k\ne j}(s_{\varepsilon,k}, s_{\varepsilon,j}) \le o(\varepsilon^2)$. Let $\bar p_\varepsilon = \{p_{\varepsilon,j},\ j \in J_\varepsilon\}$ be a family of probability vectors. Set

$$\pi_\varepsilon = \pi_\varepsilon(\bar s_\varepsilon, \bar p_\varepsilon) = \sum_{j\in J_\varepsilon} p_j\delta_{s_{\varepsilon,j}}, \quad P_{\pi_\varepsilon} = \sum_{j\in J_\varepsilon} p_j P_{\varepsilon,s_{\varepsilon,j}}.$$

Analogously to Proposition 3.10 we have

Proposition 3.11 *Let $H_0 : s = 0$, $H_1 : s \in S_\varepsilon$ and let $\pi_\varepsilon = \pi_\varepsilon(\bar{s}_\varepsilon, \bar{p}_\varepsilon)$ be the family of asymptotically semiorthogonal priors such that*

$$s_{\varepsilon,j} \in S_\varepsilon \quad \forall\, j \in J_\varepsilon, \quad \sum_j p_{\varepsilon,j}^2(\exp(\varepsilon^{-2}\|s_{\varepsilon,j}\|^2) - 1) = o(1).$$

Then $\gamma(S_\varepsilon) \to 1$, $\beta(\alpha, S_\varepsilon) \to 1 - \alpha$.

3.3.8 Minimax Distinguishability for l_p^n-Balls, $p = \infty$

Let

$$V_n = V_{n,q}(\rho_n, R_n) = \{v \in R^n : |v|_\infty \geq \rho_n,\ |v|_q \leq R_n\}, \tag{3.95}$$
$$V_n^+ = V_{n,q}^+(\rho_n, R_n) = V_n \cap R_+^n. \tag{3.96}$$

In particular, if $R_n = \infty$, we have

$$V_n = V_{n,q}(\rho_n, \infty) = \breve{D}_{n,\infty}(\rho_n), \quad V_n^+ = V_{n,q}^+(\rho_n, \infty) = \breve{D}_{n,\infty}^+(\rho_n).$$

The conditions for minimax distinguishability and nondistinguishability in the problem are of a different form with respect to Section 3.3.6.

Proposition 3.12 *For both alternatives V_n or V_n^+ and for all $0 < \rho_n \leq R_n \leq \infty$, $0 < q \leq \infty$ one has:*
(1) *Let $\limsup \rho_n/\sqrt{\log n} < 1$. Then $\gamma_n(V_n) \geq \gamma_n(V_n^+) \to 1$.*
(2) *Let $\liminf \rho_n/\sqrt{\log n} > \sqrt{2}$. Then $\gamma_n(V_n^+) \leq \gamma_n(V_n) \to 0$.*

Proof of Proposition 3.12. To obtain lower bounds we consider the smallest alternative V_n^+ with $R_n = \rho_n$, $q < \infty$, which contains only n points $v_{n,j} = \rho_n e_j$, $j = 1, \ldots, n$, where $\{e_j\}$ is the canonical basis in R^n. Consider the orthogonal priors $\pi^n = \pi(\bar{v}_n,\ \bar{p}_n)$ with $p_{n,j} = 1/n$. Clearly, $\pi^n(V_n) = 1$ and

$$\sum_j p_{n,j}^2 \exp\left(|v_{n,j}|^2\right) = n^{-1}\exp(\rho_n^2) = o(1).$$

Thus we can use Proposition 3.10.

To obtain the upper bounds let us consider the widest alternative

$$V_n = \breve{D}_{n,\infty}(\rho_n).$$

Applying supreme tests (3.36) and using Corollary 3.11 with $R_n = \rho_n$ we get

$$\gamma_n(V_{n,q}(\rho_n, R_n)) \leq \gamma_n(V_{n,q}(\rho_n, \infty), \psi_n) \leq \Phi(\sqrt{2\log n} - \rho_n) \to 0, \tag{3.97}$$

as $\rho_n - \sqrt{2\log n} \to \infty$. □

Remark. For $p = \infty$, Proposition 3.12 gives the rates $\rho_n^* = \sqrt{\log n}$. However, distinguishability conditions are of different type than for $p < \infty$: they are not of the forms $\rho_n/\rho_n^* \to 0$ or $\rho_n/\rho_n^* \to \infty$, but depend on constants which bound the ratio. The constants are different here. Later we show that the "correct" constant is in Proposition 3.12 (2) and that the upper bounds in (3.97) are asymptotically sharp.

3.4 Triviality and Classical Asymptotics

3.4.1 Minimax Triviality

Let us return to the general minimax hypothesis testing problem

$$H_0 : P \in \mathcal{P}_0, \quad H_1 : P \in \mathcal{P}_1.$$

We term the minimax hypothesis testing problem as *trivial*, if $\gamma = \gamma(1, \mathcal{P}_0, \mathcal{P}_1) = 1$ (return to Section 2.4 for a definition of $\gamma(t, \mathcal{P}_0, \mathcal{P}_1)$).

It follows from the convex properties in Lemma 2.3 and from the inequalities (2.11) that relation $\gamma = 1$ is equivalent to $\gamma(t, \mathcal{P}_0, \mathcal{P}_1) = \min(1, t)\ \forall\, t > 0$ and to $\beta(\alpha, \mathcal{P}_0, \mathcal{P}_1,) = 1 - \alpha\ \forall\, \alpha \in (0, 1)$. Minimax setting is not of interest in this case. Therefore we first need to select some trivial problems.

By Theorem 2.1 (ii) the triviality of the problem is equivalent to the existence of sequences $P_{0,m} \in [\mathcal{P}_0]$, $P_{1,m} \in [\mathcal{P}_1]$ such that $|P_{0,m} - P_{1,m}|_1 \to 0$ as $m \to \infty$.

Under the sequence Gaussian model (2.1) let us test the simple hypothesis $H_0 :\ v = 0$ against alternative $H_1 :\ v \in V,\ 0 \notin V$. The triviality of the problem is equivalent to the existence of a sequence of priors π_n such that $\pi_n(V) \to 1,\ |P_{\pi_n} - P_0|_1 \to 0$.

Triviality for finite n is possible if and only if the closure of V contains point 0; in particular, if $V \subset R^n$ is a closed set, then $\gamma < 1$. In fact, in this case, $\rho = \inf_{v \in V} |v| > 0$ and in view of (1.14) and of the monotonicity of the noncentral distribution function $G_n(t, h)$ in h one has

$$\beta(\alpha, V) \le \beta(\alpha, \breve{D}^n(\rho)) = G_n(t_{n,\alpha}, \rho^2) < 1 - \alpha.$$

Thus to obtain nontrivial problem it suffices "to remove" some neighborhood of point 0. However, it was noted in Section 1.4.2 that the infinite-dimensional case is essentially different.

3.4.2 Triviality for l_p-Balls

Let $\#(I) = \infty$. Denote by $D_{p,l}(\rho) \subset l^2$ the l_p-ball of radius ρ defined by inequality $|v|_p < \rho$. Remember that

$$|v|_p = \left(\sum_{i \in I} |v_i|^p\right)^{1/p}, \quad 0 < p < \infty; \quad |v|_\infty = \sup_{i \in I} |v_i|.$$

Denote by $S_{p,l}(\rho)$, $\breve{D}_{p,l}(\rho)$ the boundary and complement of the ball $D_{p,l}(\rho)$.

Proposition 3.13 *Under the sequence Gaussian model* (2.1) *let us test $H_0 : v = 0$ against $H_1 : v \in V$ where $V = S_{p,l}(\rho)$ or $V = \breve{D}_{p,l}(\rho)$. Then minimax hypothesis testing problems are trivial for any $\rho > 0$, $0 < p \leq \infty$.*

Proof. It suffices to consider $V = S_{p,l}(\rho)$. Let $\{e_j,\ j \in I\}$ be canonical the orthonormal basis in l^2. Consider the sequence of orthogonal collections $\bar{v}_n = (v_{n,1}, \dots v_{n,n}) : v_{n,j} = \rho e_j$ and put $p_{n,j} = 1/n,\ j = 1, \dots, n$. Consider the priors $\pi_n = \pi(\bar{v}_n, \bar{p}_n)$ and mixtures $P_{\pi_n} = P_\pi$ from Section 3.3.7. Clearly, $\pi_n(S_{p,l}(\rho)) = 1$. Using (3.94) we get

$$|P_{\pi_n} - P_0|_1 \leq n^{-1}(\exp(\rho^2) - 1) \to 0. \qquad \square$$

Corollary 3.14 *Under the functional Gaussian model* (2.5) *let us test $H_0 : s = 0$ against $H_1 : s \in S$ with $S = \breve{D}_{p,L}(r)$ or $S = S_{p,L}(r)$. Then $\gamma_\varepsilon(S) = 1$ for any $r > 0$, $\varepsilon > 0$, $1 \leq p \leq \infty$.*

Proof of Corollary 3.14. Take the collections of functions $s_j(t) = b_p^{-1} r \sin 2\pi j t$ and put $p_{n,j} = 1/n,\ j = 1, \dots, n,\ t \in (0,1)$,

$$b_p^p = \int_0^1 |\sin(2\pi j t)|^p \, dt = (2\pi)^{-1} \int_0^{2\pi} |\sin u|^p \, du.$$

These functions are orthogonal and $\|\psi_j\|_p = r$. Therefore they are contained in S for any n. Note that $\|s_j\|^2 = r^2/2b_p^2$ does not depend on n and $n^{-1}\exp(\varepsilon^{-2}\|s_j\|^2) \to 0$ as $n \to \infty$. Thus we can use Proposition 3.11. $\square$

Proposition 3.13 and Corollary 3.14 mean that it is not enough to remove small signals for the detection of *all large signals*, and one of the reasons is that there are too many orthogonal signals.

The situation is different, if we remove not the "ball-shaped," but the "ellipsoid-shaped" neighborhoods of the null-hypothesis.

3.4.3 Triviality and Classical Asymptotics for Power Norms

Remember that we call the $l_{\bar{a},p}$-ellipsoid the set $D_p(\bar{a}) \subset l^2$ defined by inequality $|v|_{\bar{a},p} < \rho$ under the norm (2.70). Let us consider an alternative corresponding to the complement of an l_p-ellipsoid

$$V(\rho) = \breve{D}_{\bar{a},p}(\rho) = \{v \in l^2 : |v|_{\bar{a},p} \geq \rho\}.$$

If $a_i = i^r$, we denote this set $V_{r,p}(\rho)$. Put

$$g_p(\bar{a}) = \begin{cases} \sum_i a_i^{4p/(4-p)}, & \text{if } 0 < p \leq 2, \\ \sum_i a_i^{2p}, & \text{if } 2 < p < \infty. \end{cases} \tag{3.98}$$

For $p = \infty$, define

$$g(\bar{a}, h) = \sum_i \exp(-h/a_i^2). \tag{3.99}$$

Theorem 3.1

(1) *Let $p < \infty$, $g_p(\bar{a}) = \infty$ or $p = \infty$, and let $g(\bar{a}, h) = \infty$ for all $h > 0$. Then the problem is trivial: $\gamma(V(\rho)) = 1$ for any $\rho > 0$.*

(2) *Let $p < \infty$, $g_p(\bar{a}) < \infty$ or $p = \infty$ and let $g(\bar{a}, h) < \infty$ for some $h > 0$. Then there are classical asymptotics in the problem: $\gamma(V(\rho)) \to 0$, as $\rho \to \infty$. In particular, let $a_i = i^r$, $-\infty < r < \infty$, $i \in \mathbf{N}$. Set*

$$r_p^* = \begin{cases} 1/4 - 1/p, & \text{if } p \le 2, \\ -1/2p, & \text{if } 2 < p < \infty, \\ 0, & \text{if } p = \infty. \end{cases} \tag{3.100}$$

Then the problem is trivial for $r \ge r_p^*$, and this is of classical asymptotics for $r < r_p^*$. Moreover, for any p and any $\rho_\varepsilon \to \infty$, there exist common tests $\psi_{\varepsilon,\rho_\varepsilon,p}$ which do not depend on r and are consistent for any alternative $V_{p,r}(\rho_\varepsilon)$ with $r < r_p^*$.

Note that Theorem 3.1 describes regions C of the classical asymptotics presented in Figures 1–8 in Section 6.4. The structure of the tests $\psi_{\varepsilon,\rho_\varepsilon,p}$ is given in the proof of the upper bounds of Theorem 3.1. [6]

Proof. Consider lower bounds. In view of (2.46) and Proposition 2.9 it suffices to construct sequences of priors π^n such that $\pi^n(V(\rho)) \to 1$ and $|P_{\pi^n} - P_0|_1 \to 0$ as $n \to \infty$.

Let $p \le 2$. Consider product priors π^n with symmetric two-point factors $\bar{\pi}_n = \{\pi_{n,i}\}$, $\pi_{n,i} = \pi(z_{n,i}, h_{n,i})$, where $h_{n,i} = 1$, $z_{n,i} = \rho a_i^{p/(4-p)}/A_n$, if $i \le n$ and $z_{n,i} = 0$, if $i > n$; here

$$A_n^p = \sum_{i=1}^n a_i^{4p/(4-p)} \to \infty, \qquad \text{as } n \to \infty.$$

The priors π^n are concentrated on $V(\rho)$ because of $\sum_i (z_{n,i} a_i)^p = \rho^p$. On the other hand,

$$\sum_i z_{i,n}^4 = \rho^4 A_n^{p-4} \to 0, \quad n \to \infty.$$

Therefore $\max_i z_i = o(1)$ and by $2\sinh^2(z^2/2) \sim z_{n,i}^4/2$ as $z \to 0$ one has

$$\|\bar{\pi}_n\|^2 = 2\sum_{i=1}^n \sinh^2(z_{n,i}^2/2) \to 0.$$

In view of Section 3.3.2 these imply triviality.

---

[6] Theorem 3.1 has been obtained in [52].

Let $2 < p < \infty$. Assume $a_n \to 0$. For large enough n consider product priors π^n with symmetric three-point factors $\pi_{n,i} = \pi(z_{n,i}, h_{n,i})$, where $z_{n,i} = 1$, $h_{n,i} = \tilde{\rho}^p a_i^p / A_n$, if $i \leq n$ and $h_{n,i} = 0$, if $i > n$; here

$$A_n = \sum_{i=1}^n a_i^{2p} \to \infty, \qquad \text{as } n \to \infty; \quad \tilde{\rho} > \rho.$$

The priors π^n are not concentrated on $V(\rho)$, but

$$E_{\pi^n}\left(\sum_i |v_i a_i|^p\right) = \sum_{i=1}^n h_{n,i} a_i^p = \tilde{\rho}^p > \rho^p, \tag{3.101}$$

$$\begin{aligned} \mathrm{Var}_{\pi^n}\left(\sum_i |v_i a_i|^p\right) &= \sum_{i=1}^n h_{n,i}(1 - h_{n,i}) a_i^{2p} \\ &\leq \sum_{i=1}^n h_{n,i} a_i^{2p} = \tilde{\rho}^p \sum_{i=1}^n a_i^{3p} / A_n \to 0. \end{aligned} \tag{3.102}$$

In fact, one can choose $m = m_n \to \infty$ such that $A_m = o(A_n)$. Then, by $a_n \to 0$, $A_n \to \infty$ one has

$$\sum_{i=1}^n a_i^{3p} / A_n \leq \max_i a_i^p A_m / A_n + \max_{i > m} a_i^p \to 0.$$

Using the Chebyshev inequality and (3.101), (3.102) one gets:

$$\begin{aligned} 1 - \pi^n(V(\rho)) &= \pi^n\left(\sum_i |v_i a_i|^p < \rho^p\right) \\ &= \pi^n\left(E_{\pi^n}\left(\sum_i |v_i a_i|^p\right) - \sum_i |v_i a_i|^p > \tilde{\rho}^p - \rho^p\right) \leq \\ &\leq \mathrm{Var}_{\pi^n}\left(\sum_i |v_i a_i|^p\right)(\tilde{\rho}^p - \rho^p)^{-2} \to 0. \end{aligned} \tag{3.103}$$

Moreover,

$$\|\bar{\pi}_n\|^2 = c \sum_{i=1}^n h_{n,i}^2 = c \tilde{\rho}^{2p} A_n^{-1} \to 0,$$

where $c = 2\sinh^2(1/2)$. In view of Section 3.3.2 these imply triviality.

Let $2 < p < \infty$, $\limsup a_n > 0$. This means one can find a subsequence $m_i \to \infty$ such that $a_{m_i} \geq B$, $i \to \infty$. Let us consider orthogonal collections $v_{n,i} = \rho e_{m_i} / a_{m_i}$, $i = 1, \ldots, n$, and orthogonal priors $\pi_n = \pi(\bar{v}_n, \bar{p}_n)$ with $p_{n,i} = 1/n$, $i = 1, \ldots, n$. These are concentrated on $V(\rho)$ and in view of Section 3.3.7 they imply triviality.

Let $p = \infty$. Let us consider the orthogonal collections $v_{n,i} = \rho e_i/a_i$, $i = 1,\dots,n$, and the orthogonal priors $\pi_n = \pi(\bar{v}_n, \bar{p}_n)$ with

$$p_{n,i} = \exp(-\rho^2/a_i^2)\left(\sum_{i=1}^n \exp(-\rho^2/a_i^2)\right)^{-1}, \quad i = 1,\dots,n.$$

They are concentrated on V and, as $n \to \infty$,

$$\sum_{i=1}^n p_{n,i}^2(\exp(|v_{n,i}|^2) - 1) \le \sum_{i=1}^n p_{n,i}^2 \exp(\rho^2/a_i^2) = \left(\sum_{i=1}^n \exp(-\rho^2/a_i^2)\right)^{-1} \to 0.$$

In view of Section 3.3.7 these imply triviality.

Consider upper bounds. Assume $\rho = \rho_\varepsilon \to \infty$, $g_p(\bar{a}) < \infty$, if $p < \infty$, or $g(\bar{a}, h) < \infty$ for some $h > 0$, if $p = \infty$.

Let $p \le 2$. Consider tests of χ^2-type (3.8) with

$$w_{\varepsilon,i} = w_i = (2g_p(\bar{a}))^{-1/2} a_i^{2p/(4-p)}.$$

In view of Corollary 3.3 it suffices to show that

$$\inf_{v \in V(\rho)} \sum_i a_i^{2p/(4-p)} v_i^2 \to \infty, \quad \text{as } \rho \to \infty. \tag{3.104}$$

Using the Hölder inequality

$$\sum_i x_i y_i \le \left(\sum_i x_i^\mu\right)^{1/\mu} \left(\sum_i y_i^\nu\right)^{1/\nu}$$

with

$$\mu = 2/(2-p), \quad \nu = 2/p, \quad x_i = a_i^{2p(2-p)/(4-p)}, \quad y_i = |v_i|^p a_i^{p^2/(4-p)},$$

we get

$$\rho^p \le \sum_i |a_i v_i|^p \le \left(\sum_i a_i^{4p/(4-p)}\right)^{1-p/2} \left(\sum_i a_i^{2p/(4-p)} v_i^2\right)^{p/2}, \tag{3.105}$$

which imply (3.104). For the case $a_i = i^r$, $r < 1/4 - 1/p$, $i \in \mathbf{N}$, one can consider tests of χ^2-type (3.8) with

$$w_i = (Ci\log^2(i+1))^{-1/2}, \quad C^2 = 2\sum_i (i\log^2(i+1))^{-1},$$

which are based on statistics

$$t_\varepsilon = t_{\varepsilon,\bar{p}} = C\sum_i (i\log(i+1))^{-1/2}(x_i^2 - 1)$$

and thresholds $T_\varepsilon = \rho_\varepsilon^2/2$. In fact, since

$$(i\log^2(i+1))^{-1/2} > i^{2rp/(4-p)}, \quad \text{for } r < 1/4 - 1/p, \quad i \in \mathbf{N},$$

using (3.105) we get, for some $B > 0$,

$$\begin{aligned}\sum_i w_i v_i^2 &= C^{-1/2}\sum_i (i\log^2(i+1))^{-1/2} v_i^2 \\ &\geq B\Big(\sum_i i^{rp}|v_i|^p\Big)^{2/p} \geq B\rho^2 \to \infty.\end{aligned}$$

Let $2 < p < \infty$. Consider tests of χ^p-type (3.17) with $w_{\varepsilon,i} = w_i = (d_p g(\bar{a}))^{-1/2} a_i^p$. Since, for all $v \in V(\rho)$,

$$(d_p g_p(\bar{a}))^{1/2} \sum_i w_i |v_i|^p = \sum_i |a_i v_i|^p \geq \rho^p \to \infty,$$

the consistency follows directly from Proposition 3.2.

For the case $a_i = i^r$, $r < -1/2p$, $i \in \mathbf{N}$, one can consider tests of χ^p-type (3.17) with

$$w_i = (Ci\log^2(i+1))^{-1/2}, \quad C = d_p \sum_i (i\log^2(i+1))^{-1},$$

which are based on statistics

$$t_\varepsilon = t_{\varepsilon,\tilde{p}} = C^{-1/2}\sum_i (i\log^2(i+1))^{-1/2}(|x_i|^p - c_p)$$

and thresholds $T_\varepsilon = B_p\rho_\varepsilon^p/(4d_p g_p(\bar{a}))^{1/2}$. Since $(i\log^2(i+1))^{-1/2} > i^{rp}$ for $r < -1/2p$, $i \in \mathbf{N}$, we have

$$C^{1/2}\sum_i w_i|v_i|^p = \sum_i (i\log^2(i+1))^{-1/2}|v_i|^p \geq \sum_i i^{rp}|v_i|^p \geq \rho^p \to \infty,$$

we apply Proposition 3.2 again.

Let $p = \infty$. Note that the assumption $g(\bar{a}, h) < \infty$ for some $h > 0$ is possible only if $a_i \to 0$ and, therefore, $a = \max_i a_i < \infty$. Moreover, $g(\bar{a}, h) \to 0$, as $h \to \infty$ by $e^{-h/a^2} < 1$. Consider supreme tests (3.27) with $T_{\varepsilon,i} = c\rho_\varepsilon/a_i$, $c \in (0,1)$. Using (3.35) one has

$$\sum_i \Phi(-T_{\varepsilon,i}) \asymp \sum_i \exp(-T_{\varepsilon,i}^2/2)/T_{\varepsilon,i} \leq g(\bar{a}, (c\rho_\varepsilon)^2/2) \to 0. \qquad (3.106)$$

Since

$$\inf_{v\in V(\rho)} \sup_i (|v_i| - T_{\varepsilon,i}) \geq \rho_\varepsilon(1-c)/a \to \infty,$$

the consistency follows directly from Proposition 3.4. For the case $a_i = i^r$, $r < 0$, let $\psi_{\varepsilon,\rho_\varepsilon,p}$ be supreme tests with $T_{\varepsilon,i} = \rho_\varepsilon^{1/2}\log(i+1)$. The consistency follows from evaluations analogous to (3.106). □

3.4.4 Triviality and Classical Asymptotics for Besov Norms

Theorem 3.2 *Let $V(\rho) = \breve{B}_{r,p,h}(\rho)$ be the complement of a Besov body of radius ρ. If $r \geq r_p^*$, then $\gamma(V(\rho)) = 1$ for any $\rho > 0$, $0 < h \leq \infty$. If $r < r_p^*$, then $\gamma(V(\rho_\varepsilon)) \to 0$, as $\rho_\varepsilon \to \infty$.*

The case $r \geq r_p^*$ corresponds to the trivial problem, whenever the case $r < r_p^*$ corresponds to the classical asymptotics. Note that Theorem 3.2 describes regions C of classical asymptotics presented in Figures 1–8 in Section 6.4.

Proof. Let us establish lower bounds. Consider a sequence $j \to \infty$ and a jth resolution space $L_j = R^n$, $n = 2^j$ for all $j > J_0$. Then V contains a complement of the l_p^n-ball

$$V_j = \breve{D}_{n,p}(\rho_n), \quad \rho_n = 2^{-rj}\rho = n^{-r}\rho.$$

The statement of Theorem 3.2 for $p < \infty, r > r_p^*$ and $p = \infty$, $r > r_\infty^* = 0$ follows from Proposition 3.9.

Let $p < \infty$, $r = r_p^*$. Denote by $\pi^j = \pi_j^n$ the product priors which are constructed in the proof of Proposition 3.9. They are asymptotically concentrated on $V_j \subset L_j$, $n = 2^j$, $\pi^j(V_j) = 1 + o(1)$, as $j \to \infty$. One can easily check that $\|\bar{\pi}_j\| = 2^j \|\pi_j\|^2 \asymp 1$ for $r = r_p^*$.

Let $j \to \infty$, $m \to \infty$. Consider the "second-order mixture"

$$\pi_{j,m} = m^{-1} \sum_{k=j+1}^{j+m} \pi^k.$$

By $V_j \subset V(\rho)$, one has

$$\pi_{j,m}(V(\rho)) \geq m^{-1} \sum_{j=j+1}^{j+m} \pi^k(V_k) = 1 + o(1).$$

The likelihood ratio is of the form

$$L_{j,m} = \frac{dP_{\pi_{j,m}}}{dP_0} = m^{-1} \sum_{j=j+1}^{j+m} L_k, \quad L_k = \frac{dP_{\pi^k}}{dP_0}.$$

Note that statistics L_k, $j < k \leq m + j$, are independent and $E_0(L_k) = 1$. Therefore, analogously to the case of semiorthogonal priors, using equality (3.68), and by $e^x - 1 \asymp x$ as $x \asymp 1$, we get

$$\begin{aligned} E_0\left(\frac{dP_{\pi_{j,m}}}{dP_0} - 1\right)^2 &= m^{-2} \sum_{k=j+1}^{j+m} E_0(L_k - 1)^2 \\ &= m^{-2} \sum_{k=j+1}^{j+m} \left(\exp \|\bar{\pi}^k\|^2 - 1\right) \asymp m^{-2} \sum_{k=j+1}^{j+m} \|\bar{\pi}^k\|^2 \to 0. \end{aligned}$$

These imply $\gamma(V(\rho)) = 1$.

To establish the upper bounds it is enough to observe the embedding: for any $\tilde{r} < r$ there exists such a constant $C > 0$ that for any h one has $\breve{B}_{r,p,h}(\rho) \subset \breve{B}_{\tilde{r},p,p}(C\rho)$. On the other hand, since the Besov norm $|\cdot|_{r,p,p}$ is equivalent to the norm $|\cdot|_{r,p}$ after the natural renumbering of coordinates, the set $\breve{B}_{\tilde{r},p,p}(C\rho)$ is equivalent (up to a constant) to the set $\breve{D}_{r,p}(\rho)$ (recall that this is the complement of the ball of the radius ρ in the power norm $|\cdot|_{r,p}$), and we can use tests $\psi_{\varepsilon,\rho_\varepsilon,p}$ from Theorem 3.1. $\square$

3.4.5 Triviality and Classical Asymptotics for a Positive Alternative

Theorem 3.1 can be extended to a "positive" alternative corresponding to the complement of an l_p-ellipsoid

$$V^+(\rho) = \{v \in l^2 : v_i \geq 0 \quad \forall i, \quad |v|_{\bar{a},p} \geq \rho\}.$$

If $a_i = i^r$, we denote this set $V^+_{r,p}(\rho)$. Put

$$g_p^+(\bar{a}) = \begin{cases} \sum_i a_i^{2p/(2-p)}, & \text{if } 0 < p \leq 1, \\ \sum_i a_i^{2p}, & \text{if } 1 < p < \infty. \end{cases} \tag{3.107}$$

For $p = \infty$, define $g(\bar{a}, h)$ by (3.99).

Theorem 3.3
(1) *Let $p < \infty$, $g_p^+(\bar{a}) = \infty$ or $p = \infty$, $g(\bar{a}, h) = \infty$ for all $h > 0$. Then the problem is trivial: $\gamma(V^+(\rho)) = 1$ for any $\rho > 0$.*
(2) *Let $p < \infty$, $g_p^+(\bar{a}) < \infty$ or $p = \infty$, $g(\bar{a}, h) < \infty$ for some $h > 0$. Then $\gamma(V^+(\rho)) \to 0$, as $\rho \to \infty$. In particular, let $a_i = i^r$. Set*

$$r^*_{p,+} = \begin{cases} 1/2 - 1/p, & \text{if } 0 < p \leq 1, \\ -1/2p, & \text{if } 1 < p \leq \infty. \end{cases} \tag{3.108}$$

Then the problem is trivial for $r \geq r^*_{p,+}$, and the problem is of classical asymptotics with some common tests $\psi_{\varepsilon,\rho_\varepsilon,p}$, if $r < r^*_{p,+}$.

Proof. Note that the case $0 < p \leq 1$ corresponds to a convex alternative and was considered in Section 2.4.3. Let $p > 1$. Then the lower bounds in Theorem 3.3 follow directly from the proof of lower bounds in Theorem 3.1 with a replacement of symmetric two- or three-point factors by non-symmetric one- or two-point factors. The upper bounds for $p \geq 2$ follow from the upper bounds in Theorem 3.1. To obtain the upper bounds for $1 < p < 2$ one can replace tests of χ^2-type by tests of χ^+-type based on statistics (3.25) and use Proposition 3.3. $\square$.

For the Besov bodies case we have an analogous statement.

Theorem 3.4 *Let $V^+(\rho) = \breve{B}_{r,p,h}(\rho) \cap l^2_+$ be a nonnegative part of the complement of a Besov body of radius ρ. If $r \geq r^*_{p,+}$, then $\gamma(V^+(\rho)) = 1$*

for any $\rho > 0$, $0 < h \le \infty$, *i.e., the problem is trivial. If* $r < r^*_{p,+}$, *then* $\gamma(V^+(\rho_\varepsilon)) \to 0$, *as* $\rho_\varepsilon \to \infty$, *i.e., we have the classical asymptotics..*

Proof of Theorem 3.4. This is analogous to the proof of Theorem 3.2. □

By comparing Theorems 3.1 and 3.3 or Theorems 3.2 and 3.4 we get the same conditions of triviality for $p \ge 2$, however, they are essentially different for $0 < p < 2$; in particular, the break point is not $p = 2$, but $p = 1$ for positive signals.

3.4.6 Nonempty and Triviality for Power Norms: Two-Sided Constraints

It follows from Theorems 3.1, 3.2 that there are two principal different situations in infinite-dimensional hypotheses testing problems.

The first situation corresponds to norms $|v|_{\bar{a},p}$ with $g_p(\bar{a}) < \infty$: it is possible to detect *all* signals which are large enough in the sense of this norm. We call this case *classical* because the same holds in the parametric case. In particular, these hold for norms $|\cdot|_{r,p}$, $|\cdot|_{r,p,h}$ with $r < r^*_p$.

The second situation corresponds to norms $|v|_{\bar{a},p}$ with $g_p(\bar{a}) = \infty$: it is impossible to detect *all* large enough signals. Some additional constraints are necessary to obtain nontrivial minimax hypotheses testing problems. In particular, these hold for norms $|\cdot|_{r,p}$, $|\cdot|_{r,p,h}$ with $r \ge r^*_p$. We call this case *nonclassical*. This is the main subject of interest in this book.

We consider the alternative under two-sided constraints in the power norms

$$V = V(\bar{a}, \bar{b}, p, q, \rho, R) = \{v \in l^2 : |v|_{a,p} \ge \rho, \ |v|_{\bar{b},q} \le R\}; \tag{3.109}$$

where $\bar{a}, \bar{b}$ are positive sequences, $0 < p, q \le \infty$. In particular, for $a_i = i^r$, $b_i = i^s$, $i \in \mathbf{N}$, we have

$$V = V(\kappa, \rho, R) = \{v \in l^2 : |v|_{r,p} \ge \rho, \ |v|_{s,q} \le R\}; \quad \kappa = (r, s, p, q), \tag{3.110}$$

assuming $-\infty < r, s < \infty$, $0 < p, q \le \infty$.

Analogously, we can consider the positive alternative

$$V^+ = \{v \in V : v_i \ge 0\} = V \cap l^2_+,$$

where V is defined by (3.109), (3.110).

The first question is: *What are* ρ, R *for* $V = V(\bar{a}, \bar{b}, p, q, \rho, R)$ *or* $V(\kappa, \rho, R)$ *to be nonempty?*

Set

$$S = \begin{cases} \left(\sum_i (a_i/b_i)^{pq/(q-p)}\right)^{(q-p)/pq}, & \text{if } p < q < \infty, \\ \left(\sum_i (a_i/b_i)^p\right)^{1/p}, & \text{if } p < q = \infty, \\ \sup_i a_i/b_i, & \text{if } p \ge q. \end{cases} \tag{3.111}$$

Proposition 3.14 *If set V is nonempty, then $SR \geq \rho$. If $SR > \rho$, then set V is nonempty (it is possible that $S = \infty$).*

Proof. In view of symmetry it is enough to consider positive alternative V^+. Let V^+ be notempty. To show that $SR \geq \rho$ we need to consider case $S < \infty$ only. Let $p < q < \infty$ (case $p < q = \infty$ is analogous). Using the Hölder inequality we get

$$\sum_i a_i^p v_i^p \leq \left(\sum_i (a_i/b_i)^{pq/(q-p)}\right)^{(q-p)/q} \left(\sum_i b_i^q v_i^q\right)^{p/q}.$$

Therefore, if $v \in V^+$, then $\rho \leq |v|_{\bar{a},p} \leq S|v|_{\bar{b},q} \leq SR$.

Let $SR > \rho$. Consider an increasing sequence of finite subsets

$$\ldots \subset J_k \subset J_{k+1} \subset \ldots \subset I; \quad \bigcup_k J_k = I.$$

Then, for large enough k, the set V^+ contains the points with coordinates

$$v_i = \rho(a_i^p/b_i^q)^{1/(q-p)} \left(\sum_{i \in J_k} (a_i/b_i)^{pq/(q-p)}\right)^{-1/p}, \quad i \in J_k; \quad v_i = 0, \quad i \notin J_k. \tag{3.112}$$

Let $\infty > p \geq q$ (case $p = \infty$ is analogous). Let $v \in V^+$. Since $v_i b_i \leq R$, we get:

$$\rho^p \leq \sum_i a_i^p v_i^p \leq \sup_i (v_i b_i)^{p-q} (a_i/b_i)^p \sum_i b_i^q v_i^q \leq \sup_i (a_i/b_i)^p R^p.$$

This implies the necessity. The sufficiency follows from the fact that the set V^+ contains the point with only one nonzero coordinate $v_i = \rho/a_i$ with i such that $\rho/a_i \leq R/b_i$. □

Remark. Let $S < \infty$ and either $p \geq q$ and the supremum in (3.111) is attained, or $p < q$ and $\sum_{i \in I} v_i^2 < \infty$ where v_i are defined analogously to (3.112) with $J_k = I$. Then one can change the nonempty condition $SR > \rho$ by $SR \geq \rho$.

Corollary 3.15 *If $p \geq q$, $s < r$ or $p < q$, $s - r \leq 1/p - 1/q$, then the set $V = V(\kappa, \rho, R)$ is nonempty for all ρ, R. If $p \geq q$, $s \geq r$, then the set V is nonempty for $R \geq \rho$. If $p < q$, $s - r > 1/p - 1/q$, then the set V is nonempty for $R \geq \rho(S(\tau))^{(p-q)/pq}$, where*

$$S(\tau) = \sum_{i=1}^{\infty} i^{-\tau}, \quad \tau = pq(s-r)/(q-p) > 1.$$

We consider the alternatives (3.110) determined by the power norms. Set (see Figures 1–8 in Section 6.4):

$$\Xi_C = \{\kappa : r < r_p^*\}, \quad \Xi_C^+ = \{\kappa : r < r_{p,+}^*\}. \tag{3.113}$$

Using Theorems 3.1, 3.2 we get

Corollary 3.16 *Let*

$$V_\varepsilon = V(\kappa, \rho_\varepsilon, R_\varepsilon), \quad \kappa \in \Xi_C \quad \text{or} \quad V_\varepsilon = V^+(\kappa, \rho_\varepsilon, R_\varepsilon), \quad \kappa \in \Xi_C^+.$$

Then under the nonempty conditions from Corollary 3.15 we have the classical asymptotics

$$\gamma(V_\varepsilon) \to 0 \;\; \text{iff} \;\; \rho_\varepsilon \to \infty; \quad \gamma(V_\varepsilon) \to 1 \;\; \text{iff} \;\; \rho_\varepsilon \to 0. \tag{3.114}$$

The second question is: *Under nonempty conditions what are (κ, ρ, R) for nontriviality (this means $\gamma(V) < 1$)? What should be the second constraint to obtain a nontrivial problem?*

Here we give the answer for the opposite question: *What are triviality conditions?* We describe the region Ξ_T in a four-dimensional space of parameters $\{\kappa\}$ such that $\gamma(V(\kappa, \rho, R)) = 1$ for all ρ, R under nonempty condition.

Note that this region is presented in Figures 1–8 in Section 6.4. This will be the basis for studying the sharp or rate asymptotics for regions with nonclassical and nontrivial asymptotics.

Set

$$\mu = s - r, \quad \lambda = s/p - r/q, \quad I = 2(s-r) - 4(s/p - r/q) + 1/q - 1/p \tag{3.115}$$

with evident modifications for $q = \infty$. Define the set Ξ_T by the inequalities (here $0 < p, q \le \infty$):

$$r \ge r_p^*, \quad \begin{cases} \mu \le 0 \text{ and } \lambda \le 0 \text{ and } I \ge 0, & \text{if } 2 > p > q\,, \\ \mu \le 1/p - 1/q \text{ and } I \le 0, & \text{if } 2 < p < q, \\ \mu \le 0 \text{ and } \lambda \le 0, & \text{if } p \ge 2,\ p > q, \\ \mu \le 1/p - 1/q, & \text{if } p \le 2,\ p \le q, \text{ or } p = q > 2, \end{cases} \tag{3.116}$$

where the values r_p^* are determined by (3.100). [7]

Theorem 3.5 *Let $\kappa \in \Xi_T$ and if $\mu = 0$, then $R > \rho_\varepsilon$. Then the problem is trivial: $\gamma(V_\varepsilon(\kappa, \rho, R)) \equiv 1$.*

Proof. Let $\kappa \in \Xi_T$, $\infty \ge p \ge q$, $r \ge 0$ (note that $s \le r$ in this case) and $R > \rho$, if $s = r$. If $r > 0$, then the set V contains the points $v_i = i^{-r} e_i \rho \in l^2$ for large enough i; since $|v_i| \to 0$, as $i \to \infty$, this implies the statement of Theorem 3.5 in this case. If $r = 0$, $\infty \ge p \ge q$, then V contains the orthogonal collections $\bar{v}_n = \{v_i = \rho e_i \in l^2,\ m+1 \le i \le n+m\}$ for large enough m. We can construct orthogonal priors with $p_{n,i} = 1/n$, $i = m+1, \dots, n+m$, concentrated on V such that

$$\sum_{i=m+1}^{m+n} p_{n,i}^2 (\exp(|v_i|^2) - 1) = n^{-1}(e^{\rho^2} - 1) = o(1).$$

[7] Triviality for $r = 0$ was studied in [92].

Therefore the statement of Theorem 3.5 follows from Proposition 3.10 in this case.

To obtain the statement of Theorem 3.5 for other cases it suffices to construct sequences $\bar{\pi}_n = \{\pi_{n,i}\} = \bar{\pi}_{n,\kappa,\rho,R}$ such that

$$\|\bar{\pi}_n\| \to 0, \quad \pi^n(V) \to 1; \quad V = V(\kappa, \rho, R), \tag{3.117}$$

where π^n is the product measure corresponding to $\bar{\pi}_n$. We use the sequences of three-point measures

$$\pi_{n,i} = (1 - h_{n,i})\delta_0 + \frac{h_{n,i}}{2}(\delta_{z_{n,i}} + \delta_{-z_{n,i}}).$$

Let $p = q < \infty$, $s \le r < 0$. Put, for $p \le 2$,

$$h_{n,i} = 1, \quad z_{n,i} = \begin{cases} 0, & \text{if } i \le m, \ i > m+n, \\ b_n i^{rp/(4-p)}, & \text{if } m+1 \le i \le m+n, \end{cases}$$

and, for $p > 2$,

$$z_{n,i} = 1, \quad h_{n,i} = \begin{cases} 0, & \text{if } i \le m, \ i > m+n, \\ a_n i^{rp}, & \text{if } m+1 \le i \le m+n, \end{cases}$$

where m is large enough and a_n, b_n are values such that

$$\sum_{i=m+1}^{n+m} i^{rp} z_{n,i}^p = b_n^p \sum_{i=m+1}^{m+n} i^{4rp/(4-p)} = \rho^p, \quad \sum_{i=m+1}^{n+m} i^{rp} h_{n,i} = a_n \sum_{i=m+1}^{m+n} i^{2rp} = \tilde{\rho}^p;$$

here $\tilde{\rho} > \rho$ and $R > \tilde{\rho} > \rho$, if $r = s$. Then we can obtain relations (3.117) from the evaluations analogous to those that were given in the proof of the lower bounds of Theorem 3.1.

Thus we need to consider $\kappa \in \Xi_T$ with $\infty > p > q$, $r < 0$ and with $p < q$. To simplify, we assume $q < \infty$; for the case $p < q = \infty$ we can use similar considerations. Set

$$\Delta = pq(4\lambda - \mu) = sq(4-p) - rp(4-q). \tag{3.118}$$

Lemma 3.3 *Let*

$$p \ne q, \quad \Delta/(q-p) > 0, \quad 0 \le pq\mu/(q-p) \le 1, \quad \lambda/(q-p) \ge 0,$$

(if $\mu = 0$, then we assume $\rho \le R$). Put

$$h_{n,i} = 1, \quad z_{n,i} = \begin{cases} \delta_n i^{-pq\lambda/(q-p)}, & \text{if } m_{n,1} \le i \le m_{n,2}, \\ 0, & \text{otherwise}, \end{cases}$$

where $m_1 = m_{n,1} \to \infty$, $m_2 = m_{n,2} \to \infty$, $\delta_n \asymp 1$, as $n \to \infty$ are such values that

$$A_n = \sum_{i=m_1}^{m_2} i^{-pq\mu/(q-p)} \asymp 1, \quad \rho^p \le \delta_n^p A_n, \quad R^q \ge \delta_n^q A_n.$$

Then relations (3.117) hold.

Note that if $\mu > 0$, then one can easily choose values A_n, δ_n which satisfy the assumptions of Lemma 3.3: since items in the sums for A_n tend to 0, one can define A_n, δ_n by the asymptotic relations $\rho^p \sim C^{-1}\delta_n^p A_n$, $R^q \sim C\delta_n^q A_n, C > 1$. If $\mu = 0$, then we can put $m_1 = m_2 = n$, $\delta_n = R$.

Proof of Lemma 3.3. By the assumptions of Lemma 3.3 one has

$$\sum_{i=m_1}^{m_2} z_{n,i}^p i^{rp} = \delta_n^p \sum_{i=m_1}^{m_2} i^{-pq\mu/(q-p)} = \delta_n^p A_n \geq \rho^p,$$

$$\sum_{i=m_1}^{m_2} z_{n,i}^q i^{sq} = \delta_n^q \sum_{i=m_1}^{m_2} i^{-pq\mu/(q-p)} = \delta_n^q A_n \leq R^q,$$

which imply $\pi^n(V(\kappa, \rho, R)) = 1$. Moreover,

$$\begin{aligned} \|\bar{\pi}_n\|^2 &\asymp \sum_{i=m_1}^{m_2} z_{n,i}^4 \delta_n^4 \sum_{i=m_1}^{m_2} i^{-(pq\mu+\Delta)/(q-p)} < \\ &< m_1^{-\Delta/(q-p)} \delta_n^4 \sum_{i=m_1}^{m_2} i^{-pq\mu/(q-p)} = O(m_1^{-\Delta/(q-p)}) \to 0. \end{aligned}$$

Lemma 3.3 is proved. □

Lemma 3.4 *Let*

$$p \neq q, \quad \Delta/(q-p) \leq 0, \quad I/(q-p) \leq 0, \quad \lambda/(q-p) \geq 0, \quad 0 < \mu/(q-p).$$

Put

$$h_{n,i} = \begin{cases} a_n i^{\Delta/(q-p)}, & \text{if } m_{n,1} \leq i \leq m_{n,2}, \\ 0, & \text{otherwise}, \end{cases}$$

$$z_{n,i} = \begin{cases} \delta_n i^{-pq\lambda/(q-p)}, & \text{if } m_{n,1} \leq i \leq m_{n,2}, \\ 0, & \text{otherwise}, \end{cases}$$

where $m_1 = m_{n,1} \to \infty$, $m_2 = m_{n,2} \to \infty$, $\delta_n \asymp 1$, $a_n \to 0$, *as* $n \to \infty$ *are values such that*

$$A_n = a_n \sum_{i=m_1}^{m_2} i^{-1-pqI/(q-p)} \asymp 1, \quad \rho^p < \delta_n^p A_n, \quad R^q > \delta_n^q A_n.$$

(one can easily choose such values). Then relations (3.117) *hold.*

Proof of Lemma 3.4. Set $d = pq/(q-p)$. Since $dI < 0$, $d\mu > 0$ one has

$$\begin{aligned} E_{\pi^n}\left(\sum_i i^{rp}|v_i|^p\right) &= \sum_{i=m_1}^{m_2} h_{n,i} z_{n,i}^p i^{rp} \\ &= a_n \delta_n^p \sum_{i=m_1}^{m_2} i^{-1-dI} = \delta_n^p A_n > \rho^p, \end{aligned}$$

$$E_{\pi^n}\left(\sum_i i^{sq}|v_i|^q\right) = \sum_{i=m_1}^{m_2} h_{n,i} z_{n,i}^q i^{sq}$$

$$= a_n \delta_n^q \sum_{i=m_1}^{m_2} i^{-1-dI} = \delta_n^q A_n < R^q,$$

$$\mathrm{Var}_{\pi^n}\left(\sum_i i^{rp}|v_i|^p\right) \asymp \mathrm{Var}_{\pi^n}\left(\sum_i i^{sq}|v_i|^q\right)$$

$$\asymp a_n \sum_{i=m_1}^{m_2} i^{-1-d(I+\mu)} < m_1^{-d\mu} A_n \to 0,$$

which by the Chebyshev inequality imply $\pi^n(V(\kappa, \rho, R)) \to 1$. Also

$$\|\bar{\pi}_n\|^2 \asymp \sum_{i=m_1}^{m_2} h_{n,i}^2 z_{n,i}^4 = a_n^2 \delta_n^4 \sum_{i=m_1}^{m_2} i^{-1-dI} = O(a_n) \to 0. \qquad \square$$

The statement of Theorem 3.5 for $\infty > p > q$, $r < 0$ and for $p < q$ follows directly from Lemmas 3.3, 3.4. In fact, observe the embedding

$$V(\kappa, \rho, R) \subset V(\kappa_1, \rho_1, R_1), \tag{3.119}$$

for $p \geq p_1$, $q \leq q_1$, $r \leq r_1$, $s \geq s_1$, $\rho \geq \rho_1$, $R \leq R_1$. In view of (3.119) one has

$$\gamma(V(\kappa, \rho, R)) \leq \gamma(V(\kappa_1, \rho_1, R_1)). \tag{3.120}$$

In view of (3.119), (3.120) it is enough to check the triviality for a large enough s "on the boundary" of the region Ξ_T.

Let $\infty > p > q$, $r < 0$. If $0 > r \geq -1/2p$, then we use Lemma 3.4, since this corresponds to $\lambda \leq 0$ and $\mu < 0$, $I \geq 0$, $\Delta \geq 0$ for $\kappa \in \Xi_T$ with large enough s. If $1/4 - 1/p \leq r \leq -1/2p$ (it is possible for $p < 2$), then we also use Lemma 3.4, since this corresponds to $I \geq 0$ and $\lambda \leq 0$, $\mu < 0$, $\Delta \geq 0$ for large enough s. Let $p < q$. If $r > 1/4 - 1/p$, then we use Lemma 3.3 since this corresponds to $0 \leq \mu \leq 1/p - 1/q$ and to $\lambda > 0$, $\Delta > 0$ for large enough s. If $r \leq 1/4 - 1/p$, then we use Lemma 3.4 again, since this corresponds to $I \leq 0$ and to $\mu > 0$, $\lambda \geq 0$, $\Delta \leq 0$ for large enough s.

Remark. One can control the required inequalities between parameters λ, μ, I, Δ by using a formal calculation. However we recommend that the reader refer to Figures 1–8 in Section 6.4 and Figures 10–13 in Section A.3.1 below because these inequalities follow easily from consideration of the (r, s)-half-planes determined by these inequalities.

Thus Theorem 3.5 is proved. $\square$

Triviality for Positive Alternative V^+

Let us consider the positive alternative

$$V^+ = V^+(\kappa, \rho, R) = V(\kappa, \rho, R) \cap l^2_+.$$

The triviality conditions are somewhat different in this case.

As above, consider the quantities $\mu = s - r$, $\lambda = s/p - r/q$ and introduce

$$I_+ = 2s(1 - 1/p) - 2r(1 - 1/q) + 1/q - 1/p.$$

Define the set Ξ_T^+ by the inequalities (here $0 < p, q \le \infty$):

$$r \ge r^*_{p,+}, \quad \begin{cases} \mu \le 0 \text{ and } \lambda \le 0 \text{ and } I_+ \ge 0, & \text{if } 1 > p > q, \\ \mu \le 1/p - 1/q \text{ and } I_+ \le 0, & \text{if } 1 < p < q, \\ \mu \le 0 \text{ and } \lambda \le 0, & \text{if } p \ge 1,\ p > q, \\ \mu \le 1/p - 1/q, & \text{if } p \le 1,\ p \le q \text{ or } p = q > 1, \end{cases} \tag{3.121}$$

where the values $r^*_{p,+}$ are defined by (3.108).

Theorem 3.6 *Let $\kappa \in \Xi_T$ and if $\mu = 0$, then $R > \rho_\varepsilon$. Then the problem is trivial: $\gamma(V_\varepsilon^+(\kappa, \rho, R)) \equiv 1$.*

Proof of Theorem 3.6. This is analogous to the proof of Theorem 3.5. The differences are as follows:

(1) Using the lemma analogous to Lemma 3.3 we consider nonsymmetric one-point measures (these correspond to Dirac masses at the points $\bar{z}_n \in V^+$), and we show that $|\bar{z}_n|^2 = \sum_i z^2_{n,i} \to 0$.

(2) Using the lemma analogous to Lemma 3.4 we consider nonsymmetric two-point measures; we show that $\sum_i h^2_{n,i} z^2_{n,i} \to 0$. We leave calculations in the proof to the reader. □

3.4.7 Triviality for Besov Norms: Two-Sided Constraints

Let us consider the constraints determined by the Besov norms (2.72) in the sequence space. Denote $\tau = (\kappa, t, h)$, $\kappa = (r, s, p, q)$ and consider the alternative

$$V = V(\tau, \rho, R) = \{v \in l^2 : \ |v|_{r,p,h} \ge \rho, \ |v|_{s,q,t} \le R\}. \tag{3.122}$$

For Besov norms we have analogous triviality conditions such as for the power norms (2.71).

Theorem 3.7 *Let $\kappa \in \Xi_T$ where the set Ξ_T is determined by (3.116) and if $\mu = 0$, then $R > \rho_\varepsilon$. Then the problem is trivial: $\gamma(V_\varepsilon(\tau, \rho, R)) \equiv 1$.*

Proof. The scheme of the proof is analogous to the case of power norms. Let $\kappa \in \Xi_T$.

First, assume that $I \ne 0$ and $R > \rho_\varepsilon$, if $s = r$. Then the considerations are analogous to the above. Let us consider a resolution $j \to \infty$. Let $\kappa \in \Xi_T$, $\infty \ge p \ge q$, $r \ge 0$ and let $R > \rho$, if $s = r$. For each $j > J_0 > 0$ the set V contains the points $v_{ij} = 2^{-rj}\rho e_{ij} \in l^2$, $i = 1, \dots, 2^j$, $|v_{ij}| = 2^{-rj}\rho$; here e_{ij} is the canonical orthonormal basis in the j-resolution subspace L_j. For $r > 0$, taking one point v_{1j} in each resolution $j \to \infty$, by $|v_{ij}| \to 0$, we obtain triviality. If $r = 0$, then, analogously to the proof of Theorem 3.5, applying Proposition 3.10 to the collection $\bar{v}_j$ which consists of $n = 2^j$ points with probabilities n^{-1}, we obtain the statement of Theorem 3.7 in this case.

Let $p = q < \infty$, $s \le r < 0$. Taking a sequences of resolutions $j_0 \to \infty$ we consider the product priors π_{j_0} with symmetric three-point factors. For $p \le 2$ and $r > 1/4 - 1/p$ we set

$$h_{j_0,ij} = 1, \quad z_{j_0,ij} = \begin{cases} 0, & \text{if } j \neq j_0, \\ 2^{-j_0(r+1/p)}\rho, & \text{if } j = j_0,\ i = 1, \dots, 2^{j_0}. \end{cases}$$

If $p > 2$ and $r > -1/2p$, we take $\tilde{\rho} > \rho$, if $s = r$, then we assume $R > \tilde{\rho} > \rho$. We set

$$z_{j_0,ij} = \tilde{\rho}, \quad h_{j_0,ij} = \begin{cases} 0, & \text{if } j \neq j_0, \\ 2^{-j_0(rp+1)}, & \text{if } j = j_0,\ i = 1, \dots, 2^{j_0}. \end{cases}$$

Then we can easily obtain relations (3.117). If $r = r_p^*$, then, by the embedding (3.120), it suffices to consider the case $s = r$. However, the considerations in this case are the same as in the proof of triviality for the one-sided constraint in Theorem 3.2.

Let $\infty > p \neq q$. The following statements are analogous to Lemmas 3.3, 3.4.

Lemma 3.5 *Let $j_0 \to \infty$,*

$$p \neq q, \quad \Delta/(q-p) > 0, \quad 0 \le pq\mu/(q-p) \le 1, \quad \lambda/(q-p) \ge 0.$$

If either $\mu = 0$ or $\mu = 1/p - 1/q$, then assume that $\rho \le R$. Put

$$h_{j_0,ij} = 1, \quad z_{j_0,ij} = \begin{cases} b_0 z_{j_0}, & \text{if } j = j_0,\ 1 \le i \le m, \\ 0, & \text{otherwise}, \end{cases}$$

where $z_{j_0} = 2^{-j_0 pq\lambda/(q-p)}$, $m = a_0 2^{j_0 pq\mu/(q-p)}$, a_0 and b_0 are values such that $\rho \le b_0 a_0^{1/p}$, $R \ge b_0 a_0^{1/q}$, and $a_0 = 1$ if $\mu = 0$ or $\mu = 1/p - 1/q$. Then relations (3.117) hold.

Proof of Lemma 3.5. We have

$$\left(\sum_j \left(2^{jrp}\sum_i |z_{j_0,ij}|^p\right)^{h/p}\right)^{p/h} = m2^{j_0 rp}(b_0 z_{j_0})^p = a_0 b_0^p,$$

$$\left(\sum_j \left(2^{jsq}\sum_i |z_{j_0,ij}|^q\right)^{t/q}\right)^{q/t} = m2^{j_0 sq}(b_0 z_{j_0})^q = a_0 b_0^q,$$

which imply $\pi_{j_0}(V) = 1$. Also

$$\|\bar{\pi}_{j_0}\|^2 \asymp m z_{j_0}^4 = a_0 2^{-j_0\Delta/(q-p)} \to 0. \qquad \square$$

Lemma 3.6 *Let* $j_0 \to \infty$,

$$p \neq q, \quad \Delta/(q-p) \le 0, \quad I/(q-p) < 0, \quad \lambda/(q-p) \ge 0, \quad 0 < \mu/(q-p).$$

Put

$$h_{j_0,ij} = \begin{cases} a_0 h_{j_0}, & \text{if } j = j_0,\ 1 \le i \le m, \\ 0, & \text{otherwise,} \end{cases}$$

$$z_{j_0,ij} = \begin{cases} b_0 z_{j_0}, & \text{if } j = j_0,\ 1 \le i \le m, \\ 0, & \text{otherwise,} \end{cases}$$

where a_0 *and* b_0 *are values such that* $\rho < b_0 a_0^{1/p}$, $R > b_0 a_0^{1/q}$, *and*

$$h_{j_0} = h_0 2^{j_0\Delta/(q-p)}, \quad z_{j_0} = 2^{-j_0 pq\lambda/(q-p)}, \quad m = 2^{j_0(1+pqI/(q-p))}/h_0,$$

where $h_0 \to 0$, $h_0 \ge 2^{j_0 pqI/(q-p)}$. *Then relations* (3.117) *hold.*

Proof of Lemma 3.6. We have

$$E_{\pi_{j_0}}\left(\sum_j (2^{jrp}\sum_i |v_{ij}|^p)^{h/p}\right)^{p/h} = m2^{j_0rp}h_{j_0}z_{j_0}^p a_0 b_0^p = a_0 b_0^p,$$

$$E_{\pi_{j_0}}\left(\sum_j (2^{jsq}\sum_i |v_{ij}|^q)^{t/q}\right)^{q/t} = m2^{j_0sq}h_{j_0}z_{j_0}^q a_0 b_0^q = a_0 b_0^q;$$

$$\begin{aligned}\operatorname{Var}_{\pi_{j_0}}\left(\sum_j (2^{jrp}\sum_i |v_{ij}|^p)^{h/p}\right)^{p/h} &\asymp m2^{2j_0rp}h_{j_0}z_{j_0}^{2p} \\ &\asymp 2^{-j_0pq\mu/(q-p)} \to 0,\end{aligned}$$

$$\begin{aligned}\operatorname{Var}_{\pi_{j_0}}\left(\sum_j (2^{jsq}\sum_i |v_{ij}|^q)^{t/q}\right)^{q/t} &\asymp m2^{2j_0sq}h_{j_0}z_{j_0}^{2q} \\ &\asymp 2^{-j_0pq\mu/(q-p)} \to 0,\end{aligned}$$

which imply $\pi_{j_0}(V) \to 1$ as $j_0 \to \infty$. Also

$$\|\bar{\pi}_{j_0}\|^2 \asymp m h_{j_0}^2 z_{j_0}^4 \asymp h_0 \to 0.$$

Lemma 3.6 is proved. □

For $I \neq 0$ and $\infty > p > q$, $r < 0$ or $p < q$ the statement of Theorem 3.7 follows directly from Lemmas 3.5, 3.6 and from the monotone property analogous to (3.119), (3.120).

Let $\kappa \in \Xi_T$, $I = 0$. It suffices to construct measures π^{j_0} on l^2, $j_0 \to \infty$, such that

$$\pi^{j_0}(V) \to 1, \quad E_0\left(\frac{dP_{\pi^{j_0}}}{dP_0} - 1\right)^2 = E_0\left(\frac{dP_{\pi^{j_0}}}{dP_0}\right)^2 - 1 \to 0, \tag{3.123}$$

where $P_{\pi^{j_0}}$ is the mixture over the measure π^{j_0} (see the proof of Theorem 3.2 above). To simplify we consider the case $p \neq q$ only.

Lemma 3.7 *Assume*

$$I = 0, \quad p \neq q, \quad \Delta/(q-p) \leq 0, \quad \lambda/(q-p) \geq 0, \quad 0 < \mu/(q-p).$$

Let us consider the product measures π_k for the sequences $\bar{\pi}_k$ with three-point factors corresponding to $\bar{h}_k$, $\bar{z}_k$ where

$$h_{k,ij} = \begin{cases} a_0 h_k, & \textit{if } j = k, \ 1 \leq i \leq 2^k, \\ 0, & \textit{otherwise}, \end{cases}$$

$$z_{k,ij} = \begin{cases} b_0 z_k, & \textit{if } j = k, \ 1 \leq i \leq 2^k, \\ 0, & \textit{otherwise}. \end{cases}$$

Here b_0 and a_0 are values such that $\rho < b_0 a_0^{1/p}$, $R > b_0 a_0^{1/q}$ and

$$h_k = 2^{k\Delta/(q-p)}, \quad z_k = 2^{-kpq\lambda/(q-p)}.$$

Put

$$\pi^{j_0} = j_0^{-1} \sum_{k=j_0}^{2j_0-1} \pi_k, \quad j_0 \to \infty.$$

Then relations (3.123) *hold.*

Proof of Lemma 3.7. Let us consider the variables

$$x_j = 2^{jrp} \sum_{i=1}^{2^j} |v_{ij}|^p, \quad y_j = 2^{jsq} \sum_{i=1}^{2^j} |v_{ij}|^q.$$

We have $\pi_k(x_j = 0,\ y_j = 0) = 1$ for $j \neq k$, and if $j = k$, then

$$\begin{aligned} E_{\pi_k}(x_k) &= a_0 2^{k(rp+1)} h_k (b_0 z_k)^p = a_0 b_0^p, \\ E_{\pi_k}(y_k) &= a_0 2^{k(sq+1)} h_k (b_0 z_k)^q = a_0 b_0^q, \\ \mathrm{Var}_{\pi_k}(x_k) &\asymp \mathrm{Var}_{\pi_k}(y_k) \asymp 2^{-kpq\mu/(q-p)}. \end{aligned}$$

By the Chebyshev inequality these relations imply that $\pi_k(V) \to 1$ as $k \to \infty$ which imply the first relation in (3.123).

To obtain the second relation note that

$$\|\bar{\pi}_k\|^2 \asymp 2^k h_k^2 z_k^4 \asymp 1$$

and

$$E_0\left(\frac{dP_{\pi^{j_0}}}{dP_0}-1\right)^2 = j_0^{-2}\sum_{j,k=j_0}^{2j_0-1} E_0\left(\frac{dP_{\pi_j}}{dP_0}-1\right)\left(\frac{dP_{\pi_k}}{dP_0}-1\right)$$
$$= j_0^{-2}\sum_{k=j_0}^{2j_0-1} E_0\left(\frac{dP_{\pi_k}}{dP_0}-1\right)^2 \to 0,$$

since

$$E_0\left(\frac{dP_{\pi_j}}{dP_0}-1\right)\left(\frac{dP_{\pi_k}}{dP_0}-1\right) = 0, \quad \text{if } k \neq j,$$

and (see (3.68))

$$E_0\left(\frac{dP_{\pi_k}}{dP_0}-1\right)^2 \leq \exp(\|\bar{\pi}_k\|^2) - 1 = O(1).$$

Lemma 3.7 and Theorem 3.7 are proved. □

3.4.8 Triviality under the Functional Gaussian Model

Using the wavelet transform and relationship (2.84) between Besov norms in functional and sequence spaces, as well as relationship (2.80) between Sobolev and Besov norms, we can partially extend Theorem 3.7 to the functional Gaussian model (2.5). Consider the set of signals $S_\varepsilon \subset L_2(0,1)$ defined by the constraints

$$\|s\|_{(\sigma,p)} \geq r_\varepsilon, \quad \|s\|_{(\eta,q)} \leq R,$$

where $\|\cdot\|_{(\sigma,p)}$, $\|\cdot\|_{(\eta,q)}$ are either Sobolev norms $\|\cdot\|_{\sigma,p}$, $\|\cdot\|_{\eta,q}$ or Besov norms $\|\cdot\|_{\sigma,p,h}$, $\|\cdot\|_{\eta,q,t}$ with any $h,\ t>0$. Let

$$\sigma > 1/p - 1, \quad \sigma \geq 0, \quad \eta > 1/q - 1, \quad \eta \geq 0. \tag{3.124}$$

Then, using an m-regular wavelet transform with a large enough m and the replacement

$$r = \sigma + 1/2 - 1/p, \quad s = \eta + 1/2 - 1/q; \quad \rho_\varepsilon = r_\varepsilon/\varepsilon, \quad R_\varepsilon = R/\varepsilon,$$

in view of (2.84), we obtain the set V_ε such that $V_{b,\varepsilon} \subset V_\varepsilon$; here

$$V_{b,\varepsilon} = V(\kappa, h, t; b\rho_\varepsilon, R_\varepsilon/b), \quad b = b(\kappa,h,t) \in (0,1).$$

Since the results of Theorem 3.7 do not depend on ρ, R (for $R > \rho$, if $\mu = 0$), we can reformulate the results.

Introduce the constraints

$$\begin{cases} \eta - 1/q \leq \sigma - 1/p \text{ and } \eta/p + 1/2p \leq \sigma/q + 1/2q, & \text{if } 2 > p > q, \\ \eta - 1/q \leq \sigma - 1/p, & \text{if } p \geq 2,\ p > q, \\ \eta \leq \sigma, & \text{if } p \leq q, \end{cases} \tag{3.125}$$

these correspond to the set Ξ_T defined by (3.116) under the additional constraints (3.124).

Theorem 3.8 *Assume* (3.124), (3.125) *and if* $\sigma = \eta$, *then* $R > Br$ *for a large enough constant* $B > 0$. *Then the problem is trivial:* $\gamma(S_\varepsilon) \equiv 1$.

3.5 Distinguishability Conditions for Smooth Signals

In this section we describe the rates in the problem of detection of a smooth signal. Let us remove from consideration signals with an L_q-norm smaller than r_ε.

The question is: *What is a family r_ε to obtain distinguishability?*

Of course, the answer depends on the degree of smoothness of the unknown signals; moreover, it depends on the norm which is used to measure the smoothness. Here we consider some special cases; a general consideration will be presented in the following chapters.

The problem is the following. Under the functional Gaussian model (2.5) let us test $H_0 : s = 0$ against $H_1 : s \in S_\varepsilon$. Let the alternatives $S_\varepsilon = S_\varepsilon(\eta, p, q; r_\varepsilon, R)$ be determined by the inequalities

$$S_\varepsilon = \{s \in L_2(0,1) : \|s\|_p \geq r_\varepsilon, \quad \|s\|_{(\eta,q)} \leq R\}. \tag{3.126}$$

Here $\|\cdot\|_{(\eta,q)}$ is either a Sobolev norm $\|\cdot\|_{\eta,q}$ or a Besov norm $\|\cdot\|_{\eta,q,t}$ with any $t \geq 1$. These mean that we consider η-smooth signals in an L_q-norm which are bounded away from zero in an L_p-norm. Here we consider the cases

$$1 \leq p \leq 2, \quad p \leq q \leq \infty \quad \text{or} \quad 2 < p = q \leq \infty; \quad \eta > 0. \tag{3.127}$$

3.5.1 Rates and Tests for Smooth Signals

Introduce the quantities

$$r_\varepsilon^* = \begin{cases} \varepsilon^{4\eta/(4\eta+1)}, & \text{if } 1 \leq p \leq 2,\ q \geq p, \\ \varepsilon^{2\eta/(2\eta+1-1/p)}, & \text{if } 2 < p = q < \infty, \\ \left(\varepsilon^2 \log(\varepsilon^{-1})\right)^{\eta/(2\eta+1)}, & \text{if } p = q = \infty, \end{cases} \tag{3.128}$$

(we will show that r_ε^* are the rates in the problem). To describe consistent tests families, first, introduce the integer-valued family

$$n = n_\varepsilon \asymp (r_\varepsilon^*)^{-1/\eta} \to \infty, \tag{3.129}$$

and consider the family of orthonormal collection $\bar{\varphi}_{\varepsilon,n}$ in $L_2(0,1)$,

$$\varphi_{\varepsilon,i} = n^{1/2} \mathbf{1}_{\delta_{i,n}}, \quad i = 1, \dots, n,$$

generated by indicator functions of the equispaced partition of $[0,1]$ to intervals $\delta_{i,n} = [(i-1)/n, i/n]$ of length $1/n$. Put

$$X_{\varepsilon,i} = \varepsilon^{-1} n^{1/2} \int_{\delta_{i,n}} dX_\varepsilon(t) = \varepsilon^{-1} n^{1/2} (X_\varepsilon(i/n) - X_\varepsilon((i-1)/n))$$

and consider the families of statistics

$$t_{\varepsilon,p} = \begin{cases} (2n)^{-1/2} \sum_{i=1}^n (X_{\varepsilon,i}^2 - 1), & \text{if } 1 \le p \le 2,\ q \ge p, \\ (d_p n)^{-1/2} \sum_{i=1}^n (|X_{\varepsilon,i}|^p - c_p), & \text{if } 2 < p = q < \infty, \\ \max_{1 \le i \le n} |X_{\varepsilon,i}|, & \text{if } p = q = \infty, \end{cases} \tag{3.130}$$

where $c_p = E|\zeta|^p$, $d_p = \mathrm{Var}|\zeta|^p$, $\zeta \sim N(0,1)$ (compare with Section 3.1.3). We use tests families based on statistics (3.130),

$$\psi_{\varepsilon,p} = \mathbf{1}_{\{t_{\varepsilon,p} > T_\varepsilon\}}. \tag{3.131}$$

Theorem 3.9 *Under the functional Gaussian noise model* (2.5) *let us test* $H_0 : s = 0$ *against* $H_1 : s \in S_\varepsilon$ *where the sets* S_ε *are determined by the constraints* (3.126), (3.127).

(1) *Let* $1 \le p < \infty$. *Then* $\gamma_\varepsilon(S_\varepsilon) \to 1$ *as* $r_\varepsilon = o(r_\varepsilon^*)$ *and* $\liminf \gamma_\varepsilon(S_\varepsilon) > 0$ *as* $r_\varepsilon = O(r_\varepsilon^*)$. *If* $r_\varepsilon / r_\varepsilon^* \to \infty$, *then there exists a family of thresholds* $T_\varepsilon \to \infty$ *such that* $\gamma_\varepsilon(\psi_{\varepsilon,p}, S_\varepsilon) \to 0$.

(2) *Let* $p = \infty$. *There exist constants* $0 < \lambda_1 \le \lambda_2$ *such that* $\gamma_\varepsilon(S_\varepsilon) \to 1$ *as* $\limsup r_\varepsilon / r_\varepsilon^* < \lambda_1$, *and* $\gamma_\varepsilon(\psi_{\varepsilon,\infty}, S_\varepsilon) \to 0$, *as* $\liminf r_\varepsilon / r_\varepsilon^* > \lambda_2$, *where the thresholds in* (3.131) *are of the form* $T_\varepsilon = \sqrt{2 \log n_\varepsilon}$.

Before to prove Theorem 3.9, let us discuss the results.

The statements of Theorem 3.9 mean that the quantities r_ε^* defined by (3.128) are rates in the problem. If $p \le 2$, $q \ge p$, then the rates and minimax consistent tests (they are of χ^2-type with increasing n) do not depend on p, q. However, there is an essential dependence of the rates and of dimension $n = n_\varepsilon(\eta)$ on the smoothness degree η. If $p = q > 2$, then there is essential dependence of the rates and of the test statistics on p, η as well.

For $p = q = \infty$ (this means we use a supreme-norm to measure the signal and smoothness), the meaning of the rates is somewhat different: this case allows a sharper division between distinguishability and nondistinguishability. Below we show that the reason is an essential difference in the asymptotic behavior of the log-likelihood ratio for the least favorable priors: this is of Gaussian type for finite p and of degenerate type (this tends to a nonrandom constant) for $p = \infty$.

If the smoothness degree η tends to infinity, then the exponent in definition (3.128) tends to one which corresponds to the classical asymptotics.

It seems of interest to compare the rates r_ε^* in hypothesis testing with the analogous rates $r_\varepsilon^{\mathrm{est}}$ in minimax estimation problems for the same degree of smoothness of unknown signals and the same losses defined by an L_p-norm (see (2.86), (2.87)):

$$r_\varepsilon^{\text{est}} \asymp \begin{cases} \varepsilon^{2\eta/(2\eta+1)}, & \text{if } p = q < \infty, \\ (\varepsilon^2 \log(\varepsilon^{-1}))^{\eta/(2\eta+1)}, & \text{if } p = q = \infty. \end{cases} \tag{3.132}$$

We see that rates in estimation problems do not depend on p and $r_\varepsilon^* = o(r_\varepsilon^{\text{est}})$ for finite p; this means that it is impossible to estimate the unknown signals whenever it is possible to detect these signals!

If $p = \infty$, then we have the same rates: $r_\varepsilon^* = r_\varepsilon^{\text{est}}$.

For the case of Sobolev and Nikol'skii norms the results of Theorem 3.9 were obtained in [49] for the functional Gaussian model.

Under the sample model these results were obtained in [46]; in [47] the results were extended to a composite null-hypothesis of independence. Minimax properties of Pearson χ^2-tests and of tests of Kolmogorov and Cramér–von Mises–Smirnov type are compared in [48].

The case $p = q = 2$ was studied in [43] the sample model and in [44] under the spectral density model.

3.5.2 Proof of the Lower Bounds

To prove the lower bounds it is sufficient to consider the case of Besov norms with $h = 1$ which corresponds to the "smallest" alternative.

Let $m < \eta \le m + 1$, $m \ge 0$, be an integer. Fix an $(m+2)$-smooth function $\varphi(t)$, $t \in R^1$, supported on the interval $(0, 1)$ and such that

$$\|\varphi\|_2 = 1; \quad C_{l,h} = \|\varphi^{(l)}\|_q < \infty, \quad l = 0, \dots, m+2.$$

Introduce an integer-valued family $n = n_\varepsilon$ which is specified below and orthonormal collections of functions

$$\varphi_{n,i}(t) = n^{1/2}\varphi(nt - i), \quad i = 0, \dots, n-1.$$

The functions $\varphi_{n,i}$ have disjoint supports in the interval $(0, 1)$ and $\|\varphi_{n,i}\|_2 = 1$. Below it is convenient to use the following relations for the norms of the functions generated by the collections $\varphi_{n,i}$.

Lemma 3.8 *Let us consider the functions*

$$g(t, \theta) = \sum_{i=1}^{n} \theta_i \varphi_{n,i}(t), \quad t \in [0, 1], \quad \theta \in R^n. \tag{3.133}$$

Then there exist constants $C_{l,q} > 0$, $l = 0, 1, \dots, m+2$, which depend on φ, and $C > 0$ which depend on φ, $q \in [1, \infty]$, $\eta > 0$ such that

$$\|g^{(l)}(\cdot, \theta)\|_q = C_{l,q} n^{1/2+l-1/q} |\theta|_q, \quad l = 0, \dots, m+2, \tag{3.134}$$

$$\|g(\cdot, \theta)\|_{\eta,q,1} \le C n^{\eta+1/2-1/q} |\theta|_q, \tag{3.135}$$

where $|\cdot|_q$ is an l^q-norm in R^n.

Proof of Lemma 3.8. Consider the case $q < \infty$; the considerations for $q = \infty$ are analogous and simpler. Since the functions $\varphi_{n,i}$ in (3.133) have disjoint supports, we have

$$\begin{aligned}\|g^{(l)}(\cdot,\theta)\|_q^q &= \sum_{i=1}^{n} |\theta_i|^q \int_0^1 |\varphi_{n,i}^{(l)}(t)|^q\,dt \\ &= |\theta|_q^q\, n^{q(l+1/2)} \int_0^1 |\varphi^{(l)}(nt-i)|^q\,dt = |\theta|_q^q n^{q(l+1/2)-1}\|\varphi^{(l)}\|_q^q,\end{aligned}$$

which yields equalities (3.134) with $C_{l,q} = \|\varphi^{(l)}\|_q$. It follows from equality (3.134) for $l = 0$ that, to obtain inequality (3.135), it suffices to evaluate the Besov semi-norm $\|g(\cdot,\theta)\|^0_{\eta,q,1}$. Set $\eta = m + \delta$ where $\delta \in (0,1]$. By definition (2.76), we need to evaluate the integral

$$A_n = \int_0^{1/2} \omega_q^{(2)}(g^{(m)}, y) y^{-1-\delta}\,dy$$

where

$$\omega_q^{(2)}(g^{(m)}, y) = \sup_{0\le x\le y} \|\Delta_x^{(2)}(g^{(m)},\cdot)\|_{L_q(0,1-2x)}.$$

Assuming $n > 2$ we set $A_n = A_{n,1} + A_{n,2}$ where the item $A_{n,1}$ corresponds to the integral over the interval $(0, 1/n)$ and $A_{n,2}$ corresponds to the integral over the interval $[1/n, 1/2]$.

First, let $y \in (0, 1/n)$. Using an integral presentation of the difference and the Hölder inequality we have

$$\begin{aligned}\Delta_x^{(2)}(g^{(m)},t) &= g^{(m)}(t+2x) - 2g^{(m)}(t+x) + g^{(m)}(t) \\ &= \int_0^x\int_0^x g^{(m+2)}(t+u+v)\,dudv; \\ |\Delta_x^{(2)}(g^{(m)},t)|^q &\le x^{2(q-1)} \int_0^x\int_0^x |g^{(m+2)}(t+u+v)|^q\,dudv;\end{aligned}$$

$$\begin{aligned}&\|\Delta_x^{(2)}(g^{(m)},t)\|^q_{L_q(0,1-2x)} \\ &\le x^{2(q-1)} \int_0^{1-2x} dx \int_0^x\int_0^x |g^{(m+2)}(t+u+v)|^q\,dudv \\ &= x^{2(q-1)} \int_0^x\int_0^x \|g^{(m+2)}\|^q_{L_q(u+v,1-2x+u+v)}\,dudv \le x^{2q}\|g^{(m+2)}\|_q^q,\end{aligned}$$

since $(u+v, 1-2x+u+v) \subset [0,1]$ for $0 \le u, v \le x$; recall that $\|g\|_q = \|g\|_{L_q(0,1)}$. This yields

$$\omega_q^{(2)}(g^{(m)}, y) = \sup_{0\le x\le y} \|\Delta_y^{(2)}(g^{(m)},\cdot)\|_{L_q(0,1-2x)} \le y^2\|g^{(m+2)}\|_q.$$

Therefore, taking into account equality (3.134) for $l = m+2$, we get

$$A_{n,1} \le \|g^{(m+2)}\|_q \int_0^{1/n} y^{1-\delta} du = \frac{\|g^{(m+2)}\|_q}{2-\delta} n^{\delta-2} = C_1 n^{\eta+1/2-1/q} |\theta|_q.$$

Let $y \in [1/n, 1/2]$. For any $x \in [0, 1/2]$, using the definition of the second difference and the triangle inequality, we get

$$\begin{aligned} \|\Delta_x^{(2)}(g^{(m)}, \cdot)\|_{L_q(0,1-2x)} &\le \|g^{(m)}\|_{L_q(2x,1)} + 2\|g^{(m)}\|_{L_q(x,1-x)} \\ &+ \|g^{(m)}\|_{L_q(0,1-2x)} \le 4\|g^{(m)}\|_q. \end{aligned}$$

Therefore, $\omega_q^{(2)}(g^{(m)}, y) \le 4\|g^{(m)}\|_q$. Using equality (3.134) for $l = m$ we have

$$A_{n,2} \le 4\|g^{(m)}\|_q \int_{1/n}^{1/2} y^{-1-\delta}\, dy = 4\|g^{(m)}\|_q \frac{n^\delta - 2^\delta}{\delta} \le C_2 n^{\eta+1/2-1/q} |\theta|_q.$$

This yields inequality (3.134). □

Return to the proof of the lower bounds. Let $1 \le p \le 2$. Without loss of generality assume $\|\cdot\|_{(\eta,q)} = \|\cdot\|_{\eta,\infty,1}$ to be a Besov norm with $q = \infty$ (this corresponds to the "smallest" alternative). Consider the product priors $\pi^\varepsilon = \pi^\varepsilon(ds)$ which correspond to the family $\bar{\pi}_\varepsilon$ of symmetric two-point factors with $h_{\varepsilon,i} = 1,\ z_{\varepsilon,i} = z_\varepsilon,\ i = 0, \dots, n-1$, and to embedding (3.70). These priors are concentrated on the functions

$$s_{\varepsilon,\bar{\xi}}(t) = \varepsilon z_\varepsilon \sum_i \xi_i \varphi_{n,i}(t), \quad \xi_i = \pm 1, \quad t \in (0,1).$$

In view of Proposition 3.7 it suffices to choose $z_\varepsilon = o(1),\ n = n_\varepsilon \to \infty$ such that:

(i) $\pi^\varepsilon(S_\varepsilon) = 1$;

(ii) $\|\bar{\pi}_\varepsilon\|^2 \asymp n_\varepsilon z_\varepsilon^4 = o(1)$ or $O(1)$, if $r_\varepsilon / r_\varepsilon^* = o(1)$ or $O(1)$.

To check (i) observe that the functions $s_{\varepsilon,\bar{\xi}}$ are of the form (3.133) with

$$\theta_i = \varepsilon z_\varepsilon \xi_i, \quad |\theta|_p = n^{1/p} \varepsilon z_\varepsilon; \quad |\theta|_\infty = \varepsilon z_\varepsilon.$$

Using Lemma 3.8 one has

$$\|s_{\varepsilon,\bar{\xi}}\|_\infty = n^{1/2} \varepsilon z_\varepsilon C_{0,\infty}; \quad \|s_{\varepsilon,\bar{\xi}}\|_p = n^{1/2} \varepsilon z_\varepsilon C_{0,p}; \tag{3.136}$$

$$\|s_{\varepsilon,\bar{\xi}}\|_{\eta,\infty,1} \le C n^{\eta+1/2} \varepsilon z_\varepsilon. \tag{3.137}$$

In view of (3.136), (3.137) we get (i), if

$$\varepsilon z_\varepsilon n^{\eta+1/2} C \le R, \quad \varepsilon z_\varepsilon n^{1/2} C_{0,p} \ge r_\varepsilon.$$

The last relations hold for some

$$n \asymp r_\varepsilon^{-1/\eta}, \quad z_\varepsilon \asymp \varepsilon^{-1} n^{-(\eta+1/2)}.$$

However, in this case, one has $nz_\varepsilon^4 \asymp (r_\varepsilon/r_\varepsilon^*)^{4+1/\eta}$ which implies (ii).

Let $2 < p = q < \infty$. Consider the product priors $\pi^\varepsilon = \pi^\varepsilon(ds)$ which correspond to the family $\bar{\pi}_\varepsilon$ of symmetric three-points factors with $h_{\varepsilon,i} = n^{-1/2}$, $z_{\varepsilon,i} = z_\varepsilon$, $i = 0, \dots, n-1$, under embedding (3.70). These priors are concentrated on the functions

$$s_{\varepsilon,\xi} = \varepsilon z_\varepsilon \sum_i \xi_i \varphi_{n,i}, \quad \xi_i = 0, \pm 1,$$

and $|\xi_i|$ are i.i.d. Bernoulli random variables in π^ε-probability,

$$\pi^\varepsilon(|\xi_i| = 1) = n^{-1/2}.$$

In view of Proposition 3.7 it is enough to choose $z_\varepsilon = o(1)$, $n = n_\varepsilon \to \infty$ such that:

(i) $\pi^\varepsilon(S_\varepsilon) \to 1$;

(ii) $\|\bar{\pi}_\varepsilon\|^2 \asymp z_\varepsilon^4 = o(1)$ or $O(1)$, if $r_\varepsilon/r_\varepsilon^* = o(1)$ or $O(1)$.

Analogously to the above, using Lemma 3.8,

$$\|s_{\varepsilon,\xi}\|_p = \varepsilon z_\varepsilon n^{1/2-1/p} C_{0,p} Y_\varepsilon, \quad \|s_{\varepsilon,\xi}\|_{\eta,p,1} \le C\varepsilon z_\varepsilon n^{\eta+1/2-1/p} Y_\varepsilon, \tag{3.138}$$

where

$$Y_\varepsilon = \left(\sum_{i=0}^{n-1} |\xi_i| \right)^{1/p}.$$

In view of (3.138), relation (i) follows from the inequalities

$$\pi^\varepsilon(\varepsilon z_\varepsilon n^{\eta+1/2-1/p} C Y_\varepsilon < R, \quad \varepsilon z_\varepsilon n^{1/2-1/p} C_{0,p} Y_\varepsilon > r_\varepsilon) \to 1. \tag{3.139}$$

Observe that

$$E_{\pi^\varepsilon}(Y_\varepsilon^p) = nh_\varepsilon = n^{1/2}, \quad \mathrm{Var}_{\pi^\varepsilon}(Y_\varepsilon^p) = nh_\varepsilon(1 - h_\varepsilon) < n^{1/2},$$

and using Chebyshev inequality one gets, for all $\zeta \in (0,1)$,

$$\pi^\varepsilon\left(Y_\varepsilon < (1+\zeta) n^{1/2p}\right) \to 1, \quad \pi^\varepsilon\left(Y_\varepsilon > (1-\zeta) n^{1/2p}\right) \to 1. \tag{3.140}$$

In view (3.140) we obtain (3.139) by taking $n = n_\varepsilon$, z_ε such that

$$n \asymp r_\varepsilon^{-1/\eta}, \quad z_\varepsilon \asymp \varepsilon^{-1} n^{-(\eta+1/2-1/2p)} \asymp (r_\varepsilon/r_\varepsilon^*)^{1+1/2\eta-1/2\eta p},$$

which yield (i) and (ii).

Let $p = q = \infty$. Consider the orthogonal priors

$$\pi^\varepsilon = n^{-1} \sum_{i=1}^n \delta_{s_{\varepsilon,i}}$$

which correspond to $p_{\varepsilon,i} = n^{-1}$ and are concentrated on the functions

$$s_{\varepsilon,i} = \varepsilon z_\varepsilon \varphi_{n,i}, \quad i = 0, \dots, n-1.$$

Using Lemma 3.8 we have

$$\|s_{\varepsilon,i}\|_\infty = n^{1/2}\varepsilon z_\varepsilon C_{0,\infty}, \quad \|s_{\varepsilon,i}\|_{\eta,\infty,1} \le Cn^{\eta+1/2}\varepsilon z_\varepsilon. \tag{3.141}$$

By Proposition 3.11 it suffices to choose z_ε, $n = n_\varepsilon \to \infty$ such that:

(i) $n^{1/2}\varepsilon z_\varepsilon C_0 > r_\varepsilon$; $n^{1/2+\eta}\varepsilon z_\varepsilon C_\eta < R$ for constants C_0, $C_\eta > 0$;

(ii) $n^{-1}\exp(z_\varepsilon^2) \to 0$, if $r_\varepsilon/r^*_{\varepsilon,\infty} < \lambda_1$ for some $\lambda_1 > 0$.

Let us take

$$n_\varepsilon \sim c_1 r_\varepsilon^{-1/\eta}, \quad z_\varepsilon \sim c_2\varepsilon^{-1}r_\varepsilon^{(2\eta+1)/2\eta}; \quad C_0^{-1}c_1^{-1/2} < c_2 < C_\eta^{-1}c_1^{-\eta-1/2}R,$$

which imply (i). To obtain (ii) it suffices that $\limsup z_\varepsilon^2/\log n < 1$. The last relation follows from

$$\limsup r_\varepsilon^{2+1/\eta}(\varepsilon^2 \log r_\varepsilon^{-1})^{-1} < (1+(2\eta)^{-1})\lambda_1^{(2\eta+1)/\eta} < \eta^{-1}c_2^{-2}. \tag{3.142}$$

The second inequality in (3.142) holds for small enough $\lambda_1 > 0$. □

Remark. We can give some simpler proof of lower bounds using the wavelet transform and relationship (2.84) between Besov norms in the functional space and in the space of wavelets coefficients. We can take a family $j = j_\varepsilon$ such that $2^j \asymp n$, and take collections $\varphi_{\varepsilon,i}$, $i = 1,\dots 2^j$ which consist of the functions from the wavelet basis in the subspace L_j. Then the required inequalities (3.137), (3.138), (3.141) follow from (2.84). However, we prefer to give a direct proof here.

3.5.3 Proof of the Upper Bounds

We need to prove the consistency of tests families based on statistics (3.130). Recall that we chose $n \asymp (r_\varepsilon^*)^{-1/\eta}$. Moreover, for $p = \infty$ we take $n \sim C_\eta(r_\varepsilon^*)^{-1/\eta}$ and choose some $C_\eta > 0$.

Let $1 \le p \le 2$. Since $q \ge p$ here, it is enough to assume the widest alternative with $q = p$. Considered tests are the χ^2-tests (3.14). Let $Pr_n s$ be a projection of $s \in L_2(0,1)$ to a linear hull L_n of $\bar\varphi_{\varepsilon,n}$. In view of Proposition 2.16 and by the choice for n one has the inequalities, for any $s \in S_\varepsilon$,

$$\|Pr_n s\|_p \ge C_1\|s\|_p - C_2Rn^{-\eta} \ge Cr_\varepsilon > 0. \tag{3.143}$$

Using (3.143) we get

$$\begin{aligned} n^{-1/4}\varepsilon^{-1}\inf_{s\in S_\varepsilon}\|Pr_n s\|_2 &\ge n^{-1/4}\varepsilon^{-1}\inf_{s\in S_\varepsilon}\|Pr_n s\|_p \\ &\ge C\varepsilon^{-1}n^{-1/4}r_\varepsilon \asymp r_\varepsilon/r_\varepsilon^* \to \infty. \end{aligned}$$

In view of Corollary 3.5 this implies the consistency.

Let $2 < p = q < \infty$. Considered tests are the χ^p-tests (3.22), (3.24).

Using (3.143) we get

$$\varepsilon^{-1}n^{1/2p-1/2}\inf_{s\in S_\varepsilon}\|Pr_n(s)\|_p \ge C\varepsilon^{-1}n^{1/2p-1/2}r_\varepsilon \asymp r_\varepsilon/r_\varepsilon^* \to \infty.$$

In view of Corollary 3.7 this implies the consistency.

Let $p = q = \infty$. Take $C_\eta = \eta + 1/2$. Considered tests are the supreme tests ψ_n defined by (3.24), (3.36). Using (3.143) we get

$$\varepsilon^{-1}(n\log n)^{-1/2} \inf_{s\in S_\varepsilon} \|Pr_n(s)\|_\infty \geq C\varepsilon^{-1}(n\log n)^{-1/2} r_\varepsilon \sim C r_\varepsilon / r^*_{\varepsilon,\infty} > \sqrt{2}$$

for large enough $\lambda_2 > 0$. By Corollary 3.12 this implies the consistency. □

3.5.4 Rates for Positive Smooth Signals

Let the set $S_\varepsilon^+ = \{s \in S_\varepsilon : s(t) \geq 0\}$ be the set of nonnegative smooth signals; here the set S_ε is defined by (3.126). We consider the cases $p = 1,\ q \geq p$ and $1 < p = q \leq \infty$.

Introduce the quantities (they are rates in the problem):

$$r_\varepsilon^{*+} = \begin{cases} \varepsilon^{2\eta/(2\eta+1-1/p)}, & \text{if } p = 1,\ q \geq p \text{ or } 1 < p = q < \infty, \\ \left(\varepsilon^2 \log(\varepsilon^{-1})\right)^{\eta/(2\eta+1)}, & \text{if } p = q = \infty. \end{cases} \tag{3.144}$$

Theorem 3.10 *Let us test $H_0 : s = 0$ against $H_1 : s \in S_\varepsilon^+$.*

(1) *Let $p = 1$ or $1 < p = q < \infty$, $\eta > 0$. Then $\gamma_\varepsilon(S_\varepsilon^+) \to 1$ as $r_\varepsilon = o(r_\varepsilon^{*+})$ and $\liminf \gamma_\varepsilon(S_\varepsilon^+) > 0$ as $r_\varepsilon = O(r_\varepsilon^{*+})$. If $r_\varepsilon / r_\varepsilon^{*+} \to \infty$, then $\gamma_\varepsilon(S_\varepsilon^+) \to 0$.*

(2) *Let $p = q = \infty$. Then there exist constants $0 < \lambda_1 \leq \lambda_2$ such that $\gamma_\varepsilon(S_\varepsilon^+) \to 1$ as $\limsup r_\varepsilon / r_\varepsilon^{*+} < \lambda_1$ and $\gamma_\varepsilon(S_\varepsilon^+) \to 0$ as $\liminf r_\varepsilon / r_\varepsilon^{*+} > \lambda_2$.*

Note that we have classical rates $r_\varepsilon^{*+} = \varepsilon$ for $p = 1$.

Proof of Theorem 3.10. This follows the scheme of the proof of Theorem 3.9 above. To obtain the lower bounds we choose a positive function φ. For $p < \infty$ we use product priors with nonsymmetric two point factors (one point factors, if $p = 1$) and positive z_ε. For $p = \infty$ we take orthogonal priors, as above. Under this choice the priors are concentrated on S_ε for $p = 1$ and $p = \infty$; if $p \in (1, \infty)$, then the constructed priors are asymptotically concentrated on S_ε.

To obtain the upper bounds, as above, we choose an equispaced partition $[0, 1]$ to $n \asymp (r_\varepsilon^{*+})^{-1/\eta}$ intervals and orthonormal collections of functions generated by indicators of intervals. For $p = 1$ we use tests based on linear statistics of the form

$$t_\varepsilon = n^{-1/2} \sum_{i=1}^n X_{\varepsilon,i} = \varepsilon^{-1} \sum_{i=1}^n \int_{\delta_{\varepsilon,i}} dX_\varepsilon(t) = \varepsilon^{-1}(X_\varepsilon(1) - X_\varepsilon(0)).$$

For $1 < p = q < 2$ we use the tests χ^+_{n,T_ε} based on a combination of linear and χ^2-statistics (3.26) and we use Corollary 3.9. We leave to the reader the calculations which lead to the rates given in Theorem 3.10.

If $p \geq 2$, then we have the same rates as in Theorem 3.9 and, by $S_\varepsilon^+ \subset S_\varepsilon$, we can use the same tests as in Theorem 3.9. □

4

Sharp Asymptotics. I

In this chapter we start to develop methods for the study of sharp asymptotics under the sequence Gaussian model. These methods are based on the study of likelihood ratio statistics in Bayesian problems and on the study of the minimax properties of tests based on these statistics.

We would like to show that for a wide class of hypothesis testing problems there exists a duality between the choice of an asymptotically best product prior and the selection of asymptotically best tests. Namely, both problems lead to the extreme problem: to find "nearest to zero point" over a specific convex set in a Hilbert space.

First we consider the simplest case: Bayesian priors are Dirac masses concentrated on points in the alternative. Here the Bayesian likelihood ratio statistics lead to tests based on linear statistics and, if an alternative set is convex, then the selection of the best tests, based on linear statistics, is dual (nonasymptotically) to the choice of the best prior. We study in detail the extreme problem for the positive alternative determined by two-sided constraints for the power norms and for the norms (2.70) defined by the sequences of exponential type with $p \leq 1,\ q \geq p$.

The reason of this study is that the analogous extreme problems we obtain in general cases. In particular, this duality is extended to symmetric two-point product priors and tests of χ^2-type. These provide a translation of results from positive alternatives to general alternatives determined by two-sided constraints for the power and Besov norms with $p \leq 2,\ q \geq p$.

The study of the relationships between orthogonal priors and supreme-tests provide a description of the regions of the asymptotics of degenerate type.

4.1 Tests Based on Linear Statistics and Convex Alternatives

In this and the next section we will study sharp asymptotics for special classes of infinite-dimensional alternatives under convex assumptions (see Section 2.4.3 above). In this case the minimax hypothesis testing problem is reduced to a special extreme problem. Our aim is to describe the key ideas of duality and analytical methods to study the extreme problem which will be generalized below for a wide class of minimax hypothesis testing problems under an asymptotic approach.

Under the sequence Gaussian model (2.1) let us test $H_0 : v = 0$ against $H_1 : v \in V \subset l^2$. One can write simple lower bounds in the problem. Denote

$$u = \inf_{v \in V} |v|. \tag{4.1}$$

For any $v \in V$ we have (see Example 2.1)

$$\beta(\alpha, V) \geq \beta(\alpha, \{v\}) = \Phi(T_\alpha - |v|), \quad \gamma(V) \geq \gamma(\{v\}) = 2\Phi(-|v|/2).$$

By taking supremum we get

$$\beta(\alpha, V) \geq \Phi(T_\alpha - u), \quad \gamma(V) \geq 2\Phi(-u/2). \tag{4.2}$$

The problem is: *Are the lower bounds* (4.2) *good enough?*

Sometimes we have a positive answer.

Consider tests based on linear statistics $\psi_{r,T} = \mathbf{1}_{\{(x,r)>T\}}$ based on the linear statistics

$$t_r = (x, r) = \sum_i x_i r_i, \quad r \in l^2; \quad |r|^2 = |r|_2^2 = (r, r) = 1,$$

and T is a nonrandom threshold. It was shown in Section 3.1.1 that

$$\alpha(\psi_{r,T}) = \Phi(-T), \quad \beta(\psi_{r,T}, V) = \sup_{v \in V} \Phi(T - (r, v)) = \Phi(T - h(r, V)),$$

where

$$h(r, V) = \inf_{v \in V} (r, v)$$

is the minimum of the projection of $v \in V$ to the direction r. Thus to find the best linear test we need to choose the direction r which maximizes $h(r, V)$. This corresponds to the maximin problem

$$h = h(V) = \sup_{r \in l^2 : |r|=1} h(r, V) = \sup_{r \in l^2 : |r|=1} \inf_{v \in V} (r, v). \tag{4.3}$$

Suppose $h > 0$. Then one can easily see that it is possible to change the constraint $|r| = 1$ by the convex constraint $|r| \leq 1$. Assume the set V is convex. Then, using the equality

$$\sup_{r \in l^2 : |r| \leq 1} (r, v) = |v|$$

and formally applying Theorem 2.2 on the minimax one has

$$h = \sup_{r\in l^2:|r|\le 1} \inf_{v\in V} (r,v) = \inf_{v\in V} \sup_{r\in l^2:|r|\le 1} (r,v) = \inf_{v\in V} |v| = u.$$

This implies

$$h = u = \inf_{v\in V} |v|. \tag{4.4}$$

The minimaxity of tests based on linear statistics follows from the evaluations of the type I and type II errors for $T = T_\alpha$ and $T = h/2$ joined with (4.2).

Let us give a direct proof of (4.4) which does not use Theorem 2.2 (it uses some topological assumptions).

Let L be a real linear space with a Hilbert structure (v,u) (this means (v,u) is a positive defined bilinear form, i.e., $|v|^2 = (v,v) > 0\ \forall\, v \ne 0$). We do not assume L be complete in the norm $|\cdot|$ or separable.

Proposition 4.1 *Let V be a convex subset of L and there exists a minimum point $v^* \in V$ such that*

$$|v^*| = \inf_{v\in V} |v| > 0. \tag{4.5}$$

Then

$$h = \sup_{r\in L:|r|=1} \inf_{v\in V} (r,v) = \inf_{v\in V} (r^*,v) = |v^*|, \tag{4.6}$$

where $r^* = v^*/|v^*|$.

Proof of Proposition 4.1. First, without any assumptions, one has the inequality

$$h(V) = \sup_{r\in L:|r|=1} \inf_{v\in V} (r,v) \le \inf_{v\in V} \sup_{r\in L:|r|=1} (r,v).$$

Next, standard consideration shows that $(r,v) \le |r||v|$ and the equality holds for $r = \lambda v$. These imply $\sup_{r\in L:|r|=1}(r,v) = |v|$ and

$$\inf_{v\in V} \sup_{r\in L:|r|=1} (r,v) = \inf_{v\in V} (v,v)/|v| = \inf_{v\in V} |v| = |v^*|.$$

Thus we need to control $h(V) \ge |v^*|$; it suffices to verify the inequality

$$\inf_{v\in V} (r^*,v) \ge |v^*|. \tag{4.7}$$

Assume that there exists $v \in V$ such that $(r^*,v) < |v^*|$; this is the same as $\lambda = (v^* - v, v^*) > 0$. Consider the interval

$$I = \{v(t) = tv + (1-t)v^* = v^* - t(v^* - v), \quad t \in [0,1]\} \subset V,$$

because of convexity of V. Then

$$|v(t)|^2 = |v^*|^2 - 2t\lambda + t^2|v^* - v|^2 < |v^*|^2$$

for small enough $t > 0$. This contradicts with a minimality of v^*. □

Proposition 4.2 *Let us test $H_0 : v = 0$, against $H_1 : v \in V$ and the set V is convex. Assume there exists $v^* \in V$, such that*

$$u = |v^*| = \inf_{v \in V} |v|. \tag{4.8}$$

Then

$$\gamma(V) = 2\Phi(-u/2), \quad \beta(\alpha, V) = \Phi(T_\alpha - u). \tag{4.9}$$

In particular, $\gamma \to 0$ and $\beta(\alpha) \to 0$ for any $\alpha \in (0,1)$, if $u = |v^*| \to \infty$.

Remark. First, note that the assumption of the existence of the minimum point $v^* \in V$ is not essential here. In fact, the minimum point $v^* \in V$ exists for a closed set V (this holds because l^2 is the complete Hilbert space, see [94], Section 2.7.1, for example). If V is not closed, then one can change V by its closure $\overline{V}$, because one can easily see that

$$\gamma(\overline{V}) = \gamma(V), \quad \beta(\alpha, \overline{V}) = \beta(\alpha, V).$$

Second, observe that the minimum point $|v^*|$ is unique. In fact, if there exists $v_1^* \in V$ such that $v^* \neq v_1^*$, $|v_1^*| = |v^*|$, then $v_2^* = (v^* + v_1^*)/2 \in V$ because of the convexity and one can easily check $|v_2| < |v^*|$.

Proof of Proposition 4.2. The lower bounds follow from (4.2). To obtain the upper bounds consider the tests based on the linear statistics $\psi_\alpha^* = \psi_{r^*, T_\alpha}$ and $\psi^* = \psi_{r^*, u/2}$. In view of the evaluations above and Proposition 4.1 one has

$$\begin{aligned}
\alpha(\psi_a^*) &= \alpha, \quad \alpha(\psi^*) = \Phi(-u/2); \\
\beta(\psi_a^*, V) &= \Phi\left(T_\alpha - \inf_{v \in V}(r^*, v)\right) = \Phi(T_\alpha - u), \\
\gamma(\psi^*, V) &= 2\Phi\left(u/2 - \inf_{v \in V}(r^*, v)\right) = 2\Phi(-u/2),
\end{aligned}$$

which implies

$$\gamma(V) \leq \gamma(\psi^*, V) = 2\Phi(-u/2), \quad \beta(\alpha, V) \leq \beta(\psi_\alpha^*, V) = \Phi(T_\alpha - u). \qquad \square$$

Sometimes one can weaken the assumption that V is a convex set.

Proposition 4.3 *Assume there exists a convex set Z in a sequence space and a coordinate-wise one-to-one mapping*

$$\phi : Z \to V; \quad v_i = \phi_i(z_i) \geq 0, \quad z \in Z, \quad \phi(Z) = V,$$

such that $\phi_i(t)$ are nonnegative convex functions. Assume there exists $z^* \neq 0$ such that

$$u = |\phi(z^*)| = \inf_{z \in Z} |\phi(z)| \tag{4.10}$$

(this is equivalent to the existence of a minimum point $v^* = \phi(z^*) \neq 0$ in the extreme problem (4.8)). Then the statements of Proposition 4.2 hold.

Remark. Analogously with the Remark to Proposition 4.2, the minimum points z^* and v^* are unique.

Proof of Proposition 4.3. Clearly, $v^* = \phi(z^*)$ is the minimum point in the extreme problem (4.8). In the proof of (4.9) convexity was required to apply Proposition 4.1 only. Therefore it suffices to control relation (4.7) which can be rewritten

$$\inf_{z\in Z}(r^*, \phi(z)) \geq |v^*| = |\phi(z^*)| = \inf_{z\in Z}|\phi(z)|,$$

where $\phi(z) = \{\phi_i(z_i)\}$, $r^* = \phi(z^*)/|\phi(z^*)|$. Analogously with the proof of (4.7) assume that there exists a $z \in Z$ such that $(r^*, \phi(z)) < |\phi(z^*)|$; this is the same as $\lambda = (\phi(z^*) - \phi(z), \phi(z^*)) > 0$. Consider the interval

$$I = \{z(t) = tz + (1-t)z^* = z^* + t(z - z^*), \quad t \in [0,1]\}$$

and note that $I \subset Z$ because of convexity of the set Z;

$$0 \leq \phi_i(z_i(t)) \leq t\phi_i(z_i) + (1-t)\phi_i(z_i^*)$$

because of convexity of functions ϕ_i. Then

$$\begin{aligned} |\phi(z(t))|^2 &= \sum_i \phi_i^2(z_i(t)) \leq \sum_i (t\phi_i(z_i) + (1-t)\phi_i(z_i^*))^2 \\ &= |\phi(z^*)|^2 - 2t\lambda + t^2|\phi(z^*) - \phi(z)|^2 < |\phi(z^*)|^2 = |v^*|^2 \end{aligned}$$

for small enough $t > 0$. This contradicts with minimality v^*. □

In view of relationship (2.8), Proposition 4.2 is extended to the case of the convex set $S \subset L_2(0,1)$ under the functional Gaussian model (2.5). We get

$$\gamma_\varepsilon(S) = 2\Phi(-u/2), \quad \beta_\varepsilon(\alpha, S) = \Phi(T_\alpha - u); \quad u = \|s^*\|/\varepsilon = \varepsilon^{-1}\inf_{s\in S}\|s\|. \tag{4.11}$$

Let us return to an n-dimensional Gaussian model. Let us consider the n-dimensional positive alternative defined by the two-sided constraints

$$V_n^+ = \{v \in R_+^n : |v|_p \geq \rho, \ |v|_q \leq R\}. \tag{4.12}$$

Let $p \leq 1$, $q \geq p$. Assume $R \geq \rho n^{1/q - 1/p}$ for V_n to be nonempty. The function $\phi(t) = t^{1/p}$, $t \geq 0$, is convex. Set $v_i = z_i^{1/p}$. Then set V_n^+ corresponds to set Z_n determined by the constraints

$$Z_n = \left\{ z \in R_+^n : \sum_{i=1}^n z_i \geq \rho^p, \quad \sum_{i=1}^n z_i^{q/p} \leq R^q \right\}.$$

Clearly, the set Z_n is convex and the minimum point z^* in the extreme problem (4.10) corresponds to the minimum point v^* in the extreme problem (4.8): $v_i^* = n^{-1/p}\rho$, $i = 1, \ldots, n$. Thus we improve Proposition 2.10.

Corollary 4.1 *Let us test $H_0 : v = 0$ against $H_1 : v \in V_n^+$, where V_n^+ is defined by the two-sided constraints* (4.12). *Then we have*

$$\beta(\alpha, V_n^+) = \Phi(T_\alpha - u_n) + o(1), \quad \gamma(\alpha, V_n^+) = 2\Phi(-u_n/2) + o(1),$$

where $u_n = n^{1/2-1/p}\rho$.

4.2 Two-Sided Constraints for the Positive Alternatives, $p \le 1$, $q \ge p$

We extend the considerations above. Consider the alternative $V \subset l_+^2$ defined by the two-sided constraints:

$$V = \{v \in l_+^2 : |v|_{\bar{a},p} \ge \rho, \ |v|_{\bar{b},q} \le R\},$$

where $\bar{a}$, $\bar{b}$ are two positive sequences. We assume, in this section, $0 < p \le 1$, $q \ge p$.

If $q \ge 1$, then V is convex and, using Proposition 4.2, we go to the extreme problem (4.8). Moreover, let $q \ge p$. Put $v_i = z_i^{1/p}$ and define the set Z by the inequalities

$$z_i \ge 0, \quad \sum_i z_i a_i^p \ge \rho^p, \quad \sum_i z_i^{q/p} b_i^q \le R^q,$$

(for $q = \infty$ the second constraint is changed by $\sup_i z_i b_i^p \le R^p$). Since the function $\phi(t) = t^{1/p}$ and the set Z is convex, we can use Proposition 4.3 to obtain the extreme problem

$$u^2 = \inf_{v \in V} \sum_i v_i^2 = \inf_{z \in Z} \sum_i z_i^{2/p}. \tag{4.13}$$

Consider the case $a_i = i^r$, $b_i = i^s$, $i \in \mathbf{N}$, and $r \ge r_{p,+}^*$. The sets $V = V(\kappa, \rho, R)$ are defined by

$$v_i \ge 0, \quad |v|_{r,p}^p = \sum_{i=1}^\infty v_i^p i^{rp} \ge \rho^p, \quad |v|_{s,q}^q = \sum_{i=1}^\infty v_i^q i^{sq} \le R^q, \tag{4.14}$$

where $\kappa = (p, q, r, s)$.

The extreme problem (4.13) is of the form: if $q < \infty$, then

$$u^2 = u^2(\kappa, \rho, R) = \inf \left\{ \sum_{i=1}^\infty v_i^2 \ : \ v_i \ge 0, \sum_{i=1}^\infty i^{rp} v_i^p \ge \rho^p, \sum_{i=1}^\infty i^{sq} v_i^q \le R^q \right\}, \tag{4.15}$$

and if $q = \infty$, then

$$u^2 = u^2(\kappa, \rho, R) = \inf \left\{ \sum_{i=1}^\infty v_i^2 \ : \ v_i \ge 0, \sum_{i=1}^\infty i^{rp} v_i^p \ge \rho^p, \ \sup_{1 \le i < \infty} i^s v_i \le R \right\}. \tag{4.16}$$

We will use the parameters $\mu = s - r$, $\lambda = s/p - r/q$ which have been introduced in (3.115). Assume below that $\mu > 1/p - 1/q$ (this yields $s > r$ for $q = p$) and $r \geq r^*_{p,+}$; recall that the quantities $r^*_{p,+}$ are defined by (3.108). This yields the inequality

$$\lambda > 1/2p - 1/2q > 0 \quad \text{for } r \geq r^*_{p,+}.$$

Remark. One can give an analytical proof of this inequality. However, this inequality and many analogous inequalities that we use below follow from geometric considerations: the point of intersection of the lines $\ell_1 = \{\mu = 1/p - 1/q\}$ and $\ell_2 = \{\lambda = 1/2p - 1/2q\}$ is $r = 1/2 - 1/p = r^*_{p,+}$, $s = 1/2 - 1/q$ whenever the slope of line ℓ_1 is larger than the slope of line ℓ_2. We recommend that the reader draw a picture. Compare with the Remark on page 122.

For $q > p$ we assume

$$R > R(\rho) = \rho(S(\tau))^{(p-q)/pq}, \tag{4.17}$$

where (see Corollary 3.15)

$$S(\tau) = S^{pq/(q-p)} = \sum_{i=1}^{\infty} i^{-\tau}, \quad \tau = pq(s-r)/(q-p) > 1,$$

and $S(\tau) < \infty$ for $s - r > 1/p - 1/q$. If $p = q$ and $s > r$, assume $\rho \leq R$. In view of Corollary 3.15 and Theorem 3.6 these imply that set V is nonempty, and that the hypothesis testing problem is nontrivial.

4.2.1 System of Equations for an Extreme Problem

Using the Lagrange multipliers rule for a convex extreme problem in terms of variables $z_i = v_i^p$ and returning to variables v_i, we obtain a system of equations for the extreme point $\{v_i^*\}$ with unknown nonnegative Lagrange multipliers A, B, C_i which are not all zero

$$2v_i^{2-p} = Api^{rp} - Bqv_i^{q-p}i^{sq} + C_i,$$

$$C_i v_i = 0, \quad A\left(\sum_i v_i^p i^{rp} - \rho^p\right) = 0, \quad B\left(\sum_i v_i^q i^{sq} - R^q\right) = 0.$$

In view of the convex properties in the problem, it suffices to find *any* solution this system; this solution provides the minimum in the problem. Recall that the uniqueness follows from the Remarks on pages 139 and 140. However, the existence of solutions is not evident.

We fill try to find the solutions with $A > 0$, $B > 0$ and, moreover, with $C_i = 0$ if $q > p$. These imply a system of equations for $\{v_i^*\}, A, B$

$$2v_i^{2-p} = (Api^{rp} - Bqv_i^{q-p}i^{sq})_+; \quad \sum_i v_i^p i^{rp} = \rho^p, \quad \sum_i v_i^q i^{sq} = R^q, \tag{4.18}$$

where $t_+ = \max(0, t)$. Change the variables $A > 0,\ B > 0$ by $v_0 > 0,\ m > 0$ and which are determined by equations $v_0^{2-p} m^{-rp} = A,\ v_0^{2-q} m^{-sq} = B$. This means

$$v_0 = (A^{sq}/B^{rp})^{1/\Delta_+}, \quad m = (A^{2-q}/B^{2-p})^{1/\Delta_+},$$

where we set

$$\Delta_+ = \Delta_+(p, q, r, s) = sq(2-p) - rp(2-q).$$

These values are correctly defined because one can check that $\Delta_+ > 0$ for $\mu > 1/p - 1/q,\ r \ge r^*_{p,+}$. Analogously to the above, this follows from consideration of the lines $\Delta_+ = 0$ and $\mu = 1/p - 1/q$ (see the Remark on page 142).

Set $v_i = v_0 u_i,\ y_i = i/m$. Then we can rewrite (4.18):

$$2u_i^{2-p} = (py_i^{rp} - qy_i^{sq} u_i^{q-p})_+; \tag{4.19}$$

$$v_0^p m^{rp} \sum_i u_i^p y_i^{rp} = \rho^p, \quad v_0^q m^{sq} \sum_i u_i^q y_i^{sq} = R^q; \tag{4.20}$$

$$|v^*|^2 = v_0^2 \sum_i u_i^2. \tag{4.21}$$

Introduce the function $u(y) = u(y; \kappa)$ which is used extensively below. This is the nonnegative solution of the equation

$$2u^{2-p} = py^{rp} - qy^{sq} u^{q-p}. \tag{4.22}$$

If $p = q$ and $s > r$, then

$$u(y) = (p(y^{rp} - y^{sp})/2)^{1/(2-p)}, \text{ if } y \in (0,1); \quad u(y) = 0, \text{ if } y \ge 1. \tag{4.23}$$

Clearly, $u(y)$ is an analytical positive function on $(0,1)$ and it is a Lipschitzian function of degree $1/2$ on $(0, \infty)$; this follows from the relation

$$u(1-x) \sim (xp^2(s-r)/2)^a, \text{ as } x > 0,\ x \to 0; \quad a = 1/(2-p) \ge 1/2.$$

Let $p < q$. Then the function $g(u) = 2u^{2-p} + qy^{sq}u^{q-p}$ is monotone and increases from 0 to ∞ for $u \ge 0$. Therefore, for any $y > 0$ there exists a unique positive solution of equation (4.22). Usually it is impossible to give a direct presentation for $u(y)$. However, one easily observes that $u(y)$ is an analytical positive function on $y > 0$ with the asymptotics

$$u(y) \sim \begin{cases} C(p) y^{rp/(2-p)}, & \text{as } y \to 0; \\ C(p,q) y^{(rp-sq)/(q-p)}, & \text{as } y \to \infty, \end{cases} \tag{4.24}$$

where

$$C(p) = (p/2)^{1/(2-p)}, \quad C(p,q) = (p/q)^{1/(q-p)}.$$

In fact, relation (4.24) is evident for $y \to 0$; for $y \to \infty$ the relation follows from $u^{2-p} = o(y^{sq}u^{q-p})$.

We rewrite (4.20), (4.21):

$$v_0^p m^{rp+1} S_{m,1} = \rho^p, \quad v_0^q m^{sq+1} S_{m,2} = R^q; \quad |v^*|^2 = v_0^2 m S_{m,0}, \qquad (4.25)$$

where

$$S_{m,1} = \frac{1}{m} \sum_i u_i^p y_i^{rp}, \quad S_{m,2} = \frac{1}{m} \sum_i u_i^q y_i^{sq}, \quad S_{m,0} = \frac{1}{m} \sum_i u_i^2; \qquad (4.26)$$

here and below $u_i = u(y_i)$, $y_i = i/m$. This yields the equation for the quantity m:

$$H(m) = \rho/R \quad \text{where} \quad H(m) = \left(\sum_i u_i^p i^{rp} \right)^{1/p} \left(\sum_i u_i^q i^{sq} \right)^{-1/q} \qquad (4.27)$$

and the quantity v_0 is defined by any equations in (4.20).

4.2.2 Solution of an Extreme Problem: Case $p < q < \infty$

Consider equation (4.27). Let $m \to 0$. Since

$$y_i \to \infty, \quad \tau = pq(s-r)/(q-p) > 1,$$

using (4.24), one has

$$\begin{aligned} \sum_{i=1}^{\infty} u_i^p i^{rp} &\sim C^p(p,q) m^{p(sq-rp)/(q-p)} \sum_{i=1}^{\infty} i^{-\tau} \\ &\sim C^p(p,q) S(\tau) m^{p(sq-rp)/(q-p)}, \\ \sum_{i=1}^{\infty} u_i^q i^{sq} &\sim C^q(p,q) m^{q(sq-rp)/q-p)} \sum_{i=1}^{\infty} i^{-\tau} \\ &\sim C^q(p,q) S(\tau) m^{q(sq-rp)/q-p)}. \end{aligned}$$

These imply

$$H(m) \sim (S(\tau))^{(q-p)/pq}, \quad m \to 0.$$

Analogous evaluations and the continuity of $u(y)$ in $y > 0$ show that the series converge uniformly over $m \in K$ for any compact $K \subset R_+^1$, and $H(m)$ is continuous in $m > 0$. Let $m \to \infty$. Then one can easily see that $H(m) \to 0$ (later we will give detailed considerations). Therefore, the image of the function $H(m)$, when $m > 0$, contains the interval $(0, (S(\tau))^{(q-p)/pq})$ for the ratio ρ/R under constraint (4.17). Thus there exists a solution $m > 0$ in (4.27) and $|v^*| > 0$ for all $R > R(\rho)$ in the extreme problem (4.15). Note that the function $H(m)$ is monotone. It is difficult to control this fact directly, however, it follows from the uniqueness of the solution of (4.27). In fact, if $m_1 \neq m_2$, then the vectors $u_1 = \{u(i/m_1)\}$ and $u_2 = \{u(i/m_2)\}$ could not be collinear. This follows from the analytical properties of $u(y)$ and from the asymptotics (4.24); we leave the verification to the reader.

It is difficult to give a direct description of the solutions and we consider an asymptotic approach. Assume $\rho = \rho_\varepsilon$, $R = R_\varepsilon$, and $\rho/R \to 0$. This implies $m \to \infty$ and we can change the sums (4.26) by integrals.

First, assume $r > r^*_{p,+}$. Then

$$S_{m,1} \sim \int_0^\infty u^p(y) y^{rp}\, dy = c_1^p = c_1^p(\kappa), \tag{4.28}$$

$$S_{m,2} \sim \int_0^\infty u^q(y) y^{sq}\, dy = c_2^q = c_2^q(\kappa); \tag{4.29}$$

$$S_{m,0} \sim \int_0^\infty u^2(y)\, dy = c_0 = c_0(\kappa), \tag{4.30}$$

where $u(y)$ is the function defined by (4.22).

Using the assumption on r, s, p, q and asymptotics (4.24) one can check that the integrals in (4.28)–(4.30) are finite. This implies asymptotic relations for v_0, m:

$$c_1 v_0 m^{r+1/p} \sim \rho, \quad c_2 v_0 m^{s+1/q} \sim R, \quad |v^*|^2 \sim c_0 v_0^2 m. \tag{4.31}$$

Using (4.31) one has

$$m \sim d_1 (R/\rho)^\eta, \quad v_0 \sim d_2 (\rho^{s+1/q}/R^{r+1/p})^\eta, \quad \eta = (s - r + q^{-1} - p^{-1})^{-1}, \tag{4.32}$$

where $\eta > 0$ (this is equivalent to $\mu = p(s-r) > 1 - p/q$), and

$$d_1 = (c_1/c_2)^\eta, \quad d_2 = (c_2^{r+1/p}/c_1^{s+1/q})^\eta.$$

Therefore, we get

$$|v^*|^2 \sim d_3 (\rho^{2s-1+2/q}/R^{2r-1+2/p})^\eta; \quad d_3 = c_0 d_1 d_2^2. \tag{4.33}$$

Observe that

$$|v^*|^2 \sim d_3 \rho^2 (\rho/R)^{(2r-1+2/p)\eta},$$

and, if $\rho/R \to 0$, then the distinguishability (i.e., $|v^*| \to \infty$) is possible for $\rho \to \infty$, $R \to \infty$ only; moreover, we need

$$\rho \gg R^\delta, \quad \delta = (2r - 1 + 2/p)/(2s - 1 + 2/q) \in (0, 1).$$

Let $r = r^*_{p,+} = 1/2 - 1/p$. Then we get (4.29), but the integrals in (4.28), (4.30) diverge at point 0 because $u^p(y) y^{rp} \asymp u^2(y) \asymp y^{-1}$, as $y \to 0$. In this case, we have

$$S_{m,1} \sim C^p \log m, \quad S_0 \sim C^2 \log m; \quad C = C(p).$$

This implies asymptotic relations for v_0, m:

$$C v_0 m^{r+1/p} (\log m)^{1/p} \sim \rho, \quad c_2 v_0 m^{s+1/q} \sim R; \quad |v^*|^2 \sim C^2 v_0^2 m \log m. \tag{4.34}$$

Using (4.34) one has

$$m \sim d_1(R/\rho)^\eta(\log(R/\rho))^{\eta/p},$$
$$v_0 \sim d_2(\rho^{s+1/q}/R^{r+1/p})^\eta(\log(R/\rho))^{-\eta(sq+1)/pq},$$

where

$$d_1 = (Cc_2^{-1}\eta^{1/p})^\eta, \quad d_2 = (c_2^{q(rp+1)}/C^{p(sq+1)}\eta^{sq+1})^{\eta/pq}.$$

Therefore, we get

$$|v^*|^2 \sim d_3(\rho^{2s-1+2/q}/R^{2r-1+2/p})^\eta(\log(R/\rho))^{1-(2sq-q+2)\eta/pq}, \tag{4.35}$$

where $d_3 = \eta C^2 d_1 d_2^2$.

4.2.3 Solution of an Extreme Problem: Case $p = q$

Assume $p = q$. Then $s > r$, function $u(y)$ is determined by (4.23), and we get

$$H^p(m) = \frac{\sum_{1\le i<m}((i/m)^{rp} - (i/m)^{sp})^{1/(2-p)} i^{rp}}{\sum_{1\le i<m}((i/m)^{rp} - (i/m)^{sp})^{1/(2-p)} i^{sp}} = (\rho/R)^p. \tag{4.36}$$

Clearly, $H(2) = 1$ and $H(m) \to 0$ as $m \to \infty$; the function $H(m)$ is continuous in $m > 1$ (it suffices to control integers m). These imply the existence of positive solutions for $\rho \le R$ and $|v^*| > 0$.

It is difficult to study the solutions exactly. Consider an asymptotic setting. Assume $\rho/R \to 0$. Then $m \to \infty$ and we can change the sums in (4.26) by integrals.

First, let $r > r^*_{p,+}$. Then

$$S_{m,1} \sim C^p\int_0^1 (y^{rp} - y^{sp})^{p/(2-p)} y^{rp}\,dy = c_1^p = c_1^p(\kappa), \tag{4.37}$$

$$S_{m,2} \sim C^p\int_0^1 (y^{rp} - y^{sp})^{p/(2-p)} y^{sp}\,dy = c_2^p = c_2^p(\kappa), \tag{4.38}$$

$$S_{m,0} \sim C^2\int_0^1 (y^{rp} - y^{sp})^{2/(2-p)}\,dy = c_0 = c_0(\kappa), \tag{4.39}$$

where $C = C(p) = (p/2)^{1/(2-p)}$. One can rewrite the right-hand sides of (4.37)–(4.39) in terms of the Euler beta-function

$$c_1^p = AB(a,c), \quad c_2^p = AB(b,c), \quad c_0 = DB(a,d),$$

where

$$B(x,y) = \int_0^1 t^{x-1}(1-t)^{y-1}\,dt; \quad A = \frac{C^p}{p(s-r)}, \quad D = \frac{C^2}{p(s-r)},$$

$$a = \frac{2rp + 2 - p}{p(2-p)(s-r)}, \quad b = \frac{(2-p)(1+ps) + rp^2}{p(2-p)(s-r)}, \quad c = \frac{2}{2-p}, \quad d = \frac{4-p}{2-p}.$$

Since $a > 0$ for $r > r^*_{p,+}$ and $b > 0$ for $r \ge r^*_{p,+}$, $c > 0,\ d > 0$, the integrals in (4.37)–(4.39) are finite.

These imply asymptotic relations for $v_0,\ m,\ |v^*|^2$ analogous with (4.31):

$$c_1 v_0 m^{r+1/p} \sim \rho, \quad c_2 v_0 m^{s+1/p} \sim R, \quad |v^*|^2 \sim c_0 v_0^2 m. \tag{4.40}$$

Using (4.40) we have (4.32)–(4.33) with $\eta = (s-r)^{-1} > 0,\ p = q$.

Let $r = r^*_{p,+} = 1/2 - 1/p$. Then $a = 0$ and the integrals in (4.37), (4.39) diverge. Analogously to the above, we have

$$S_{m,i} \sim C^{a(i)} \log m, \quad i = 0, 1; \quad a(0) = 2, \quad a(1) = p.$$

This leads to relations (4.34)–(4.35) with $\eta = (s-r)^{-1}$.

4.2.4 Solution of an Extreme Problem: Case $p < q = \infty$

Note that the constraint $\mu > 1/p - 1/q$ is of the form $s > r + 1/p$ in this case. Using the Lagrange multipliers rule for extreme problem (4.16) we obtain a system of equations for the extreme sequence $\{v_i^*\}$ with unknown non-negative Lagrange multipliers A, C_i, D_i which are not all zero,

$$2v_i^{2-p} = Api^{rp} + C_i - D_i,$$

$$C_i v_i = 0, \quad A\left(\sum_i v_i^p i^{rp} - \rho^p\right) = 0, \quad D_i(v_i - Ri^{-s}) = 0.$$

One can exclude the variables C_i, D_i:

$$v_i = \begin{cases} (Ap/2)^{1/(2-p)} i^{rp/(2-p)}, & \text{if } (Ap/2)^{1/(2-p)} i^{rp/(2-p)} \le Ri^{-s}, \\ Ri^{-s}, & \text{if } (Ap/2)^{1/(2-p)} i^{rp/(2-p)} > Ri^{-s}. \end{cases} \tag{4.41}$$

Introduce the variables $m > 0,\ v_0 > 0$ defined by relations

$$v_0^{2-p} m^{-rp} = Ap/2, \quad v_0 = Rm^{-s},$$

and the function $u(y),\ y > 0$,

$$u(y) = \begin{cases} y^{rp/(2-p)}, & \text{if } y < 1, \\ y^{-s}, & \text{if } y \ge 1. \end{cases} \tag{4.42}$$

One can rewrite (4.41): $v_i = v_0 u(y_i),\ y_i = i/m;\ v_0 = Rm^{-s}$, and

$$v_0^p m^{rp} \left(\sum_{1 \le i < m} y_i^{2rp/(2-p)} + \sum_{m \le i < \infty} y_i^{p(r-s)} \right) = \rho^p. \tag{4.43}$$

Thus we have

$$H(m) = m^{-pD} \sum_{1 \le i < m} i^{2rp/(2-p)} + \sum_{m \le i < \infty} i^{p(r-s)} = (\rho/R)^p, \tag{4.44}$$

where $D = s + rp/(2-p) > 0$. If $m < 1$, then $H(m) = S(\tau)$; if $m \to \infty$, then $H(m) \to 0$ and $H(m)$ is continuous in $m > 0$. Therefore, a positive solution exists and $|v^*| > 0$ for any $R \geq \rho S^{-1/p}(\tau)$ (remember that it is nonempty condition (4.17) for $q = \infty$).

Let $\rho/R \to 0$. Then $m \to \infty$ and we can change the sums by integrals in (4.43). If $r > r^*_{p,+}$, then we get the relations

$$v_0^p m^{rp+1}\left(\int_0^1 y^{2rp/(2-p)}\,dy + \int_1^\infty y^{p(r-s)}\,dy\right) \tag{4.45}$$
$$= v_0^p m^{rp+1}\left(\frac{2-p}{2rp+2-p} + \frac{1}{p(s-r)-1}\right) \sim \rho^p; \quad v_0 = Rm^{-s};$$

$$|v^*|^2 = \sum_i v_i^2 \sim v_0^2 m\left(\int_0^1 y^{2rp/(2-p)}dy + \int_1^\infty y^{-2s}dy\right) \tag{4.46}$$
$$= v_0^2 m\left(\frac{2-p}{2rp+2-p} + \frac{1}{2s-1}\right).$$

These relations imply

$$m \sim d_1(R/\rho)^\eta; \quad |v^*|^2 \sim d_3(\rho^{2s-1}/R^{2r-1+2/p})^\eta,$$

where $\eta = (s - r - p^{-1})^{-1}$ and we leave for reader to write the direct relations for constants d_1, d_3.

If $r = r^*_{p,+}$, then the first integrals in (4.45)–(4.46) diverge. In this case, we have

$$v_0^p m^{rp+1}\log m \sim \rho^p, \quad v_0 = Rm^{-s}; \quad |v^*|^2 \sim v_0^2 m \log m, \tag{4.47}$$

which imply

$$m \sim (R/\rho)^\eta(\eta\log(R/\rho))^{\eta/p};$$
$$|v^*|^2 \sim (\rho^{2s-1}/R^{2r-1+2/p})^\eta(\eta\log(R/\rho))^{1-(2s-1)\eta/p}.$$

Observe that relations (4.47) correspond to limits (4.31), (4.34) as $q \to \infty$.

Let us combine the main results above.

Theorem 4.1 *Let $p \leq 1$, $p \leq q \leq \infty$,*

$$V_\varepsilon = \{v \in l_+^2 : |v|_{r,p} \geq \rho_\varepsilon, |v|_{s,q} \leq R_\varepsilon\}, \quad r \geq r^*_{p,+} = 1/2 - 1/p,$$

and let $\rho_\varepsilon, R_\varepsilon$ be such that $V_\varepsilon = V(\rho_\varepsilon, R_\varepsilon)$ is nonempty.

If $s - r \leq 1/p - 1/q$, then $\gamma(V_\varepsilon) \equiv 1$. If $s - r > 1/p - 1/q$, then, as $R_\varepsilon/\rho_\varepsilon \to \infty$, one has nonasymptotic Gaussian relations for the minimax errors

$$\beta(\alpha, V_\varepsilon) = \Phi(T_\alpha - u_\varepsilon), \quad \gamma(V_\varepsilon) = 2\Phi(-u_\varepsilon/2),$$

where the quantities $u_\varepsilon = u_\varepsilon(\rho_\varepsilon, R_\varepsilon)$ are defined by the relation

$$u_\varepsilon^2 \sim \begin{cases} c_0 v_0^2 m, & \text{if } r > r^*_{p,+}, \\ c_0 v_0^2 m \log m, & \text{if } r = r^*_{p,+}, \end{cases} \tag{4.48}$$

where $m = m_\varepsilon$, $v_0 = v_{0,\varepsilon}$ *satisfy*

$$\rho_\varepsilon \sim \begin{cases} c_1 v_0 m^{r+1/p}, & \text{if } r > r^*_{p,+}, \\ c_1 v_0 m^{r+1/p}(\log m)^{1/p}, & \text{if } r = r^*_{p,+}, \end{cases} \quad R_\varepsilon \sim c_2 v_0 m^{s+1/q}; \quad (4.49)$$

where $c_l = c_l(\kappa)$, $l = 0, 1, 2$, *are positive functions in* $\kappa = (r, s, p, q)$ *which are continuous in* r, s, p, $1/q$ *over the region* $0 < p \le 1$, $p \le q$, $r > r^*_{p,+}$, $s - r > 1/p - 1/q$. *In this case, minimax tests are based on the linear statistics*

$$t_\varepsilon = \sum_i r_{\varepsilon,i} x_i, \quad r^2_{\varepsilon,i} = u^2(i/m) \left(\sum_i u^2(i/m) \right)^{-1},$$

where, if $q < \infty$, *the function* $u(y)$, $y > 0$, *is the solution of the equation*

$$2u^{2-p} = p y^{rp} - q y^{sq} u^{q-p},$$

and, if $q = \infty$,

$$u(y) = \begin{cases} y^{rp/(2-p)}, & \text{if } y < 1, \\ y^{-s}, & \text{if } y \ge 1. \end{cases}$$

Note that for $r > r^*_{p,+}$, $q < \infty$, direct formulas for $c_l = c_l(\kappa), l = 0, 1, 2$, are given by relations (4.28)–(4.30). If $r > r^*_{p,+}$, $q = \infty$, then $c_2 = 1$ and the relations for c_1, c_0 are presented in (4.45), (4.46). If $r = r^*_{p,+}$, then one can find these quantities from (4.34), (4.47).

Remarks.

(1) It is not difficult to change the relative accuracy (i.e., asymptotic equivalences $u_\varepsilon \sim \ldots$) by absolute accuracy (i.e., by the relations of the type $u_\varepsilon = \ldots + o(1)$) in the asymptotic relations above.

(2) One can easily check that analogous relations hold for the case when the norms $|v|_{r,p}$, $|v|_{s,q}$ are changed by the norms $|v|_{\bar a,p}$, $|v|_{\bar b,q}$ with $a_i \sim i^r$, $b_i \sim i^s$ (one needs to change the equalities by asymptotic equivalences). The difference is that the equalities $r^2_{\varepsilon,i} = u^2(i/m)/\sum_i u^2(i/m)$ are changed by the asymptotic relations $r^2_{\varepsilon,i} \sim u^2(i/m)/\sum_i u^2(i/m)$ which are uniform over i.

(3) It follows from relations (4.48), (4.49) that, if $r > r^*_{p,+}$, then

$$u^2_\varepsilon(\rho_\varepsilon, R_\varepsilon) \sim d\rho_\varepsilon^A R_\varepsilon^{-B}, \quad (4.50)$$

where

$$A = \frac{2/q - 1 + 2s}{s - r + 1/q - 1/p}, \quad B = \frac{2/p - 1 + 2r}{s - r + 1/q - 1/p}, \quad (4.51)$$

and, if $r = r^*_{p,+}$, then

$$u^2_\varepsilon(\rho_\varepsilon, R_\varepsilon) \sim d\rho^2_\varepsilon (\log(R_\varepsilon/\rho_\varepsilon))^{(p-2)/p}, \quad (4.52)$$

where $d = d(p, q, r, s) > 0$.

(4) Let $R_\varepsilon \to \infty$. If the assumption $R_\varepsilon/\rho_\varepsilon \to \infty$ does not hold, then $u_\varepsilon \to \infty$ and $\gamma(V_\varepsilon) = o(1)$. In fact, let $R_\varepsilon/\rho_\varepsilon = O(1)$. Then V_ε could be an empty set (see Corollary 3.15) which formally corresponds to $\gamma(V_\varepsilon) = 0$ and $u_\varepsilon = \infty$. If V_ε is nonempty set, then it easily follows, from (4.50)–(4.52), that one can change ρ_ε by smaller $\tilde{\rho}_\varepsilon = o(\rho_\varepsilon)$ such that $\tilde{u}_\varepsilon = u(\tilde{\rho}_\varepsilon, R_\varepsilon) \to \infty$. By $R_\varepsilon/\tilde{\rho}_\varepsilon \to \infty$, we can apply Theorem 4.1 to the sets $\tilde{V}_\varepsilon = V(\tilde{\rho}_\varepsilon, R_\varepsilon)$. By $V_\varepsilon \subset \tilde{V}_\varepsilon$, we have the inequality $\gamma(V_\varepsilon) \le \gamma(\tilde{V}_\varepsilon) \to 0$.

These considerations also show that the relations of Theorem 4.1 hold true under the assumptions $R_\varepsilon \to \infty$ and $u_\varepsilon \asymp 1$.

4.2.5 Constraints of Exponential Type with $p \le 1$, $q \ge p$

Consider the case $a_j = e^{rj}, b_j = e^{sj}$, $j \in \mathbf{N}$ (we change the notation i to j avoid confusion with $\sqrt{-1}$). In view of Theorem 3.3 for $r < 0$ we have classical asymptotics without the second constraint; therefore, assume $r \ge 0$. In view of Proposition 3.14, the nonempty condition is of the form $s > r$ or $s = r$ and $R \ge \rho$, if $p \ge q$. We assume $p \le 1$, $q \ge p$ which imply convexity of the set Z in the extreme problem (4.13).

If $r \ge 0$, $s \le r$, then there is triviality in the hypothesis testing problem. In fact, if $r > 0$, then the infimum is zero in the extreme problem (4.13), since the set $V = V(\kappa, \rho, R)$ contains the points with only one nonzero coordinate $v_j = \rho e^{-jr} e_j$ where $\{e_j\}$ is a standard basis in l^2, and $v_j \to 0$, as $j \to \infty$. As above, we denote $\kappa = (p, q, r, s)$. If $r = 0$, then V contains all the points ρe_j which imply triviality in view of Proposition 3.10.

Therefore, we consider the case

$$r \ge 0, \quad s > r; \quad 0 < p \le 1, \quad q \ge p.$$

Analogously to the consideration in Section 4.2.1 we obtain the system of equations for the coordinates of minimum point in (4.13):

$$2v_j^{2-p} = (Ape^{rpj} - Bqv_j^{q-p}e^{sqj})_+; \quad \sum_{j=1}^{\infty} v_j^p e^{rpj} = \rho^p, \quad \sum_{j=1}^{\infty} v_j^q e^{sqj} = R^q. \tag{4.53}$$

If $q < \infty$, then we change the variables $A > 0$, $B > 0$ by variables $v_0 > 0$, $j_0 > 0$ defined by the equations

$$v_0^{2-p} e^{-rpj_0} = A, \quad v_0^{2-q} e^{-sqj_0} = B.$$

If $q = \infty$, then

$$v_0^{2-p} e^{-rpj_0} = Ap/2, \quad v_0 = Re^{-sj_0}.$$

Using the first equations in (4.53) we get

$$v_j = v_0 u_j; \quad u_j = u(y_j); \quad y_j = e^{j-j_0}, \tag{4.54}$$

where the function $u(y)$ is defined by (4.22) for $q < \infty$, and by (4.42) for $q = \infty$. Setting

$$C_1^p(j_0) = \sum_{j=1}^{\infty} u^p(y_j)y_j^{rp}, \quad C_2^q(j_0) = \sum_{j=1}^{\infty} u^q(y_j)y_j^{sq}; \quad C_0(j_0) = \sum_{j=1}^{\infty} u^2(y_j),$$

we can rewrite other equations in (4.53): if $q < \infty$, then

$$v_0^p e^{rpj_0} C_1^p(j_0) = \rho^p, \quad v_0^q e^{sqj_0} C_2^q(j_0) = R^q; \tag{4.55}$$

$$u^2(\rho, R) = |v^*|^2 = v_0^2 C_0(j_0).$$

If $q = \infty$, then the second equation in (4.55) is changed by $v_0 e^{sj_0} = R$.
These yield the equation for the quantities j_0,

$$R/\rho = \Psi(j_0), \quad \text{where} \quad \Psi(j_0) = e^{j_0(s-r)} C_2(j_0)/C_1(j_0). \tag{4.56}$$

One can see that the functions $C_0(j_0)$, $e^{rj_0}C_1(j_0)$, $e^{sj_0}C_2(j_0)$ are positive and continuous. The function $\Psi(j_0)$ in (4.56) is strictly increasing in j_0 (in the opposite case there exist more than one extreme points in the extreme problem which contradict to convexity) and tends to ∞ as $j_0 \to \infty$. Therefore, at least for large enough $x = R/\rho$, there exists an inverse function $J(x) = j_0(x) = \Psi^{-1}(x)$ which determines the solution of equations (4.55).

Let us study the asymptotics of the solutions as $x = R/\rho \to \infty$ (this implies $j_0 \to \infty$). Let $r > 0$. Using asymptotics (4.24) we get, for $q < \infty$, that there exists $b = b(\kappa) > 0$ such that

$$C_k(j_0) = c_k(j_0) + o(e^{-bj_0}), \quad \text{as } j_0 \to \infty, \quad k = 0, 1, 2.$$

Here the functions $c_k(j_0) = c_k(j_0, \kappa)$ and are defined by the series

$$c_1^p(j_0) = \sum_{l=-\infty}^{\infty} u^p(e^{l-j_0}) e^{rp(l-j_0)}, \tag{4.57}$$

$$c_2^q(j_0) = \sum_{l=-\infty}^{\infty} u^q(e^{l-j_0}) e^{sq(l-j_0)},$$

$$c_0(j_0) = \sum_{l=-\infty}^{\infty} u^2(e^{l-j_0}), \tag{4.58}$$

and the series converge. For $q = \infty$ we have $c_2(j_0) \equiv 1$. For the special case $p = q$, using the direct form (4.23) of $u(y)$, one can easily rewrite the series.

Clearly, *the functions* $c_k(j_0)$ *are* 1*-periodic in* j_0, i.e.,

$$c_k(j_0) = c_k(j_0 + 1) \quad \forall\, j_0 \in R; \quad k = 0, 1, 2.$$

Thus we get asymptotic relations for the quantities $|v^*| = u(\rho, R)$:

$$u^2(\rho, R) \sim c_0(j_0) v_0^2, \tag{4.59}$$

where the quantities j_0, v_0 are determined by the asymptotic relations

$$\rho \sim c_1(j_0)v_0 e^{rj_0}, \qquad R \sim c_2(j_0)v_0 e^{sj_0}. \tag{4.60}$$

These yield the relation

$$|v^*|^2 = u^2(\rho, R) \sim d(j_0)(\rho^s/R^r)^{2/(s-r)}; \tag{4.61}$$

the function

$$d(j_0) = c_0(j_0)(c_2^r(j_0)/c_1^s(j_0))^{2/(s-r)}$$

is positive continuous 1-periodic in j_0, and j_0 is determined by equation

$$x = R/\rho \sim \psi(j_0), \quad \text{where} \quad \psi(j_0) = e^{(s-r)j_0}c_2(j_0)/c_1(j_0). \tag{4.62}$$

Note that the function $\Psi(j_0)$ is strictly increasing and

$$\psi(j_0) \sim \Psi(j_0), \quad \log\psi(j_0) = \log\Psi(j_0) + o(e^{-bj_0}), \quad \text{as } j_0 \to \infty.$$

One also has $\psi(j_0+1) = e^{s-r}\psi(j_0)$ and the function $\psi(j_0)$ is increasing in j_0. However, it is not clear that it is strictly increasing which is needed to define a continuous inverse function $\psi^{-1}(x)$ to solve the equation in (4.62). To obtain this property it suffices to show that the function $\Psi^{-1}(e^y)$ (which is the inverse function to the function $\log\Psi(j_0)$) is Lipschitzian in y of order $1/2$ for large enough y. This fact follows from

Lemma 4.1 *There exist $C > 0$, $B > 0$, $\delta > 0$ such that $(J(x(1+t)) - J(x))^2 \le C|t|$ for $x > B$, $|t| < \delta$, where, recall, $J(x) = j_0(x) = \Psi^{-1}(x)$.*

Proof. Let B be such that the functions $C_l(x)$, $l = 0, 1, 2$, are bounded and bounded away from 0 for $x > B$. Then it follows from (4.56) for $x = R/\rho$ that there exists $C_1 > 0$ such that

$$|J(x) - (\log x)/(s-r)| \le C_1, \quad \text{if } x > B. \tag{4.63}$$

Therefore it suffices to assume that the quantity $J(x)$ is large enough and the quantity $|J(x(1+2t)) - J(x)|$ is bounded.

Fix R, ρ such that $x = R/\rho > 2B$ and consider the function

$$f(t) = u^2(\rho, (1+t)R)/u^2(\rho, R), \quad |t| \le 1/2; \quad f(0) = 1.$$

It follows from the monotonicity of $u^2(\rho, R)$ in R that $f(t)$ decreases in t. It follows from Proposition 2.8 that $f(t)$ is a convex function, and it follows from (4.55), (4.63) that $f(t)$ is bounded and bounded away from 0 in $|t| \le 1/2$ uniformly over $x > 2B$. By convexity one can easily get that $f(t)$ is Lipschitzian, i.e., there exists $C_2 > 0$ such that $|f(t+y) - f(t)| \le C_2|y|$ for all $|t| \le 1/3$, $|y| < 1/6$, $x > 2B$, which yields

$$f(t) \ge 1 - C_2|t|. \tag{4.64}$$

On the other hand, let $w \in l^2$ and $w_1 \in l^2$ be extreme points in the problems which correspond to the quantities (ρ, R) and $(\rho, (1+2t)R)$, i.e.,

$$|w|^2 = u^2(\rho, R), \quad |w_1|^2 = u^2(\rho, (1+2t)R) = f(2t)u^2(\rho, R),$$

where the points $w = vu$, $w_1 = v_1 u_1$ are determined by (4.54) and correspond to the quantities v, v_1 and to the vectors

$$u = \{u(e^{j-j_0})\}, \quad u_1 = \{u(e^{j-j_1})\}, \quad j > 0; \quad j_0 = J(x), \quad j_1 = J(x(1+2t)).$$

Set $w_2 = (w + w_1)/2$. By convexity, the point w_2 is contained in the set $V(\kappa, \rho, (1+t)R)$. Therefore,

$$u^2(\rho, (1+t)R) = u^2(\rho, R) f(t) \le |w_2|^2.$$

Since $f(t)$ is Lipschitzian, there exists $C_3 > 0$ such that

$$u^2(\rho, R)(1 - C_2|t|) \le u^2(\rho, R) f(t) \le |w_2|^2 = (|w|^2 + |w_1|^2)/2$$
$$-(|w - w_1|/2)^2 \le u^2(\rho, R)(1 + C_3|t| - (|w - w_1|/2)^2/u^2(\rho, R)). \quad (4.65)$$

Therefore, for some $C_4 > 0$, we get

$$|w - w_1|^2/u^2(\rho, R) = |u - cu_1|^2/|u|^2 \le C_4|t|, \quad c = v_1/v,$$

and this suffices to control the inequality

$$|u - cu_1|^2/|u|^2 \ge b(j_1 - j_0)^2, \quad b > 0. \quad (4.66)$$

To verify (4.66) we need to establish lower bounds for $|u - cu_1|^2/|u|^2$. Clearly, the qualities $|u|^2$ are bounded and bounded away from 0 for large enough j and $|u - cu_1|^2 \ge |\tilde{u} - c\tilde{u}_1|^2$, where $\tilde{u}$, $\tilde{u}_1$ are any subvectors of the vectors u, u_1. For this reason, it suffices to obtain lower bounds for the quantities $|\tilde{u} - c\tilde{u}_1|^2/|\tilde{u}|^2$, where $\tilde{u}$, $\tilde{u}_1$ are some two-dimensional subvectors. Take $j = j_0 \pm a$, with bounded and large enough a, and

$$\tilde{u} = (u(e^{-a}), u(e^{a})), \quad \tilde{u}_1 = (u(e^{j_0 - j_1 - a}), u(e^{j_0 - j_1 + a})).$$

Note that

$$|\tilde{u} - c\tilde{u}_1|^2/|\tilde{u}|^2 \ge \inf_c |\tilde{u} - c\tilde{u}_1|^2/|\tilde{u}|^2 = |\tilde{u} - P_{\tilde{u}_1}(\tilde{u})|^2/|\tilde{u}|^2 = 1 - \mathrm{cor}^2(\tilde{u}, \tilde{u}_1),$$

where $P_{\tilde{u}_1}(\tilde{u})$ is the projection of vector $\tilde{u}$ in direction $\tilde{u}_1$ and $\mathrm{cor}(\tilde{u}, \tilde{u}_1) = (\tilde{u}, \tilde{u}_1)/|\tilde{u}||\tilde{u}_1|$. Since the quantities $|\tilde{u}|$, $|\tilde{u}_1|$ are bounded, it suffices to verify that there exists $b > 0$ such that

$$T = |\tilde{u}|^2|\tilde{u}_1|^2(1 - \mathrm{cor}^2(\tilde{u}, \tilde{u}_1)) \ge b(j_0 - j_1)^2. \quad (4.67)$$

To establish inequality (4.67) note that, taking the derivative in (4.22) by $y = e^{j-j_0}$, we get an asymptotic relation for the derivative of $u(e^{j-j_0})$ in j:

$$u'_j \sim \begin{cases} B_1 u(y), & \text{as } y = e^{j-j_0} \to 0, \\ B_2 u(y), & \text{as } y = e^{j-j_0} \to \infty. \end{cases}$$

Here $B_1 = rp/(2-p)$, $B_2 = (rp - sq)/(q-p)$. Define the functions $c_0(j)$ by the equality $u(e^j) = c_0(j)C(p)e^{B_1 j}$. Note that $c_0(j - j_0) \sim 1$, as $j - j_0 \to -\infty$ and

$$c'_0(j - j_0) = (e^{-B_1(j-j_0)}u(e^{j-j_0}))'/C(p) = o(1), \quad \text{as } j - j_0 \to -\infty.$$

Analogously, setting $c_1(j) = u(e^j)e^{-B_2 j}/C(p,q)$, we have $c_1'(j-j_0) = o(1)$, as $j - j_0 \to \infty$. Set $\Delta j = j_0 - j_1$. Direct calculation gives

$$T = De^{-2aB_0}\left(c_0(-a)c_1(\Delta j + a)e^{B_2\Delta j} - c_1(a)c_0(\Delta j - a)e^{B_1\Delta j}\right)^2,$$

where $B_0 = \Delta_+/(q-p)(2-p)$, $D = C(p)^2C(p,q)^2$.

Note that either $B_1\Delta j > 0$ or $B_2\Delta j > 0$. Let $B_1\Delta j > 0$ (the case $B_2\Delta j > 0$ is considered in an analogous way). Recall that we can assume Δj to be bounded. Using the asymptotics above we can take $a > 0$ such that, for all $\theta_0, \theta_1 \in (0,1)$ and large enough j, the following inequalities are fulfilled:

$$c_0(-a)c_1'(a + \theta_1\Delta j) - c_1(a)c_0'(-a + \theta_0\Delta j) < B_0\Delta j/4,$$
$$c_0(a)c_1(\Delta j + a) > 1/2.$$

Using inequality $e^x > 1 + x$ we have $T > De^{-2aB_0}B_0(j_0 - j_1)^2/16$. □

Thus the function $d(j_0)$ in (4.61) is of the form $d(j_0) = d_1(\log(R/\rho))$ and $d_1(t)$ is a continuous periodic function in $t = \log(R/\rho)$ of period $s - r$.

Let $r = 0$. Then the series (4.57), (4.58) diverge and

$$C_1^p(j_0) \sim C^p j_0, \quad C_0(j_0) \sim C^2 j_0; \quad C = C(p).$$

In this case, we get

$$u^2(R,\rho) = |v^*|^2 \sim C^2 v_0^2 j_0; \quad \rho \sim C v_0 j_0^{1/p}, \quad R \sim c_2(j_0)v_0 e^{sj_0}. \tag{4.68}$$

The equation for j_0 is of the form

$$sj_0 = \log(R/\rho) + p^{-1}\log j_0 + \log(C/c_2(j_0)),$$

which yields $j_0 \sim s^{-1}\log(R/\rho)$ as $R/\rho \to \infty$. Therefore, we get the relations

$$|v^*|^2 \sim \rho^2 j_0^{1-2/p} \sim \rho^2(s^{-1}\log(R/\rho))^{1-2/p}, \tag{4.69}$$

and a periodic component is not essential in the asymptotics.

Analogously to the Remarks to Theorem 4.1 one can replace relative accuracy by absolute accuracy in the main relations above.

4.3 Sharp Asymptotics of Gaussian Type: Product Priors

In this section we obtain the sharp asymptotics for alternatives under constraints determined by power norms and Besov norms with $p \le 2, q \ge 2$. In this case, product priors with symmetric two-point factors are asymptotically least favorable and tests of the χ^2-type are asymptotically minimax. These results are based on asymptotic normality in the Bayesian problems and on duality between the problems: to find asymptotically least favorable priors and to find asymptotically best tests of χ^2-type. In view of this duality we pass to extreme problems analogous that which have been studied in Section 4.2 and we use the results above.

4.3.1 Asymptotic Normality for Product Priors

To obtain sharp asymptotics we need a more detailed study of the Bayesian likelihood ratio. Here we give the study for product priors with two-point and symmetric three-point factors (these are sufficient for our aims).

Let us consider a family of product priors π^ε which corresponds to a family of factors $\bar{\pi}_\varepsilon = \{\pi_{\varepsilon,i}\}$. Our goal is to establish the relation

$$L_\varepsilon = \frac{dP_{\pi^\varepsilon}}{dP_0} = \exp(-\|\bar{\pi}_\varepsilon\|^2/2 + \|\bar{\pi}_\varepsilon\|t_\varepsilon + \delta_\varepsilon), \tag{4.70}$$

where $\|\bar{\pi}_\varepsilon\|$ is defined by (3.72),

$$t_\varepsilon \to \eta \sim N(0,1), \quad \delta_\varepsilon \to 0, \tag{4.71}$$

in P_0-probability.

Note that for $\|\bar{\pi}_\varepsilon\| = O(1)$, the relations (4.70), (4.71) imply the relations for the Bayesian problem

$$\beta(\alpha; P_0, P_{\pi^\varepsilon}) = \Phi(T_\alpha - \|\bar{\pi}_\varepsilon\|) + o(1), \quad \gamma(P_0, P_{\pi^\varepsilon}) = 2\Phi(-\|\bar{\pi}_\varepsilon\|/2) + o(1). \tag{4.72}$$

Proof of (4.72). It suffices to consider the case $\|\bar{\pi}_\varepsilon\| \asymp 1$ because if $\|\bar{\pi}_\varepsilon\| \to 0$, then relations (4.72) follow from Proposition 2.12 and (3.68). Assume $\|\bar{\pi}_\varepsilon\| \to u > 0$, without loss of generality. By the Neyman–Pearson Lemma 2.1

$$\beta(\alpha; P_0, P_{\pi^\varepsilon}) = E_{P_{\pi^\varepsilon}}(1 - \psi_{\varepsilon,\alpha}) = E_{P_0} e^{l_\varepsilon}(1 - \psi_{\varepsilon,\alpha}), \quad l_\varepsilon = \log L_\varepsilon,$$

where

$$1 - \psi_{\varepsilon,\alpha} = \begin{cases} 1, & \text{if } l_\varepsilon < t_{\varepsilon,\alpha}, \\ 0, & \text{if } l_\varepsilon > t_{\varepsilon,\alpha}, \end{cases}$$

and $t_{\varepsilon,\alpha}$ is a $(1-\alpha)$-quantile of statistic l_ε under P_0-probability. Because of (4.70), (4.71) one has $t_{\varepsilon,\alpha} \to -u^2/2 + uT_\alpha$, which implies the relation for $\beta(\alpha)$:

$$E_{P_0}((e^{l_\varepsilon}(1 - \psi_{\varepsilon,\alpha})) \to E\left(\exp(-u^2/2 + u\eta)\mathbf{1}_{\{\eta < T_\alpha\}}\right) = \Phi(T_\alpha - u)$$

(we can pass to limit under expectation since the expected function is bounded and continuous with the exception of the sets of P_0-probability tending to 0). The relation for γ follows from the relation for $\beta(\alpha)$. □

Assume $\pi^\varepsilon(V_\varepsilon) \to 1$. Then relations (4.70), (4.71) imply the lower bounds

$$\beta(\alpha; V_\varepsilon) \geq \Phi(T_\alpha - \|\bar{\pi}_\varepsilon\|) + o(1), \quad \gamma(V_\varepsilon) \geq 2\Phi(-\|\bar{\pi}_\varepsilon\|/2) + o(1). \tag{4.73}$$

In fact, if $\|\bar{\pi}_\varepsilon\| = O(1)$, then (4.73) follow from (4.72) and Proposition 2.11. If $\|\bar{\pi}_\varepsilon\| \to \infty$, then the lower bounds (4.73) are trivial.

For our purpose it is sufficient to establish (4.70), (4.71) for families of product priors π^ε with two-point and symmetric three-point factors $\{\pi_{\varepsilon,i}\}$, i.e.,

$$\pi_{\varepsilon,i} = (1 - h_{\varepsilon,i})\delta_0 + h_{\varepsilon,i}\delta_{z_{\varepsilon,i}}$$

or

$$\pi_{\varepsilon,i} = (1 - h_{\varepsilon,i})\delta_0 + h_{\varepsilon,i}(\delta_{z_{\varepsilon,i}} + \delta_{-z_{\varepsilon,i}})/2.$$

Denote

$$u_{\varepsilon,i} = \|\pi_{\varepsilon,i}\|, \quad u_\varepsilon^2 = \|\pi_\varepsilon\|^2 = \sum_i u_{\varepsilon,i}^2.$$

Introduce the assumption:

A.1. $w_\varepsilon = \sup_i u_{\varepsilon,i} = o(1)$; $u_\varepsilon^2 \asymp 1$.

For any $\delta > 0$, denote

$$J_{\varepsilon,\delta} = \{i \in I : \ e^{2z_{\varepsilon,i}^2} u_{\varepsilon,i} > \delta\}, \qquad u_{\varepsilon,\delta}^2 = \sum_{i \in J_{\varepsilon,\delta}} u_{\varepsilon,i}^2.$$

A.2. $u_{\varepsilon,\delta}^2 = o(1)$ *for any* $\delta > 0$.

A.3. *There exists* $\delta > 0$ *such that*

$$\sum_i u_{\varepsilon,i}^2 \exp(\delta z_{\varepsilon,i}^2) = O(u_\varepsilon^2).$$

A.4. $\sup_i |z_{\varepsilon,i}| = O(1)$.

Clearly, assumption A.4 implies A.3, assumption A.2 implies $w_\varepsilon = o(1)$ in assumptions A.1. Note that assumptions A.1 and A.3 imply A.2. In fact, under A.1, for any $\delta_1 > 0$,

$$\exp(2z_{\varepsilon,i}^2)w_\varepsilon \geq \exp(2z_{\varepsilon,i}^2)u_{\varepsilon,i} > \delta_1, \quad \text{if } \ i \in J_{\varepsilon,\delta_1}.$$

Therefore, for any $\delta > 0$, we have $\exp(\delta z_{\varepsilon,i}^2) \geq (\delta_1/w_\varepsilon)^{\delta/2}$, which yields, under A.1, A.3,

$$\sum_{i \in J_{\varepsilon,\delta_1}} u_{\varepsilon,i}^2 \leq (w_\varepsilon/\delta_1)^{\delta/2} \sum_i \exp(\delta z_{\varepsilon,i}^2) u_{\varepsilon,i}^2 = O(w_\varepsilon^{\delta/2} u_\varepsilon^2) = o(1). \qquad (4.74)$$

Moreover, it follows from A.2 that there exists a family $\delta_\varepsilon \to 0$ such that $u_{\varepsilon,\delta_\varepsilon}^2 \to 0$.

Proposition 4.4 *Assume either that* $u_\varepsilon = o(1)$ *or* $u_\varepsilon \asymp 1$ *and* A.2. *Then* (4.70), (4.71) *hold.*

Proof. If $u_\varepsilon \to 0$, then it follows from evaluation in (3.68), that $L_\varepsilon \to 1$ under P_0-probability, which yields (4.70), (4.71) in this case. Therefore, it suffices to assume $u_\varepsilon \asymp 1$ and A.2.

Consider the symmetric three-point factors $\pi_{\varepsilon,i} = \pi(h_i, z_i)$ (the case of two-point factors is considered analogously). Denote

$$g_{\varepsilon,i}(t) = h_i \xi(t, z_i),$$

where, as above,

$$\xi(t,z) = e^{-z^2/2}\cosh(tz) - 1.$$

Using A.2 take a family $\delta_\varepsilon \to 0$ such that $u^2_{\varepsilon,\delta_\varepsilon} \to 0$. Denote

$$I_{\varepsilon,1} = J_{\varepsilon,\delta_\varepsilon}, \quad I_{\varepsilon,2} = I \backslash I_{\varepsilon,1}.$$

Consider the priors π_1^ε, π_2^ε with the factors $\pi(h_i, z_i)$ for $i \in I_{\varepsilon,1}$ and $i \in I_{\varepsilon,2}$. Set $P_{\varepsilon,k} = P_{\pi_k^\varepsilon}$, $k = 1,2$. Then we have

$$L_\varepsilon = \prod_i (1 + g_{\varepsilon,i}(x_i)) = L_{\varepsilon,1}L_{\varepsilon,2}, \quad L_{\varepsilon,k} = \frac{dP_{\varepsilon,k}}{dP_0} = \prod_{i\in I_{\varepsilon,k}} (1 + g_{\varepsilon,i}(x_i)). \tag{4.75}$$

Analogously to the evaluations in Section 3.3.2, using A.2, we get

$$\begin{aligned} E_{P_0}(L_{\varepsilon,1} - 1)^2 &= \prod_{i\in I_{\varepsilon,1}} \int\int \exp(uv)\pi_{\varepsilon,i}(du)\pi_{\varepsilon,i}(dv) - 1 \\ &\le \exp\left(\sum_{i\in I_{\varepsilon,1}} \|\pi_{\varepsilon,i}\|^2\right) - 1 \sim u^2_{\varepsilon,\delta_\varepsilon} \to 0. \end{aligned}$$

Therefore $L_{\varepsilon,1} \to 1$ in P_0-probability and we can change the priors π^ε by π_2^ε, the statistics L_ε by $L_{\varepsilon,2}$, and the values u_ε^2 by

$$u^2_{\varepsilon,2} = \sum_{i\in I_{\varepsilon,2}} u^2_{\varepsilon,i} = u_\varepsilon^2 + O(u^2_{\varepsilon,\delta_\varepsilon}).$$

Thus we consider $i \in I_{\varepsilon,2}$ only, i.e., assume below that $e^{2z^2_{\varepsilon,i}}u_{\varepsilon,i} \le \delta_\varepsilon \to 0$.

Using (3.85) we observe

$$\begin{aligned} \sum_i E^4_{P_0} g_{\varepsilon,i}(x) &\le 60\sum_i h^4_{\varepsilon,i}\sinh^4(z^2_{\varepsilon,i}/2)e^{4z^2_{\varepsilon,i}} \\ &\le 15\sup_i(u^2_{\varepsilon,i}e^{4z^2_{\varepsilon,i}})\sum_i u^2_{\varepsilon,i} \le 15u_\varepsilon^2\delta_\varepsilon^2. \end{aligned} \tag{4.76}$$

Introduce the set

$$X_{\varepsilon,\tau} = \left\{x : \sup_i |g_{\varepsilon,i}(x_i)| < \tau\right\}$$

and take a positive family $\tau_\varepsilon \to 0$ such that $\tau_\varepsilon^{-2}\delta_\varepsilon u_\varepsilon \to 0$ (this is possible since $u_\varepsilon \asymp 1$). Using (4.76) and the Chebyshev inequality let us evaluate the P_0-probability of the complement of $X_{\varepsilon,\tau_\varepsilon}$:

$$P_0(\overline{X_{\varepsilon,\tau_\varepsilon}}) \le \sum_i P_0(|g_{\varepsilon,i}(x_i)| \ge \tau_\varepsilon)$$

$$\leq \tau_\varepsilon^{-4} \sum_i E_{P_0} g_{\varepsilon,i}(x_i)^4 \leq \tau_\varepsilon^{-4} u_\varepsilon^2 \delta_\varepsilon^2 \to 0.$$

Thus it suffices to consider $x \in X_{\varepsilon,\tau_\varepsilon}$. In this case, we have the expansion

$$\log(1 + g_{\varepsilon,i}(x_i)) = g_{\varepsilon,i}(x_i) - g_{\varepsilon,i}^2(x_i)/2 + \zeta_{\varepsilon,i} g_{\varepsilon,i}^2(x_i), \tag{4.77}$$

where $|\zeta_{\varepsilon,i}| \leq \zeta_\varepsilon \asymp \tau_\varepsilon \to 0$ as $\tau_\varepsilon \to 0$. Using (4.75), (4.77) we get

$$l_{\varepsilon,2} = \log L_{\varepsilon,2} = t_{\varepsilon,1} - t_{\varepsilon,2}/2 + t_{\varepsilon,3},$$

where

$$t_{\varepsilon,1} = \sum_i g_{\varepsilon,i}(x_i), \quad t_{\varepsilon,2} = \sum_i g_{\varepsilon,i}^2(x_i), \quad t_{\varepsilon,3} = \sum_i \zeta_{\varepsilon,i} g_{\varepsilon,i}^2(x_i). \tag{4.78}$$

Using (3.84) analogously with (4.76) we get

$$E_{P_0}|t_{\varepsilon,3}| \leq \zeta_\varepsilon \sum_i E_{P_0} g_{\varepsilon,i}^2(x_i) = \zeta_\varepsilon u_{\varepsilon,2}^2 \to 0;$$

$$E_{P_0} t_{\varepsilon,2} = u_{\varepsilon,2}^2; \quad \mathrm{Var}_{P_0} t_{\varepsilon,2} = \sum_i \mathrm{Var}_{P_0} g_{\varepsilon,i}^2(x_i) \leq \sum_i E_{P_0} g_{\varepsilon,i}^4(x_i) \to 0,$$

because of (4.76). These imply $t_{\varepsilon,3} \to 0$, $t_{\varepsilon,2} - u_\varepsilon^2 \to 0$ in P_0-probability. Finally, note that $l_\varepsilon = t_{\varepsilon,1}/u_\varepsilon \to \eta \sim N(0,1)$ under P_0-probability, because

$$E_{P_0}(g_{\varepsilon,i}(x_i)) = 0, \quad \sum_i \mathrm{Var}_{P_0}(g_{\varepsilon,i}(x_i)) = u_{\varepsilon,2}^2 \sim u_\varepsilon^2,$$

and by (4.76) the Lyapunov ratio tends to 0:

$$R_\varepsilon = \frac{\left(\sum_i E_{P_0}(g_{\varepsilon,i}^4(x))\right)^{1/2}}{\sum_i \mathrm{Var}_{P_0}(g_{\varepsilon,i}(x))} \leq B \frac{u_\varepsilon \delta_\varepsilon}{u_\varepsilon^2} \to 0. \qquad \square$$

We will need to consider the case $u_\varepsilon \to \infty$ and to evaluate the accuracy of the Gaussian approximation below. Rough estimations are sufficient for our aims.

Introduce the assumptions which modify A.1, A.2:

A.1a. $\liminf u_\varepsilon > 0$, $w_\varepsilon = o(1)$, *and* $u_\varepsilon = o(w_\varepsilon^{-a})$ *for any* $a > 0$.

A.2a. There exist $a > 0$, $b > 0$ such that for $\delta_\varepsilon = w_\varepsilon^a$ one has $u_{\varepsilon,\delta_\varepsilon} = o(w_\varepsilon^b)$.

It follows from evaluations (4.74) that A.1a and A.3 yield assumption A.2a with any $a \in (0,1)$, $b \in (0, (1-a)\delta/2)$, where $\delta > 0$ is the parameter from A.3.

Corollary 4.2

(1) *Assume* A.2 *and* $u_\varepsilon \asymp 1$. *Then the statistics*

$$t_\varepsilon = u_\varepsilon^{-1} \sum_i g_i(x_i), \quad \lambda_\varepsilon = (l_\varepsilon + u_\varepsilon^2/2)/u_\varepsilon$$

are asymptotically $(0,1)$*-Gaussian and* $t_\varepsilon - \lambda_\varepsilon \to 0$ *in* P_0*-probability; recall that* $l_\varepsilon = \log L_\varepsilon$ *is the Bayesian log-likelihood ratio.*

(2) *Assume* A.1a *and* A.2a. *Then there exist constants* $d = d(a,b) > 0$, $d_1 = d_1(a,b) > 0$, $d_2 = d_2(a,b) > 0$, *such that*

$$\sup_t |P_0(t_\varepsilon < t) - \Phi(t)| = o(w_\varepsilon^d); \quad P_0(|t_\varepsilon - \lambda_\varepsilon| > w_\varepsilon^{d_1}) = o(w_\varepsilon^{d_2}).$$

Proof of Corollary 4.2. This follows directly from the von Bahr–Essen inequality and from evaluations in the proof of Proposition 4.4. □

Remark. Let us consider product priors with n symmetric common two-point factors $\pi_i = \pi(z_n, 1) = (\delta_{z_n} + \delta_{-z_n})/2$, $i = 1, \dots, n \to \infty$ (here $h_i = 1$; other factors are δ_0). They are concentrated on the 2^n points $v = (\pm z_n, \dots, \pm z_n) \in R^n$; $z_n = n^{-1/2}\rho_n$; $|v|_2 = \rho_n$. Comparing with Proposition 2.15 we obtain the same asymptotics in the Bayesian problem as for spherical priors concentrated on $S^{n-1}(\rho_n)$. Thus these product priors provide the same sharp asymptotics that one has for spherical priors.

4.3.2 Symmetric Two-Point Factors and Tests of χ^2*-Type*

Under the sequence Gaussian model (2.1) let us test $H_0 : v = 0$ against $H_1 : v \in V_\varepsilon$. Consider the product priors π^ε with symmetric two-point factors ($h_i = 1$):

$$\pi_{\varepsilon,i} = \pi(z_{\varepsilon,i}, 1) = (\delta_{z_{\varepsilon,i}} + \delta_{-z_{\varepsilon,i}})/2, \tag{4.79}$$

where $\bar{z}_\varepsilon = \{z_{\varepsilon,i}\}$ is a family of nonnegative sequences. Assume

$$\sup_i z_{\varepsilon,i} = o(1), \quad \text{if } \sum_i z_{\varepsilon,i}^4 = O(1). \tag{4.80}$$

This assumption implies A.1 and A.4; therefore assumptions A.2 and A.3 hold as well.

Using (3.87) and (4.80) one has

$$\|\bar{\pi}_\varepsilon\|^2 \sim \sum_i z_{\varepsilon,i}^4/2 = u_\varepsilon^2(\bar{z}_\varepsilon).$$

Let $\pi^\varepsilon(V_\varepsilon) \to 1$. If $u_\varepsilon^2(\bar{z}_\varepsilon) \to 0$, then $\gamma(V_\varepsilon) \to 0$ (this follows from Proposition 2.12 and (3.68)) and if (4.80) holds, then

$$\beta(V_\varepsilon, \alpha) \geq \Phi(T_\alpha - u_\varepsilon(\bar{z}_\varepsilon)) + o(1), \quad \gamma(V_\varepsilon) \geq 2\Phi(-u_\varepsilon(\bar{z}_\varepsilon)/2) + o(1).$$

Thus to find the asymptotically best product priors with symmetric two-point factors (4.79) we get the problem: to minimize

$$u_\varepsilon^2(\bar{z}_\varepsilon) = \sum_i z_{\varepsilon,i}^4/2$$

under the asymptotic constraint $\pi^\varepsilon(V_\varepsilon) \to 1$.

Assume the sets V_ε to be *sign-symmetric: if* $v = \{v_i\} \in V_\varepsilon$, *then* $\tilde{v} = \{\pm v_i\} \in V_\varepsilon$ *for all changes of signs of coordinates.*

Note that the set under the constraints which are determined by power and Besov norms are sign-symmetric. For sign-symmetric sets V_ε the relation $\pi^\varepsilon(V_\varepsilon) = 1$ is equivalent to $\bar{z}_\varepsilon \in V_\varepsilon$ and we can rewrite the extreme problem

$$u_\varepsilon^2 = \frac{1}{2} \inf_{\bar{z}_\varepsilon \in V_\varepsilon} \sum_i z_{\varepsilon,i}^4. \tag{4.81}$$

If there exists an asymptotically extremal family $\bar{z}_\varepsilon \in V_\varepsilon$, such that

$$\frac{1}{2} \sum_i z_{\varepsilon,i}^4 \sim u_\varepsilon^2 + o(1)$$

and (4.80) holds, then

$$\beta(\alpha, V_\varepsilon) \geq \Phi(T_\alpha - u_\varepsilon) + o(1), \quad \gamma(V_\varepsilon) \geq 2\Phi(-u_\varepsilon/2) + o(1). \tag{4.82}$$

On the other hand, let us consider tests of χ^2-type $\psi_{\varepsilon,w_\varepsilon;T_\varepsilon}$ defined by (3.8) where

$$w_\varepsilon \in W = \left\{ w \, : \, w_i \geq 0, \, \sum_i w_i^2 = 1/2 \right\}.$$

Denote, as above

$$h_\varepsilon(w_\varepsilon) = \inf_{v \in V_\varepsilon} \sum_i w_{\varepsilon,i} v_i^2.$$

Suppose that if $h_\varepsilon(w_\varepsilon) = O(1)$, then $\sup_i w_{\varepsilon,i} = o(1)$. Then, using Corollary 3.3, one has

$$\beta(\alpha, V_\varepsilon) \leq \Phi(T_\alpha - h_\varepsilon(w_\varepsilon)) + o(1), \quad \gamma(V_\varepsilon) \leq 2\Phi(-h_\varepsilon(w_\varepsilon)/2) + o(1). \tag{4.83}$$

Therefore, to find the asymptotically best test families of χ^2-type we go to the maximin problem:

$$h_\varepsilon = \sup_{w_\varepsilon \in W} \inf_{v \in V_\varepsilon} \sum_i w_{\varepsilon,i} v_i^2. \tag{4.84}$$

Let $w_\varepsilon \in W$ be an asymptotically extremal family, such that

$$h_\varepsilon(w_\varepsilon) \sim h_\varepsilon + o(1).$$

If $h_\varepsilon \to \infty$, then $\gamma_\varepsilon(V_\varepsilon, \psi_{\varepsilon,w_\varepsilon;h_\varepsilon/2}) \to 0$ (this follows from Corollary 3.3 as well) and if the extreme family w_ε is such that

$$\sup_i w_{\varepsilon,i} = o(1), \quad \text{as } h_\varepsilon = O(1),$$

then (4.83) hold with $h_\varepsilon(w_\varepsilon) = h_\varepsilon$.

Set $u_i = v_i^2/\sqrt{2}$ and let $U_\varepsilon \subset l_+^1$ be an image of the set $V_\varepsilon \subset l^2$ under the mapping: $v \to u$. Let $r_{\varepsilon,i} = \sqrt{2}w_{\varepsilon,i}$. Then one can rewrite the maximin problem (4.84) and the extreme problem (4.81):

$$h_\varepsilon = \sup_{r \in l_+^2 : |r|=1} \inf_{u \in U_\varepsilon} \sum_i r_{\varepsilon,i} u_i; \quad u_\varepsilon = \inf_{u \in U_\varepsilon} |u|. \tag{4.85}$$

However, these problems are the same as the problems in (4.3), (4.1) (the additional constraint $r \in l_+^2$ in the maximin problem is not essential since $u \in l_+^1$). Therefore, using the considerations from Section 4.1, we obtain the following statement:

Proposition 4.5 *Let sets V_ε be sign-symmetric and let sets U_ε be convex. Assume that there exist minimum points $u_\varepsilon^* \in U_\varepsilon$ in the right-hand side extreme problem (4.85), $u_\varepsilon = |u_\varepsilon^*| > 0$, such that*

$$\sup_i u_{\varepsilon,i}^* = o(1), \qquad \text{if } u_\varepsilon \asymp 1. \tag{4.86}$$

Then $h_\varepsilon = u_\varepsilon$ in (4.85). These yield the sharp asymptotics of Gaussian type,

$$\beta(\alpha, V_\varepsilon) = \Phi(T_\alpha - u_\varepsilon) + o(1), \quad \gamma(V_\varepsilon) = 2\Phi(-u_\varepsilon/2) + o(1), \tag{4.87}$$

and χ^2-type tests with $w_\varepsilon = u_\varepsilon^*/(|u_\varepsilon^*|\sqrt{2})$ are asymptotically minimax.

Moreover, analogously with Proposition 4.5, we can weaken the assumption of convexity for U_ε:

Proposition 4.6 *Assume there exists a family of convex sets G_ε in a sequence space and a coordinate-wise mapping $\phi_\varepsilon : G_\varepsilon \to U_\varepsilon$, $u_i = \phi_{\varepsilon,i}(g_i)$, such that $\phi_\varepsilon(G_\varepsilon) = U_\varepsilon$ and $\phi_{\varepsilon,i}(t)$ are convex functions $\phi_{\varepsilon,i}(t) \geq 0$. Assume there exist $g_\varepsilon^* \in G_\varepsilon$ such that*

$$|\phi_\varepsilon(g_\varepsilon^*)| = \inf_{g \in G_\varepsilon} |\phi_\varepsilon(g)| > 0,$$

and $u_\varepsilon^* = \phi_\varepsilon(g_\varepsilon^*)$ satisfy (4.86). Then the statements of Proposition 4.5 hold.

If the derivatives of the functions $\phi_{\varepsilon,i}$ satisfy $\phi'_{\varepsilon,i}(t) \neq 0$ for $\phi_{\varepsilon,i}(t) \neq 0$, and $\phi_{\varepsilon,i}(t) = 0$ iff $t = 0$, then minimum points g_ε^*, u_ε^* are unique. In particular, consider the n-dimensional two-sided constraints

$$V_n = \{v \in R^n : |v|_p \geq \rho, \ |v|_q \leq R\}; \quad p \leq 2, \ p \leq q, \tag{4.88}$$

and assume $\rho n^{1/q-1/p} \leq R$ for V_n to be nonempty. Setting $\phi(g) = g^{2/p}$, $g \geq 0$, we obtain the convex set G_n. By symmetry of the problem the coordinates of the minimum point in the extreme problem are of the form $v_i^* = n^{-1/p}\rho$, $i = 1, \ldots, n$. Thus we improve Corollary 3.13.

Corollary 4.3 *Let us test $H_0 : v = 0$ against $H_1 : v \in V_n$, where V_n is defined by the two-sided constraints (4.88), $\rho = \rho_n \to \infty$, $R = R_n \geq$*

$\rho n^{1/q-1/p}$; $n \to \infty$. Then we have the sharp asymptotics of Gaussian type

$$\beta(\alpha, V_n) = \Phi(T_\alpha - u_n) + o(1), \quad \gamma(\alpha, V_n) = 2\Phi(-u_n/2) + o(1),$$

where $u_n^2 = n^{1-4/p}\rho_n^4/2$.

4.3.3 The Power Norms: $p \leq 2,\ q \geq p$

Consider the alternative V_ε defined by two ellipsoidal constraints

$$V_\varepsilon = \{v \in l^2 : |v|_{\bar{a},p} \geq \rho_\varepsilon, \ |v|_{\bar{b},q} \leq R_\varepsilon\}, \tag{4.89}$$

where $\bar{a}$, $\bar{b}$ are two positive sequences. Then V_ε is sign-symmetric and U_ε is defined by constraints

$$u \in l_+^1 : \ |u|_{\bar{a}^2,p/2} \geq \rho_\varepsilon^2/\sqrt{2}, \quad |u|_{\bar{b}^2,q/2} \leq R_\varepsilon^2/\sqrt{2}.$$

If $p \leq 2$, $q \geq 2$, then U_ε is convex. Moreover, if $p \leq 2$, $q \geq p$, then using convex functions $\phi(g) = g^{2/p}$, $g \geq 0$, we obtain the convex sets $G_\varepsilon \subset l_+^{2/p}$ defined by the constraints

$$G_\varepsilon = \left\{g \in l_+^{2/p} : |g|_{\bar{a}^p,1} \geq 2^{-p/4}\rho_\varepsilon^p, \ |g|_{\bar{b}^q,q/p} \leq 2^{-q/4}R_\varepsilon^q\right\}, \tag{4.90}$$

where $\bar{a}^p = \{a_i^p\}$, $\bar{b}^q = \{b_i^q\}$. Thus we obtain the extreme problem

$$u_\varepsilon^2 = \inf_{u \in U_\varepsilon} \sum_i u_i^2 = \inf_{g \in G_\varepsilon} \sum_i g_i^{4/p}. \tag{4.91}$$

However, this is the same as the extreme problems (4.13) with the changes

$$\tilde{p} = p/2,\ \tilde{q} = q/2,\ \tilde{\rho}_\varepsilon = 2^{-1/2}\rho_\varepsilon^2,\ \tilde{R}_\varepsilon = 2^{-1/2}R_\varepsilon^2,\ \tilde{a}_i = a_i^2\ \tilde{b}_i = b_i^2.$$

Consider the case $a_i = i^r$, $b_i = i^s$. The original extreme problem is

$$u_\varepsilon^2 = \frac{1}{2}\inf \sum_i z_i^4 : z_i \geq 0; \quad \sum_i (i^r z_i)^p \geq \rho_\varepsilon^p, \quad \sum_i (i^s z_i)^q \leq R_\varepsilon^q; \tag{4.92}$$

in terms of variables $u_i = z_i^2/\sqrt{2}$ this is of the form

$$u_\varepsilon^2 = \inf \sum_i u_i^2 : u_i \geq 0; \quad \sum_i (i^{2r}u_i)^{p/2} \geq 2^{-p/4}\rho_\varepsilon^p,$$

$$\sum_i (i^{2s}u_i)^{q/2} \leq 2^{-q/4}R_\varepsilon^q; \tag{4.93}$$

if $q = \infty$, then the constraint (4.93) is changed by

$$\sup_i i^{2s}u_i \leq 2^{-1/2}R_\varepsilon^2.$$

Denote, as above (see (3.118))

$$\Delta = \Delta(p, q, r, s) = sq(4-p) - rp(4-q).$$

Make the change of parameters

$$\tilde{p} = p/2, \quad \tilde{q} = q/2; \quad \tilde{r} = 2r; \quad \tilde{s} = 2s; \quad \tilde{\rho}_\varepsilon = 2^{-1/2}\rho_\varepsilon^2,$$
$$\tilde{R}_\varepsilon = 2^{-1/2}R_\varepsilon^2; \quad \tilde{\Delta} = \Delta(p,q,r,s)/2 = \Delta_+(\tilde{p},\tilde{q},\tilde{r},\tilde{s}). \tag{4.94}$$

After this change we obtain the extreme problem (4.15) which has been studied in Section 4.2. Therefore the solution of the extreme problem (4.92) corresponds to the solution of problem (4.15) after the change of parameters above.

We can solve problem (4.92) directly by repeating the considerations from Section 4.2. In this case, we have the relation for coordinates $z_{\varepsilon,i}$ of the extreme point $z_{\varepsilon,i} = z_0 z(i/\tilde{m})$ where the function $z(y),\ y > 0$, is defined by the equations: if $q < \infty$, then (compare with (4.22))

$$2z^{4-p} = py^{rp} - qy^{sq}z^{q-p}, \tag{4.95}$$

if $p = q,\ s > r$, then (compare with (4.23))

$$z(y) = (p(y^{rp} - y^{sp})/2)^{1/(4-p)}, \quad \text{if } y \in (0,1); \quad z(y) = 0, \quad \text{if } y \geq 1, \tag{4.96}$$

and if $q = \infty$, then (compare with (4.42))

$$z(y) = \begin{cases} y^{rp/(4-p)}, & \text{if } 0 < y < 1, \\ y^{-s}, & \text{if } y \geq 1. \end{cases} \tag{4.97}$$

On the other hand, using the change of parameters noted above, we obtain the equations for coordinates $u_{\varepsilon,i}$ of the extreme point in problem (4.93) (see (4.18)):

$$4u_i^{2-p/2} = (Api^{rp}/2 - Bqu_i^{q/2-p/2}i^{sq})_+, \tag{4.98}$$

where

$$u_i = u_{\varepsilon,i} = v_0 u(i/m) = z_{\varepsilon,i}^2/\sqrt{2}, \quad z_{\varepsilon,i} = z_0 z(i/\tilde{m}). \tag{4.99}$$

Let us choose m, v_0 such that to obtain equation (4.95) on $z = z(\tilde{y})$, $\tilde{y} = i/\tilde{m}$ from (4.98), (4.99). One can set

$$m = \tilde{m}2^{(p-q)/\Delta}; \ v_0 = z_0^2 2^{pq(s-r)/2\Delta}; \ v_0^{2-p/2}m^{-rp} = A, \ v_0^{2-q/2}m^{-sq} = B.$$

This leads to the relation

$$y = \tilde{y}\, 2^{-(p-q)/\Delta}, \quad u(y) = z^2(\tilde{y})2^{-2(sq-rp)/\Delta}.$$

The values c_l, $l = 0,1,2$, are changed by

$$\tilde{c}_0 = c_0 2^h, \quad \tilde{c}_1 = c_1^{1/2}2^{h/p}; \quad \tilde{c}_2 = c_2^{1/2}2^{h/q},$$

where $h = (pq(s-r) + p - q)/\Delta$.

Thus we can rewrite Theorem 4.1 (below we omit the tilde).

Theorem 4.2 *Let $p \le 2$, $q \ge p$,*

$$V_\varepsilon = \{v \in l^2 : |v|_{r,p} \ge \rho_\varepsilon, \ |v|_{s,q} \le R_\varepsilon\}, \quad r \ge r_p^* = 1/4 - 1/p,$$

and let $\rho_\varepsilon, R_\varepsilon$ be such that V_ε is nonempty. Then:

(1) If $s - r \le 1/p - 1/q$, then $\gamma(V_\varepsilon) \equiv 1$, i.e., the hypothesis testing problem is trivial.

(2) If $s - r > 1/p - 1/q$, then, as $R_\varepsilon \to \infty$, one has sharp asymptotics (4.87) of the Gaussian type, and the quantities u_ε are defined by the relation

$$u_\varepsilon^2 \sim \begin{cases} c_0 z_0^4 m, & \text{if } r > r_p^*, \\ C^4(p) z_0^4 m \log m/2, & \text{if } r = r_p^*, \end{cases} \tag{4.100}$$

where $C(p) = (p/2)^{1/(4-p)}$ and $m = m_\varepsilon$, $z_0 = z_{0,\varepsilon}$ satisfy

$$\rho_\varepsilon \sim \begin{cases} c_1 z_0 m^{r+1/p}, & \text{if } r > r_p^*, \\ C(p) z_0 m^{1/4} (\log m)^{1/p}, & \text{if } r = r_p^*, \end{cases} \qquad R_\varepsilon \sim c_2 z_0 m^{s+1/q}.$$

Here the functions $c_l = c_l(\kappa)$, $l = 0, 1, 2$, $\kappa = (r, s, p, q)$, are positive continuous in $(r, s, p, 1/q)$ over the region

$$0 < p \le 2, \quad p \le q, \quad r > r_p^*, \quad s - r > 1/p - 1/q.$$

The tests of χ^2-type, $\psi_\varepsilon = \psi_{\varepsilon, w_\varepsilon, T_\varepsilon}$ with $T_\varepsilon = T_\alpha$ and $T_\varepsilon = u_\varepsilon/2$, are asymptotically minimax, where

$$w_{\varepsilon,i} = z^2(i/m) \left(2 \sum_i z^4(i/m) \right)^{-1/2},$$

and, when $q < \infty$, the function $z(y)$, $y > 0$, is the solution of equations (4.95) or is of the forms (4.97). The coordinates of the minimum point are

$$z_{\varepsilon,i} = z_0 z(i/m), \quad u_{\varepsilon,i} = 2^{-1/2} z_{\varepsilon,i}^2.$$

Note that we exclude the assumption $R_\varepsilon/\rho_\varepsilon \to \infty$ of Theorem 4.1 in view of Remark 4 to Theorem 4.1.

For $r = 0$ the sharp asymptotics for power norms with $p = q = 2$ have been obtained by Ermakov [22]; the results for $p = q < 2$ have been studied in [51, 52]; the general case $r = 0$, $p \le 2$, $q \ge p$ was studied in [92].

It follows from Theorem 4.2 that, under given constraints on κ and for $r > r_p^*$, $s - r > 1/p - 1/q$, as $R_\varepsilon/\rho_\varepsilon \to \infty$, we have

$$u_\varepsilon^2 = u_\varepsilon^2(\kappa, \rho_\varepsilon, R_\varepsilon) \sim d(\kappa) \rho_\varepsilon^{A(\kappa)} R_\varepsilon^{-B(\kappa)},$$

where $d(\kappa) > 0$ and

$$A(\kappa) = \frac{4/q - 1 + 4s}{s - r + 1/q - 1/p}, \qquad B(\kappa) = \frac{4/p - 1 + 4r}{s - r + 1/q - 1/p}. \tag{4.101}$$

Jointly with Theorem 3.1, we obtain regions of different types of asymptotics which are presented in Figures 1 and 4 in Section 6.4 below. Direct relations for the function $c_l(\kappa)$ are given in (4.104) below.

Proof of Theorem 4.2. We need to control the assumption (4.86) only: $\sup_i z_{\varepsilon,i} = o(1)$, if $u_\varepsilon \asymp 1$ under the assumptions of (2).

One can rewrite relation (4.24) which describe the asymptotics of the function $z(y)$ for $p \neq q$:

$$z(y) \sim \begin{cases} C(p)y^{rp/(4-p)}, & \text{as } y \to 0, \\ C(p,q)y^{(rp-sq)/(q-p)}, & \text{as } y \to \infty, \end{cases} \tag{4.102}$$

where $C(p) = (p/2)^{1/(4-p)}$ and $C(p,q) = (p/q)^{1/(q-p)}$ (if $q = \infty$, then the analogous asymptotics hold). Using these relations for $p \neq q$ and (4.96) for $p = q$ one has

$$\sup_i z_{\varepsilon,i} \asymp \begin{cases} z_0, & \text{if } r \geq 0, \\ z_0 m^{-rp/(4-p)}, & \text{if } r_p^* \leq r < 0. \end{cases}$$

Let $u_\varepsilon \asymp 1$. Then, by $R_\varepsilon/\rho_\varepsilon \to \infty$ (see Remark 4 to Theorem 4.1) and by $s - r + 1/q - 1/p > 0$, one has: if $r > r_p^*$, then $4z_0 \asymp m^{-1/4}$, $R_\varepsilon/\rho_\varepsilon \asymp m^{s-r+1/q-1/p}$,4 therefore, $m \to \infty$ and $\sup_i z_{\varepsilon,i} = o(1)$. If $r = r_p^*$, then

$$z_0 \asymp (m \log m)^{-1/4}, \quad R_\varepsilon/\rho_\varepsilon \asymp m^{s-r+1/q-1/p}(\log m)^{-1/p}.$$

These imply

$$m \to \infty, \quad z_0 \to 0, \quad z_0 m^{-rp/(4-p)} \asymp \begin{cases} m^{-p(r-r_p^*)/(4-p)}, & \text{if } r > r_p^* \\ (\log m)^{-1/4}, & \text{if } r = r_p^* \end{cases} \to 0. \tag{4.103}$$

Therefore $\sup_i z_{\varepsilon,i} = o(1)$ as well. □

Note that for $r > r_p^*$, $q < \infty$ the functions $c_l = c_l(\kappa)$ are defined as integrals

$$c_1^p = \int_0^\infty z^p(y)y^{rp}\,dy, \tag{4.104}$$

$$c_2^q = \int_0^\infty z^q(y)y^{sq}\,dy, \tag{4.105}$$

$$c_0 = \frac{1}{2}\int_0^\infty z^4(y)\,dy, \tag{4.106}$$

where $z(y)$ is the function defined by (4.95) and integrals are finite. If $q = \infty$, then $z(y)$ is defined by (4.97) $c_2 = 1$ and c_1, c_0 are defined by (4.104) and (4.106). For $r = r_p^*$ the integrals (4.104) and (4.106) diverge. These imply logarithmic factors in (4.100).

One can easily check that analogous relations hold for this case, when the constraints on $|v|_{r,p}, |v|_{s,q}$ are changed by analogous constraints on $|v|_{\bar{a},p}, |v|_{\bar{b},q}$ with $a_i \sim i^r$, $b_i \sim i^s$. The equalities

$$w_{\varepsilon,i} = z^2(i/m)\left(2\sum_i z^4(i/m)\right)^{-1/2}$$

are changed by the asymptotic relations

$$w_{\varepsilon,i} \sim z^2(i/m)\left(2\sum_i z^4(i/m)\right)^{-1/2},$$

which are uniform over i.

One can translate the results to the functional Gaussian model (2.5). In particular, return to Examples 3.1, 3.3 (smooth signals in the L_2-norm). Here $S = \tilde{S}_{\sigma,2}(R)$ is the set of periodic σ-smooth functions in the L_2-norm of period 1 and $S_\varepsilon = \{s \in S : \ \|s\|_2 \geq \eta_\varepsilon\}$. By passing to the sequence Gaussian model (2.1) under a standard Fourier basis, we obtain the set $V_\varepsilon = V_\varepsilon \subset l^2$ of the considered type with

$$p = q = 2, \quad a_i = 1, \quad \rho_\varepsilon = r_\varepsilon/\varepsilon, \quad R_\varepsilon = \pi^{-\eta}R/\varepsilon, \quad b_i \sim i^\eta.$$

One can check that we obtain the same rates as in Examples 3.1, 3.3, however, additionally, we obtain sharp asymptotics of the Gaussian type here.

Let us consider the case of exponential semi-axes: $a_j = e^{rj}$, $b_j = e^{sj}$. Then, under the change (4.94), we get the same extreme problem as in Section 4.2.5. Certainly we can rewrite the results using the change (4.94), however, it is more convenient to write the results directly by repeating the considerations above. Thus, for $p \leq 2$, $q \geq p$, we obtain the asymptotics of the solution of the extreme problem

$$u_\varepsilon^2 = \frac{1}{2}\inf_{z \in V_\varepsilon} \sum_i z_j^4 \tag{4.107}$$

for the sets

$$V_\varepsilon = \left\{ z \in l^2 : \ \sum_j |z_j|^p e^{rpj} \geq \rho_\varepsilon^p, \ \sum_j |z_j|^q e^{sqj} \leq R_\varepsilon^q \right\} \tag{4.108}$$

with $p \leq 2$, $q \geq p$; $r \geq 0$ (if $q = \infty$, then the second constraint is changed by $\sup_j |v|e^{sj} \leq R_\varepsilon$) and $\rho_\varepsilon \leq R_\varepsilon$ for $p = q$, $s = r$.

Proposition 4.7

(1) If $s - r \leq 0$, then $u_\varepsilon = 0$, i.e., the hypothesis testing problem is trivial.

(2) If $s - r > 0$, $r \geq 0$, then, as $R_\varepsilon/\rho_\varepsilon \to \infty$, the quantities u_ε are defined by the relation

$$u_\varepsilon^2 \sim \begin{cases} c_0(j_0)z_0^4, & \text{if } r > 0, \\ \frac{1}{2}C^4(p)z_0^4 j_0, & \text{if } r = 0, \end{cases} \tag{4.109}$$

where $j_0 = j_{0,\varepsilon}$, $z_0 = z_{0,\varepsilon}$ satisfy

$$\rho_\varepsilon \sim \begin{cases} c_1(j_0)z_0 e^{rj_0}, & \text{if } r > 0, \\ C(p)z_0 j_0^{1/p}, & \text{if } r = 0, \end{cases} \qquad R_\varepsilon \sim c_2(j_0)z_0 e^{sj_0}. \tag{4.110}$$

Here $c_k(j_0) = c_k(j_0, \kappa)$, $k = 0, 1, 2$, *are positive functions in* $\kappa = (r, s, p, q)$ *and* 1*-periodic functions in* j_0. *The coordinates of minimum points* $z_\varepsilon = \{z_{\varepsilon,j}\}$ *in the extreme problem* (4.107), (4.108) *are of the form* $z_{\varepsilon,j} = z_0 z(e^{j-j_0})$, *where* $z(y)$, $y > 0$, *is the solution of equation* (4.95) *with* $y = e^{j-j_0}$ *for* $q < \infty$ *and is defined by* (4.97) *for* $q = \infty$.

One can check that the analogous asymptotics hold when $a_j \sim e^{rj}, b_j \sim e^{sj}$: the equalities $z_{\varepsilon,j} = z_0 z(e^{j-j_0})$ are changed by asymptotic relations

$$z_{\varepsilon,j} \sim z_0 z(e^{j-j_0}),$$

which are uniform over j.

If $r > 0$, then the 1-periodic functions $c_k(j_0) = c_k(j_0, \kappa)$, $k = 0, 1, 2$ are defined by series, for $q < \infty$,

$$c_1^p(j_0) = \sum_{l=-\infty}^{\infty} z^p(e^{l-j_0}) e^{rp(l-j_0)},$$

$$c_2^q(j_0) = \sum_{l=-\infty}^{\infty} z^q(e^{l-j_0}) e^{sq(l-j_0)},$$

$$c_0(j_0) = \frac{1}{2} \sum_{l=-\infty}^{\infty} z^4(e^{l-j_0}),$$

and $c_2(j_0) \equiv 1$ for $q = \infty$. Here $z(y)$ is the function defined by (4.95), (4.97), and the series converge.

If $r = 0$, then the series for $c_0(j_0)$, $c_1(j_0)$ diverge. These imply factors j_0, $j_0^{1/p}$ in (4.109), (4.110) with coefficients $c_1 = 1$, $c_0 = 1/2$.

Using Proposition 4.7 we obtain the rate asymptotics in the hypothesis testing problem; unfortunately, we have sharp asymptotics for $r = 0$ since the assumption (4.86) only holds in this case.

Corollary 4.4 *Assume* $p \le 2$, $q \ge p$.

(1) *Let* $s > r > 0$. *Then* $\gamma_\varepsilon(V_\varepsilon) \to 0$, *as* $\rho_\varepsilon / R_\varepsilon^{r/s} \to \infty$ *and* $\gamma_\varepsilon(V_\varepsilon) \to 1$, *as* $\rho_\varepsilon / R_\varepsilon^{r/s} \to 0$.

(2) *Let* $s > 0$, $r = 0$; $R_\varepsilon / \rho_\varepsilon \to \infty$. *Then we have sharp asymptotics of the Gaussian type* (4.87) *where (compare with* (4.69))

$$u_\varepsilon^2 \sim \rho_\varepsilon^4 (s^{-1} \log(R_\varepsilon / \rho_\varepsilon))^{1-4/p} / 2. \tag{4.111}$$

Proof. Let $r > 0$. Using (4.110) we get $u_\varepsilon \asymp z_0^2 \asymp (\rho_\varepsilon^s / R_\varepsilon^r)^{2/(s-r)}$. Therefore $u_\varepsilon \to 0$, as $\rho_\varepsilon / R_\varepsilon^{r/s} \to 0$ and $u_\varepsilon \to \infty$, as $\rho_\varepsilon / R_\varepsilon^{r/s} \to \infty$. Thus the statements of (1) follow from Proposition 2.12 and from Corollary 3.3.

To obtain sharp asymptotics in (2), since $R_\varepsilon / \rho_\varepsilon \to \infty$, using (4.110), we get $j_0 \to \infty$; using (4.109), we get $z_0 \asymp u_\varepsilon^{1/2} j_0^{-1/4} \to 0$, as $u_\varepsilon = O(1)$. Using relations (4.96), (4.102) one has $\sup_j z_{\varepsilon,j} \asymp z_0 = o(1)$, if $u_\varepsilon \asymp 1$. Thus (4.86) holds. The relation for u_ε follows from evaluations analogous to those which yield (4.69). □

Analogously to Remark (1) to Theorem 4.1 one can replace the relative accuracy by absolute accuracy in the above main relations.

One can extend the results to the functional Gaussian model (2.5). In particular, return to Examples 3.2, 3.4 (analytical signals). Here $S = \tilde{S}(\mu, H)$ is the set of 1-periodic real-valued functions $s(t)$ which have an analytical extension $s(z)$, $z = x + iy$, to the region in the complex plane $\{|y| \leq \mu/\pi\}$ such that (3.49) holds. By passing to the sequence Gaussian model (2.1) under the standard Fourier basis these correspond to the set $V_\varepsilon = V_\varepsilon \subset l^2$ of the considered type with

$$p = q = 2, \quad a_j = 1, \quad \rho_\varepsilon = r_\varepsilon/\varepsilon, \quad R_\varepsilon = H/\varepsilon, \quad b_j \sim e^{\mu j}.$$

One can check that we get the same sharp asymptotics as in Examples 3.2, 3.4.

4.3.4 Besov Norms: $p \leq 2$, $q \geq p$

Consider the Besov bodies cases

$$V_\varepsilon = \{v \in l^2 : |v|_{r,p,h} \geq \rho_\varepsilon, \ |v|_{s,q,t} \leq R_\varepsilon\}; \tag{4.112}$$

here the norms are defined by (2.72). To simplify the notation assume $J_0 = 0$, $K = 0$. Then we can rewrite the constraints in (4.112):

$$\sum_{j=1}^{\infty} 2^{hrj} \left(\sum_{i=1}^{2^j} |v_{ij}|^p \right)^{h/p} \geq \rho_\varepsilon^h, \quad \sum_{j=1}^{\infty} 2^{tsj} \left(\sum_{i=1}^{2^j} |v_{ij}|^q \right)^{t/q} \leq R_\varepsilon^t, \tag{4.113}$$

for q, $t < \infty$ with a modification of the second constraint

$$\sup_{1 \leq j < \infty} 2^{sj} \left(\sum_{i=1}^{2^j} |v_{ij}|^q \right)^{1/q} \leq R_\varepsilon, \quad \text{if } q < \infty, \quad t = \infty, \tag{4.114}$$

$$\sup_{1 \leq j < \infty} \sup_{1 \leq i \leq 2^j} 2^{sj} |v_{ij}| \leq \rho_\varepsilon, \quad \text{if } q = t = \infty, \tag{4.115}$$

(one can easily write the second constraint for $t < q = \infty$).

It is clear that V_ε is sign-symmetric and if $p \leq 2$, $q \geq p$, then using the change of variables $u_{ij} = v_{ij}^2/\sqrt{2}$, $g_{ij} = u_{ij}^{p/2}$, which correspond to the convex functions $\phi(g) = g^{2/p}$, $g \geq 0$, we obtain the sets $G_\varepsilon = \{g\} \subset l_+^1$ defined by the constraints

$$\sum_{j=1}^{\infty} 2^{hrj} \left(\sum_{i=1}^{2^j} 2^{p/4} g_{ij} \right)^{h/p} \geq \rho_\varepsilon^h, \quad \sum_{j=1}^{\infty} 2^{tsj} \left(\sum_{i=1}^{2^j} 2^{q/4} g_{ij}^{q/p} \right)^{t/q} \leq R_\varepsilon^t, \tag{4.116}$$

for q, $t < \infty$; for the cases $q < \infty$, $t = \infty$, and $q = t = \infty$, the second constraint is modified analogously to (4.114), (4.115).

Now observe that if $h \le p \le 2,\ t \ge q \ge p$, then the sets G_ε are convex and we can use the methods above. Thus we obtain an extreme problem

$$u_\varepsilon^2 = \frac{1}{2}\inf_{z\in V_\varepsilon}\sum_{j=1}^{\infty}\sum_{i=1}^{2^j} z_{ij}^4 = \inf_{g\in G_\varepsilon}\sum_{j=1}^{\infty}\sum_{i=1}^{2^j} g_{ij}^{4/p}. \tag{4.117}$$

It is clear from the symmetry that coordinates $z_{ij} = z_j$ of the extreme point do not depend on $i = 1,\dots,2^j$ for any j. We can rewrite (4.117) in terms of the unknown z_j:

$$u_\varepsilon^2 = \frac{1}{2}\inf\sum_{j=1}^{\infty} 2^j z_j^4 \,:\, \sum_{j=1}^{\infty} 2^{j(rh+h/p)}|z_j|^h \ge \rho_\varepsilon^h,\ \ \sum_{j=1}^{\infty} 2^{j(st+t/q)}|z_j|^t \le R_\varepsilon^t, \tag{4.118}$$

for $q,\ t < \infty$ with modification of the second constraint

$$\sup_{1\le j<\infty} 2^{j(s+1/q)}|z_j| \le R_\varepsilon, \quad \text{if } q \le \infty,\ t = \infty. \tag{4.119}$$

Changing the variables $x_j = 2^{j/4} z_j$, we rewrite (4.118), (4.119)

$$u_\varepsilon^2 = \frac{1}{2}\inf\sum_{j=0}^{\infty} x_j^4 \tag{4.120}$$

under the constraints

$$\sum_{j=1}^{\infty} 2^{jh(r+1/p-1/4)}|x_j|^h \ge \rho_\varepsilon^h,\quad \sum_{j=1}^{\infty} 2^{jt(s+1/q-1/4)}|x_j|^t \le R_\varepsilon^t; \tag{4.121}$$

for $q,\ t < \infty$ with modification of the second constraint

$$\sup_{1\le j<\infty} 2^{j(s+1/q-1/4)}|x_j| \le R_\varepsilon, \quad \text{if } q \le \infty,\ t = \infty. \tag{4.122}$$

However, this is the same extreme problem as for the ellipsoidal case with exponential semi-axes $a_j = e^{rj},\ b_j = e^{sj}$ after the change

$$\tilde p = h,\quad \tilde q = t,\quad \tilde r = (r + 1/p - 1/4)\log 2,\quad \tilde s = (s + 1/q - 1/4)\log 2. \tag{4.123}$$

Thus we can use Proposition 4.7 which describes the solution of the extreme problem (4.121), (4.122) and obtain asymptotics in original extreme problem. Certainly we can rewrite the results using the change (4.123); however, it is more convenient to write these directly by repeating the considerations above.

Proposition 4.8 *For extreme problem* (4.118), (4.119) *with* $h \le p,\ t \ge q$, *and* $r \ge r_p^* = 1/4 - 1/p$ *one has:*

(1) *If* $s - r \le 1/p - 1/q$, *then* $u_\varepsilon = 0$, *i.e., the hypothesis testing problem is trivial.*

(2) *If $s-r>1/p-1/q$, $r\ge r_p^*$, then, as $R_\varepsilon/\rho_\varepsilon\to\infty$, the quantities u_ε are determined by the relation*

$$u_\varepsilon^2 \sim \begin{cases} c_0(j_0)2^{j_0}z_0^4, & \text{if } r>r_p^*,\\ C^4(h)2^{j_0-1}z_0^4 j_0, & \text{if } r=r_p^*,\end{cases} \tag{4.124}$$

where $j_0=j_{0,\varepsilon}$, $z_0=z_{0,\varepsilon}$ satisfy

$$\rho_\varepsilon \sim \begin{cases} c_1(j_0)z_0 2^{(r+1/p)j_0}, & \text{if } r>r_p^*,\\ C(h)z_0 2^{(r+1/p)j_0}j_0^{1/h}, & \text{if } r=r_p^*,\end{cases} \qquad R_\varepsilon\sim c_2(j_0)z_0 2^{(s+1/q)j_0}. \tag{4.125}$$

Here $c_k(j_0)=c_k(j_0,\kappa,h,t)>0$, $k=0,1,2$, are 1-periodic functions in j_0 determined by the relations

$$c_1^h(j_0)=\sum_{l=-\infty}^{\infty} z^h(2^{l-j_0})2^{(l-j_0)(rh+h/p)},$$

$$c_2^t(j_0)=\sum_{l=-\infty}^{\infty} z^t(2^{l-j_0})2^{(l-j_0)(st+t/q)};$$

$$c_0(j_0)=\frac{1}{2}\sum_{l=-\infty}^{\infty} z^4(2^{l-j_0})2^{l-j_0},$$

and $c_2=1$, if $t=\infty$. Here the function $z(y)$, $y=2^{j-j_0}>0$, is the solution of the equations that are analogous to (4.95), (4.97): if $t<\infty$, then

$$2z^{4-h}=hy^{r_1h}-ty^{s_1t}z^{t-h},\quad r_1=r+1/p-1/h,\quad s_1=s+1/q-1/t,$$

and, for $t=\infty$, we have $r_1h+(4-h)(s+1/q)>0$ and

$$z(y)=\begin{cases} y^{r_1/(4-h)}, & \text{if } 0<y<1,\\ y^{-(s+1/q)}, & \text{if } y\ge 1.\end{cases}$$

Moreover, if $r=r_p^*$, then (compare with (4.111))

$$u_\varepsilon^2\sim\frac{1}{2}\rho^4(s_*^{-1}\log_2(R/\rho))^{1-4/h},\quad s_*=s+1/q-1/4, \tag{4.126}$$

and a periodical component is not essential in the asymptotics.

Coordinates of the minimum point in the extreme problem (4.118) are of the form $z_{\varepsilon,j}=z_0z(2^{j-j_0})$.

Denote $\kappa=(p,q,r,s)$, $\tau=(\kappa,h,t)$. It follows from (4.124), (4.125), and Lemma 4.1 that under the assumptions above on τ and for $r>r_p^*$, $s-r>1/p-1/q$, as $R_\varepsilon\to\infty$, we have

$$u_\varepsilon^2=u_\varepsilon^2(\tau,\rho_\varepsilon,R_\varepsilon)\sim d(j_0,\tau)\rho_\varepsilon^{A(\kappa)}R_\varepsilon^{-B(\kappa)}, \tag{4.127}$$

where (compare with (4.101))

$$A(\kappa)=\frac{4/q-1+4s}{s-r+1/q-1/q},\quad B(\kappa)=\frac{4/p-1+4r}{s-r+1/q-1/p}, \tag{4.128}$$

and $d(j_0, \tau) = d_1(\log_2(R_\varepsilon/\rho_\varepsilon), \tau)$ where $d_1(x, \tau)$ is a periodic continuous function in $x = \log_2(R_\varepsilon/\rho_\varepsilon)$ of the period $(s - r + 1/q - 1/p)$.

Analogously to Remark (1) to Theorem 4.1 one can replace the relative accuracy by absolute accuracy in the above main relations.

Observe the following. Up to constants, the asymptotics of the solutions of the extreme problem for Besov norm, given in Proposition 4.8, do not depend on h, t for $h \le p$, $t \ge q$, and are the same that for the norms of power type which have been described in Theorem 4.2 (one can put $m = 2^{j_0}$). In particular, assumption (4.86) holds (this was proved in the proof of Theorem 4.2). Therefore, we get

Theorem 4.3 *Let $h \le p$, $t \ge q$, $r \ge r_p^*$, and the sets V_ε are defined by the Besov norms* (4.112). *Then:*

(1) *If $s - r \le 1/p - 1/q$, then $\gamma(V_\varepsilon) \equiv 1$, i.e., the hypothesis testing problem is trivial.*

(2) *If $s - r > 1/p - 1/q$, then, as $R_\varepsilon/\rho_\varepsilon \to \infty$, one has sharp asymptotics of the Gaussian type* (4.87) *and the quantities u_ε are defined by* (4.126), (4.127).

Remark. The methods above are not applicable to obtaining sharp asymptotics for $h > p$ or $t < q$ since the sets G_ε are not convex. However, we have the same rates of testing. In fact, in view of the inequalities

$$|v|_{r,p,h} \le |v|_{r,p,h'}, \quad |v|_{s,q,t} \ge |v|_{s,q,t'} \quad \text{for } h \le h', \quad t \ge t',$$

we have the smaller sets V_ε or G_ε in this case. Therefore, it suffices to obtain the same rates in the lower bounds. It suffices to consider the case $u_\varepsilon \asymp 1$. In this case, one can easily check that for all h, t the sets V_ε contain n dimensional l_q^n-balls of radius R_ε with l_p^n-balls of radius ρ_ε removed in an $n = 2^j$-subspace corresponding to j-resolution, $j \asymp j_0$. Therefore the lower bounds follow from Corollaries 3.13 or 4.3.

Thus, Theorem 4.3 jointly with Theorem 3.2, leads to regions of different types of asymptotics presented in Figures 1 and 4 in Section 6.4 for Besov norms as well as for the power norms above.

4.3.5 Smooth Signals: Case $p \le 2$, $q \ge p$

Using the wavelet transform and the relationship (2.84) between Besov norms in functional and sequence spaces, as well as relationship (2.80) between Sobolev and Besov norms, we can extend the rate asymptotics which follow from Theorem 4.3 and the Remark above, to the functional Gaussian model (2.5).

In fact, analogously with Section 3.4.8, consider the set of signals $S_\varepsilon \subset L_2[0, 1]$ defined by constraints (2.91):

$$\|s\|_{(\sigma,p)} \ge r_\varepsilon, \quad \|s\|_{(\eta,q)} \le R, \tag{4.129}$$

where $\|\cdot\|_{(\sigma,p)}$, $\|\cdot\|_{(\eta,q)}$ are either Sobolev norms $\|\cdot\|_{\sigma,p}$, $\|\cdot\|_{\eta,q}$ or Besov norms $\|\cdot\|_{\sigma,p,h}$, $\|\cdot\|_{\eta,q,t}$ with any $h,\ t > 0$. Assume

$$\sigma \geq 0, \quad \sigma > 1/p - 1, \quad \eta \geq 0, \quad \eta > 1/q - 1.$$

Then, using the m-regular wavelet transform for large enough m and the change

$$r = \sigma + 1/2 - 1/p, \quad s = \eta + 1/2 - 1/q; \quad \rho_\varepsilon = r_\varepsilon/\varepsilon, \quad R_\varepsilon = R/\varepsilon,$$

by (2.84), we obtain set V_ε such that

$$V_{b,\varepsilon} \subset V_\varepsilon \subset V_{b^{-1},\varepsilon}, \tag{4.130}$$

where $V_{b,\varepsilon} = V(\kappa, h, t; b\rho_\varepsilon, R_\varepsilon/b)$ and $b = b(\kappa, h, t) \in (0,1)$. In view of the relations from Proposition 4.8, we get the inequalities

$$cu_{b^{-1},\varepsilon} \leq u_\varepsilon \leq c^{-1}u_{b,\varepsilon}, \quad c = c(b,\kappa) \in (0,1), \tag{4.131}$$

where $u_{b,\varepsilon} = u(\kappa, b\rho_\varepsilon, R_\varepsilon/b)$, $u_\varepsilon = u_{1,\varepsilon}$ and c is a constant. Since $V_{b,\varepsilon}$ are the sets from Theorem 4.3 with radii that are different in constant factors, we get the inequalities

$$\Phi(cu_\varepsilon - T_\alpha) + o(1) \leq \beta_\varepsilon(\alpha, S_\varepsilon) \leq \Phi(c^{-1}u_\varepsilon - T_\alpha) + o(1), \tag{4.132}$$

$$2\Phi(cu_\varepsilon) + o(1) \leq \gamma_\varepsilon(S_\varepsilon) \leq 2\Phi(c^{-1}u_\varepsilon) + o(1). \tag{4.133}$$

Recall that if $\eta - \sigma \leq 0$, then the problem is trivial: $\gamma(S_\varepsilon) \equiv 1$ (see Theorem 3.8 above).

Theorem 4.4 *Under the functional Gaussian model* (2.5) *let the sets* S_ε *be defined by* (4.129) *and let*

$$p \leq 2, \quad q \geq p, \quad \eta > \sigma > 1/p - 1, \quad \sigma \geq 0.$$

Then, as $r_\varepsilon \to 0$, *one has* (4.132), (4.133) *with some constant* c *and with the quantities* u_ε *determined by asymptotic relations*

$$u_\varepsilon^2 \asymp nz_0^4; \quad r_\varepsilon \asymp \varepsilon z_0 n^{(\sigma+1/2)}, \quad R \asymp \varepsilon z_0 n^{\eta+1/2}. \tag{4.134}$$

These imply rate asymptotics (2.63), (2.64) *with the rate*

$$r_\varepsilon^* \asymp \varepsilon^{(\eta-\sigma)/(\eta+1/4)}. \tag{4.135}$$

If $u_\varepsilon \to \infty$, *then the asymptotically consistent tests could be of* χ^2*-type. These are based on statistics*

$$t_{\varepsilon,w_\varepsilon} = \sum_j \sum_{i=1}^{2^j} w_{\varepsilon,j}(x_{ij}^2 - 1), \quad x_{ij} = \varepsilon^{-1} \int_0^1 \varphi_{ij}(t)\, dX_\varepsilon(t), \tag{4.136}$$

where the sequence $w_\varepsilon = \{w_{\varepsilon,j}\}$ *is defined by Proposition 4.8 with* $n = 2^{j_0}$ *and* $\{\varphi_{ij}\}$ *is an* m*-regular wavelet basis in* $L_2(0,1)$, $m > \eta + 1$.

The rates (4.135) extend the rates from Theorem 3.9 for $\sigma = 0,\ 1 < p \leq 2,\ q \geq p$; formally the case $p = 1$ is not covered by Theorem 4.4; however the same rates follow from Theorem 3.9.

4.4 Sharp Asymptotics: Asymptotic Degeneracy

In this section we obtain sharp asymptotics for alternatives determined by the power norms and rates for Besov norms with $p > q,\ sq \leq rp$. In this case orthogonal priors are asymptotically least favorable and supreme-tests are asymptotically minimax. The degeneracy means that the Bayesian likelihood ratio tends to nonrandom constant in this case.

4.4.1 *Asymptotic Degeneracy in Bayesian Problems*

Here we study the asymptotic distribution of the Bayesian likelihood ratio for orthogonal priors (see Section 3.3.7).

Let $\bar{v}_\varepsilon = \{v_{\varepsilon,j}\},\ j \in J_\varepsilon$, be a family of collections of orthogonal vectors $v_{\varepsilon,j} \in l^2$ and let the sets V_ε consist of the points $v_{\varepsilon,j},\ j \in J_\varepsilon$, from these collections. Let us test $H_0 : v = 0$ against $H_1 : v \in V_\varepsilon$. Denote $u_{\varepsilon,j} = |v_{\varepsilon,j}|$ and consider the family of priors

$$\pi_\varepsilon = \pi(\bar{v}_\varepsilon, \bar{p}_\varepsilon) = \sum_{j\in J_\varepsilon} p_{\varepsilon,j}\delta_{v_{\varepsilon,j}}$$

for a family of probability vectors $\bar{p}_\varepsilon = \{p_{\varepsilon,j},\ j \in J_\varepsilon\}$, i.e.,

$$p_{\varepsilon,j} \geq 0, \quad \sum_{j\in J_\varepsilon} p_{\varepsilon,j} = 1.$$

The likelihood ratio is of the form

$$L_\varepsilon(x) = \frac{dP_{\pi_\varepsilon}}{dP_0}(x) = \sum_{j\in J_\varepsilon} p_{\varepsilon,j}\zeta(\eta_{\varepsilon,j}, u_{\varepsilon,j}),$$

where

$$\zeta(\eta, u) = \exp(-u^2/2 + u\eta), \quad \eta_{\varepsilon,j} = (x, v_{\varepsilon,j})/u_{\varepsilon,j} \text{ are i.i.d. } N(0,1)$$

in P_0-probability. The items in the sum are independent, however we cannot use the Central Limit Theorem here (one can check that the Lyapunov ratio does not tend to 0). This is essential because the real limit distribution is often not Gaussian, but degenerate: this means

$$L_\varepsilon = C_\varepsilon + o(1) \quad \text{in } P_0\text{-probability}, \tag{4.137}$$

where $C_\varepsilon \in [0,1]$ is a nonrandom family.

Proposition 4.9 *Assume* (4.137). *Then one has asymptotic relations for* $\beta_\varepsilon(\alpha) = \beta(\alpha; P_0, P_{\pi_\varepsilon}),\ \gamma_\varepsilon = \gamma(P_0, P_{\pi_\varepsilon})$ *in the Bayesian problem:*

$$\beta_\varepsilon(\alpha) = (1-\alpha)(C_\varepsilon + o(1)), \quad \gamma_\varepsilon = C_\varepsilon(1 + o(1)), \tag{4.138}$$

i.e., we have the asymptotic degeneracy in the Bayesian problem. Since $\pi_\varepsilon(V_\varepsilon) = 1$, *this implies*

$$\beta(\alpha, V_\varepsilon) \geq (1-\alpha)(C_\varepsilon + o(1)), \quad \gamma(V_\varepsilon) \geq C_\varepsilon(1 + o(1)). \tag{4.139}$$

Proof. It suffices to establish the first relation in (4.138). Let $\psi_{\varepsilon,\alpha}$ be the Neyman–Pearson test for testing $H_0 : P = P_0$ against $H_1 : P = P_{\pi_\varepsilon}$ which is based on the likelihood ratio L_ε; recall that

$$\alpha(\psi_{\varepsilon,\alpha}, P_0) = E_{P_0}\psi_{\varepsilon,\alpha} = \alpha; \quad \beta(\alpha) = E_{P_1}(1 - \psi_{\varepsilon,\alpha}) = E_{P_0}(1 - \psi_{\varepsilon,\alpha})L_\varepsilon.$$

Since $L_\varepsilon = C_\varepsilon + o(1)$, one can see that the statistics $L_\varepsilon(x)$ are uniformly bounded over the set $\{x : \psi_{\varepsilon,\alpha}(x) < 1\}$, and we can pass to limits under the expectations

$$\beta(\alpha) = E_{P_0}(1-\psi_{\varepsilon,\alpha})L_\varepsilon = (C_\varepsilon + o(1))E_{P_0}(1-\psi_{\varepsilon,\alpha}) = (1-\alpha)(C_\varepsilon + o(1)). \quad \square$$

Now we give conditions for relation (4.137) to hold true.

Proposition 4.10 *Denote* $u_\varepsilon = \inf_{j\in J_\varepsilon} u_{\varepsilon,j}$. *Let* D_ε *be a family such that*

$$w_\varepsilon = u_\varepsilon - D_\varepsilon \to \infty, \quad \sum_{j\in J_\varepsilon} \exp(-w_{\varepsilon,j}^2/2) \asymp 1; \quad w_{\varepsilon,j} = u_{\varepsilon,j} - D_\varepsilon.$$

Consider the family of probability vectors $\bar{p}_\varepsilon$,

$$p_{\varepsilon,j} = \exp(-w_{\varepsilon,j}^2/2)\left(\sum_{j\in J_\varepsilon} \exp(-w_{\varepsilon,j}^2/2)\right)^{-1}, \quad j \in J_\varepsilon.$$

Then $L_\varepsilon = \Phi(-D_\varepsilon) + o(1)$ *in* P_0*-probability. These imply* (4.138), (4.139) *with* $C_\varepsilon = \Phi(-D_\varepsilon)$.

Proof. Introduce the w-truncated statistics

$$\zeta_w(\eta, u) = \zeta(\eta, u)\mathbf{1}_{\{\eta<w\}}, \quad L_{\varepsilon,w} = \sum_j p_{\varepsilon,j}\zeta_{w_{\varepsilon,j}}(\eta_{\varepsilon,j}, u_{\varepsilon,j})$$

(here and below in the proof the suprema and sums are taken over $j \in J_\varepsilon$). Observe that

$$\begin{aligned} P_0(L_\varepsilon \neq L_{\varepsilon,w}) &\le \sum_j P_0(\eta_j \ge w_{\varepsilon,j}) \\ &= \sum_j \Phi(-w_{\varepsilon,j}) \asymp \sum_j \exp(-w_{\varepsilon,j}^2/2)/w_{\varepsilon,j} \to 0; \end{aligned}$$

here and below we use the well-known relation (see (3.35)):

$$\Phi(-x) \sim \exp(-x^2/2)/x\sqrt{2\pi}, \quad x \to \infty. \tag{4.140}$$

Therefore it suffices to consider the truncated statistics and to prove that

$$L_{\varepsilon,w} = \Phi(-D_\varepsilon) + o(1) \tag{4.141}$$

in P_0-probability. Let $D_\varepsilon \le O(1)$. Then $u_\varepsilon = \inf_j u_{\varepsilon,j} \to \infty$ and equality (4.141) follows from the relations

$$E(L_{\varepsilon,w}) = \Phi(-D_\varepsilon), \quad \mathrm{Var}(L_{\varepsilon,w}) \to 0. \tag{4.142}$$

If $D_\varepsilon \to \infty$, then (4.141) follows from the first equality in (4.142).

Let us prove (4.142) assuming $u_\varepsilon \to \infty$ for the second relation. Using the equalities

$$E\zeta_w(\eta, u) = \Phi(w-u), \quad E\zeta_w^2(\eta, u) = \exp(u^2)\Phi(w-2u), \tag{4.143}$$

we obtain the first relation in (4.142) directly. Next,

$$\begin{aligned}\operatorname{Var}(L_{\varepsilon,w}) &\le \sum_i p_{\varepsilon,j}^2 E\zeta_{w_{\varepsilon,j}}^2(\eta_{\varepsilon,j}, u_{\varepsilon,j}) \\ &\asymp \sum_j p_{\varepsilon,j}\exp(u_{\varepsilon,j}^2 - w_{\varepsilon,j}^2/2)\Phi(w_{\varepsilon,j} - 2u_{\varepsilon,j}) \\ &\le \sup_j \exp(u_{\varepsilon,j}^2 - w_{\varepsilon,j}^2/2)\Phi(w_{\varepsilon,j} - 2u_{\varepsilon,j}).\end{aligned}$$

By considering differently the cases $w_{\varepsilon,j} > 3u_{\varepsilon,j}/2$ and $w_{\varepsilon,j} \le 3u_{\varepsilon,j}/2$, we get

$$\begin{aligned}&\exp(u_{\varepsilon,j}^2 - w_{\varepsilon,j}^2/2)\Phi(w_{\varepsilon,j} - 2u_{\varepsilon,j}) \\ &\le \begin{cases}\exp(u_{\varepsilon,j}^2 - w_{\varepsilon,j}^2/2), & \text{if } w_{\varepsilon,j} > 3u_{\varepsilon,j}/2, \\ Bu_{\varepsilon,j}^{-1}\exp(-(u_{\varepsilon,j} - w_{\varepsilon,j})^2), & \text{if } w_{\varepsilon,j} \le 3u_{\varepsilon,j}/2,\end{cases} = o(1),\end{aligned}$$

which implies the second relation in (4.142). □

Remark. One can extend Proposition 4.10 to the case when the priors are asymptotically semiorthogonal. This follows from the proof of Lemma 7.2 below.

Let us compare Proposition 4.10 with Corollary 3.10 which give upper bounds based on the properties of supreme-tests. Observing the analogy between H_ε and D_ε, between $T_{\varepsilon,j}$ and $w_{\varepsilon,j}$, we combine these statements in the following:

Corollary 4.5 *Let a real-valued family D_ε be given.*

(1) *Let $\bar{r}_\varepsilon = \{r_{\varepsilon,j}\}$ be a family of collections $r_{\varepsilon,j} \in l^2$, $|r_{\varepsilon,j}| = 1$ such that*

$$\inf_{v\in V_\varepsilon} \sup_j (|(r_{\varepsilon,j}, v)| - w_{\varepsilon,j}) \ge D_\varepsilon,$$

where $\bar{w}_\varepsilon = \{w_{\varepsilon,j}\}$ is a family of real-valued collections such that

$$\inf_j w_{\varepsilon,j} \to \infty; \quad \sum_j \exp(-w_{\varepsilon,j}^2/2) \asymp 1. \tag{4.144}$$

In particular, let the set V_ε be such that

$$\inf_{v\in V_\varepsilon} \sup_{i\in I} (|v_i| - w_{\varepsilon,i}) \ge D_\varepsilon,$$

where v_i are coordinates of the point $v \in l^2$. Then one has

$$\beta(\alpha, V_\varepsilon) \le (1-\alpha)(\Phi(-D_\varepsilon) + o(1)), \quad \gamma(V_\varepsilon) \le \Phi(-D_\varepsilon) + o(1).$$

(2) *Let e_i, $i \in I$, be the standard basis in l^2, the sets V_ε contain all points from the collections $\bar{v}_\varepsilon = \{u_{\varepsilon,i}e_i\}$, $i \in I_\varepsilon \subset I$, and relation* (4.144) *holds with $w_{\varepsilon,i} = u_{\varepsilon,i} - D_\varepsilon$; the sums and the infima are taken over $i \in I_\varepsilon$. Then*

$$\beta(\alpha, V_\varepsilon) \geq (1-\alpha)(\Phi(-D_\varepsilon) + o(1)), \quad \gamma(V_\varepsilon) \geq \Phi(-D_\varepsilon) + o(1).$$

In particular, let $I = \mathbf{N}$, $u_{\varepsilon,i} = u_\varepsilon$ for $1 \leq i \leq n_\varepsilon \to \infty$, $D_\varepsilon = u_\varepsilon - \sqrt{2\log n_\varepsilon}$. Set

$$w_{\varepsilon,i} = \begin{cases} \sqrt{2\log n_\varepsilon}, & \text{if } i \leq n_\varepsilon, \\ \infty, & \text{if } i > n_\varepsilon. \end{cases}$$

Then

$$\beta(\alpha, V_\varepsilon) \geq (1-\alpha)\Phi(\sqrt{2\log n_\varepsilon} - u_\varepsilon) + o(1), \ \gamma(V_\varepsilon) \geq \Phi(\sqrt{2\log n_\varepsilon} - u_\varepsilon) + o(1). \tag{4.145}$$

We can improve Proposition 3.12. By taking

$$u_{\varepsilon,i} = \rho_n, \ 1 \leq i \leq n; \quad D_\varepsilon = \rho_n - \sqrt{2\log n},$$

we get [1]

Proposition 4.11 *For both alternatives V_n or V_n^+, defined by* (3.95) *or* (3.96) *and for all $0 < \rho_n \leq R_n \leq \infty$, $0 < q \leq \infty$, one has*

$$\beta_n(\alpha) = (1-\alpha)\Phi(\sqrt{2\log n} - \rho_n) + o(1), \quad \gamma_n = \Phi(\sqrt{2\log n} - \rho_n) + o(1).$$

Thus, we have the asymptotics of degenerate type in the problem.

4.4.2 Degenerate Asymptotics for Power Norms

Consider the alternatives

$$V_\varepsilon = V_\varepsilon(\kappa, \rho_\varepsilon, R_\varepsilon) = \{v \in l^2 : \ |v|_{r,p} \geq \rho_\varepsilon, \ |v|_{s,q} \leq R_\varepsilon\},$$

for the region of parameters $\kappa = (r, s, p, q)$,

$$\Xi_D = \{\kappa : \ s > r \geq 0, \ \lambda = s/p - r/q \leq 0\}. \tag{4.146}$$

Our aim is to show that, as $R_\varepsilon \to \infty$, $R_\varepsilon/\rho_\varepsilon \to \infty$, we have degenerate asymptotics for $\kappa \in \Xi_D$. Degenerate asymptotics are characterized by the relations

$$\beta(\alpha, V_\varepsilon) = (1-\alpha)\Phi(-D_\varepsilon) + o(1), \tag{4.147}$$

where

$$\begin{aligned} D_\varepsilon = D_\varepsilon(\kappa, \rho_\varepsilon, R_\varepsilon) &= n_\varepsilon^{-r}\rho_\varepsilon - \sqrt{2\log n_\varepsilon}, \\ n_\varepsilon = n_\varepsilon(\kappa, \rho_\varepsilon, R_\varepsilon) &= (R_\varepsilon/\rho_\varepsilon)^{1/(s-r)} \to \infty. \end{aligned} \tag{4.148}$$

[1] This result was obtained in [51]; see also [52].

It follows from the definition of the set Ξ_D that the equality $r = 0$ is possible for $p = \infty$ only and one has either $p > q$ or $p = q = \infty$.

We can write the direct relations

$$\beta(\alpha, V_\varepsilon) = (1-\alpha)\Phi\left(\sqrt{\frac{2\log(R_\varepsilon/\rho_\varepsilon)}{s-r}} - \rho_\varepsilon^{s/(s-r)} R_\varepsilon^{-r/(s-r)}\right) + o(1). \tag{4.149}$$

Introduce the rates $\rho_\varepsilon^* = \rho_\varepsilon^*(\kappa, R_\varepsilon, \rho_\varepsilon)$ by the relation

$$\rho_\varepsilon^* = R_\varepsilon^{r/s}\left(\log(R_\varepsilon/\rho_\varepsilon^*)\right)^{(s-r)/2s}, \tag{4.150}$$

and set $\Lambda = \Lambda(\kappa) = (2/(s-r))^{(s-r)/2s}$. This corresponds to

$$\Lambda^{s/(s-r)}\rho_\varepsilon^* \sim n_\varepsilon^r\sqrt{2\log n_\varepsilon}, \quad n_\varepsilon = n_\varepsilon(\kappa, \rho_\varepsilon^*, R_\varepsilon).$$

Relation (4.149) allows the frontier between distinguishability and nondistinguishability, for any $\alpha \in (0,1)$,

$$\beta(\alpha, V_\varepsilon) \to 0, \quad \text{if } \liminf \rho_\varepsilon/\rho_\varepsilon^* > \Lambda \tag{4.151}$$

and

$$\beta(\alpha, V_\varepsilon) \to 1-\alpha, \quad \text{if } \limsup \rho_\varepsilon/\rho_\varepsilon^* < \Lambda. \tag{4.152}$$

One can write a small different direct relation for ρ_ε^*,

$$\rho_\varepsilon^* = R_\varepsilon^{r/s}(\log(R_\varepsilon))^{(s-r)/2s} \tag{4.153}$$

which corresponds to a different relation for the constant $\Lambda = (2/s)^{(s-r)/2s}$ in relations (4.151)–(4.152).

It is important that there exists a *common asymptotically minimax test procedure which does not depend on the parameter* $\kappa \in \Xi_D$. This is of the form

$$\psi_{\varepsilon,\alpha} = (1-\alpha)\mathbf{1}_{X_\varepsilon} + \alpha \tag{4.154}$$

and based on the thresholding

$$X_\varepsilon = \left\{\max_{1\le i\le N_\varepsilon} |x_i| > \sqrt{2\log N_\varepsilon}\right\} \bigcup \left\{\sup_{N_\varepsilon < i < \infty} |x_i|/T_i > 1\right\} \tag{4.155}$$

with $T_i = \sqrt{2\log i + 2\log\log i}$ and $N_\varepsilon \asymp \log\log R_\varepsilon \to \infty$.

Theorem 4.5 *Let* $\kappa \in \Xi_D$ *and let* $R_\varepsilon/\rho_\varepsilon \to \infty$, $R_\varepsilon \to \infty$. *Then:*

(1) $\beta(\alpha, V_\varepsilon) \ge (1-\alpha)\Phi(-D_\varepsilon) + o(1)$.

(2) *For the test family* (4.154), (4.155) *one has* $\alpha(\psi_{\varepsilon,\alpha}) = \alpha + o(1)$, *and uniformly over* $\kappa \in K$, *for any compact* $K \subset \Xi_D$,

$$\beta(\psi_{\varepsilon,\alpha}, V_\varepsilon) \le (1-\alpha)\Phi(-D_\varepsilon) + o(1).$$

Here the quantities $D_\varepsilon = D_\varepsilon(\kappa, \rho_\varepsilon, R_\varepsilon)$ *are defined by* (4.148).

These relations imply the sharp asymptotics (4.147) of degenerate type in the problem and asymptotic minimaxity of the tests (4.154), (4.155). [2]

Proof of the Lower Bounds

The lower bounds of Theorem 4.5 follow from the relation: if $\kappa \in \Xi_D$ and $n_\varepsilon = (R_\varepsilon/\rho_\varepsilon)^{1/(s-r)}$, then

$$\beta(\alpha, V_\varepsilon) \geq (1-\alpha)\Phi(\sqrt{2\log n_\varepsilon} - n_\varepsilon^{-r}\rho_\varepsilon) + o(1). \tag{4.156}$$

To prove (4.156) we can assume

$$n_\varepsilon^{-r}\rho_\varepsilon = O(\sqrt{2\log n_\varepsilon}) \tag{4.157}$$

(if $n_\varepsilon^{-r}\rho_\varepsilon/\sqrt{2\log n_\varepsilon} \to \infty$, then the lower bounds are trivial). Introduce the family of orthogonal collections $\bar{v}_\varepsilon = \{v_k = u_{\varepsilon,k}e_k,\ n_1 \leq k \leq n\}$, where e_k is the standard orthonormal basis in l^2 and

$$n = n_\varepsilon, \quad n_1 = n_{1,\varepsilon} = n(1 - 1/\log n), \quad u_{\varepsilon,k} = k^{-r}\rho_\varepsilon.$$

Clearly, $\bar{v}_\varepsilon \subset V_\varepsilon$ and, using Corollary 4.5 (2), we obtain the inequality

$$\beta(\alpha, V_\varepsilon) \geq (1-\alpha)\Phi(-D_\varepsilon) + o(1), \tag{4.158}$$

where D_ε are quantities such that

$$\min_{n_1 \leq k \leq n}(u_{\varepsilon,k} - D_\varepsilon) \to \infty, \qquad \sum_{k=n_1}^{n} \exp(-(u_{\varepsilon,k} - D_\varepsilon)^2/2) \asymp 1. \tag{4.159}$$

Put $D_\varepsilon = n_\varepsilon^{-r}\rho_\varepsilon - \sqrt{2\log n_\varepsilon} + \delta_\varepsilon$. Then relations (4.157)–(4.158) imply (4.156), if we could choose a $\delta_\varepsilon \to 0$ such that (4.159) holds. Clearly, the possibility follows from the relations that for any $\delta > 0$,

$$\sum_{n_1 \leq k \leq n} \exp(-(u_{\varepsilon,k} - D_\varepsilon - \delta)^2/2) \to \infty, \tag{4.160}$$

$$\sum_{n_1 \leq k \leq n} \exp(-(u_{\varepsilon,k} - D_\varepsilon + \delta)^2/2) \to 0. \tag{4.161}$$

Since $u_{\varepsilon,k} - D_\varepsilon = \sqrt{2\log n} + o(1)$ uniformly over $n_1 \leq k \leq n$, one can easily check relations (4.160), (4.161). Relation (4.156) and Theorem 4.5 (1) are proved. □

Proof of the Upper Bounds

Let us consider the tests (4.154), (4.155). Since

$$\alpha(\psi_{\varepsilon,\alpha}) = \alpha + (1-\alpha)P_0(X_\varepsilon), \quad \beta(\psi_{\varepsilon,\alpha}, v) = (1-\alpha)P_v(\bar{X}_\varepsilon),$$

---

[2] This result was obtained in [62].

we need to show that, uniformly over $\kappa \in K \subset \Xi_D$,

$$P_0(X_\varepsilon) \to 0, \quad \sup_{v \in V_\varepsilon} P_v(\overline{X_\varepsilon}) \leq \Phi\left(\sqrt{2\log n_\varepsilon} - n_\varepsilon^{-r}\rho_\varepsilon\right) + o(1), \quad (4.162)$$

where $\overline{X_\varepsilon}$ is a complement of X_ε,

$$V_\varepsilon = V_\varepsilon(\kappa, \rho_\varepsilon, R_\varepsilon), \quad n = n_\varepsilon(r, s, \rho_\varepsilon, R_\varepsilon) = (R_\varepsilon/\rho_\varepsilon)^{1/(s-r)} \to \infty,$$

since $s > r \geq 0$ for $\kappa \in \Xi_D$ ($r > 0$ for $p < \infty$). Without loss of generality one can assume $n_\varepsilon > N_\varepsilon$ (in the opposite case one can make ρ_ε smaller to obtain this inequality).

Let us consider the properties of the thresholding (4.155). Using relation (4.140) we have the first relation in (4.162):

$$\begin{aligned} P_0(X_\varepsilon) &\leq 2N_\varepsilon \Phi(-\sqrt{2\log N_\varepsilon}) + 2\sum_{i=N_\varepsilon}^{\infty} \Phi(-T_i) \\ &\asymp (\log N_\varepsilon)^{-1/2} + \sum_{i=N_\varepsilon}^{\infty} i^{-1}(\log i)^{-3/2} \to 0. \end{aligned}$$

Let $v \in V_\varepsilon$ and let

$$\tilde{T}_i = \begin{cases} \sqrt{2\log N_\varepsilon}, & \text{if } i \leq N_\varepsilon, \\ T_i, & \text{if } i > N_\varepsilon. \end{cases}$$

Then, since T_i increases in i and by $n_\varepsilon > N_\varepsilon$, we get

$$P_v(\overline{X_\varepsilon}) \leq \inf_i (\Phi(\tilde{T}_i - |v_i|) - \Phi(-\tilde{T}_i - |v_i|)) \leq \min_{i \leq n} \Phi(T_n - |v_i|)$$

and, since $T_n = \sqrt{2\log n} + o(1)$, the second relation in (4.162) follows from

Lemma 4.2 *Let*

$$v \in V_\varepsilon = V_\varepsilon(\kappa, R_\varepsilon, \rho_\varepsilon), \quad n = n_\varepsilon(\kappa, R_\varepsilon) - (R_\varepsilon/\rho_\varepsilon)^{1/(s-r)},$$

and let $s > r \geq 0$, $p \geq q$, $sq - rp \leq 0$ (note that these assumptions hold for $\kappa \in \Xi_D$). Then

$$\inf_{v \in V_\varepsilon} \max_{i \leq n} |v_i| \geq \rho_\varepsilon n^{-r}.$$

Proof of Lemma 4.2. To simplify, assume $q < p < \infty$ (the case $\infty = p \geq q$ is simpler). First, note that

$$\rho_\varepsilon^p \leq \sum_i (i^r|v_i|)^p \leq \sup_i (i^{rp-sq}|v_i|^{p-q}) \sum_i (i^s|v_i|)^q \leq \sup_i i^{rp-sq}|v_i|^{p-q} R_\varepsilon^q,$$

which imply

$$\sup_i i^{rp-sq}|v_i|^{p-q} \geq \rho_\varepsilon^p / R_\varepsilon^q. \quad (4.163)$$

Next, by $|v|_{s,q} \leq R_\varepsilon$ we have $|v_i| \leq R_\varepsilon i^{-s}$ and, by definition of n, we have for $i > n$,

$$i^{rp-sq}|v_i|^{p-q} \leq i^{p(r-s)} R_\varepsilon^{p-q} < \rho_\varepsilon^p / R_\varepsilon^q.$$

Therefore inequality (4.163) is strong and the supremum is attained in a $i \le n$. These relations imply

$$n^{rp-sq} \max_{i\le n} |v_i|^{p-q} \ge \max_{i\le n} i^{rp-sq} |v_i|^{p-q} > \rho_\varepsilon^p / R_\varepsilon^q.$$

Thus we have the inequality of Lemma 4.2,

$$\max_{i\le n} |v_i| \ge n^{(sq-rp)/(p-q)} (\rho_\varepsilon / R_\varepsilon)^{q/(p-q)} \rho_\varepsilon = n^{-r} \rho_\varepsilon.$$

Lemma 4.2 and Theorem 4.5 are proved. □

4.4.3 Degenerate Asymptotics for Besov Norms

Consider the alternatives determined by Besov norms in the sequence space

$$V_\varepsilon = V_\varepsilon(\tau, \rho_\varepsilon, R_\varepsilon) = \{v \in l^2 : \ |v|_{r,p,h} \ge \rho_\varepsilon, \ |v|_{s,q,t} \le R_\varepsilon\}; \quad \tau = (\kappa, h, t),$$

for the region of parameters $\kappa \in \Xi_D$ of degenerate asymptotics with additional constraints

$$\lambda = s/p - r/q < 0, \quad \text{if } pt > qh.$$

Recall that $\lambda \le 0$, $0 \le r < s$ for $\kappa \in \Xi_D$. Our aim is to show that we have two-sided bounds of the type

$$(1-\alpha)\Phi(-D_{\varepsilon,c_1}) + o(1) \le \beta(\alpha, V_\varepsilon) \le (1-\alpha)\Phi(-D_{\varepsilon,c_2}) + o(1), \quad (4.164)$$

where $c_1(\tau) \ge 1 \ge c_2(\tau) > 0$ and

$$D_{\varepsilon,c} = D_{\varepsilon,c}(\tau, \rho_\varepsilon, R_\varepsilon) = c n_\varepsilon^{-r} \rho_\varepsilon - \sqrt{2 \log n_\varepsilon}, \quad (4.165)$$

$$n_\varepsilon = n_\varepsilon(r, s, \rho_\varepsilon, R_\varepsilon) = (R_\varepsilon / \rho_\varepsilon)^{1/(s-r)}. \quad (4.166)$$

Note that we can take $c_1 = 1$, if $(R_\varepsilon/\rho_\varepsilon)^{s-r} = 2^{j_0}$ with an integer j_0 or $r = 0$ (the last equality is possible for $p = \infty$). For other cases the constant $c_1 > 1$ is defined below. Also, if $pt \le qh$, then we can take $c_2 = 1$. If $pt > qh$, then the constant $c_2 \in (0,1)$ is defined below.

Relation (4.164) leads to the distinguishability conditions

$$\beta(\alpha, V_\varepsilon) \to 0, \quad \text{if } \liminf \rho_\varepsilon / \rho_\varepsilon^* > \Lambda_1 \quad (4.167)$$

and

$$\beta(\alpha, V_\varepsilon) \to 1 - \alpha, \quad \text{if } \limsup \rho_\varepsilon / \rho_\varepsilon^* < \Lambda_2 \quad (4.168)$$

with the same rates (critical radii) ρ_ε^* which were defined by (4.153); however, if $c_1 > c_2$, then the constants Λ_1, Λ_2 are different.

There exist minimax consistent test procedures which do not depend on $\kappa \in \Xi_D$, t, h,

$$\psi_{\varepsilon,\alpha} = (1-\alpha)\mathbf{1}_{X_\varepsilon} + \alpha \quad (4.169)$$

which are based on the thresholding

$$X_\varepsilon = \left\{ \max_{1\le j\le J_\varepsilon} \max_{1\le l\le 2^j} |x_{lj}| > \sqrt{2CJ_\varepsilon} \right\} \bigcup \left\{ \sup_{J_\varepsilon<j<\infty} \max_{1\le l\le 2^j} |x_{lj}|/T_j > 1 \right\}, \tag{4.170}$$

where $C = \log 2,\ T_j = \sqrt{2C(j+\log j)},\ J_\varepsilon \asymp \log\log R_\varepsilon \to \infty$.

Theorem 4.6 *Let $\kappa \in \Xi_D$ and if $a = qh/pt < 1$, then $\lambda < 0$; let $R_\varepsilon/\rho_\varepsilon \to \infty$, $R_\varepsilon \to \infty$. Then:*

(1) *There exists a constant $c_1 \ge 1$, such that*

$$\beta(\alpha, V_\varepsilon) \ge (1-\alpha)\Phi(-D_{\varepsilon,c_1}) + o(1).$$

If either $j_0 = (s-r)^{-1}\log_2(R_\varepsilon/\rho_\varepsilon)$ is an integer or $r = 0$, then one can take $c_1 = 1$.

(2) *There exists a constant $0 < c_2 \le 1$ such that for the test family (4.169), (4.170) one has $\alpha(\psi_{\varepsilon,\alpha}) = \alpha + o(1)$ and, uniformly over $\kappa \in K$ for any compact $K \subset \Xi_D$, one has*

$$\beta(\psi_{\varepsilon,\alpha}, V_\varepsilon) \le (1-\alpha)\Phi(-D_{\varepsilon,c_2}) + o(1).$$

If $a = qh/pt \ge 1$, then one can take $c_2 = 1$.

Here the quantities $D_{\varepsilon,c} = D_{\varepsilon,c}(\tau, \rho_\varepsilon, R_\varepsilon)$ are defined by (4.165) with the quantities $n = n_\varepsilon(r, r, \rho_\varepsilon, R_\varepsilon)$ defined by relation (4.166).

The relations of Theorem 4.6 imply the rate asymptotics (4.167), (4.168) of degenerate type and minimax consistency of the tests (4.169), (4.170). [3]

Proof of the Lower Bounds

Let $j_0 = j_{0,\varepsilon}$ be the integral part of the quantity $(s-r)^{-1}\log_2(R_\varepsilon/\rho_\varepsilon)$ and let $\delta = \delta_\varepsilon$ be the fractional part of this. Set $\tilde n_\varepsilon = 2^{j_0} = n_\varepsilon 2^{-\delta}$ where n_ε is defined by relation (4.166). The lower bounds of Theorem 4.6 follow from the relation, if $\kappa \in \Xi_D$,

$$\beta(\alpha, V_\varepsilon(\tau, R_\varepsilon, \rho_\varepsilon)) \ge (1-\alpha)\Phi(\sqrt{2\log \tilde n_\varepsilon} - \tilde n_\varepsilon^{-r}\rho_\varepsilon) + o(1). \tag{4.171}$$

By $n_\varepsilon \to \infty$ we have $\sqrt{2\log n_\varepsilon} = \sqrt{2\log \tilde n_\varepsilon} + o(1)$ and relation (4.171) yields the needed relation with $c_{1,\varepsilon} = 2^{r\delta}$. Therefore, we can take $c_1 = 1$ for $r\delta = 0$.

To prove (4.171) let us consider the resolution j_0 and the family of orthogonal collections $\bar v_\varepsilon = \{v_i = u_{\varepsilon,j_0} e_{j_0,i},\ 1 \le i \le 2^{j_0}\}$, where $e_{j,i},\ 1 \le i \le 2^j$, is a standard orthonormal basis in j-resolution and $u_{\varepsilon,j_0} = 2^{-j_0 r}\rho_\varepsilon = \tilde n^{-r}\rho_\varepsilon$. Clearly, all points from the collection $\bar v_\varepsilon$ are contained in the set V_ε, and using Proposition 4.11 (with the change ρ_ε by $\tilde\rho_\varepsilon = 2^{-j_0 r}\rho_\varepsilon$) we get the inequality

$$\beta(\alpha, V_\varepsilon) \ge (1-\alpha)\Phi(\sqrt{2\log \tilde n_\varepsilon} - \tilde n^{-r}\rho_\varepsilon) + o(1).$$

---

[3] This result was obtained in [62].

These considerations show that we can take $c_1 = 2^r$ in (4.171) which do not depend on R_ε, ρ_ε. Relation (4.171) and Theorem 4.6 (1) are proved. □

Proof of the Upper Bounds

The consideration in this case is analogous to the above: we need the relations

$$P_0(X_\varepsilon) \to 0, \quad \sup_{v \in V_\varepsilon} P_v(\overline{X_\varepsilon}) \leq \Phi(\sqrt{2\log n_\varepsilon} - c_2 n_\varepsilon^{-r} \rho_\varepsilon) + o(1). \quad (4.172)$$

The first relation in (4.172) is obtained analogously to the above. To obtain the second relation we use the considerations analogous to the proof of the upper bounds in Section 4.4.2. For simplicity, assume $p, q, t, h < \infty$. Without loss of generality one can assume $n_\varepsilon > \log R_\varepsilon$ (in the opposite case one can make ρ_ε smaller to obtain this inequality).

It suffices to prove the following:

Lemma 4.3 *Let $v \in V_\varepsilon = V_\varepsilon(\tau, \rho_\varepsilon, R_\varepsilon)$, and let $s > r > 0$, $p > q$, $sq - rp \leq 0$, and $sq - rp < 0$, if $a = qh/pt < 1$. Then there exists a constant $d \in (0,1]$, $c = d^{s/(s-r)}$ such that*

$$\inf_{v \in V_\varepsilon} \max_{j \leq j_0} \max_{1 \leq i \leq 2^j} |v_{ij}| \geq c\rho_\varepsilon n^{-r}.$$

Here j_0 is defined by the relation $d2^{j_0(s-r)} = R_\varepsilon/\rho = n^{s-r}$ where $n = n_\varepsilon$ is defined by (4.166). *Moreover, if $a = qh/pt \geq 1$, then $c = d = 1$*

Proof of Lemma 4.3. Let $v \in V_\varepsilon$. For a positive sequence $\{d_j\}$ (which is determined concretely below) we have

$$\begin{aligned}\rho_\varepsilon^h &\leq \sum_{j=1}^{\infty} 2^{jrh} \left(\sum_{i=1}^{2^j} |v_{ij}|^p \right)^{h/p} \\ &\leq \left(\sup_{1 \leq j < \infty} 2^{(rp-sq)j} d_j^{-1} \max_{1 \leq i \leq 2^j} |v_{ij}|^{p-q} R_\varepsilon^q \right)^{h/p} \sum_{j=1}^{\infty} d_j^{h/p} x_j^a,\end{aligned}$$

where

$$x_j = \left(2^{jsq} \sum_{i=1}^{2^j} |v_{ij}|^q / R_\varepsilon^q \right)^{t/q}; \quad \sum_{j=1}^{\infty} x_j \leq 1$$

(the last inequality follows from $|v|_{s,q,t} \leq R_\varepsilon$). This implies, for any j_0,

$$\sup_j 2^{(rp-sq)(j-j_0)} d_j^{-1} \max_i |v_{ij}|^{p-q} \geq 2^{-(rp-sq)j_0} \rho_\varepsilon^p R_\varepsilon^{-q} \left(\sum_j d_j^{h/p} x_j^a \right)^{-p/h}.$$

Let $a \geq 1$. Set $d_j = 1$, $d = 1$ and, by $\sum_j x_j^a \leq 1$, we have

$$\sup_j 2^{(rp-sq)(j-j_0)} \max_i |v_{ij}|^{p-q} \geq 2^{-(rp-sq)j_0} \rho_\varepsilon^p / R_\varepsilon^q = (2^{-rj_0}\rho_\varepsilon)^{p-q},$$

and by

$$\sup_j 2^{sj} \max_i |v_{ij}| \leq |v|_{s,q,t} \leq R_\varepsilon, \tag{4.173}$$

analogously to the proof of Lemma 4.2, we have

$$\sup_{j>j_0} 2^{(rp-sq)(j-j_0)/(p-q)} \max_i |v_{ij}| < 2^{-rj_0}\rho_\varepsilon.$$

These relations imply the required inequality with $c = 1$.

Let $a < 1$, $\lambda < 0$. For any integer j, $-\infty < j < \infty$, set

$$d_j = \begin{cases} 2^{(rp-sq)(j-j_0)}, & \text{if } j \leq j_0, \\ 2^{p(r-s)(j-j_0)}, & \text{if } j > j_0. \end{cases}$$

Applying the Hölder inequality one has

$$\begin{aligned} \sum_{j=1}^{\infty} d_j^{h/p} x_j^a &\leq \left(\sum_{j=1}^{\infty} x_j\right)^a \left(\sum_{j=1}^{\infty} d_j^{h/(p(1-a))}\right)^{1-a} \\ &< \left(\sum_{j\leq j_0} 2^{(j-j_0)h(rp-sq)/p(1-a)} + \sum_{j>j_0} 2^{(j-j_0)h(r-s)/(1-a)}\right)^{1-a} \\ &= b(j_0). \end{aligned}$$

The sums in the final parentheses are taken over integers j, $-\infty < j < \infty$. The last equality defines the function $b(j_0)$. This is periodic function in j_0 of period 1. Set

$$b = \max_{j_0 \in [0,1]} b(j_0) > 1, \quad d = b^{-1/h} \in (0,1).$$

Let us take $\tilde{n}_\varepsilon = 2^{j_0}$ according to the relation $d\tilde{n}_\varepsilon^{s-r} = R_\varepsilon/\rho_\varepsilon$ which corresponds to $\tilde{n}_\varepsilon = nd^{-1/(s-r)}$. Then we get

$$\max\left\{\max_{j\leq j_0}\max_i |v_{ij}|, \ \sup_{j>j_0} 2^{(j-j_0)s} \max_i |v_{ij}|\right\} > d2^{-j_0 r}\rho_\varepsilon,$$

and, by (4.173), we have

$$\sup_{j>j_0} 2^{(j-j_0)s} \max_i |v_{ij}| \leq 2^{-sj_0} R_\varepsilon = d2^{-rj_0}\rho_\varepsilon.$$

Thus we get

$$\max_{j\leq j_0}\max_i |v_{ij}| > d2^{-rj_0}\rho_\varepsilon = cn^{-r}\rho_\varepsilon, \quad c = d^{s/(s-r)}.$$

Lemma 4.3 and Theorem 4.6 are proved. □

4.4.4 Degenerate Asymptotics for Smooth Signals

Using the wavelet transform and relationship (2.84) between Besov norms in functional and sequence spaces, as well as relationship (2.80) between Sobolev and Besov norms, we can extend the rate asymptotics of the degenerate type which follow from Theorem 4.6 to the functional Gaussian model (2.5).

Let the set of signals $S_\varepsilon \subset L_2(0,1)$ be determined by constraints (4.129). Analogously with Section 4.3.5 we get

Theorem 4.7 *Assume*

$$\sigma \geq \frac{1}{p} - \frac{1}{2}, \quad \sigma \geq 0, \quad \eta - \sigma > \frac{1}{q} - \frac{1}{p}, \quad p > q, \quad \frac{\eta}{p} - \frac{\sigma}{q} \leq \frac{1}{2q} - \frac{1}{2p}; \tag{4.174}$$

if $pt > qh$, *then* $\eta/p - \sigma/q < 1/2q - 1/2p$ *(these correspond to* $\kappa \in \Xi_D$ *and if* $pt > qh$, *then* $\lambda = s/p - r/q < 0$*); let* $r_\varepsilon \to 0$, $\varepsilon \to 0$. *Then:*

(1) *There exists a constant* $c_1 > 0$ *such that*

$$\beta(\alpha, S_\varepsilon) \geq (1-\alpha)\Phi(-D_{\varepsilon,c_1}) + o(1).$$

(2) *There exists a constant* $c_2 > 0$ *such that for the test family* (4.169), (4.170) *with* x_{ij} *defined by* (4.136) *one has* $\alpha(\psi_{\varepsilon,\alpha}) = \alpha + o(1)$, *and*

$$\beta(\psi_{\varepsilon,\alpha}, S_\varepsilon) \leq (1-\alpha)\Phi(-D_{\varepsilon,c_2}) + o(1).$$

Here the quantities $D_{\varepsilon,c}$ *are determined by the relations*

$$D_{\varepsilon,c} = c n_\varepsilon^{-(\sigma+1/2-1/p)} r_\varepsilon/\varepsilon - \sqrt{2\log n_\varepsilon},$$
$$R/r_\varepsilon = n_\varepsilon^{\eta-\sigma+1/p-1/q}.$$

These relations imply the rate asymptotics analogous to (4.167), (4.168) *with the rates*

$$r_\varepsilon^* = \left(\varepsilon\sqrt{\log(\varepsilon^{-1})}\right)^{(\eta-\sigma+1/p-1/q)/(\eta+1/2-1/q)} \tag{4.175}$$

and some constants $\Lambda_1 > 0$, $\Lambda_2 > 0$, *as well as asymptotic consistency for the test family* (4.169), (4.170), *as* $\liminf r_\varepsilon/r_\varepsilon^* > \Lambda_1$.

Remember that the test family (4.169), (4.170) does not depend on the parameters σ, η, p, q which determine alternative S_ε.

Observe that the rates r_ε^* defined by (4.175) in region (4.174) are the same as the rates r_ε^{est} determined by (2.88) in the estimation problem.

For the case $\sigma = 0$ these rates were obtained in [91].

It was shown by Lepski [76] and by Lepski and Tsybakov [80] that $\Lambda_1 = \Lambda_2$ in the relations (4.167), (4.168) for the alternative defined by the Hölder norm with the L_∞-ball removed (this corresponds to the case $\sigma = 0$, $\eta > 0$, and $p = q = \infty$).

5

Sharp Asymptotics. II

In this chapter we continue the study of hypothesis testing problems for alternatives determined by power and Besov norms. This is based on duality which corresponds to Hilbert structure introduced in Section 3.3.3. To obtain the duality we study the minimax properties of tests based on the approximation of log-likelihood statistics for product priors with two- or three-point factors jointly with thresholding.

Therefore to find the asymptotically least favorable priors of product type we find an extreme problem: to minimize the Hilbert norm on the space of the sequences of measures. For a wide enough class of alternatives the extreme problem is modified and separated: we show that extreme sequences consist of two- or three-point measures in the problems of interest.

5.1 Tests Based on Log-Likelihood Statistics and Thresholding

5.1.1 Test Procedures

Under the sequence Gaussian model (2.1) let us test the simple hypothesis $H_0 : v = 0$ against the alternatives $H_1 : v \in V_\varepsilon \subset l^2$. We would like to describe test procedures which provide asymptotic minimaxity for a wide enough class of alternatives determined by power and Besov norms.

Consider a family of the sequences of two-point or symmetric three-point measures (see Section 4.3.1),

$$\bar{\pi}_\varepsilon = \{\pi_{\varepsilon,i}(h_{\varepsilon,i}, z_{\varepsilon,i})\}, \quad z_{\varepsilon,i} > 0, \quad h_{\varepsilon,i} \in (0,1], \tag{5.1}$$

where

$$\pi(h,z) = (1-h)\delta_0 + h\delta_z \quad \text{or} \quad \pi(h,z) = (1-h)\delta_0 + h(\delta_z + \delta_{-z})/2.$$

The family $\bar{\pi}_\varepsilon$ define the product priors π^ε and it is natural to study the test procedures based on log-likelihood statistics which are of the form (see (4.75)):

$$l_\varepsilon = \log \frac{dP_{\pi^\varepsilon}}{dP_0}(x) = \sum_i \log(1 + g_{\varepsilon,i}(x_i)), \quad g_{\varepsilon,i}(t) = g(t; h_{\varepsilon,i}, z_{\varepsilon,i}), \quad t \in R^1,$$

where for the two-point case we denote

$$g(t;h,z) = h\nu(t,z) = h(e^{-z^2/2+tz} - 1),$$

and for the three-point case

$$g(t;h,z) = h\xi(t,z) = h(e^{-z^2/2}\cosh(tz) - 1).$$

We consider test statistics of some different form. Put

$$t_\varepsilon^{(1)} = t_{\varepsilon;\bar{h}_\varepsilon,\bar{z}_\varepsilon}^{(1)}(x) = \sum_i g_{\varepsilon,i}(x_i).$$

Thus, we change the log-likelihood statistics by the sums $t_\varepsilon^{(1)}$ of the first items in the expansion $\log(1+g) \approx g$. It was shown in the proof of Proposition 4.4 that the difference between l_ε and $t_{\varepsilon,1}$ is asymptotically nonrandom under the general assumptions A.1, A.2 and, at least, under null-hypothesis.

In addition, take a family $\bar{Q}_\varepsilon = \{Q_{\varepsilon,i}\}$ such that

$$\sum_i \Phi(-Q_{\varepsilon,i}) = o(1) \tag{5.2}$$

(we construct the family $\bar{Q}_\varepsilon$ below). We consider tests $\psi_{\varepsilon,S} = \psi_{\varepsilon,S;\bar{\pi}_\varepsilon,\bar{Q}_\varepsilon}$ based on the normalized statistics $t_\varepsilon = t_\varepsilon^{(1)}/u_\varepsilon$, i.e.,

$$t_\varepsilon = t_{\varepsilon,\bar{h}_\varepsilon,\bar{z}_\varepsilon} = u_\varepsilon^{-1} \sum_i g_{\varepsilon,i}(x_i), \tag{5.3}$$

jointly with the thresholding analogous to (3.27).

Namely, for symmetric three-point measures, we set

$$\psi_{\varepsilon,S} = \mathbf{1}_{X_{\varepsilon,S}}; \quad X_{\varepsilon,S} = \{t_\varepsilon > S\} \bigcup \left\{ \sup_i |x_i|/Q_{\varepsilon,i} \geq 1 \right\}, \tag{5.4}$$

and we change the inequality $|x_i|/Q_{\varepsilon,i} \geq 1$ by $x_i/Q_{\varepsilon,i} \geq 1$ for the two-point case.

For given $\alpha \in (0,1)$ we denote $\psi_{\varepsilon,\alpha} = \psi_{\varepsilon,T_\alpha}$ (i.e., we consider $S = T_\alpha$). Recall that T_α is a $(1-\alpha)$-quantile of the standard Gaussian distribution,

$$u_\varepsilon^2 = \|\bar{\pi}_\varepsilon\|^2 = \sum_i \|\pi_{\varepsilon,i}\|^2, \quad u_{\varepsilon,i} = \|\pi_{\varepsilon,i}\|,$$

where the norms are defined in Section 3.3.3.

We can easily evaluate the type I errors of the test family (5.4) under the assumptions which have been introduced in Section 4.3.1.

Proposition 5.1

(1) *Assume* A.1, A.2, *and* (5.2). *Then* $\alpha(\psi_{\varepsilon,S}) \to \Phi(-S)$ *for any* S. *In particular, for any* $\alpha \in (0,1)$ *and* $S = T_\alpha$, *one has* $\alpha(\psi_{\varepsilon,\alpha}) = \alpha + o(1)$.

(2) *Assume* A.1a, A.2a, *and*

$$\sum_i \Phi(-Q_{\varepsilon,i}) = \delta_{\varepsilon,Q} = o(1). \tag{5.5}$$

Then

$$\alpha(\psi_{\varepsilon,S}) \le \Phi(-S) + 2\delta_{\varepsilon,Q} + o(w_\varepsilon^c) \quad \forall\, c > 0$$

uniformly over S.

Recall that we denote $w_\varepsilon = \sup_i u_{\varepsilon,i}$.

Proof. Assume A.1, A.2, and (5.2). Since $\alpha(\psi_{\varepsilon,S}) = P_0(X_{\varepsilon,S})$, one has

$$P_0(t_\varepsilon > S) \le \alpha(\psi_{\varepsilon,S}) < P_0(t_\varepsilon > S) + P_0(\sup_i |x_i|/Q_{\varepsilon,i} \ge 1).$$

Using Corollary 4.2 (1) we get $P_0(t_\varepsilon > S) = \Phi(-S) + o(1)$. Also we have

$$P_0(\sup_i |x_i|/Q_{\varepsilon,i} \ge 1) \le \sum_i P_0(|x_i| \ge Q_{\varepsilon,i}) = 2\sum_i \Phi(-Q_{\varepsilon,i}) \to 0.$$

The proof of the second statement is based on Corollary 4.2 (2). □

To describe type II errors of the test family (5.4) we would like to establish asymptotic $N(h_\varepsilon(v), \sigma_\varepsilon^2(v))$-normality of the statistics t_ε under alternatives, where

$$\begin{aligned} h_\varepsilon(v) &= E_v(t_\varepsilon) = u_\varepsilon^{-1} \sum_i E_{v_i}(g_{\varepsilon,i}), \\ \sigma_\varepsilon^2(v) &= \mathrm{Var}_v(t_\varepsilon) = u_\varepsilon^{-2} \sum_i \mathrm{Var}_{v_i}(g_{\varepsilon,i}). \end{aligned}$$

However if the quantities $z_{\varepsilon,i}$ are too large, then the items $g_{\varepsilon,i}$ have "too heavy tails". It follows from the general Central Limit Theorem that we need "to truncate the tails" in this case.

Introduce the threshold families $T_{\varepsilon,i} = T_{\varepsilon,i}(h_{\varepsilon,i}, z_{\varepsilon,i}) \ge 0$ which are defined by the relations

$$h\nu(T,z) = 1 \quad \text{or} \quad h\xi(T,z) = 1.$$

For the two-point case, we have

$$T = T(h, z) = \frac{z}{2} + \frac{\log(1 + h^{-1})}{z},$$

and for the three-point case,

$$T = T(h, z) = \frac{z}{2} + \frac{\log(1 + h^{-1})}{z} + O(z^{-1}), \quad \text{as } z \to \infty \text{ or } h \to 0. \tag{5.6}$$

Using the relations above one easily gets the inequalities

$$Tz \geq \log(1 + h^{-1}) \geq \log 2. \tag{5.7}$$

Moreover,

$$\begin{aligned} \exp(z^2/2 - Tz) \leq h \leq 4\exp(z^2/2 - Tz); \\ h_{\varepsilon,i} \sim d\exp(z_{\varepsilon,i}^2/2 - T_{\varepsilon,i} z_{\varepsilon,i}), \quad \text{as } h_{\varepsilon,i} \to 0, \end{aligned} \tag{5.8}$$

where $d = 1$ for the two-point case and $d = 2$ for the three-point case.

The inequality (5.7) yields $T_{\varepsilon,i} \to \infty$ as $z_{\varepsilon,i} \to 0$.

Note that under assumptions A.1, A.3 one has

$$\sup_i u_{\varepsilon,i}^2 \exp(\delta_1 z_{\varepsilon,i}^2) = o(1) \quad \forall\, \delta_1 \in (0, \delta),$$

where $\delta > 0$ is the parameter from A.3. Also, using (5.8) we have, for large enough $z_{\varepsilon,i} \geq B > 0$,

$$\begin{aligned} u_{\varepsilon,i}^2 \exp(\delta_1 z_{\varepsilon,i}^2) &\asymp h_{\varepsilon,i}^2 \exp(z_{\varepsilon,i}^2(1 + \delta_1)) \\ &\asymp \exp(2z_{\varepsilon,i}(z_{\varepsilon,i}(1 + \delta_1/2) - T_{\varepsilon,i})) \to 0. \end{aligned} \tag{5.9}$$

This yields

$$T_{\varepsilon,i} > z_{\varepsilon,i}(1 + \delta_1/2), \quad T_{\varepsilon,i} \to \infty, \quad \text{as } z_{\varepsilon,i} \asymp 1.$$

Moreover, under assumption A.1 we have, if $z_{\varepsilon,i} > 1$, then

$$\log w_\varepsilon \geq \log u_{\varepsilon,i} = z_{\varepsilon,i}^2 - z_{\varepsilon,i} T_{\varepsilon,i} + O(1).$$

Using the inequality $x + y \geq \sqrt{2xy}$; $x, y > 0$, we get

$$T_{\varepsilon,i} \geq z_{\varepsilon,i} + \frac{\log w_\varepsilon^{-1}}{z_{\varepsilon,i}} + O(1) \geq \sqrt{2\log w_\varepsilon^{-1}} + O(1).$$

If $z_{\varepsilon,i} \leq 1$, then $u_{\varepsilon,i}^2 \asymp h_{\varepsilon,i}^2 z_{\varepsilon,i}^d$, where $d = 2$ for the two-point case and $d = 4$ for the three-point case. By the inequality $z^{-1} \geq 1 + \log z^{-1}$, $z > 0$, we get

$$\begin{aligned} T_{\varepsilon,i} &\geq z_{\varepsilon,i}^{-1} \log(1 + h_{\varepsilon,i}^{-1}) > \log h_{\varepsilon,i}^{-1} + \log z_{\varepsilon,i}^{-1} \log(1 + h_{\varepsilon,i}^{-1}) \\ &\geq \log h_{\varepsilon,i}^{-1} + \log(2) \log z_{\varepsilon,i}^{-1} > c(\log h_{\varepsilon,i}^{-1} + 2\log z_{\varepsilon,i}^{-1}) \\ &= c\log u_{\varepsilon,i}^{-1} + O(1) \geq c\log w_\varepsilon^{-1} + O(1), \quad c > 0. \end{aligned}$$

In any case, under assumptions A.1, A.3, we have

$$T_{\varepsilon,i} > z_{\varepsilon,i}(1+\delta_1/2), \quad T_\varepsilon = \inf_i T_{\varepsilon,i} \geq \sqrt{2\log w_\varepsilon^{-1}} + O(1) \to \infty. \tag{5.10}$$

The same holds under assumptions A.1a, A.3 as well. In fact, if

$$u_{\varepsilon,i}^2 \exp(\delta z_{\varepsilon,i}^2) \leq B u_\varepsilon^2 = o(w_\varepsilon^{-a}),$$

then, for any $\delta_1 < \delta$, we have

$$u_{\varepsilon,i}^2 \exp(\delta_1 z_{\varepsilon,i}^2) \leq \left(w_\varepsilon^b u_{\varepsilon,i}^2 \exp(\delta z_{\varepsilon,i}^2)\right)^{\delta_1/\delta} = o(w_\varepsilon^c),$$

where $b = 2(\delta/\delta_1 - 1) > 0$, $c = (b-a)\delta_1/\delta > 0$ for small enough $a > 0$. Other evaluations are the same.

We assume *the family* $\bar{Q}_\varepsilon$ *to be such that there exist* $\delta_0 > 0$, $\eta_0 > 0$, $\tau \in (0,1)$, *such that*

$$Q_{\varepsilon,i} \leq T_{\varepsilon,i} - \delta_0 z_{\varepsilon,i} + o(1); \quad Q_{\varepsilon,i} \leq \tau T_{\varepsilon,i}, \quad \text{if } z_{\varepsilon,i} \leq \eta_0 T_{\varepsilon,i}. \tag{5.11}$$

It is clear that if (5.11) holds for $\delta = \delta_0 > 0$, $\eta = \eta_0 > 0$, then it holds for any $\delta \in (0,\delta_0)$, $\eta \in (0,\eta_0)$ as well.

Proposition 5.2

(1) *Under assumptions* A.1, A.3 *there exists the family* $\bar{Q}_\varepsilon$ *which satisfies* (5.2), (5.11) *with any* $\delta_0 \in (0,\delta/2)$ *and any* $\eta_0 \in (0,1/4)$; *here* $\delta > 0$ *is the parameter from* A.3.

(2) *Under assumptions* A.1a, A.3 *there exists the family* $\bar{Q}_\varepsilon$ *which satisfies* (5.5), (5.11) *with any* $\delta_0 \in (0,\delta/2)$ *and any* $\eta_0 \in (0,1/4)$.

Proof. Let us prove statement (1). Define the family $\bar{Q}_\varepsilon$ by the relation

$$\exp(-Q_{\varepsilon,i}^2/2) \asymp u_{\varepsilon,i}^2 \exp(\delta_1 z_{\varepsilon,i}^2), \quad \delta_1 \in (0,\delta).$$

Under assumptions A.1, A.3, one has

$$\sup_i u_{\varepsilon,i}^2 \exp(\delta_1 z_{\varepsilon,i}^2) \to 0, \quad \sum_i u_{\varepsilon,i}^2 \exp(\delta_1 z_{\varepsilon,i}^2) = o(1).$$

Therefore using (4.140), we have

$$\inf_i Q_{\varepsilon,i} \to \infty; \quad \sum_i \Phi(-Q_{\varepsilon,i}) \asymp \sum_i \exp(-Q_{\varepsilon,i}^2/2)/Q_{\varepsilon,i} \to 0. \tag{5.12}$$

It follows from (5.9) that for $z_{\varepsilon,i} > B > 0$ one has

$$\begin{aligned} Q_{\varepsilon,i}^2 &= 4z_{\varepsilon,i}T_{\varepsilon,i} - 2z_{\varepsilon,i}^2(2+\delta_1) + O(1) = (T_{\varepsilon,i} - \delta_1 z_{\varepsilon,i}/2)^2 \\ &\quad -(T_{\varepsilon,i} - z_{\varepsilon,i}(2+\delta_1/2))^2 + O(1) \leq (T_{\varepsilon,i} - \delta_1 z_{\varepsilon,i}/2)^2 + O(1). \end{aligned}$$

These inequalities yield the first inequality (5.11) for $z_{\varepsilon,i} > B$ and the second inequality for $B < z_{\varepsilon,i} < \eta_0 T_{\varepsilon,i}$ with $\eta_0 = (\tau/2)^2 \in (0,1/4)$.

If $z_{\varepsilon,i} \leq B$, using (5.7), (5.8) we get

$$u_{\varepsilon,i}^2 \exp(\delta z_{\varepsilon,i}^2) \asymp h_{\varepsilon,i}^2 z_{\varepsilon,i}^d \geq B T_{\varepsilon,i}^{-d} \exp(z_{\varepsilon,i}^2 - 2z_{\varepsilon,i}T_{\varepsilon,i}),$$

where $d = 2$ for the two-point case and $d = 4$ for the three-point case. Therefore

$$Q_{\varepsilon,i}^2 \le 4z_{\varepsilon,i}T_{\varepsilon,i} + 2d\log T_{\varepsilon,i} + O(1); \quad Q_{\varepsilon,i} = o(T_{\varepsilon,i}).$$

These relations yield (5.11).

To prove statement (2) under A.1a, A.3, define the family $\bar{Q}_\varepsilon$ by the relation

$$\exp(-Q_{\varepsilon,i}^2/2) \asymp u_{\varepsilon,i}^{2+\zeta}\exp(\delta z_{\varepsilon,i}^2),$$

where $\zeta > 0$ is small enough. For any $\zeta_1 \in (0,\zeta)$ these yield

$$\sum_i \exp(-Q_{\varepsilon,i}^2/2) \le Bw_\varepsilon^\zeta \sum_i u_{\varepsilon,i}^2 \exp(\delta z_{\varepsilon,i}^2) = O(w_\varepsilon^\zeta u_\varepsilon^2) = o(w_\varepsilon^{\zeta_1}).$$

Other evaluations are analogous to the proof of statement (1). □

Assuming (5.11), take $\eta \in (0,\eta_0)$ such that $3\eta/2 + \tau < 1$ and denote

$$I_{\varepsilon,\eta} = \{i : z_{\varepsilon,i} > \eta T_{\varepsilon,i}\}. \tag{5.13}$$

Set

$$h_{\varepsilon,T}(v) = h_{1,\varepsilon}(v) + h_{2,\varepsilon}(v) = \sum_{i\notin I_{\varepsilon,\eta}} h_{\varepsilon,i}(v_i) + \sum_{i\in I_{\varepsilon,\eta}} h_{\varepsilon,i,T}(v_i), \tag{5.14}$$

$$u_\varepsilon^2 = u_{1,\varepsilon}^2 + u_{2,\varepsilon}^2 = \sum_{i\notin I_{\varepsilon,\eta}} u_{\varepsilon,i}^2 + \sum_{i\in I_{\varepsilon,\eta}} u_{\varepsilon,i}^2, \tag{5.15}$$

where, for the two-point case we have

$$u_{\varepsilon,i}^2 = h_{\varepsilon,i}^2(\exp(z_{\varepsilon,i}^2) - 1), \quad h_{\varepsilon,i}(v_i) = u_\varepsilon^{-1}h_{\varepsilon,i}(\exp(v_i z_{\varepsilon,i}) - 1); \tag{5.16}$$

$$h_{\varepsilon,i,T}(v_i) = u_\varepsilon^{-1}h_{\varepsilon,i}(\exp(v_i z_{\varepsilon,i}) - 1)\Phi(T_{\varepsilon,i} - z_{\varepsilon,i} - |v_i|), \tag{5.17}$$

and for the three-point case we have

$$u_{\varepsilon,i}^2 = 2h_{\varepsilon,i}^2\sinh^2(z_{\varepsilon,i}^2/2), \quad h_{\varepsilon,i}(v_i) = u_\varepsilon^{-1}2h_{\varepsilon,i}\sinh^2(v_i z_{\varepsilon,i}/2); \tag{5.18}$$

$$h_{\varepsilon,i,T}(v_i) = u_\varepsilon^{-1}2h_{\varepsilon,i}\sinh^2(v_i z_{\varepsilon,i}/2)\Phi(T_{\varepsilon,i} - z_{\varepsilon,i} - |v_i|). \tag{5.19}$$

Recall that assumptions A.1 or A.1a and A.3 yield relation (5.10). Using (5.10) and (5.13) we get

$$\begin{aligned}\sum_{i\in I_{\varepsilon,\eta}} u_{\varepsilon,i}^2 &\le \left(\sup_{i\in I_{\varepsilon,\eta}} \exp(-\delta z_{\varepsilon,i}^2)\right)\sum_i u_{\varepsilon,i}^2\exp(\delta z_{\varepsilon,i}^2)\\ &\le \exp(-\delta\eta^2 T_\varepsilon^2)\sum_i u_{\varepsilon,i}^2\exp(\delta z_{\varepsilon,i}^2) = o(w_\varepsilon^a).\end{aligned}$$

Therefore under assumptions A.1 or A.1a and A.3, for small enough $a \in (0,\delta\eta^2)$, we have

$$u_{1,\varepsilon}^2 = u_\varepsilon^2 + o(w_\varepsilon^a), \quad u_{2,\varepsilon}^2 = o(w_\varepsilon^a). \tag{5.20}$$

Proposition 5.3 *Assume either* A.1 *or* A.1a *and* A.3. *Let the thresholds family* $Q_{\varepsilon,i}$ *satisfy* (5.2) *and* (5.11). *Then, uniformly over* $v \in l^2$ *for any* S, *one has*

$$\beta(\psi_{\varepsilon,S}, v) \le \Phi(S - h_{\varepsilon,T}(v)) + o(1). \tag{5.21}$$

Moreover, take a family $C_\varepsilon \to \infty$, $C_\varepsilon = o(T_\varepsilon)$ *and set* $H_{\varepsilon,i} = Q_{\varepsilon,i} + C_\varepsilon$. *Introduce the sets*

$$V_{\varepsilon,1} = \left\{ v \in l^2 : \sup_i (|v_i| - H_{\varepsilon,i}) > 0 \text{ or } \sum_i \Phi(|v_i| - Q_{\varepsilon,i}) > C_\varepsilon \right\}$$

and $V_{\varepsilon,2} = l^2 \backslash V_{\varepsilon,1}$. *Then* $\beta(\psi_{\varepsilon,S}, v) \to 0$ *uniformly on* S, $v \in V_{\varepsilon,1}$. *Also*

$$\beta(\psi_{\varepsilon,S}, v) \le \Phi(S - h_{\varepsilon,T}(v)) + \delta_\varepsilon(v); \quad \sup_{v \in V_{\varepsilon,2}} \delta_\varepsilon(v) = o(1). \tag{5.22}$$

Under assumptions A.1a, A.3, *relations* (5.21), (5.22) *are uniform over* S *such that* $S \le o(w_\varepsilon^{-b})$ *for any* $b > 0$.

Proof of Proposition 5.3. This is given in the Appendix, Section A.2. [1]

5.1.2 Reduction to the Nontruncated Case

The results (5.22) are formulated in terms of the truncation: we do not use $h_{\varepsilon,i}(v_i)$ defined by (5.16), (5.18), but the values $h_{\varepsilon,i,T}(v_i)$ defined by (5.17), (5.19). This is not a defect our method of the proof: in Chapter 8 we study some problems where the truncation is of importance.

However, in many hypothesis testing problems of interest (in particular, for alternatives V_ε determined by power or Besov norms), we can pass to the relation analogous to (5.22) but without the truncation: we can change the values $h_{\varepsilon,i,T}(v_i)$ by $h_{\varepsilon,i}(v_i)$ and the values $h_{\varepsilon,T}(v)$ by

$$h_\varepsilon(v) = \sum_i h_{\varepsilon,i}(v_i) \tag{5.23}$$

in relations (5.21), (5.22). If this change is possible then, using Proposition 5.3, we obtain the upper bounds for the type II errors: [2]

$$\beta(\psi_{\varepsilon,S}, V_\varepsilon) \le \Phi(S - h_\varepsilon(V_\varepsilon^H)) + o(1), \quad h_\varepsilon(V_\varepsilon^H) = \inf_{v \in V_\varepsilon^H} h_\varepsilon(v), \tag{5.24}$$

where we denote the set $V_{\varepsilon,2}$ from Proposition 5.3 by V_ε^H, assuming

$$H_{\varepsilon,i} = Q_{\varepsilon,i} + C_\varepsilon, \quad C_\varepsilon \to \infty, \quad C_\varepsilon = o(T_\varepsilon),$$

---

[1] The proof in Section A.2 is a modification of the proof which is given in [55], [62] under some different assumptions.

[2] These relations proved in [55], [62] under additional assumptions.

i.e.,

$$V_\varepsilon^H = \left\{ v \in V_\varepsilon : |v_i| \le H_{\varepsilon,i} \ \forall i; \ \sum_i \Phi(|v_i| - Q_{\varepsilon,i}) \le C_\varepsilon \right\}. \tag{5.25}$$

Certainly this change is possible in the case when the sets $I_{\varepsilon,\eta}$, defined by (5.13), are empty; particularly, this holds under assumption A.4. Thus we have relation (5.24) in this case. Unfortunately this is not sufficient for our aims: we will see below that for alternatives determined by the power or Besov norms with $p > q$ we need to study the cases when $z_{\varepsilon,i} \to \infty$, as $i \to \infty$.

We change v by "smaller" $\tilde{v}$ with $\tilde{v}_i \le |v_i|$ for the two-point case and $|\tilde{v}_i| \le |v_i|$ for the three-point case; we change $h_{\varepsilon,T}(v)$ by $h_\varepsilon(\tilde{v})$ for $v \in V_\varepsilon^H$ in these cases. We assume the alternative is positive, i.e., $v \in l_+^2$ for the case of two-point measures, when we use relations (5.16) and (5.17). We consider the three-point case below, since considerations for the two-point case are analogous and simpler.

We will use two variants of the change. The first variant is used for Besov norms. Set

$$\tilde{v} = \{\tilde{v}_i\}, \quad \tilde{v}_i = \begin{cases} v_i, & \text{if } i \notin I_{\varepsilon,\eta}, \\ 0, & \text{if } i \in I_{\varepsilon,\eta}. \end{cases} \tag{5.26}$$

Then $h_\varepsilon(\tilde{v}) = h_{\varepsilon,T}(\tilde{v}) \le h_{\varepsilon,T}(v)$, and we change the upper bounds (5.24) since

$$\beta(\psi_{\varepsilon,S}, V_\varepsilon) \le \Phi(S - h_\varepsilon(\tilde{V}_\varepsilon^H)) + o(1), \quad \tilde{V}_\varepsilon^H = \{\tilde{v} : v \in V_\varepsilon^H\}. \tag{5.27}$$

The second variant is used for power norms. Taking a positive family $\zeta_\varepsilon \to 0$, we set

$$\tilde{v}_i = \begin{cases} v_i, & \text{if } i \notin I_{\varepsilon,\eta}, \\ (1 - 2\zeta_\varepsilon)v_i, & \text{if } i \in I_{\varepsilon,\eta}. \end{cases} \tag{5.28}$$

Proposition 5.4 *Assume* A.1 *or* A.1a *and* A.3. *Let $\zeta_\varepsilon \to 0$ be a family such that $T_\varepsilon\zeta_\varepsilon \to \infty$ and take $C_\varepsilon \to \infty$ such that $C_\varepsilon = o(T_\varepsilon\zeta_\varepsilon)$. Assume that for $\delta_\varepsilon = 1 - \zeta_\varepsilon$ one has: if $i \in I_{\varepsilon,\eta}$, then*

$$Q_{\varepsilon,i} \le T_{\varepsilon,i} - \delta_\varepsilon z_{\varepsilon,i}/2, \quad T_{\varepsilon,i} \ge z_{\varepsilon,i}(1 + \delta_\varepsilon/2), \tag{5.29}$$

the set $I_{\varepsilon,\eta}$ is defined by (5.13). *Furthermore, under* A.1a *assume $u_\varepsilon \to \infty$ and the test thresholds in* (5.4) *are such that $S = S_\varepsilon = O(u_\varepsilon)$. Let $\tilde{v}_\varepsilon$ be defined by* (5.28). *Then relation* (5.27) *holds.*

Proof. If the test thresholds $S = O(u_\varepsilon)$, then one can easily see that the change $h_{\varepsilon,T}(v)$ by $h_\varepsilon(\tilde{v})$ is possible, if

$$h_{\varepsilon,T}(v) \ge h_\varepsilon(\tilde{v})(1 - o(u_\varepsilon^{-1})) + o(u_\varepsilon^{-1}) \quad \text{uniformly over } v \in V_\varepsilon^H.$$

This relation follows from

$$\sum_{i \in I_{\varepsilon,\eta}} h_{\varepsilon,i,T}(v_i) \ge \sum_{i \in I_{\varepsilon,\eta}} h_{\varepsilon,i}(\tilde{v}_i)(1 - o(u_\varepsilon^{-1})) + o(u_\varepsilon^{-1}). \tag{5.30}$$

Here, as above,

$$h_{\varepsilon,i,T}(v_i) = h_{\varepsilon,i}(v_i)\Phi(-Y_{\varepsilon,i}), \quad Y_{\varepsilon,i} = z_{\varepsilon,i} + |v_i| - T_{\varepsilon,i}.$$

Let us control (5.30). Taking a small enough value $b > 0$ and a family $B_\varepsilon \to \infty$ such that $B_\varepsilon = o(T_\varepsilon \zeta_\varepsilon)$, introduce the sets

$$\begin{aligned} I^0_{\varepsilon,\eta,B_\varepsilon} &= \{i \in I_{\varepsilon,\eta} : Y_{\varepsilon,i} < -bT_\varepsilon\},\\ I^1_{\varepsilon,\eta,B_\varepsilon} &= \{i \in I_{\varepsilon,\eta} : -bT_\varepsilon \le Y_{\varepsilon,i} < -B_\varepsilon\},\\ I^2_{\varepsilon,\eta,B_\varepsilon} &= \{i \in I_{\varepsilon,\eta} : Y_{\varepsilon,i} \ge -B_\varepsilon\}.\end{aligned}$$

Denote

$$h_{\varepsilon,k}(\tilde v) = \sum_{i\in I^k_{\varepsilon,\eta,B_\varepsilon}} h_{\varepsilon,i}(\tilde v_i), \quad k = 0,1,2.$$

Let $i \in I^0_{\varepsilon,\eta,B_\varepsilon}$. Since $h_{\varepsilon,i}(\tilde v_i) < h_{\varepsilon,i}(v_i)$, relation (5.30), for the sum over $I^0_{\varepsilon,\eta,B_\varepsilon}$, follows from A.1 or A.1a and from relations (4.140), (5.10):

$$\Phi(Y_{\varepsilon,i}) = o(\exp(-b^2T_\varepsilon^2/2)) = o(w_\varepsilon^{b_2}), \quad 0 < b_2 < b^2.$$

Let $i \in I^1_{\varepsilon,\eta,B_\varepsilon}$. Then it follows from (5.10) that $z_{\varepsilon,i}|v_i| > aT_\varepsilon^2$ for some $a > 0$ and, using the assumptions on ζ_ε, we have

$$h_{\varepsilon,i}(\tilde v_i) \asymp h_{\varepsilon,i}(v_i)\exp(-2\zeta_\varepsilon z_{\varepsilon,i}|v_i|) = o(h_{\varepsilon,i}(v_i)).$$

Since $\Phi(Y_{\varepsilon,i}) \le \Phi(-B_\varepsilon) = o(1)$, we have $h_{\varepsilon,i,T}(v_i) \sim h_{\varepsilon,i}(v_i)$.

Therefore relation (5.30) holds for the sum over $I^1_{\varepsilon,\eta,B_\varepsilon}$. Thus it suffices to establish the relation

$$h_{\varepsilon,2}(\tilde v) = o(u_\varepsilon^{-1}) \quad \text{uniformly over } v \in V_\varepsilon^H. \tag{5.31}$$

Since

$$\sum_i \Phi(|v_i| - Q_{\varepsilon,i}) \le C_\varepsilon, \quad v \in V_\varepsilon^H,$$

to establish (5.31) it suffices to check that, uniformly over $v \in V_\varepsilon^H$, $i \in I^2_{\varepsilon,\eta,B_\varepsilon}$,

$$u_\varepsilon h_{\varepsilon,i}(\tilde v_i) \asymp \exp(z_{\varepsilon,i}(|\tilde v_i| + z_{\varepsilon,i}/2 - T_{\varepsilon,i})) = o(\Phi(|v_i| - Q_{\varepsilon,i})/C_\varepsilon). \tag{5.32}$$

The first equivalence in (5.32) follows from (5.8) and (5.16), (5.18). Let us control the second relation.

Using (5.10), for $v \in V_\varepsilon^H$, $i \in I^2_{\varepsilon,\eta,B_\varepsilon}$, we have

$$Q_{\varepsilon,i} + C_\varepsilon \ge |v_i| \ge T_{\varepsilon,i} - z_{\varepsilon,i} - B_\varepsilon \ge \delta_\varepsilon z_{\varepsilon,i}/2 - B_\varepsilon, \tag{5.33}$$

which yields

$$-((1+\zeta_\varepsilon)z_{\varepsilon,i}/2 + B_\varepsilon) \le |v_i| - Q_{\varepsilon,i} \le C_\varepsilon. \tag{5.34}$$

Note that the function $\exp(\kappa_\varepsilon z^2)/z^2$, $\kappa_\varepsilon > 0$, is monotone increasing in z for large z. Therefore, using (4.140) we see that, to obtain (5.32), it suffices to verify the inequality

$$2z_{\varepsilon,i}(|v_i|(1-2\zeta_\varepsilon)+z_{\varepsilon,i}/2-T_{\varepsilon,i})\le -(|v_i|-Q_{\varepsilon,i})^2-\kappa_\varepsilon z_i^2,$$

where $\kappa_\varepsilon = 4(\eta T_\varepsilon)^{-2}\log(\eta T_\varepsilon)=o(\zeta_\varepsilon)$, which is equivalent to

$$(|v_i|+z_{\varepsilon,i}(1-2\zeta_\varepsilon)-Q_{\varepsilon,i})^2\le 2z_{\varepsilon,i}(T_{\varepsilon,i}-Q_{\varepsilon,i}+2\zeta_\varepsilon Q_{\varepsilon,i})-z_{\varepsilon,i}^2(4\zeta_\varepsilon-4\zeta_\varepsilon^2+\kappa_\varepsilon). \tag{5.35}$$

By (5.34) the squared value on the left-hand side of (5.35) is positive. By (5.34), (5.29) the left-hand side of (5.35) is no more than $z_{\varepsilon,i}^2(1-2\zeta_\varepsilon+o(\zeta_\varepsilon))^2$, and the right-hand side of (5.35) is no less than $z_{\varepsilon,i}^2(1-3\zeta_\varepsilon+o(\zeta_\varepsilon))$. These yield (5.35). □

5.2 Extreme Problem in the Space of Sequences of Measures

Using the results of Sections 4.3.1, 5.1 we can obtain the asymptotic lower and upper bounds for minimax errors in terms of the specific extreme problem. [3]

5.2.1 *Statement of the Problem*

Proposition 4.4 yields to relations (4.72) for the Bayesian problems. Therefore, to obtain asymptotically the best lower bounds in minimax problems, we need to minimize the norms $\|\bar\pi_\varepsilon\|$ under the asymptotic constraint $\pi^\varepsilon(V_\varepsilon)\to 1$.

Let us formulate the upper bounds. Let $\bar\pi_\varepsilon=\{\pi(h_{\varepsilon,i},z_{\varepsilon,i})\}$ be the family which consists of two-point or of symmetric three-point measures, such that either assumptions A.1, A.4 hold and or assumptions A.1, A.3 hold and there are maps $v\to\tilde v$ of type (5.26) or (5.28) such that (5.31) holds in the last case. Then we have the upper bounds (5.24), (5.27) which are determined by the values

$$h_\varepsilon(v)=h(v,\bar r_\varepsilon)=(\bar r_\varepsilon,\bar\delta_v);\quad \bar r_\varepsilon=\|\bar\pi_\varepsilon\|^{-1}\bar\pi_\varepsilon,\quad \|\bar r_\varepsilon\|=1; \tag{5.36}$$

$$h(V_\varepsilon,\bar r_\varepsilon)=\inf_{v\in V_\varepsilon}h(v,\bar r_\varepsilon),\quad h(\tilde V_\varepsilon,\bar r_\varepsilon)=\inf_{v\in\tilde V_\varepsilon}h(v,\bar r_\varepsilon). \tag{5.37}$$

Here $\bar\delta_v=\{\delta_{v_i}\}$ is the sequence which consists of Dirac masses concentrated on the coordinates of $v=\{v_i\}\in l^2$. Given a set $V\subset l^2$ we denote

$$\Delta(V)=\{\bar\delta_v,\ v\in V\}\subset\bar\Pi\subset L;$$

recall that $\bar\Pi$ is the set of sequences which consists of probability measures and L is the space with the Hilbert structure as defined in Section 3.3.3.

---

[3]The considerations in this section follows [55].

Let $K \subset \bar{\Pi}$, $\bar{r} \in L$. Set

$$h(K,\bar{r}) = \inf_{\bar{\pi}\in K} (\bar{\pi},\bar{r}),$$

these are the quantities of the same structure as in (5.37) for $K = \Delta(V_\varepsilon)$ or $K = \Delta(\tilde{V}_\varepsilon)$.

Let $\bar{\Pi}_\varepsilon$ be a convex subset $\bar{\Pi}$ such that $\Delta(V_\varepsilon) \subset \bar{\Pi}_\varepsilon$ or $\Delta(\tilde{V}_\varepsilon) \subset \bar{\Pi}_\varepsilon$ (certainly we can change V_ε by V_ε^H but we omit it here). By the inequalities

$$h(\Delta(V_\varepsilon),\bar{r}_\varepsilon) \geq h(\bar{\Pi}_\varepsilon,\bar{r}_\varepsilon), \quad h(\Delta(\tilde{V}_\varepsilon),\bar{r}_\varepsilon) \geq h(\bar{\Pi}_\varepsilon,\bar{r}_\varepsilon),$$

we get, from (5.24) and (5.27),

$$\beta(\alpha, V_\varepsilon) \leq \Phi(T_\alpha - h(\bar{\Pi}_\varepsilon,\bar{r}_\varepsilon)) + o(1). \tag{5.38}$$

Moreover, we get

$$\gamma(V_\varepsilon) \leq 2\Phi(-h(\bar{\Pi}_\varepsilon,\bar{r}_\varepsilon)/2) + o(1).$$

In fact, the last inequality is evident for $h_\varepsilon = h(\bar{\Pi}_\varepsilon,\bar{r}_\varepsilon) \to 0$. If $h_\varepsilon \to \infty$, then this follows from (5.38): using Propositions 5.1, 5.3 we can take a family $\alpha = \alpha_\varepsilon \to 0$, such that $T_\alpha \to \infty$, $T_\alpha = o(h_\varepsilon)$ and $\alpha(\psi_{\varepsilon,\alpha}) \to 0$, $\beta(\psi_{\varepsilon,\alpha}, V_\varepsilon) \to 0$ by (5.38); if $h_\varepsilon \asymp 1$, then we use (5.38) with $T = h_\varepsilon/2$.

Thus the asymptotically best tests correspond to the family $\bar{r}_\varepsilon$ which maximizes the function $h(\bar{\Pi}_\varepsilon,\bar{r})$ over $\bar{r}$ such that $\|\bar{r}\| = 1$. This leads to the maximin problem

$$h(\bar{\Pi}_\varepsilon) = \sup_{\|\bar{r}\|=1} \inf_{\bar{\pi}\in\bar{\Pi}_\varepsilon} (\bar{\pi},\bar{r}). \tag{5.39}$$

Next the considerations are analogous to those in Sections 4.1, 4.3.2. The maximin problems (5.39) are the same as in Proposition 4.1. We consider the extreme problem

$$u(\bar{\Pi}_\varepsilon) = \inf_{\bar{\pi}\in\bar{\Pi}_\varepsilon} \|\bar{\pi}\|, \tag{5.40}$$

and if there exists the solution $\bar{\pi}_\varepsilon$ of the extreme problem (5.40) such that $u_\varepsilon = u(\bar{\Pi}_\varepsilon) = \|\bar{\pi}_\varepsilon\|$ then, using Proposition 4.1, we get

$$h(\bar{\Pi}_\varepsilon) = u(\bar{\Pi}_\varepsilon) = \inf_{\bar{\pi}\in\bar{\Pi}_\varepsilon} (\bar{\pi},\bar{r}_\varepsilon) = u_\varepsilon, \quad \bar{r}_\varepsilon = \|\bar{\pi}_\varepsilon\|^{-1}\bar{\pi}_\varepsilon.$$

These lead to the inequality

$$\inf_{v\in V} (\bar{r}_\varepsilon,\bar{\delta}_v) = \|\bar{\pi}_\varepsilon\|^{-1} \inf_{v\in V} (\bar{\pi}_\varepsilon,\bar{\delta}_v) \geq \|\bar{\pi}_\varepsilon\|, \tag{5.41}$$

where $\bar{\pi}_\varepsilon$ is the extreme sequence in (5.40).

Thus the results of Sections 5.1, 4.3.1 (see Proposition 4.4 and relation (4.73)) imply

Proposition 5.5 *Let $\bar{\Pi}_\varepsilon \subset L$ be convex sets such that $\Delta(V_\varepsilon) \subset \bar{\Pi}_\varepsilon$ or $\Delta(\tilde{V}_\varepsilon) \subset \bar{\Pi}_\varepsilon$. Here the sets $\tilde{V}_\varepsilon$ correspond to the change, either* (5.26) *or* (5.28). *For the change* (5.28) *with $\delta = \delta_\varepsilon = 1 - 2\zeta_\varepsilon$ and $\zeta_\varepsilon = o(1)$ we* *assume ζ_ε satisfies the assumption of Proposition 5.4. As was noted above, we can change the set V_ε by the set V_ε^H defined by* (5.25).

(1) *Let $\bar{\pi}_\varepsilon$ be the solution of the extreme problems* (5.40). *Let the family $\bar{\pi}_\varepsilon = \{\pi(h_{\varepsilon,i}, z_{\varepsilon,i})\}$ consist of two-point or of symmetric three-point measures and satisfies A.1, A.3. Then*

$$\beta(\alpha, V_\varepsilon) \le \Phi(T_\alpha - u_\varepsilon) + o(1); \quad \gamma(V_\varepsilon) \le 2\Phi(-u_\varepsilon/2) + o(1). \tag{5.42}$$

(2) *Let there exist a sequence $\tilde{\pi}_\varepsilon$ which consists of two-point or of symmetric three-point measures and satisfies A.1, A.2. Assume*

$$\|\tilde{\pi}_\varepsilon\| = u_\varepsilon + o(1), \quad \tilde{\pi}^\varepsilon(V_\varepsilon) \to 1,$$

where $\tilde{\pi}^\varepsilon$ is the product prior which corresponds to the sequence $\tilde{\pi}_\varepsilon$. Then

$$\beta(\alpha, V_\varepsilon) \ge \Phi(T_\alpha - u_\varepsilon) + o(1); \quad \gamma(V_\varepsilon) \ge 2\Phi(-u_\varepsilon/2) + o(1). \tag{5.43}$$

Thus relations (5.42), (5.43) yield to the sharp asymptotics of Gaussian type in the hypothesis testing problem in terms of solutions of the extreme problems (5.40):

$$\beta(\alpha, V_\varepsilon) = \Phi(T_\alpha - u_\varepsilon) + o(1); \quad \gamma(V_\varepsilon) = 2\Phi(-u_\varepsilon/2) + o(1). \tag{5.44}$$

5.2.2 Moment Constraints

To construct the extreme problems (5.40) we need to take the convex sets $\bar{\Pi}_\varepsilon \subset \bar{\Pi}$ such that $\bar{\Pi}_\varepsilon \supset \Delta(V_\varepsilon)$ or $\bar{\Pi}_\varepsilon \supset \Delta(\tilde{V}_\varepsilon)$. Certainly, the smallest convex sets $\bar{\Pi}_\varepsilon$ are the convex hulls of $\Delta(V_\varepsilon)$ or of $\Delta(\tilde{V}_\varepsilon)$. However, it is difficult to study extreme problems in this case. We consider some different sets $\bar{\Pi}_\varepsilon$ in the problems of interest.

We assume the sets $V_\varepsilon = V_\varepsilon(H_{1,\varepsilon}, H_{2,\varepsilon}) \subset l^2$ to be determined by constraints of a *separable symmetric type*:

$$V_\varepsilon = \{v \in l^2 : G_1(v) \ge H_{1,\varepsilon},\ G_2(v) \le H_{2,\varepsilon}\} \quad \text{or} \quad V_\varepsilon^+ = V_\varepsilon \cap l_+^2, \tag{5.45}$$

(the sets $V_\varepsilon^+ = V_\varepsilon^+(H_{1,\varepsilon}, H_{2,\varepsilon})$ correspond to the positive alternatives $v \in V_\varepsilon$), where $H_{1,\varepsilon}$, $H_{2,\varepsilon}$ are families of thresholds.

We assume the constraint functions to be of the form

$$G_1(v) = F_1(\bar{f}_1(v)), \quad G_2(v) = F_2(\bar{f}_2(v)).$$

Here F_1, F_2 are functionals defined in the space of positive sequences $\bar{x}$, with a possible value $F_l(\bar{x}) = \infty$ but $F_l(\bar{0}) = 0$, $l = 1, 2$. We assume F_1, F_2 to be monotone; this means $F_l(\bar{x}_1) \le F_l(\bar{x}_2)$, if $0 \le x_{1,i} \le x_{2,i}\ \forall\, i \in I$, and F_1 to be concave, F_2 to be convex. Moreover, $\bar{f}_1(v)$ and $\bar{f}_2(v)$ are separated sequences of the nonnegative functions

$$\bar{f}_1(v) = \{f_{i,1}(v_i)\}, \quad \bar{f}_2(v) = \{f_{i,2}(v_i)\}, \quad i \in I,$$

the functions $f_{i,l}(t) = f_l(t)$ *do not depend on* i *and are symmetric nonnegative functions in the real line*

$$f_l(0) = 0, \quad f_l(t) = f_l(-t) \geq 0, \quad t \in R^1, \quad l = 1, 2.$$

These correspond to problems of our interest (see (1.54), (2.89), (2.90)). In fact, for alternative (2.89) determined by power norms (2.71) with $p, q < \infty$ we put, $H_{1,\varepsilon} = \rho_\varepsilon^p$, $H_{2,\varepsilon} = R_\varepsilon^q$,

$$f_1(t) = |t|^p, \quad f_2(t) = |t|^q; \quad F_1(\bar{f}) = \sum_i i^{rp} f_i, \quad F_2(\bar{f}) = \sum_i i^{sq} f_i. \tag{5.46}$$

The functionals F_1, F_2 are linear in this case. For $q = \infty$ we put

$$H_{2,\varepsilon} = R_\varepsilon, \quad f_2(t) = |t|; \quad F_2(\bar{f}) = \sup_i i^s f_i. \tag{5.47}$$

We do not consider the case $p = \infty$ since it has been studied in Section 4.4.2.

For the alternative (2.90) determined by Besov norms (2.72) with $p, q, h, t < \infty$ on the sequence space of a pyramidal structure we have, $H_{1,\varepsilon} = \rho_\varepsilon^h$, $H_{2,\varepsilon} = R_\varepsilon^t$,

$$\bar{f}_l(v) = \{f_l(v_{ij});\ i = 1, \ldots, 2^j,\ j \geq 1\}, \quad l = 1, 2;$$

(remember that we assume $K = 0$, $J_0 = 0$ in the definition (2.72) of the Besov norm in the sequence space). Here, as above,

$$f_1(t) = |t|^p, \quad f_2(t) = |t|^q.$$

The functionals F_1, F_2 are of the form

$$F_1(\bar{f}) = \sum_j F_{1,j}(\bar{f}_j), \quad F_2(\bar{f}) = \sum_j F_{2,j}(\bar{f}_j), \tag{5.48}$$

where $\bar{f}_j = \{f_{ij},\ i = 1, \ldots, 2^j\}$, $l = 1, 2$, and

$$F_{1,j}(\bar{f}_j) = 2^{jrh} \left(\sum_{i=1}^{2^j} f_{ij} \right)^{h/p}, \quad F_{2,j}(\bar{f}_j) = 2^{jst} \left(\sum_{i=1}^{2^j} f_{ij} \right)^{t/q}. \tag{5.49}$$

Analogously, if q or $t = \infty$, then sums are changed by suprema: if $q < \infty$, $t = \infty$, then $H_{2,\varepsilon} = R_\varepsilon^q$,

$$f_2(t) = |t|^q, \quad F_2(\bar{f}) = \sup_j F_{2,j}(\bar{f}_j) = \sup_j 2^{jsq} \sum_i f_{ij}, \tag{5.50}$$

and if $q = t = \infty$, then $H_{2,\varepsilon} = R_\varepsilon$,

$$f_2(t) = |t|, \quad F_2(\bar{f}) = \sup_j F_{2,j}(\bar{f}_j) = \sup_j 2^{js} \left(\max_i f_{ij} \right). \tag{5.51}$$

We will study the sharp asymptotics for a Besov norm in the cases $h \leq p < \infty$, $q \leq t \leq \infty$; the functional F_1 is concave and the functional F_2 is convex under these constraints.

Let us extend functions $f_l(t)$, $t \in R^1$, $l = 1, 2$, *to functionals*

$$\varphi_l(r) = \int_{R^1} f_l(t) r(dt)$$

defined in the space $\mathcal{L}$ *of signed measures* r *in the real line, with possible values* $\varphi_l(r) = \infty$.

We have the relations

$$\varphi_l(\delta_x) = f_l(x); \quad \varphi_l(r + a\delta_0) = \varphi_l(r) \; \forall r \in \mathcal{L}, \quad a \in R^1, \quad l = 1, 2. \quad (5.52)$$

Therefore we can consider the functionals $\varphi_l(r)$ in the factorspace $\tilde{\mathcal{L}} = \mathcal{L}/\mathcal{L}_0$ (recall that $\mathcal{L}_0$ is generated by masses δ_0 at point 0).

Analogously, *let us extend functionals* $G_1(v)$, $G_2(v)$ *to spaces* L, $\tilde{L}$ *which consist of the sequences* $\bar{r}$ *of measures* $r_i \in \mathcal{L}$ *and* $r_i \in \tilde{\mathcal{L}}$,

$$\mathcal{G}_1(\bar{r}) = F_1(\bar{\varphi}_1(\bar{r})), \quad \mathcal{G}_2(\bar{r}) = F_2(\bar{\varphi}_2(\bar{r})), \quad (5.53)$$

where $\bar{\varphi}_l(\bar{r}) = \{\varphi_l(r_i)\}$. Since φ_l is linear, *the functional* $\mathcal{G}_1(\bar{r})$ *is concave and the functional* $\mathcal{G}_2(\bar{r})$ *is convex on* L, $\tilde{L}$.

We use these extensions for power and Besov norms with $p, q < \infty$,

$$\varphi_1(r) = \int |t|^p r(dt), \quad \varphi_2(r) = \int |t|^q r(dt), \quad (5.54)$$

if r is a probability measure, then it is the pth or qth absolute moment of r.

Let $q = \infty$. Then we use a different extension. Denote

$$|r| = \inf\{t > 0 : r([-t, t]) = r(R^1)\} \quad (5.55)$$

as the half-length of the minimum symmetric interval which contains the support of the measure r. Put

$$\varphi_2(r) = |r|.$$

The functional $\varphi_2(r)$ satisfies (5.52). It is not convex but *quasi-convex: this means that for any* $a \geq 0$ *the set* $\{r \in \mathcal{L} : |r| \leq a\}$ *is convex.*

For the functional F_2, defined by (5.47) or (5.51), we set $\mathcal{G}_2(\bar{r}) = F_2(\bar{\varphi}_2(\bar{r}))$. One can easily check that the functional $\mathcal{G}_2(\bar{r})$ is quasi-convex on L, $\tilde{L}$.

For positive alternatives let Π_+ be the set of probability measures supported on positive half-line $R^1_+ = \{x \geq 0\}$ and let $\bar{\Pi}_+$ be the set of sequences $\{\pi_i\}$ with $\pi_i \in \Pi_+$.

We consider the sets [4]

$$\bar{\Pi}_\varepsilon = \bar{\Pi}(H_{1,\varepsilon}, H_{2,\varepsilon}) = \{\bar{\pi} \in \bar{\Pi} : \ \mathcal{G}_1(\bar{\pi}) \geq H_{1,\varepsilon}, \ \mathcal{G}_2(\bar{\pi}) \leq H_{2,\varepsilon}\}; \quad (5.56)$$

$$\bar{\Pi}_\varepsilon = \bar{\Pi}_{\varepsilon,+} = \bar{\Pi}_+(H_{1,\varepsilon}, H_{2,\varepsilon}) = \bar{\Pi}(H_{1,\varepsilon}, H_{2,\varepsilon}) \cap \bar{\Pi}_+. \quad (5.57)$$

By above, *these sets are convex.*

We denote by $u_\varepsilon = u(H_{1,\varepsilon}, H_{2,\varepsilon})$ the infimum in the extreme problems (5.40) for the sets $\bar{\Pi}_\varepsilon$ defined by (5.56) or (5.57); sometimes we will use notations $u_{\varepsilon,+}$ to emphasize the nonnegativity.

It is clear that $\Delta(V_\varepsilon) \subset \bar{\Pi}_\varepsilon$. Under the assumptions of Proposition 5.5 we get the asymptotics (5.44).

Consider the sets $\tilde{V}_\varepsilon$ under mappings (5.26), (5.28). For the case of the mapping (5.28), and for power norms with $p < \infty$, one can see that

$$F_1(\tilde{v}) \geq (1 - \zeta_\varepsilon)^p F_1(v), \quad F_2(\tilde{v}) \leq F_2(v).$$

Therefore, if $V_\varepsilon = V(H_{1,\varepsilon}, H_{2,\varepsilon})$ then, for some $B > 0$, one has

$$\Delta(\tilde{V}_\varepsilon) \subset \Delta(V_\varepsilon(\tilde{H}_{1,\varepsilon}, H_{2,\varepsilon})) \subset \bar{\Pi}(\tilde{H}_{1,\varepsilon}, H_{2,\varepsilon}), \quad \tilde{H}_{1,\varepsilon} = H_{1,\varepsilon}(1 - B\zeta_\varepsilon). \quad (5.58)$$

For the mapping (5.26) and for Besov norms with $h \leq p < \infty$, we consider the functionals (5.48), (5.49). It follows from monotony that $F_{1,2}(\tilde{v}) \leq F_{1,2}(v)$ and $\tilde{v}$ satisfies the second constraint in (5.45). However, we cannot evaluate directly how much the first constraint should be changed.

Let us introduce the assumption which is used and controlled for alternatives determined by Besov norms. In the mapping (5.26) the set $I_{\varepsilon,\eta}$ is determined by families $h_{\varepsilon,ij}$, $z_{\varepsilon,ij}$. Let these families depend on the index j only,

$$h_{\varepsilon,ij} = h_{\varepsilon,j}, \quad z_{\varepsilon,ij} = z_{\varepsilon,j}.$$

This yields

$$T_{\varepsilon,ij} = T_{\varepsilon,j}, \quad I_{\varepsilon,\eta} = \{(i,j) \ : \ j \in J_{\varepsilon,\eta}, \ i = 1, \dots 2^j\},$$

where

$$J_{\varepsilon,\eta} = \{j : \ z_{\varepsilon,j} > \eta T_{\varepsilon,j}\}. \quad (5.59)$$

Relation (5.2) is of the form

$$\sum_j 2^j \Phi(-Q_{\varepsilon,j}) = o(1). \quad (5.60)$$

A.5. *Given the families $C_\varepsilon \to \infty$ and $Q_{\varepsilon,j}$ which satisfy* (5.60) *and* (5.11) *with the change i by j, assume the following relation holds:*

$$\sup_{v \in V_\varepsilon^H(H_{1,\varepsilon}, H_{2,\varepsilon})} \sum_{j \in J_{\varepsilon,\eta}} F_{1,j}(\bar{f}_{1,j}(v_j)) = \zeta_\varepsilon H_{1,\varepsilon}, \quad \zeta_\varepsilon = o(1),$$

[4] The sets of analogous structure were considered in [17], [18] in the minimax estimation problem.

where the set $V_\varepsilon^H = V_\varepsilon^H(H_{1,\varepsilon}, H_{2,\varepsilon})$ *is determined analogously to* (5.25)*:*
$V_\varepsilon^H(H_{1,\varepsilon}, H_{2,\varepsilon}) = \{v \in V_\varepsilon(H_{1,\varepsilon}, H_{2,\varepsilon}) \; : \; |v_{ij}| \le H_{\varepsilon,j}\}, \quad H_{\varepsilon,j} = Q_{\varepsilon,j} + C_\varepsilon,$
and

$$F_{1,j}(\bar{f}_{1,j}(v_j)) = 2^{jrh}\left(\sum_i |v_{ij}|^p\right)^{h/p}, \quad H_{1,\varepsilon} = \rho_\varepsilon^h.$$

Assumption A.5 yields the relations: if $v \in V_\varepsilon^H(H_{1,\varepsilon}, H_{2,\varepsilon})$, then

$$\begin{aligned} G_1(\tilde{v}) &= \sum_{j \notin J_{\varepsilon,\eta}} F_{1,j}(\bar{f}_{1,j}(v_j)) = \\ &= \sum_j F_{1,j}(\bar{f}_{1,j}(v_j)) - \sum_{j \in J_{\varepsilon,\eta}} F_{1,j}(\bar{f}_{1,j}(v_j)) \ge H_{1,\varepsilon}(1-\zeta_\varepsilon). \end{aligned}$$

Therefore under assumption A.5 we get relation (5.58) for Besov norms as well.

Remark. The assumption $u_\varepsilon \asymp 1$ in Proposition 5.5 is not necessary for the asymptotics (5.44). In fact, suppose we had established the upper bounds (5.42) and the lower bounds (5.43) for all $H_{1,\varepsilon}, H_{2,\varepsilon} \to \infty$ such that $u_\varepsilon \asymp 1$. Let $u_\varepsilon \to 0$. Then the upper bounds (5.42) are of the form

$$\beta(V_\varepsilon, \alpha) \le 1 - \alpha + o(1), \quad \gamma(V_\varepsilon) \le 1 + o(1),$$

which are evident and trivial. The lower bounds $\beta(V_\varepsilon, \alpha) \ge 1 - \alpha + o(1)$, $\gamma(V_\varepsilon) \ge 1 + o(1)$ follow from Proposition 4.4 as well.

Let $u_\varepsilon \to \infty$. Then the lower bounds (5.43) are of the form $\beta(V_\varepsilon, \alpha) \ge o(1)$, $\gamma(V_\varepsilon) \ge o(1)$, which are evident and trivial. To obtain the upper bounds let us change the values $H_{1,\varepsilon}$ to smaller values $\tilde{H}_{1,\varepsilon} < H_{1,\varepsilon}$ such that $u_\varepsilon \sim u$ with large u (this is possible because of the monotonicity and continuity). Then $V_\varepsilon(H_{1,\varepsilon}, H_{2,\varepsilon}) \subset \tilde{V}_\varepsilon = V(\tilde{H}_{1,\varepsilon}, H_{2,\varepsilon})$ which yields

$$\beta(\alpha, V_\varepsilon) \le \beta(\alpha, \tilde{V}_\varepsilon), \quad \gamma(V_\varepsilon) \le \gamma(\tilde{V}_\varepsilon); \quad \beta(\psi, V_\varepsilon) \le \beta(\psi, \tilde{V}_\varepsilon) \;\; \forall\, \psi \in \Psi.$$

Thus we obtain the upper bounds (5.42) with arbitrarily large $u_\varepsilon \sim u$ which yields $\beta(\alpha, \tilde{V}_\varepsilon)$, $\gamma(\tilde{V}_\varepsilon)$ to be arbitrary small. This yields $\beta(\alpha, V_\varepsilon) \to 0$, $\gamma(V_\varepsilon) \to 0$, which corresponds to the upper bounds (5.42) with $u_\varepsilon \to \infty$.

5.2.3 Convex Properties of the Extreme Problem

First, note that *the function* $f(\bar{r}) = \|\bar{r}\|^2$ *is strictly convex in the linear space* $\tilde{L}$. In fact, one can easily check that

$$p\|\bar{r}_1\|^2 + (1-p)\|\bar{r}_2\|^2 - \|p\bar{r}_1 + (1-p)\bar{r}_2\|^2 = p(1-p)\|\bar{r}_1 - \bar{r}_2\|^2, \quad (5.61)$$

the strict convexity holds because the bilinear form $(\bar{r}_1, \bar{r}_2)$ is positively defined (see Section 3.3.3). This yields *the convexity of the extreme problems* (5.40) *with* (5.56) *or* (5.57) under the assumptions of Section 5.2.2.

Let r be the measure on the real line. Denote by $\hat{r}$ the measure defined by $r(A) = \hat{r}(-A)$ for all Borelian $A \subset R^1$. We call the measure r on the real line *symmetric*, if $r = \hat{r}$. Clearly, for the case (5.56) the sets Π_ε are symmetric in the sense: if $\pi \in \Pi_\varepsilon$, then $\hat{\pi} \in \Pi_\varepsilon$.

Note that *the function* $\|r\|$ *and functionals* φ_l *are symmetric* in the sense:

$$\|r\| = \|\hat{r}\|, \quad \varphi_l(r) = \varphi_l(\hat{r}), \quad l = 1,2,$$

which yield the symmetry of functionals $\|\bar{r}\|$, $\mathcal{G}_l$, $l = 1,2$, and of the set $\bar{\Pi}_\varepsilon$ in the same sense.

These considerations imply the following properties of the extreme problem (5.40) with the sets $\bar{\Pi}_\varepsilon$ determined by (5.56) or (5.57). Introduce the assumption:

A.6. *Let $H_{1,\varepsilon}$, $H_{2,\varepsilon}$ be positive families. Assume that for any families $b_{\varepsilon,1} \asymp 1$, $b_{\varepsilon,2} \asymp 1$, one has*

$$u_\varepsilon(b_{\varepsilon,1}H_{1,\varepsilon}, b_{\varepsilon,2}H_{2,\varepsilon}) \asymp u_\varepsilon(H_{1,\varepsilon}, H_{2,\varepsilon}). \tag{5.62}$$

Let $R^2_+ = \{(x,y) \in R^2,\ x > 0,\ y > 0\}$ be the positive corner in the plane.

Proposition 5.6 *Under the assumptions of Section 5.3.2 we have:*

(1) *The function $u^2_\varepsilon(H_1, H_2)$ is convex in $(H_1, H_2) \in R^2_+$ and $u^2_\varepsilon(0, H_2) = 0$ for any $H_2 > 0$. This increases in H_1 and decreases in H_2. If there exists a solution $\bar{\pi}_\varepsilon$ of the extreme problem (5.40), then this is unique and $\mathcal{G}_1(\bar{\pi}_\varepsilon) = H_1$. All measures $\pi_{\varepsilon,i}$, $i \in I$, are symmetric for the case (5.56). In particular, these hold for power and Besov norms with $h \le p < \infty$, $t \ge q$.*

(2) *Let $u_\varepsilon(H_1, \infty) = 0$ (this means we consider the extreme problem without the second constraint; this equality holds for $r \ge r^*_p$ or $r \ge r^*_{p,+}$ for positive alternatives determined by the pover or Besov norms, the values r^*_p, $r^*_{p,+}$ are defined by (3.100), (3.108)). Let there exist the solution $\bar{\pi}_\varepsilon$ of extreme problems (5.40) for $H_1 > 0$. Then $\mathcal{G}_2(\bar{\pi}_\varepsilon) = H_2$.*

(3) *For any $t \in (0,1)$ and $H_1 > 0$, $H_2 > 0$ one has the inequality*

$$u_\varepsilon(tH_1, tH_2) \le t u_\varepsilon(H_1, H_2).$$

(4) *Assume A.6. Then the functions $f_\varepsilon(b_1, b_2) = u^2_\varepsilon(b_1 H_{1,\varepsilon}, b_2 H_{2,\varepsilon})/U^2_\varepsilon$ are uniformly continuous and Lipschitzian over any compact $K \subset R^2_+$ uniformly in $(b_1, b_2) \in K$ and for small enough $\varepsilon > 0$. In particular, these yield the relation: for any positive family $\zeta_\varepsilon = o(1)$, if $\tilde{H}_{1,\varepsilon} = H_{1,\varepsilon}(1 + O(\zeta_\varepsilon))$, $\tilde{H}_{2,\varepsilon} = H_{2,\varepsilon}(1 + O(\zeta_\varepsilon))$, then*

$$\tilde{u}_\varepsilon = u_\varepsilon(\tilde{H}_{1,\varepsilon}, \tilde{H}_{2,\varepsilon}) = u_\varepsilon(1 + O(\zeta_\varepsilon)). \tag{5.63}$$

Proof. We give the proof for the sets $\bar{\Pi}_\varepsilon$ defined by (5.56) with remarks for other cases. The convexity of the function $u^2_\varepsilon(H_1, H_2)$ follows from the convexity of the function $\|\bar{\pi}\|^2$ and from Proposition 2.8. Monotone properties follow from the natural embedding properties of the sets $\bar{\Pi}_\varepsilon(H_1, H_2)$:

$$\bar{\Pi}_\varepsilon(H_1, H_2) \subset \bar{\Pi}_\varepsilon(T_1, T_2), \quad \text{if } H_1 \geq T_1, \quad H_2 \leq T_2.$$

If $H_1 = 0$, then $\bar{0} \in \bar{\Pi}_\varepsilon$ (this corresponds to the sequence $\bar{\delta}_0$) and $\|\bar{0}\| = 0$. If there exist two different solutions

$$\bar{\pi}_{\varepsilon,1} \in \bar{\Pi}_\varepsilon, \quad \bar{\pi}_{\varepsilon,2} \in \bar{\Pi}_\varepsilon, \quad u_\varepsilon = \|\bar{\pi}_{\varepsilon,1}\| = \|\bar{\pi}_{\varepsilon,2}\|,$$

then $\bar{\pi}_\varepsilon = (\bar{\pi}_{\varepsilon,1} + \bar{\pi}_{\varepsilon,2})/2 \in \bar{\Pi}_\varepsilon$ and it follows, from (5.61), that

$$\|\bar{\pi}_\varepsilon\|^2 < (\|\bar{\pi}_{\varepsilon,1}\|^2 + \|\bar{\pi}_{\varepsilon,2}\|^2)/2 = u_\varepsilon^2,$$

which is a contradiction. Analogously, since the function $\|\bar{\pi}\|$ and the set $\bar{\Pi}_\varepsilon$ are symmetric, the extreme sequence of measures $\bar{\pi}_\varepsilon$ consists of symmetric measures (this is the unique point in the proof which does not hold for the positive case (5.57)).

The case $H_2 = \infty$ corresponds to the problem without the second constraint. The equality $u_\varepsilon(H_1, \infty) = 0$ for $r \geq r_p^*$ has been established in the proofs of the lower bounds of Theorems 3.1, 3.2. Moreover, it follows from the proofs that there exists sequences $H_{2,n} \to \infty$ and $\bar{\pi}_n \in \bar{\Pi}(H_1, H_{2,n})$ such that $\|\bar{\pi}_n\| \to 0$ as $n \to \infty$. In the case (5.57) with $r \geq r_{p,+}^*$ we use Theorems 3.3, 3.4.

Let $\mathcal{G}_2(\bar{\pi}_\varepsilon) = \tilde{H}_2 < H_2$. Taking n and $\bar{\pi}_n \in \bar{\Pi}(H_1, H_{2,n})$ such that $\delta_n = \|\bar{\pi}_n\| < u_\varepsilon$, consider the sequence

$$\bar{\pi}_{n,h} = h\bar{\pi}_n + (1-h)\bar{\pi}_\varepsilon, \quad h \in (0,1).$$

By the convexity of $\mathcal{G}_2(\bar{r})$ one has

$$\mathcal{G}_2(\bar{\pi}_{n,h}) \leq h\mathcal{G}_2(\bar{\pi}_n) + (1-h)\mathcal{G}_2(\bar{\pi}_\varepsilon) = hH_{2,n} + (1-h)\tilde{H}_2 \leq H_2$$

for $H_{2,n} \leq H_2$ or for $h \leq (H_2 - \tilde{H}_2)/(H_{2,n} - \tilde{H}_2) = h_0$. Analogously, by the convexity of $\|\bar{r}\|^2$ and the concavity of $\mathcal{G}_1(\bar{r})$, for any $h \in (0,1)$, one has

$$\|\bar{\pi}_{n,h}\|^2 < u_\varepsilon^2, \quad \mathcal{G}_1(\bar{\pi}_{n,h}) \geq H_1.$$

Therefore, if $h \leq h_0$, then $\bar{\pi}_{n,h} \in \bar{\Pi}(H_1, H_2)$ and $\|\bar{\pi}_{n,h}\|^2 < u_\varepsilon^2$, which is a contradiction.

To prove statement (3) note that if $\bar{\pi} \in \bar{\Pi}(H_1, H_2)$ then, by $\mathcal{G}_1(\bar{0}) = 0$, $\mathcal{G}_2(\bar{0}) = 0$, it follows from convex assumptions, that, for any $t \in (0,1)$,

$$\mathcal{G}_1(t\bar{\pi}) \geq t\mathcal{G}_1(\bar{\pi}), \quad \mathcal{G}_2(t\bar{\pi}) \leq t\mathcal{G}_2(\bar{\pi}),$$

i.e., $t\bar{\pi} \in \bar{\Pi}(tH_1, tH_2)$. Since $\|t\bar{\pi}\| = t\|\bar{\pi}\|$ these yield statement (3).

To prove statement (4) we use the following:

Lemma 5.1 *Let f be a convex function defined on the convex compact $X \subset R^k$. Suppose*

$$\sup_{x,y \in X} |f(x) - f(y)| \leq D,$$

and let $K \subset X$ be a compact such that the distance δ between K and the boundary of X is positive. Then f is Lipschitzian on K with the constant $C = D/\delta$.

Proof of Lemma 5.1. Let $x,\ y \in K,\ f(x) < f(y)$. Consider the line $x + t\ell,\ t \geq 0,\ |\ell| = 1$ such that $y = x + t_0\ell;\ t_0 = |x - y|$. Let $t_1 \geq t_0 + \delta$ be such that the point $z = x + t_1\ell$ belongs to the boundary of X. It follows from convexity that $f(y) \leq (1-p)f(x) + pf(z),\ p = t_0/t_1$. This yields

$$\frac{f(y) - f(x)}{|y - x|} \leq \frac{f(z) - f(x)}{|z - x|} \leq \frac{D}{\delta}. \qquad \square$$

Return to the proof of statement (4). Let $K \subset R^2_+$ be a compact and let K_δ be a δ-neighborhood of K. Choose $\delta > 0$ small enough such that $K_\delta \subset R^2_+$. Let $X \subset R^2_+$ be a closure of the convex hull of K_δ. Then under assumption A.6 there exist $D > 0,\ \varepsilon_0 > 0$ such that

$$\sup_{b,c \in X} |u^2_\varepsilon(b_1 H_{1,\varepsilon}, b_2 H_{2,\varepsilon}) - u^2_\varepsilon(c_1 H_{1,\varepsilon}, c_2 H_{2,\varepsilon})| \leq D u^2_\varepsilon$$

for $\varepsilon < \varepsilon_0$. Since we have a common Lipschitz constant, it follows from Lemma 5.1 that the function $u^2(b_1 H_{1,\varepsilon}, b_2 H_{2,\varepsilon})/u^2_\varepsilon$ is uniform continuous in $b = (b_1, b_2) \in K$ and in $0 < \varepsilon < \varepsilon_0$. $\square$

Note that, in view of (5.58), under assumptions A.5, A.6, we can change the set $\bar{\Pi}_\varepsilon = \bar{\Pi}_\varepsilon(H_{1,\varepsilon}, H_{2,\varepsilon})$ by $\tilde{\Pi}_\varepsilon = \bar{\Pi}_\varepsilon(\tilde{H}_{1,\varepsilon}, H_{2,\varepsilon})$ with $\tilde{H}_{1,\varepsilon} \sim H_{1,\varepsilon}$ which leads to $u_\varepsilon = u(\bar{\Pi}_\varepsilon) \sim \tilde{u}_\varepsilon = u(\tilde{\Pi}_\varepsilon)$ in the extreme problem (5.40). We also can change u_ε to $\tilde{u}_\varepsilon$ in Proposition 5.5.

In view of the consideration above for alternatives of interest determined by (5.45)–(5.51) our aim below is to study the extreme problem (5.40) with the sets $\bar{\Pi}_\varepsilon,\ \bar{\Pi}_{\varepsilon,+}$ determined by (5.56)–(5.57) for $u_\varepsilon \asymp 1,\ H_{2,\varepsilon} \to \infty$. In view of Proposition 5.5 it suffices to show that there exist solutions of the extreme problem which consist of the symmetric three-point measures $\pi_{\varepsilon,i} = \pi(h_{\varepsilon,i}, z_{\varepsilon,i})$ (or of the two-point measures for the positive case) and to study the properties of the extreme families $h_{\varepsilon,i}, z_{\varepsilon,i}$, particularly, to control assumptions A.1–A.6.

5.3 Separation of the Problem

Later in this chapter we will consider the cases which correspond to power or Besov norms.

It is convenient to separate the extreme problems. Let us consider problem (5.40) with the sets $\bar{\Pi}_\varepsilon$ determined by (5.56):

$$u^2(H_1, H_2) = \inf\left\{\|\bar{\pi}\|^2 :\ \mathcal{G}_1(\bar{\pi}) \geq H_1,\ \mathcal{G}_2(\bar{\pi}) \leq H_2;\ \pi_i \in \Pi\right\}, \qquad (5.64)$$

where $\mathcal{G}_l(\{\pi_i\}) = F_l(\{\varphi_l(\pi_i)\}),\ l = 1, 2$.

5.3.1 *Case $p = q$*

Let

$$f_1(t) = f_2(t) = |t|^p, \quad \varphi_1(\pi) = \varphi_2(\pi) = E_\pi f(t).$$

These correspond to $0 < p = q < \infty$ and $\varphi_1(\pi) = \varphi_2(\pi) = E_\pi|t|^p$. Fix $x \geq 0$, denote

$$\Pi(x,p) = \{\pi \in \Pi : E_\pi|t|^p = x\}, \quad u^2(x,p) = \inf_{\pi\in\Pi(x,p)} \|\pi\|^2. \tag{5.65}$$

Note that *the extreme problem* (5.65) *is convex* (this means that the function under the infimum is convex and the set for minimization is convex). It follows from Proposition 2.8 that the function $u^2(x,p)$ is convex in $x \geq 0$.

We call the extreme problem (5.65) *one-dimensional.* Introduce the set of nonnegative sequences

$$\Lambda(H_1, H_2) = \{\bar{x} : F_1(\bar{x}) \geq H_1, \ F_2(\bar{x}) \leq H_2\}. \tag{5.66}$$

The set $\Lambda(H_1, H_2; p)$ is convex because F_2 is convex and F_1 is concave. Set

$$\bar{\Pi}(\bar{x}, p) = \{\bar{\pi} \in \bar{\Pi} : \pi_i \in \Pi(x_i, p); \ i \in I\}. \tag{5.67}$$

The set $\bar{\Pi}(H_1, H_2)$ consists of sequences $\bar{\pi} \in \bar{\Pi}(\bar{x}, p)$ with $\bar{x} \in \Lambda(H_1, H_2)$:

$$\bar{\Pi}(H_1, H_2) = \bigcup_{\bar{x}\in\Lambda(H_1,H_2)} \bar{\Pi}(\bar{x}, p).$$

Since the function $\|\bar{\pi}\|^2$ is additive, one has

$$\inf_{\bar{\pi}\in\bar{\Pi}(\bar{x},p)} \|\bar{\pi}\|^2 = \sum_i \inf_{\pi_i\in\Pi(x_i,p)} \|\pi_i\|^2 = \sum_i u^2(x_i, p).$$

Therefore we get the relation

$$u^2(H_1, H_2) = \inf_{\bar{x}\in\Lambda(H_1,H_2)} u^2(\bar{x}, p); \quad u^2(\bar{x}, p) = \sum_i u^2(x_i, p). \tag{5.68}$$

We call relation (5.68) the *separation of the extreme problem for $p = q$.* Note that *the extreme problem* (5.68) *is convex.*

Analogous separation is possible for positive alternatives as well:

$$u_+^2(H_1, H_2) = \inf_{\bar{x}\in\Lambda(H_1,H_2)} \sum_i u_+^2(x_i, p) \tag{5.69}$$

where

$$u_+^2(x,p) = \inf_{\pi\in\Pi_+(x,p)} \|\pi\|^2, \quad \Pi_+(x,p) = \Pi(x,p) \cap \Pi_+. \tag{5.70}$$

5.3.2 *Case $p \neq q$*

The separation is possible for $p \neq q$ as well. Recall that if $p, q < \infty$, then

$$f_1(t) = |t|^p, \quad f_2(t) = |t|^q, \quad \varphi_l(\pi) = E_\pi f_l(t), \quad l = 1, 2.$$

If $q = \infty$, then $\varphi_2(\pi) = |\pi|$ where $|\pi|$ is defined by (5.55). Fix $x, y \geq 0$, introduce the set $\Pi(x, y; p, q)$: if $p, q < \infty$, then

$$\Pi(x, y; p, q) = \{\pi \in \Pi : E_\pi |t|^p \geq x, \ E_\pi |t|^q \leq y\},$$

and if $p < \infty, \ q = \infty$, then

$$\Pi(x, y; p, \infty) = \{\pi \in \Pi : E_\pi |t|^p \geq x, \ |\pi| \leq y\}.$$

We call the *one-dimensional extreme problem* the minimization problem

$$u^2(x, y; p, q) = \inf_{\pi \in \Pi(x,y;p,q)} \|\pi\|^2. \tag{5.71}$$

Since the set $\Pi(x, y; p, q)$ and the function $\|\pi\|^2$ are convex, it follows from Proposition 2.8 that the function $u^2(x, y; p, q)$ is convex in $(x, y) \in R_+^2$.

Introduce the set of pairs of nonnegative sequences

$$\Lambda(H_1, H_2) = \{(\bar{x}, \bar{y}) \ : \ F_1(\bar{x}) \geq H_1, \ F_2(\bar{y}) \leq H_2\}. \tag{5.72}$$

The set $\Lambda(H_1, H_2)$ is convex because F_2 is convex and F_1 is concave. Set

$$\bar{\Pi}(\bar{x}, \bar{y}; p, q) = \{\bar{\pi} \in \bar{\Pi} \ : \ \pi_i \in \Pi(x_i, y_i; p, q); \ i \in I\}. \tag{5.73}$$

Since the functionals F_1, F_2 are monotone, one has

$$\bar{\Pi}(H_1, H_2) = \bigcup_{(\bar{x},\bar{y}) \in \Lambda(H_1,H_2)} \bar{\Pi}(\bar{x}, \bar{y}; p, q),$$

which yield

$$\inf_{\bar{\pi} \in \bar{\Pi}(\bar{x},\bar{y};p,q)} \|\bar{\pi}\|^2 = \sum_i \inf_{\pi_i \in \Pi(x_i,y_i;p,q)} \|\pi_i\|^2 = \sum_i u^2(x_i, y_i; p, q).$$

Therefore, we get the relation

$$u^2(H_1, H_2) = \inf_{(\bar{x},\bar{y}) \in \Lambda(H_1,H_2)} u^2(\bar{x}, \bar{y}; p, q) \tag{5.74}$$

where $u^2(\bar{x}, \bar{y}; p, q) = \sum_i u^2(x_i, y_i; p, q)$ and *the extreme problem* (5.74) *is convex.*

To study the extreme problem (5.74) we can use classical methods based on the Lagrange multipliers rule. By the structure of the constraints (5.72) the Lagrange function for problem (5.74) is of the form

$$L(\bar{x}, \bar{y}, A, B; H_1, H_2) = u^2(\bar{x}, \bar{y}; p, q) - A(F_1(\bar{x}) - H_1) + B(F_2(\bar{y}) - H_2)$$

where $A \geq 0, \ B \geq 0$. The extreme problem (5.74) is equivalent to minimization of the Lagrange function

$$u^2(H_1, H_2) = \inf_{\bar{x},\bar{y}} L(\bar{x}, \bar{y}, A, B; H_1, H_2) \tag{5.75}$$

subject to $A \geq 0, \ B \geq 0; \ x_i \geq 0, \ y_i \geq 0 \ \forall \ i \in I$.

Consider the case of power norms with $p,\ q < \infty$,

$$F_1(\bar{x}) = \sum_{i=1}^{\infty} i^{rp} x_i, \quad F_2(\bar{y}) = \sum_{i=1}^{\infty} i^{sq} y_i.$$

Then problem (5.75) is separated to a sequence of Lagrange extreme problems for fixed A, B:

$$\inf_{x \geq 0, y \geq 0} \{u^2(x, y; p, q) - A i^{rp} x + B i^{sp} y\}. \tag{5.76}$$

Let the infima in (5.76) be attained in (x_i, y_i), where $x_i = x_i(A, B) \geq 0,\ y_i = y_i(A, B) \geq 0$. Then the quantities A, B are determined by the equations and constraints

$$A(F_1(\bar{x}) - H_1) = 0, \quad B(F_2(\bar{x}) - H_2), \quad A \geq 0, \quad B \geq 0. \tag{5.77}$$

The case $A = 0$ corresponds to the extreme problem (5.64) without the first constraint ($H_1 = 0$) and $u^2(0, H_2) = 0$ by Proposition 5.6. The case $B = 0$ corresponds to $H_2 = \infty$ and $u^2(H_1, \infty) = 0$ for the problems of interest with $r \geq r_p^*$. Therefore, for the study of the problems (5.76), (5.77), we consider the case $A > 0,\ B > 0$ and we rewrite (5.77) in the form of the equations

$$F_1(\bar{x}) = H_1, \quad F_2(\bar{y}) = H_2. \tag{5.78}$$

If $p < \infty,\ q = \infty$, then $F_2(\bar{y}) = \sup_i i^s y_i$ and analogous considerations lead to a sequence of Lagrange extreme problems

$$\inf u^2(x, y; p, \infty) - A i^{rp} x, \quad \text{subject to } x \geq 0, \quad 0 \leq i^s y \leq H_2, \tag{5.79}$$

and to the equation (for $r \geq r_p^*$):

$$F_1(\bar{x}) = H_1. \tag{5.80}$$

Consider the case of Besov norms (2.72) with $h \leq p,\ t \geq q$ assuming $t < \infty$; $J = K = 0$. In this case the sequences $\bar{x},\ \bar{y}$ are of pyramidal structure: $\bar{x} = \{\bar{x}_j,\ j \geq 1\}$; $\bar{x}_j = \{x_{ij},\ i = 1, \ldots, 2^j\}$, $x_{ij} \geq 0$, analogously for $\bar{y}$ and

$$F_1(\bar{x}) = \sum_{j=1}^{\infty} 2^{jrh} F_{1,j}(\bar{x}_j), \quad F_2(\bar{y}) = \sum_{j=1}^{\infty} 2^{jst} F_{1,j}(\bar{y}_j)$$

where

$$F_{1,j}(\bar{x}_j) = \left(\sum_{i=1}^{2^j} x_{ij} \right)^{h/p}, \quad F_{2,j}(\bar{y}_j) = \left(\sum_{i=1}^{2^j} y_{ij} \right)^{t/q}.$$

In this case we can separate problem (5.64) in a different way. Set

$$u_j^2(a, b; p, q) = \inf \left\{ \sum_{i=1}^{2^j} \|\pi_i\|^2 : \sum_{i=1}^{2^j} E_{\pi_i} |t|^p \geq a, \sum_{i=1}^{2^j} E_{\pi_i} |t|^p \leq b \right\} \tag{5.81}$$

(the infimum is taken over collections $\{\pi_1, \dots, \pi_{2^j}\}$ which consist of probability measures π_i). Then

$$u^2(H_1, H_2) = \inf \sum_{j=1}^{\infty} u_j^2(a_j, b_j; p, q), \tag{5.82}$$

where the infimum is taken over the set of sequences $\bar{a} = \{a_j\}$, $\bar{b} = \{b_j\}$, $a_j \geq 0$, $b_j \geq 0$, such that

$$\sum_{j=1}^{\infty} 2^{jhr} a_j^{h/p} \geq H_1, \quad \sum_{j=1}^{\infty} 2^{jts} b_j^{t/p} \leq H_2. \tag{5.83}$$

Certainly the extreme problem (5.81) is separated to a one-dimensional problem as well. On the other hand, this problem is symmetric with respect to all permutations of the indexes i. Because of the convexity of the problem, it follows from Proposition 5.6 that the extreme collection $\bar{\pi}_j = \{\pi_{ij}\}$ is unique which yields that the measures π_{ij} do not depend on i, i.e., $\pi_{ij} = \pi_j$. Therefore π_j is the extreme measure in a one-dimensional problem with $x_j = 2^{-j}a$, $y_j = 2^{-j}b$ and we get

$$u_j^2(a, b; p, q) = 2^j u^2(x_j, y_j; p, q).$$

Thus, we can rewrite the extreme problem (5.82) in terms of sequences of the nonpyramidal structure $\bar{x} = \{x_j\}$, $\bar{y} = \{y_j\}$, $x_j \geq 0$, $y_j \geq 0$,

$$u^2(H_1, H_2) = \inf \left\{ \sum_{j=1}^{\infty} 2^j u^2(x_j, y_j; p, q) : \Phi_1(\bar{x}) \geq H_1, \, \Phi_2(\bar{y}) \leq H_2 \right\}, \tag{5.84}$$

where the infimum is taken over sequences $(\bar{x}, \bar{y})$ under the constraints which are determined by the functions

$$\Phi_1(\bar{x}) = \sum_{j=1}^{\infty} 2^{j(rh+h/p)} x_j^{h/p}, \quad \Phi_2(\bar{y}) = \sum_{j=1}^{\infty} 2^{j(st+t/p)} y_j^{t/p}.$$

The Lagrange function for problem (5.84) is of the form

$$\sum_{j=1}^{\infty} 2^j u^2(x_j, y_j; p, q) - A(\Phi_1(\bar{x}) - H_1) + B(\Phi_2(\bar{y}) - H_2).$$

We can assume $A > 0$, $B > 0$ for $r \geq r_p^*$ by Proposition 5.6, which leads to the sequence of Lagrange extreme problems

$$\inf_{x \geq 0, y \geq 0} \{2^j u^2(x, y; p, q) - A 2^{j(rh+h/p)} x^{h/p} + B 2^{j(st+t/q)} y^{t/q}\} \tag{5.85}$$

and to the equations

$$\Phi_1(\bar{x}) = H_1, \quad \Phi_2(\bar{y}) = H_2. \tag{5.86}$$

If $t = \infty$, then we can rewrite (5.85) in the form

$$\inf \left\{ 2^j u^2(x, y; p, q) - A 2^{j(hr+h/p)} x^{h/p} \right\} \tag{5.87}$$

subject to

$$x \geq 0, \quad 0 \leq y \leq H_2 \cdot \begin{cases} 2^{-j(sq+1)}, & \text{if } q < \infty, \\ 2^{-js}, & \text{if } q = \infty. \end{cases}$$

Equations (5.86) are changed by

$$\Phi_1(\bar{x}) = H_1. \tag{5.88}$$

Analogous separation is possible for the positive alternatives

$$u_+^2(H_1, H_2) = \inf_{(\bar{x},\bar{y}) \in \Lambda(H_1,H_2)} \sum_i u_+^2(x_i, y_i; p, q), \tag{5.89}$$

where

$$u_+^2(x, y, p, q) = \inf_{\pi \in \Pi_+(x,y;p,q)} \|\pi\|^2. \tag{5.90}$$

The extreme problem (5.90) is convex as well. For power and Besov norms we get analogous Lagrange extreme problems and equations.

Thus, we reduce the extreme problems (5.64), defined in the set of sequences of probability measures, to the one-dimensional problems (5.65), (5.70) or (5.71), (5.90), and to the extreme problems of the type (5.68), (5.69) or (5.75), (5.90) which are defined in the space of real-valued sequences or to a sequence of Lagrange extreme problems of the type (5.76), (5.85) and to an equation of the type (5.78), (5.86) in the variables $A > 0, B > 0$.

In view of convexity, *any solution of equations* (5.78) *or* (5.86) *provides the minimum in the extreme problem under consideration.*

The plan of study below is as follows.

In the next section we study one-dimensional problems (5.65), (5.70) or (5.71), (5.90). We show that the solutions are two-point or symmetric three-point measures.

In Chapter 6 we study Lagrange extreme problems and the equations for the power and Besov norms.

5.4 Solution of One-Dimensional Problems

The method of study is as follows. By using the specific change of probability measures $\pi \to \nu$ we reduce the extreme problems (5.65), (5.70) or (5.71), (5.90) to the minimization problem of the type

$$\inf_{\nu \in \Xi} \int_0^\infty \int_0^\infty h(u, v) \nu(du) \nu(dv), \tag{5.91}$$

where $\Xi \subset \Pi$ and, for any fixed $v \geq 0$, the function $h(u,v) = h(v,u)$ is a symmetric convex function in the variable $u \geq 0$; recall that Π is the set which consists of probability measures on $(R^1, \mathcal{B})$. We show that the extreme measure ν^* in (5.91) is the Dirac mass at point $u \in R^1$. Returning to the original problem, we obtain the extreme measure π^*. [5]

5.4.1 Lemma on Convexity

Let $\nu \in \Pi$ be a measure supported on $[0, \infty)$. Set

$$u(\nu) = \int_0^\infty u\, \nu(du),$$

and let the set $\Xi \subset \Pi$ be such that if $\nu \in \Xi$, then $\delta_{u(\nu)} \in \Xi$. In view of the convexity, using Jensen's inequality twice, one has

$$\int_0^\infty \int_0^\infty h(u,v)\nu(du)\nu(dv) \geq h(u(\nu), u(\nu)), \tag{5.92}$$

and we have the equality in (5.92) for the Dirac masses $\nu = \delta_u,\ u \geq 0$. Therefore we go to the problem: minimize $h(u) = h(u,u)$ over the set $U = \{u : \delta_u \in \Xi\}$.

For $u > 0$, define the functions

$$f_{p,\alpha}(u) = 2u^{-\alpha p} \sinh^2(u^\alpha/2), \qquad f_p(u) = 2\sinh^2(u^{1/p}/2); \tag{5.93}$$

$$f_{p,\alpha,+}(u) = u^{-\alpha p}(\exp(u^\alpha) - 1), \quad f_{p,+}(u) = \exp(u^{1/p}) - 1. \tag{5.94}$$

Denote by $z_p,\ z_+(p)$ the positive solution of the equations

$$p\tanh(u^2/2) = u^2 \ \text{ for } p > 2; \quad p(1 - \exp(-u^2)) = u^2 \ \text{ for } p > 1. \tag{5.95}$$

The following lemma is an analytical basis for the solution of the extreme problems.

Lemma 5.2

(1) *The function $f_p(u)$ is strictly convex in $u \geq 0$ for $p \leq 2$ and the function $f_{p,\alpha}(u)$ is strictly convex in $u \geq 0$ for $\alpha \geq 1/2,\ p \geq 2$ and for $\alpha < 0,\ p \leq 2$. The function $f_p(u)$ increases to $u > 0$ for any $p > 0$.*

If $\alpha < 0$ and $p < 2$, then $f_{p,\alpha}(u)$ decreases from ∞ (if $p = 2$, then from $1/2$) to 0, when u increases from 0 to ∞.

If $\alpha > 0,\ p > 2$, then there exists a unique minimum of $f_{p,\alpha}(u)$ at point $u_{p,\alpha} = z_p^{2/\alpha}$.

If $\alpha > 0,\ p = 2$, then the function $f_{p,\alpha}(u)$ increases in $u > 0$ and

$$\inf_{u \geq 0} f_{p,\alpha}(u) = \lim_{u \to 0} f_{p,\alpha}(u) = 1/2.$$

[5] The extreme problem (5.65) has been studied in [51] for $p > 2$; the general extreme problem (5.71) has been solved in [93].

(2) *The function* $f_{p,+}(u)$ *is strictly convex in* $u \geq 0$ *for* $p \leq 1$ *and the function* $f_{p,\alpha,+}(u)$ *is strictly convex in* $u \geq 0$ *for* $\alpha \geq 1,\ p \geq 1$ *and for* $\alpha < 0,\ p \leq 1$. *The function* $f_{p,+}(u)$ *increases to* $u > 0$ *for any* $p > 0$.

If $\alpha < 0$ *and* $p < 1$, *then* $f_{p,\alpha,+}(u)$ *decreases from* ∞ (*if* $p = 1$, *then from* 1) *to* 0, *when* u *increases from* 0 *to* ∞.

If $\alpha > 0,\ p > 1$, *then there exists unique minimum of* $f_{p,\alpha}(u)$ *at point* $u_{p,\alpha,+} = z_+^{2/\alpha}(p)$.

If $\alpha > 0,\ p = 1$, *then the function* $f_{p,\alpha}(u)$ *increases on* $u > 0$ *and*

$$\inf_{u \geq 0} f_{p,\alpha,+}(u) = \lim_{u \to 0} f_{p,\alpha,+}(u) = 1.$$

This lemma directly implies

Corollary 5.1 *Let*

$$\begin{aligned}
h_{p,\alpha}(u,v) &= 2(uv)^{-\alpha p} \sinh^2((uv)^\alpha/2), & h_{p,\alpha}(u) = h_{p,\alpha}(u,u);\\
h_p(u,v) &= 2\sinh^2((uv)^{1/p}/2, & h_p(u) = h_p(u,u);\\
h_{p,\alpha,+}(u,v) &= (uv)^{-\alpha p}(\exp((uv)^\alpha) - 1), & h_{p,\alpha,+}(u) = h_{p,\alpha,+}(u,u);\\
h_{p,+}(u,v) &= (\exp((uv)^{1/p}) - 1), & h_{p,+}(u) = h_{p,+}(u,u).
\end{aligned}$$

Then the function $h_{p,\alpha}(u,v)$ *is convex in* $u > 0$ *for any* $v > 0$ *and in* $v > 0$ *for any* $u > 0$, *if* $\alpha \geq 1/2,\ p \geq 2$ *and for* $\alpha < 0,\ p \leq 2$. *The same holds for the function* $h_p(u,v)$ *with* $p \leq 2$. *The function* $h_p(u)$ *increases in* $u > 0$ *for any* $p > 0$. *If* $\alpha > 0,\ p > 2$, *then the function* $h_{p,\alpha}(u)$ *has a unique minimum point* $z_{p,\alpha} = z_p^{1/\alpha}$.

The same holds for the functions $h_{p,\alpha,+}(u,v)$, $h_{p,\alpha,+}(u)$ *after the change of the constraints* $p > 2,\ p \leq 2$ *by* $p > 1,\ p \leq 1$ *and* $\alpha \geq 1/2$ *by* $\alpha \geq 1$.

Proof of Lemma. We consider the functions $f_{p,\alpha}$, f_p only; the study of the functions $f_{p,\alpha,+}$, $f_{p,+}$ follows the same procedure.

The first derivatives are of the forms

$$f'_{p,\alpha}(u) = \alpha u^{-(\alpha p+1)} \left(u^\alpha \sinh(u^\alpha) - 2p \sinh^2(u^\alpha/2)\right);$$
$$f'_p(u) = p^{-1} u^{1/p-1} \sinh(u^{1/p}).$$

Since $\tanh(x) < x$ for $x > 0$, these yield

$$f'_p(u) > 0\ \forall\, u > 0,\ p > 0; \quad f'_{p,\alpha}(u) < 0\ \forall\, u > 0,\ \alpha < 0,\ p \leq 2.$$

If $\alpha > 0,\ p > 2$, then the minimum point of $f_{p,\alpha}(u)$ is determined by the equation $f'_{p,\alpha}(u) = 0$. After replacement of u^α by u^2 this is equivalent to (5.95). If $\alpha \geq 1/2,\ p > 2$, then the existence and uniqueness of a minimum point follow from convexity, which is proved later, and the inequality $z_p > 0$, where z_p is the solution of the first equality (5.95). If $0 < \alpha < 1/2,\ p > 2$, then we get the same, since the change of variable $v = u^\alpha$ is monotonously increasing.

Since the function $\tanh y/y$ is monotonously decreasing for $y > 0$, these yield $f'_{2,\alpha}(u) > 0$, $\alpha > 0$, $u > 0$, and the function $f_{2,\alpha}(u)$, $\alpha > 0$, $u > 0$, is monotonously increasing.

To prove the convexity we need to show $f''_p(u) > 0$; $f''_{p,\alpha}(u) > 0$. As for $f'_p(u)$, we get directly

$$f''_p(u) = p^{-2}u^{2/p-2}\cosh(u^{1/p})\left(1 + (1-p)u^{-1/p}\tanh(u^{1/p})\right).$$

Since $x^{-1}\tanh x < 1$ for all $x > 0$, the inequality $f''_p(u) > 0$ follows from $1 + a(1-p) > 0$ for any $a \in (0,1)$, $p \le 2$.

Consider a power series for the second derivative $f''_{p,\alpha}(u)$:

$$\begin{aligned} f''_{p,\alpha}(u) &= \alpha u^{-(\alpha p+2)}\Big((\alpha - 2p\alpha - 1)u^\alpha \sinh(u^\alpha) \\ &\quad + (\alpha u^{2\alpha} + p(p\alpha+1))\cosh(u^\alpha) - p(p\alpha+1)\Big) \\ &= u^{-(\alpha p+2)}\sum_{n=1}^{\infty} a_n u^{2n\alpha}, \end{aligned}$$

where

$$a_n = \frac{\alpha^2}{(2n)!}\left(4n^2 - 4np + p^2 + \frac{p-2n}{\alpha}\right).$$

The negative items $a_n < 0$ are possible for $p/2 < n < (p+\alpha^{-1})/2$, if $\alpha > 0$, and for $(p+\alpha^{-1})/2 < n < p/2$, if $\alpha < 0$.

Let $\alpha < 0$, $p \le 2$. Then the negative items correspond to $n < 1$. However, we do not have the items with $n < 1$ in the series. Thus, $f_{p,\alpha}$ is convex for $\alpha < 0$, $p \le 2$.

Let $\alpha \ge 1/2$, $p \ge 2$. Then $a_n < 0$ is possible only for n with

$$p/2 < n < (p+\alpha^{-1})/2 < p/2 + 1.$$

Therefore, a unique negative item is possible for $n > p/2 \ge 1$. Let us show that this item is compensated for by the neighboring items

$$\begin{aligned} S &= a_{n-1}u^{2(n-1)\alpha} + a_n u^{2n\alpha} + a_{n+1}u^{2(n+1)\alpha} \\ &= u^{2(n-1)\alpha}\left(a_{n-1} + a_n u^{2\alpha} + a_{n+1}u^{4\alpha}\right) \ge 0, \end{aligned}$$

where $a_{n-1} > 0$, $a_n < 0$, $a_{n+1} > 0$. It suffices to say that the determinant of the quadratic form in parentheses is nonpositive,

$$D = a_n^2 - 4a_{n-1}a_{n+1} \le 0.$$

We have

$$D = \frac{2\alpha^4}{(2n+2)!\,(2n)!}\left((n+1)(2n+1)\beta^2\gamma^2 - 4n(2n-1)(\gamma^2-4)(\beta^2-4)\right),$$

where $\gamma = 2n - p - \alpha^{-1}$, $\beta = 2n - p$. The inequality $D \le 0$ follows from

$$\frac{4n(2n-1)}{(n+1)(2n+1)}(\gamma^2-4)(\beta^2-4) - (\beta\gamma)^2 \ge 0.$$

To check the last inequality note that

$$4n(2n-1)\big((n+1)(2n+1)\big)^{-1} \geq 8/5$$

for the integer $n > 1$. Denote $2n - p - 1/2\alpha = l$. Then

$$|l| < 1/2\alpha, \quad \beta = l + 1/2\alpha, \quad \gamma = l - 1/2\alpha.$$

Therefore, for $\alpha \geq 1/2$ we get $(\gamma^2 - 4)(\beta^2 - 4) > 0$ and

$$\begin{aligned} &4n(2n-1)(\gamma^2-4)(\beta^2-4)/\left((n+1)(2n+1)\right) - (\beta\gamma)^2 \\ &> 16 - 4(\beta^2+\gamma^2) = 16 - 8l^2 - 2/\alpha^2 > 16 - 4/\alpha^2 \geq 0, \end{aligned}$$

which yield the required relation.

Thus we have proved the convexity of $f_{p,\alpha}(u)$ in the case $\alpha \geq 1/2$, $p \geq 2$ and the existence and uniqueness of the minimum point $u_{p,\alpha} = z_p^{2/\alpha}$ of $f_{p,\alpha}(u)$ in the case $\alpha > 0$, $p > 2$. □

5.4.2 Case $p = q$

We begin with problem (5.65); problem (5.70) is considered in an analogous way and we present only the results.

It is convenient to denote $x = \eta^p$. If $x = 0$, then $u^2(p,x) = \|\delta_0\|^2 = 0$, therefore we can assume $x > 0$. Let $\hat{\Pi}(x,p) \subset \Pi(x,p)$ be the set of symmetric probability measures in $\Pi(x,p)$. In view of Proposition 5.6 (1) we can find the solution of the extreme problem (5.65) in the set $\hat{\Pi}(x,p)$. Since

$$\left(e^{uv} + e^{-uv} - 2\right)/2 = 2\sinh^2(uv/2),$$

we can rewrite (5.65) in the form

$$u^2(x,p) = \inf_{\pi \in \hat{\Pi}(x,p)} \int_0^\infty \int_0^\infty 2\sinh^2(uv/2)\pi(du)\pi(dv). \tag{5.96}$$

First, let $p \leq 2$. Change the variable $|u|^p = y$ and consider the corresponding change of measure $\pi(du) \to \nu(dy)$:

$$\nu(y \in [a,b]) = \pi(|u| \in [a^{1/p}, b^{1/p}]); \quad 0 \leq a \leq b < \infty. \tag{5.97}$$

In terms of new measures and variables, we can rewrite (5.96):

$$\|\pi\|^2 = W_p(\nu) = \int_0^\infty \int_0^\infty 2\sinh^2((yz)^{1/p}/2)\nu(dy)\nu(dz); \tag{5.98}$$

$$E_\pi(|u|^p) = E_\nu(y). \tag{5.99}$$

The kernel in (5.98) is of the form $h_p(y,z)$. Denote $E_\nu(y) = u(\nu)$ and recall that $u(\nu) = x$ for $\pi \in \hat{\Pi}(x,p)$. The kernel is convex in view of Corollary 5.1. In view of (5.92) we get

$$W_p(\nu) \geq W_p(\delta_{u(\nu)}) = 2\sinh^2(x^{2/p}/2) = 2\sinh^2(\eta^2/2). \tag{5.100}$$

Thus, if $p \le 2$, then the extreme measure is $\nu = \delta_x$.

Next, let $p > 2$. Any symmetric probability measure $\pi \in \hat{\Pi}(x,p)$ can be presented in the form

$$\pi = (1-h)\delta_0 + h(r+\hat{r})/2, \quad h \in (0,1), \quad r(R^1_+) = \hat{r}(R^1_-) = 1, \quad (5.101)$$

where r is a probability measure concentrated on the positive half-line $R^1_+ = \{x > 0\}$ and $\hat{r}$ is symmetric to r and concentrated on the negative half-line $R^1_- = \{x < 0\}$. Note that the relation $E_\pi |u|^p = x$ is equivalent to $hE_r u^p = x$.

Introduce the probability measure μ on R^1_+ by the relation

$$\frac{d\mu}{dr}(t) = \frac{ht^p}{x}. \quad (5.102)$$

For $\pi \in \hat{\Pi}(x,p)$ we get

$$\int_0^\infty \mu(dt) = \int_0^\infty ht^p r(dt)/x = 1, \quad 1 \ge hr(R^1_+) = x\int_0^\infty t^{-p}\mu(dt). \quad (5.103)$$

Therefore, we get one-to-one correspondence between the set $\hat{\Pi}(x,p)$ and the set $\Xi(x,p)$ which consists of probability measures μ concentrated on R^1_+ such that $E_\mu(t^{-p}) \le 1/x$. Observe the relation for value h in terms of measure μ:

$$h = xE_\mu(t^{-p}).$$

In particular, if $E_\mu(t^{-p}) = x^{-1}$, then π does not have a δ-component at point 0.

We can rewrite (5.65) in the form

$$\begin{aligned} u^2(x,p) &= \inf_{\mu \in \Xi(x,p)} U_{x,p}(\mu); \\ U_{x,p}(\mu) &= x^2 \int_0^\infty\int_0^\infty 2(uv)^{-p}\sinh^2(uv/2)\mu(du)\mu(dv). \end{aligned} \quad (5.104)$$

The kernel in (5.104) is of the form $h_{p,1}(u,v)$. This is convex in view of Corollary 5.1. Analogously, in view of (5.92), we get

$$U_{x,p}(\mu) \ge U_{x,p}(\delta_{u(\mu)}).$$

Moreover, since the function t^{-p} is convex in R^1_+, we get

$$E_{\delta_{u(\mu)}}(t^{-p}) \le E_\mu(t^{-p}).$$

Therefore, if $\mu \in \Xi(x,p)$, then $\delta_{u(\mu)} \in \Xi(x,p)$ and $u^{-p}(\mu) \le 1/x$.

Thus, to minimize (5.104) it suffices to minimize $h_{p,1}(u) = h_{p,1}(u,u)$ over $u \ge x^{1/p} = \eta$. Since $h_{p,1}(u)$ decreases to $0 < u \le z_p$ and increases to $u \ge z_p$, the minimum point z is $z = \max\{\eta,\ z_p\}$. Therefore, we get

Theorem 5.1 *The solution of the extreme problem* (5.65) *is of the form:*

(1) *If $p \le 2$, then $\pi = \pi(h,z)$ is a symmetric two-point measure, concentrated on the points $\pm\eta$:*

$$\pi = \pi(1,z) = (\delta_z + \delta_{-z})/2; \quad z = \eta; \quad u^2(x,p) = 2\sinh^2(z^2/2) \ge z^4/2. \tag{5.105}$$

(2) *If $p > 2$, then $\pi = \pi(h,z)$ is a symmetric three-point measure*

$$\pi = \pi(h,z) = (1-h)\delta_0 + h(\delta_z + \delta_{-z})/2, \tag{5.106}$$

where

$$h = \min\{1, (\eta/z_p)^p\}, \quad z = \max\{\eta,\ z_p\}; \quad u^2(x,p) = 2h^2\sinh^2(z^2/2).$$

In any case we have

$$E_\pi|t|^p = x = hz^p, \quad u^2(x,p) \ge 2h^2\sinh^2(z_p^2/2), \quad h = \min\{1, (\eta/z_p)^p\}.$$

Using Theorem 5.1 we can rewrite the extreme problem (5.68) in terms of variables $z = z(x),\ h = h(x)$.

For problem (5.70) the analogous considerations are based on the properties of the functions $h_{p,+}$ for $p \le 1$ and $h_{p,1,+}$ for $p > 1$ from Corollary 5.1. These yield the following results:

Theorem 5.2 *The solution of the extreme problem* (5.70) *is of the form:*
(1) *If $p \le 1$, then π is the Dirac mass at point η,*

$$\pi = \pi(1,z) = \delta_z, \quad z = \eta; \quad u_+^2(x,p) = \|\pi\|^2 = e^{z^2} - 1. \tag{5.107}$$

(2) *If $p > 1$, then π is a two-point measure*

$$\pi = \pi(h,z) = (1-h)\delta_0 + h\delta_z, \tag{5.108}$$

where $z = \max\{\eta,\ z_+(p)\}$, the quantity $z_+(p)$ is the solution of equation (5.95), *and*

$$h = \min\{1, (\eta/z_+(p))^p\}; \quad u_+^2(x,p) = \|\pi\|^2 = h^2(e^{z^2} - 1).$$

In any case we have

$$E_\pi t^p = x = hz^p, \quad u^2(x,p) \ge h^2(e^{z_+^2(p)} - 1), \quad h = \min\{1, (\eta/z_+(p))^p\}.$$

5.4.3 Case $p \ne q$

We begin with problem (5.71); for problem (5.90) we present the results only.

As above, it is convenient to denote $x = \eta^p$ and $y = \zeta^q$ for $q < \infty$; $y = \zeta$ for $q = \infty$. We assume $p, q < \infty$; the case $p < q = \infty$ is more easily considered in an analogous way. Denote

$$\hat\Pi_0(x,y;p,q) = \{\pi \in \hat\Pi(x,y;p,q) \ :\ E_\pi|t|^p = x\}$$

(recall that $\hat{\Pi}(x,y;p,q)$ consists of symmetric measures in $\Pi(x,y;p,q)$). We can change $\Pi(x,y;p,q)$ in (5.71) to $\hat{\Pi}_0(x,y;p,q)$ without loss of generality, since if $\pi \in \Pi(x,y;p,q)$ and $E_\pi|t|^p = \tilde{x} > x$, we can change π to a symmetric measure

$$\pi_h = (1-h)\delta_0 + h(\pi + \hat{\pi})/2, \quad h = x/\tilde{x} \in (0,1),$$

such that $\pi_h \in \hat{\Pi}_0(x,y;p,q)$ and $\|\pi_h\|^2 < \|\pi\|^2$.

In view of the inequality

$$\eta = (E_\pi|u|^p)^{1/p} \le (E_\pi|u|^q)^{1/q} \le \zeta, \quad p \le q;$$

we assume $\eta \le \zeta$, when $p \le q$ (in the opposite case the set $\Pi(x,y;p,q)$ is empty and we put $u^2(x,y;p,q) = \infty$).

First, let $p < q$, $p \le 2$. Analogously to the case $p = q$ consider a change of variables and measures $t = |u|^p$ and $\pi \to \nu$. We get (5.98) and the constraint $\pi \in \Pi_0(x,y;p,q)$ corresponds to

$$E_\nu(t) = \eta^p; \qquad E_\nu(t^{q/p}) \le \zeta^q,$$

(if $q = \infty$, the last inequality is changed by $|\nu| \le \zeta^p$). By passing $\nu \to \delta_u$ with $u = E_\nu(t)$, in view of the convexity of the function $t^{q/p}$, we get

$$E_{\delta_u}(t) = u \ge \eta^p, \ E_{\delta_u}(t^{q/p}) = u^{q/p} \le E_\nu(t^{q/p}) \le \zeta^q;$$
$$W_p(\nu) > W_p(\delta_u) \ge 2\sinh^2(\eta^2/2),$$

(the second inequality is analogous to (5.100)). Therefore the extreme measure ν is supported at point x and the extreme measure $\pi \in \Pi_0(x,y;p,q)$ is of the form (5.105).

Next, we consider the cases $p > 2$ and either $p > q$ or $q - p \ge 1$. Let $\pi \in \Pi_0(x,y;p,q)$ be defined by (5.101). By using the change $\pi \to \mu$ defined by (5.102) analogously with (5.103) we get

$$E_\mu(t^{-p}) \le 1/x, \quad y \ge E_\pi(|t|^q) = xE_\mu(|t|^{q-p}),$$

and there is one-to-one correspondence between the set $\hat{\Pi}_0(x,y;p,q)$ and the set $\Xi(x,y;p,q)$ of probability measures μ which are concentrated on R^1_+ under the constraints

$$E_\mu(t^{-p}) \le 1/x, \quad E_\mu(t^{q-p}) \le y/x.$$

We can rewrite the extreme problem (5.71):

$$u^2(x,y;p,q) = \inf_{\mu \in \Xi(x,y;p,q)} U_{x,p}(\mu);$$

the functional $U_{x,p}(\mu)$ is determined by (5.104) and corresponds to the kernel $h_{p,1}(u,v)$. In considered cases the functions t^{-p}, t^{q-p} are convex in $t > 0$, the kernel $h_{p,1}(u,v)$ is convex on $u > 0$ and in $v > 0$ in view of Corollary 5.1. Therefore, if $\mu \in \Xi(x,y;p,q)$ and $u = E_\mu(t)$, then $\delta_u \in \Xi(x,y;p,q)$ and

$$U_{x,p}(\mu) \ge U_{x,p}(\delta_u) = x^2 h_{p,1}(u).$$

Thus, the extreme measure μ is concentrated on point z which minimizes function $h_{p,1}(u)$ over the set

$$U = \{u \geq 0 : \ u^{-p} \leq 1/x, \ u^{q-p} \leq y/x\}, \tag{5.109}$$

and the extreme measure π is of the form (5.106).

Let $p > 2$, $q-p \geq 1$. Then the set U is a finite interval $[\eta, (\zeta^q/\eta^p)^{1/(q-p)}]$. In view of Corollary 5.1 point z_p gives the absolute minimum for the convex function $h_{p,1}(u)$. Therefore we get

$$\begin{cases} z = z_p, \ h = (\eta/z_p)^p, & \text{if } \eta \leq z_p \leq (\zeta^q/\eta^p)^{1/(q-p)}, \\ z = \eta, \ h = 1, & \text{if } \ z_p < \eta, \\ z = (\zeta^q/\eta^p)^{1/(q-p)}, \ h = (\eta/\zeta)^{pq/(q-p)}, & \text{if } (\zeta^q/\eta^p)^{1/(q-p)} < z_p, \end{cases} \tag{5.110}$$

where the last case corresponds to the relations

$$E_\pi(|t|^p) = hz^p = \eta^p = x; \quad E_\pi(|t|^q) = hz^q = \zeta^q = y. \tag{5.111}$$

Let $p > 2$, $p > q$. Set $a = \max(\eta, (\eta^p/\zeta^q)^{1/(p-q)})$. Then the set U is the half-line $[a, \infty)$ and we get

$$\begin{cases} z = z_p, \ h = (\eta/z_p)^p, & \text{if } a \leq z_p, \\ z = \eta, \ h = 1, & \text{if } z_p < \eta \leq \zeta, \\ z = (\eta^p/\zeta^q)^{1/(p-q)}, \ h = (\zeta/\eta)^{pq/(p-q)}, & \text{if } z_p < (\eta^p/\zeta^q)^{1/(p-q)}, \ \zeta \leq \eta, \end{cases} \tag{5.112}$$

where the last case corresponds to (5.111).

Finally, we consider the cases: either $q < p \leq 2$ or $p > 2$, $0 < q - p < 1$. In these cases we use the change $\pi \to r$, $r \to \mu$ defined by (5.101), (5.102) and the transform $\mu \to \nu$ which corresponds to the change of variables $t = u^{q-p}$; these are analogous to (5.97). We have

$$E_\mu(u^{-p}) = E_\nu(t^{-p/(q-p)}), \quad E_\mu(u^{q-p}) = E_\nu(t); \quad U_{x,p}(\mu) = W_p(\nu),$$

where $U_{x,p}(\mu)$ is defined by (5.104). The function $W_p(\nu)$ is constructed analogously to (5.98); this is based on $U_{x,p}(\mu)$ and on the change $\mu \to \nu$:

$$W_p(\nu) = x^2 \int_0^\infty \int_0^\infty 2(st)^{-p/(q-p)} \sinh^2((st)^{1/(q-p)}/2)\nu(ds)\nu(dt).$$

This corresponds to the kernel $h_{p,\alpha}(s,t)$ with $\alpha = 1/(q-p)$. We need to minimize $W_p(\nu)$ over the set of probability measures ν concentrated on R^1_+ under the constraints

$$E_\nu(t) \leq y/x, \qquad E_\nu(t^{-p/(q-p)}) \leq 1/x. \tag{5.113}$$

If $p > 2$, $0 < q-p < 1$, then $-p/(q-p) < 0$, $\alpha > 1/2$, and if $q < p \leq 2$, then $-p/(q-p) > 1$, $\alpha < 0$. Therefore the functions t, $t^{-p/(q-p)}$ are convex in R^1_+ and the kernel $h_{p,\alpha}(s,t)$ is convex in $s > 0$ and in $t > 0$ in view of Corollary 5.1, and we can pass to the δ_t-measures, where $t = E_\nu(t)$. By the convexity noted above the moments of δ_t-measures satisfy (5.113). Thus we get the problem: to minimize $h_{p,\alpha}(t)$ over the set $T = \{t \leq y/x, \ t^{-p/(q-p)} \leq 1/x\}$.

Let $q < p \leq 2$. Then the problem corresponds to the maximization of $h_{p,1}(u)$ over half-line $U = [\max(\eta, (\eta^p/\zeta^q)^{1/(p-q)}), \infty)$. Since $h_{p,1}(u)$ decreases in u for $p \leq 2$, an extreme measure π is of the form (5.106) with

$$\begin{cases} z = \eta, \ h = 1, & \text{if } \zeta > \eta, \\ z = (\eta^p/\zeta^q)^{1/(p-q)}, \ h = (\zeta/\eta)^{pq/(p-q)}, & \text{if } \zeta \leq \eta, \end{cases}$$

where the last case corresponds to (5.111).

Let $p > 2$, $0 < q-p < 1$. Then the problem corresponds to the maximization of $h_{p,1}(u)$ over the interval U determined by (5.109) and we extend the relations (5.110) to this case.

Let us combine the results above.

Theorem 5.3 *Let $p,\ q < \infty$, $x = \eta^p$, $y = \zeta^q$ or $y = \zeta$ for $q = \infty$, and $\eta \leq \zeta$ for $p \leq q$. The solution of the extreme problem* (5.71) *is the symmetric three-point measure $\pi(h, z)$ of the form* (5.106) *with three possible relations for parameters h, z:* (i) $\quad z = \eta, \quad h = 1$;

(ii) $\quad z = z_p, \quad hz^p = \eta^p \quad$ *(for $p > 2$)*;

(iii) $\quad hz^p = \eta^p, \quad hz^q = \zeta^q \quad$ *(or $z = \zeta$ for $q = \infty$)*.

The relationship between p, q, η, ζ is the following:

(1) *Let $p \leq 2$ and $p \leq q$. Then relation* (i) *holds.*

(2) *Let $q < p \leq 2$. Then* (i) *holds for $\eta < \zeta$ and* (iii) *holds for $\eta \geq \zeta$.*

(3) *Let $2 < p < q < \infty$. Then* (i) *holds for $\eta > z_p$,* (ii) *holds for $\eta \leq z_p \leq (\zeta^q/\eta^p)^{1/(q-p)}$, and* (iii) *holds for $\eta \leq z_p$, $(\zeta^q/\eta^p)^{1/(q-p)} \leq z_p$.*

If $p > 2$, $q = \infty$, then (i) *holds for $\eta > z_p$,* (ii) *holds for $\eta \leq z_p \leq \zeta$, and* (iii) *holds for $\eta \leq \zeta \leq z_p$.*

(4) *Let $\infty > p > 2$ and $p > q$. Then* (i) *holds for $z_p < \eta < \zeta$,* (ii) *holds for $\eta \geq \zeta$ and $(\eta/z_p)^p \leq (\zeta/z_p)^q$ or $\zeta > \eta$ and $\eta \leq z_p$, and* (iii) *holds for $\eta \geq \zeta$ and $(\eta/z_p)^p > (\zeta/z_p)^q$.*

If $\eta > \zeta$ for $p = q < \infty$ or $p < q \leq \infty$, then the set $\Pi(x, y; p, q)$ is empty and we put $u^2(x, y, p, q) = \infty$.

In all cases we have the equality and inequality

$$E_{\pi(h,z)}|u|^p = hz^p = \eta^p, \quad \begin{cases} E_{\pi(h,z)}|u|^q = hz^q \leq \zeta^q, & \text{if } q < \infty, \\ |\pi(h, z)| = z \leq \zeta, & \text{if } q = \infty. \end{cases} \tag{5.114}$$

Relation (i) corresponds to a two-point measure. Relation (iii) yields, for all $\zeta > 0$, $\eta > 0$, and one has

$$z = \left(\frac{\eta^p}{\zeta^q}\right)^{\frac{1}{(p-q)}}, \quad h = \left(\frac{\zeta}{\eta}\right)^{\frac{pq}{(p-q)}}, \quad \text{if } q < \infty; \quad z = \zeta, \quad h = \left(\frac{\zeta}{\eta}\right)^p, \quad \text{if } q = \infty. \tag{5.115}$$

Relation (ii) is possible for $p > 2$ only.

Note that the function $u(0, y; p, q) = 0 \, \forall \, y \geq 0$ because the set $\Pi(0, y; p, q)$ contains the measure δ_0. On the other hand, the function $u(x, 0; p, q) = \infty \ \ \forall x > 0$ since the set $\Pi((x, 0; p, q) = \emptyset$.

Consider problem (5.90) which corresponds to the positive alternatives. We use analogous arguments based on the properties of the functions $h_{p,+}$, $h_{p,\alpha,+}$ from Corollary 5.1. We formulate only the results.

Theorem 5.4 *Let p, $q < \infty$, $x = \eta^p$, $y = \zeta^q$ or $y = \zeta$ for $q = \infty$ and $\eta \le \zeta$ for $p \le q$. The solution of the extreme problem* (5.90) *is the two-point measure $\pi(h, z)$ of the form* (5.108) *with three possible relations for parameters h, z:*

(i) $z = \eta, \quad h = 1;$

(ii) $z = z_+(p), \quad hz^p = \eta^p$ *(for $p > 1$);*

(iii) $hz^p = \eta^p, \quad hz^q = \zeta^q$ *(or $z = \zeta$ for $q = \infty$).*

Here the quantity $z_+(p)$ is the solution of equation (5.95).

The relationship between p, q, η, ζ is the following:

(1) *Let $p \le 1$ and $p \le q$. Then relation* (i) *holds.*

(2) *Let $q < p \le 1$. Then* (i) *holds for $\eta < \zeta$ and* (iii) *holds for $\eta \ge \zeta$.*

(3) *Let $1 < p < q < \infty$. Then* (i) *holds for $\eta > z_+(p)$,* (ii) *holds for $\eta \le z_+(p) \le (\zeta^q/\eta^p)^{1/(q-p)}$, and* (iii) *holds for $\eta \le z_+(p)$, $(\zeta^q/\eta^p)^{1/(q-p)} \le z_+(p)$.*

If $p > 1$, $q = \infty$, then (i) *holds for $\eta > z_+(p)$,* (ii) *holds for $\eta \le z_+(p) \le \zeta$, and* (iii) *holds for $\eta \le \zeta \le z_+(p)$.*

(4) *Let $\infty > p > 2$ and $p > q$. Then* (i) *holds for $z_+(p) < \eta < \zeta$,* (ii) *holds for $\eta \ge \zeta$ and $(\eta/z_p)^p \le (\zeta/z_+(p))^q$ or $\zeta > \eta$ and $\eta \le z_+(p)$, and* (iii) *holds for $\eta \ge \zeta$ and $(\eta/z_+(p))^p > (\zeta/z_+(p))^q$.*

If $\eta > \zeta$ for $p = q < \infty$ or $p < q \le \infty$, then the set $\Pi_+(x, y; p, q)$ is empty and we put $u_+^2(x, y, p, q) = \infty$.

In all cases we have the equality and inequality (5.114). Relation (i) corresponds to a Dirac mass at point η. Relation (iii) implies (5.115). Relation (ii) is possible for $p > 1$ only.

5.4.4 Reduction of the Extreme Problems (5.76), (5.85)

Let us return to the extreme problems (5.76), (5.85) from Section 5.3.2. We would like to show that there exists a unique minimum point $w_{\min} = (x_{\min}, y_{\min})$ providing a minimum in the problems and to give some constraints on this point.

Introduce the sets $\Lambda_0(p, q)$, $\Lambda_1(p, q) \subset R_+^2$ which consist of $x > 0$, $y > 0$ under the constraints

$$\Lambda_0(p,q) = \begin{cases} x < y^{p/q}, & \text{if } \ p < q, \\ x > y^{p/q}, & \text{if } \ p > q, \end{cases} \quad x > y\, z_p^{p-q}, \text{ if } p > 2; \tag{5.116}$$

$$\Lambda_1(p,q) = \{(x,y) : x^q = y^p;\ x > yz_p^{p-q}, \quad \text{if } p > 2,\ p > q\}. \tag{5.117}$$

Set

$$\Lambda(p,q) = \Lambda_0(p,q) \cup \Lambda_1(p,q), \quad \Lambda^*(p,q) = \begin{cases} \bar{\Lambda}(p,q), & \text{if } p > q \text{ or } p > 2, \\ \Lambda_1(p,q), & \text{if } p < q,\ p \le 2, \end{cases} \tag{5.118}$$

where $\bar{\Lambda}$ is the closer of the set Λ in $R^2_+ \cup (0,0)$ (i.e., we add the point $(0,0)$ and, for $p > 2$, a part of the sets which correspond to the equalities $x = yz_p^{p-q}$, $x > 0$). Relations (iii) provide one-to-one correspondence between the sets $\Lambda_0(p,q)$, $\Lambda_1(p,q)$ of the variables (x,y) and the sets $\Theta_0(p,q)$, $\Theta_1(p,q)$ of the variables (h,z) which determine the extreme measure. Namely, the $\Theta_0(p,q)$ is determined by the constraints

$$z > 0,\ h \in (0,1);\ \ z < z_p, \text{ if } 2 < p < q \text{ or } z > z_p, \text{ if } p > 2,\ p > q, \tag{5.119}$$

and the set $\Theta_1(p,q)$ consists of (h,z) : $h = 1$, $z > 0$ for $p \le 2$ or $2 < p < q$ and $z > z_p$ for $p > 2$, $p > q$. Analogously, set

$$\Theta(p,q) = \Theta_0(p,q) \cup \Theta_1(p,q), \quad \Theta^*(p,q) = \bar{\Theta}(p,q), \text{ if } p > q \text{ or } p > 2, \tag{5.120}$$

where $\bar{\Theta}$ is the closer of the set Θ in R_2. The set $\Theta^*(p,q)$ consists of (z,h) : $z \ge 0$, $h \in [0,1]$; $z \ge z_p$ for $p > q$, $p > 2$ and $z \le z_p$ for $q > p > 2$. There is a one-to-one correspondence between $\Lambda^*(p,q)$ and $\Theta^*(p,q)$ except the point $(0,0)$ which corresponds to $h = 0$ and any $z \ge 0$ or to $z = 0$ and any $h \in [0,1]$.

Consider a slightly different extreme problem

$$u_0(x,y;p,q) = \inf\{\|\pi\| \ :\ E_\pi|t|^p = x,\ E_\pi|t|^q = y\}$$

(certainly if $p < q$, then we assume $x^q \le y^p$ and if $p > q$, then we assume $x^q \ge y^p$ for the set under the constraints to be nonempty; we set $u_0(x,y;p,q) = \infty$ for the opposite cases). Clearly, $u_0(x,y;p,q) \ge u(x,y;p,q)$ since the set under the constraints is smaller: we change the inequality to the equality. It follows from the above that the first equality does not change the problem. On the other hand, if y is too large, then the second equality does not hold, which yields $u_0(x,y;p,q) > u(x,y;p,q)$. It follows from Theorem 5.3 that the second equality holds for $(x,y) \in \Lambda^*(p,q)$.

It follows from Proposition 2.8 that the function $u^2(x,y;p,q)$ is strictly convex in $\bar{\Lambda}(p,q)$ for $p > 2$ or $p > q$; note that the function $2\sinh^2(z^2/2)$ is strictly convex in $z \in R^1$.

Let us show that there exists a unique minimum point $(x,y) \in \Lambda(p,q)$ in problems (5.76), (5.85). This follows from

Lemma 5.3 *For any $A > 0$, $B > 0$, $d \in (0,1]$, $D \ge 1$ there exists a unique minimum point $w_{\min} = (x_{\min}, y_{\min}) \in \Lambda^*(p,q)$ of the extreme problem*

$$\inf_{(x,y)\in R^2_+} L(x,y); \quad L(x,y) = u^2(x,y;p,q) - Ax^d + By^D \tag{5.121}$$

which corresponds to $\theta_{\min} = (h_{\min}, z_{\min}) \in \Theta^*(p,q)$.

The infimum is attained in the set $\Lambda(p,q)$; i.e., $\theta_{\min} \in \Theta(p,q)$.

Moreover, let $2 < p < q$ and the infumum over the set $\bar{\Theta}_0(p,q)$ is attained in point $w_0 \neq (z_p^p, z_p^q)$ (i.e., $\theta_0 \neq (1, z_p)$). Then $w_0 = w_{\min}$.

Proof. First, note that

$$\inf_{(x,y)\in R_+^2} L(x,y) = \inf_{(x,y)\in\Lambda^*(p,q)} L(x,y). \tag{5.122}$$

In fact, it suffices to assume $u^2(x,y;p,q) < \infty$. Let $(x,y) \in R_+^2$ and $(x,y) \notin \Lambda^*(p,q)$. Then we have either relation (i) or (ii) and the quantity $u^2(x,y;p,q)$ does not depend on y. Therefore, by making y smaller, such that $(x,y) \in \Lambda^*(p,q)$ (this corresponds to relation (iii)), we do not change the value $u^2(x,y;p,q)$ but we make the quantity $L(x,y)$ smaller.

If $p \leq 2$, $q > p$, then $\Lambda^*(p,q) = \bar{\Lambda}_1(p,q)$ and the extreme problem in the right-hand side of (5.122) is of the form

$$\inf_{(x,y)\in\bar{\Lambda}_1(p,q)} L(x,y) = \inf_{z\geq 0} F(1,z); \tag{5.123}$$

$$F(h,z) = 2h^2\sinh^2(z^2/2) - Ah^d z^{dp} + Bh^D z^{Dq}. \tag{5.124}$$

By making the change of variables $u = z^{dp}$ in the function

$$F(1,z) = 2\sinh^2(z^2/2) - Az^{dp} + Bz^{Dq},$$

we get the function

$$G(u) = 2\sinh^2(u^{2/dp}/2) - Au + Bu^{Dq/dp}$$

which is strictly convex in $u \geq 0$ by Lemma 5.2. One can easily see that this function decreases for small u and tends to infinity as $u \to \infty$. Therefore the infimum in (5.123) is unique and this is attained in positive $z_{\min}$ such that $(x_{\min}, y_{\min}) \in \Lambda_1(p,q)$. This yields the statement of Lemma 5.3 for this case.

Let $p > 2$ or $p > q$. Consider the extreme problem

$$\inf_{(x,y)\in\bar{\Lambda}(p,q)} L(x,y) = \inf_{(h,z)\in\bar{\Theta}_0(p,q)} F(h,z); \tag{5.125}$$

(if $p > q$, then it is the same as the extreme problem in the right-hand side of (5.122)). Let us show that there exists a unique point $w_0 = (x_0, y_0)$ which provides a minimum in (5.125) and $w_0 \in \Lambda(p,q)$ (in terms of the variables (h,z) these correspond to $\theta_0 = (h_0, z_0) \in \Theta(p,q)$).

The uniqueness follows from the strict convexity of $u^2(x,y;p,q)$ on the convex set $\bar{\Lambda}(p,q)$ which yields the strict convexity of the minimized function $L(x,y)$. Therefore, we need to prove that the minimum in (5.125) is attained; the minimum point $w_0 \neq (0,0)$ and $z_0 > z_p$ for $p > q$, $p > 2$ or $z_0 < z_p$ when $h_0 < 1$ for $2 < p < q$.

Let us show that point $(0,0)$ is not a minimum point. Recall that $u^2(0,0;p,q) = 0$ which yields $L(0,0) = 0$. It suffices to show that the function $F(h,z)$ has a negative value at some point $(z_0, h_0) \in \Theta_0(p,q)$.

First, let $d < 1$ or $D > 1$. Take $(h, z_0) \in \Theta_0(p,q)$ and consider the function $f(t) = F(th, z_0)$, $t \in (0,1]$, which is of the form

$$f(t) = -A_1 t^d + B_1 t^D + C_1 t^2 \sim -A_1 t^d, \qquad \text{as } t \to 0,$$

where $A_1 > 0$, $B_1 > 0$, $C_1 > 0$. Thus $F(h, z_0)$ could be negative for any $z = z_0 > 0$ and for a small $h > 0$. Next, let $D = d = 1$. Then one can use the consideration above by taking $(h, z_0) \in \Theta_0(p,q)$ such that $Az_0^p > Bz_0^q$ (i.e., $z_0 > (A/B)^{1/(p-q)}$ for $p > q$ and $z_0 < (B/A)^{1/(q-p)}$ for $p < q$). This yields $A_1 > B_1$.

Let us show that the minimum in the right-hand side of (5.125) is attained. Note that the function $F(h, z)$ is continuous on the closed set $\bar{\Theta}_0(p,q)$. If $p < q$, then $\bar{\Theta}_0(p,q)$ is compact, which yield the infimum is attained on $\bar{\Theta}_0(p,q)$. This is negative which yields $z_0 > 0$, $h_0 > 0$.

Let $p > q$. Then $\bar{\Theta}_0(p,q)$ is not compact. Let us show that

$$\liminf F(h_n, z_n) \geq 0 \quad \forall\, z_n \to \infty, \quad (h_n, z_n) \in \Theta_0(p,q). \tag{5.126}$$

Let $F_0 = F(h_0, z_0) < 0$, where $(h_0, z_0) \in \Theta_0(p,q)$. Relation (5.126) yields that there exists $z_1 > z_0$ such that

$$\inf_{(z,h) \in \bar{\Theta}_0(p,q):\ z > z_1} F(h, z) > F_0.$$

and we can take the infimum over the compact set $\{(h, z) \in \bar{\Theta}_0(p,q) : z \leq z_1\}$ and the infimum is attained.

To verify (5.126) note that if $h_n = o(z_n^{-p})$, then the negative item in the right-hand side of (5.124) tends to 0. Let $h_n \geq bz_n^{-p}$ for some $b > 0$. Then the order of the first item in the right-hand side of (5.124) is no less than $z_n^{-2p}\exp(z_n^2)$ and the order of the negative item is no more than z_n^{dp}. Therefore, $F(h_n, z_n) \to \infty$ as $z_n \to \infty$ which yield to (5.126). These yield that the infimum is attained for $p > q$.

Let us show that for $p > 2$, $p > q$ the minimum point θ_0 of extreme problem (5.125) satisfies $z_0 > z_p$. First, let $h = h_0 \in (0,1)$. It suffices to show that the equality

$$h\frac{\partial F}{\partial h}(h, z_p) = 4h^2 \sinh^2(z_p^2/2) - Adh^d z_p^{dp} + DBh^D z_p^{Dq} = 0. \tag{5.127}$$

yields the inequality

$$z_p \frac{\partial F}{\partial z}(h, z_p) = \frac{4z_p^2 h^2 \sinh^2(z_p^2/2)}{\tanh(z_p^2/2)} - Adph^d z_p^{dp} + DqBh^D z_p^{Dq} < 0. \tag{5.128}$$

In fact, equality (5.127) corresponds to the minimum of the function $F(h, z_p)$ in $h \in (0,1)$ for fixed $z = z_p$. By inequality (5.128) this yields that, by taking some $z > z_p$, one can decrease the quality $F(h_0, z_p)$.

To verify (5.128) under (5.127) one can rewrite (5.127), (5.128):

$$Adph^d z_p^{dp} = 4ph^2 \sinh^2(z_p^2/2) + DpBh^D z_p^{Dq}, \tag{5.129}$$

$$Adph^d z_p^{dp} > \frac{4z_p^2 h^2 \sinh^2(z_p^2/2)}{\tanh(z_p^2/2)}, + DqBh^D z_p^{Dq} \tag{5.130}$$

and (5.130) follows from $p > q$ and from the equality $p \tanh(z_p^2/2) = z_p^2$ which determines the point z_p. Thus $\theta_{\min} = (h_{\min}, z_p)$, $h_{\min} = h_0 \in (0,1)$ is impossible.

Let $h_0 = 1$. Analogously to the above this yields to the inequalities:

$$\begin{aligned} \frac{\partial F}{\partial h}(1, z_p) &= 4\sinh^2(z_p^2/2) - Adz^{dp} + DBz^{Dq} \le 0; \\ Adpz_p^{dp} &\ge 4p\sinh^2(z_p^2/2) + DpBz_p^{Dq}; \\ z_p \frac{\partial F}{\partial z}(1, z_p) &= \frac{4z_p^2 \sinh^2(z_p^2/2)}{\tanh(z_p^2/2)} - Adpz_p^{dp} + DqBz_p^{Dq} < 0. \end{aligned}$$

It follows from the last inequality that, by making z some larger than z_p, we can decrease the value $F(1, z_p)$. Thus $\theta_{\min} = (1, z_p)$ is also impossible. These yield the statement of Lemma 5.3 for $p > q$.

Let $2 < p < q$. Let us show that the minimum point θ_0 of the extreme problem (5.125) satisfies $z_0 < z_p$, when $h_0 < 1$. Analogously to the case $p > q$ we need to verify the inequality

$$z_p \frac{\partial F}{\partial z}(h, z_p) > 0$$

which is opposite to inequality (5.128) under equality (5.127):

$$h \frac{\partial F}{\partial h}(h, z_p) = 0.$$

Since $p < q$, we use arguments analogous to the above. Therefore we need $h_0 = 1$ for a possible minimum point with $z_0 = z_p$. For $2 < p < q$ we need to consider the infimum over the sets $\bar{\Lambda}_2(p,q)$, $\bar{\Theta}_2(p,q)$ also; here

$$\Lambda_2(p,q) = \Lambda^*(p,q) \backslash \bar{\Lambda}_0(p,q) \quad \text{or} \quad \Theta_2(p,q) = \Theta^*(p,q) \backslash \bar{\Theta}_0(p,q),$$

which correspond to $h = 1$, $z \ge z_p$. The function $F(1, z)$ is continuous in $z \ge z_p$ and $F(1,z) \to \infty$ as $z \to \infty$. Therefore the infimum is attained over the set $\bar{\Theta}_2(p,q)$ as well. Since $\bar{\Lambda}_2(p,q) \subset \Lambda(p,q)$, $\bar{\Theta}_2(p,q) \subset \Theta(p,q)$, these yield the existence of the minimum point $w_{\min} \in \Lambda(p,q)$ in this case.

To complete the proof of Lemma 5.3 for $2 < p < q$, let us verify that if $w_0 \ne (z_p^p, z_p^q)$, then this point provides a global miminum and the minimum point $w_{\min}$ is unique for the problem in the right-hand side of (5.122).

First, let $L(w_0) \ge L(w)$ for some point $w = (x, y) \in \Lambda_1$ which corresponds to $h = 1$, $z > z_p$. Consider the segment

$$W_1 = \{(1-t)w_0 + tw, \, t \in [0,1]\},$$

and note that there exists $t_0 \in (0,1)$ such that

$$W_{t_0} = \{(1-t)w_0 + tw, \, t \in [0, t_0]\} \subset \bar{\Lambda}_0(p,q).$$

The function $L(x,y)$ is convex on the set W_1 and this is strictly convex on the set W_{t_0}. Therefore there exists a point $w_0^* \in W_{t_0}$ such that $L(w_0^*) < L(w_0)$ which contradicts that the point w_0 provides a minimum of the function $L(x,y)$ on $\bar{\Lambda}_0(p,q)$.

Next, let there exist two minimum points $w_1 \in \Lambda(p,q)$, $w_2 \in \Lambda(p,q)$. By the above we need to consider the case $w_1 \in \Lambda_1(p,q)$, $w_2 \in \Lambda_1(p,q)$. Then the segment $W = \{(1-t)w_1 + tw_2,\ t \in [0,1]\}$ is a minimum set by convexity of the function $L(x,y)$. However, the set $\Lambda_1(p,q)$ is not linear and the segment W contains the point $w \notin \Lambda^*(p,q)$ which contradicts that all minimum points are contained in $\Lambda^*(p,q)$. □

It follows from Lemma 5.3 that we can change the extreme problem (5.121) by

$$\inf_{(h,z)\in\Theta(p,q)} F(h,z), \quad \Theta(p,q) = \Theta_0(p,q) \cup \Theta_1(p,q), \tag{5.131}$$

and find a minimum point in (5.131) for $p > 2$ or $p > q$ by two steps. The first step is to try to find a minimum point over the set $\Theta_0(p,q)$. This leads to the system of equations

$$\frac{\partial F}{\partial z}(h,z) = 0, \quad \frac{\partial F}{\partial h}(h,z) = 0, \tag{5.132}$$

where the function $F(h,z)$ is defined by (5.124). If we find solution (5.132) such that $(h,z) \in \Theta_0(p,q)$, then it is a minimum point in the problem (moreover, we can admit $h = 1$). If solution (5.132) does not satisfy these constraints (below is typical situation where there exists the solution with $h > 1$ only), then the second step is to find a minimum point over the set $\Theta_1(p,q)$:

$$\inf_{z>0} F(1,z); \quad z > z_p \quad \text{for } p > 2,\ p > q. \tag{5.133}$$

This leads to the equation

$$\frac{\partial F}{\partial z}(1,z) = 0 \tag{5.134}$$

and we need to select solution (5.134) which corresponds to a minimum of $F(1,z)$.

It follows from Lemma 5.3 that *in this way we necessarily find the solution of the extreme problems* (5.131) *and* (5.121).

We will realize this scheme of asymptotic study of the extreme problems (5.76), (5.85) in the following chapters.

Note that the analogous results are fulfilled for the analogous Lagrange extreme problems which correspond to positive alternative. The main difference is that the inequalities $p > 2$, $p \le 2$ are changed by $p > 1$, $p \le 1$. Since we will not use these later, we leave it to the reader to formulate them.

5.5 Sharp Asymptotics for l^n-Balls

Let us return to the n-dimensional alternatives under the constraints defined by the l_p^n-norms

$$V_n = V_n(\rho_n, R_n) = \{v \in R^n : |v|_p \geq \rho_n,\ |v|_q \leq R_n\}; \quad V_n^+ = V_n \cap R_+^n. \tag{5.135}$$

Assume the nonempty conditions

$$\rho_n \leq \begin{cases} R_n, & \text{for } p > q, \\ R_n n^{-1/q+1/p}, & \text{for } p \leq q < \infty, \\ R_n n^{1/p}, & \text{for } q = \infty. \end{cases} \tag{5.136}$$

For the cases $p \leq 2,\ q \geq p$ (or $p \leq 1,\ q \geq p$ for the positive case) these problems were studied in Sections 4.1, 4.3.2 and we have obtained asymptotics of the Gaussian type (see Corollaries 4.1, 4.3). These considerations were based on the convexity of the sets V_n or Z_n. However, these do not hold in other cases.

To study the remaining cases we use the methods developed above. We improve the nonempty assumption (5.136) in the cases $p > q$ or $p > 2$ (or $p > 1$ for the positive case): assume that there exists a constant $c \in (0,1)$ such that

$$\rho_n/R_n \leq (1-c)n^{(1/p-1/q)_+}, \quad t_+ = \max(0,t). \tag{5.137}$$

All limits in this section are considered as $n \to \infty$.

5.5.1 *One-Sided Constraint*

We begin with the one-sided constraint

$$V_n = V_n(\rho_n) = \breve{D}^n(\rho) = \left\{ v \in R^n : \sum_{i=1}^n |v_i|^p \geq \rho_n^p \right\}; \quad p > 2, \tag{5.138}$$

or with the one-sided constraint for the positive case

$$V_n = V_n^+(\rho_n) = V_n(\rho_n) \cap R_+^n, \quad p > 1, \tag{5.139}$$

(these correspond to $R_n = \infty$ in (5.135)); for the cases $p \leq 2$ or $p \leq 1$ the hypothesis testing problems were considered in Section 2.4.3. Assume $\rho_n \to \infty$.

Let us consider the first case: $V_n = V_n(\rho_n),\ p > 2$. The extreme problem (5.40), under the constraints (5.56), is of the form

$$u_n^2 = \inf\left\{ \sum_{i=1}^n \|\pi_i\|^2 : \sum_{i=1}^n E_{\pi_i}|u|^p \geq \rho_n^p \right\}.$$

This extreme problem is separable, convex, and symmetric and its solution $\bar{\pi}_n = (\pi_n, \dots, \pi_n)$ consists of identical measures where $\pi_n = \pi(p, x_n)$ is the solution of the one-dimensional extreme problem (5.65) with $x = x_n = \rho_n^p/n$. It follows from Theorem 5.1 that the π_n is a symmetric three-point measure $\pi_n = \pi(h_n, z_n)$ with $nh_n z_n^p = \rho_n^p$. If $\eta = \rho_n n^{-1/p} > z_p$, then we get relation (i) and

$$h_n = 1, \quad z_n = \rho_n n^{-1/p}, \quad u_n^2 = u_n^2(V_n) = 2n \sinh^2(z_n^2/2). \tag{5.140}$$

and if $\rho_n n^{-1/p} \le z_p$, then we get relation (ii) and

$$z_n = z_p, \quad h_n = n^{-1}(\rho_n/z_p)^p, \tag{5.141}$$

$$u_n^2 = u_n^2(V_n) = 2n^{-1}(\rho_n/z_p)^{2p} \sinh^2(z_p^2/2) \asymp n^{-1}\rho_n^{2p}. \tag{5.142}$$

Note that relations (5.140) yield

$$u_n^2 > 2n \sinh^2(z_p^2/2) \asymp n \to \infty.$$

Therefore, if $u_n^2 = o(n)$, then we necessarily get (5.141).

Assume that $u_n \asymp 1$ which yields $z_n = z_p$ and this is equivalent to $h_n \asymp n^{-1/2}$, $\rho_n \asymp n^{1/2p}$. Then we get A.1, A.4 and, using Proposition 5.5, we obtain the upper bounds (5.42) with the change of the quantities u_ε by the quantities u_n, $n \to \infty$, determined by (5.142).

However, the measure $\pi^n = \pi_n \times \ldots \times \pi_n$ is not concentrated on V_n. Let us change ρ_n to the values $\tilde{\rho}_n = \rho_n(1+\delta)$ with small $\delta > 0$ and consider the measure $\tilde{\pi}_n = \pi(\tilde{h}_n, z_p)$ which corresponds to $\tilde{\rho}_n$. Then $\tilde{\pi}^n$ is asymptotically concentrated on V_n:

$$\tilde{\pi}^n(V_n(\rho_n)) \to 1, \quad \text{as } n \to \infty.$$

In fact, put

$$G(v) = \sum_{i=1}^n |v_i|^p,$$

which corresponds to (5.45) with $H_{2,\varepsilon} = \infty$. By definition of the product measure $\tilde{\pi}^n$ we have (compare with Section 3.3.6)

$$\begin{aligned} E_{\tilde{\pi}^n}(G(v)) &= n\tilde{h}_n z_p^p = \tilde{\rho}_n^p = \rho_n^p(1+\delta)^p; \\ \mathrm{Var}_{\tilde{\pi}^n}(G(v)) &\le \sum_{i=1}^n E_{\tilde{\pi}_n}|v_i|^{2p} = n\tilde{h}_n z_p^{2p} = \tilde{\rho}_n^p z_p^p, \end{aligned}$$

and using the Chebyshev inequality we get, for any $\delta > 0$,

$$\begin{aligned} 1 - \tilde{\pi}^n(V_n(\rho_n)) &= \tilde{\pi}^n(R^n \setminus V_n(\rho_n)) = \tilde{\pi}^n(G(v) < \rho_n^p) \\ &\le \tilde{\pi}^n(|E_{\tilde{\pi}^n}(G(v)) - G(v)| > \rho_n^p((1+\delta)^p - 1) \\ &\le \mathrm{Var}_{\tilde{\pi}^n}(G(v))/\rho_n^{2p}((1+\delta)^p - 1)^2 \asymp \rho_n^{-p} \to 0, \end{aligned}$$

moreover, we can take $\delta = \delta_n \to 0$ such that $\delta\rho_n^p \to \infty$. On the other hand, as $\delta = \delta_n \to 0$,

$$\tilde{u}_n^2 = 2n^{-1}(\tilde{\rho}_n/z_p)^{2p} \sinh^2(z_p^2/2) = u_n^2(1+\delta)^{2p} \sim u_n^2.$$

Therefore using Proposition 5.5 we get the lower bounds (5.43) with the change of u_ε by u_n defined by (5.142).

The case of a positive alternative $V_n = V_n^+(\rho_n)$ with $p > 1$ is considered in an analogous way. If $u_n \asymp 1$, then we get (ii) and

$$z_n = z_+(p), \quad h_n = n^{-1}(\rho_n/z_+(p))^p, \tag{5.143}$$

$$u_n^2 = u_n^2(V_n) = n^{-1}(\rho_n/z_+(p))^{2p}(\exp(z_+^2(p)) - 1) \asymp n^{-1}\rho_n^{2p}. \tag{5.144}$$

This is based on Theorem 5.4 and we obtain the sharp asymptotics (5.42), (5.43) with the change of u_ε to u_n defined by (5.144).

The considerations above, jointly with the Remark from Section 5.2.3, yield

Proposition 5.7 *For the alternatives defined by* (5.138), (5.139) *we have the sharp asymptotics of Gaussian type*

$$\beta(V_n, \alpha) = \Phi(T_\alpha - u_n) + o(1), \quad \gamma(V_n) = 2\Phi(-u_n/2) + o(1), \quad n \to \infty,$$

with the values u_n defined by (5.142), (5.144).

Proposition 5.7 gives the same distinguishability conditions as in Section 3.3.6 for $V_n = V_n(\rho_n)$ with $p > 2$ or for $V_n = V_n^+(\rho_n)$ with $p > 1$:

$$\gamma_n \to 1 \quad \text{iff} \quad \rho_n n^{-1/2p} \to 0; \quad \gamma_n \to 0 \quad \text{iff} \quad \rho_n n^{-1/2p} \to \infty.$$

5.5.2 *Two-Sided Constraints*

Let us return to the general case (5.135) assuming $R_n \to \infty$.

Let $q < \infty$. The sets V_n correspond to (5.45) with $\varepsilon = n$,

$$G_1(v) = \sum_{i=1}^n |v_i|^p, \quad G_2(v) = \sum_{i=1}^n |v_i|^q, \quad H_{1,n} = \rho_n^p, \quad H_{2,n} = R_n^q.$$

The extreme problem (5.40) under constraints (5.56) is of the form

$$u_n^2 = \inf\left\{\sum_{i=1}^n \|\pi_i\|^2 : \sum_{i=1}^n E_{\pi_i}|u|^p \geq \rho_n^p, \ \sum_{i=1}^n E_{\pi_i}|u|^q \leq R_n^q\right\}.$$

This extreme problem is separable, convex and symmetric and its solution $\bar{\pi}_n = (\pi_n, \dots, \pi_n)$ consists of identical measures where π_n is the solution of the one-dimensional extreme problem (5.71) with

$$x = x_n = \rho_n^p/n = \eta^p, \quad y = y_n = \zeta^q = R_n^q/n. \tag{5.145}$$

If $q = \infty$, then, analogously, we have $y = \zeta = R_n$.

It follows from Theorem 5.3 that π_n is the symmetric three-point measure $\pi_n = \pi(h_n, z_n)$ with the parameters $h_n = h(\rho_n, R_n)$, $z_n = z(\rho_n, R_n)$

defined by (i)–(iii) and by the relations in Theorem 5.3 according to (5.145), and

$$u_n^2 = 2nh_n^2 \sinh^2(z_n^2/2). \tag{5.146}$$

Let $p \le 2$. Then relation (i) holds for $q \ge p$ which corresponds to the results of Section 4.3.2 and Corollary 4.3. Relation (i) holds for $p > q$, $\rho_n \le R_n n^{1/p-1/q}$ as well: this means that the second constraint is not essential for a large enough R_n. However, if $2 \ge p > q$ and

$$R_n n^{1/p-1/q} \le \rho_n \le R_n, \tag{5.147}$$

then we get (iii). The second inequality in (5.147) corresponds to the condition that the set V_n, defined by (5.135), is nonempty. Assume $u_n \asymp 1$. Then $z_n = o(1)$ under (i). Recall that this holds for $p \le 2$, $q \ge p$.

Let $p > 2$. Then (i) is possible for $z_n = \eta_n > z_p$ which yields $u_n \to \infty$. Assume $u_n \asymp 1$. Then we get either (ii) or (iii). Note that it follows from Theorem 5.3 that $z_n \le z_p$ for $p \le q$ and $z_n \ge z_p$ for $p > q$.

In particular, for $p > 2$, $q = \infty$ we get $\rho_n n^{-1/p} \le R_n$ because of the nonempty condition. If $z_p < \rho_n n^{-1/p}$, then we get (i) and $u_n \to \infty$; if $\rho_n n^{-1/p} \le z_p \le R_n$, then we get (ii) and $z_n = z_p$, $h_n = (\rho_n/z_p)^p n^{-1} \le 1$; and the second constraint is not essential. Finally, if $R_n \le z_p$, then we get (iii) with $z = R_n$, $h_n = (\rho_n/R_n)^p n^{-1}$.

Thus under the assumption $u_n \asymp 1$ we get $z_n = O(1)$ for all $p \le q$, however, if $p > q$, then it is possible that $z_n \to \infty$; this corresponds to (iii) and

$$z_n = (\rho_n^p/R_n^q)^{1/(p-q)}, \quad h_n = n^{-1}(R_n/\rho_n)^{pq/(p-q)} \ge n^{-1};$$

the last inequality corresponds to the condition that the set V_n is nonempty. Using the considerations above we get that, if $q < \infty$, then

$$H_{1,n} = \rho_n^p = nh_n z_n^p \to \infty; \quad H_{2,n} = R_n^q \to \infty, \quad H_{2,n} \ge nh_n z_n^q, \tag{5.148}$$

and if, moreover, $u_n \asymp 1$, $z_n = O(1)$, then, under (ii) or (iii), one has

$$nh_n \to \infty, \quad nh_n z_n^q \to \infty. \tag{5.149}$$

Assume $z_n = O(1)$, $u_n \asymp 1$. Then A.1, A.4 hold and, using Proposition 5.5, we obtain the upper bounds (5.42) with the change of u_ε to u_n and $n \to \infty$.

Under (i) the measure π^n is supported on V_n which yields to the lower bounds (5.43). Under (ii) or (iii) we change the thresholds ρ_n, R_n to

$$\tilde{\rho}_n = \rho_n(1+\delta), \quad \tilde{R}_n = R_n(1-\delta), \quad \delta \in (0, c/2),$$

where c is the constant from (5.137); this entails that the set $\tilde{V}_n = V_n(\tilde{\rho}_n, \tilde{R}_n)$ is contained in V_n and is nonempty by (5.137) (note that for $q = \infty$ we need only one change, $\rho_n \to \tilde{\rho}_n$). Change the measure $\pi_n = \pi(h_n, z_n)$ to $\tilde{\pi}_n = \pi(\tilde{h}_n, \tilde{z}_n)$, where the values $\tilde{h}_n = h_n(\tilde{\rho}_n, \tilde{R}_n)$, $\tilde{z}_n = z_n(\tilde{\rho}_n, \tilde{R}_n)$

are defined by Theorem 5.3 and (5.145). It suffices to check that, for any $\delta \in (0, c/2)$,

$$\tilde{\pi}^n(V_n) = \tilde{\pi}^n(G_1(v) \geq H_{1,n},\ G_2(v) \leq H_{2,n}) \to 1,$$

which follows from

$$\tilde{\pi}^n(G_1(v) < H_{1,n}) \to 0, \quad \tilde{\pi}^n(G_2(v) > H_{2,n}) \to 0, \tag{5.150}$$

(when $q = \infty$, the second inequality in (5.150) is changed to $\tilde{z}_n \leq R_n$, which is fulfilled for a small enough $\delta > 0$ under (5.137)).

One can easily see *the continuity property*: if $z_n = O(1)$, then one has $\tilde{h}_n \asymp h_n$, $\tilde{z}_n \asymp z_n$, $\tilde{u}_n \asymp u_n$ for any $\delta \in (0,1)$, and $\tilde{h}_n \sim h_n$, $\tilde{z}_n \sim z_n$, $\tilde{u}_n \sim u_n$, as $\delta = \delta_n \to 0$.

Using the continuity property, (5.148), and (5.149) we easily get

$$\begin{aligned}
E_{\tilde{\pi}^n}(G_1(v)) &= n\tilde{h}_n\tilde{z}_n^p = (1+\delta)^p H_{1,n} \to \infty,\\
\mathrm{Var}_{\tilde{\pi}^n}(G_1(v)) &\leq n\tilde{h}_n\tilde{z}_n^{2p} \asymp (E_{\tilde{\pi}^n}(G_1(v)))^2/n\tilde{h}_n;\\
E_{\tilde{\pi}^n}(G_2(v)) &= n\tilde{h}_n\tilde{z}_n^q \leq (1-\delta)^q H_{2,n} \to \infty,\\
\mathrm{Var}_{\tilde{\pi}^n}(G_2(v)) &\leq n\tilde{h}_n\tilde{z}_n^{2q} \asymp (E_{\tilde{\pi}^n}(G_2(v)))^2/n\tilde{h}_n.
\end{aligned}$$

The evaluations above and the Chebyshev inequality yield (5.150). Thus, we get the lower bounds (5.43) under (ii) and (iii) as well.

By repeating the considerations from the Remark in Section 5.2.3 we can exclude the assumption $u_n \asymp 1$ for the upper and lower bounds (5.42), (5.43) to be fulfilled.

The positive alternatives V_n^+ are considered in an analogous way: using Theorem 5.4 we change the inequality $p > 2$ to $p > 1$ and the values z_p to $z_+(p)$.

We combine the results in the following:

Proposition 5.8 *For the alternatives defined by* (5.135), (5.137) *let the values* z_n, h_n *be defined by Theorems 5.3, 5.4 with* x_n, y_n *defined by the relations* (5.145). *Let* $q \leq p$ *or* $p > q$ *and let* $z_n = O(1)$. *Then we have the sharp asymptotics of Gaussian type*

$$\beta(V_n, \alpha) = \Phi(T_\alpha - u_n) + o(1), \qquad \gamma(V_n) = 2\Phi(-u_n/2) + o(1); \tag{5.151}$$

$$\beta(V_n^+, \alpha) = \Phi(T_\alpha - u_{n,+}) + o(1),\ \gamma(V_n^+) = 2\Phi(-u_{n,+}/2) + o(1), \tag{5.152}$$

as $n \to \infty$ *with the values*

$$u_n^2 = 2nh_n^2 \sinh^2(z_n^2/2), \quad u_{n,+}^2 = nh_n^2(\exp(z_n^2/2) - 1). \tag{5.153}$$

If $nh_n^2 \to \infty$ *or* $z_n \to 0$, *then we can change* (5.153) *by*

$$u_n^2 = nh_n^2 z_n^4/2, \quad u_{n,+}^2 = nh_n^2 z_n^2. \tag{5.154}$$

In particular, let $q = \infty$. If $p \leq 2$ (or $p \leq 1$ for the positive alternative), then the second constraint is not essential under the nonempty condition (5.136) and we get the distinguishability conditions from Corollaries 4.1,

4.3. For $q = \infty$ and $p > 2$ (or $p > 1$ for the positive alternative) we get distinguishability conditions of a different form:

Corollary 5.2 *Let $q = \infty$ and $p > 2$. Then*

$$\gamma(V_n) \to 1, \quad \textit{iff} \begin{cases} n^{-1}\rho_n^{2p} \to 0, & \textit{for } \liminf R_n > 1, \\ n^{-1}\rho_n^{2p} R_n^{4-2p} \to 0, & \textit{for } R_n = O(1), \end{cases}$$

$$\gamma(V_n) \to 0, \quad \textit{iff} \begin{cases} n^{-1}\rho_n^{2p} \to \infty, & \textit{for } \liminf R_n > 1, \\ n^{-1}\rho_n^{2p} R_n^{4-2p} \to \infty, & \textit{for } R_n = O(1), \end{cases}$$

Let $q = \infty$ and $p > 1$. Then

$$\gamma(V_{n,+}) \to 1, \quad \textit{iff} \begin{cases} n^{-1}\rho_n^{2p} \to 0, & \textit{for } \liminf R_n > 1, \\ n^{-1}\rho_n^{2p} R_n^{2-2p} \to 0, & \textit{for } R_n = O(1), \end{cases}$$

$$\gamma(V_{n,+}) \to 0, \quad \textit{iff} \begin{cases} n^{-1}\rho_n^{2p} \to \infty, & \textit{for } \liminf R_n > 1, \\ n^{-1}\rho_n^{2p} R_n^{2-2p} \to \infty, & \textit{for } R_n = O(1). \end{cases}$$

Note that we exclude the case $z_n \to \infty$ which is possible for $p > q$. This case will be considered in Chapter 8.

Example 5.1 *Radii of the power form.* Let the radius of the ball in the second constraint be of the form

$$R_n = An^a, \quad a > 0, \quad A > 0; \quad a \geq 1/2q \quad \text{for } p > q.$$

Assume ρ_n to be such that the nonempty condition (5.136) holds and htat the extended nonempty condition (5.137) holds for $p > q$ or $p > 2$. Using the above results one gets $\beta_n(\alpha) \to 1 - \alpha$, if $a < 1/q - 1/4$ for either $p \leq q$, $p \leq 2$ or $2 < p \leq q$. In the other cases let us try to determine the asymptotics ρ_n such that $u_n \asymp 1$, $z_n = O(1)$.

Denote by $\tau = (p, q, a, A, C)$ the collection of parameters in the problem and define the sets of collections

$$\begin{aligned} \Xi_1 &= \left\{\tau : p \leq 2 \text{ and either } a > \frac{1}{q} - \frac{1}{4} \text{ or } a = \frac{1}{q} - \frac{1}{4}, \ A \geq C\right\}, \\ \Xi_2 &= \left\{\tau : p > 2 \text{ and either } a > \frac{1}{2q} \text{ or } a = \frac{1}{2q}, \ C^p \leq A^q z_p^{p-q}\right\}, \\ \Xi_3 &= \Xi_{31} \cup \Xi_{32} \cup \Xi_{33}, \end{aligned}$$

where

$$\begin{aligned} \Xi_{31} &= \left\{\tau : q > p > 2 \text{ and either } \frac{1}{q} - \frac{1}{4} < a < \frac{1}{2q} \text{ or } a = \frac{1}{q} - \frac{1}{4}, \ C \leq A\right\}, \\ \Xi_{32} &= \left\{\tau : 2 \geq p > q \text{ and either } \frac{1}{2q} \leq a < \frac{1}{q} - \frac{1}{4} \text{ or } a = \frac{1}{2q}, \ C > A\right\}, \\ \Xi_{33} &= \left\{\tau : p > 2, \ a = \frac{1}{2q}, \ C^p > A^q z_p^{p-q}\right\}; \end{aligned}$$

if $\tau \in \Xi_{31}$ and $a = 1/q - 1/4$, then we assume additionally that $q < 4$ for $a > 0$.

Let the radii ρ_n be of the form $\rho_n = Cn^d$, $C \in (0, \infty)$ with

$$d = \begin{cases} 1/p - 1/4, & \text{if } \tau \in \Xi_1, \\ 1/2p, & \text{if } \tau \in \Xi_2, \\ (p - q + 2aq(2-p))/2p(2-q), & \text{if } \tau \in \Xi_3. \end{cases}$$

Then, using Proposition 5.8, we obtain the parameters h_n, z_n:

$$\begin{cases} z_n = Cn^{-1/4}, & h_n = 1, & \text{if } \tau \in \Xi_1, \\ z_n = z_p, & h_n = (C/z_p)^p n^{-1/2}, & \text{if } \tau \in \Xi_2, \\ z_n = n^{-\delta}(C^p/A^q)^{1/(p-q)}, & h_n = n^{\gamma-1}(A/C)^{pq/(p-q)}, & \text{if } \tau \in \Xi_3, \end{cases}$$

where, for $\tau \in \Xi_3$,

$$\gamma = \frac{q(4a-1)}{2(2-q)} \in [1/2, 1], \qquad \delta = \frac{2aq-1}{2(2-q)} \geq 0,$$

and we get the asymptotics (5.151) with

$$u_n^2 \sim \begin{cases} C^4/2, & \text{if } \tau \in \Xi_1, \\ 2(C/z_p)^{2p} \sinh^2(z_p^2/2), & \text{if } \tau \in \Xi_2, \\ C^{2p(2-q)/(p-q)} A^{-2q(2-p)/(p-q)}/2, & \text{if } \tau \in \Xi_3, \quad a \neq 1/2q, \\ 2(A/C)^{2pq/(p-q)} \sinh^2((C^p/A^q)^{2/(p-q)}/2), & \text{if } \tau \in \Xi_3, \quad a = 1/2q. \end{cases}$$

Note that we do not consider the cases $p > q$, $0 < a < 1/2q$. One can see that relation (iii) holds in these cases and if $z_n = O(1)$, then $u_n \to 0$. Therefore we need to study the case $z_n \to \infty$ for $p > q$, $0 < a < 1/2q$. This case will be considered later in Section 8.3.2.

6

Gaussian Asymptotics for Power and Besov Norms

In this chapter we consider the alternatives V_ε under constraints determined by the power norms (2.71),

$$V_\varepsilon = V(\kappa, \rho_\varepsilon, R_\varepsilon) = \{v \in l^2 : \ |v|_{r,p} \geq \rho_\varepsilon, \ |v|_{s,q} \leq R_\varepsilon\}; \tag{6.1}$$

and by the Besov norms (2.72)

$$V_\varepsilon = V(\tau, \rho_\varepsilon, R_\varepsilon) = \{v \in l^2 : \ |v|_{r,p,h} \geq \rho_\varepsilon, \ |v|_{s,q,t} \leq R_\varepsilon\}; \tag{6.2}$$

(compare with (3.110), (3.122)); we denote $\kappa = (r, s, p, q)$, $\tau = (\kappa, h, t)$. We consider the general cases

$$-\infty < r, s < \infty, \quad 0 < p, h < \infty, \quad 0 < q, t \leq \infty,$$

for the rate asymptotics problem and add some additional constraints on h, t for the sharp asymptotics problem. We assume the nonempty conditions from Corollary 3.15 and we suppose $\rho_\varepsilon \to \infty$, $R_\varepsilon \to \infty$.

These problems were studied above for the cases $\kappa \in \Xi_C$ (the set Ξ_C is defined by (3.113) and corresponds to the asymptotics of classical type (3.114)), for $\kappa \in \Xi_T$ (the set Ξ_T is defined by (3.116) and corresponds to the triviality; see Sections 3.4.6, 3.4.7), and for $\kappa \in \Xi_D$ (the set Ξ_D is defined by (4.146) and corresponds to the asymptotics of degenerate type; see Section 4.4). Moreover the case $p \leq 2$, $q \geq p$ have been studied in Sections 4.3.3, 4.3.4 and we have obtained the asymptotics of Gaussian type.

Here we extend these results. We consider the case $\kappa \in \Xi_G$ where

$$\Xi_G = \{\kappa : r \geq r_p^*, \ \kappa \notin \Xi_T, \ \kappa \notin \Xi_D\}. \tag{6.3}$$

We describe the regions Ξ_{G_1}, Ξ_{G_2} of the principal types and the regions Ξ_{G_3}–Ξ_{G_5} of frontier types of Gaussian asymptotics in the problems. Using the wavelet transform we translate the rate asymptotics to alternatives determined by Besov norms in the functional space (2.78) under a functional Gaussian model.

For the regions of the principal types, we show that the rate asymptotics for the Besov norms are the same as that for the power norms and do not depend on the additional parameters t, h. These provide the translations of the results to alternatives determined by Sobolev norms in the functional space.

6.1 Extreme Problems

We would like to apply Proposition 5.5 and to obtain the asymptotics (5.44) in the problems under consideration. The main point is the study of the extreme problem (5.40) with the sets $\bar{\Pi}_\varepsilon$ specified for the problems of interest. In view of the considerations from Section 5.2.2, we can rewrite the extreme problem (5.40) in the form (5.64) where $H_l = H_{l,\varepsilon}$ and the functionals $\mathcal{G}_l(\bar{\pi})$ are determined by the functionals $G_l(v)$, $l = 1, 2$, according to (5.53), (5.54). This has been discussed in Sections 5.2.2, 5.3 and we will combine the results.

First, let us consider the problem for power norms. If $q < \infty$, then the set V_ε is of the form (5.45) with

$$G_1(v) = \sum_i i^{rp}|v_i|^p, \quad H_{1,\varepsilon} = \rho_\varepsilon^p; \quad G_2(v) = \sum_i i^{sq}|v_i|^q, \quad H_{2,\varepsilon} = R_\varepsilon^q.$$

If $q = \infty$, then

$$G_2(v) = \sup_i i^s|v_i|, \quad H_{2,\varepsilon} = R_\varepsilon.$$

Recall that the considerations from Section 5.2 lead to the extreme problem

$$u_\varepsilon^2 = u_\varepsilon^2(\kappa; \rho_\varepsilon, R_\varepsilon) = \inf \sum_{i=1}^{\infty} \|\pi_i\|^2 \tag{6.4}$$

subject to, if $q < \infty$,

$$\pi_i \in \Pi, \quad \sum_{i=1}^{\infty} i^{rp} E_{\pi_i}|t|^p \geq \rho_\varepsilon^p, \quad \sum_{j=1}^{\infty} i^{sq} E_{\pi_i}|t|^q \leq R_\varepsilon^q; \tag{6.5}$$

if $q = \infty$, then the last constraint in (6.5) is changed to $\sup_i i^s|\pi_i| \leq R_\varepsilon$.

Let $p = q < \infty$. Taking into account the separation (5.68) we reduce the extreme problem (6.4), (6.5) to the following:

$$u_\varepsilon^2 = u_\varepsilon^2(\rho_\varepsilon, R_\varepsilon) = \inf \sum_i u^2(x_i, p) \;:\; \sum_i i^{rp} x_i \geq \rho_\varepsilon^p, \quad \sum_i i^{rp} x_i \leq R_\varepsilon^p, \tag{6.6}$$

where $u^2(x,p)$ is the solution of the one-dimensional problem (5.65) determined by Theorem 5.1.

Let $p \neq q$. Using the separation (5.74) we have reduced the extreme problem (6.4), (6.5) to the following:

$$u_\varepsilon^2 = \inf \sum_i u^2(x_i, y_i, p, q) \;:\; \sum_i i^{rp} x_i \geq \rho_\varepsilon^p, \quad \begin{cases} \sum_i i^{sq} y_i \leq R_\varepsilon^q, & \text{if } q < \infty, \\ \sup_i i^s y_i \leq R_\varepsilon, & \text{if } q = \infty, \end{cases} \tag{6.7}$$

where $u^2(x,y,p,q)$ is the solution of the one-dimensional problems (5.71) which are determined by Theorem 5.3.

Recall the extreme problem for Besov norms. To simplify, assume

$$p,\ h,\ q,\ t < \infty, \quad p \neq q; \quad p > h, \quad q < t. \tag{6.8}$$

Recall that the constraints $h \leq p$, $q \leq t$ yield convex properties of the problem and are essential for the study of sharp asymptotics. The equalities $p = h$, $q = t$, or $q, t = \infty$ correspond to more simple considerations.

The considerations from Section 5.2 lead to the extreme problem

$$u_\varepsilon^2 = u_\varepsilon^2(\tau; \rho_\varepsilon, R_\varepsilon) = \inf \sum_{j=1}^{\infty} \sum_{i=1}^{2^j} \|\pi_{ij}\|^2 \tag{6.9}$$

subject to

$$\pi_{ij} \in \Pi, \quad \sum_{j=1}^{\infty} 2^{jhr} F_{1,j}(\bar{\pi}_j) \geq \rho_\varepsilon^h, \quad \sum_{j=1}^{\infty} 2^{jts} F_{2,j}(\bar{\pi}_j) \leq R_\varepsilon^t, \tag{6.10}$$

where we denote $\bar{\pi}_j = \{\pi_{ij},\ i = 1, \dots, 2^j\}$,

$$F_{1,j}(\bar{\pi}_j) = \left(\sum_{i=1}^{2^j} E_{\pi_{ij}} |t|^p \right)^{h/p}, \quad F_{2,j}(\bar{\pi}_j) = \left(\sum_{i=1}^{2^j} E_{\pi_{ij}} |t|^q \right)^{t/q}.$$

Using the separation (5.74) we have reduced the extreme problem to the following (compare with (5.84)):

$$u_\varepsilon^2 = \inf \left\{ \sum_{j=1}^{\infty} 2^j u^2(x_j, y_j; p, q) \;:\; \Phi_1(\bar{x}) \geq \rho_\varepsilon^h,\ \Phi_2(\bar{y}) \leq R_\varepsilon^t \right\}, \tag{6.11}$$

where the infimum is taken over the sequences $(\bar{x}, \bar{y})$ and the constraints are determined by the functions

$$\Phi_1(\bar{x}) = \sum_{j=1}^{\infty} 2^{j(rh+h/p)} x_j^{h/p}, \quad \Phi_2(\bar{y}) = \sum_{j=1}^{\infty} 2^{j(st+t/p)} y_j^{t/p}.$$

We will study the extreme problems (6.6), (6.7), (6.11) and will show that there exist the extreme sequences of measures in the problems which satisfy assumptions A.1, A.3, A.6. Moreover, if $p \le q$, then assumption A.4 is fulfilled as well.

In the case $p > q$, to obtain the upper bounds, for the reasons noted in Section 5.1.2, we need to modify the sets V_ε by changing (5.26), (5.28). For the power norms and for the change (5.28) this leads to the embedding

$$\tilde{V}_\varepsilon^H(\rho_\varepsilon, R_\varepsilon) \subset V_\varepsilon^H(\tilde{\rho}_\varepsilon, R_\varepsilon) \quad \text{for } \tilde{\rho}_{l,\varepsilon} = \rho_{l,\varepsilon}(1-\zeta_\varepsilon), \quad \zeta_\varepsilon = o(1).$$

For the Besov norms we control assumption A.5. This leads to an analogous embedding under the change (5.26) for Besov norms as well. Therefore, we need to consider the extreme problem for some wider sets by changing the quantities ρ_ε in the extreme problems to $\tilde{\rho}_{l,\varepsilon}$. If assumption A.6 is fulfilled, it follows from Proposition 5.6 that the quantities $\tilde{u}_\varepsilon$, in the modified extreme problem, are of the same asymptotics as the previous one, i.e., $\tilde{u}_\varepsilon \sim u_\varepsilon$.

Typically the relation $\pi^\varepsilon(V_\varepsilon) \to 1$ is not fulfilled for product measures which correspond to the extreme sequences $\bar{\pi}_\varepsilon$ in the extreme problems. However, we will show that it is possible to modify the extreme problems and extreme sequences such that for the modified sequences $\tilde{\pi}_\varepsilon$ one has $\tilde{\pi}^\varepsilon(V_\varepsilon) \to 1$ and $\|\tilde{\pi}_\varepsilon\| \sim \|\bar{\pi}_\varepsilon\| = u_\varepsilon$. Modified extreme problems correspond to some smaller sets $\tilde{V}_\varepsilon = V_\varepsilon(\tilde{\rho}_\varepsilon, \tilde{R}_\varepsilon)$ with

$$\tilde{\rho}_\varepsilon = \rho_\varepsilon(1+\zeta_\varepsilon), \quad \tilde{R}_\varepsilon = R_\varepsilon(1-\zeta_\varepsilon), \quad \zeta_\varepsilon = o(1)$$

(compare with Section 5.5). The modification of extreme sequences is based on the "truncation of tails." These lead to the required lower bounds.

The scheme of study of extreme problems is presented in Sections 6.6, 6.7, 6.9 below; technical details of the study and of the verification of the required properties of extreme sequences are presented in the Appendix, Sections A.3, A.4.

However, we start with descriptions and discussions of the results.

6.2 Principal Types of Gaussian Asymptotics

Gaussian asymptotics are characterized by the relations (see (5.44))

$$\beta(\alpha, V_\varepsilon) = \Phi(T_\alpha - u_\varepsilon) + o(1); \quad \gamma(V_\varepsilon) = 2\Phi(-u_\varepsilon/2) + o(1). \tag{6.12}$$

The quantities $u_\varepsilon = u_\varepsilon(\kappa, \rho_\varepsilon, R_\varepsilon)$ for power norms or $u_\varepsilon = u_\varepsilon(\tau, \rho_\varepsilon, R_\varepsilon)$ for Besov norms characterize the minimax distinguishability in the hypothesis testing problem for the alternatives V_ε defined by (6.1) and (6.2). The quantities u_ε are of the form

$$u_\varepsilon^2 = \sum_i \|\pi_{\varepsilon,i}\|^2 = \sum_i 2h_{\varepsilon,i}^2 \sinh^2(z_{\varepsilon,i}^2/2), \tag{6.13}$$

where $\pi_{\varepsilon,i} = \pi(h_{\varepsilon,i}, z_{\varepsilon,i})$ is the family which consists of symmetric three-point measures and corresponds to the extreme family $(h_{\varepsilon,i},\ z_{\varepsilon,i})$. However properties of these families and of asymptotics of the function u_ε depend essentially on the point $\kappa \in \Xi_G$; for Besov norms they depend on (t, h) too.

There are two principal and three "frontier" types of Gaussian asymptotics. We start with the description of the principal types G_1 and G_2 of asymptotics which seem to be the most important and most interesting.

6.2.1 *Principal Types of Gaussian Asymptotics for Power Norms*

There are two principal types of the function u_ε:

$$u_\varepsilon^2(\kappa, \rho_\varepsilon, R_\varepsilon) \sim d(\kappa)\rho_\varepsilon^{A_l(\kappa)} R_\varepsilon^{-B_l(\kappa)}, \quad l = 1, 2, \tag{6.14}$$

where $d(\kappa) > 0$. For type G_1 one has (compare with (4.101)),

$$A_1(\kappa) = \frac{4/q - 1 + 4s}{s - r + 1/q - 1/p}, \qquad B_1(\kappa) = \frac{4/p - 1 + 4r}{s - r + 1/q - 1/p}, \tag{6.15}$$

and for type G_2 one has

$$A_2(\kappa) = \frac{1/q + 2s}{s/p - r/q}, \qquad B_2(\kappa) = \frac{1/p + 2r}{s/p - r/q}, \tag{6.16}$$

with a natural extension to $q = \infty$.

Define the function $I = I(\kappa)$ which determines the frontier between the principal types of asymptotics

$$I = I(\kappa) = 2(s - r) - 4(s/p - r/q) + 1/q - 1/p. \tag{6.17}$$

Introduce the sets

$$\begin{aligned} \Xi_{G_1} &= \left\{\kappa \in \Xi_G : r > r_p^* \text{ and either } I < 0 \text{ or } p = q = 2\right\}, \\ \Xi_{G_2} &= \left\{\kappa \in \Xi_G : r > r_p^*,\ I > 0\right\}, \end{aligned}$$

(this means we do not consider the cases $I = 0$ and $r = r_p^*$).

Enough cumbersome relations (6.14)–(6.16) correspond to the solution of simple enough equations for the values $z_0 = z_{0,\varepsilon}(\kappa)$, $m = m_{0,\varepsilon}(\kappa)$ or $h_0 = h_{0,\varepsilon}(\kappa)$, $n = n_{0,\varepsilon}(\kappa)$.

Theorem 6.1 *Let $R_\varepsilon \to \infty$. Then for the alternatives* (6.1) *the relations* (6.12) *hold. Moreover, the quantities u_ε in* (6.12) *could be determined by the relations: if $\kappa \in \Xi_{G_1}$, then*

$$u_\varepsilon^2 \sim c_0(\kappa) m z_0^4, \tag{6.18}$$

where

$$\rho_\varepsilon \sim c_1(\kappa) m^{r+1/p} z_0, \quad R_\varepsilon \sim c_2(\kappa) \begin{cases} m^{s+1/q} z_0, & \text{if } q < \infty, \\ m^s z_0, & \text{if } q = \infty, \end{cases} \tag{6.19}$$

and if $\kappa \in \Xi_{G_2}$, *then*

$$u_\varepsilon^2 \sim c_0(\kappa) n h_0^2, \tag{6.20}$$

where

$$\rho_\varepsilon \sim c_1(\kappa) n^{r+1/p} h_0^{1/p}, \quad R_\varepsilon \sim c_2(\kappa) \begin{cases} n^{s+1/q} h_0^{1/q}, & \text{if } q < \infty, \\ n^s, & \text{if } q = \infty. \end{cases} \tag{6.21}$$

Here $c_0(\kappa)$, $c_1(\kappa)$, $c_2(\kappa)$ *are positive functions bounded away from* 0 *and from* ∞ *over any compact* $K \subset \Xi_{G_k}$, $k = 1, 2$.

Let ρ_ε *be such that* $u_\varepsilon \asymp 1$ *in* (6.18), (6.20). *Then the upper bounds are provided by the test family* $\psi_{\varepsilon,\alpha}$ *from Section 5.1 of the type* (5.4) *which are based on the extreme families* $h_{\varepsilon,i}$, $z_{\varepsilon,i}$ *in* (6.13) *and correspond to the extreme problem with* $\tilde{\rho}_\varepsilon = \rho_\varepsilon$ *for* $p \le q$ *and* $\tilde{\rho}_\varepsilon = \rho_\varepsilon(1 - \zeta_\varepsilon)$ *for* $p > q$; *here* $\zeta_\varepsilon \to 0$ *is a positive family such that* $\zeta_\varepsilon \sqrt{\log R_\varepsilon} \to \infty$. *Moreover, we can use a thresholding of the form*

$$Q_{\varepsilon,i} = \sqrt{2 \log i + c \log\log(i+1) + 4 \log\log R_\varepsilon}; \quad c > 1. \tag{6.22}$$

This means that these test families are asymptotically minimax.

It follows from simple calculations that relations (6.18), (6.19) yield (6.14), (6.15) and relations (6.20), (6.21) yield (6.14), (6.16).

We give a description of the extreme families $h_{\varepsilon,i}$, $z_{\varepsilon,i}$ in (6.13) and the function $c_l(\kappa)$, $l = 0, 1, 2$, in relations (6.18)– (6.21) in terms of special functions which are determined as solutions of the special transcendental equations. Direct description can be given for special cases.

Particularly, for $p \le 2, q \ge p$, we have the asymptotics of the type G_1 and $h_{\varepsilon,i} \equiv 1$ which corresponds to symmetric two-point measures. The values $z_{\varepsilon,i} = z_0 z(i/m)$ are defined by the function $z(y)$ which is the solution of the equation (4.95) for $p < q$ or is of the form (4.96) for $p = q \le 2$. The functions $c_l(\kappa)$, $l = 0, 1, 2$, in (6.18), (6.19) are defined by relations (4.104). These correspond to the results that were obtained in Section 4.3.

If $2 < p = q < \infty$, then we have asymptotics of type G_2. In this case $z_{\varepsilon,i} \equiv z_p$ and $h_{\varepsilon,i}$ are the families $h_{\varepsilon,i} \in [0,1]$ defined by

$$h_{\varepsilon,i} = h_0(x^{rp} - x^{sp})_+, \quad x = i/n, \tag{6.23}$$

the values $u_\varepsilon(\kappa)$, h_0, n are of the form relations (6.20), (6.21) with

$$c_0(\kappa) \sim 2 \sinh^2(z_p^2/2) \int_0^1 (x^{rp} - x^{sp})^2 dx, \tag{6.24}$$

$$c_1^p(\kappa) \sim z_p^p \int_0^1 (x^{rp} - x^{sp}) x^{rp} dx, \tag{6.25}$$

$$c_2^p(\kappa) \sim z_p^p \int_0^1 (x^{rp} - x^{sp}) x^{sp} dx. \tag{6.26}$$

Recall that for $p > 2$ we denote by z_p the positive solution of the equation

$$p \tanh(z^2/2) = z^2.$$

The values z_p minimize the functions $f_p(z) = z^{-p} \sinh(z^2/2)$ in $z > 0$. If $p \leq 2$, then we put, formally, $z_p = 0$. [1]

6.2.2 *Principal Types of Gaussian Asymptotics for Besov Norms*

The results from Section 6.2.1 are extended to Besov norms in the sequence space l^2 of a pyramidal structure. Assume

$$0 < h \leq p < \infty, \quad 0 < q \leq t \leq \infty, \tag{6.27}$$

(these constraints yield convex properties of the problem and are essential for the study of sharp asymptotics).

Note that if $p = h,\ q = t$, then Besov norms are of the form

$$|\cdot|_{r,p,p} = |\cdot|_{\bar{a},p}, \quad |\cdot|_{s,q,t} = |\cdot|_{\bar{b},q},$$

with piecewise constant sequences of dyadic form

$$a_i = 2^{rj}, \quad b_i = 2^{sj}, \quad 2^{j-1} \leq i < 2^j, \quad j = 1, 2, \ldots,$$

that is, $a_i \asymp i^r,\ b_i \asymp i^s$. Therefore the rate asymptotics for this case are the same as for the power norm; the sharp asymptotics can be studied analogously. We will not study this case separately.

Then we have the analogous relations (6.18), (6.19) or (6.20), (6.21) with $n = 2^{j_0}$ or $m = 2^{j_1}$, where $j_0 = j_{\varepsilon,0},\ j_1 = j_{\varepsilon,1}$ are real-valued families, and the positive functions $c_k = c_k(\kappa)$ are replaced by the positive functions $c_k = c_k(\tau, j^*)$ which are 1-periodic in j^*; here j^* is either j_0 or j_1. These relations yield relations (6.14), (6.15) for type G_1 (compare with (4.127), (4.128)) and relations (6.14), (6.16) for type G_2 with a replacement of the positive function $d(\kappa)$ by the positive function $d(\tau, j^*)$ which is 1-periodic in j^*. For type G_1 this function is of the form $d_1(\tau, \log_2(R_\varepsilon/\rho_\varepsilon))$ where $d_1(\tau, x)$ is of period $(s - r + 1/q - 1/p)$ in $x = \log_2(R_\varepsilon/\rho_\varepsilon)$. For type G_2 this function is of the form $d_2(\tau, \log_2(R_\varepsilon^q/\rho_\varepsilon^p))$ where $d_2(\tau, x)$ is of period $(sq - rp)$ in $x = \log_2(R_\varepsilon^q/\rho_\varepsilon^p)$. Namely, we have

[1]The results for $r = 0,\ p = q = 2$ were obtained in [22]; the cases $p \leq 2,\ q \geq p,\ 2 < p = q < \infty$ and $p = \infty$ correspond to [51], [52].

These results are extended to the sample model in the [52], [54]. For composite null-hypothesis of independence, analogous results are presented in [53]. The case $p = q = 2$ was studied in [26] where some composite null-hypotheses were considered. For $p = q = 2$ the extension on spectral density model is given in [23].

The results for $r = 0,\ p < \infty,\ p \neq q$ correspond to [92], [93].

Theorem 6.2 *Consider the alternatives (6.2) for $\kappa \in \Xi_{G_1}$ or $\kappa \in \Xi_{G_2}$ and $R_\varepsilon \to \infty$.*

(1) Assume (6.27). Then relations (6.12) hold. Moreover, for $\kappa \in \Xi_{G_1}$, one has

$$u_\varepsilon^2 \sim c_0(\tau, j_1)2^{j_1} z_0^4, \tag{6.28}$$

where the quantities z_0, j_1 are determined by the relations

$$\rho_\varepsilon \sim c_1(\tau, j_1)2^{j_1(r+1/p)} z_0, \quad R_\varepsilon \sim c_2(\tau, j_1)2^{j_1(s+1/q)} z_0. \tag{6.29}$$

For $\kappa \in \Xi_{G_2}$, one has

$$u_\varepsilon^2 \sim c_0(\tau, j_0)2^{j_0} h_0^2, \tag{6.30}$$

where the quantities h_0, j_0 are determined by the relations

$$\rho_\varepsilon \sim c_1(\tau, j_0)2^{j_0(r+1/p)} h_0^{1/p}, \quad R_\varepsilon \sim c_2(\tau, j_0)2^{j_0(s+1/q)} h_0^{1/q}, \tag{6.31}$$

with a natural extension to $q = \infty$.

Here the functions $c_l(\tau, j^*) > 0$, $l = 0, 1, 2$, are 1-periodic in j^* and bounded away from 0 and from ∞ over any compact.

Let ρ_ε be such that $u_\varepsilon \asymp 1$ in (6.28), (6.30). Then the upper bounds are provided by the tests $\psi_{\varepsilon,\alpha}$ from Section 5.1 of the type (5.4). They are based on the extreme families $h_{\varepsilon,ij} = h_{\varepsilon,j}$, $z_{\varepsilon,ij} = z_{\varepsilon,j}$ which do not depend on $i = 1, \dots, 2^j$ for any j and correspond to the extreme problem with $\tilde{\rho}_\varepsilon = \rho_\varepsilon$ for $p \le q$ and $\tilde{\rho}_\varepsilon = \rho_\varepsilon(1 - \zeta_\varepsilon)$ for $p > q$; here $\zeta_\varepsilon \to 0$ is some positive family. Moreover, we can use a thresholding which does not depend on i and is of the form

$$Q_{\varepsilon,ij} = Q_{\varepsilon,j} = \sqrt{2(j \log 2 + c \log j + 2 \log\log R_\varepsilon)}, \quad c > 1/2. \tag{6.32}$$

This means that these tests are asymptotically minimax.

(2) Without the constraints (6.27) for any $h > 0$, $t > 0$ one has the lower bounds

$$\beta(\alpha, V_\varepsilon) \ge \Phi(T_\alpha - b(\tau)u_\varepsilon) + o(1), \ \gamma(V_\varepsilon) \ge 2\Phi(-b(\tau)u_\varepsilon/2) + o(1),$$

where $b(\tau)$ is a positive function and u_ε is determined by the relations analogous to (6.28), (6.29) or to (6.30), (6.31) with some $c_k = c_k(\tau) > 0$.

In the proof below we obtain relations for the functions $c_l(\tau, j^*)$, $l = 0, 1, 2$, of the form

$$c_l^{a_l}(\tau, j^*) = \sum_{-\infty}^{\infty} \omega_{\tau,l}(j - j^*), \quad l = 0, 1, 2; \quad a_0 = 1, \quad a_1 = h, \quad a_2 = t, \tag{6.33}$$

where $\omega_{\tau,l}(v)$, $v \in R^1$, are continuous functions in v and the series in (6.33) converge; here the quantity j^* is either j_0 or j_1. These entail that the functions $c_l(\tau, j^*)$, $l = 0, 1, 2$, are 1-periodic in j^*.

Using Theorems 6.1, 6.2 we get distinguishability conditions for the power and Besov norms. They are formulated in terms of the rates $\rho_\varepsilon^* =$

$\rho^*(\kappa, R_\varepsilon)$ and correspond to the relation $u_\varepsilon(\tau, \rho^*_\varepsilon, R_\varepsilon) \asymp 1$. The constraints (6.27) are not essential for rate asymptotics in view of statement (2) of Theorem 6.2 and the natural embedding properties.

Corollary 6.1 *Let $R_\varepsilon \to \infty$, $\kappa \in \Xi_{G_1} \cup \Xi_{G_2}$, $t > 0$, $h > 0$. Then*

$$\gamma(V_\varepsilon) \to 0, \quad \text{iff } \rho_\varepsilon/\rho^*_\varepsilon \to \infty, \tag{6.34}$$

$$\gamma(V_\varepsilon) \to 1, \quad \text{iff } \rho_\varepsilon/\rho^*_\varepsilon \to 0, \tag{6.35}$$

where, for $\kappa \in \Xi_{G_l}$, $l = 1, 2$,

$$\rho^*_\varepsilon = \rho^*(\kappa, R_\varepsilon) = R_\varepsilon^{B_l/A_l}. \tag{6.36}$$

Recall that the quantities A_l, B_l, $l = 1, 2$, are defined by (6.15), (6.16).

Using the wavelet transform and relations (2.84) between Besov norms in the sequence space l^2 and in the function space $L_2(0,1)$, we can translate the rate results of Corollary 6.1 to the functional Gaussian model (2.5) and to alternatives which are determined by Besov norms in the functional space $L_2(0,1)$

$$S_\varepsilon = S_\varepsilon(\tilde{\tau}; r_\varepsilon, R) = \{s \in L_2(0,1) : \ \|s\|_{\sigma,p,h} \geq r_\varepsilon, \ \|s\|_{\eta,q,t} \leq R\},$$

(at least for $\sigma, \eta \geq 0, p, q, h, t \geq 1$); here $\tilde{\tau} = (\tilde{\kappa}, h, t)$ and $\tilde{\kappa} = (\sigma, \eta, p, q)$. In fact, let

$$V_\varepsilon = V_\varepsilon(\tilde{\tau}, \rho_\varepsilon, R_\varepsilon) \subset l^2, \quad \rho_\varepsilon = r_\varepsilon/\varepsilon, \quad R_\varepsilon = R/\varepsilon,$$

be the set which consists of sequences $v = \{v_{ij}\}$, $s \in S_\varepsilon$ of normalized wavelet coefficients $v_{ij} = \varepsilon^{-1}(s, \varphi_{ij})$ where $\{\varphi_{ij}\}$ is a regular wavelet basis (see Section 2.7.3). It follows from (2.84) that there exist constants $0 < B_1 \leq B_2 < \infty$ such that

$$V_{\varepsilon,1} = V(\tau, B_1\rho_\varepsilon, B_2R_\varepsilon) \subset V_\varepsilon \subset V(\tau, B_2\rho_\varepsilon, B_1R_\varepsilon) = V_{\varepsilon,2} \tag{6.37}$$

where τ corresponds to

$$r = \sigma + 1/2 - 1/p, \quad s = \eta + 1/2 - 1/q. \tag{6.38}$$

Moreover, let us consider the alternatives determined by Sobolev norms, with $p, q \geq 1$,

$$S_\varepsilon = S_\varepsilon(\tilde{\kappa}; r_\varepsilon, R) = \{s \in L_2(0,1) : \ \|s\|_{\sigma,p} \geq r_\varepsilon, \ \|s\|_{\eta,q} \leq R\}.$$

By relations (2.80) one can take the constants $0 < b_1 \leq b_2 < \infty$ such that

$$S_{\varepsilon,1} = S(\tau_1, b_1r_\varepsilon, b_2R) \subset S_\varepsilon \subset S(\tau_2, b_2r_\varepsilon, b_1R) = S_{\varepsilon,2}, \tag{6.39}$$

where $\tau_1 = (\kappa, \infty, 1)$, $\tau_2 = (\kappa, 1, \infty)$. The alternative S_ε under the functional Gaussian model (2.5) is equivalent to the alternative V_ε under the Gaussian model in the sequence space of pyramidal structure and $\gamma_\varepsilon(S_\varepsilon) = \gamma(V_\varepsilon)$. The rate asymptotics in Corollary 6.1 do not depend on constants in the radii and on parameters h, t. Therefore, using the embedding (6.37), (6.39) and relations (6.38), we get [2]

Corollary 6.2 *Let $\sigma, \eta \geq 0$, $p, q, h, t \geq 1$ be such that $\kappa \in \Xi_{G_1} \cup \Xi_{G_2}$ under the transform* (6.38). *Then*

$$\gamma(S_\varepsilon) \to 0, \quad \textit{iff } r_\varepsilon / r_\varepsilon^* \to \infty,$$
$$\gamma(S_\varepsilon) \to 1, \quad \textit{iff } r_\varepsilon / r_\varepsilon^* \to 0.$$

Here, for $\kappa \in \Xi_{G_l}$, one has

$$r_\varepsilon^* = r_\varepsilon^*(\tilde{\kappa}) = \varepsilon^{C_l}, \quad l = 1, 2,$$

where

$$C_1 = \frac{4(\eta - \sigma)}{4\eta + 1}, \quad C_2 = \frac{2(\eta - \sigma) + p^{-1} - q^{-1}}{2\eta + 1 - q^{-1}}.$$

6.3 Frontier Log-Types of Gaussian Asymptotics

Let us consider the "frontier" sets

$$\begin{aligned}
\Xi_{G_3} &= \{r > r_p^*,\ I(\kappa) = 0 \text{ without } p = q = 2\}, \\
\Xi_{G_4} &= \{r = r_p^*,\ s > s_{pq}^*,\ \text{either } \ p < 2 \ \text{ or } \ p = 2,\ p \leq q\}, \\
\Xi_{G_5} &= \{r = r_p^*,\ s > s_{pq}^*,\ \text{either } \ p > 2 \ \text{ or } \ p = 2,\ p > q\}.
\end{aligned}$$

The set Ξ_{G_3} is the frontier between Ξ_{G_1} and Ξ_{G_2}, the set Ξ_{G_4} is the frontier between Ξ_{G_1} and Ξ_C, and the set Ξ_{G_5} is the frontier between Ξ_{G_2} and Ξ_C.

6.3.1 Frontier Types of Asymptotics for Power Norms

For frontier cases we have relation (6.12) with

$$u_\varepsilon^2 \sim d(\kappa) \begin{cases} \rho_\varepsilon^{A_2(\kappa)} / (R_\varepsilon^{B_2(\kappa)} \log(R_\varepsilon/\rho_\varepsilon)), & \text{if } \kappa \in \Xi_{G_3}, \\ \rho_\varepsilon^4 (\log(R_\varepsilon/\rho_\varepsilon))^{(p-4)/p}, & \text{if } \kappa \in \Xi_{G_4}, \\ \rho_\varepsilon^{2p} / \log(R_\varepsilon/\rho_\varepsilon), & \text{if } \kappa \in \Xi_{G_5}. \end{cases} \tag{6.40}$$

Here the functions $d(\kappa)$ are positive and bounded away from 0 and ∞ over any compact $K \subset \Xi_{G_k}$, $k = 3, 4, 5$.

Note, that $A_1(\kappa) = A_2(\kappa)$, $B_1(\kappa) = B_2(\kappa)$ for $\kappa \in \Xi_{G_3}$. The asymptotics are near to classical in regions G_4, G_5 (the difference is in the log-factor only). Relations (6.40) correspond to the following versions of relations (6.18)–(6.21).

---

[2] The case $p = 2$, $\sigma = 0$, $q < 2$, $\eta > 0$ was studied in [61], [79]. The rates for the functional Gaussian model and $\sigma = 0$ correspond to [91] (up to log log-factors and under some additional constraints).

Theorem 6.3 *Let us consider the alternative* (6.1) *with* $\kappa \in \Xi_{G_k}$, $k = 3,4,5$. *Then we have* (6.12) *where the quantities* $u_\varepsilon = u_\varepsilon(\kappa)$ *are determined by the following. Let* $\kappa \in \Xi_{G_3}$.

$$u_\varepsilon^2 \sim c_0(\kappa) n h_0^2 \log h_0^{-1}, \tag{6.41}$$

where

$$\begin{gathered} \rho_\varepsilon = c_1(\kappa) n^{1/p+r} (h_0 \log h_0^{-1})^{1/p}, \\ R_\varepsilon = c_2(\kappa) \begin{cases} n^{1/q+s}(h_0 \log h_0^{-1})^{1/q}, & \text{if } q < \infty, \\ n^s, & \text{if } q = \infty. \end{cases} \end{gathered} \tag{6.42}$$

Let $\kappa \in \Xi_{G_4}$. *Then*

$$u_\varepsilon^2 \sim c_0(\kappa) m z_0^4 \log m, \tag{6.43}$$

where

$$\rho_\varepsilon = c_1(\kappa) m^{1/4} z_0 (\log m)^{1/p}, \quad R_\varepsilon = c_2(\kappa) \begin{cases} m^{s+1/q} z_0, & \text{if } q < \infty, \\ m^s z_0, & \text{if } q = \infty. \end{cases} \tag{6.44}$$

Let $\kappa \in \Xi_{G_5}$. *Then*

$$u_\varepsilon^2 \sim c_0(\kappa) n h_0^2 \log n, \tag{6.45}$$

where

$$\rho_\varepsilon \sim c_1(\kappa) n^{1/2p} (h_0 \log n)^{1/p}, \; R_\varepsilon \sim c_2(\kappa) \begin{cases} n^{1/q+s} h_0^{1/q}, & \text{if } q < \infty, \\ n^s, & \text{if } q = \infty. \end{cases} \tag{6.46}$$

Let $u_\varepsilon \asymp 1$. *Then the upper bounds are provided by the tests of the type* (5.4) *which are based on the extreme families* $h_{\varepsilon,i}$, $z_{\varepsilon,i}$ *in* (6.13) *corresponding to the extreme problem with* $\tilde\rho_\varepsilon = \rho_\varepsilon$ *for* $p \le q$ *and* $\tilde\rho_\varepsilon = \rho_\varepsilon(1-\zeta_\varepsilon)$ *for* $p > q$; *here* $\zeta_\varepsilon \to 0$ *is a positive family. We can use the thresholding which is defined by* (6.22).

Direct relations for the functions $c_l(\kappa)$, $l = 0,1,2$, and for the extreme families $h_{\varepsilon,i}$, $z_{\varepsilon,i}$ will be presented in the proof.

6.3.2 *Frontier Types of Asymptotics for Besov Norms*

Define the function

$$\zeta = \zeta(\tau) = 2(st - rh) + t(rh+1)/q - h(st+1)/p. \tag{6.47}$$

Consider the alternative (6.2) with $\kappa \in \Xi_{G_k}$, $k = 3, 4, 5$ assuming (6.27) and $\rho_\varepsilon \to \infty$, $R_\varepsilon/\rho_\varepsilon \to \infty$. Then we have (6.12) with

$$u_\varepsilon^2 \sim d(\tau) \begin{cases} \rho_\varepsilon^{A_2(\tau)} R_\varepsilon^{-B_2(\tau)} (\log R_\varepsilon/\rho_\varepsilon)^{-\zeta/ht\lambda}, & \text{if } \kappa \in \Xi_{G_3}, \\ \rho_\varepsilon^4 (\log R_\varepsilon/\rho_\varepsilon)^{(h-4)/h}, & \text{if } \kappa \in \Xi_{G_4}, \\ \rho_\varepsilon^{2p} (\log R_\varepsilon/\rho_\varepsilon)^{(h-2p)/h}, & \text{if } \kappa \in \Xi_{G_5}. \end{cases} \tag{6.48}$$

Recall that $\lambda = s/p - r/q$. Here $d(\tau)$ is a positive function which is bounded away from 0 and ∞ over any compact $\mathcal{T} = \{(\kappa, h, t)\}$ such that $\kappa \in K$ where $K \subset \Xi_{G_k}$, $k = 3, 4, 5$, is a compact.

Note that the asymptotics are near to classical in regions G_4, G_5 (the difference is in the log-factor only).

Relations (6.48) correspond to the following versions of relations (6.28)–(6.31).

Theorem 6.4 *Assume* (6.27). *Let* $\kappa \in \Xi_{G_3}$. *Then*

$$u_\varepsilon^2 \sim c_0(\tau) 2^{j_0} h_0^2 \log(h_0^{-1}), \tag{6.49}$$

where the quantities h_0, j_0 *are determined by the relations*

$$\begin{cases} \rho_\varepsilon \sim c_1(\tau) 2^{j_0(1/p+r)} h_0^{1/p} (\log h_0^{-1})^{1/h}, \\ R_\varepsilon \sim c_2(\tau) 2^{j_0(1/q+s)} h_0^{1/q} (\log h_0^{-1})^{1/t}. \end{cases} \tag{6.50}$$

Let $\kappa \in \Xi_{G_4}$. *Then*

$$u_\varepsilon^2 \sim c_0(\tau) 2^{j_1} z_0^4 j_1 \tag{6.51}$$

where the quantities z_0, j_1 *are determined by the relations*

$$\rho_\varepsilon = c_1(\tau) 2^{j_1/4} z_0 j_1^{1/h}, \quad R_\varepsilon = c_2(\tau, j_1) 2^{j_1(1/q+s)} z_0. \tag{6.52}$$

Let $\kappa \in \Xi_{G_5}$. *Then*

$$u_\varepsilon^2 \sim c_0(\tau) 2^{j_0} h_0^2 j_0, \tag{6.53}$$

where the quantities h_0, j_0 *are determined by the relations*

$$\rho_\varepsilon = c_1(\tau) 2^{j_0/2p} h_0^{1/p} j_0^{1/h}, \quad R_\varepsilon = c_2(\tau, j_0) 2^{j_0(1/q+s)} h_0^{1/q}. \tag{6.54}$$

Here $c_l(\tau)$, $l = 0, 1, 2$, $c_2(\tau, j_k)$ *are positive functions and* $c_2(\tau, j_k)$ *is* 1*-periodic in* j_k, $k = 0, 1$.

Let $u_\varepsilon \asymp 1$. *Then the upper bounds are provided by the tests of the type* (5.4) *which are based on the extreme families* $h_{\varepsilon,i}$, $z_{\varepsilon,i}$ *in* (6.9) *corresponding to the extreme problem with* $\tilde{\rho}_\varepsilon = \rho_\varepsilon$ *for* $p \le q$ *and* $\tilde{\rho}_\varepsilon = \rho_\varepsilon(1 - \zeta_\varepsilon)$ *for* $p > q$; *here* $\zeta_\varepsilon \to 0$ *is some positive family. Moreover, we can use the thresholding of the form* (6.32).

Direct relations for the functions $c_l(\tau)$, $l = 0, 1, 2$, $c_2(\tau, j_k)$, $k = 0, 1$, and for extreme families $h_{\varepsilon,i}$, $z_{\varepsilon,i}$ will be presented in the proofs.

These sharp asymptotics imply the rate asymptotics (6.34), (6.35) with the rates of the following form: if $\kappa \in \Xi_{G_3}$, then

$$\rho_\varepsilon^*(\tau) = R_\varepsilon^{B_2(\kappa)/A_2(\kappa)} (\log R_\varepsilon)^{\zeta/ht\lambda A_2(\kappa)}, \tag{6.55}$$

if $\kappa \in \Xi_{G_4}$, then

$$\rho_\varepsilon^*(\tau) = (\log R_\varepsilon)^{(4-h)/4h}; \tag{6.56}$$

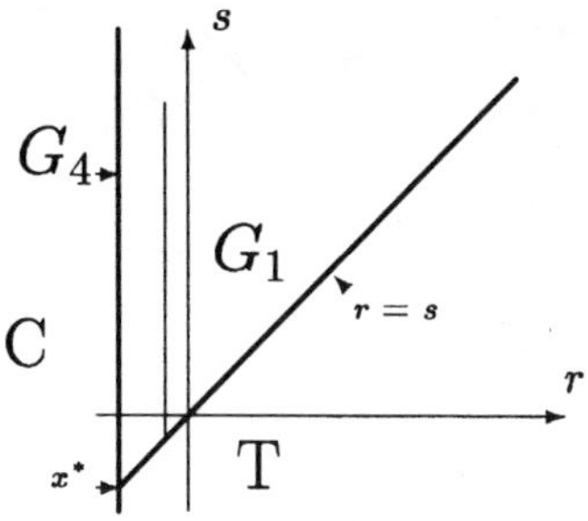

Fig. 1. $p = q \leq 2$.

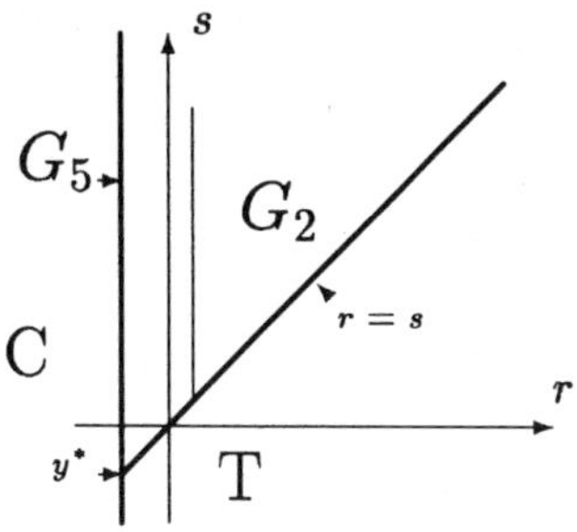

Fig. 2. $2 < p = q < \infty$.

and if $\kappa \in \Xi_{G_5}$, then

$$\rho_\varepsilon^*(\tau) = (\log R_\varepsilon)^{(2p-h)/2ph}; \tag{6.57}$$

Analogously to Corollary 6.2 we can translate these rates to Besov norms in functional space. However, because the rates depend on h, t, after the translation to Sobolev norms we get different upper and lower bounds (the difference is in the exponent of the logarithmic factors).

6.4 Graphical Presentation

In Figures 1–8 we describe the partition of the planes of parameters $\{r, s\}$ to regions of the asymptotics of different types for the fixed parameters p, q in the power norms. Recall that T stands for trivial type, C for classical, and G_1, G_2 for the principal types of Gaussian asymptotics. The same partition holds for Besov norms in the sequence space as well. Under the functional Gaussian model (1.36) for Sobolev and Besov norms we have similar partitions (at least for $\sigma \geq 0,\ \eta \geq 0,\ p \geq 1,\ q \geq 1$) after the translation (1.53). These correspond to moving the origin of coordinates to the point $(1/2 - 1/p,\ 1/2 - 1/q)$ in the illustrations. This point is the beginning of the vertical half-line $\{\sigma = 0,\ \eta > 0\}$ which is presented in the illustrations (this half-line corresponds to the Sobolev or Besov ball of smoothness $\eta > 0$ with the L_p-balls removed).

The partition is defined in terms of the quantities λ, μ, I which are linear functions in r, s for fixed p, q:

$$\lambda = s/p - r/q, \quad \mu = s - r, \quad I = 2(s - r) - 4(s/p - r/q) + 1/q - 1/p.$$

Using the translation (1.53) one can rewrite these functions in terms of the variables $\sigma \geq 0,\ \eta \geq 0$ which determine the smoothness

$$\lambda = (\eta + 1/2)/p - (\sigma + 1/2)/q, \quad \mu = \eta - \sigma + 1/p - 1/q,$$
$$I = 2(\eta - \sigma) - 4(\eta/p - \sigma/q) + 1/q - 1/p.$$

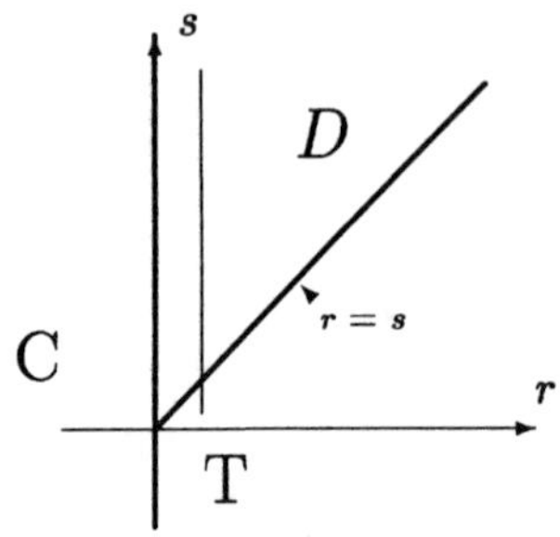

Fig. 3. $q \leq p = \infty$.

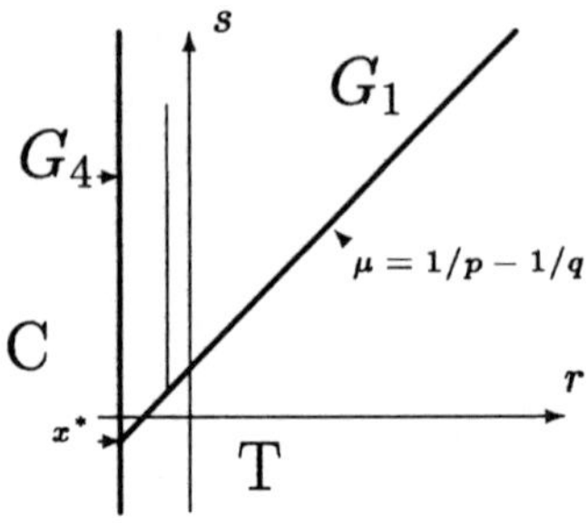

Fig. 4. $p \leq 2,\ p < q \leq \infty$.

We denote by $x^* = x^*_{p,q}$ and $y^* = y^*_{p,q}$ the points at the plane $\{r, s\}$ with coordinates

$$x^* = (1/4 - 1/p,\ 1/4 - 1/q), \quad y^* = (-1/2p,\ -1/2q).$$

In Figures 1–2 we show the partitions for finite $p = q$. The classical asymptotics C correspond to $r < r^*_p$ with $r^*_p = 1/4 - 1/p$, if $p \leq 2$ and $r^*_p = -1/2p$, if $p > 2$. If $r \geq r^*_p$, then we have a trivial type T for $s \leq r$, of course. If $s > r > r^*_p$, then we have Gaussian asymptotics of type G_1 for $p \leq 2$, and of type G_2 for $p > 2$. The frontier $r = r^*_p$, $s > r^*_p$ between C and either G_1 or G_2 corresponds to either G_4 or G_5.

The case $q \leq p = \infty$ is presented in Figure 3. Region C of the classical asymptotics corresponds to $r < 0$ and the Gaussian asymptotics G are changed to the degenerate asymptotics D in this case.

The following figures correspond to $p < \infty$, $p \neq q$.

The case $p \leq 2$, $p < q \leq \infty$ (see Figure 4) is similar to $p = q \leq 2$: we have regions C, T, G_1 with some translation of the frontier between the regions of trivial and Gaussian types; the frontier between types C and G_1 corresponds to G_4 as well. Thus, if $r = 0$, then we have a positive interval $(0,\ (q-p)/pq]$ which corresponds to the trivial type.

Let us consider the other case $p \neq q$.

The case $2 < p < q \leq \infty$ is presented in Figures 5–6. The frontier between regions G and T is not linear in this case: the break point is x^*. We have regions G_1 and G_2 of the principal types of Gaussian asymptotics and we have type G_3 in the frontier half-line $I = 0$ from point x^*. The frontier between G_2 and C corresponds to type G_5. The difference between cases $p < 4$ and $p > 4$ is the position of point x^*. Note that if $r = 1/2 - 1/p$ (vertical half-line), then we have the interval $0 < \eta < (q-p)/2q(p-2)$ which corresponds to type G_1 and the half-line $\eta > (q-p)/2q(p-2)$ which corresponds to type G_2.

The most interesting cases seem to be $p > q$ (see Figures 7–8).

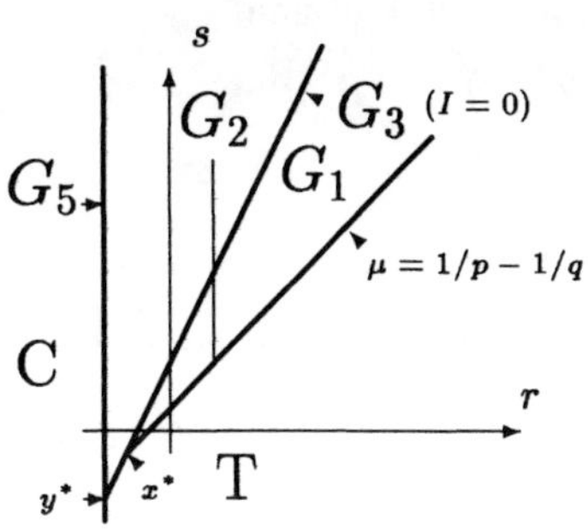

Fig. 5. $2 < p < q \leq \infty$; $p \leq 4$.

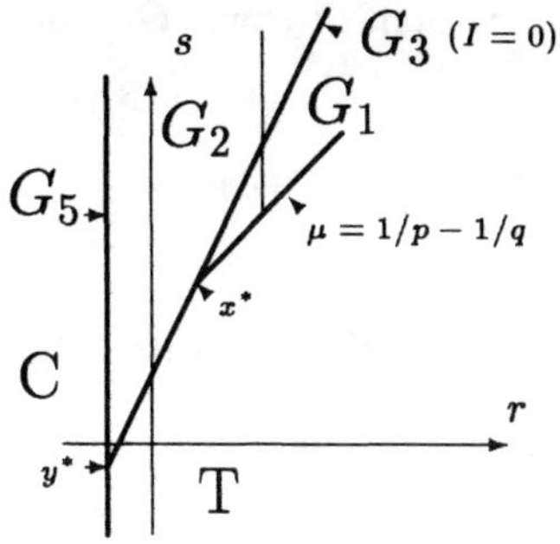

Fig. 6. $2 < p < q \leq \infty$; $p > 4$.

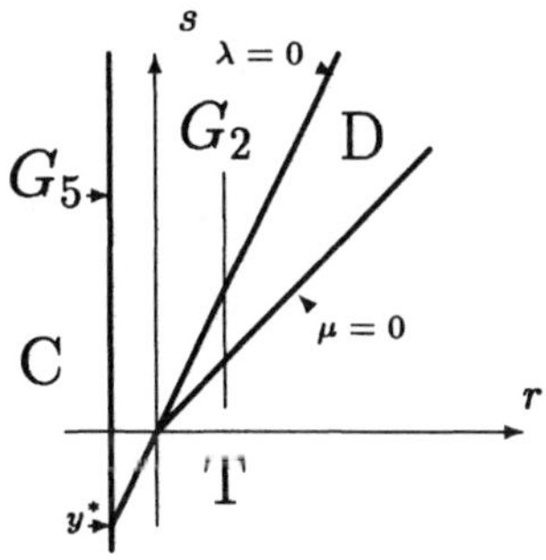

Fig. 7. $p > q$, $p \geq 2$.

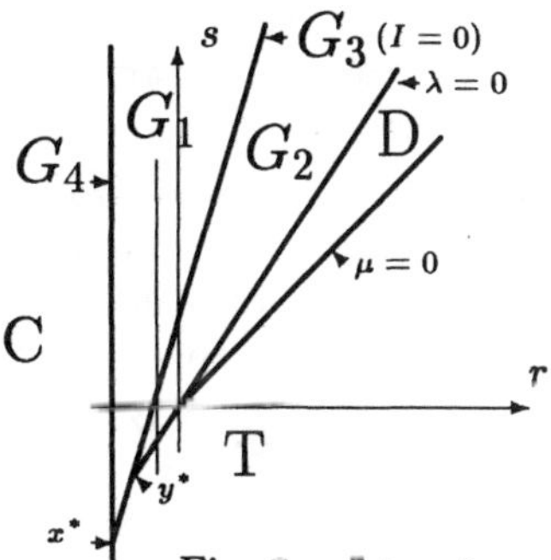

Fig. 8. $2 > p > q$.

We have regions D of degenerate type here. If $p \geq 2$, then the principal Gaussian type is G_2; the frontier type G_5 (which corresponds to the frontier between G_2 and C for $r = -1/2p$) is presented as well. For $r = 1/2 - 1/p$ (vertical half-line) we have the interval

$$I_T \,:\, 0 < \eta \leq (p-q)/pq$$

which corresponds to the type T, the interval

$$I_D \,:\, (p-q)/pq < \eta \leq (p-q)/2q$$

which corresponds to type D and half-line $\eta > (p-q)/2q$ which corresponds to type G_2.

If $p < 2$, then all types of asymptotics are presented (with the exception of the frontier G_5-type). The frontier of region T has break points x^*, y^* and $(0,0)$. For $r = 1/2 - 1/p$ (vertical half-line) we have the interval

$$I_T \,:\, 0 < \eta \leq (p-q)/2q$$

which corresponds to type T, the interval

$$I_{G_2} \,:\, (p-q)/2q < \eta < (p-q)/2q(2-p)$$

which corresponds to type G_2 (for $p > 1$), and the half-line

$$\eta > (p-q)/2q(2-p)$$

which corresponds to type G_1.

6.5 Remarks on the Proofs of Theorems 6.1–6.4

To prove Theorems 6.1–6.4 we study the extreme problems (5.40) specified in Section 6.1, and verify the assumptions of Proposition 5.5 on the extreme sequences $\bar{\pi}_\varepsilon$. In addition to the statements of Proposition 5.5 we specify thresholding $Q_{\varepsilon,i}$ or $Q_{\varepsilon,j}$ which does not depend on κ, ρ_ε (recall that we have established the existence of the needed thresholding in Proposition 5.2 only). We will show, in the study of the extreme problems, that the thresholding satisfies (5.11) (relation (5.2) follows from (4.140) and $R_\varepsilon \to \infty$). We give some remarks, which we use in the proofs and applications of the results, below.

1. It follows from the asymptotics for u_ε in Theorems 6.1–6.4 that assumption A.6. in Proposition 5.6 (4) is fulfilled. Therefore we do not need to control it after the study of the extreme problems.

2. We assume, in the proofs below, that

$$u_\varepsilon = u_\varepsilon(\kappa, \rho_\varepsilon, R_\varepsilon) \asymp 1 \quad \text{or} \quad u_\varepsilon(\tau, \rho_\varepsilon, R_\varepsilon) \asymp 1.$$

Given a family $R_\varepsilon \to \infty$, we consider these as assumptions on the quantities $\rho_\varepsilon = \rho_\varepsilon(\kappa)$ or $\rho_\varepsilon = \rho_\varepsilon(\tau)$ for the quantities u_ε defined by relations (6.18)–(6.21), (6.28)-=(6.31), (6.41)–(6.46), (6.49)–(6.54). Analogously to the Remark in Section 5.2.3, this assumption is not essential for the asymptotics (6.12). In particular, if $u_\varepsilon(\tau, \rho_\varepsilon) \to \infty$, then the lower bounds are trivial. To obtain the upper bounds, by making ρ_ε smaller, for any large $U > 0$, we can change ρ_ε to $\tilde{\rho}_\varepsilon < \rho_\varepsilon$ such that $u_\varepsilon(\kappa, \tilde{\rho}_\varepsilon, R_\varepsilon) \sim U$. Since

$$\beta(\alpha, V_\varepsilon(\kappa, \rho_\varepsilon, R_\varepsilon)) \le \beta(\alpha, V_\varepsilon(\kappa, \tilde{\rho}_\varepsilon, R_\varepsilon)),$$

by using Theorems 6.1–6.4 with $u_\varepsilon \asymp 1$, we obtain the quantities $\beta(\alpha, V_\varepsilon(\kappa, \tilde{\rho}_\varepsilon, R_\varepsilon))$ arbitrarily small for large enough U. This means that $\beta(\alpha, V_\varepsilon) \to 0$.

3. Assuming $R_\varepsilon \to \infty$, $u_\varepsilon \asymp 1$, for the power norm case, we get the following relations from (6.18)–(6.21) and (6.41)–(6.46).

Let $\kappa \in \Xi_{G_l}$, $l = 1, 4$. Then

$$z_0 m^{-\lambda/(p-q)} \to 0, \text{ if } p > q; \; z_0 m^{-rp/(4-p)} \to 0, \text{ if } p \le 2; \; z_0 \to 0, \; m \to \infty,$$

and

$$\log m_\varepsilon \asymp \log z_0^{-1} \asymp \log R_\varepsilon.$$

If $\kappa \in \Xi_{G_l}$, $l = 2, 3, 5$, then

$$n^{-rp} h_0 \to 0, \quad \text{if } p > 2; \quad h_0 \to 0, \; n \to \infty; \quad \log n_\varepsilon \asymp \log h_0^{-1} \asymp \log R_\varepsilon.$$

These relations are extended to the cases

$$\liminf u_\varepsilon > 0; \quad \begin{cases} \log u_\varepsilon = o(\log R_\varepsilon), & \text{for } \kappa \in \Xi_{G_l}, \quad l = 1,2, \\ \log u_\varepsilon = o(\log\log R_\varepsilon), & \text{for } \kappa \in \Xi_{G_l}, \quad l = 3,4,5. \end{cases} \tag{6.58}$$

For Besov norms, analogous statements follow from the asymptotics (6.28)–(6.31), (6.49)–(6.54) with the modifications $n = 2^{j_0}, \ m = 2^{j_1}$.

4. To study the adaptive problems in Chapter 7 we need a small extension of the upper bounds of Theorem 6.1 to the case $u_\varepsilon^2 \asymp \log\log R_\varepsilon$ and $\alpha = \alpha_\varepsilon \to 0$ such that $T_{\alpha_\varepsilon} = O(u_\varepsilon)$. Following the proof of Theorem 6.1, which will be presented below, one can show that, if

$$\liminf u_\varepsilon > 0, \quad u_\varepsilon = o(R_\varepsilon^a) \quad \text{for any } a > 0, \tag{6.59}$$

(i. e., $\log u_\varepsilon = o(\log R_\varepsilon)$), then, for some $b > 0$,

$$w_\varepsilon = \inf_i u_{\varepsilon,i} = o(R_\varepsilon^{-b}), \quad T_\varepsilon = \inf_i T_{\varepsilon,i} \asymp \sqrt{\log R_\varepsilon}. \tag{6.60}$$

Recall that the thresholds $T_{\varepsilon,i}$ are defined in Section 5.1.1, and the assumptions of Proposition 5.5 and relations (5.11) are fulfilled; also for the thresholds (6.22) in relation (5.5) we have

$$\delta_{\varepsilon,Q} = \sum_i \Phi(-Q_{\varepsilon,i}) = O((\log R_\varepsilon)^{-2}). \tag{6.61}$$

For the case $p > q$, to obtain the upper bounds, we use the change (5.28). The assumptions of Proposition 5.4 are fulfilled with $\zeta_\varepsilon = o(1)$ such that

$$\zeta_\varepsilon \sqrt{\log R_\varepsilon} / \log u_\varepsilon \to \infty.$$

In particular, we can take

$$\zeta_\varepsilon \asymp (\log R_\varepsilon)^{-b}, \quad b \in (0, 1/2). \tag{6.62}$$

Since $\delta_{\varepsilon,Q} = O(\log R_\varepsilon)^{-2})$ for the thresholding (6.22), it follows from Proposition 5.1 that $\alpha(\psi_{\varepsilon,\alpha}) = \alpha + o\left((\log R_\varepsilon)^{-1/2}\right)$ uniformly over $\alpha \in (0,1)$.

Moreover, it follows from Proposition 5.6 (4) that, to provide the upper bounds using the test families (5.4), we can change ρ_ε to $\tilde{\rho}_\varepsilon = \rho_\varepsilon(1 - \zeta_\varepsilon)$ and we have $\tilde{u}_\varepsilon = u_\varepsilon(\tilde{\rho}_\varepsilon, R_\varepsilon) = u_\varepsilon + o(1)$ under (6.62) and for

$$u_\varepsilon = o((\log R_\varepsilon)^{b_1}), \quad b_1 > 0.$$

In particular, if $u_\varepsilon^2 \asymp \log\log R_\varepsilon$, then, uniformly over $\alpha \in (0,1)$ such that $T_\alpha = O(u_\varepsilon)$, we get

$$\alpha(\psi_{\varepsilon,\alpha}) = \alpha + o\left((\log R_\varepsilon)^{-1/2}\right), \quad \beta(\psi_{\varepsilon,\alpha}, V_\varepsilon) \le \Phi(T_\alpha - u_\varepsilon) + o(1). \tag{6.63}$$

5. It follows from the proof below that the function $d(\kappa)$ in (6.14) is a continuous Lipschitz function with the exception, possibly, of some three-dimensional submanifolds in Ξ_{G_k}, $k = 1, 2$; namely, we except the submanifolds defined by the equalities $p = q, \ p = 2, \ \lambda = 0, \ \Delta = 0$, where

the functions $\lambda(\kappa)$, $\Delta(\kappa)$ are defined by (3.115), (3.118). Moreover, it follows from the proof that, under assumptions (6.59), the relative accuracy of relations (6.18)–(6.21) is $1 + o(R_\varepsilon^{-\delta})$. This means

$$u_\varepsilon^2 = c_0(\kappa) m z_0^4 (1 + o(R_\varepsilon^{-\delta})) \quad \text{or} \quad u_\varepsilon^2 = c_0(\kappa) n h_0^2 (1 + o(R_\varepsilon^{-\delta}));$$

relations (6.19), (6.21) are modified in an analogous way. Here $\delta = \delta(\kappa)$ is positive and bounded away from 0 uniformly over any compact $K \subset \Xi_{G_k}$, $k = 1, 2$. Moreover, $c_l(\kappa)$, $l = 0, 1, 2$, are positive functions, and these functions are Lipschitz on K with the exception, possibly, of the three-dimensional submanifolds in Ξ_{G_k}, $k = 1, 2$, noted above.

Analogous statements hold true for Besov norms as well with evident modifications.

6. Using relations (6.18)–(6.21), the remark on the accuracy and Lipschitz properties of the functions $c_l(\kappa)$, $l = 0, 1, 2$, we get the following relations. Let $\kappa \in \Xi_{G_1}$. Set

$$\phi_1(\kappa) = (s + 1/q - 1/4)^{-1}.$$

Then, for some $\delta > 0$, one has

$$m = d_0(\kappa)(R_\varepsilon / u^{1/2})^{\phi_1(\kappa)} (1 + o(R_\varepsilon^{-\delta})), \tag{6.64}$$
$$z_0 = d_1(\kappa) u^{1/2} (u^{1/2}/R_\varepsilon)^{\phi_1(\kappa)/4} (1 + o(R_\varepsilon^{-\delta})); \tag{6.65}$$

moreover,

$$d_0(\kappa) = (c_0(\kappa)^{1/4}/c_2(\kappa))^{\phi_1(\kappa)}, \quad d_1(\kappa) = (c_0(\kappa) d_0(\kappa))^{-1/4}.$$

Let $\kappa \in \Xi_{G_2}$. Set

$$\phi_2(\kappa) = (s + 1/2q)^{-1}.$$

Then, for some $\delta > 0$, one has

$$n = d_0(\kappa)(R_\varepsilon / u^{1/q})^{\phi_2(\kappa)} (1 + o(R_\varepsilon^{-\delta})), \tag{6.66}$$
$$h_0 = d_1(\kappa) u (u^{1/q}/R_\varepsilon)^{\phi_2(\kappa)/2} (1 + o(R_\varepsilon^{-\delta})); \tag{6.67}$$

moreover,

$$d_0(\kappa) = (c_0(\kappa)^{1/2q}/c_2(\kappa))^{\phi_2(\kappa)}, \quad d_1(\kappa) = (c_0(\kappa) d_0(\kappa))^{-1/2}.$$

These entail that the functions $d_0(\kappa)$, $d_1(\kappa)$ are Lipschitz over any compact $K \subset \Xi_{G_l}$, $l = 1, 2$, with the exception, possibly, of the three-dimensional submanifolds in Ξ_{G_k}, $k = 1, 2$, noted above, these functions are positive and bounded away from 0 over K.

The functions $\phi_l(\kappa)$ are positive and bounded away from 0 over any compact $K \subset \Xi_{G_l}$, $l = 1, 2$, as well.

Using relations (6.64)–(6.67) under the assumption (6.59) we get that there exist $\delta_l = \delta_l(\kappa) > 0$, $l = 1, 2$, such that

$$m R_\varepsilon^{-\delta_1} \to \infty, \quad z_0 = o(R_\varepsilon^{-\delta_2}), \quad \text{if } \kappa \in \Xi_{G_1},$$
$$n R_\varepsilon^{-\delta_1} \to \infty, \quad h_0 = o(R_\varepsilon^{-\delta_2}), \quad \text{if } \kappa \in \Xi_{G_2}.$$

Moreover, if $\kappa \in \Xi_{G_1}$, then there exists $\delta = \delta(\kappa) > 0$ such that

$$z_0 m^{-\lambda/(p-q)} = o(m^{-\delta}), \quad \text{if } p > q; \quad z_0 m^{-rp/(4-p)_+} = o(m^{-\delta}), \quad \text{if } p \le 2,$$

and if $\kappa \in \Xi_{G_2}$, then there exists $\delta = \delta(\kappa) > 0$ such that

$$n^{-rp_+} h_0 = o(n^{-\delta}), \quad \text{if } p > 2.$$

7. For Besov norms, to obtain relations (6.14), (6.15) for type G_1 and (6.14), (6.16) for type G_2 with the periodic functions $d_1(\tau, x)$, $x = \log_2(R_\varepsilon/\rho_\varepsilon)$ or $d_2(\tau, x)$, $x = \log_2(R_\varepsilon^q/\rho_\varepsilon^p)$, we need the statement analogous to Lemma 4.1. The proof of this statement follows a scheme analogous to the proof of Lemma 4.1 and is based on the convex properties from Proposition 5.6, and on the properties of the extreme sequences of measures in extreme problems. We omit it for simplification

6.6 Proof of Theorems 6.1 and 6.3 for $p \le 2,\ q \ge p$, and $p = q$

Certainly we can take the reduction of the extreme problem (6.6), (6.7) to the sequence of Lagrange extreme problems (5.76), (5.79). However, we prefer to study (6.6), (6.7) directly in these cases.

6.6.1 *Case* $p \le 2,\ q \ge p$

Recall that this case has been studied in Section 4.3.3. We consider this case here to compare the methods from Section 4.3.3 and the general methods based on the results of Chapter 5.

In this case, the set Ξ_G is defined by the inequalities

$$r \ge r_p^* = 1/p - 1/4, \quad s - r > 1/p - 1/q,$$

and consists of that part of Ξ_{G_1} which corresponds to the strict inequality $r > r_p^*$ and from that part of Ξ_{G_4} which corresponds to the equality $r = r_p^*$.

First, let $p = q \le 2$. Put $x = z^p$. We have, from Theorem 5.1, $u^2(x, p) = 2\sinh^2(z^2/2)$. Therefore, we can rewrite the extreme problem (6.6):

$$\inf_{\bar{z}} \sum_i 2\sinh^2(z_i^2/2) : \sum_i i^{rp} z_i^p \ge \rho_\varepsilon^p, \quad \sum_i i^{sp} z_i^p \le R_\varepsilon^p, \quad z_i \ge 0. \quad (6.68)$$

Let $p \le 2,\ q > p$. Set $x = z^p,\ y = \zeta^q$, if $q < \infty$ and $y = \zeta$, if $q = \infty$. It was shown in Proposition 5.5 that we can pass to problem (5.64) in the form (5.74) for the sequences of measures $\bar{\pi}$ and for the real-valued sequences $\bar{x},\ \bar{y}$. Recall that, if $x > y^{p/q}$, then $u^2(x, y, p, q) = \infty$, and if $x \le y^{p/q}$, then it follows from Theorem 5.3 that $u^2(x, y, p, q) = 2\sinh^2(z^2/2)$ and the solutions of one-dimensional problems are symmetric two-point measures with $h = 1$, $z = x^{1/p}$. Therefore, we can rewrite the extreme problem (6.7):

$$\inf_{\bar z} \sum_i 2\sinh^2(z_i^2/2) \ : \ \sum_i i^{rp} z_i^p \ge \rho_\varepsilon^p, \quad \begin{cases} \sum_i i^{sq} z_i^q \le R_\varepsilon^q, & \text{if } q < \infty, \\ \sup_i i^s z_i \le R_\varepsilon, & \text{if } q = \infty. \end{cases} \tag{6.69}$$

The extreme problem (6.69) corresponds to (6.68) for $p = q$.

Consider a slightly different extreme problem

$$\tilde u_\varepsilon^2 = \inf_{\bar z} \frac{1}{2} \sum_i z_i^4 \ : \ z_i \ge 0, \ \sum_i i^{rp} z_i^p \ge \rho_\varepsilon^p, \quad \begin{cases} \sum_i i^{sq} z_i^q \le R_\varepsilon^q, & \text{if } q < \infty, \\ \sup_i i^s z_i \le R_\varepsilon, & \text{if } q = \infty. \end{cases} \tag{6.70}$$

Suppose the extreme sequence $z_{\varepsilon,i}$ in problem (6.70) is such that $z_\varepsilon = \sup_i z_{\varepsilon,i} \to 0$. Since $2\sinh^2(z^2/2) \sim z^4/2$, as $z \to 0$, this entails $\tilde u_\varepsilon \sim u_\varepsilon$. Thus, it suffices to study the extreme problem (6.70) and to control $z_\varepsilon \to 0$.

However, the extreme problem (6.70) with $q < \infty$ is the same as (4.92) and the solution is presented in Theorem 4.2. These correspond to the asymptotics (6.18), (6.19) for $r > r_p^*$ and to (6.43), (6.44) for $r = r_p^*$. The relation $z_\varepsilon \to 0$ and assumption A.1 have been controlled in the proof of Theorem 4.2. The relation $z_\varepsilon \to 0$ yields A.4. The product measure π^ε which corresponds to the family $\pi_{\varepsilon,i} = \pi(1, z_{\varepsilon,i})$ is supported on V_ε. Thus we can apply Proposition 5.5 and obtain the required sharp asymptotics.

Certainly we get the same results as in Theorem 4.2 where these asymptotics have been obtained. Moreover, Theorem 4.2 gives simpler tests of χ^2-type than that which follow from Proposition 5.5 (in particular, thresholding is not required).

6.6.2 Case $p = q > 2$

Put $h = \min\{1, (z/z_p)^p\}$. We have, from Theorem 5.1,

$$u^2(x,p) \ge 2h^2 \sinh^2(z_p^2/2); \quad u^2(x,p) = 2h^2 \sinh^2(z_p^2/2), \quad \text{if } h < 1. \tag{6.71}$$

Consider the extreme problem

$$u_{\varepsilon,1}^2 = \inf_{\bar h} 2\sinh^2 \frac{z_p^2}{2} \sum_i h_i^2 \ : \ z_p^p \sum_i i^{rp} h_i \ge \rho_\varepsilon^p, \quad z_p^p \sum_i i^{sp} h_i \le R_\varepsilon^p. \tag{6.72}$$

Suppose the extreme sequence $h_{\varepsilon,i}$ in problem (6.72) is such that $h_\varepsilon = \sup_i h_{\varepsilon,i} < 1$. Then it follows from (6.71) that $u_{\varepsilon,1} = u_\varepsilon$. Therefore it is enough to study problem (6.72) and to control $h_\varepsilon < 1$. We will show that $h_\varepsilon \to 0$ below.

Let us put $h_i = z_i^2$. Then the extreme problem (6.72) corresponds to (6.70) up to the factor $2\sinh^2(z_p^2/2)$ and to the replacement

$$\tilde p = \tilde q = 2, \quad \tilde r = rp/2, \quad \tilde s = sp/2, \quad \tilde \rho_\varepsilon = (\rho_\varepsilon/z_p)^{p/2}, \quad \tilde R_\varepsilon = (R_\varepsilon/z_p)^{p/2}.$$

Thus we get the following results. The values h_i which minimize (6.72) are determined, by positive parameters $h_0 = h_{\varepsilon,0},\ n = n_\varepsilon$, by the relations

$$h_i = h_0((i/n)^{rp} - (i/n)^{sp})_+.$$

This corresponds to relations (6.23). The values $n = n_\varepsilon,\ h_0 = h_{0,\varepsilon},\ u_\varepsilon^2$ are determined by the equalities

$$\rho_\varepsilon^p = z_p^p h_0 n^{rp+1} \left(n^{-1} \sum_{1 \le i \le n} ((i/n)^{rp} - (i/n)^{sp})(i/n)^{rp} \right), \quad (6.73)$$

$$R_\varepsilon^p = z_p^p h_0 n^{sp+1} \left(n^{-1} \sum_{1 \le i \le n} ((i/n)^{rp} - (i/n)^{sp})(i/n)^{sp} \right), \quad (6.74)$$

$$u_\varepsilon^2 = 2\sinh^2(z_p^2/2) n h_0^2 \left(n^{-1} \sum_{1 \le i \le n} ((i/n)^{rp} - (i/n)^{sp})^2 \right). \quad (6.75)$$

Let $s > r > r_p^* = -1/2p$ which corresponds to $\kappa \in \Xi_{G_2}$. Suppose $R_\varepsilon/\rho_\varepsilon \to \infty$. Then $n \to \infty$ and, by replacing the normalized sums in (6.73)–(6.75) by the integrals, we get relations (6.20), (6.21) with the values $c_l = c_l(\kappa)$ determined by (6.24)–(6.26). Moreover, if $R_\varepsilon \to \infty,\ u_\varepsilon \asymp 1$, then $R_\varepsilon/\rho_\varepsilon \to \infty,\ h_0 \to 0,\ n \to \infty$ and $h_\varepsilon = \sup_i h_{\varepsilon,i} \to 0$ (recall that these follow from the study of the extreme problem (6.70) in Theorem 4.2).

Let $r = r_p^* = -1/2p,\ s > r_p^*$ which corresponds to $\kappa \in \Xi_{G_5}$. Then the normalized sum in (6.74) tends to integral (6.26), however, integrals (6.24)–(6.25) diverge. The normalized sums in (6.73) and in (6.75) are of the rate

$$n^{-1} \sum_{1 \le i \le n} ((i/n)^{-1/2} - (i/n)^{sp})(i/n)^{-1/2} \sim \log n,$$

$$n^{-1} \sum_{1 \le i \le n} ((i/n)^{-1/2} - (i/n)^{sp})^2 \sim \log n. \quad (6.76)$$

Suppose $R_\varepsilon/\rho_\varepsilon \to \infty$. Then $n \to \infty$ and we obtain relations (6.43)–(6.44) with $c_0(\kappa) = c_1(\kappa) = 1$ and $c_2(\kappa)$ defined by (6.26). It is easy to check that if $R_\varepsilon \to \infty,\ u_\varepsilon \asymp 1$, then $h_0 \to 0,\ n \to \infty\ R_\varepsilon/\rho_\varepsilon \to \infty$ and $h_\varepsilon \to 0$.

Thus we get the solution of the extreme problem (6.71) and assumptions A.1, A.4 are fulfilled as well (this follows from $\sup_i z_i = z_p$). These yield the upper bounds from Proposition 5.5.

We need to control the assumption (5.11) for the thresholding (6.22). Let us consider the case $r > r_p^*$ (for $r = r_p^*$ the study is analogous). Note that because of (5.6) we have $T_{\varepsilon,i} = \infty$, if $i \ge n$, which yields (5.11). Let $1 \le i < n$. Then in view of Remark 6 in Section 6.5 we have

$$h_{\varepsilon,i} \le h_0 n^{-rp+} \le n^{-\delta}, \quad \delta > 0,$$

and using (5.6), we get

$$T_{\varepsilon,i} \geq z_p^{-1}\log(h_{\varepsilon,i}^{-1}) + O(1) \asymp \log n.$$

Recall that $\log(R_\varepsilon) \asymp \log n$ by Remark 3 in Section 6.5. Therefore we get

$$Q_{\varepsilon,i} \leq B\sqrt{\log n} = o(T_{\varepsilon,i}),$$

which yields (5.11).

To obtain the lower bounds we need to construct the families $\tilde{\pi}_{\varepsilon,i}$ such that

$$\tilde{\pi}^\varepsilon(V_\varepsilon) \to 1, \quad \|\tilde{\pi}_\varepsilon\| \sim u_\varepsilon.$$

It suffices to assume $u_\varepsilon \asymp 1$ which yields $h_0 \asymp n^{-1/2}$ for $r > r_p^*$ and $h_0 \asymp (n\log n)^{-1/2}$ for $r = r_p^*$. Recall that it follows from the asymptotics (6.18)–(6.19) and (6.43)–(6.44) which have been proved previously, that assumption A.6 is fulfilled.

Take a positive family $\delta_\varepsilon \to 0$ and put $\tilde{\rho}_\varepsilon = (1+\delta_\varepsilon)\rho_\varepsilon$, $\tilde{R}_\varepsilon = (1-\delta_\varepsilon)R_\varepsilon$. Let us consider the solutions $\{\tilde{\pi}_{\varepsilon,i} = \pi(\tilde{h}_{\varepsilon,i}, z_p)\}$ of the extreme problem (6.72) with replacement $\rho_\varepsilon \to \tilde{\rho}_\varepsilon$, $R_\varepsilon \to \tilde{R}_\varepsilon$. It follows from Proposition 5.6 (4) that $\|\tilde{\pi}_\varepsilon\| \sim u_\varepsilon$. On the other hand (compare with Section 5.5.2), we have the inequality

$$\begin{aligned}\tilde{\pi}^\varepsilon(V_\varepsilon) &= \tilde{\pi}^\varepsilon\left(G_1(v) \geq H_{\varepsilon,1},\ G_2(v) \leq H_{\varepsilon,2}\right) \\ &\geq 1 - \tilde{\pi}^\varepsilon\left(G_1(v) < H_{\varepsilon,1}\right) - \tilde{\pi}^\varepsilon\left(G_2(v) > H_{\varepsilon,2}\right).\end{aligned}$$

In view of the Chebyshev inequality, to obtain the relation $\tilde{\pi}^\varepsilon(V_\varepsilon) \to 1$, it suffices to control the relations, for some $\delta_\varepsilon \to 0$,

$$E_{\tilde{\pi}^\varepsilon}(G_1(v)) = (1+\delta_\varepsilon)^p H_{\varepsilon,1}, \quad E_{\tilde{\pi}^\varepsilon}(G_2(v)) = (1-\delta_\varepsilon)^p, H_{\varepsilon,2} \tag{6.77}$$

and

$$\mathrm{Var}_{\tilde{\pi}^\varepsilon}(G_1(v)) = o(H_{\varepsilon,1}^2\delta_\varepsilon^2), \quad \mathrm{Var}_{\tilde{\pi}^\varepsilon}(G_2(v)) = o(H_{\varepsilon,2}^2\delta_\varepsilon^2). \tag{6.78}$$

Relations (6.77) follow from the construction of the sequence $\tilde{\pi}_\varepsilon$. Let us verify (6.78). Using direct calculation we have:

$$\begin{aligned}\mathrm{Var}_{\tilde{\pi}^\varepsilon}(G_2(v)) &= (z_p)^{2p}\sum_{i=1}^n \tilde{h}_i(1-\tilde{h}_i)i^{2sp} \asymp n^{1+2sp}h_0\int_{1/n}^1 x^{(2s+r)p}\,dx, \\ \mathrm{Var}_{\tilde{\pi}^\varepsilon}(G_1(v)) &= (z_p)^{2p}\sum_{i=1}^n \tilde{h}_i(1-\tilde{h}_i)i^{2rp} \asymp n^{1+2rp}h_0\int_{1/n}^1 x^{3rp}\,dx.\end{aligned}$$

Note that

$$\begin{aligned}H_{\varepsilon,2} \asymp h_0 n^{1+sp}, \quad H_{\varepsilon,1} \asymp h_0 n^{1+rp}, \quad &\text{if } r > r_p^*; \\ H_{\varepsilon,1} \asymp h_0 n^{1+rp}\log n, \quad &\text{if } r = r_p^*.\end{aligned}$$

Therefore, relations (6.78) follow from evaluations of the integrals. For the first integral,one has

$$\int_{1/n}^{1} x^{(2s+r)p} dx \asymp \begin{cases} 1, & \text{if } (2s+r)p+1>0, \\ \log n, & \text{if } (2s+r)p+1=0, \\ n^{-(2s+r)p-1}, & \text{if } (2s+r)p+1<0, \end{cases}$$

and by $(2s+r)p+1 > -1/2$ for $s > r$ we have the required relation for $\mathrm{Var}_{\tilde{\pi}^\varepsilon}(G_2(v))$ with $\delta_\varepsilon = (\log\log n)^{-1}$.

If $r > r_p^*$, then $3rp+1 > -1/2$ and, by using analogous estimation for the integral of the function x^{3rp}, we get the required relation for $\mathrm{Var}_{\tilde{\pi}^\varepsilon}(G_1(v))$ with the same δ_ε. If $r = r_p^*$, then $3rp+1 = -1/2$ and we have the required relations

$$H_1 \asymp \log^{1/2} n, \quad \mathrm{Var}_{\tilde{\pi}^\varepsilon}(G_1(v)) \asymp \log^{-1/2} n.$$

Thus Theorems 6.1, 6.3 are proved for the case $p = q > 2$.

6.7 Extreme Problem for Power Norms: $p \neq q$

In Section 5.3.2 we have reduced the extreme problem (6.7) (which corresponds to (6.4), (6.5)) to the sequence of the Lagrange extreme problems (5.76):

$$\inf_{x\geq 0, y \geq 0} \{u^2(x,y;p,q) - Ai^{rp}x + Bi^{sp}y\} \tag{6.79}$$

and to the equations for minimum sequences in (6.79):

$$\sum_{i=1}^{\infty} i^{rp}x_i = \rho_\varepsilon^p, \quad \sum_{i=1}^{\infty} i^{sq}y_i = R_\varepsilon^q, \tag{6.80}$$

(we can assume $A > 0$, $B > 0$). If $q = \infty$, then (6.79), (6.80) are replaced by

$$\inf\{u^2(x,y;p,\infty) - Ai^{rp}x\}, \quad \text{subject to } x \geq 0, \quad 0 \leq i^s y \leq R_\varepsilon, \tag{6.81}$$

and

$$\sum_{i=1}^{\infty} i^{rp}x_i = \rho_\varepsilon^p. \tag{6.82}$$

Here $u^2(x,y,p,q)$ are the solutions of the one-dimensional extreme problems from Theorem 5.3. By Lemma 5.3 the infimum is attained and could be taken over $(x_i, y_i) \in \Lambda(p,q)$; the set $\Lambda(p,q)$ is defined by (5.118).

We can rewrite the relations in terms of the parameters (z,h) of extreme measure in the one-dimensional extreme problem

$$x_i = h_i z_i^p, \quad y_i = \begin{cases} h_i z_i^q, & \text{if } q < \infty, \\ z_i, & \text{if } q = \infty, \end{cases} \quad u^2(x_i, y_i; p, q) = 2h_i^2 \sinh^2(z_i^2/2).$$

Therefore we can reduce the extreme problem to the following: if $q < \infty$, then

$$\inf F(h_i, z_i) = \inf\left\{2h_i^2 \sinh^2(z_i^2/2) - Ai^{rp}h_i z_i^p + Bi^{sp}h_i z_i^q\right\}, \qquad (6.83)$$

$$\sum_i i^{rp}h_i z_i^p = \rho_\varepsilon^p, \quad \sum_i i^{sq}h_i z_i^q = R_\varepsilon^q, \qquad (6.84)$$

the infimum in (6.83) is taken over $(h, z) \in \Theta(p, q)$; the set $\Theta(p, q)$ corresponds to $\Lambda(p, q)$ and is defined in Section 5.4.4. If $q = \infty$, then the Lagrange extreme problem is of the form

$$\inf F(h_i, z_i) = \inf\{2h_i^2 \sinh^2(z_i^2/2) - Ai^{rp}hz^p \ : \ h \in (0, 1], \ 0 < i^s z_i \le R_\varepsilon\}, \qquad (6.85)$$

and we have the first equation in (6.84) only.

6.7.1 Equations System for $q < \infty$

It was noted in Section 5.4.4 that we can find the minimum point in problem (6.83) for any fixed i in two steps. The first step is the study of the extreme problem (6.83) without the constraint $h_i \le 1$. These lead to the system of equations (5.132) where the function $F(h, z)$ is defined by (5.124) and $D = d = 1$. This is of the form

$$4h_i \sinh^2 \frac{z_i^2}{2} = Ai^{rp}z_i^p - Bi^{sq}z_i^q, \qquad (6.86)$$

$$4h_i \sinh^2 \frac{z_i^2}{2}\left(\frac{z_i^2}{\tanh(z_i^2/2)}\right) = Api^{rp}z_i^p - Bqi^{sq}z_i^q, \qquad (6.87)$$

and the parameters $A > 0$, $B > 0$ are determined by equations (6.84).

If we can find the solution $h_i > 0$, $z_i > 0$ of (6.86), (6.87), such that $h_i \le 1$ and $z_i < z_p$ for $2 < p < q$ or $z_i > z_p$ for $2 < p$, $q < p$, then this provides the minimum in the problem. If there does not exist a solution with $h_i \le 1$, then we consider the extreme problems (5.133) which correspond to $h_i = 1$

$$\inf_{z_i > 0} \left\{2\sinh^2(z_i^2/2) - Ai^{rp}z_i^p + Bi^{sp}z_i^q\right\}. \qquad (6.88)$$

Certainly we need to find the solution of (6.88) with $z_i > z_p$ for $2 < p$, $q < p$; however, we will show that in the opposite case there exists the solution of (6.86), (6.87) with $h_i < 1$ for all i under the assumptions of Theorems 6.1–6.4. This leads to equation (5.134)

$$2z_i^2 \sinh(z_i^2) = Api^{rp}z_i^p - Bqi^{sq}z_i^q. \qquad (6.89)$$

By the above, if we go to the second step, then there exists the unique minimum point $z_i > 0$ in problem (6.88) which provides the minimum in problem (6.83) and satisfies (6.89). Sometimes equation (6.89) has two roots, but only one of them (maximal) corresponds to the minimum point in (6.88).

In what follows we study the systems (6.86), (6.87) and equations (6.84) under several assumptions, either on $A = A_\varepsilon$, $B = B_\varepsilon$ or on some other parameters determined by A, B. We find these parameters from equations (6.84) and then we control the required assumptions. Thus we get the sequences $\bar{h}, \bar{z}$ which provide the minimum in the Lagrange extreme problems and the minimum in the extreme problem (5.64),

$$u_\varepsilon^2 = 2\sum_i h_i^2 \sinh^2(z_i^2/2). \tag{6.90}$$

We realize this outline in the Appendix, Section A.3.

6.7.2 Equations System for $q = \infty$

For the case $p < q = \infty$, to find the minimum points in (6.85) and to satisfy the first equation in (6.84), we obtain the system of equations

$$4h_i \sinh^2 \frac{z_i^2}{2} = Ai^{rp} z_i^p - C_i, \tag{6.91}$$

$$4h_i^2 \sinh^2 \frac{z_i^2}{2}\left(\frac{z_i^2}{\tanh(z_i^2/2)}\right) = Api^{rp} h_i z_i^p - B_i i^s z_i, \tag{6.92}$$

and the constraints

$$\sum_i h_i z_i^p i^{rp} = \rho_\varepsilon^p, \quad \sup_i z_i i^s \le R_\varepsilon, \quad h_i \le 1. \tag{6.93}$$

Here $A = A_\varepsilon > 0$ and the unknown parameters $B_i = B_{\varepsilon,i} \ge 0$, $C_i = C_{\varepsilon,i} \ge 0$ control the inequalities $z_i i^s \le R_\varepsilon$, $h_i \le 1$, i.e., if $C_i > 0$, then $h_i = 1$, if $B_i > 0$, then $z_i = i^{-s} R_\varepsilon$ (certainly we find the solutions (6.91), (6.92) assuming $h_i > 0$, $z_i > 0$; $z_i \le z_p$ for $p > 2$).

First, we try to find the solutions h_i, z_i of (6.91), (6.92) assuming $B_i = C_i = 0$. It is possible for $p > 2$ and we obtain the relations

$$z_i = z_p, \quad h_i = AC(p) i^{rp}; \quad C(p) = \psi_p(z_p), \tag{6.94}$$

where

$$\psi_p(z) = z^p / 4\sinh^2(z^2/2). \tag{6.95}$$

If relations (6.94) lead to

$$h_i \le 1, \quad i^{-s} R_\varepsilon \ge z_i = z_p, \tag{6.96}$$

then these relations determine the solutions of (6.91), (6.92). Let either $p \le 2$ or $p > 2$ and relations (6.96) are not fulfilled for h_i, z_i defined by (6.94). If $p \le 2$, then we set $h_i = 1$, $B_i = 0$. Assuming $z_i = o(1)$ and using the relations $\sinh x \sim \tanh x \sim x$, as $x \to 0$, we obtain the equation and the constraint

$$z_i = (Api^{rp}/2)^{1/(4-p)} \le i^{-s} R_\varepsilon \tag{6.97}$$

(note that the inequality $C_i \ge 0$ holds, since the left-hand side of (6.92) is asymptotically equivalent to twice the left-hand side of (6.91), as $z_i \to 0$ and $p \le 2$). If $p > 2$ and (6.96) are not fulfilled for h_i, z_i defined by (6.94), then, setting $z_i = i^{-s} R_\varepsilon$, $C_i = 0$, we obtain the equation and the constraint

$$h_i = A i^{rp} \psi_p(z_i) \le 1 \tag{6.98}$$

(the inequality $B_i > 0$ holds by $p > 2$).

We realize this outline in the Appendix, Section A.3.7

6.8 Properties of the Extreme Sequences for Power Norms

It was noted above that the solution of the extreme problem (5.40) is the sequence $\bar{\pi}_\varepsilon = \{\pi(h_{\varepsilon,i}, z_{\varepsilon,i})\}$ which consists of symmetric three-point measures. The sequences $h_{\varepsilon,i} = h_{\varepsilon,i}(\kappa, \rho_\varepsilon, R_\varepsilon)$ and $z_{\varepsilon,i} = z_{\varepsilon,i}(\kappa, \rho_\varepsilon, R_\varepsilon)$ are such that assumption A.4 holds for $p \le q$. If $p > q$, then the assumptions of Proposition 5.4 are fulfilled uniformly over a compact $K \subset \Xi_{G_1}$ or $K \subset \Xi_{G_2}$. Recall that $w_\varepsilon = o(R_\varepsilon^{-a})$ for some $a > 0$, see (6.60); for $p > q$ we consider the change (5.28) with $\zeta_\varepsilon = o(1)$ which satisfies (6.62). Moreover, it will be shown in Section A.3 that we can construct the priors $\pi_1^\varepsilon = \pi_1^\varepsilon(\kappa)$ of the product type (these correspond to the sequences $\bar{\pi}_{\varepsilon,1}(\kappa) = \{\pi(h_{\varepsilon,i,1}, z_{\varepsilon,i,1})\}$ under the "rejecting of tails" and a small correction of the parameters ρ_ε, R_ε) such that assumption A.1a and A.3 are fulfilled as well, and $\pi_1^\varepsilon(V_\varepsilon) \to 1$, $\|\bar{\pi}_{\varepsilon,1}(\kappa)\| = \|\bar{\pi}_\varepsilon(\kappa)\| + o(1)$ uniformly over a compact set $K \subset \Xi_{G_1}$ or $K \subset \Xi_{G_2}$.

The sequences $h_{\varepsilon,i} = h_{\varepsilon,i}(\kappa, \rho_\varepsilon, R_\varepsilon)$ and $z_{\varepsilon,i} = z_{\varepsilon,i}(\kappa, \rho_\varepsilon, R_\varepsilon)$ are determined by the families $n = n_\varepsilon(\kappa, \rho_\varepsilon, R_\varepsilon)$, $h_0 = h_{0,\varepsilon}(\kappa, \rho_\varepsilon, R_\varepsilon)$ such that relations (6.18), (6.19) are fulfilled, or by $m = m_\varepsilon(\kappa, \rho_\varepsilon, R_\varepsilon)$, $z_0 = z_{0,\varepsilon}(\kappa, \rho_\varepsilon, R_\varepsilon)$ such that relations (6.20), (6.21) are fulfilled. The sequences $h_{\varepsilon,i,1}$, $z_{\varepsilon,i,1}$ are of the same structure.

To apply the results for the adaptive problems in the next chapter we need some additional properties of solutions. To simplify, we consider the case $q < \infty$. Denote

$$\Xi_{G_0} = \{\kappa \in \Xi_{G_1} \cup \Xi_{G_2} : p \ne q,\ p \ne 2,\ \lambda \ne 0,\ \Delta \ne 0\},$$

where

$$\Delta = \Delta(\kappa) = sq(4-p) - rp(4-q), \quad \lambda = \lambda(\kappa) = s/p - r/q,$$

and set

$$\Xi_{G_{01}} = \Xi_{G_0} \cap \Xi_{G_1}, \quad \Xi_{G_{02}} = \Xi_{G_0} \cap \Xi_{G_2}.$$

6.8.1 Continuity Properties

For $L_1 > 0$, $L_2 > 0$, and $\kappa_0 \in \Xi_{G_{02}}$ (or $\kappa_0 \in \Xi_{G_{01}}$, respectively) set $L = (L_1, L_2)$ and, for a given R_ε, let $\Delta_\varepsilon(\kappa_0, L)$ be any nonempty subset which consists of all pairs $(\kappa, \rho_\varepsilon)$, $\kappa \in \Xi_{G_0}$, such that

$$d(\kappa, \kappa_0) = |r - r_0| + |s - s_0| + |p - p_0| + |q - q_0| < L_1, \tag{6.99}$$

$$|\tilde{n}/n - 1| + |\tilde{h}_0/h_0 - 1| < L_2 \quad \text{or} \quad |\tilde{m}/m - 1| + |\tilde{z}_0/z_0 - 1| < L_2. \tag{6.100}$$

Here $n = n_\varepsilon(\kappa_0, \rho_\varepsilon, R_\varepsilon)$, $h_0 = h_{\varepsilon,0}(\kappa_0, \rho_\varepsilon, R_\varepsilon)$ and $m = m_\varepsilon(\kappa_0, \rho_\varepsilon, R_\varepsilon)$, $z_0 = z_{\varepsilon,0}(\kappa_0, \rho_\varepsilon, R_\varepsilon)$ are the quantities noted above for κ_0, ρ_ε, R_ε; also $\tilde{n}$, $\tilde{h}_0$ (or $\tilde{m}$ and $\tilde{z}_0$) are the analogous quantities for κ, ρ_ε, R_ε. Let

$$\bar{h}_0 = \bar{h}_\varepsilon(\kappa_0, \rho_\varepsilon, R_\varepsilon), \quad \bar{z}_0 = \bar{z}_\varepsilon(\kappa_0, \rho_\varepsilon, R_\varepsilon),$$

be the sequences $h_i(\kappa_0, n, h_0)$, $z_i(\kappa_0, n, h_0)$ or $h_i(\kappa_0, m, z_0)$, $z_i(\kappa_0, m, z_0)$, respectively, for the extreme sequences $\bar{\pi}_\varepsilon(\kappa_0, \rho_\varepsilon, R_\varepsilon)$. Also let $\bar{h}^* = \bar{h}^*_L(\kappa_0)$, $\bar{z}^* = \bar{z}^*_L(\kappa_0)$ be the sequences which are determined by the relations

$$h_i^* = \sup_{(\kappa, \tilde{n}, \tilde{h}_0) \in \Delta_\varepsilon(\kappa_0, L)} h_i(\kappa, \tilde{n}, \tilde{h}_0), \quad \hat{z}_i = \sup_{(\kappa, \tilde{n}, \tilde{h}_0) \in \Delta_\varepsilon(\kappa_0, L)} z_i(\kappa, \tilde{n}, \tilde{h}_0), \tag{6.101}$$

or, respectively,

$$h_i^* = \sup_{(\kappa, \tilde{m}, \tilde{z}_0) \in \Delta_\varepsilon(\kappa_0, L)} h_i(\kappa, \tilde{m}, \tilde{z}_0), \quad \hat{z}_i = \sup_{(\kappa, \tilde{m}, \tilde{z}_0) \in \Delta_\varepsilon(\kappa_0, L)} z_i(\kappa, \tilde{m}, \tilde{z}_0), \tag{6.102}$$

and

$$z_i^* = \begin{cases} \hat{z}_i, & \text{if } i \in I_\varepsilon^*, \\ 0, & \text{if } i \notin I_\varepsilon^*, \end{cases} \quad I_\varepsilon^* = \{i : \hat{z}_i \le BQ_{\varepsilon,i}\}. \tag{6.103}$$

The quantities $Q_{\varepsilon,i}$ are determined by (6.22) and B is a positive constant. Consider the quantities $u_L = u_{L,\varepsilon}$, $u_0 = u_{0,\varepsilon}$ of the form

$$u_L = u(\bar{h}^*, \bar{z}^*) = \sum_i u_i^2(h_i^*, z_i^*), \quad u_0 = u(\bar{h}_0, \bar{z}_0) = \sum_i u_i^2(h_{0,i}, z_{0,i}),$$

where, as above, $u_i^2(h_i, z_i) = 2h_i^2 \sinh^2(z_i^2/2)$. Fix a compact $K \subset \Xi_{G_0}$.

Proposition 6.1 *Assume* $\liminf u_0 > 0$, $u_0 = o(R_\varepsilon^a)$ *for all* $a > 0$. *Then for any* $B > 0$ *in* (6.103) *one has:*

(1) *Let* $L_1 = L_{\varepsilon,1} = o((\log\log R_\varepsilon)^{-1})$, $L_2 = L_{\varepsilon,2} = o(1)$. *Then the sequences* $\bar{h}^*_L(\kappa_0)$, $\bar{z}^*_L(\kappa_0)$ *satisfy the assumptions of Proposition 5.4. In particular, assumption* A.1a *holds with* $w_\varepsilon = o(R_\varepsilon^{-b})$ *for small enough* $b > 0$. *Moreover, if* $p < q$ *or* $2 > p > q$, $\lambda < 0$ *for* $\kappa = \kappa_0$, *then assumption* A.4 *holds. If* $p > q$ *and* $\lambda > 0$, *then we can use the change* (5.28) *with* $\zeta_\varepsilon^* \to 0$ *such that* $\zeta_\varepsilon^* \sqrt{\log R_\varepsilon} \to \infty$.

(2) *There exists* $C_1 > 0$, $C_2 > 0$ *(which depend on a compact* K*) such that, for* $L_1 = L_{\varepsilon,1} = o((\log\log R_\varepsilon)^{-C_1})$, $L_2 = L_{\varepsilon,2} = o((\log\log R_\varepsilon)^{-C_1})$,

one has

$$u_L = u_0(1 + O((\log\log R_\varepsilon)^{C_1}(L_2 + L_1) + (\log R)^{-C_2})).$$

Proof of Proposition 6.1. This is given in the Appendix, Section A.3.8.

6.8.2 Correlation Properties

Denote by $\rho_\varepsilon(\kappa_1, \kappa_2)$ the correlation coefficients

$$\begin{aligned}\rho_\varepsilon(\kappa_1, \kappa_2) &= \frac{(\bar{\pi}_\varepsilon(\kappa_1), \bar{\pi}_\varepsilon(\kappa_2))}{\|\bar{\pi}_\varepsilon(\kappa_1)\| \, \|\bar{\pi}_\varepsilon(\kappa_2)\|} \\ &= \frac{\sum_i h_{1,i}h_{2,i}\sinh^2(z_{1,i}z_{2,i}/2)}{\sqrt{\sum_i h_{1,i}^2 \sinh^2(z_{1,i}^2/2)}\sqrt{\sum_i h_{2,i}^2 \sinh^2(z_{2,i}^2/2)}},\end{aligned}$$

where $\bar{\pi}_{\kappa_l,\varepsilon}$ is the sequence of the three-point measures corresponding to the sequences $\bar{h}_\varepsilon(\kappa_l) = \{h_{l,i}\}$, $\bar{z}_\varepsilon(\kappa_l) = \{z_{l,i}\}$, $l = 1, 2$. Let $n_l = n_{l,\varepsilon}$ or $m_l = m_{l,\varepsilon}$ be the quantities corresponding to these sequences. Let

$$\tilde{\rho}_\varepsilon(\kappa_1, \kappa_2) = \frac{(\bar{\pi}_{\varepsilon,1}(\kappa_1), \bar{\pi}_{\varepsilon,1}(\kappa_2))}{\|\bar{\pi}_{\varepsilon,1}(\kappa_1)\| \, \|\bar{\pi}_{\varepsilon,1}(\kappa_2)\|}$$

be the analogous correlation coefficients for the sequences which correspond to priors $\pi_1^\varepsilon(\kappa)$.

Let $K_1 \subset \Xi_{G_{0,1}}$ or $K_2 \subset \Xi_{G_{0,2}}$ be a compact.

Proposition 6.2 *Let $\kappa_l \in K_1$ (or $\kappa_l \in K_2$) and let*

$$b \le u_l = \|\bar{\pi}_\varepsilon(\kappa_l)\| = o(R_\varepsilon^a) \quad \forall\, a > 0, \quad l = 1, 2,$$

for some $b > 0$. Then there exist positive δ_1, δ_2, L_0, B (which could depend on a compact K_1 or K_2) such that for any $L < L_0$, $d(\kappa_1, \kappa_2) < L$ one has: if $\kappa_l \in K_2$, $l = 1, 2$, and $n_1 \le n_2$, then

$$\rho_\varepsilon(\kappa_1, \kappa_2) \le B\left(\left(\frac{n_1}{n_2}\right)^{\delta_1} + R_\varepsilon^{-\delta_2}\right),$$

if $\kappa_l \in K_1$, $l = 1, 2$, and $m_1 \le m_2$, then

$$\rho_\varepsilon(\kappa_1, \kappa_2) \le B\left(\left(\frac{m_1}{m_2}\right)^{\delta_1} + R_\varepsilon^{-\delta_2}\right).$$

The same relations hold for $\tilde{\rho}_\varepsilon(\kappa_1, \kappa_2)$.

Proof of Proposition 6.2. This is given in the Appendix, Section A.3.8.

6.9 Extreme Problem for Besov Norms

We study the case $p > 2$ or $p > q$ (the case $p \le 2$, $p \le q$ was considered in Section 4.3.4). The scheme of the study repeats in the principle the scheme for the power norms, however, the number of variants we need to consider is much larger.

In Section 5.3.2 we have reduced the extreme problem (6.11) (which corresponds to (6.9), (6.10)) to the sequence of Lagrange extreme problems (5.85):

$$\inf_{x_j \geq 0, y_j \geq 0} L(x_j, y_j; A, B), \tag{6.104}$$

where

$$L(x_j, y_j; A, B) = 2^j u^2(x_j, y_j; p, q) - A 2^{j(hr+h/p)} x_j^{h/p} + B 2^{j(ts+t/q)} y_j^{t/q},$$

and to the equations for minimum sequences in (6.104):

$$\sum_{j=1}^{\infty} 2^{j(hr+h/p)} x_j^{h/p} = \rho_\varepsilon^h, \quad \sum_{j=1}^{\infty} 2^{j(ts+t/p)} y_j^{t/p} = R_\varepsilon^t, \tag{6.105}$$

(we can assume $A > 0$, $B > 0$). Here $u(x, y; p, q)$ is the solution of the one-dimensional problem

$$u(x, y; p, q) = \inf \{ \|\pi\| \, : \, E_\pi |t|^p \geq x, \; E_\pi |t|^q \leq y \} \tag{6.106}$$

which was described in Section 5.4, Theorem 5.3. By Lemma 5.3 we can assume $(x_j, y_j) \subset \Lambda(p, q) \; \forall \, (i, j)$ where the set $\Lambda(p, q)$ is defined by (5.116)–(5.118) and corresponds to the set $\Theta(p, q)$ which is described in Section 5.4.4 and consists of the variables (h, z), $h \in (0, 1]$, $z > 0$, such that

$$x = hz^p, \quad y = hz^q; \quad u^2(x, y; p, q) = 2h^2 \sinh^2(z^2/2); \quad (x, y) \in \Lambda(p, q).$$

We can rewrite the Lagrange extreme problems (6.104) in terms of the variables h_j, z_j:

$$\inf_{(h_j, z_j) \in \Theta(p,q)} \{ 2^{j+1} h_j^2 \sinh^2(z_j^2/2) - A 2^{j(hr+h/p)} h_j^{h/p} z_j^h + B 2^{j(ts+t/q)} h_j^{t/q} z_j^t \}, \tag{6.107}$$

and by Lemma 5.3, for any $A > 0$, $B > 0$, $j \geq 1$, there exists a unique $(h_j, z_j) \in \Theta(p, q)$ which provides the minimum in (6.107). The parameters A, B are determined by equations (6.105) which correspond to

$$\sum_{j=1}^{\infty} 2^{j(hr+h/p)} h_j^{h/p} z_j^h = \rho_\varepsilon^h, \quad \sum_{j=1}^{\infty} 2^{j(ts+t/p)} h_j^{t/q} z_j^t = R_\varepsilon^t. \tag{6.108}$$

Then the solution of the extreme problem (6.9) is of the form

$$u_\varepsilon^2 = \sum_{j=1}^{\infty} 2^{j+1} h_j^2 \sinh^2(z_j^2/2). \tag{6.109}$$

We study the extreme problems (6.107) in two steps according to Section 5.4.4. The first step is to study the extreme problem (6.107) without the constraint $h_i \le 1$. This leads to the system of equations (5.132) which is of the form

$$2^{j+2}h_j^2 \sinh^2 \frac{z_j^2}{2} = \frac{Ah}{p} 2^{j(rh+h/p)} h_j^{h/p} z_j^h - \frac{Bt}{q} 2^{j(st+t/q)} h_j^{t/q} z_j^t, \qquad (6.110)$$

$$2^{j+2}h_j^2 \sinh^2 \frac{z_j^2}{2} \left(\frac{z_j^2}{\tanh(z_j^2/2)} \right) = Ah2^{j(rh+h/p)} h_j^{h/p} z_j^h - Bt2^{j(st+t/q)} h_j^{t/q} z_j^t. \qquad (6.111)$$

If we can find the solution $h_j > 0$, $z_j > 0$ of (6.110), (6.111) such that $h_j \le 1$ and $z_j < z_p$ for $2 < p < q$ or $z_j > z_p$ for $2 < p$, $q < p$, then this provides the minimum in the problem. If such a solution does not exist, then we consider the extreme problems (5.133) which correspond to $h_j = 1$:

$$\inf_{z_j > 0} 2^{j+1} \sinh^2(z_j^2/2) - A2^{j(hr+h/p)} z_j^h + B2^{j(ts+t/q)} z_j^t \qquad (6.112)$$

and $z_j > z_p$ for $2 < p$, $q < p$. This leads to the equation

$$2^{j+1} z_j^2 \sinh(z_j^2) = Ah2^{j(rh+h/p)} z_j^h - Bt2^{j(st+t/q)} z_j^t \qquad (6.113)$$

and we select the root of (6.113) which provides the minimum in (6.112).

In the proofs below we study the solution of the system (6.110), (6.111), or of equations (6.113) under several assumptions, either on $A = A_\varepsilon$, $B = B_\varepsilon$, or on other parameters which are determined by A, B. Then we find these parameters by solving (6.108) and we check these assumptions.

We realize this outline in the Appendix, Section A.4.

To study the adaptive setting in the following chapter it is convenient to combine some rate properties of the extreme sequences

$$z_{\varepsilon,j} = z_j(\tau, \rho_\varepsilon, R_\varepsilon), \quad h_{\varepsilon,j} = h_j(\tau, \rho_\varepsilon, R_\varepsilon),$$

which determine the extreme sequences measures $\bar{\pi}_{\varepsilon,j} = \bar{\pi}_j(\tau, \rho_\varepsilon, R_\varepsilon)$ in the extreme problem (6.11).

Proposition 6.3 *Assume* $0 < h < p < \infty$, $0 < q < t < \infty$, $p \ne q$.

I. *Let* $\kappa \in \Xi_{G_1}$. *Denote*

$$\mathcal{J}_\varepsilon^0 = \begin{cases} \{j : j > j_1(1+\eta)\}, & \text{if } q < p < 2, \\ \{j : j \le j_1(1-\eta)\}, & \text{if } q > p > 2, \end{cases}$$

where $\eta > 0$ *is small enough. Then there exists* $\delta > 0$ *such that*

$$\sup_{j \notin \mathcal{J}_\varepsilon^0} z_{\varepsilon,j} 2^{\delta \max(j,j_1)} = o(1).$$

II. *Let* $\kappa \in \Xi_{G_2}$ *and* $q > p > 2$. *Then, for any* $\eta > 0$, *there exists* $A > 0$ *and* $\delta > 0$, *such that*

$$z_{\varepsilon,j} \le \begin{cases} A, & \text{if } j \le j_0(1+\eta), \\ 2^{-\delta(j-j_0)}, & \text{if } j > j_0(1+\eta). \end{cases}$$

III. *Let $\kappa \in \Xi_{G_2}$ and $p > q$. Then there exists $A > 0$ such that*

$$z_{\varepsilon,j} \leq \begin{cases} A, & \text{if } j \leq j_0, \\ A\sqrt{1+j-j_0}, & \text{if } j > j_0. \end{cases}$$

IV. *Denote*

$$j_\varepsilon^* = \begin{cases} j_1, & \text{if } \tau \in \Xi_{G_1}, \\ j_0, & \text{if } \tau \in \Xi_{G_2}. \end{cases}$$

Then, for any $\tau = (\kappa, h, t)$ with $\kappa \in \Xi_{G_1} \cup \Xi_{G_2}$, there exist $a = a(\tau) > 0$ and $b = b(\tau) > 0$, such that for ε small enough,

$$\frac{u_{\varepsilon,j}}{u_\varepsilon} \leq a 2^{-b|j-j_\varepsilon^*|}, \tag{6.114}$$

where

$$u_{\varepsilon,j}^2 = u_{\varepsilon,j}^2(\tau) = 2^j h_{\varepsilon,j}^2 \sinh^2(z_{\varepsilon,j}^2/2), \quad u_\varepsilon^2 = u_\varepsilon^2(\tau) = \sum_{j=1}^{\infty} u_{\varepsilon,j}^2(\tau).$$

Proof of Proposition 6.3. This is given in the Appendix, Section A.4.7.

7
Adaptation for Power and Besov Norms

An important point in the results of Section 4.4 is that the asymptotically minimax families of tests for power norms or minimax consistent families of tests for Besov norms do not depend on the parameters κ, R_ε, ρ_ε for $\kappa \in \Xi_D$. This means that there exists a *common* family of tests which is asymptotically minimax (or minimax consistent) *for all* $\kappa \in \Xi_D$ uniformly over any compact $K \subset \operatorname{Int} \Xi_D$ (recall that these tests are based on the thresholding). Analogously one can construct a common family of minimax consistent tests for $K \subset \operatorname{Int} \Xi_C$. This means that these tests should be asymptotically minimax (or minimax consistent) for the "union" alternatives

$$V_\varepsilon(K) = \bigcup_{\kappa \in K} V(\kappa, \rho_\varepsilon(\kappa), R_\varepsilon).$$

Certainly, radii of the removing sets should depend on κ and be large enough to obtain minimax consistent tests: for a given family $R_\varepsilon \to \infty$ we need to consider the alternatives $V_\varepsilon(\kappa) = V(\kappa, \rho_\varepsilon(\kappa), R_\varepsilon)$ with

$$\lim \inf_{\kappa \in K} \rho_\varepsilon(\kappa)/\rho_\varepsilon^*(\kappa) > \Lambda(\kappa)$$

only; here $\rho_\varepsilon^*(\kappa) = \rho^*(\kappa, R_\varepsilon)$ are the rates (critical radii) and the constants $\Lambda(\kappa)$ are defined in Section 4.4.2 (see (4.150)).

However, this does not hold for $K \subset \Xi_G$: there is an essential dependence of the structure of asymptotically minimax (or consistent) families of tests on $\kappa \in \Xi_G$. Thus we have the problem: to construct a common family of tests which has good minimax (or consistent) properties *for all* $\kappa \in K \subset \Xi_G$

with wide enough subsets K. Of course, we need

$$\lim \inf_{\kappa \in K} \rho_\varepsilon(\kappa)/\rho_\varepsilon^*(\kappa) > 0 \quad \text{or} \quad \lim \inf_{\kappa \in K} \rho_\varepsilon(\kappa)/\rho_\varepsilon^*(\kappa) \to \infty, \tag{7.1}$$

where the rates $\rho_\varepsilon^*(\kappa) = \rho^*(\kappa, R_\varepsilon)$ are determined by (6.36) for the regions of the main types of Gaussian asymptotics. One has the same question for Besov norms in the sequence space, and for Sobolev and Besov norms in the functional space. This problem is of importance from a practical point of view since, as usual, a statistician has no information on norms to measure a smoothness of alternatives and a distance from the null-hypothesis.

Following Spokoiny [90], [91] who started the study of the problem, we call this setting *adaptive*. Asymptotics of minimax total error probabilities $\gamma(V_\varepsilon(K))$ or type II error probabilities $\beta(\alpha, V_\varepsilon(K))$ for "union" alternatives $V_\varepsilon(K)$ we also call *adaptive* (sharp or rate).

Spokoiny [90] has considered the case when the quantities $p = 2,\ q \geq p,\ \sigma = 0$ (L_p-balls removed) are known and the smoothness parameter $\eta > 0$ is unknown and belongs to an interval (η_0, η_1). It was shown in [90] that one needs to increase the rates up to some power of $\log\log\varepsilon^{-1}$. For increased rates "adaptive" tests $\psi_\varepsilon = \psi_{\varepsilon,p}$ have been constructed in [91] for the case of alternative types of Besov bodies with the L_p-balls removed. The proposed tests are minimax consistent uniformly over any compact set

$$K \subset (\Xi_{G_1} \cup \Xi_{G_2}) \cap \{\sigma = 0,\ p = \text{const},\ I \neq 0\}).$$

We would like to obtain sharp adaptive asymptotics for the power norms and sharp adaptive rates for the case of Besov norms for the principal regions Ξ_{G_l}, $l = 1, 2$, of the Gaussian asymptotics. Certainly these results jointed with the wavelet transform give adaptive rates for Sobolev or Besov norms under the functional Gaussian model. To simplify, we do not consider cases when κ belongs to the boundary of the sets $\Xi_{G_{1,2}}$ and to some submanifolds (particularly, $\lambda = 0,\ \Delta = 0,\ p = q,\ p = 2$ in these regions.)

7.1 Adaptive Setting

7.1.1 *Idea of the Study*

As was noted above, test procedures for Gaussian asymptotics depend essentially on the parameter κ which determines the norms in the constraints. In the adaptive problem we can consider $\kappa \in K \subset R^4$ as an unknown parameter. On the other hand, it was noted that for Gaussian asymptotics we reduce the hypotheses testing problems of interest to signal detection problems: an unknown "signal" $\bar{\pi}_\varepsilon = \bar{\pi}_\varepsilon(\kappa)$ corresponds to the asymptotically least favorable prior $\pi^\varepsilon = \pi^\varepsilon(\kappa)$ of the product type and the likelihood ratio is asymptotically Gaussian,

$$\frac{dP_{\pi^\varepsilon(\kappa)}}{dP_0} = \exp(-\|\bar{\pi}_\varepsilon(\kappa)\|/2 + t_{\varepsilon,\kappa}\|\bar{\pi}_\varepsilon(\kappa)\| + \eta_\varepsilon),$$

where $t_{\varepsilon,\kappa} \to \xi \sim N(0,1)$, $\eta_\varepsilon \to 0$ in P_0-probability. Thus, roughly, we can consider an adaptive problem as a parametric signal detection problem with an unknown "signal" $\bar{\pi}_\varepsilon = \bar{\pi}_\varepsilon(\kappa)$ in the Hilbert space $\bar{\Pi}$ which consists not of real-valued sequences, but of sequences of measures on the real line and the Hilbert structure is determined in Section 3.3.3.

We try to use the maximum likelihood method in the problem. However, the properties of the maximum likelihood method are not standard here. The problem is analogous to the following one.

Let us observe an unknown signal s in the white Gaussian noise of level $\varepsilon = 1$ under model (1.36). Then we want to test the null-hypothesis H_0 : $s = 0$ against the parametric alternative $H_1 : s = s_\theta,\ \theta \in \Theta_T$, where Θ_T is the interval (θ_0, T) and the signal s is the sine with an unknown frequency θ,

$$s_\theta(t) = \sqrt{2}\rho_T \sin((t-1/2)\theta + \pi/4); \quad t \in [0,1], \quad \theta_0 > 0, \quad T \to \infty.$$

By the choice of the argument of the sine, we have $\|s_\theta(\cdot)\|_2 = \rho_T$ for all θ.

To obtain the lower bounds in this problem let us take the collection of orthogonal signals

$$s_k(t) = s_{\theta_k}(t), \quad \theta_k = k/2\pi, \quad k = 1,\dots,M \sim T/2\pi,$$

and consider the prior π^M concentrated on this collection

$$\pi^M = M^{-1} \sum_{k=1}^{M} \delta_{s_k(t)}.$$

Using the results from Corollary 4.5 we get

$$\begin{aligned}\beta(\alpha, \Theta_T) \ &\geq \ \beta(\alpha, \pi^M) = (1-\alpha)\Phi(\sqrt{2\log M} - \rho_T) + o(1)\\ &= \ (1-\alpha)\Phi(\sqrt{2\log T} - \rho_T) + o(1).\end{aligned}$$

Therefore we need to consider $\rho_T \geq \sqrt{2\log T} + O(1) \to \infty$.

To obtain the upper bounds assume $\rho_T \asymp \sqrt{2\log T}$ and consider the tests $\psi_\theta = \mathbf{1}_{\{t_\theta > H\}}$ which are based on the statistics

$$t_\theta = \rho_T^{-1} \int_0^1 s_\theta(t)\, dX(t). \tag{7.2}$$

The statistics t_θ are standard Gaussian under the null-hypothesis and for each θ we have $\alpha(\psi_\theta) = \Phi(-H)$. Each test ψ_θ is optimal for the simple alternative θ. For a different alternative θ_1 we have $\beta(\phi_\theta, \theta_1) = \Phi(H - \eta\rho_T)$ where $\eta = (s_\theta, s_{\theta_1})/\|s_\theta\|\|s_{\theta_1}\|$ is the correlation between the signals s_θ and s_{θ_1}. Using simple calculations we get

$$\eta = \frac{2}{\theta - \theta_1} \sin\left(\frac{\theta - \theta_1}{2}\right) \geq 1 - b(\theta - \theta_1)^2$$

for some $b > 0$. Thus the test ψ_θ is asymptotically optimal for the alternatives θ_1 such that $(\theta - \theta_1)^2 = o(\rho_T^{-1})$.

Now let us use a discrete version of the maximum likelihood method. Take the collection

$$\theta_k = k(\log T)^{-1}, \quad k = 1, \dots, M_1 \sim T \log T; \quad \theta_{M_1} > T.$$

Set

$$H = \sqrt{2 \log M_1} = \sqrt{2 \log T} + o(1),$$

and consider combinations of the tests based on the statistics (7.2),

$$\psi_T = \max_{1 \le k \le M_1} \psi_{\theta_k} = \mathbf{1}_{\mathcal{X}_T}, \quad \mathcal{X}_T = \left\{ \max_{1 \le k \le M_1} t_{\theta_k} > H \right\}.$$

Setting $\Delta_k = [\theta_{k-1}, \theta_k]$ and using relation (3.35), we have

$$\begin{aligned} \alpha(\psi_T) &\le \sum_{k=1}^{M_1} \alpha(\psi_{\theta_k}) \le M_1 \Phi(-H) \to 0, \\ \beta(\psi_T, \Theta) &\le \max_{1 \le k \le M_1} \beta(\psi_{\theta_k}, \Delta_k) \le \Phi(H - \rho_T) + o(1) \\ &= \Phi(\sqrt{2 \log T} - \rho_T) + o(1). \end{aligned}$$

Now, the asymptotically minimax tests $\psi_{T,\alpha}$ could be obtained by randomization: $\psi_{T,\alpha} = \alpha + (1 - \alpha)\psi_T$.

Thus the considerations are analogous to those which lead to the asymptotics of degenerate type. We will apply this scheme to the adaptive problem which is, roughly, parametric with an unknown parameter κ.

7.1.2 Statement of the Problem

Recall that we use the parameters

$$\Delta = sq(4 - p) - rp(4 - q), \quad \lambda = s/p - r/q,$$
$$I = 2(s - r) - 4(s/p - r/q) + 1/p - 1/q.$$

Denote, as in Section 6.8,

$$\Xi_{G_0} = \{\kappa \in \Xi_{G_1} \cup \Xi_{G_2} : \ p \ne q, \ p \ne 2, \ \lambda \ne 0, \ \Delta \ne 0\}$$

and set

$$\Xi_{G_{01}} = \Xi_{G_0} \cap \Xi_{G_1}, \quad \Xi_{G_{02}} = \Xi_{G_0} \cap \Xi_{G_2}.$$

Here $\kappa = (p, q, r, s)$ is a four-dimensional parameter which determines the asymptotics in Chapter 6. Assume that a compact K is given, $K \subset \Xi_{G_0}$. For the power norms let a family of functions $\rho_\varepsilon(\kappa)$, $\kappa \in K$, be given. Set

$$V_\varepsilon(K) = \bigcup_{\kappa \in K} V(\kappa, \rho_\varepsilon(\kappa), R_\varepsilon).$$

Here, as above, the sets $V_\varepsilon(\kappa) = V(\kappa, \rho_\varepsilon(\kappa), R_\varepsilon) \subset l_2$ are determined by the constraints

$$|v|_{r,p} \geq \rho_\varepsilon(\kappa), \quad |v|_{s,q} \leq R_\varepsilon,$$

however, we suppose that the radii ρ_ε could depend on the parameter κ. We test the null-hypothesis $H_0 : v = 0$ against the "union" alternative $H_1 : v \in V_\varepsilon(K)$.

For Besov norms we denote $\tau = (\kappa, h, t, R)$ and let $\Gamma \subset K \times D^{(3)}$ where $D^{(3)}$ is a subset of the set $\{(h, t, R) : 0 < a \leq h, t, R \leq \infty\}$. We want to apply our results to the functional Gaussian model (2.5) directly. For this reason we specify $R_\varepsilon = R/\varepsilon$, $\rho_\varepsilon(\tau) = r_\varepsilon(\tau)/\varepsilon$ where $R > 0$ is fixed, $r_\varepsilon(\tau)$ is a positive family of function, and $\varepsilon \to 0$. Set

$$V_\varepsilon(\Gamma) = \bigcup_{\tau \in \Gamma} V_\varepsilon(\tau, r_\varepsilon(\tau)),$$

where the sets $V_\varepsilon = V_\varepsilon(\tau, r_\varepsilon) \subset L_2$ are determined by constraints

$$|v|_{r,p,h} \geq \rho_\varepsilon(\tau) = r_\varepsilon(\tau)/\varepsilon, \quad |v|_{s,q,t} \leq R_\varepsilon = R/\varepsilon.$$

We test the null-hypothesis $H_0 : v = 0$ against the alternative $H_1 : v \in V_\varepsilon(\Gamma)$.

7.1.3 *Main Results*

To describe the sharp adaptive asymptotics for power norms let us consider the family of functions $u(\kappa, \rho_\varepsilon(\kappa), R_\varepsilon)$ and put

$$u_\varepsilon(K) = \inf_{\kappa \in K} u(\kappa, \rho_\varepsilon(\kappa), R_\varepsilon), \quad H_\varepsilon = u_\varepsilon(K) - \sqrt{2 \log\log R_\varepsilon}.$$

Theorem 7.1 *Assume $K \subset \Xi_{G_0}$ is compact. Then:*
(1) *Upper bounds:*

$$\beta(\alpha, V_\varepsilon(K)) \leq (1 - \alpha)\Phi(-H_\varepsilon) + o(1).$$

(2) *Assume for any $\delta > 0$ that there exists an open set $\Delta = \Delta(\delta) \subset K$ such that $u_\varepsilon(\kappa) < u_\varepsilon(K) + \delta$ for $\kappa \in \Delta$. Then the following lower bounds hold:*

$$\beta(\alpha, V_\varepsilon(K)) \geq (1 - \alpha)\Phi(-H_\varepsilon) + o(1).$$

Theorem 7.1 shows that under the assumption of (2),

$$\beta(\alpha, V_\varepsilon(K)) = (1 - \alpha)\Phi(-H_\varepsilon) + o(1).$$

Remark. The assumption in (2) means that the infimum $u_\varepsilon(K)$ is "essential." We can use a weaker assumption: there exists an interval in K which is not "tangent" to the hypersurfaces

$$\{\phi_l(\kappa) = \text{const}; \ \kappa \in \Xi_{G_l}\}, \quad l = 1, 2,$$

where the functions $\phi_l(\kappa)$ are defined by (7.44), (7.50) later (see also Remark 6 in Section 6.5). More exactly, it suffices to assume that the image $\phi_l(K)$ contains a nontrivial interval. However, it is possible to show that for any finite collection $\{\kappa_i\}$ we can provide better lower bounds.

Let us define the *adaptive rate functions* $\rho^*_{\varepsilon,\mathrm{ad}}(\kappa)$ by the relation

$$u(\kappa, \rho^*_{\varepsilon,\mathrm{ad}}(\kappa), R_\varepsilon) = \sqrt{2\log\log R_\varepsilon} + O(1).$$

Using (6.14) one has, for $\kappa \in \Xi_{G_l}$, $l = 1, 2$,

$$\rho^*_{\varepsilon,\mathrm{ad}}(\kappa) \sim R_\varepsilon^{B_l(\kappa)/A_l(\kappa)}((2\log\log R_\varepsilon)/d_l(\kappa))^{1/A_l(\kappa)}, \tag{7.3}$$

where $A_l(\kappa), B_l(\kappa)$, $l = 1, 2$, are defined by (6.15), (6.16).

The difference the adaptive rate in (7.3) with the nonadaptive rate (6.36) is the factor $(\log\log R_\varepsilon)^{1/A_l(\kappa)}$. One can interpret this factor as *payment for adaptation.*

Using Theorem 7.1 we get the following relations. Let

$$\liminf_{\varepsilon\to 0} \inf_{\kappa\in K} \rho_\varepsilon(\kappa)/\rho^*_{\varepsilon,\mathrm{ad}}(\kappa) > 1. \tag{7.4}$$

Then $\beta(\alpha, V_\varepsilon(K)) \to 0$. On the other hand, let there exist an open set $\Delta \subset K$ such that

$$\limsup_{\varepsilon\to 0} \sup_{\kappa\in\Delta} \rho_\varepsilon(\kappa)/\rho^*_{\varepsilon,\mathrm{ad}}(\kappa) < 1. \tag{7.5}$$

Then $\beta(\alpha, V_\varepsilon(K)) \to 1 - \alpha$.

For Besov norms we denote $u_\varepsilon(\tau, r_\varepsilon(\tau)) = u(\tau, r_\varepsilon(\tau)/\varepsilon, R/\varepsilon)$ (recall that we consider radii of the form $\rho_\varepsilon = r_\varepsilon/\varepsilon$, $R_\varepsilon = R/\varepsilon$ which correspond to the functional Gaussian model (2.5)). For the set Γ from Section 7.1.2 we denote

$$K = K(\Gamma) = \{\kappa(\tau) : \tau \in \Gamma\}.$$

Let $K \subset \Xi_{G_1} \cup \Xi_{G_2}$ and let K be a compact. We put

$$u_\varepsilon(\Gamma) = \inf_{\tau\in\Gamma} u_\varepsilon(\tau, r_\varepsilon(\tau)).$$

Theorem 7.2 *There exist constants $C = C(\Gamma)$, $c = c(\Gamma)$, $C \geq c > 0$ such that:*

(1) *Let*

$$\liminf_{\varepsilon\to 0} u_\varepsilon^2(\Gamma)/\log\log\varepsilon^{-1} > C.$$

Then there exists a family of tests ψ_ε such that

$$\alpha(\psi_\varepsilon) = o(1), \quad \beta(\psi_\varepsilon, V_\varepsilon(\Gamma)) = o(1).$$

These relations imply $\beta(\alpha, V_\varepsilon(\Gamma)) \to 0$ for all $\alpha \in (0, 1)$.

(2) *Let there exist an open set $\Delta \subset K$ and functions $\kappa \to (h, t, R) : \Delta \to D^{(3)}$, such that*

$$\Gamma_0 = \{\tau(\kappa) = (\kappa, h(\kappa), t(\kappa), R(\kappa)),\ \kappa \in \Delta\} \subset \Gamma$$

and

$$\limsup_{\varepsilon\to 0}\sup_{\tau\in\Gamma_0} u_\varepsilon^2(\tau, r_\varepsilon(\tau))/\log\log\varepsilon^{-1} < c.$$

Then $\beta(\alpha, V_\varepsilon(\Gamma)) \to (1-\alpha)$ *for all* $\alpha \in (0,1)$.

Using Theorem 7.2 we obtain the relations analogous to (7.4), (7.5) with the adaptive rate s (7.3). [1]

Corollary 7.1 *For Besov norms there exist constants*

$$\Lambda_1 = \Lambda_1(\Gamma), \quad \Lambda_2 = \Lambda_2(\Gamma); \quad \Lambda_1 > \Lambda_2 > 0,$$

such that

$$\text{if } \liminf_{\varepsilon\to 0}\inf_{\tau\in\Gamma} r_\varepsilon(\tau)/r^*_{\varepsilon,\mathrm{ad}}(\kappa) > \Lambda_1, \text{ then } \beta(\alpha, V_\varepsilon(\Gamma)) \to 0, \tag{7.6}$$

$$\text{if } \limsup_{\varepsilon\to 0}\sup_{\tau\in\Delta} r_\varepsilon(\tau)/r^*_{\varepsilon,\mathrm{ad}}(\kappa) < \Lambda_2, \text{ then } \beta(\alpha, V_\varepsilon(\Gamma)) \to 1-\alpha, \tag{7.7}$$

where $\kappa = \kappa(\tau)$ *are the corresponding coordinates of vector parameter* τ, *and* Δ *is the subset* Γ *such that* $\kappa(\Delta) = \{\kappa(\tau) : \tau \in \Delta\}$ *is the open subset* $\Xi_{G_1} \cap \Xi_{G_2}$.

7.2 Lower Bounds

To obtain the lower bounds we follow the Bayesian approach. Let π^ε be a sequence of probability measures (priors) on the space l^2 and let P_{π^ε} be a sequence of mixtures

$$P_{\pi^\varepsilon}(A) = \int P_v(A)\pi^\varepsilon(dv).$$

Consider the Bayesian hypothesis testing problem: $H_0 : P = P_0$; $H_1 : P = P_{\pi^\varepsilon}$. Assume that the L_2-distance between P_0 and P_{π^ε} tends to 0

$$E_{P_0}\left(\frac{dP_{\pi^\varepsilon}}{dP_0} - 1\right)^2 \to 0.$$

Then it follows from Proposition 2.12 that, for all $\alpha \in (0,1)$,

$$\beta(\alpha, P_0, P_{\pi^\varepsilon}) \to 1-\alpha, \quad \gamma(P_0, P_{\pi^\varepsilon}) \to 1. \tag{7.8}$$

Assume that the P_0-distributions of the Bayesian likelihood ratios are asymptotically degenerate: for some (nonrandom) sequence C_ε, in P_0-probability,

$$dP_{\pi^\varepsilon}/dP_0 = C_\varepsilon + o(1). \tag{7.9}$$

[1] This result for $\sigma = 0$, $p = \text{const}$ corresponds to the result by Spokoiny [91].

Relation (7.9) yields (compare with formulas (4.138), (4.139))

$$\gamma(P_0, P_{\pi^\varepsilon}) \geq C_\varepsilon + o(1), \quad \beta(\alpha, P_0, P_{\pi^\varepsilon}) \geq (1-\alpha)\, C_\varepsilon + o(1). \tag{7.10}$$

If, moreover,

$$\pi^\varepsilon(V_\varepsilon) \to 1, \tag{7.11}$$

then the lower bounds hold: under (7.8) we get

$$\beta(\alpha, V_\varepsilon) \to 1-\alpha, \quad \gamma(V_\varepsilon) \to 1, \tag{7.12}$$

and under (7.10) we get

$$\gamma(V_\varepsilon) \geq C_\varepsilon + o(1), \quad \beta(\alpha, V_\varepsilon) \geq (1-\alpha)\, C_\varepsilon + o(1). \tag{7.13}$$

7.2.1 Asymptotically Orthogonal Priors

We consider priors which are finite mixtures

$$\pi^\varepsilon = \sum_{l=1}^{M_\varepsilon} p_{\varepsilon,l}\pi_l^\varepsilon, \quad M_\varepsilon \to \infty, \tag{7.14}$$

where $p_{\varepsilon,l} \geq 0,\ \sum_l p_{\varepsilon,l} = 1$. Clearly, if

$$\min_{1\leq l\leq M_\varepsilon} \pi_l^\varepsilon(V_\varepsilon) = 1 - o(1), \tag{7.15}$$

then the mixture π^ε is asymptotically concentrated on V_ε, i.e., (7.11) holds.

We considered, in Sections 3.3.7, 4.4.1, the case when $\pi_l^\varepsilon = \delta_{v_{\varepsilon,l}}$, where $\{v_{\varepsilon,l}\} \in l^2$ is an orthogonal collection in l^2, and δ_v is a Dirac mass at a point $v \in l^2$. It was shown in Proposition 3.10 that if $v_{\varepsilon,l} \in V_\varepsilon$ and

$$\sum_l p_{\varepsilon,l}^2 \exp(u_{\varepsilon,l}^2) \to 0, \quad u_{\varepsilon,l} = |v_{\varepsilon,l}|, \tag{7.16}$$

then $\gamma(V_\varepsilon) \to 1,\ \beta(\alpha, V_\varepsilon) \to 1-\alpha$. Moreover, denote

$$u_\varepsilon = \min_l u_{\varepsilon,l}, \quad \omega_{\varepsilon,l} = u_{\varepsilon,l} - D_\varepsilon.$$

It was shown in Proposition 4.10 that if the quantities D_ε are such that $u_\varepsilon - D_\varepsilon \to \infty$ and

$$\sum_l \exp(-\omega_{\varepsilon,l}^2/2) \asymp 1, \quad p_{\varepsilon,l} = \exp(-\omega_{\varepsilon,l}^2/2)\left(\sum_l \exp(-\omega_{\varepsilon,l}^2/2)\right)^{-1}, \tag{7.17}$$

then we get asymptotics of the type (7.9) with $C_\varepsilon = \Phi(-D_\varepsilon)$. In particular (see Corollary 4.5 (2)), if $u_{\varepsilon,l} = u_\varepsilon,\ l = 1, \dots, M_\varepsilon \to \infty$, then $C_\varepsilon = \Phi(\sqrt{2\log M_\varepsilon} - u_\varepsilon)$.

We would like to extend these considerations to the priors of product type

$$\pi_l^\varepsilon = \prod_{i\in I} \pi_{\varepsilon,l,i} \tag{7.18}$$

which correspond to sequences of the probability measures $\bar{\pi}_{\varepsilon,l} = \{\pi_{\varepsilon,l,i}, i \in I\}$ on the real line R^1. Here i is either a positive integer for the power norms or pyramidal index $i = (\iota, j) \in \mathcal{J}$ for the Besov norms, and $\mathcal{J}$ is the index set $\mathcal{J} = \{(\iota, j),\ 1 \le \iota \le 2^j,\ j \ge 1\}$. The orthogonality is assumed in the sense of the scalar product in Section 3.3.3.

Using the relations (compare with (3.66)):

$$\begin{aligned}
E_{P_0}\left(\frac{dP_{\pi_k^\varepsilon}}{dP_0}\frac{dP_{\pi_l^\varepsilon}}{dP_0}\right) &= \prod_i E_{P_0}\left(\frac{dP_{\pi_{\varepsilon,k,i}}}{dP_0}\frac{dP_{\pi_{\varepsilon,l,i}}}{dP_0}\right) \\
&= \prod_i \int\int e^{(uv)}\pi_{\varepsilon,k,i}(du)\pi_{\varepsilon,l,i}(dv) \\
&= \prod_i (1 + (\pi_{\varepsilon,k,i}, \pi_{\varepsilon,l,i})) \\
&\le \exp\left(\sum_i (\pi_{\varepsilon,k,i}, \pi_{\varepsilon,l,i})\right) = \exp(\bar{\pi}_{\varepsilon,k}, \bar{\pi}_{\varepsilon,l}),
\end{aligned}$$

analogously with Section 3.3.7, we get

$$\begin{aligned}
E_{P_0}\left(\frac{dP_{\pi^\varepsilon}}{dP_0} - 1\right)^2 &= \sum_{l,k=1}^{M_\varepsilon} p_{\varepsilon,k}p_{\varepsilon,l}E_{P_0}\left(\left(\frac{dP_{\pi_l^\varepsilon}}{dP_0} - 1\right)\left(\frac{dP_{\pi_k^\varepsilon}}{dP_0} - 1\right)\right) \\
&\le \sum_{l,k=1}^{M_\varepsilon} p_{\varepsilon,k}p_{\varepsilon,l}(\exp(\bar{\pi}_{\varepsilon,l}, \bar{\pi}_{\varepsilon,k}) - 1).
\end{aligned}$$

These yield the following nondistinguishability conditions.

Lemma 7.1 *Assume*

$$\sup_{1\le l<k\le M_\varepsilon} (\bar{\pi}_{\varepsilon,l}, \bar{\pi}_{\varepsilon,k}) = o(1); \quad \sum_{l=1}^{M_\varepsilon} p_{\varepsilon,l}^2 \exp(\|\bar{\pi}_{\varepsilon,l}\|^2) = o(1).$$

Then (7.8) *holds. If, moreover,* (7.15) *is fulfilled, then* (7.12) *holds.*

7.2.2 Asymptotic Sharp Lower Bounds

To obtain the asymptotic sharp lower bounds for the priors (7.14) let us consider the likelihood ratio statistics

$$L_\varepsilon = dP_{\pi^\varepsilon}/dP_0 = \sum_{l=1}^{M} p_{\varepsilon,l}L_{\varepsilon,l}; \quad L_{\varepsilon,l} = dP_{\pi_l^\varepsilon}/dP_0, \quad l_{\varepsilon,l} = \log\left(dP_{\pi_l^\varepsilon}/dP_0\right).$$

Introduce the following assumption.

Let there be given families of collections

$$u_{\varepsilon,l} > 0, \quad \omega_{\varepsilon,l} > 0,\ 1 \le l < k \le M = M_\varepsilon,$$

a family $\rho_{\varepsilon,kl} = \rho_{\varepsilon,lk}$, $1 \le l,k \le M$, $l \ne k$, a family D_ε, and a quantity $b > 0$ such that

$$\min_{1\le l\le M} \omega_{\varepsilon,l} \to \infty; \quad b \le u_{\varepsilon,l} \le \omega_{\varepsilon,l} + D_\varepsilon + o(1), \quad \sum_{l=1}^{M} \exp(-\omega_{\varepsilon,l}^2/2) \asymp 1. \tag{7.19}$$

Set

$$p_{\varepsilon,l} = \exp(-\omega_{\varepsilon,l}^2/2)\left(\sum_{l=1}^{M} \exp(-\omega_{\varepsilon,l}^2/2)\right)^{-1}. \tag{7.20}$$

Consider the statistics

$$\lambda_{\varepsilon,l} = (l_{\varepsilon,l} + u_{\varepsilon,l}^2/2)/u_{\varepsilon,l}.$$

Seting $r_{\varepsilon,kl} = \rho_{\varepsilon,lk} u_{\varepsilon,k} u_{\varepsilon,l}$, assume

$$\max_{1\le l<k\le M} r_{\varepsilon,kl} = o(1). \tag{7.21}$$

The assumptions (7.19), (7.21) yield

$$\max_{1\le l<k\le M} \rho_{\varepsilon,kl} = o(1).$$

Denote

$$\begin{aligned} \Delta_{\varepsilon,l} &= \sup_{x\in R^1} |P_0(\lambda_{\varepsilon,l} < x) - \Phi(x)|, \\ \Delta_{\varepsilon,kl} &= \sup_{(x,y)\in R^2} |P_0(\lambda_{\varepsilon,k} < x,\ \lambda_{\varepsilon,l} < y) - \Phi_{\rho_{\varepsilon,kl}}(x,y)|, \end{aligned}$$

where $\Phi_\rho(x,y)$ is the two-dimensional distribution function of the Gaussian random variables (ν_1,ν_2) with

$$E(\nu_k) = 0, \quad \mathrm{Var}(\nu_k) = 1, \quad k = 1,2, \quad \mathrm{Cov}(\nu_1,\nu_2) = \rho.$$

Assume

$$\Delta_\varepsilon = \sum_{1\le l\le M} \Delta_{\varepsilon,l} + \sum_{1\le l<k\le M} \Delta_{\varepsilon,lk} = o(1). \tag{7.22}$$

Let

$$L_{\varepsilon,l,w} = \exp(l_{\varepsilon,l}) \mathbf{1}_{\{\lambda_{\varepsilon,l} < \omega_{\varepsilon,l}\}}$$

be the truncated statistics $L_{\varepsilon,l} = \exp(l_{\varepsilon,l})$.

The following lemma extends Proposition 4.10:

Lemma 7.2 *Under the assumptions* (7.19)–(7.22) *one has*

$$\beta(\alpha, P_0, P_{\pi^\varepsilon}) \ge (1-\alpha)\Phi(-D_\varepsilon) + o(1), \quad \gamma(P_0, P_{\pi^\varepsilon}) \ge \Phi(-D_\varepsilon) + o(1). \tag{7.23}$$

If, moreover, (7.15) *is fulfilled, then* (7.13) *hold with* $C_\varepsilon = \Phi(-D_\varepsilon)$.

In particular, we get

Corollary 7.2 *Assume* (7.21), (7.22) *with*

$$p_{\varepsilon,l} = 1/M, \quad \omega_{\varepsilon,l} = \omega_\varepsilon = \sqrt{2\log M}, \quad u_{\varepsilon,l} = u_\varepsilon + o(1).$$

Then the inequalities (7.23) *hold with* $D_\varepsilon = u_\varepsilon - \sqrt{2\log M}$. *If, moreover,* (7.15) *is fulfilled, then* (7.13) *holds with* $C_\varepsilon = \Phi(-D_\varepsilon)$.

Proof of Lemma 7.2. In view of Proposition 4.9, it suffices to show that, in P_0-probability,

$$L_\varepsilon = \sum_{l=1}^{M} p_{\varepsilon,l} L_{\varepsilon,l} = C_\varepsilon + o(1), \quad C_\varepsilon \geq \Phi(-D_\varepsilon). \tag{7.24}$$

Note that we can substitute $L_{\varepsilon,l}$ for $L_{\varepsilon,l,w}$. In fact, let

$$L_\varepsilon = \sum_{l=1}^{M} p_{\varepsilon,l} L_{\varepsilon,l}, \quad L_{\varepsilon,w} = \sum_{l=1}^{M} p_{\varepsilon,l} L_{\varepsilon,l,w},$$

(the index w corresponds to $\omega_{\varepsilon,l}$-truncation). We get

$$P_0(L_\varepsilon \neq L_{\varepsilon,w}) \leq \sum_{l=1}^{M} P_0(\lambda_{\varepsilon,l} \geq \omega_{\varepsilon,l}) \leq \Delta_\varepsilon + \sum_{l=1}^{M} \Phi(-\omega_{\varepsilon,l}) = o(1);$$

the last relation follows from the last relation in (7.19):

$$\sum_{l=1}^{M} \Phi(-\omega_{\varepsilon,l}) \asymp \sum_{l=1}^{M} \exp(-\omega_{\varepsilon,l}^2/2)/\omega_{\varepsilon,l} \leq B\omega_\varepsilon^{-1} \to 0.$$

To obtain (7.24) with this change it suffices to check that

$$\sum_{l=1}^{M} p_{\varepsilon,l} E_0 L_{\varepsilon,l,w} \geq \Phi(-D_\varepsilon) + o(1), \quad \sum_{l=1}^{M} p_{\varepsilon,l}^2 E_0 L_{\varepsilon,l,w}^2 = o(1), \tag{7.25}$$

$$\sum_{1\leq l<k\leq M} p_{\varepsilon,l} p_{\varepsilon,k} \operatorname{Cov}_0(L_{\varepsilon,l,w}, L_{\varepsilon,k,w}) = o(1). \tag{7.26}$$

In fact, using (7.25), (7.26) we get

$$\begin{aligned} E_0 L_{\varepsilon,w} &\geq \Phi(-D_\varepsilon) + o(1), \\ \operatorname{Var}_0(L_{\varepsilon,w}) &\leq \sum_{l=1}^{M} p_{\varepsilon,l}^2 E_0 L_{\varepsilon,l,w}^2 \\ &\quad + 2 \sum_{1\leq l<k\leq M} p_{\varepsilon,l} p_{\varepsilon,k} \operatorname{Cov}_0(L_{\varepsilon,l,w}, L_{\varepsilon,k,w}) = o(1), \end{aligned}$$

and using the Chebyshev inequality we get (7.24).

Let us check (7.25), (7.26). Observe the equalities for the moments of the bounded random variables $0 \leq X < H$, $Y < H$ with distribution functions

$F_X(x) = P(X < x)$, $F_Y(y) = P(Y < y)$, and $F(x, y) = P(X < x, Y < y)$:

$$\int_0^H x\, dF_X(x) = \int_0^H (1 - F_X(x))\, dx, \tag{7.27}$$

$$\int_0^H x^2\, dF_X(x) = 2\int_0^H x(1 - F_X(x))\, dx, \tag{7.28}$$

$$\int_0^H \int_0^H xy\, dF(x, y) = \int_0^H \int_0^H (1 - F_X(x) - F_Y(y), + F(x, y))\, dx\, dy, \tag{7.29}$$

where the differential $dF_{X,Y}(x, y)$ means $P(dx, dy)$ for the measure P which corresponds to the random variables (X, Y).

Equalities (7.27) and (7.28) follow from the integration by parts formula. To obtain equality (7.29), using (7.27) and the following equalities for conditional distributions:

$$dF(x, y) = dF(x|y)\, dF_Y(y); \quad F(x, y) = F_X(x) F(y|X < x),$$

$$dF_Y(y) F(x|y) = F_X(x)\, dF(y|X < x) \quad \text{a. s.},$$

we get

$$\begin{aligned} \int_0^H \int_0^H xy\, dF(x, y) &= \int_0^H y\, dF_Y(y) \int_0^H x\, dF(x|y) \\ &= \int_0^H y\, dF_Y(y) \int_0^H (1 - F(x|y))\, dx \\ &= H \int_0^H (1 - F_Y(y))\, dy \\ &\quad - \int_0^H F_X(x)\, dx \int_0^H y\, dF(y|X < x) \\ &= H \int_0^H (1 - F_Y(y))\, dy - H \int_0^H F_X(x)\, dx \\ &\quad + \int_0^H \int_0^H F(x, y))\, dx\, dy. \end{aligned}$$

Let X_k, Y_k be the bounded random variables, $0 \le X_k,\ Y_k < H$, with the distribution functions

$$F_{X_k}(x) = P(X_k < x), \quad F_k(x, y) = P(X_k < x, Y_k < y), \quad k = 1, 2.$$

Using equalities (7.27), (7.28), (7.29) and noting

$$F_{X_k}(x) = F_k(x, H), \quad F_{Y_k}(y) = F_k(H, y),$$

we obtain the inequalities for the differences of the moments

$$|EX_1 - EX_2| \le H \sup_x |F_{X_1}(x) - F_{X_2}(x)|, \tag{7.30}$$

$$|EX_1^2 - EX_2^2| \le 2H^2 \sup_x |F_{X_1}(x) - F_{X_2}(x)|; \tag{7.31}$$

$$|EX_1Y_1 - EX_2Y_2| \le 3H^2 \sup_{x,y} |F_1(x,y) - F_2(x,y)|. \tag{7.32}$$

Let ν_l, $l = 1, \dots, M$, be Gaussian random variables with

$$E(\nu_l) = 0, \quad \mathrm{Var}(\nu_l) = 1, \quad \mathrm{Cov}(\nu_l, \nu_k) = \rho_{\varepsilon,lk}, \quad l,k = 1,\dots,M.$$

Set

$$X_l = \exp(-u_{\varepsilon,l}^2/2 + u_{\varepsilon,l}\nu_l), \quad H_{\varepsilon,l} = \exp(-u_{\varepsilon,l}^2/2 + u_{\varepsilon,l}\omega_{\varepsilon,l}),$$

and consider the truncated random variables $X_{l,w} = X_l \mathbf{1}_{\{\nu_l < \omega_{\varepsilon,l}\}}$. Using (7.30)–(7.32) we get

$$E_0 L_{\varepsilon,l,w} = EX_{l,w} + O(H_{\varepsilon,l}\Delta_{\varepsilon,l}), \quad E_0 L_{\varepsilon,l,M}^2 = EX_{l,w}^2 + O(H_{\varepsilon,l}^2\Delta_{\varepsilon,l}),$$

$$\mathrm{Cov}_0(L_{\varepsilon,l,w}, L_{\varepsilon,k,w}) = \mathrm{Cov}(X_{l,w}, X_{k,w}) + O(H_{\varepsilon,l}H_{\varepsilon,k}\Delta_{\varepsilon,lk}).$$

Note that

$$p_{\varepsilon,l}H_{\varepsilon,l} \asymp \exp(-(u_{\varepsilon,l} - \omega_{\varepsilon,l})^2/2) \le 1.$$

Jointly with (7.22) this yields that we can change $L_{\varepsilon,l,w}$ to $X_{l,w}$ in (7.25), (7.26) and the sums of the remainder terms are $O(\Delta_\varepsilon) = o(1)$.

For the variables $X_{l,w}$ one has

$$EX_{l,w} = P(\nu_l + u_{\varepsilon,l} < \omega_{\varepsilon,l}) = \Phi(\omega_{\varepsilon,l} - u_{\varepsilon,l}) \ge \Phi(-D_\varepsilon) + o(1), \tag{7.33}$$

which implies the first relation in (7.25).

To obtain the second relation in (7.25) it suffices to show that, uniformly over l,

$$p_{\varepsilon,l}EX_{l,w}^2 \asymp \exp(u_{\varepsilon,l}^2 - \omega_{\varepsilon,l}^2/2)\Phi(\omega_{\varepsilon,l} - 2u_{\varepsilon,l}) = o(1).$$

Consider differently the cases $\omega_{\varepsilon,l} > 3u_{\varepsilon,l}/2$ and $\omega_{\varepsilon,l} \le 3u_{\varepsilon,l}/2$. We get in these cases, respectively,

$$\begin{aligned} &\exp(u_{\varepsilon,l}^2 - \omega_{\varepsilon,l}^2/2)\Phi(\omega_{\varepsilon,l} - 2u_{\varepsilon,l}) \\ &\le \begin{cases} \exp(u_{\varepsilon,l}^2 - \omega_{\varepsilon,l}^2/2) \le \exp(-\omega_{\varepsilon,l}^2/18) & = o(1), \\ B\omega_{\varepsilon,l}^{-1}\exp(-(\omega_{\varepsilon,l} - u_{\varepsilon,l})^2) & = o(1). \end{cases} \end{aligned}$$

Finally, to verify (7.26), note the inequality

$$\sup_{x,y} |\Phi_\rho(x+a, y+b) - \Phi(x)\Phi(y)| \le B(|\rho| + |a| + |b|) \tag{7.34}$$

which holds for some $B > 0$ and small enough $|\rho|$. To obtain (7.34), note that

$$\begin{aligned} &\sup_{x,y} |\Phi_\rho(x+a, y+b) - \Phi(x)\Phi(y)| \\ &\le \sup_{x,y} |\Phi_\rho(x+a, y+b) - \Phi(x+a)\Phi(y+a)| \end{aligned} \tag{7.35}$$

$$+\sup_{x,y}|\Phi(x+a)\Phi(y+a)-\Phi(x)\Phi(y)|. \tag{7.36}$$

Item (7.36) is controlled by the variation distance (2.23) between the two-dimensional Gaussian measures $P_{(a,b)}$ and $P_{(0,0)}$ with the unit covariation matrix and mean vectors (a,b) and $(0,0)$ and we can use inequality (2.26). Item (7.35) is controlled by the variation distance between the two-dimensional Gaussian measures P_0 and P_ρ with zero mean vectors. and the unit variances and correlations 0 and ρ. To control the variation distance $|P_0-P_\rho|_1$ it suffices to calculate the L_2-distance $|P_0-P_\rho|_2$. Simple calculation gives

$$|P_0-P_\rho|_2^2=E_{P_0}\left(\frac{dP_\rho}{dP_0}\right)^2-1=\frac{1}{\sqrt{1-6\rho^2+\rho^4}}-1\sim 3\rho^2,\quad \rho\to 0.$$

These yield (7.34).

For the variables $X_{l,w}$, $X_{k,w}$, $k\neq l$, we have the equality

$$EX_{l,w}X_{k,w}=\exp(\rho_{\varepsilon,lk}u_{\varepsilon,l}u_{\varepsilon,k})\Phi_{\rho_{\varepsilon,lk}}(x_\varepsilon,y_\varepsilon);$$

$$x_\varepsilon=\omega_{\varepsilon,l}-u_{\varepsilon,l}-u_{\varepsilon,k}\rho_{\varepsilon,lk},\quad y_\varepsilon=\omega_{\varepsilon,k}-u_{\varepsilon,k}-u_{\varepsilon,l}\rho_{\varepsilon,lk}.$$

Therefore, using (7.34) under (7.21) for $k\neq l$, we get

$$EX_{l,w}X_{k,w}=\Phi(\omega_{\varepsilon,l}-u_{\varepsilon,l})\Phi(\omega_{\varepsilon,l}-u_{\varepsilon,l})+o(1).$$

Jointly with (7.33) this yields $\operatorname{Cov}(X_{l,w},X_{k,w})=o(1)$ uniformly over $l\neq k$. □

Let us consider product priors π_l^ε of the type (7.18) which correspond to the factors $\pi_{\varepsilon,l,i}=\pi(z_{\varepsilon,l,i},h_{\varepsilon,l,i})$ with symmetric three-point measures

$$\pi(z,h)=(1-h)\delta_0+\frac{h}{2}(\delta_z+\delta_{-z}) \tag{7.37}$$

(or two-point measures, if $h_{\varepsilon,l,i}=1$). We take, in Lemma 7.2,

$$u_{\varepsilon,l}^2=\|\bar\pi_{\varepsilon,l}\|^2=\sum_i u_{\varepsilon,l,i}^2,\quad u_{\varepsilon,l,i}^2=\|\pi_{\varepsilon,l,i}\|^2=2h_{\varepsilon,l,i}^2\sinh^2\frac{z_{\varepsilon,l,i}^2}{2},$$

and $\rho_{\varepsilon,lk}=r_{\varepsilon,lk}/u_{\varepsilon,l}u_{\varepsilon,k}$, $k\neq l$, where

$$r_{\varepsilon,lk}=(\bar\pi_{\varepsilon,l},\bar\pi_{\varepsilon,k})=2\sum_i h_{\varepsilon,l,i}h_{\varepsilon,k,i}\sinh^2(z_{\varepsilon,l,i}z_{\varepsilon,k,i}/2).$$

Recall that the log-likelihood ratio is of the form

$$l_{\varepsilon,l}=\log\left(dP_{\pi_l^\varepsilon}/dP_0\right)=\sum_i\log(1+g_{\varepsilon,l,i}),$$

where

$$g_{\varepsilon,l,i}(x)=h_{\varepsilon,l,i}\xi(x,z_{\varepsilon,l,i}),\quad \xi(x,z)=e^{-z^2/2}\cosh zx-1.$$

Set

$$t_{\varepsilon,l} = u_{\varepsilon,l}^{-1} \sum_i g_{\varepsilon,l,i}(x_i).$$

It follows from (3.84) that

$$E_0(t_{\varepsilon,l}) = 0, \quad \mathrm{Var}_0(t_{\varepsilon,l}) = 1, \quad \mathrm{Cov}_0(t_{\varepsilon,l}, t_{\varepsilon,k}) = \rho_{\varepsilon,lk}, \quad k \neq l.$$

Let the sequences $\bar{\pi}_{\varepsilon,l}$ satisfy assumptions A.1a, A.3 from Section 4.3.1 uniformly over l, i.e., for any $a > 0$,

$$w_\varepsilon = \sup_{i,l} u_{\varepsilon,l,i} = o(1); \quad u_\varepsilon = \max_l u_{\varepsilon,l} = o(w_\varepsilon^{-a}), \tag{7.38}$$

and there exists $\delta > 0$ such that, uniformly over l,

$$\sum_i u_{\varepsilon,i,l}^2 \exp(\delta z_{\varepsilon,i,l}^2) = O(u_{\varepsilon,l}^2). \tag{7.39}$$

It was noted in Section 4.3.1 that this yields assumption A.2a uniformly over l; namely, for any $a \in (0,1)$, $b \in (0,(1-a)\delta/2)$ for $\delta_\varepsilon = w_\varepsilon^a$ one has $u_{\varepsilon,\delta_\varepsilon} = o(w_\varepsilon^b)$ where $\delta > 0$ is the parameter from (7.39).

Lemma 7.3 *Assume $M = o(w_\varepsilon^{-a})$ for any $a > 0$ and* (7.21), (7.38), (7.39) *are fulfilled. Then relations* (7.22) *hold.*

Proof. Relations (7.22) for one-dimensional distributions directly follow from Corollary 4.2. Relation (7.22) for two-dimensional distributions follows from the approximation of the statistics $\lambda_{\varepsilon,l}$ by $t_{\varepsilon,l}$ given in the corollary and from the two-dimensional version of the Bahr–Essen inequality

$$\Delta_{\varepsilon,lk} \leq A \frac{\sum_i E_0|g_{\varepsilon,l,i}|^3 + \sum_i E_0|g_{\varepsilon,k,i}|^3}{\left((1-|\rho_{\varepsilon,lk}|)\left(\sum_i E_0 g_{\varepsilon,l,i}^2 + \sum_i E_0 g_{\varepsilon,k,i}^2\right)\right)^{3/2}};$$

analogously to the proof of Proposition 4.4, the evaluations for the third moments follow from evaluations for the fourth moments and lead to the fact that the right-hand side is $o(w_{\varepsilon,l}^c)$ for some $c > 0$. □

7.2.3 Lower Bounds for Power Norms

To obtain the lower bounds of Theorem 7.1 (2) we can assume that

$$u_\varepsilon(K) = \sqrt{2 \log\log R_\varepsilon} + O(1). \tag{7.40}$$

Fix $\delta_0 > 0$. Our aim is to construct a family of collections of the product priors π_l^ε, $l = 1, \ldots, M = M_\varepsilon$, with $\sqrt{2\log M_\varepsilon} = \sqrt{2\log\log R_\varepsilon} + o(1)$ corresponding to sequences of the symmetric three-point factors $\bar{\pi}_{\varepsilon,l}$ which satisfy the assumptions of Lemma 7.3, and are such that $\pi_l^\varepsilon(V_\varepsilon) = 1 + o(1)$ uniformly over l and $u_{\varepsilon,l} = \|\bar{\pi}_{\varepsilon,l}\| \geq u_\varepsilon(K) - \delta_0$. Then, using Corollary 7.2, we obtain the required lower bounds.

Let $\Delta \subset K$ be a set such that for any $\kappa \in \Delta$ one has

$$u_\varepsilon(K) \le u_\varepsilon(\kappa) < u_\varepsilon(K) + \delta. \tag{7.41}$$

Let us take the collection of the points $\kappa_{\varepsilon,l} \in \Delta$, $l = 1, \dots, M$, and let $\bar{\pi}_{\varepsilon,l} = \tilde{\pi}_\varepsilon(\kappa_{\varepsilon,l}, \rho_\varepsilon(\kappa_{\varepsilon,l}), R_\varepsilon)$ be the sequences which are constructed in the proof of the lower bounds of Theorem 6.1 (see Sections A.3.3, A.3.6). It was noted in Section 6.8 that these collections satisfy the assumptions of Proposition 5.5 uniformly over $\kappa \in K$; in particular, assumptions (7.38), (7.39) hold with $w_\varepsilon = o(R^{-a})$ for any $a > 0$ and

$$\pi_l^\varepsilon(V_\varepsilon(K)) \ge \pi_l^\varepsilon(V(\kappa_l, \rho_\varepsilon(\kappa_{\varepsilon,l}), R_\varepsilon) = 1 - o(1)$$

which yield (7.15). Therefore, we need to find the collections $\kappa_{\varepsilon,l} \in \Delta$, $l = 1, \dots, M$, which satisfy the assumption (7.21). To verify these we can use Proposition 6.2 in Section 6.8.2.

First assume $\Delta \subset \Xi_{G_{01}}$. It follows from the study of the extreme problem in Chapter 6 (see Section 6.8) that there exist values

$$m_l = m(\kappa_{\varepsilon,l}, \rho_\varepsilon(\kappa_{\varepsilon,l}), R_\varepsilon), \quad z_{0,l} = z_0(\kappa_{\varepsilon,l}, \rho_\varepsilon(\kappa_{\varepsilon,l}), R_\varepsilon)$$

such that, uniformly over l,

$$u_{\varepsilon,l}^2 \asymp m_l z_{0,l}^4, \quad m_l^{s+1/q} z_{0,l} \asymp R_\varepsilon. \tag{7.42}$$

These relations imply that uniformly over l one has (compare with (6.64))

$$m_l \asymp \left(R_\varepsilon / u_{\varepsilon,l}^{1/2}\right)^{\phi_1(\kappa_{\varepsilon,l})}, \tag{7.43}$$

where

$$\phi_1(\kappa) = (s + 1/q - 1/4)^{-1} > 0, \quad \kappa \in \Xi_{G_{01}}. \tag{7.44}$$

Let $k \ne l$, $m_{\varepsilon,l} \le m_{\varepsilon,k}$. It follows from Proposition 6.2 that

$$(\bar{\pi}_{\varepsilon,l}, \bar{\pi}_{\varepsilon,k}) \le B u_{\varepsilon,l} u_{\varepsilon,k} \left(\left(\frac{m_{\varepsilon,l}}{m_{\varepsilon,k}} \right)^{\delta_1} + R_\varepsilon^{-\delta_2} \right), \quad \delta_1 > 0, \quad \delta_2 > 0. \tag{7.45}$$

Put

$$M_\varepsilon \asymp \log R_\varepsilon / \log\log R_\varepsilon, \quad \delta_\varepsilon = (\log\log R_\varepsilon)^b / \log R_\varepsilon, \quad b \in (0,1).$$

By $M_\varepsilon \delta_\varepsilon = o(1)$, one can construct collections

$$\Delta_\varepsilon = \{\kappa_{\varepsilon,l}, \ l = 1, \dots, M_\varepsilon\} \subset \Delta$$

such that

$$\max_{l \ne k} |\phi_1(\kappa_{\varepsilon,l}) - \phi_1(\kappa_{\varepsilon,k})| = o(1), \quad \min_{l \ne k} |\phi_1(\kappa_{\varepsilon,l}) - \phi_1(\kappa_{\varepsilon,k})| > 2\delta_\varepsilon.$$

By choice M_ε we have $\sqrt{2 \log M_\varepsilon} = \sqrt{2 \log\log R_\varepsilon} + o(1)$. It follows from (7.40), (7.41) and (7.43) that for some $B > 0$, $c > 0$ and any $B_1 > 0$ one has, for large enough R_ε, uniformly over $k \neq l$,

$$m_{\varepsilon,l}/m_{\varepsilon,k} \leq B(\log\log R_\varepsilon)^c R_\varepsilon^{-|\phi_1(\kappa_{\varepsilon,l})-\phi_1(\kappa_{\varepsilon,k})|} \leq BR_\varepsilon^{-\delta_\varepsilon} \leq B(\log R_\varepsilon)^{-B_1}. \tag{7.46}$$

Using (7.45), (7.46) we have, for any $B_2 > 0$,

$$\max_{1\leq l<k\leq M_\varepsilon} r_{\varepsilon,lk} \leq B(\log R_\varepsilon)^{-B_2}\log\log R_\varepsilon = o(1). \tag{7.47}$$

Assume $\Delta \subset \Xi_{G_{02}}$. Analogously to the above (see Section 6.8) there exist the values

$$n_l = n(\kappa_{\varepsilon,l}, \rho_\varepsilon(\kappa_{\varepsilon,l}), R_\varepsilon), \quad h_{0,l} = h_0(\kappa_{\varepsilon,l}, \rho_\varepsilon(\kappa_{\varepsilon,l}), R_\varepsilon),$$

such that, uniformly over l,

$$u_{\varepsilon,l}^2 \asymp n_l h_{0,l}^2, \quad n_l^{s+1/q} h_{0,l}^{1/q} \asymp R_\varepsilon. \tag{7.48}$$

These relations imply (compare with (6.66)) that uniformly over l one has

$$n_l \asymp \left(R_\varepsilon/u_{\varepsilon,l}^{1/q}\right)^{\phi_2(\kappa)}, \tag{7.49}$$

where

$$\phi_2(\kappa) = (s + 1/2q)^{-1} > 0, \quad \kappa \in \Xi_{G_{02}}. \tag{7.50}$$

Let $k \neq l$, $m_{\varepsilon,l} \leq m_{\varepsilon,k}$. It follows from Proposition 6.2 that

$$(\bar\pi_{\varepsilon,l}, \bar\pi_{\varepsilon,k}) \leq Bu_{\varepsilon,l}u_{\varepsilon,k}\left(\left(\frac{n_{\varepsilon,l}}{n_{\varepsilon,k}}\right)^{\delta_1} + R_\varepsilon^{-\delta_2}\right), \quad \delta_1 > 0, \quad \delta_2 > 0.$$

The next considerations repeat the case $\Delta \subset \Xi_{G_{01}}$. Theorem 7.1 (2) is proved.

7.2.4 Nondistinguishability Conditions for Besov Norms

To obtain the lower bounds of Theorem 7.2 we can assume that

$$c_0 \leq u_\varepsilon^2(\tau, r_\varepsilon(\tau)) \leq c\log\log\varepsilon^{-1} \tag{7.51}$$

for small enough $c > 0$, $c_0 > 0$ and all $\tau \in \Delta$. We consider the cases $\Delta \subset K \times D^{(3)}$ where $K \subset \Xi_{G_1}$ or $K \subset \Xi_{G_2}$.

Taking a quantity $B > 1$ let us construct a family of sequences of the three-point measures which are concentrated on a resolution I_{j^*},

$$\bar\pi_\varepsilon = \bar\pi_\varepsilon(\tau, r_\varepsilon, B) = \{\pi_{\varepsilon,i}, \ i = (\iota, j) \in \mathcal{J}\}, \quad \tau = (\kappa, h, t)$$

(recall that $\mathcal{J}$ is the index set of pyramical structure, $\mathcal{J} = \{(\iota, j), \ 1 \leq \iota \leq 2^j, \ j \geq 1\}$). This means that

$$\pi_{\varepsilon,\iota,j} = \begin{cases} \delta_0, & \text{if } j \neq j^*, \\ (1-h_j)\delta_0 + h_j(\delta_{z_j} + \delta_{-z_j})/2, & \text{if } j = j^*, \end{cases} \quad 1 \leq \iota \leq 2^j,$$

where

$$\begin{cases} j^* = j_0, & h_{j^*} = h_0, \quad z_{j^*} = 1, \quad \text{if } \kappa \in \Xi_{G_2}, \\ j^* = j_1, & h_{j^*} = 1, \quad z_{j^*} = z_0, \quad \text{if } \kappa \in \Xi_{G_1}. \end{cases}$$

Here $j_0 = j_0(\tau, r_\varepsilon, B) \to \infty$, $j_1 = j_1(\tau, r_\varepsilon, B) \to \infty$ and the positive values $h_0 = h_0(\tau, r_\varepsilon, B) = o(1)$, $z_0 = h_0(\tau, r_\varepsilon, B) = o(1)$ are determined by the relations (later in this section asymptotic relations are uniform over any compact in Ξ_{G_1} or Ξ_{G_2})

$$\begin{cases} 2^{j_1(rp+1)} z_0^p \sim (Br_\varepsilon/\varepsilon)^p, & 2^{j_1(sq+1)} z_0^q \sim (R/B\varepsilon)^q, \quad \text{if } \kappa \in \Xi_{G_1}, \\ 2^{j_0(rp+1)} h_0 \sim (Br_\varepsilon/\varepsilon)^p, & 2^{j_0(sq+1)} h_0 \sim (R/B\varepsilon)^q, \quad \text{if } \kappa \in \Xi_{G_2}. \end{cases}$$

Under this choice one has

$$\|\bar{\pi}_\varepsilon(\tau, r_\varepsilon, B)\|^2 \asymp u_\varepsilon^2(\tau, r_\varepsilon) \asymp \begin{cases} 2^{j_1} z_0^4, & \text{if } \kappa \in \Xi_{G_1}, \\ 2^{j_0} h_0^2, & \text{if } \kappa \in \Xi_{G_2}. \end{cases}$$

In view of (7.51) these relations yield

$$\begin{cases} j_1 \sim \phi_1(\kappa) \log_2 \varepsilon^{-1}, & \text{if } \kappa \in \Xi_{G_1}, \\ j_0 \sim \phi_2(\kappa) \log_2 \varepsilon^{-1}, & \text{if } \kappa \in \Xi_{G_2}. \end{cases} \tag{7.52}$$

These relations are analogous to (7.42), (7.43) and to (7.48), (7.49) with $m = 2^{j_1}$ or $n = 2^{j_0}$. Note that the cross-section of the sets $V_\varepsilon(\tau, r_\varepsilon)$ by the 2^{j^*}-dimensional linear subspace L_{j^*} corresponding to the j^*th resolution is the l^q-ball of radius $2^{-j^* s} R/\varepsilon$ with the l^p-ball of radius $2^{-j^* r} r_\varepsilon/\varepsilon$ removed. Analogously to the considerations in Section 5.5.2, for any $B > 1$, one has

$$\pi^\varepsilon(V_\varepsilon(\tau, r_\varepsilon(\tau))) \to 1. \tag{7.53}$$

Set

$$M_\varepsilon \sim \log \varepsilon^{-1} / \log\log \varepsilon^{-1}, \quad \delta_\varepsilon = (\log\log \varepsilon^{-1})^b / \log \varepsilon^{-1}; \quad b \in (0,1).$$

By $M_\varepsilon \delta_\varepsilon = o(1)$, one can construct the collections

$$\Gamma_\varepsilon = \{\tau_{\varepsilon,1}, \dots, \tau_{\varepsilon,M}\} \subset \Gamma, \quad \tau_{\varepsilon,l} = (\kappa_{\varepsilon,l}, h(\kappa_{\varepsilon,l}), t(\kappa_{\varepsilon,l}), R(\kappa_{\varepsilon,l})), \quad \kappa_{\varepsilon,l} \in \Delta,$$

such that

$$\min_{1 \le l < k \le M_\varepsilon} |\phi(\kappa_{\varepsilon,l}) - \phi(\kappa_{\varepsilon,k})| > 2\delta_\varepsilon,$$

where $\phi(\kappa) = \phi_1(\kappa)$ for $\Delta \subset \Xi_{G_1}$ or $\phi(\kappa) = \phi_2(\kappa)$ for $\Delta \subset \Xi_{G_2}$. Put $\bar{\pi}_{\varepsilon,l} = \bar{\pi}_\varepsilon(\tau_{\varepsilon,l}, r_\varepsilon(\tau_{\varepsilon,l}), B)$ with

$$j_l = \begin{cases} j_1(\tau_{\varepsilon,l}, r_\varepsilon(\tau_{\varepsilon,l})), & \text{if } \kappa \in \Xi_{G_1}. \\ j_2((\tau_{\varepsilon,l}, r_\varepsilon(\tau_{\varepsilon,l})), & \text{if } \kappa \in \Xi_{G_2}. \end{cases}$$

By the choice of M_ε, δ_ε, using (7.52), we have, for some $B > 0$,

$$\min_{1 \le l < k \le M_\varepsilon} |j_l - j_k| \ge B(\log\log \varepsilon^{-1})^b \to \infty.$$

Since measures in $\bar{\pi}_{\varepsilon,l}$ are concentrated on one level j_l, this relation implies

$$(\bar{\pi}_{\varepsilon,l}, \bar{\pi}_{\varepsilon,k}) = 0, \quad 1 \le l < k \le M_\varepsilon,$$

and one can choose $c > 0$ in (7.51) such that, for small enough $\delta > 0$,

$$\|\bar{\pi}_{\varepsilon,l}\|^2 \leq (1-\delta)\log M_\varepsilon, \quad 1 \leq l \leq M_\varepsilon.$$

Using (7.53) and Lemma 7.1 with $p_{\varepsilon,l} = M_\varepsilon^{-1}$, $l = 1, \dots, M_\varepsilon$, we get

$$\sum_{l=1}^{M_\varepsilon} p_{\varepsilon,l}^2 \exp(\|\bar{\pi}_{\varepsilon,l}\|) \leq M_\varepsilon^{-\delta} = o(1).$$

Thus we have the nondistinguishability conditions of Theorem 7.2 (2).

7.3 Upper Bounds for Power Norms

7.3.1 Methods of Constructions

We would like to construct asymptotically minimax families of tests $\psi_\varepsilon = \psi_{\varepsilon,\alpha}$ such that

$$\alpha(\psi_{\varepsilon,\alpha}) \leq \alpha + o(1); \quad \beta(\psi_{\varepsilon,\alpha}, V_\varepsilon(K)) \leq (1-\alpha)\Phi(-D_\varepsilon) + o(1), \tag{7.54}$$

where $D_\varepsilon = u_\varepsilon(K) - \sqrt{2\log\log R_\varepsilon}$. It suffices to find a family ψ_ε such that

$$\alpha(\psi_\varepsilon) \to 0; \quad \beta(\psi_\varepsilon, V_\varepsilon(K)) \leq \Phi(-D_\varepsilon) + o(1). \tag{7.55}$$

In fact, let us consider the randomized test families $\psi_{\varepsilon,\alpha} = \alpha + (1-\alpha)\psi_\varepsilon$. Then the relations (7.55) imply (7.54)

The test families ψ_ε that we propose are, combinations of collections of tests $\{\psi_{\varepsilon,l},\ 1 \leq l \leq M\}$, i.e.,

$$\psi_\varepsilon(x) = \max_{1 \leq l \leq M} \psi_{\varepsilon,l}(x), \quad M = M_\varepsilon \to \infty. \tag{7.56}$$

These correspond to the "Bonferroni method": one rejects the null-hypothesis if and only if this is rejected by some test $\psi_{\varepsilon,l}$ in the collection.

For the tests (7.56) one has

$$\alpha(\psi_\varepsilon) \leq \sum_{1 \leq l \leq M} \alpha(\psi_{\varepsilon,l}); \quad \beta(\psi_\varepsilon, v) \leq \min_{1 \leq l \leq M} \beta(\psi_{\varepsilon,l}, v), \tag{7.57}$$

and to obtain (7.55) it suffices to construct collections of tests $\psi_{\varepsilon,l}$ and partitions of the set K to subsets

$$K_l = K_{\varepsilon,l} \subset K, \quad \bigcup_{1 \leq l \leq M} K_l = K,$$

such that, uniformly over l, $1 \leq l \leq M$,

$$\alpha(\psi_{\varepsilon,l}) = o(M^{-1}); \tag{7.58}$$

$$\beta(\psi_{\varepsilon,l}, V_\varepsilon(K_l)) \leq \Phi(-D_\varepsilon) + o(1). \tag{7.59}$$

We consider the tests $\psi_{\varepsilon,l} = \psi(\bar{h}_{\varepsilon,l}, \bar{z}_{\varepsilon,l})$ of the type (compare with (5.4))

$$\psi(\bar{h}_{\varepsilon,l}, \bar{z}_{\varepsilon,l}) = \mathbf{1}_{\left\{t_{\bar{h}_{\varepsilon,l},\bar{z}_{\varepsilon,l}} > \sqrt{2\log M}\right\}} \cup X_\varepsilon \tag{7.60}$$

which are based on the statistics

$$t_{\bar{h},\bar{z}}(x) = \|\bar{\pi}\|^{-1} \sum_i h_i \xi(x_i, z_i).$$

Here $\bar{\pi} = \bar{\pi}(\bar{h}, \bar{z}) = \{\pi_i(h_i, z_i)\}$ is a sequence which consists of the symmetric three-point measures (7.37). The set X_ε corresponds to the threshold procedure

$$X_\varepsilon = \left\{x : \sup_i |x_i|/Q_{\varepsilon,i} \geq 1\right\},$$

where

$$Q_{\varepsilon,i} = \sqrt{2\log i + c\log\log(i+1) + 4\log\log R_\varepsilon}, \quad c > 1,$$

(compare with (6.22)). These test families are determined by collections

$$\bar{h}_{\varepsilon,l} = \{h_{\varepsilon,l,i}\}, \quad h_{\varepsilon,l,i} \in [0,1], \quad \bar{z}_{\varepsilon,l} = \{z_{\varepsilon,l,i}\}, \quad z_{\varepsilon,l,i} \geq 0; \quad 1 \leq l \leq M. \tag{7.61}$$

Let $M = M_\varepsilon$ be such that

$$\sqrt{2\log M} = \sqrt{2\log\log R_\varepsilon} + o(1) \tag{7.62}$$

and the collections (7.61) satisfy assumptions A.1a, A.3 with $w_\varepsilon = w_{\varepsilon,l} = o(R_\varepsilon^{-a})$ for some $a > 0$ uniformly over $1 \leq l \leq M$. Then, using Propositions 5.1 (2) for $R = \sqrt{2\log M}$ and (6.61), we get (7.58).

Let the assumptions of Proposition 5.4 be fulfilled uniformly over $1 \leq l \leq M$ with $\zeta_\varepsilon = \zeta_{\varepsilon,l} = o(1/u_\varepsilon)$. Then under the change (5.28) we get the inequality

$$\begin{aligned}\beta(\psi_{\varepsilon,l}, V_\varepsilon(K_l)) &\leq \Phi\left(\sqrt{2\log M_\varepsilon} - \inf_{v \in V_\varepsilon^H(K_l)} h_{\varepsilon,l}(\tilde{v})\right) + o(1) \\ &\leq \Phi\left(\sqrt{2\log\log R_\varepsilon} - \inf_{v \in V_{\varepsilon,1}^H(K_l)} h_{\varepsilon,l}(v)\right) + o(1). \end{aligned} \tag{7.63}$$

Here $\tilde{v}$ is defined by (5.28) and we denote, as above,

$$h_{\varepsilon,l}(v) = 2u_{\varepsilon,l}^{-1} \sum_i h_{\varepsilon,l,i} \sinh^2(v_i z_{\varepsilon,i,l}/2), \quad u_{\varepsilon,l}^2 = 2\sum_i h_{\varepsilon,l,i}^2 \sinh^2(z_{\varepsilon,i,l}^2/2);$$

$$V_{\varepsilon,1}^H(K_l) = \bigcup_{\kappa \in K_l} V_\varepsilon^H(\kappa, \rho'_\varepsilon(\kappa), R_\varepsilon), \quad \rho'_\varepsilon(\kappa) = (1 - 2\zeta_\varepsilon)\rho_\varepsilon(\kappa). \tag{7.64}$$

The sets $V_\varepsilon^H(\kappa, \rho'_\varepsilon(\kappa), R_\varepsilon)$ are defined analogously to (5.25) with C_ε is some family such that $C_\varepsilon \to \infty$, $C_\varepsilon = o(\sqrt{\log R_\varepsilon})$ (we take $C_\varepsilon \asymp \sqrt{\log\log R_\varepsilon}$). The second inequality in (7.63) is fulfilled because if $v \in V(\kappa, \rho_\varepsilon(\kappa), R_\varepsilon)$, then $\tilde{v} \in V(\kappa, \rho'_\varepsilon(\kappa), R_\varepsilon)$.

Therefore we need to construct a family of collections (7.61) and a family of partitions

$$K_l \subset K, \qquad \bigcup_{1\le l\le M} K_l = K,$$

such that the assumptions above are fulfilled and the following inequalities hold:

$$\min_{1\le l\le M} \inf_{v\in V^H_{\varepsilon,1}(K_l)} h_{\varepsilon,l}(v) \ge u_\varepsilon(K) + o(1). \tag{7.65}$$

7.3.2 The Construction of Collections of Tests

We can assume that the function $u(\kappa, \rho_\varepsilon(\kappa), R_\varepsilon)$ is constant in $\kappa \in K$ and

$$u(\kappa, \rho_\varepsilon(\kappa), R_\varepsilon) = \sqrt{2\log\log R_\varepsilon} + D_\varepsilon \quad \forall\, \kappa \in K, \quad D_\varepsilon = O(1). \tag{7.66}$$

In fact, let

$$u(\kappa, \rho_\varepsilon(\kappa), R_\varepsilon) > u_\varepsilon(K) = \inf_{\kappa\in K} u(\kappa, \rho_\varepsilon(\kappa), R_\varepsilon)$$

for some $\kappa \in K$. Then, by making $\rho_\varepsilon(\kappa)$ smaller and by the function $u(\kappa, \rho_\varepsilon, R_\varepsilon)$ being monotone and continuous in ρ_ε (see Proposition 5.6 (3)), we can get the case $u(\tau, \rho_\varepsilon(\kappa), R_\varepsilon) = u_\varepsilon(K)$ and obtain wider alternatives $V(\kappa, \rho_\varepsilon(\kappa), R_\varepsilon)$. We use an analogous argument for $D_\varepsilon \to \infty$. If $D_\varepsilon \to -\infty$, then the upper bounds are trivial.

Let us consider partitions of K to $M = M_\varepsilon$ cells K_l of the following type. Let $K = K^{(1)} \cup K^{(2)}$ where $K^{(k)} = K \cap \Xi_{G_{0k}}$, $k = 1, 2$. Let $C_1 = C_1(K)$ be the parameter from Proposition 6.1. These partitions correspond to partition of the region of parameters ($\kappa = (p, q, r, s)$) to the balls of the radius

$$\delta_{\varepsilon,1} \asymp (\log\log R_\varepsilon)^{-C_1-b}, \quad b > 1,$$

with respect to the distance $d(\kappa, \kappa_0)$ defined by (6.99), jointly with the partition of the region by levels of the function $\phi_k(\kappa)$, i.e., to the sets $K_j^{(k)}$, such that

$$(j-1)\delta_{\varepsilon,2} < \phi_k(\kappa) \le j\delta_{\varepsilon,2}, \quad j = 1, \dots, J_1 \asymp \delta_{\varepsilon,2}^{-1}$$

with the size

$$\delta_{\varepsilon,2} \asymp \frac{1}{(\log\log R_\varepsilon)^{b_1+C_1}\log R_\varepsilon}, \quad 1/2 < b_1 < b - 1/2.$$

Therefore, for any $\kappa \in K_l$, $\kappa_0 \in K_l$, $K_l \subset K^{(k)}$, $k = 1, 2$, $l = 1, \dots, M_\varepsilon$, one has

$$d(\kappa, \kappa_0) \le 2\delta_{\varepsilon,1}, \quad |\phi_k(\kappa) - \phi_k(\kappa_0)| \le \delta_{\varepsilon,2}.$$

Recall that the functions $\phi_1(\kappa), \phi_2(\kappa)$ are defined by (7.44), (7.50). The number of cells M_ε is of the order

$$M_\varepsilon \asymp \log R_\varepsilon (\log\log R_\varepsilon)^c,$$

for some $c > 0$ which yields (7.62).

Consider the sequences

$$\bar{h}_\varepsilon(\kappa) = \bar{h}_\varepsilon(\kappa, \rho_\varepsilon(\kappa), R_\varepsilon), \quad \bar{z}_\varepsilon(\kappa) = \bar{z}_\varepsilon(\kappa, \rho_\varepsilon(\kappa), R_\varepsilon),$$

and the quantities

$$m(\kappa) = m_\varepsilon(\kappa, \rho_\varepsilon(\kappa), R_\varepsilon), \quad z_0(\kappa) = z_{\varepsilon,0}(\kappa, \rho_\varepsilon(\kappa), R_\varepsilon); \quad \kappa \in K^{(1)},$$

or

$$n(\kappa) = n_\varepsilon(\kappa, \rho_\varepsilon(\kappa), R_\varepsilon), \quad h_0(\kappa) = h_{\varepsilon,0}(\kappa, \rho_\varepsilon(\kappa), R_\varepsilon); \quad \kappa \in K^{(2)},$$

which determine the solution of the extreme problems (6.4), (6.5).

Let $\kappa_l = \kappa_{\varepsilon,l} \in K_l$, $l = 1, \dots, M$, and let $\Delta_{l,L} = \Delta_\varepsilon(\kappa_l, L)$ be the sets which consist of the pairs $(\kappa, \rho_\varepsilon)$ determined by constraints analogous to (6.99), (6.100). Let us show that if $\kappa \in K_l$, then the pair $(\kappa, \rho_\varepsilon(\kappa)) \in \Delta_{l,L}$ for $L = (L_1, L_2)$,

$$L_1 \asymp \delta_{\varepsilon,1} \asymp (\log\log R_\varepsilon)^{-(b+C_1)}, \quad L_2 = O((\log\log R_\varepsilon)^{-(b_1+C_1)}.$$

In fact, by this construction we get (6.99) for $\kappa_0 = \kappa_l$ and for all $\kappa \in K_l$. Let us verify the inequalities (6.100). Let $\kappa_1 \in K_l$, $\kappa_2 \in K_l$. Denote $m_\varepsilon^{(k)} = m(\kappa_k)$, $z_{\varepsilon,0}^{(k)} = z_0(\kappa_k)$ or $n_\varepsilon^{(k)} = n(\kappa_k)$, $h_{\varepsilon,0}^{(k)} = h_0(\kappa_k)$, $k = 1, 2$, and recall that $u(\kappa, \rho_\varepsilon(\kappa), R_\varepsilon) = u_\varepsilon(K) = u_\varepsilon$ by (7.66). Using relations (6.64)–(6.67) we get, for some $\delta > 0$,

$$\frac{m_\varepsilon^{(1)}}{m_\varepsilon^{(2)}} = \frac{d_0(\kappa_1)}{d_0(\kappa_2)} u_\varepsilon^{(\phi_1(\kappa_2)-\phi_1(\kappa_1))/2} R_\varepsilon^{\phi_1(\kappa_1)-\phi_1(\kappa_2)}(1 + o(R_\varepsilon^{-\delta})),$$

$$\frac{z_{0,\varepsilon}^{(1)}}{z_{0,\varepsilon}^{(2)}} = \frac{d_1(\kappa_1)}{d_1(\kappa_2)} u_\varepsilon^{(\phi_1(\kappa_1)-\phi_1(\kappa_2))/8} R_\varepsilon^{(\phi_1(\kappa_2)-\phi_1(\kappa_1))/4}(1 + o(R_\varepsilon^{-\delta}));$$

or

$$\frac{n_\varepsilon^{(1)}}{n_\varepsilon^{(2)}} = \frac{d_0(\kappa_1)}{d_0(\kappa_2)} u_\varepsilon^{\phi_2(\kappa_2)/q_2-\phi_2(\kappa_1)/q_1} R_\varepsilon^{\phi_2(\kappa_1)-\phi_2(\kappa_2)}(1 + o(R_\varepsilon^{-\delta})),$$

$$\frac{h_{0,\varepsilon}^{(1)}}{h_{0,\varepsilon}^{(2)}} = \frac{d_1(\kappa_1)}{d_1(\kappa_2)} u_\varepsilon^{\phi_2(\kappa_1)/2q_1-\phi_2(\kappa_2)/2q_2} R_\varepsilon^{(\phi_2(\kappa_2)-\phi_2(\kappa_1))/2}(1 + o(R_\varepsilon^{-\delta})).$$

Since the functions $d_0(\kappa), d_1(\kappa)$ are positive Lipschitz and bounded away from 0 over any compact $K \subset \Xi_{G_k}$, $k = 1, 2$ (see Remarks 5, 6 in Section 6.5), the first ratio in the right-hand side of these relations is

$$1 + O(\delta_{\varepsilon,1}) = 1 + O\left((\log\log R_\varepsilon)^{-b-C_1}\right).$$

Analogously, because of (7.66), the second factor is

$$1 + O(\delta_{\varepsilon,1}\log u_\varepsilon)) = 1 + O\left((\log\log R_\varepsilon)^{1/2-b-C_1}\right).$$

The logarithm of the third factor is of the rate

$$\delta_{\varepsilon,2}\log R_\varepsilon \asymp (\log\log R_\varepsilon)^{-b_1} = O\left((\log\log R_\varepsilon)^{-(b_1+C_1)}\right).$$

Thus we get the constraints (6.100) with

$$L_2 = o\left((\log\log R_\varepsilon)^{-(b_1+C_1)}\right).$$

Fix $B > 2$. Analogously to (6.101), (6.102), (6.103), define the sequences $\bar{h}_{\varepsilon,l} = \{h_{\varepsilon,l,i}\}$, $\bar{z}_{\varepsilon,l} = \{z_{\varepsilon,l,i}\}$ by the relations

$$h_{\varepsilon,l,i} = \sup_{\kappa\in K_l} h_{\varepsilon,i}(\kappa,\rho'_\varepsilon(\kappa),R_\varepsilon), \hat{z}_{\varepsilon,l,i} = \sup_{\kappa\in K_l} z_{\varepsilon,i}(\kappa,\rho'_\varepsilon(\kappa),R_\varepsilon),$$

$$z_{\varepsilon,l,i} = \begin{cases} \hat{z}_{\varepsilon,l,i} & \text{if } \hat{z}_{\varepsilon,l,i} \le BQ_{\varepsilon,i}, \\ 0, & \text{if } \hat{z}_{\varepsilon,l,i} > BQ_{\varepsilon,i}. \end{cases}$$

We can use Proposition 6.1 which yields that the sequences $\bar{h}_{\varepsilon,l}$, $\bar{z}_{\varepsilon,l}$ satisfy the assumptions of Proposition 5.4 uniformly over $1 \le l \le M$ with $\zeta_\varepsilon = \zeta_{\varepsilon,l} = o(1/u_\varepsilon)$. By taking the parameters b, b_1 in the definition of $\delta_{\varepsilon,1} \asymp L_1$, $\delta_{\varepsilon,2}$ large enough, one has $u_{\varepsilon,l} = u_\varepsilon(K) + o(1)$.

We need to verify the inequalities (7.65). For $\kappa = (r,s,p,q) \in K$ let us denote $h_{\varepsilon,i} = h_{\varepsilon,i}(\kappa,\rho'_\varepsilon(\kappa),R_\varepsilon)$, $z_{\varepsilon,i} = z_{\varepsilon,i}(\kappa,\rho'_\varepsilon(\kappa),R_\varepsilon)$. Let $v = \{v_i\} \in V_\varepsilon^H(\kappa,\rho'_\varepsilon(\kappa))$.

Recall that the families $\bar{h}_\varepsilon, \bar{z}_\varepsilon$ correspond to the sequences $\bar{\pi}_\varepsilon$ which is the solution of the extreme problem (5.40) of the form (6.4), (6.5) for the case of the power norms with changed $\rho'_\varepsilon(\kappa) = \rho_\varepsilon(\kappa)(1-2\zeta_\varepsilon)$, that is, $\|\bar{\pi}_\varepsilon\| = u'_\varepsilon = u(\kappa,\rho'_\varepsilon(\kappa),R_\varepsilon)$. Therefore (compare with Section 5.2.1) it follows from Proposition 4.1 that for $\bar{r}_\varepsilon = (u'_\varepsilon)^{-1}\bar{\pi}_\varepsilon$ we have

$$(u'_\varepsilon)^{-1} 2\sum_i h_{\varepsilon,i}\sinh^2(v_i z_{\varepsilon,i}/2) = (\bar{\delta}_v,\bar{r}_\varepsilon) \ge \inf_{\bar{\pi}\in\bar{\Pi}_\varepsilon}(\bar{\pi},\bar{r}_\varepsilon) = u'_\varepsilon.$$

By $u_\varepsilon = u(\kappa,\rho_\varepsilon(\kappa),R_\varepsilon) = u(\kappa,\rho'_\varepsilon(\kappa),R_\varepsilon) + o(1)$, we have the inequality

$$2\sum_i h_{\varepsilon,i}\sinh^2(v_i z_{\varepsilon,i}/2) \ge u^2(\kappa,\rho'_\varepsilon(\kappa),R_\varepsilon) = u_\varepsilon^2 + o(u_\varepsilon).$$

Let us show that, for small enough $\delta > 0$,

$$\delta_\varepsilon = 2\sum_{i:\hat{z}_{\varepsilon,l,i}>BQ_{\varepsilon,i}} h_{\varepsilon,i}\sinh^2(v_i z_{\varepsilon,i}/2) = o(R_\varepsilon^{-\delta}). \tag{7.67}$$

In fact, if $p < q$ or $\lambda(\kappa) \le 0$, then the index set $I_\varepsilon = \{i : \hat{z}_{\varepsilon,l,i} > BQ_{\varepsilon,i}\}$ is empty for small enough ε. Let $p > q$ and let $\lambda(\kappa) > 0$. It follows from the proof of Proposition 6.1 that the index set I_ε is contained in the index set $I_{\varepsilon,1} = \{i : z_{\varepsilon,i} > B_1\sqrt{\log n}\}$ and $z_{\varepsilon,i} > B_1 Q_{\varepsilon,i}$ for any $B_1 \in (2, B)$ for $i \in I_{\varepsilon,1}$; also $x_i = i/n \to \infty$, $h_{\varepsilon,i} \to 0$ for $i \in I_{\varepsilon,1}$. Moreover, since

$$|v_i| \le Q_{\varepsilon,i} + C_\varepsilon = Q_{\varepsilon,i}(1 + o(1)), \quad \text{if } v \in V_\varepsilon^H(\kappa, \rho'_\varepsilon(\kappa)), \quad z_{\varepsilon,i} > B_1 Q_{\varepsilon,i},$$

we get $z_{\varepsilon,i}|v_i| \le b_2 z_{\varepsilon,i}^2$ for any $b_2 > 1/B_1 \in (1/B, 1/2)$. Using the asymptotics of the solutions of equations (6.86), (6.87), presented in Proposition A.2 (see Section A.3.1), we have for $i \in I_{\varepsilon,1}$,

$$z_{\varepsilon,i} \asymp (i/n)^{pq\lambda/(p-q)}, \quad h_{\varepsilon,i} \asymp h_0 (i/n)^c \exp(-z_{\varepsilon,i}^2),$$

where $c = c(\kappa)$ is bounded on K. Taking B_1, d, b_2, $b_3 \in (0, 1 - b_2)$ such that $B_1^2 b_3 > 1 - d$ and since $h_0 = o(n^{-d})$ for any $d \in (0, 1/2)$, we get the evaluations

$$\begin{aligned} \delta_\varepsilon &\le C h_0 \sum_{i \in I_{\varepsilon,1}} (i/n)^c \exp(-z_{\varepsilon,i}^2 + z_{\varepsilon,i}|v_i|) \\ &\le C n^{-d+1} \int_{b_3 B_1^2 \log n}^{\infty} e^{-y} y^\gamma \, dy \asymp n^{1-d-B_1^2 b_3} (\log n)^{-\gamma}. \end{aligned}$$

Since $\log n \asymp \log R_\varepsilon$ these yield (7.67). Relation (7.67) yields

$$\frac{2}{u_{\varepsilon,l}} \sum_{i:\hat{z}_{\varepsilon,i,l} \le BQ_{\varepsilon,i}} h_{\varepsilon,i} \sinh^2(v_i z_{\varepsilon,i}/2) \ge u_\varepsilon + o(1) = u_\varepsilon(K) + o(1).$$

By construction of the families $\bar{h}_{\varepsilon,l}$, $\bar{z}_{\varepsilon,l}$ we get, for any $v \in l^2$,

$$h_{\varepsilon,l}(v) \ge \frac{2}{u_{\varepsilon,l}} \sum_{i:\hat{z}_{\varepsilon,i,l} \le BQ_{\varepsilon,i}} h_{\varepsilon,i} \sinh^2(v_i z_{\varepsilon,i}/2).$$

These yield (7.65). The upper bounds of Theorem 7.1 (1) are proved.

7.4 Upper Bounds for Besov Norms

7.4.1 Test Procedure

For Besov norms we can assume $u_\varepsilon(\Gamma)/\sqrt{\log\log \varepsilon^{-1}} > C + o(1)$ for large enough $C = C(\Gamma) > 0$. Under this assumption we need to construct a family of tests ψ_ε such that

$$\alpha(\psi_\varepsilon) \to 0, \quad \beta(\psi_\varepsilon, V_\varepsilon(\Gamma)) \to 0. \tag{7.68}$$

We assume later without loss of generality that, uniformly over $\tau \in \Gamma$,

$$u_\varepsilon(\tau, r_\varepsilon(\tau)) \sim C\sqrt{\log\log \varepsilon^{-1}}. \tag{7.69}$$

We could use methods analogous to the previous section which are based on partitions of the set Γ to small enough cells $\{\Gamma_l\}$ and on using the combinations of tests for all sells. However we prefer to give direct constructions of the consistent test family in the problem.

First, consider supreme tests of the form

$$\psi_{\varepsilon,0} = \mathbf{1}_{X_{\varepsilon,0}}, \quad X_{\varepsilon,0} = \left\{ x : \sup_j \max_{1 \le \iota \le 2^j} (|x_{\iota j}|/T_{\varepsilon,j}) > 1 \right\}, \tag{7.70}$$

where

$$T_{\varepsilon,j} = \begin{cases} \sqrt{2CJ_{\varepsilon,0}}, & \text{if } j \le J_{\varepsilon,0}, \\ \sqrt{2(Cj + \log j)}, & \text{if } j > J_{\varepsilon,0}, \end{cases} \quad J_{\varepsilon,0} \asymp \log\log\varepsilon^{-1}, \quad C = \log 2.$$

Note that these tests provide distinguishability for region D of degenerate type (see Sections 4.4.3, 4.4.4 and Theorems 4.6, 4.7).

Next, for $j \ge J_{\varepsilon,0}$, let us consider statistics of χ^2-type (see Section 3.1.2)

$$l_j = 2^{-(j+1)/2} \sum_{\iota=1}^{2^j} (x_{\iota j}^2 - 1)$$

and note that (compare with (3.12))

$$E_0(l_j) = 0, \quad E_0(l_j^2) = 1, \tag{7.71}$$

$$E_v(l_j) = 2^{-(j+1)/2} \sum_{\iota=1}^{2^j} v_{\iota j}^2, \quad \mathrm{Var}_v(l_j) = 1 + 2^{-(j-1)} \sum_{\iota=1}^{2^j} v_{\iota j}^2. \tag{7.72}$$

Put

$$\psi_{\varepsilon,1} = \mathbf{1}_{X_{\varepsilon,1}}, \quad X_{\varepsilon,1} = \left\{ x : \sup_{j \ge J_{\varepsilon,0}} l_j/T_j > 1 \right\}, \quad T_j = 2\sqrt{\log j}, \tag{7.73}$$

i.e., we take collection of χ^2-tests.

Finally, fix a small enough value $c \in (0, (\log 2)/4)$ and put

$$K = K(j) = (\log j)/2, \qquad K(c,j) = K + cj. \tag{7.74}$$

For resolutions j such that

$$J_{\varepsilon,0} \le j \le J_{\varepsilon,1} \asymp (\log\log\varepsilon^{-1}) \log\varepsilon^{-1},$$

let us consider the collections of statistics

$$l_{j,k} = \sum_{\iota=1}^{2^j} l_{\iota;j,k} = (2^{j+1} \sinh^2(z_{j,k}^2/2))^{-1/2} \sum_{\iota=1}^{2^j} \xi(x_{\iota j}, z_{j,k}), \quad 1 \le k \le K(c,j),$$

where

$$z_{j,k} = \begin{cases} \exp(k-1)/\sqrt{j}, & \text{for } 1 \le k \le K+1, \\ \sqrt{k-K}, & \text{for } K+1 < k \le K(c,j). \end{cases}$$

These correspond to the normalized sequences of measures

$$\bar{\pi}_{j,k}^* = \bar{\pi}_j^*(z_{j,k}) = \{\pi_{\iota j}^*(z_{j,k}),\ 1 \le \iota \le 2^j\},$$

which are concentrated on the subspace L_j,

$$\pi^*_{\iota j}(z) = \pi(h_{z,j}, z), \quad h_{z,j} = (2^{j+1}\sinh^2(z^2/2))^{-1/2}; \quad \|\bar{\pi}^*_j(z)\| = 1. \tag{7.75}$$

Recall that

$$\xi(x,z) = \exp(-z^2/2)\cosh(xz) - 1$$

and (see relations (A.3), (A.4))

$$E_v(\xi(x,z)) = 2\sinh^2(zv/2),, \tag{7.76}$$

$$\mathrm{Var}_v(\xi(x,z)) = 2\sinh^2(z^2/2) + (\exp(z^2)-1)\sinh^2(zv). \tag{7.77}$$

Put

$$\psi_{\varepsilon,j,k} = \mathbf{1}_{\{l_{j,k}>t_j\}}, \quad t_j = \sqrt{5\log j}.$$

We consider the tests

$$\psi_\varepsilon = \max\left\{\psi_{\varepsilon,0},\ \psi_{\varepsilon,1},\ \max_{J_{\varepsilon,0}\le j\le J_{\varepsilon,1}}\ \max_{1\le k\le K(c,j)} \psi_{\varepsilon;j,k}\right\}. \tag{7.78}$$

Remark 1. The structure of the tests (7.78) is analogous to the adaptive tests which have been proposed by Spokoiny [91]. Particularly, the tests $\psi_{\varepsilon,0}$, $\psi_{\varepsilon,1}$ are analogous to the tests $\phi_{n,\infty}$, $\phi_{n,2}$ with $n^{-1/2} = \varepsilon$ in [91]. The main difference is that we use the tests

$$\max_{1\le k\le K(c,j)} \psi_{\varepsilon;j,k}$$

in place of the tests $\phi_{n,p,j}$ and $\max_k \phi_{n,p,\lambda_k,j}$ which are based on the statistics

$$\sum_\iota |x_{\iota j}|^p, \quad \sum_\iota |x_{\iota j}|^p \mathbf{1}_{\{|x_{\iota j}|>\lambda_k\}}$$

for fixed $p > 2$.

Remark 2. It follows from the later considerations that we can use the tests $\psi_{\varepsilon,0}$, $\psi_{\varepsilon,1}$ only if it is known that $\kappa \in \Xi_{G_1}$.

Also, it follows from the Remark to Proposition A.9, that, if it is known that $s > r$ for $p > q$ or $s + 1/q > r + 1/p$ for $p \le q$, then we can also consider only the finite number $j \le J_{\varepsilon,1}$ in the tests $\psi_{\varepsilon,0}$, $\psi_{\varepsilon,1}$.

Remark 3. Analogous adaptive test procedure based on Pearson χ^2-tests is presented in [57] under the sample model. Analogous results are given in [32] under the regression function model.

7.4.2 Type I Errors

To evaluate the type I error probabilities note that

$$\alpha(\psi_\varepsilon) \le \alpha(\psi_{\varepsilon,0}) + \alpha(\psi_{\varepsilon,1}) + \sum_{J_{\varepsilon,0}\le j\le J_{\varepsilon,1}}\ \sum_{1\le k\le K(c,j)} \alpha(\psi_{\varepsilon;j,k}). \tag{7.79}$$

First, one can see that

$$\begin{aligned}\alpha(\psi_{\varepsilon,0}) &\le 2^{J_{\varepsilon,0}+2}\Phi(-T_{\varepsilon,J_{\varepsilon,0}}) + \sum_{j>J_{\varepsilon,0}} 2^{j+1}\Phi(-T_{\varepsilon,j}) \\ &\asymp J_{\varepsilon,0}^{-1/2} + \sum_{j>J_{\varepsilon,0}} \frac{1}{j^{3/2}} = o(1). \end{aligned} \tag{7.80}$$

Next, using Gaussian approximation, the Bahr–Essen inequality, and relations (7.71), one can easily check that

$$\alpha(\psi_{\varepsilon,1}) \le \sum_{j\ge J_{\varepsilon,0}} \left(\Phi\left(-2\sqrt{\log j}\right) + O\left(2^{-j/2}\right)\right) = o(1). \tag{7.81}$$

Finally, note that

$$E_0(l_{j,k}) = 0, \quad \mathrm{Var}_0(l_{j,k}) = \|\bar{\pi}^*_{j,k}\|^2 = 1.$$

By the choice of constant c we have $d = 1 - 4c/\log 2 > 0$. Therefore, using (3.85), we get

$$R_{j,k} = \sum_{\iota=1}^{2^j} E_0 l^4_{\iota;j,k} = O\left(2^{-j}\exp(4cj)\right) = O(2^{-jd}),$$

which yields the asymptotic normality of statistics $l_{j,k}$ and the evaluations of the accuracy in the Bahr–Essen inequality. Using this inequality, we get

$$\begin{aligned}\sum_{J_{\varepsilon,0}\le j\le J_{\varepsilon,1}} \sum_{1\le k\le K(c,j)} \alpha(\psi_\varepsilon; j,k) &\le \sum_{J_{\varepsilon,0}\le j\le J_{\varepsilon,1}} \sum_{1\le k\le K(c,j)} P_0(l_{j,k} > t_j) \\ &\le \sum_{J_{\varepsilon,0}\le j\le J_{\varepsilon,1}} \sum_{1\le k\le K(c,j)} (\Phi(-t_j) + O(R_{j,k})) = o(1). \end{aligned} \tag{7.82}$$

Relations (7.79)–(7.82) yield $\alpha(\psi_\varepsilon) = o(1)$.

7.4.3 Type II Errors for Particular Tests

To obtain the second relation in (7.68) under assumption (7.69) we need, for any families $\tau = \tau_\varepsilon \in \Gamma$, $v = v_\varepsilon \in V_\varepsilon = V_\varepsilon(\tau_\varepsilon, r_\varepsilon(\tau_\varepsilon))$, to select the tests $\psi_{\varepsilon,l}$ under the maximum in (7.78), i.e., to select the numbers $l = l_\varepsilon(v_\varepsilon)$, $l = 0, 1$ or $l = (j,k)$, $j = j_\varepsilon$, $k = k_\varepsilon$, $J_{\varepsilon,0} \le j \le J_{\varepsilon,1}$, $1 \le k \le K(c,j)$, such that

$$\beta(\psi_{\varepsilon,l}, v_\varepsilon) \to 0. \tag{7.83}$$

First, taking any $\delta > 0$, let us consider the families $v_\varepsilon \in l^2$ such that

$$\limsup_{\iota,j} |v_{\varepsilon,\iota,j}|/T_{\varepsilon,j} \ge 1 + \delta \tag{7.84}$$

(we assume later without loss of generality that there exist all limits under consideration as $\varepsilon \to 0$). One can easily see that for all these families we get

$$\beta(\psi_{\varepsilon,0}, v_\varepsilon) \le \inf_{\iota,j} \Phi(T_{\varepsilon,j} - |v_{\varepsilon,\iota,j}|) = o(1).$$

Next, let us consider the families $v_\varepsilon \in l^2$ such that

$$\lim \sup_{j \ge J_{\varepsilon,0}} H_j(v_\varepsilon)/T_j \ge 1 + \delta$$

where

$$H_j(v) = E_v l_j = 2^{-(j+1)/2} \sum_{\iota=1}^{2^j} v_{\iota j}^2. \tag{7.85}$$

Using the Chebyshev inequality and relations (7.71) one can easily see that for all these families

$$\beta(\psi_{\varepsilon,1}, v_\varepsilon) \le \inf_{j \ge J_{\varepsilon,0}} \frac{1 + 2^{-(j-3)/2} H_j(v_\varepsilon)}{(H_j(v_\varepsilon) - T_j)^2} \le 4 \frac{H_j(v_\varepsilon)(\delta+1)^2}{2^{j/2} H_j^2(v_\varepsilon)\delta^2} = o(1).$$

Finally, for any (j,k) such that $J_{\varepsilon,0} \le j \le J_{\varepsilon,1}$, $1 \le k \le K(c,j)$, let us consider the families $v_\varepsilon \in l^2$ such that

$$\lim \sup_{\iota,j} |v_{\varepsilon,\iota,j}|/T_{\varepsilon,j} \le 1 + \delta, \quad \lim H_{j,k}(v_\varepsilon)/t_j \ge 1 + \delta,$$

where (see (7.75), (7.76)),

$$H_{j,k}(v) = E_v(l_{j,k}) = (\bar{\pi}^*_{j,k}, \bar{\delta}_{v_j}) = 2h_{j,k} \sum_\iota \sinh^2(v_{\iota j} z_{j,k}/2); \quad h_{j,k} = h_{z_{j,k},j},$$

(here and below v_j and $\bar{\delta}_{v_j}$ correspond to the resolution j).

Using (7.77), equality

$$\exp(x^2) \quad 1 = 2\exp(x^2/2)\sinh(x^2/2),$$

and inequality

$$\sinh^2 x \le 4\exp(|x|)\sinh^2(x/2),$$

one can see that for all these families,

$$\begin{aligned} \mathrm{Var}_v(l_{j,k}) &\le 1 + 4h_{j,k}\sinh(z_{j,k}^2/2)\exp(z_{j,k}T_{\varepsilon,j} + z_{j,k}^2/2)H_{j,k}(v) \\ &= 1 + 2^{-(j-3)/2}\exp(z_{j,k}T_{\varepsilon,j} + z_{j,k}^2/2)H_{j,k}(v). \end{aligned}$$

Taking a small enough $c > 0$ in (7.74) (for $0 < c \le 0,1$), for a small enough $\varepsilon > 0$ one has the inequality

$$z_{j,k}T_{\varepsilon,j} + z_{j,k}^2/2 \le ((2\log 2c)^{1/2}(1+o(1)) + c/2)j < (\log 2/2)j.$$

Using this inequality and the Chebyshev inequality we get

$$\beta(\psi_{\varepsilon;j,k}(v_\varepsilon)) \le \mathrm{Var}_v(l_{j,k})/(H_{j,k}(v_\varepsilon) - t_j)^2 = o(1)$$

for all families v_ε under consideration.

Thus we have shown that type II error probabilities of the supreme test $\psi_{\varepsilon,0}$ are vanishing for alternatives under the constraint (7.84). If the constraint (7.84) is not valid, then we need to use tests χ^2-tests $\psi_{\varepsilon,1}$ based on the statistics l_j, $j \geq J_{\varepsilon,0}$, or the tests $\psi_{\varepsilon;j,k}$ for $J_{\varepsilon,0} \leq j \leq J_{\varepsilon,1}$. Note that the properties of the tests $\psi_{\varepsilon,1}$ do not depend on the coordinates v_j with $j < J_{\varepsilon,0}$ and the properties of the collection $\psi_{\varepsilon;j,k}$ do not depend on the coordinates v_j with $j < J_{\varepsilon,0}$ and $j > J_{\varepsilon,1}$. Therefore, it suffices to verify that, for any sequence $v = v_\varepsilon \in V_\varepsilon$ such that (7.84) is not valid, either one of the two following inequalities hold

$$\lim \sup_{j \geq J_{\varepsilon,0}} H_j(v_\varepsilon)/T_j \geq B; \quad \lim \sup_{J_{\varepsilon,0} \leq j \leq J_{\varepsilon,1}} H_{j,k}(v_\varepsilon)/t_j \geq B,$$

for some $B > 1$.

We cannot prove these inequalities directly and we use the following construction. Fixing a family of subsets $\mathcal{J}_\varepsilon \subset \mathcal{J}$ we set

$$\tilde{v} = \tilde{v}(\mathcal{J}_\varepsilon); \quad \tilde{v}_{\iota j} = \begin{cases} v_{\iota j}, & \text{if } j \in \mathcal{J}_\varepsilon, \\ 0, & \text{if } j \notin \mathcal{J}_\varepsilon \, . \end{cases}$$

Note that the set

$$X_\varepsilon = \{x \, : \, \psi_\varepsilon(x) = 0\}$$

is convex and symmetric. It follows from Lemma 2.4 in Section 2.4.4 that we have the inequality

$$\beta(\psi_\varepsilon(v)) \leq \beta(\psi_\varepsilon(\tilde{v})).$$

The evaluations above show that, to obtain (7.83), it suffices to establish the following statement.

Lemma 7.4 *Under the assumptions above there exists a constant $C = C(\Gamma) > 0$ in (7.69) and $B > 1$ such that, for all families $\tau_\varepsilon \in \Gamma$, all families $v_\varepsilon \in V_\varepsilon = V_\varepsilon(\tau_\varepsilon, r_\varepsilon(\tau_\varepsilon))$, such that*

$$\limsup_{\iota j} |v_{\iota j}|/T_{\varepsilon,j} \leq B, \tag{7.86}$$

there exist $\delta_\varepsilon = o(1)$ and sets $\mathcal{J}_\varepsilon$ such that $\tilde{v}_\varepsilon = \tilde{v}_\varepsilon(\mathcal{J}_\varepsilon) \in \tilde{V}_\varepsilon = V_\varepsilon(\tau_\varepsilon, \tilde{r}_\varepsilon(\tau_\varepsilon))$ with $\tilde{r}_\varepsilon = (1-\delta_\varepsilon)r_\varepsilon$ and either

$$\limsup_j H_j(\tilde{v}_\varepsilon)/T_j = \lim \sup_{j \in \mathcal{J}_\varepsilon} H_j(v_\varepsilon)/T_j \geq B, \quad \mathcal{J}_\varepsilon \subset \{j \geq J_{\varepsilon,0}\}, \tag{7.87}$$

or

$$\limsup_j \sup_{1 \leq k \leq K(c,j)} H_{j,k}(\tilde{v}_\varepsilon)/t_j = \lim \sup_{j \in \mathcal{J}_\varepsilon} \sup_{1 \leq k \leq K(c,j)} H_{j,k}(v_\varepsilon)/t_j \geq B,$$
$$\mathcal{J}_\varepsilon \subset \{J_{\varepsilon,0} \leq j \leq J_{\varepsilon,1}\}. \tag{7.88}$$

Proof of Lemma 7.4. This is given in the Appendix, Section A.5.

8

High-Dimensional Signal Detection

In this chapter we would like to demonstrate some new phenomena in minimax nonparametric hypothesis testing problems.

The main part of this chapter is concerned with the nonstandard asymptotics of error probabilities. Recall that we have got the asymptotics of trivial, classical degenerate, and various Gaussian types in the problems. The main types of Gaussian asymptotics were obtained under some constraints on the asymptotics of the extreme families $\bar{h}_\varepsilon, \bar{z}_\varepsilon$ in the extreme problems (5.39) on the spaces of sequences of measures. In particular we assumed, in the studies, that the extreme sequences $z_{\varepsilon,i}$ satisfy assumption A.3 which means that these are not "too large." Fortunately, all the required assumptions are fulfilled for the power and Besov norms with any fixed $\kappa \in \Xi_{G_1} \cup \Xi_{G_2}$. However, they may not be fulfilled for a family $\kappa_\varepsilon \in \Xi_{G_2}$ with $p > q$ and $\lambda_\varepsilon = \lambda(\kappa_\varepsilon) \to 0$ (these correspond to the frontier between regions Ξ_{G_2} and Ξ_D). What has happened if they are not fulfilled?

It is difficult to study this problem in general. Therefore, we consider a particular case which corresponds to the hypothesis testing problems for l^n-balls under the n-dimensional Gaussian model $X = v + \xi$, $v \in R^n$. This problem has been considered in Example 2.2 and in Sections 2.4.3, 5.5 where we tested the null-hypothesis $H_0 : v = 0$ against the alternative $H_1 : v \in V_n$ of the type

$$V_n = V_n(\rho_n, R_n) = \{v \in R^n \, : \, |v|_p \geq \rho_n, \; |v|_q \leq R_n\} \qquad (8.1)$$

under the nonempty conditions (5.136) (which are improved up to (5.137) in the cases $p > q$ or $p > 2$). Remember that we have obtained asymptotics of degenerate type for $p = \infty$ and of Gaussian type for the cases $p \leq q$

and $q < p < \infty$ with bounded values $z_n = (\rho_n/R_n)^{1/(p-q)} = O(1)$. Here we consider the case $q < p < \infty$ with $z_n \to \infty$. We show that new types of asymptotics could arise in this case (namely, asymptotics of a specific infinite divisible type). [1]

In Section 8.5 we study new analytical phenomena which correspond to testing the composite hull-hypothesis (this corresponds to a small ball in the l^p- or L_p-norm). We show that distinguishability conditions depend essentially on the evenness of p. [2]

8.1 The Bayesian Signal Detection Problem

Let us consider the Bayesian hypothesis testing problem: we test the null-hypothesis $H_0 : P = P_{n,0}$ on an unknown probability measure generated by a random vector X against the Bayesian alternative $H_{\pi^n} : P = P_{\pi^n}$; here, as above, $P_{n,v}$ is the n-dimensional Gaussian measure $N(v, I_n)$, $v \in R^n$ with mean vector v and unit covariation matrix, which corresponds to the observation (see (1.4))

$$X = v + \xi, \quad \xi \sim N(0, I_n),$$

and $P_{\pi^n} = \int P_{n,v}\pi^n(dv)$ is the mixture for a product prior

$$\pi^n(dv) = \pi_n(dv_1) \times \ldots \times \pi_n(dv_n)$$

with common factors of symmetric three-point type:

$$\pi_n = \pi(h_n, z_n) = (1 - h_n)\delta_0 + \frac{h_n}{2}(\delta_{z_n} + \delta_{-z_n}).$$

Then one has

$$P_{\pi^n}(dx_1...dx_n) = \prod_{i=1}^{n} P_{\pi_n}(dx_i), \quad P_{\pi_n}(dx) = (1-h_n)P_0 + h_n(P_{z_n} + P_{-z_n})/2, \tag{8.2}$$

where $P_z = N(z, 1)$ is one-dimensional Gaussian measure with mean z and variation 1.

Introduce the notation

$$g_n(t) = h_n\xi(t, z_n), \quad w_n(t) = \log(1 + g_n(t)) = \log(1 + h_n\xi(t, z_n)), \quad t \in R^1,$$

where, as above, $\xi(t, z) = \exp(-z^2/2)\cosh(tz) - 1$.

The Bayesian log-likelihood ratio $l_n = \log(dP_{\pi^n}/dP_0)$ is of the form

$$l_n = l_{n,\pi^n} = \sum_{i=1}^{n} w_{n,i}, \quad w_{n,i} = w_n(X_i), \tag{8.3}$$

---

[1]These materials are based on [56], [58].

[2]These materials are based on [60].

and the items $w_{n,i}$ are i.i.d. under null-hypothesis. The Bayesian tests ψ_{α,π^n} are of the form $\psi_{\alpha,\pi^n} = \mathbf{1}_{\{l_n > T_{n,\alpha}\}}$ where $T_{n,\alpha}$ is a $(1-\alpha)$-quantile of the statistic l_n under $P_{n,0}$-probability.

At first glance, one can use the Central Limit Theorem and obtain the Gaussian limit distribution for the statistics l_n. However, this is not possible if items $w_{n,i}$ have "too heavy tails" and one needs to use the general limit theorems which lead to infinitely divisible distributions.

8.1.1 Limit Theorems and Infinitely Divisible Distributions

Recall (see [86], Chapter 2) that the random variable X of infinitely divisible type is characterized by a specific form of the characteristic function $\varphi(z) = E\exp(izX)$. The following presentation is called the *Levi formula*:

$$\log \varphi(z) = iz\gamma - \frac{\sigma^2 z^2}{2} + \int_{\breve{R}} \left(e^{izt} - 1 - \frac{izt}{1+t^2} \right) dL(t), \tag{8.4}$$

where the integration set is $\breve{R} = R^1 \setminus \{0\}$, $\gamma \in R^1$, the quantity $\sigma \geq 0$ corresponds to the Gaussian component, and the function $L(t)$ is called the *Levi spectral function*: this is nondecreasing on the intervals $(-\infty, 0)$ and $(0, \infty)$, this satisfies $L(-\infty) = 0$, $L(\infty) = 0$ and

$$\int_{-\delta}^{0} t^2\, dL(t) + \int_{0}^{\delta} t^2\, dL(t) < \infty \quad \forall\, \delta > 0.$$

The first and second items in (8.4) correspond to the Gaussian random variable; in particular, if $L(t) \equiv 0$, then $X \sim N(\gamma, \sigma^2)$. The integral item in (8.4) corresponds to a limit of weighted sums of the centered independent Poisson random variable.

One can present properties of the random variable X of infinitely divisible type in terms of the presentation (8.4). In particular, for any $a > 0$, this has the finite absolute moment

$$E|X|^a < \infty \quad \text{iff} \quad \int_{|t|\geq 1} |t|^a\, dL(t) < \infty.$$

Also let

$$l_X = \inf\{u : P(X \geq u) = 1\}$$

be the essential lower bound of the random variable X. Then

$$l_X > -\infty \quad \text{iff} \quad \sigma = 0, \quad L(t) = 0 \;\forall\, t < 0, \quad \int_{+0}^{1} t\, dL(t) < \infty.$$

If $l_X > -\infty$, then

$$l_X = \gamma - \int_{+0}^{\infty} \frac{t}{1+t^2}\, dL(t). \tag{8.5}$$

Let $L^a(t)$ be an absolute continuous component of $L(t)$. Then the distribution function of X is absolutely continuous, if either $\sigma > 0$ or

$$\int_{\breve{R}} dL^a(t) = \infty. \tag{8.6}$$

Infinitely divisible distributions form the general class of limit distributions for sums $\sum_{i=1}^{k_n} w_{n,i}$ of the i.i.d. random variables $w_{n,i}$, $i = 1, \dots, k_n$, in the series scheme under the assumption

$$\max_i P(|w_{n,i}| \geq t) \to 0 \quad \text{as } n \to \infty \quad \forall\, t > 0. \tag{8.7}$$

One can obtain the Levi spectral function $L(t)$ at a point t of continuity of the function $L(t)$ from the limit equality

$$\sum_i F_{n,i}(t) \to L(t) \quad \text{for } t < 0, \quad \sum_i (F_{n,i}(t) - 1) \to L(t) \quad \text{for } t > 0, \tag{8.8}$$

where $F_{n,i}(t) = P(w_{n,i} < t)$ is the distribution function of the random variable $w_{n,i}$. To obtain the Gaussian component σ^2 one can use the limit relation

$$\sigma^2 = \lim_{\varepsilon \to 0} \lim_{n \to \infty} \sum_i \left(\int_{|t|<\varepsilon} t^2\, dF_{n,i}(t) - \left(\int_{|t|<\varepsilon} t\, dF_{n,i}(t) \right)^2 \right). \tag{8.9}$$

Finally, one can obtain the constant γ from the relation

$$\sum_i \int_{|t|<\eta} t\, dF_{n,i}(t) \to \gamma + \int_{|t|<\eta} t\, dG(t) - \int_{|t|>\eta} \frac{dG(t)}{t}, \tag{8.10}$$

where

$$dG(t) = \frac{t^2}{1+t^2}\, dL(t)$$

and $\eta > 0$ is such that the function G (or L) is continuous at the points $\pm\eta$. The function $G(t)$ is called the *Levi–Hynchin spectral function.*

All these relations are contained in [86], Chapter 4.

Below we denote by $L^0(t)$, $L^1(t)$ the spectral function of the studied statistics under the null-hypothesis and under the Bayesian alternative, respectively.

8.1.2 Some Remarks on the Study of Log-Likelihood Ratios

To study the limit distribution of the statistics l_n defined by (8.3) we assume that $z_n \to \infty$, $h_n \to 0$. First, note that the random variables X_i in (8.3) are i.i.d. standard Gaussian under the hull-hypothesis, i.e., they have the distribution function $F_0(x) = \Phi(x)$. It follows from (8.2) that X_i are i.i.d. under the Bayesian alternative with the measure

$$P_{\pi_n} = P_0 + \frac{h_n}{2}\left(P_{z_n} + P_{-z_n} - 2P_0\right). \tag{8.11}$$

This corresponds to the distribution function

$$F_1(x) = \Phi(x) + \Delta_n \Phi(x),$$

where

$$\Delta_n \Phi(x) = \frac{h_n}{2}(\Phi(x - z_n) + \Phi(x + z_n) - 2\Phi(x)), \qquad \frac{d\Delta_n \Phi}{d\Phi}(x) = g_n(x). \tag{8.12}$$

Recall that $\Phi(x)$ stands for the distribution function of the standard Gaussian law.

Next, note that

$$g_n(x) \geq -h_n \to 0 \quad \forall\, x \in R^1.$$

Therefore, under the null-hypothesis and under the Bayesian alternative for large enough n, we have

$$\sum_{i=1}^{n} F_{n,i}(t) = nP(w_n(X) < t) = 0 \quad \forall\, t < 0, \tag{8.13}$$

where $P = P_0$ or $P = P_{\pi_n}$ and $F_{n,i}$ is the P-distribution function of i.i.d. random variables $w_n(X_i)$. Thus, for the limit spectral functions under the null-hypothesis and under the Bayesian alternative we get

$$L^0(t) = L^1(t) = 0 \quad \forall\, t < 0.$$

Let us consider the thresholds T_n such that

$$g_n(T_n) = h_n \xi(T_n, z_n) = 1,$$

which are the same as that we have considered in Section 5.1.1. In particular, we can rewrite relations (5.6), (5.8),

$$T_n = \frac{z_n}{2} + \frac{\log(1 + h_n^{-1})}{z_n} + O(z_n^{-1}) \to \infty, \quad h_n \sim 2\exp(z_n^2/2 - T_n z_n). \tag{8.14}$$

Note the inequalities

$$\begin{aligned} g_n(x) &= h_n(e^{-z_n^2/2}\cosh(xz_n) - 1) \geq \frac{h_n}{2}\left(e^{-z_n^2/2+|x|z_n} - 2\right) \\ &= \frac{h_n}{2} e^{-z_n^2/2+|x|z_n}\left(1 - 2e^{z_n^2/2-|x|z_n}\right); \\ g_n(x) &\leq h_n e^{-z_n^2/2}\cosh(xz_n) = \frac{h_n}{2} e^{-z_n^2/2+|x|z_n}\left(1 + e^{-2|x|z_n}\right), \end{aligned}$$

which yield the relations

$$\frac{h_n}{2} e^{-z_n^2/2+|x|z_n}(1 - 2\eta_n) \leq g_n(x) \leq \frac{h_n}{2} e^{-z_n^2/2+|x|z_n}(1 + \eta_n), \tag{8.15}$$

where $\eta_n = o(1)$ uniformly over $|x| > z_n/2 + \xi_n$, $z_n \xi_n \to \infty$.

Taking into account relations (8.14), (8.15) and since $h_n \to 0$, one can find a positive sequence $\delta_n \to 0$ such that $\delta_n z_n \to \infty$ and for $|x| = y + T_n$ one has

$$\sup_{|y|<\delta_n} |g_n(x) - e^{yz_n}| \le \zeta_n, \quad \sup_{|y|<\delta_n} |w_n(x) - \log(1+e^{yz_n})| \le \zeta_n, \quad \zeta_n = o(1), \tag{8.16}$$

and

$$g_n(x) \sim e^{yz_n}, \quad w_n(x) \sim \log(1+e^{yz_n}) \quad \text{uniformly over} \quad y > -\delta_n. \tag{8.17}$$

Also for any $t > 0$ let us define the thresholds $T_n(t) > 0$ from the equality $w_n(T_n(t)) = t$. It follows from (8.15)–(8.17) that

$$T_n(t) = T_n + \frac{\log(e^t - 1)}{z_n} + o(z_n^{-1}). \tag{8.18}$$

Recall the well-known relations (see (4.140))

$$\Phi(-x) \sim \frac{e^{-x^2/2}}{\sqrt{2\pi}x}, \quad \Phi(-x+a) \sim \Phi(-x)e^{ax}, \quad \text{as } x \to \infty, \quad a \to 0. \tag{8.19}$$

Since $w_n(x)$ increases in $|x|$, under the hull-hypothesis for any $t > 0$ we get

$$\sum_{i=1}^n (1 - F_{n,i}(t)) = nP(|\eta| > T_n(t)) = 2n\Phi(-T_n(t)), \quad \eta \sim N(0,1). \tag{8.20}$$

Introduce the quantities

$$\tau_n = \frac{T_n}{z_n} = \frac{1}{2} + \frac{\log(1+h_n^{-1})}{z_n^2} + O(z_n^{-2}) \tag{8.21}$$

and assume without loss of generality that $\tau_n \to \tau \in [1/2, \infty]$. Also consider the quantities

$$u_n^2 = \frac{n}{2}h_n^2 e^{z_n^2}\Phi(T_n - 2z_n), \quad c_n = 2n\Phi(-T_n), \quad \lambda_n = nh_n\Phi(z_n - T_n). \tag{8.22}$$

Using (8.14) and (8.19) one gets the relations between these quantities:

$$c_n = o(u_n^2), \text{ if } \tau \ge 2; \qquad c_n \sim \frac{(2-\tau)}{\tau}u_n^2, \text{ if } \tau < 2; \tag{8.23}$$

$$c_n \sim \lambda_n(1-\tau^{-1}), \text{ if } \tau > 1; \quad c_n = o(\lambda_n), \text{ if } \tau \le 1. \tag{8.24}$$

Note the following:

Lemma 8.1 *Let $nh_n \to \infty$ and H_n be a sequence such that $n\Phi(-H_n) \asymp 1$ (this yields $H_n = \sqrt{2\log n} + o(1)$). Put $\lambda_n^* = nh_n\Phi(z_n - H_n)$. Then:*
(1) Let $1/2 \le \tau < \infty$ and $\lambda_n \to \infty$. Then $\lambda_n^* \to \infty$.
(2) Let $\tau_n = 1$ and $\lambda_n \asymp 1$. Then $\lambda_n^* = \lambda_n + o(1)$.

Proof. Set $d_n = T_n - z_n$, $r_n = T_n - H_n$. If $r_n \geq 0$, then $\lambda_n^* \geq \lambda_n$. Therefore, to prove statement (1) it suffices to suppose that $r_n < 0$. If $H_n - z_n \leq B$ for some B, then $\lambda_n^* \to \infty$ by $nh_n \to \infty$.

Let $H_n - z_n \to \infty$. Then, using (8.14), (8.19) we get

$$\lambda_n^* \asymp \frac{nh_n e^{-(H_n - z_n)^2/2}}{H_n - z_n} \asymp \frac{ne^{-H_n^2/2 - r_n z_n}}{H_n - z_n} \asymp \frac{H_n e^{-r_n z_n}}{H_n - z_n}. \tag{8.25}$$

If $-r_n z_n \to \infty$, then by (8.25) we have $\lambda_n^* \to \infty$. Let $-r_n z_n = O(1)$. If $d_n < B$ for some B, then $\lambda_n^* = nh_n \Phi(r_n - d_n) \to \infty$. In particular, this holds for $\tau < 1$ since $d_n = z_n(\tau_n - 1) \to -\infty$ if $\tau_n \to \tau < 1$. Let $d_n \to \infty$. Since $d_n r_n = (\tau_n - 1) r_n z_n = O(1)$ by $\tau_n = O(1)$, using (8.19), we have

$$\lambda_n^* = nh_n \Phi(r_n - d_n) \sim nh_n \Phi(-d_n) \exp(d_n r_n) \asymp \lambda_n.$$

These yield statement (1).

Let $\tau = 1$, $\lambda_n \asymp 1$. This yields $d_n \to \infty$, $d_n = o(T_n)$. Using (8.14), (8.19) we get $\lambda_n \asymp nT_n \Phi(-T_n)/d_n$. This yields

$$n\Phi(-T_n) \asymp \frac{d_n}{T_n} \to 0, \quad n \exp(-T_n^2/2) \asymp d_n \to \infty.$$

These relations imply $H_n < T_n < \sqrt{2 \log n} = H_n + o(1)$ which yields $0 < r_n = o(1)$. Using (8.19) we get

$$n\Phi(-T_n) = n\Phi(-H_n - r_n) \sim n\Phi(-H_n) \exp(-H_n r_n) \asymp \exp(-H_n r_n)$$

which leads to $\exp(H_n r_n) \asymp T_n/d_n \to \infty$. Therefore, one has

$$r_n \sim H_n^{-1} \log(T_n/d_n) \sim T_n^{-1} \log(T_n/d_n).$$

Analogously to the above, we have

$$\lambda_n^* = nh_n \Phi(-d_n + r_n) \sim \lambda_n \exp(r_n d_n) \sim \lambda_n$$

by $r_n d_n \asymp (d_n/T_n) \log(T_n/d_n) \to 0$. □

To study the limit distributions of the Bayesian log-likelihood ratio we consider the different cases $\tau \geq 2$, $\tau \in (1, 2)$, and $\tau \in [1, 1/2]$.

8.1.3 Limit Distributions for $\tau \geq 2$

We would like to show that the Gaussian asymptotics hold in this case as well, but for $\tau = 2$ the qualities u_n^2 defined by (8.22) are different from $u_n^2 = 2nh_n^2 \sinh^2(z_n^2/2) \sim nh_n^2 e^{z_n^2}/2$, $z_n \to \infty$ which were used previously.

Consider the case $u_n \asymp 1$; to simplify, we consider the case $\tau_n \to \tau < \infty$. By (8.23) this yields $c_n = o(1)$. We start with limit distributions under the null-hypothesis. Let us show that the statistics l_n defined by (8.3) are asymptotically Gaussian $N(-u_n^2/2, u_n^2)$ (this means $(l_n + u_n^2/2)/u_n \to \eta \sim N(0, 1)$ in $P_{n,0}$-probability).

Using (8.8), (8.13) and (8.20) we get $L(t) = L^0(t) = 0$. Therefore the limit distribution is Gaussian. To obtain the parameters of the limit distribution, according to a general theory [86], we can pass to truncated random variables in the sums. Let us take a positive sequence $t_n \to 0$ such that

$$-\log t_n = o(z_n), \quad c_n t_n^{-\tau_n} \to 0.$$

By (8.18), (8.19) this yields

$$T_n(t_n) = T_n + z_n^{-1} \log t_n + o(z_n^{-1}), \quad 2n\Phi(-T_n(t_n)) \sim c_n t_n^{-\tau_n} \to 0.$$

Thus we can pass to the truncated random variables

$$\hat{w}_{n,i} = w_{n,i} \mathbf{1}_{\{w_{n,i} < t_n\}} = w_n(X_i) \mathbf{1}_{\{|X_i| < T_n(t_n)\}}.$$

It follows from the inequalities $\log(1 - h_n) \leq \hat{w}_{n,i} \leq t_n$ that

$$\sup_i |\hat{w}_{n,i}| = o(1).$$

Consider the sums $\hat{l}_n = \sum_{i=1}^n \hat{w}_{n,i}$. We have

$$P_0(\hat{l}_n \neq l_n) \leq \sum_{i=1}^n P_0(\hat{w}_{n,i} \neq w_{n,i}) = 2n\Phi(-T_n(t_n)) \to 0.$$

Therefore the statistics $\hat{l}_n$ have the same limit distribution and its parameters are determined by the means and variances of $\hat{l}_n$. It suffices to verify that

$$E_{n,0}\hat{l}_n = -u_n^2/2 + o(1), \quad \text{Var}_{n,0}\hat{l}_n = u_n^2 + o(1). \tag{8.26}$$

Using the relation

$$\log(1+g) = g - g^2(1/2 + o(1)), \quad \text{as} \quad g \to 0,$$

we have the presentation

$$\hat{w}_n(X_i) = \hat{g}_n(X_i) - \hat{g}_n^2(X_i)(1/2 + o(1)) = \hat{g}_n(X_i)(1 + o(1)),$$

where

$$\hat{g}_n(X_i) = h_n \hat{\xi}(X, z_n), \quad \hat{\xi}(X, z_n) = \xi(X, z_n) \mathbf{1}_{\{|X| < T_n(t_n)\}}.$$

Observe the following relations

$$\begin{aligned} E_0(\hat{\xi}(X, z_n)) &\sim -\Phi(z_n - T_n(t_n)), & (8.27)\\ E_0(\hat{\xi}^2(X, z_n)) &\sim \exp(z_n^2)\Phi(T_n(t_n) - 2z_n)/2, & (8.28)\\ nh_n\Phi(z_n - T_n(t_n)) &= o(1), & (8.29) \end{aligned}$$

(here and later E_0 is the expectation with respect to the distribution function $F_0(x) = \Phi(x)$). In fact, relations (8.27), (8.28) follow from (A.11), (A.12); relation (8.29) follows from (8.22), (8.24):

$$nh_n\Phi(z_n - T_n(t_n)) \sim nh_n\Phi(z_n - T_n)t_n^{1-\tau_n} = \lambda_n t_n^{1-\tau_n} \asymp c_n t_n^{1-\tau_n} = o(1).$$

Using (8.27), and since $\hat{w}_n(X_i) \sim \hat{g}_n(X_i)$, we get

$$\begin{aligned} E_{n,0}(\hat{l}_n) &= nE_0(\hat{w}_n(X)) = nE_0(\hat{g}_n(X)) - nE_0(\hat{g}_n^2(X))(1/2 + o(1)) \\ &= -nh_n\Phi(z_n - T_n(t_n))(1 + o(1)) \\ &\quad -\frac{nh_n^2}{4}e^{z_n^2}\Phi(T_n(t_n) - 2z_n)(1 + o(1)) = -u_n^2/2 + o(1), \\ \mathrm{Var}_{n,0}(\hat{l}_n) &= nE_0(\hat{w}_n^2(X)) - n(E_0(\hat{w}_n(X)))^2 \sim nh_n^2E_0(\hat{\xi}^2(X, z_n)) \sim u_n^2. \end{aligned}$$

Let us consider the limit distribution under the Bayesian alternative P_{π^n}. We need to show that l_n are asymptotically Gaussian $N(u_n^2/2, u_n^2)$ in P_{π^n}-probability, in particular, that $L(t) = L^1(t) = 0$. One can verify this by using analogous evaluations and relation (8.12). However this follows directly from the third Le Cam lemma (see [30], [88]).

These yield asymptotics of Gaussian type for the type II error probabilities $\beta(\alpha, \pi^n) = \beta(\alpha, P_0, P_{\pi^n})$ in the Bayesian problem.

Proposition 8.1 *Let $\tau \geq 2$. Then*

$$\beta(\alpha, \pi^n) = \Phi(T_\alpha - u_n) + o(1).$$

If $u_n \asymp 1$, then the statistics l_n are asymptotically Gaussian $N(-u_n^2/2, u_n^2)$ in $P_{n,0}$-probability and they are asymptotically Gaussian $N(u_n^2/2, u_n^2)$ in P_{π^n}-probability. In particular, we have the following distinguishability conditions:

$$\beta(\alpha, \pi^n) \to 1 - \alpha \quad \textit{iff} \quad u_n \to 0; \quad \beta(\alpha, \pi^n) \to 0 \quad \textit{iff} \quad u_n \to \infty.$$

To complete the proof of Proposition 8.1 it suffices to verify that $\beta(\alpha, \pi^n) \to 1-\alpha$ as $u_n \to 0$ and $\beta(\alpha, \pi^n) \to 0$ as $u_n \to \infty$. This follows from evaluations of the mean values and variances. We leave these to the reader.

Remark. The considerations above show that for $\tau \geq 2$ the statistics $t_n = u_n^{-1}h_n \sum_{i=1}^n \xi(x_i, z_n)$ are asymptotically Gaussian $N(0, 1)$ in $P_{n,0}$-probability.

8.1.4 Limit Distributions for $\tau \in (1, 2)$

We would like to show that the limit distributions of l_n have no Gaussian components and have nontrivial Levi spectral functions L^0 and L^1 under the null-hypothesis and under the Bayesian alternative. We consider the case $c_n \to c \in (0, \infty)$.

We start with limit distributions under the null-hypothesis. It was shown previously that $L^0(t) = 0$ for all $t < 0$. Let us calculate $L^0(t)$, $t > 0$. Using (8.20), (8.19), (8.18), we get,

$$\sum_{i=1}^n (1 - F_{n,i}(t)) = 2n\Phi(-T_n(t)) = c_n(e^t - 1)^{-\tau_n} \to c(e^t - 1)^{-\tau} = -L^0(t). \tag{8.30}$$

Recall that $F_{n,i}$ is a distribution function of $w_n(X_i)$ under the null hypothesis here.

Let us show that the limit distribution has no Gaussian component. In view of (8.9) it suffices to verify that

$$\lim_{t\to 0}\lim_{n\to\infty} n\int_{|w_n|<t} w_n^2(x)\,d\Phi(x) \to 0. \tag{8.31}$$

Since $w_n^2(x) \le Bh_n^2\xi^2(x,z_n)$ the last relation follows from

$$\lim_{t\to 0}\lim_{n\to\infty} nh_n^2\int_{|x|<T_n(t)} \xi^2(x,z_n)\,d\Phi(x) \to 0. \tag{8.32}$$

It follows from (A.12) and (8.19), (8.22) that

$$\begin{aligned} nh_n^2\int_{|x|<T_n(t)} \xi^2(x,z_n)\,d\Phi(x) &\sim \frac{n}{2}h_n^2 e^{z_n^2}\Phi(T_n(t)-2z_n) \\ &\sim u_n^2(e^t-1)^{2-\tau_n} \asymp (e^t-1)^{2-\tau}, \quad n\to\infty, \end{aligned}$$

which yields (8.32).

Let us calculate a constant $\gamma=\gamma_0$ under the null-hypothesis. Since

$$\int_1^\infty t\,dL^0(t) < \infty,$$

we can rewrite equality (8.10) in the form

$$\begin{aligned} \gamma_0 &= \lim_{n\to\infty} n\int_{|w_n|<t} w_n(x)\,d\Phi(x) - \int_0^t \frac{x^3}{1+x^2}dL^0(x) + \int_t^\infty \frac{x}{1+x^2}dL^0(x) \\ &= \lim_{n\to\infty} n\int_{|x|<T_n(t)} w_n(x)\,d\Phi(x) - \int_0^\infty \frac{x^3}{1+x^2}dL^0(x) + \int_t^\infty x\,dL^0(x). \end{aligned}$$

To estimate

$$A_n = \int_{|w_n|<t} w_n(x)\,d\Phi(x),$$

let us take a sequence

$$\delta_n \to 0, \quad z_n\delta_n \to \infty,$$

such that (8.16) holds; moreover we can assume that $e^{T_n\delta_n}\zeta_n = o(1)$ which yields

$$n\zeta_n\Phi(\delta_n - T_n) = o(1). \tag{8.33}$$

Let us write

$$\begin{aligned} A_n &= n\int_{|x|<T_n(t)} w_n(x)\,d\Phi(x) = n\int_{|x|<T_n(t)} (w_n(x)-h_n\xi(x,z_n))\,d\Phi(x) \\ &+nh_n\int_{|x|<T_n(t)} \xi(x,z_n)\,d\Phi(x) = B_n + C_n. \end{aligned}$$

By (8.27), (8.18), (8.19), (8.24), the second item is of the asymptotics

$$C_n \sim -nh_n\Phi(z_n - T_n(t)) \sim -\lambda_n(e^t-1)^{1-\tau_n} \sim -\frac{c\tau}{\tau-1}(e^t-1)^{1-\tau}.$$

Also set $B_n = B_{n,1} + B_{n,2}$ where $B_{n,1}$ corresponds to the integral over $|x| < T_n - \delta_n$ and $B_{n,2}$ corresponds to the integral over

$$\Delta_n(t) = \{x : T_n - \delta_n < |x| < T_n(t)\}.$$

Since

$$w_n(x) - h_n\xi(x, z_n) \sim -h_n^2\xi^2(x, z_n)/2 \quad \text{for } |x| < T_n - \delta_n,$$

analogously to evaluations of the variance, we get $B_{n,1} \to 0$.

To evaluate $B_{n,2}$ set

$$y = |x| - T_n, \quad h(t) = \log(e^t - 1), \quad u_n(x) = w_n(x) - h_n\xi(x, z_n).$$

Using (8.16), (8.33), and making the change of variables

$$u = z_n y, \quad v = \log(e^u + 1),$$

we get

$$\begin{aligned} n\int_{\Delta_n(t)} u_n(x)\,d\Phi(x) &\sim 2n\int_{-\delta_n}^{h(t)/z_n} (\log(1+e^{z_n y}) - e^{z_n y})\,d\Phi(T_n + y) \\ &\sim c\tau\int_{-\infty}^{h(t)} (\log(1+e^u) - e^u)e^{-\tau u}\,du \\ &= c\tau\int_0^t (v - e^v + 1)(e^v - 1)^{-\tau-1}e^v\,dv. \end{aligned}$$

Using (8.30) and making the change of variables $z = e^v - 1$ we can rewrite the last integral in the form

$$\begin{aligned} & c\tau\int_0^\infty (v - e^v + 1)(e^v-1)^{-\tau-1}e^v\,dv - \int_t^\infty v\,dL^0(v) + c\tau\int_{(e^t-1)}^\infty z^{-\tau}\,dz \\ &= c\tau\int_0^\infty (v - e^v + 1)(e^v-1)^{-\tau-1}e^v\,dv - \int_t^\infty v\,dL^0(v) + \frac{c\tau}{\tau-1}(e^t-1)^{1-\tau}. \end{aligned}$$

Therefore we get

$$\begin{aligned} \gamma_0 &= c\tau\int_0^\infty (x - e^x + 1)(e^x-1)^{-\tau-1}e^x\,dx - \int_0^\infty \frac{x^3}{1+x^2}\,dL^0(x) \\ &= \int_0^\infty (1 + \frac{x}{1+x^2} - e^x)\,dL^0(x). \end{aligned} \tag{8.34}$$

Thus $l_n \to \zeta_0$ under the null-hypothesis where ζ_0 is a random variable of infinite divisible type with the Levi spectral function (8.30), with the constant (8.34), and without the Gaussian component. We leave it for the reader to verify that

$$E_{n,0}(l_n) \to E(\zeta_0) = \int_0^\infty (x - e^x + 1)\, dL^0(x), \tag{8.35}$$

$$\mathrm{Var}_{n,0}(l_n) \to \mathrm{Var}(\zeta_0) = \int_0^\infty x^2\, dL^0(x); \quad E(\exp(\zeta_0)) = 1. \tag{8.36}$$

The last equality means that ζ_0 satisfies the log-likelihood ratio property.

Also using the evaluations above one gets: if $c_n \to 0$, then $l_n \to 0$ under null-hypothesis.

Let us consider the limit distribution under the Bayesian alternative. Let $F_{n,i}$ be a distribution function of $w_n(X_i)$ under the Bayesian alternative. For all $i = 1, \dots, n$ this is of the form

$$F_{n,i}(t) = P_{\pi_n}(w_n(x) < t) = P_{\pi_n}(|x| < T_n(t)) = \Phi(T_n(t)) + \Delta_n\Phi(T_n(t)),$$

where the measure P_{π_n} is defined by (8.11) and $\Delta_n\Phi$ is defined by (8.12). We get

$$\sum_{i=1}^n (1 - F_{n,i}(t)) = 2n(\Phi(-T_n(t)) + \Delta_n\Phi(-T_n(t)).$$

Analogously to the evaluations above we get

$$\begin{aligned} 2n\Delta_n\Phi(-T_n(t)) &\sim nh_n\Phi(z_n - T_n(t)) \sim nh_n\Phi(z_n - T_n)(e^t - 1)^{1-\tau_n} \\ &\sim \frac{c_n\tau_n}{\tau_n - 1}(e^t - 1)^{1-\tau_n} \to \frac{c\tau}{\tau - 1}(e^t - 1)^{1-\tau}. \end{aligned}$$

Thus we get the relation for the Levi spectral function under the Bayesian alternative

$$L^1(t) = L^0(t) + L^\Delta(t); \quad L^\Delta(t) = -\frac{c\tau}{\tau - 1}(e^t - 1)^{1-\tau}. \tag{8.37}$$

Note that the function $L^\Delta(t)$ is a Levi spectral function as well. This yields that the limit distribution of l_n is the distribution of the sum $\zeta_1 = \zeta_0 + \zeta_\Delta$, where ζ_0, ζ_Δ are independent and infinitely divisible random variables with the Levi spectral functions L^0 and L^Δ.

Let us show that the limit distribution has no Gaussian component. It suffices to verify that

$$\lim_{t\to 0}\lim_{n\to\infty} n \int_{|w_n|<t} w_n^2(x)\, d\Delta_n\Phi(x) = 0. \tag{8.38}$$

Since $w_n^2(x) \asymp h_n^2\xi^2(x, z_n)$ we have, from (8.12),

$$\begin{aligned} n\int_{|w_n|<t} w_n^2(x) d\Delta_n\Phi(x) &= n\int_{|w_n|<t} w_n^2(x) h_n\xi(x, z_n)\, d\Phi(x) \\ &\le Btnh_n^2 \int_{|x|<T_n(t)} \xi^2(x, z_n)\, d\Phi(x), \end{aligned}$$

and (8.38) follows from the analogous evaluations for the null-hypothesis. Let us calculate γ^1. Using (8.12) we get $\gamma^1 = \gamma^0 + \gamma^\Delta$ where

$$\gamma^\Delta = \lim_{n\to\infty} n \int_{|w_n|<t} w_n(x) d\Delta_n \Phi(x) - \int_0^t \frac{x^3}{1+x^2} dL^\Delta(x) + \int_t^\infty \frac{x}{1+x^2} dL^\Delta(x) = \lim_{n\to\infty} nh_n \int_{|x|<T_n(t)} w_n(x)\xi(x,z_n)\, d\Phi(x) + \int_0^\infty \frac{x}{1+x^2} dL^\Delta(x) - \int_0^t x\, dL^\Delta(x)$$

and the integrals are finite.

Analogously to the calculations for γ^0 one can show that

$$\lim_{n\to\infty} h_n n \int_{|x|<T_n(t)} w_n(x)\xi(x,z_n)\, d\Phi(x) = \int_0^t x\, dL^\Delta(x).$$

We leave it for the reader to repeat the calculations: the integrals over the set $|x| < T_n - \delta_n$ tend to 0; the integrals over the set $\Delta_n(t)$ are studied following the same scheme as item $B_{n,2}$ above. This yields

$$\gamma^\Delta = \int_0^\infty \frac{x}{1+x^2} dL^\Delta(x). \tag{8.39}$$

It follows from (8.5) that the random variable ζ_Δ is positively supported, i.e., $P(\zeta_\Delta \geq 0) = 1$. Note also that the random variables ζ_0, ζ_Δ have finite absolute moments of any degree and, by (8.6), these have absolutely continuous distribution functions.

Denote by $F^0_{c,\tau}(t)$, $F^1_{c,\tau}(t)$ the distribution functions of the random variables ζ_0 and $\zeta_1 = \zeta_0 + \zeta_\Delta$,

$$F^0_{c,\tau}(t) = P(\zeta_0 < t), \quad F^1_{c,\tau}(t) = P(\zeta_1 < t), \tag{8.40}$$

where ζ_0 and ζ_Δ are independent infinitely divisible random variables with the Levi spectral functions L^0, L^Δ determined by (8.30), (8.37), with the constants γ^0, γ^Δ determined by (8.34), (8.39), and with no Gaussian components.

Proposition 8.2 *Let $\tau_n \to \tau \in (1,2)$.*

(1) If $c_n \to 0$, then $\beta(\alpha,\pi^n) \to 1-\alpha$ for any $\alpha \in (0,1)$.

(2) If $c_n \to \infty$, then $\beta(\alpha,\pi^n) \to 0$ for any $\alpha \in (0,1)$.

(3) Let $c_n \to c \in (0,\infty)$. Then $l_n \to \zeta_0$ under the null-hypothesis and $l_n \to \zeta_1 = \zeta_0 + \zeta_\Delta$ under the Bayesian alternative. Therefore

$$\beta(\alpha,\pi^n) = F^1_{c,\tau}(T^0_\alpha(c,\tau)) + o(1),$$

where $T^0_\alpha(c,\tau)$ is a $(1-\alpha)$-quantile of the distribution function $F^0_{c,\tau}(t)$, i.e.,

$$F^0_{c,\tau}(T^0_\alpha(c,\tau)) = P(\zeta_0 < T^0_\alpha(c,\tau)) = 1-\alpha.$$

Proof. The evaluations above show $l_n \to 0$ in $P_{n,0}$-probability as $c_n \to 0$ which yields statement (1). Statement (3) follows from the Neyman–Pearson Lemma 2.1. Therefore we need to prove statement (2) only.

Let us consider the tests

$$\psi_{n,\alpha} = \mathbf{1}_{\mathcal{X}_{n,\alpha}}, \quad \mathcal{X}_{n,\alpha} = \left\{ x \in R^n : \max_{1 \le i \le n} |x_i| > H_{n,\alpha} \right\}, \tag{8.41}$$

where $H_n = H_{n,\alpha}$ is a sequence such that $(1 - 2\Phi(-H_{n,\alpha}))^n = 1 - \alpha$ which yields $n\Phi(-H_n) \asymp 1$ and (compare with Section 3.1.4)

$$\alpha(\psi_{n,\alpha}) = P_{n,0}(\mathcal{X}_{n,\alpha}) = \alpha.$$

It suffices to show that

$$\beta(\psi_{n,\alpha}, P_{\pi^n}) \to 0, \quad \text{if } c_n \to \infty.$$

Using (8.11), (8.2) we have

$$\begin{aligned}
\beta_n &= \beta(\psi_{n,\alpha}, P_{\pi^n}) = \prod_{i=1}^{n} P_{\pi_n}(|X_i| \le H_n) \\
&= (1 - 2\Phi(-H_n) + h_n(-\Phi(z_n - H_n) - \Phi(-z_n - H_n) + 2\Phi(-H_n)))^n \\
&= (1 - \alpha) \exp(-nh_n\Phi(z_n - H_n)(1 + o(1))),
\end{aligned}$$

which yields $\beta_n \to 0$, if

$$\lambda_n^* = nh_n\Phi(z_n - H_n) \to \infty. \tag{8.42}$$

However, since $c_n \asymp \lambda_n \to \infty$ by (8.24), the relation (8.42) follows from Lemma 8.1. □

Remark. Proposition 8.2 yields the following distinguishability conditions for $\tau \in (1, 2)$: $\beta(\psi_{n,\alpha}, P_{\pi^n}) \to 0$ iff $c_n \to \infty$. On the other hand, it follows from (8.23) that $c_n \to \infty$ iff $u_n \to \infty$ in this case. Therefore we can extend the distinguishability condition: $\beta(\phi_{n,\alpha}, P_{\pi^n}) \to 0$ iff $u_n \to \infty$ for $\tau \ge 2$ from Proposition 8.1 to the case $\tau > 1$.

8.1.5 Limit Distributions for $\tau \le 1$

Let $\lambda_n \to \lambda \in (0, \infty)$. We would like to show that $l_n \to -\lambda$ under the null-hypothesis. This means that the likelihood ratio is asymptotically constant

$$L_n = \exp(l_n) \to \exp(-\lambda) \quad \text{in } P_{n,0}\text{-probability}$$

which yields (compare with Section 4.4.1) that

$$\beta(\alpha, \pi^n) \to (1 - \alpha)\exp(-\lambda), \quad \gamma(P_{n,0}, P_{\pi^n}) = \gamma(\pi^n) \to \exp(-\lambda). \tag{8.43}$$

To verify the relation $l_n \to -\lambda$ note that $c_n = o(\lambda_n) \to 0$ by (8.24) and, analogously to (8.30), for any $t > 0$,

$$nP_0(|w_n(X)| \ge t) = 2n\Phi(-T_n(t)) \sim c_n(e^t - 1)^{-\tau_n} \to 0. \tag{8.44}$$

Consider the truncated items $\hat{w}_{n,i} = \hat{w}_n(x_i)$ and the truncated statistics $\hat{l}_n$:

$$\hat{w}_{n,i} = w_{n,i}\mathbf{1}_{\{|w_{n,i}|<t\}}, \quad \hat{l}_n = \sum_{i=1}^{n} \hat{w}_n.$$

It follows from (8.44) that $P_{n,0}(\hat{l}_n \neq l_n) \to 0$ and it suffices to verify that

$$E_{n,0}(\hat{l}_n) \to -\lambda, \quad \mathrm{Var}_{n,0}(\hat{l}_n) \to 0. \tag{8.45}$$

Since

$$w_n(x) = h_n\xi(x, z_n) + O(h_n^2\xi^2(x, z_n)), \quad w_n^2(x) = O(h_n^2\xi^2(x, z_n)),$$

we get

$$\begin{aligned} E_{n,0}(\hat{l}_n) &= nh_n \int_{\Delta_n(t)} \xi(x, z_n)d\Phi(x) + O\left(nh_n^2 \int_{\Delta_n(t)} \xi^2(x, z_n)d\Phi(x)\right) \\ &= A_n + O(B_n), \\ \mathrm{Var}_{n,0}(\hat{l}_n) &= O\left(nh_n^2 \int_{\Delta_n(t)} \xi^2(x, z_n)d\Phi(x)\right) = O(B_n), \end{aligned}$$

where $\Delta_n(t) = \{|x| < T_n(t)\}$. By (8.27), (8.28) we have

$$A_n \sim -nh_n\Phi(z_n - T_n(t)), \quad B_n \asymp nh_n^2 \exp(z_n^2)\Phi(T_n(t) - 2z_n).$$

Denote $d_n = T_n - z_n$. If $d_n \le B$ for some $B < \infty$, then using (8.18) we have

$$T_n(t) - z_n = d_n + o(1), \quad \Phi(z_n - T_n(t)) \sim \Phi(z_n - T_n),$$

which yields

$$A_n \sim -nh_n\Phi(-d_n) = -\lambda_n.$$

Let $d_n \to \infty$, which is possible for $\tau_n \to 1$ and $d_n = o(z_n)$ in this case. Using (8.18), (8.19) we get

$$\Phi(z_n - T_n(t)) \sim \Phi(-d_n)(e^t - 1)^{-d_n/z_n} \sim \Phi(-d_n)$$

which yields $A_n \sim -\lambda_n$ as well. Analogously,

$$\Phi(T_n(t) - 2z_n) \sim \Phi(T_n - 2z_n)(e^t - 1)^{2-\tau} \asymp \Phi(T_n - 2z_n)$$

which yields $B_n \asymp u_n^2 \asymp c_n = o(1)$ by (8.23), (8.24). Thus we get (8.45).

Analogous calculations show that $l_n \to 0$ as $\lambda_n \to 0$. Let $\lambda_n \to \infty$. By Lemma 8.1 this yields $\lambda_n^* \to \infty$. It was shown in the proof of statement (2) of Proposition 8.2 that $\beta(\psi_{n,\alpha}, P_{\pi^n}) \to 0$ for the tests (8.41) in this case. Thus we get

Proposition 8.3 *Let $\tau_n \to \tau \le 1$, $\lambda_n \to \lambda \in [0, \infty]$. Then relations* (8.43) *hold.*

8.1.6 Graphic Representation

To interpret the result above it is convenient to consider the dependence of the distinguishability conditions on variables $z_n \to \infty$, $h_n \to 0$. Introduce the variables

$$x_n = \frac{z_n^2}{2\log n}; \quad y_n = \frac{\log nh_n}{\log n}, \quad y_n < 1. \tag{8.46}$$

Using the variables x_n, y_n and relation (8.21) we can write

$$z_n^2 = 2x_n \log n, \quad \log h_n = (y_n - 1)\log n, \quad \tau_n = \frac{1}{2}\left(1 + \frac{1-y_n}{x_n}\right) + o(1). \tag{8.47}$$

Using (8.14), (8.22)–(8.24), and (8.47) one has the following relations. If $\tau \in [2,\infty)$ and $u_n \asymp 1$, then

$$x_n \sim x = \frac{1}{4(\tau-1)}, \quad y_n \sim y = \frac{2\tau-3}{4(\tau-1)}, \tag{8.48}$$

$$x + y = 1/2; \quad 0 < x \le \frac{1}{4}, \quad \frac{1}{4} \le y < \frac{1}{2}. \tag{8.49}$$

If $\tau \in (1,2)$ and $c_n \to c \asymp 1$, then

$$x_n \sim x = \tau^{-2}, \quad y_n \sim y = (1-\tau^{-1})^2; \tag{8.50}$$

$$x^{1/2} + y^{1/2} = 1, \quad 1/4 < x < 1, \quad 0 < y < 1/4. \tag{8.51}$$

At last, if $\tau \in (1/2, 1]$, $\lambda_n \to \lambda \in (0,\infty)$, then one has

$$x_n \sim x = \frac{1}{2\tau-1}, \quad y_n \sim y \equiv 0; \quad x \ge 1. \tag{8.52}$$

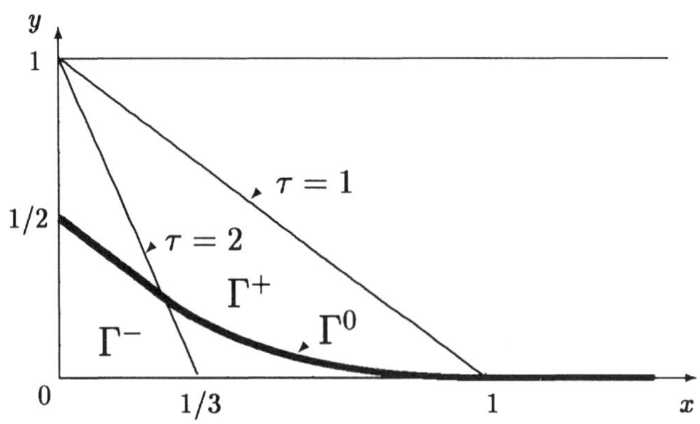

Fig. 9.

Relations (8.49), (8.51), (8.52) determine the three-piece curve

$$\Gamma^0 = \{y = f(x)\}$$

which divides the set $\{x > 0, 0 < y \leq 1\}$ into the regions

$$\Gamma^+ = \{y > f(x)\}, \quad \Gamma^- = \{y < f(x)\}.$$

In Figure 9 the partition and the lines $\tau = 1$ and $\tau = 2$ are presented. Region Γ^+ corresponds to the distinguishability: $\beta(\alpha, \pi^n) \to 0$, and region Γ^- corresponds to the nondistinguishability: $\beta(\alpha, \pi^n) \to 1 - \alpha$. Curve Γ^0 contains the cases when $\beta(\alpha, \pi^n)$ is bounded away from 0 and from $1 - \alpha$. Each point of this curve corresponds to the family of the Gaussian distributions (first piece), of the infinite divisible distributions with parameter $\tau = x^{-1/2}$ (second piece), and of the degenerate distributions (third piece).

8.2 Multichannel Signal Detection Problems

The Bayesian problem in Section 8.1 is of independent interest. The product prior π^n corresponds to a random vector of the form $v_t = (z_n t_1, \ldots, z_n t_n)$ where t_i are i.i.d. random variables with the possible values 0, ± 1, and

$$\pi_n(t_i = 0) = 1 - h_n, \quad \pi_n(t_i = \pm 1) = h_n/2.$$

One can interpret this as an output of the n-channels signal detection system which can receive (or not receive) a signal $\pm z_n$ from each channel in the Gaussian noise ξ_i background, and the number of channels which contain a signal is a random binomial $B_n(h_n)$. Therefore we can call these problems, *multichannel signal detection problems with a random number of signals*.

It is clear that the average number of signals is $k_n = nh_n$. Let us consider an analogous *multichannel signal detection problems with a nonrandom number of signals* $k_n = k$, $1 \leq k \leq n$. This corresponds to the minimax signal detection problem under the n-dimensional Gaussian model $X = v + \xi$, $v \in R^n$, with null-hypothesis $H_0 : v = 0$ and alternative $H_1 : v \in V_n = V_n(z, k)$; the set $V_n(z, k)$ consists of $M = 2^k \binom{n}{k}$ vectors of the type

$$v = (v_1, \ldots, v_n) : v_i \in \{0, \pm z\}, \quad \sum_{i=1}^{n} |v_i| = zk. \tag{8.53}$$

Clearly, the set $V_n(z, k)$ is invariant under group G which consists of orthogonal transforms corresponding to all the permutations and changes of signs of coordinates; moreover, the set $V_n(z, k)$ is an orbit of any point $v_0 \in V_n(z, k)$ under action of the group G, i.e., $V_n(z, k) = G(v_0)$.

The least favorable prior is the uniform discrete measure on $V_n(z, k)$:

$$\pi_{k,z}^n = 2^{-k} \binom{n}{k}^{-1} \sum_{v \in V_n(z,k)} \delta_v, \tag{8.54}$$

where δ_v is the Dirac mass at the point $v \in R^n$. In fact, this prior is G-invariant, the Bayesian tests ψ_α have G-invariant admissible sets $\mathcal{X}_\alpha$ (see Propositions 2.4, 2.6), and the $\beta(\psi_\alpha, v) = P_v(\mathcal{X}_\alpha)$ does not depend on $v \in V_n(z,k)$ and equals $\beta(\psi_\alpha, V_n(z,k))$. Therefore we can use Theorem 2.3.

For the case $k = 1$ this Bayesian problem has been studied in Sections 3.3.7, 4.4.1 and we have asymptotics of degenerate type. For $k = n$ these problems correspond to a product prior with symmetric two-point factors with $h_{n,i} = 1$, and was studied in Sections 3.3.5, 4.3.2.

8.2.1 Reduction to the Bayesian Problem

It is difficult to study the asymptotics for $1 < k_n < n$ directly. However we can use the results for the Bayesian problem in Section 8.1. Denote $\beta_n(\alpha, k, z) = \beta(\alpha, V_n(z,k))$ and let $\beta(\alpha, \pi^n) = \beta(\alpha, P_{n,0}, P_{\pi^n})$ be the Bayesian minimum type II errors that were studied in Section 8.1.

Theorem 8.1

(1) *Let* $k_n = nh_n \to \infty$, $z_n \to \infty$. *Then* $\beta_n(\alpha, k_n, z_n) = \beta(\alpha, \pi^n) + o(1)$.

(2) *Let* $k_n = O(1)$. *Then*

$$\beta_n(\alpha, k_n, z_n) = (1-\alpha)(\Phi(\sqrt{2\log n} - z_n))^{k_n} + o(1).$$

(3) *Let us consider the test* $\psi_{n,\alpha}$ *determined by* (8.41).

(3a) *If* $k_n = O(1)$, *then*

$$\beta_n(\psi_{n,\alpha}, V_n(z_n, k_n)) \le (1-\alpha)(\Phi(\sqrt{2\log n} - z_n))^{k_n} + o(1).$$

(3b) *If* $k_n = nh_n \to \infty$ *and* $\tau < \infty$, *then*

$$\beta_n(\psi_{n,\alpha}, V_n(z_n, k_n)) \le (1-\alpha)\exp(-\lambda_n^*) + o(1).$$

(3c) *If* $k_n = nh_n \to \infty$, $\tau \in (1,\infty)$, *and* $c_n = 2n\Phi(-T_n) \to \infty$, *then*

$$\beta_n(\psi_{n,\alpha}, V_n(z_n, k_n)) \to 0.$$

Theorem 8.1 shows that Propositions 8.1–8.3 in Section 8.1 which deal with the Bayesian alternatives are carried over to the case of minimax problem with $V_n = V_n(z_n, k_n)$ with the same z_n and with $k_n = nh_n \to \infty$. Relations (8.46), (8.50), and (8.52) hold with the change of nh_n to k_n, of the interval $(-\infty, 1)$ of y_n to $(0,1)$ because $k_n \ge 1$, and of the interval $(0,\infty)$ of x_n to $(0,1)$ because $\beta_n(\alpha, V_n) \to 0$ for $x_n \ge 1$.

By the Remark in Section 8.1.4 for $\tau > 1$ we get the distinguishability conditions: $\beta_n(\alpha, V_n) \to 0$ iff $u_n \to \infty$; here u_n is determined by (8.22) for $z_n \to \infty$, or by (5.146) for $z_n = O(1)$ (the last follows from the results of Section 5.5.2). Let $\tau = 1$. Then we can present distinguishability conditions in the form: $\beta_n(\alpha, V_n) \to 0$ iff $\lambda_n^* \to \infty$; here $\lambda_n^* = k_n\Phi(z_n - H_n)$, H_n is any sequence such that $n\Phi(-H_n) \asymp 1$. If $k_n = nh_n \to \infty$, then this follows from Lemma 8.1; if $k_n = O(1)$, then this follows from $H_n = \sqrt{2\log n} + o(1)$.

Theorem 8.1 implies that the tests $\psi_{n,\alpha}$ are asymptotically minimax for any $k_n = O(1)$ or $k_n \to \infty$, $\tau \le 1$. If $\tau \in (1,2)$ and $c_n \to \infty$, then these tests provide minimax distinguishability (recall that $c_n \to \infty$ is a necessary condition for minimax distinguishability according to Proposition 8.2).

If $k_n = nh_n \asymp 1$, then the Bayesian and minimax alternatives are essentially different. In fact, $\beta_n(\alpha, V_n(z_n, k_n)) \to 0$ as $z_n - \sqrt{2\log n} \to \infty$ for any $k_n \ge 1$. On the other hand, it follows from Proposition 8.3, for the Bayesian alternatives, that if $\tau < 1$ and $nh_n \to \lambda > 0$, then $\lambda_n \to \lambda$ and $\beta(\alpha; P_{n,0}, P_{\pi^n}) \to (1-\alpha)\exp(-\lambda)$, and that it is separated away from 0 for arbitrary large z_n.

The reason for this is the following. Recall that the prior π^n is supported on the collection V_{n,z_n} of vectors with coordinates $t_i z_n$, $t_i \in \{0, \pm 1\}$. If $z_n - \sqrt{\log 2n} \to \infty$, then tests (8.41) provide the second type error probabilities tending to 0 for all alternatives $v \in V_{n,z_n}$, $v \ne 0$ (see Corollary 3.11). Therefore the second type error probabilities for the Bayesian alternatives are specified by the probability $\pi^n(v = 0) = (1-h_n)^n$ which is the same as π^n-probabilities of the events

$$\Im_{n,0} = \left\{ \sum_{i=1}^{n} |t_i| = 0 \right\},$$

where $|t_i|$ are i.i.d. Bernoulli, $\pi_n(|t_i| = 1) = h_n$, $\pi_n(|t_i| = 0) = 1 - h_n$. If $nh_n \to \lambda$, then the random variables $\sum_{i=1}^n |t_i|$ tend to the Poisson random variable with parameter λ in π^n-probability, which implies the relation

$$\pi^n(\Im_{n,0}) \to \exp(-\lambda).$$

8.2.2 Proof of Theorem 8.1

Proof of Statement (1)

Observe the following *asymptotic continuity properties* for the Bayesian alternatives (which follow from Propositions 8.1–8.3 for $z_n \to \infty$, $h_n \to 0$, and from the results of Section 5.5 in other cases). Let either $\tau > 1$ and

$$h_n^* \sim h_n, \quad z_n^* = z_n + o(z_n^{-1}),$$

or $\tau = 1$, $d_n = T_n - z_n$,

$$h_n^* \sim h_n, \quad z_n^* = \begin{cases} z_n + o(d_n^{-1}), & \text{if } d_n \to \infty, \\ z_n + o(1), & \text{if } d_n < B, \\ z_n + O(1), & \text{if } d_n \to -\infty, \end{cases}$$

for some $B > 0$. Then, for any $\alpha \in (0,1)$, one has

$$\beta(\alpha, \pi^n(z_n^*, h_n^*)) = \beta(\alpha, \pi^n(z_n, h_n)) + o(1).$$

Let $k_n \to \infty$, $h_n = k_n/n$, and let δ_n be a sequence such that $\delta_n \to 0$, $k_n\delta_n^2 \to \infty$. Put, for $r = 1, 2$,

$$k_{n,r}^- = \max\{0, [k_n(1 - r\delta_n)]\}, \quad k_{n,r}^+ = \min\{n, [k_n(1 + r\delta_n)]\},$$

where $[t]$ denotes the integer part of t. Let us consider the sets

$$V_n^+ = \{v \in V_{n,z_n} : k_n \le k_n(v) \le k_{n,2}^+\},$$

$$V_n^- = \{v \in V_{n,z_n} : k_{n,2}^- \le k_n(v) \le k_n\}.$$

Here $k_n(v) = z_n^{-1} \sum_{i=1}^n |v_i|$ and

$$V_{n,z} = \{v = (v_1, \dots, v_n) \in R^n : v_i = zt_i, \ t_i \in \{-1, 0, +1\}\}.$$

Also put $h_n^+ = k_{n,1}^+/n$, $h_n^- = k_{n,1}^-/n$, and consider the product measures

$$\pi_+^n = \pi^n(z_n, h_n^+), \quad \pi_-^n = \pi^n(z_n, h_n^-)$$

and the conditional measures $\pi_{+,*}^n = \pi^+$, $\pi_{-,*}^n = \pi^-$ with respect to the sets V_n^+ and V_n^-:

$$\pi_{+,*}^n(A) = \pi_+^n(A \cap V_n^+)/\pi_+^n(V_n^+), \quad \pi_{-,*}^n(A) = \pi_-^n(A \cap V_n^-)/\pi_-^n(V_n^-).$$

The asymptotic continuity properties imply that

$$\beta(\alpha, \pi^n(z_n, h_n)) = \beta_n(\alpha, \pi_+^n) + o(1) = \beta(\alpha, \pi_-^n) + o(1). \tag{8.55}$$

Recall that the random variables $k_n = k_n(v)$ are the binomials $B_n(h_n^+)$ and $B_n(h_n^-)$ with respect to the measures π_+^n and π_-^n which are supported by V_{n,z_n}. Using the Chebyshev inequality one can easily check that

$$\pi_+^n(V_n^+) \to 1, \quad \pi_-^n(V_n^-) \to 1,$$

which imply the relations for the L_1-distance

$$|\pi_+^n - \pi^+|_1 \to 0, \quad |\pi_-^n - \pi^-|_1 \to 0$$

(compare with (2.49)) and

$$\beta(\alpha, \pi_+^n) = \beta(\alpha, \pi^+) + o(1), \quad \beta(\alpha, \pi_-^n) = \beta(\alpha, \pi^-) + o(1) \tag{8.56}$$

(this follows from inequality (2.32) and Proposition 2.1).

Let $\pi^0 = \pi_{z_n,k_n}^n$ be the uniform discrete measure (8.54) which is the least favorable prior for the alternative $V_n(z_n, k_n)$. This means that

$$\beta_n(\alpha, V_n(z_n, k_n)) = \beta(\alpha, \pi^0). \tag{8.57}$$

Let us show that

$$\beta(\alpha, \pi^+) \le \beta(\alpha, \pi^0) \le \beta(\alpha, \pi^-). \tag{8.58}$$

Relations (8.55)–(8.57) and inequalities (8.58) imply statement (1) of Theorem 8.1.

To prove inequalities (8.58) let us consider the admissible sets $X^0 = X^0_{n,\alpha}$ and $X^- = X^-_{n,\alpha}$ of the Bayesian tests $\psi_{n,\alpha}$ for the Bayesian alternatives P_{n,π^0} and P_{n,π^-},

$$X^0 = \left\{x \in R^n : \frac{dP_{n,\pi^0}}{dP_{n,0}}(x) < T^0\right\}, \quad X^- = \left\{x \in R^n : \frac{dP_{n,\pi^-}}{dP_{n,0}}(x) < T^-\right\},$$

where $T^0 = T^0_{n,\alpha}$ and $T^- = T^-_{n,\alpha}$ are the thresholds such that

$$P_{n,0}(X^0) = P_{n,0}(X^-) = 1 - \alpha$$

(recall that the Bayesian tests are nonrandomized almost surely according to Proposition 2.4). According to the optimality of the Bayesian tests one has the inequalities

$$\beta(\alpha, \pi^+) \leq \int_{V_n^+} P_{n,v}(X^0)\, \pi^+(dv), \tag{8.59}$$

$$\beta(\alpha, \pi^0) \leq \int_{V_n(z_n,k_n)} P_{n,v}(X^-)\, \pi^0(dv). \tag{8.60}$$

It follows from Proposition 2.6 that the sets X^0 and X^- are invariant with respect to all permutations and changes of signs of coordinates, and the probabilities $P_{n,v}(X^0)$, $P_{n,v}(X^-)$ depend on $k_n(v)$ for $v \in V_{n,z_n}$ only. In particular, one has

$$\beta(\alpha, \pi^0) = P_{n,v^0}(X^0), \quad P_{n,v}(X^-) = P_{n,v^0}(X^-), \tag{8.61}$$

for any $v \in V_n(z_n, k_n)$ where $v^0 = (\underbrace{b, \dots, b}_{k}, 0, \dots, 0)$ and $k = k_n$, $b = z_n$.

Let us show that for any $v_+ \in V_n^+$ and for any $v_- \in V_n^-$ the following inequalities hold:

$$P_{n,v_+}(X^0) \leq P_{n,v^0}(X^0), \quad P_{n,v^0}(X^-) \leq P_{n,v_-}(X^-). \tag{8.62}$$

It suffices to consider the case

$$v_+ = (\underbrace{b, \dots, b}_{k_+}, 0, \dots, 0), \quad v_- = (\underbrace{b, \dots, b}_{k_-}, 0, \dots, 0),$$

where $k_+ > k > k_-$. Recall that the sets X^0 and X^- are convex, because they are the admissible sets of Bayesian tests (see Proposition 2.5), and these sets are symmetric with respect to all permutations and changes of signs of coordinates (since the priors are symmetric; see Proposition 2.6). Applying Lemma 2.4 one obtains inequalities (8.62).

Using inequalities (8.62) and equalities (8.61) one has

$$\int_{V_n^+} P_{n,v}(X^0)\pi^+(dv) \leq \max_{v \in V_n^+} P_{n,v}(X^0) \leq P_{n,v^0}(X^0) = \beta(\alpha, \pi^0),$$

$$\int_{V_n(z_n,k_n)} P_{n,v_0}(X^-)\pi^0(dv_0) = P_{n,v^0}(X^-) \leq \min_{v \in V_n^-} P_{n,v}(X^-) \leq \beta(\alpha, \pi^-).$$

These relations and (8.59), (8.60) imply inequalities (8.58). □

Proof of Statement (2)

It suffices to consider the case $z_n = \sqrt{2\log n} + O(1)$, $k_n = k \geq 1$. The least favorable prior $\pi_n = \pi^n_{k,z_n}$ is the uniform discrete measure on the set $V_n(z_n, k)$ and the likelihood ratio $L_n = dP_{n,\pi_n}/dP_{n,0}$ is of the form

$$L_n = (C_n^k)^{-1} \sum_A \prod_{i \in A} (1 + \xi(x_i, z_n)) \tag{8.63}$$

where the sum is taken over all subsets $A \subset \{1, \dots, n\}$ such that $\#(A) = k$. Set

$$X_n = \{|x_i| < H_n,\ i = 1, \dots, n\}, \quad H_n = \sqrt{2\log n}.$$

It suffices to show that

$$L_n - (\Phi(H_n - z_n))^k \to 0 \tag{8.64}$$

in $P_{n,0}$-probability. Since $P_{n,0}(X_n) \to 1$, relation (8.64) is equivalent to

$$\hat{L}_n - (\Phi(H_n - z_n))^k \to 0; \quad \hat{L}_n = (C_n^k)^{-1} \sum_A \prod_{i \in A} (1 + \hat{\xi}(x_i, z_n)), \tag{8.65}$$

where $\hat{\xi}(x_i, z_n) = \xi(x_i, z_n) \mathbb{1}_{\{|x_i| < H_n\}}$. Relation (8.65) follows from

$$E_{n,0}(\hat{L}_n) = (\Phi(H_n - z_n))^k + o(1), \tag{8.66}$$

and

$$\mathrm{Var}_{n,0}(\hat{L}_n) = o(1). \tag{8.67}$$

Let us verify (8.66) and (8.67). Using the relation

$$E_{n,0}(\hat{\xi}(x_i, z_n)) \sim -\Phi(z_n - H_n) = \Phi(H_n - z_n) - 1$$

(see (A.11) in the Appendix, Section A.2.1), we get (8.66),

$$E_{n,0}(\hat{L}_n) = (\Phi(H_n - z_n) + o(1))^k.$$

To obtain (8.67), let us consider the presentations

$$E_{n,0}(\hat{L}_n^2) = \sum_{l=0}^{k} D_l, \quad D_l = (C_n^k)^{-2} \sum_{A,B,C} E_{n,0}(U(A, B, C)), \tag{8.68}$$

where the sum is taken over all disjoint subsets

$$A, B, C \subset \{1, \dots, n\}, \quad \#A = l, \quad \#B = \#C = k - l,$$

and

$$U(A, B, C) = \prod_{i \in A} (1 + \hat{\xi}(x_i, z_n))^2 \prod_{i \in B \cup C} (1 + \hat{\xi}(x_i, z_n)).$$

One easily gets $D_0 < (E_{n,0}(\hat{L}_n))^2$, which yields

$$\mathrm{Var}_{n,0}(\hat{L}_n) < \sum_{l=1}^{k} D_l.$$

To evaluate D_l for $l \geq 1$ we use the relation

$$E_{n,0}(\hat{\xi}^2(x_i, z_n)) \sim 2\sinh^2(z_n^2/2)\Phi(H_n - 2z_n) \sim \frac{1}{2}\Phi(H_n - 2z_n)\exp(z_n^2)$$

(see (A.12) in the Appendix, Section A.2.1). This yields that the items in the sum in (8.68) are bounded by

$$\left(B + \Phi(H_n - 2z_n)\exp(z_n^2)\right)^l. \tag{8.69}$$

Note that the quantities in (8.69) are of the order $\left(n/\sqrt{\log n}\right)^l$. In fact, by (8.19), one has

$$\Phi(H_n - 2z_n)\exp(z_n^2) \asymp \frac{\exp(-(z_n - H_n)^2 + H_n^2/2)}{2z_n - H_n} \asymp (n/\sqrt{\log n}).$$

The number of items in the sum (8.68) is of the order $n^{l+2(k-l)} = n^{2k-l}$ and by $C_n^k \asymp n^k$ one has

$$\sum_{l=1}^{k} D_l \asymp \sum_{l=1}^{k} (\log n)^{-l/2} = o(1),$$

which implies (8.67) and (8.64). □

Proof of Statements (3)

Let us calculate the type II error probability of the test (8.41) for any $v \in V_n(z_n, k_n)$ defined by (8.53). We have

$$\begin{aligned}\beta_n(\psi_{n,\alpha}, v) &= \prod_{i=1}^{n}(\Phi(H_{n,\alpha} - v_i) - \Phi(-H_{n,\alpha} - v_i)) \\ &= (1-\alpha)^{1-k_n/n}(\Phi(H_{n,\alpha} - z_n) - \Phi(-z_n - H_{n,\alpha}))^{k_n}. \quad (8.70)\end{aligned}$$

By $H_{n,\alpha} = \sqrt{2\log n} + o(1)$, this implies statement (3a).

Let $k_n = nh_n \to \infty$ and $\tau < \infty$. Then $k_n = o(n)$ and, by (8.70), we get

$$\beta_n(\psi_{n,\alpha}, v) \leq (1-\alpha)\exp(-\lambda_n^*) + o(1), \quad \lambda_n^* = k_n\Phi(z_n - H_{n,\alpha}).$$

This yields statement (3b). Note that if $\tau_n \to \tau \in (1, \infty)$, $c_n \to \infty$, then, by (8.24), we have

$$\lambda_n = k_n\Phi(z_n - T_n) \asymp c_n\tau_n/(\tau_n - 1) \to \infty.$$

Therefore statements (3c) follows from Lemma 8.1. □

8.3 Minimax Signal Detection for l^n-Balls

Return to the minimax hypothesis testing problem $H_0 : v = 0$ against the alternative $H_1 : v \in V_n$ for the set $V_n \subset R^n$ determined by (8.1). By the results of Section 5.5.2, henceforth in this section, we assume $\infty > p > q$, $z_n \to \infty$, which correspond to

$$nh_n z_n^p = \rho_n^p, \quad nh_n z_n^q = R_n^q. \tag{8.71}$$

This is the same as

$$z_n = (\rho_n^p / R_n^q)^{1/(p-q)}, \quad nh_n = (R_n/\rho_n)^{pq/(p-q)} \geq 1,$$

(the latter inequality follows from the nonempty conditions (5.136)).

8.3.1 Main Results

Consider the families T_n, τ_n, u_n, c_n, λ_n, λ_n^* which were determined in Section 8.1 and assume without loss of generality that $\tau_n \to \tau \in [1/2, \infty]$. We would like to show that in the problems under consideration we have the same results as for the multichannel signal detection problems in Section 8.2. Denote, as above, $\pi_n = \pi(z_n, h_n)$.

Theorem 8.2 Lower bounds. *Assume that $nh_n/\sqrt{\log n \log\log n} \to \infty$ or that $k_n = nh_n$ is an integer. If $k_n \to \infty$, then*

$$\beta_n(\alpha, V_n) \geq \beta(\alpha, \pi^n) + o(1),$$

where the asymptotics of $\beta(\alpha, \pi^n)$ are determined by Propositions 8.1–8.3. If $k_n = nh_n$ is an integer, then

$$\beta_n(\alpha, V_n) \geq \beta_n(\alpha, k_n, z_n) + o(1),$$

where the asymptotics of $\beta_n(\alpha, k_n, z_n)$ are determined by Theorem 8.1.

Proof of Theorem 8.2. This easily follows from Theorem 8.1 and will be given in Section 8.3.3.

Let us describe the upper bounds. First, let $\tau_n \to \tau \geq 2$. Let us consider the statistics

$$t_n = h_n \sum_{i=1}^{n} \xi(X_i, z_n),$$

and the test sequence

$$\psi_{n,\alpha} = \mathbf{1}_{\{t_n > u_n T_\alpha\} \cup \{\max_i |x_i| \geq Q_n\}}; \tag{8.72}$$

where a sequence Q_n is such that $n\Phi(-Q_n) \to 0$; this will be specified in the proof.

Theorem 8.3 *Let $\tau_n \to \tau \geq 2$. For the sequence of tests* (8.72) *one has*

$$\alpha(\psi_{n,\alpha}) = \alpha + o(1), \quad \beta_n(\psi_{n,\alpha}, V_n) \leq \Phi(T_\alpha - u_n) + o(1).$$

This means that the sequence of tests $\psi_{n,\alpha}$ is asymptotically minimax.

Proof of Theorem 8.3. This will be given in Section 8.4.1.

Let $\tau_n \to \tau \in (1,2)$. To describe the upper bounds consider the small changed quantities $\tilde{z}_n$, $\tilde{h}_n$. Consider the small decreased quantities $\tilde{\rho}_n = \rho_n(1 - z_n^{-3})$ and put $\tilde{z}_n = z_n(\tilde{\rho}_n, R_n)$, $\tilde{h}_n = h_n(\tilde{\rho}_n, R_n)$ which are determined by relations (8.71) by replacing ρ_n by $\tilde{\rho}_n$. It follows from the asymptotic continuity properties that

$$\tilde{c}_n = c_n(\tilde{h}_n, \tilde{z}_n) \sim c_n, \quad \tilde{\tau}_n = \tau_n(\tilde{h}_n, \tilde{z}_n) \sim \tau_n.$$

Let $l_n = l_n(\tilde{h}_n, \tilde{z}_n)$ be the log-likelihood statistics.

Theorem 8.4 *Let $\tau_n \to \tau \in (1,2)$. Consider the sequence of tests*

$$\psi_{n,\alpha} = \mathbf{1}_{\{l_n > T_{n,\alpha}\} \cup \{\max_i |x_i| > Q_n\}}, \tag{8.73}$$

where Q_n is a sequence (to be specified in the proof) such that $n\Phi(-Q_n) \to 0$ and that $T_{n,\alpha} = T_\alpha^0(c,\tau)$ is a $(1-\alpha)$-quantile of the distribution function $F_{c,\tau}^0$ defined by (8.40). *Then*

$$\alpha(\psi_{n,\alpha}) = \alpha + o(1), \quad \beta_n(\psi_{n,\alpha}, V_n) \leq \beta(\alpha, \pi^n) + o(1),$$

the quantities $\beta(\alpha, \pi^n)$ are determined by Proposition 8.2. This means that the sequence of tests $\psi_{n,\alpha}$ is asymptotically minimax.

Proof of Theorem 8.4. This will be given in Section 8.4.2.

Let us consider the case $\tau_n \to \tau \leq 1$. Then the results are of the following form:

Theorem 8.5 *Consider the sequence of supreme-tests* (8.41).

(1) *If $\tau_n \to \tau = 1$, $nh_n \to \infty$, then*

$$\beta_n(\psi_{n,\alpha}, V_n) \leq \beta(\alpha, \pi^n) + o(1) = (1-\alpha)\exp(-\lambda_n) + o(1).$$

(2) *If $\tau_n \to \tau = 1$, $nh_n = O(1)$, then*

$$\beta_n(\psi_{n,\alpha}, V_n) \leq (1-\alpha)(\Phi(\sqrt{2\log n} - z_n))^{nh_n} + o(1).$$

(3) *Let one of the following assumptions hold:*
(i) $\tau < 1$; (ii) $\tau = 1$ *and* $\lambda_n \to \infty$; (iii) $\tau \in (1, \infty)$ *and* $c_n \to \infty$.
Then $\beta_n(\psi_{n,\alpha}, V_n) \to 0$ *for any* $\alpha \in (0,1)$.

Proof of Theorem 8.5. This will be given in Section 8.4.3.

Note that statements (1) and (2) of Theorem 8.5 mean that for $\tau = 1$ the tests (8.41) are asymptotically minimax, if $nh_n/\sqrt{\log n \log\log n} \to \infty$ or if nh_n is an integer. Statement (3) of Theorems 8.5, 8.2 means that, for $\tau < 2$, if there exists a consistent sequence of tests, then the tests (8.41) are consistent.

8.3.2 Examples

Let us return to the examples with $R_n = An^a$ which were considered in Section 5.5.2 for the cases $p \le q$ and $p > q$, $a \ge 1/2q$. Let us consider the cases $p > q$, $0 < a < 1/2q$. We would like to select ρ_n such that $\beta_n(\alpha, V_n)$ is bounded away from 0 and from $1-\alpha$. It was noted in Section 5.5.2 that $z_n \to \infty$ in this case.

Take a sequence $C_n \asymp 1$, and put

$$\rho_n = A^{q/p}(C_n \log n)^{1/2-q/2p} n^{aq/p}; \tag{8.74}$$

$$D_n = D(C_n, aq) = \left(\frac{C_n^{1/2}}{2} + \frac{1-aq}{C_n^{1/2}}\right). \tag{8.75}$$

Using (8.71) we get $z_n^2 = C_n \log n$,

$$h_n = n^{aq-1}A^q(C_n \log n)^{-q/2}, \quad \tau_n = \frac{1}{2} + \frac{1-aq}{C_n} + o(1), \tag{8.76}$$

$$T_n = D_n\sqrt{\log n} + \frac{q(\log\log n + \log(C_n/A^2)) + \log 4}{2\sqrt{C_n \log n}} + o\left(\frac{1}{\sqrt{\log n}}\right). \tag{8.77}$$

Example 8.1 Let $1/4q < a < 1/2q$. Put

$$C = C(aq) = 1 - 2aq, \quad C_n = C + q\frac{\log\log n}{\log n} + \frac{H_n}{\log n}.$$

Using (8.76), (8.77) we get $\tau \in (2, \infty)$, and if $H_n = O(1)$, then we get the Gaussian asymptotics with

$$2u_n^2 \sim A^{2q}C^{-q}\exp(H_n);$$

if $H_n \to -\infty$, then $\beta_n(\alpha, V_n) \to 1-\alpha$, and if $H_n \to \infty$, then $\beta_n(\alpha, V_n) \to 0$.

Example 8.2 Let $a = a_n \to 1/4q$; namely, set $a_n = 1/4q + d/\sqrt{\log n}$ where $-\infty < d < \infty$. Put $C_n \asymp 1$,

$$R_n = An^{1/4q+d/\sqrt{\log n}}, \quad \rho_n = A^{q/p}(C_n \log n)^{1/2-q/2p} n^{1/4p+dq/p\sqrt{\log n}}. \tag{8.78}$$

Using (8.71) we get $z_n^2 = C_n \log n$,

$$h_n = n^{-3/4+dq/\sqrt{\log n}}A^q(C_n \log n)^{-q/2}, \quad \tau_n = \frac{1}{2} + \frac{3}{4C_n} + o(1), \tag{8.79}$$

$$T_n = D_n\sqrt{\log n} - \frac{dq}{\sqrt{C_n}} + o(1), \quad D_n = D(C_n, 1/4). \tag{8.80}$$

Put

$$C_n = \frac{1}{2} - \frac{2dq}{\sqrt{\log n}} + q\frac{\log\log n}{\log n} + \frac{H_n}{\log n}.$$

Using (8.79), (8.80) we get $\tau = 2$, and if $H_n = O(1)$, then we get the Gaussian asymptotics with

$$u_n^2 \sim 2^{q-1} A^{2q} \exp(H_n)\Phi(2\sqrt{2}dq).$$

In particular, if $H_n \to -\infty$, then $\beta_n(\alpha, V_n) \to 1-\alpha$, and if $H_n \to \infty$, then $\beta_n(\alpha, V_n) \to 0$.

Example 8.3 Let $0 < a < 1/4q$. Put

$$t = (aq)^{1/2}, \quad C = C(t) = 2(1-t)^2,$$

(this corresponds to $D(C, aq) = \sqrt{2}$ in (8.75)). Let

$$C_n = C + s\frac{\log\log n}{\log n} + \frac{H_n}{\log n}, \qquad s = \frac{(1-t)^2 + q(1-t)}{t}.$$

If $H_n = O(1)$, then using (8.76), (8.77) we get $\tau = (1-t)^{-1} \in (1,2)$ and

$$c_n \sim \frac{1}{\sqrt{\pi}}(A\tau)^{q\tau} 2^{-\tau(1+q/2)} \exp(H_n\tau(\tau-1)/2),$$

if $H_n \to -\infty$, then $\beta_n(\alpha, V_n) \to 1-\alpha$, and if $H_n \to \infty$, then $\beta_n(\alpha, V_n) \to 0$.

Example 8.4 Consider the case of small $a = d/\log n$. Let $R_n = A(\log n)^d$. Put

$$\rho_n = A^{q/p} C_n^{1/2-q/2p} (\log n)^{1/2-q/2p+dq/p}.$$

Similarly to (8.76), (8.77) we get $z_n^2 = C_n \log n$,

$$h_n = \frac{1}{n}(\log n)^{q(d-1/2)}(A^2/C_n)^{q/2}, \quad \tau_n = \frac{1}{2} + \frac{1}{C_n} + o(1), \tag{8.81}$$

$$T_n = \left(\frac{\sqrt{C_n}}{2} + \frac{1}{\sqrt{C_n}}\right)\sqrt{\log n} + O\left(\frac{\log\log n}{\sqrt{\log n}}\right). \tag{8.82}$$

Assume $d > 1/2 + 1/2q$ (this implies that $nh_n/\sqrt{\log n \log\log n} \to \infty$). Put

$$C_n = 2 - 2s\sqrt{\frac{\log\log n}{\log n}} + \frac{2\log\log\log n}{s\sqrt{\log n \log\log n}} + \frac{H_n}{\sqrt{\log n \log\log n}},$$

where $s = \sqrt{2q(2d-1)}$. Let $H_n = O(1)$. Using (8.81), (8.82) we get $\tau = 1$,

$$\beta_n(\alpha, V_n) = (1-\alpha)e^{-\lambda_n} + o(1), \quad \lambda_n \sim \frac{A^q}{s\sqrt{\pi}2^{q/2}} \exp(sH_n/4).$$

If $H_n \to -\infty$, then $\beta_n(\alpha, V_n) \to 1-\alpha$, and if $H_n \to \infty$, then $\beta_n(\alpha, V_n) \to 0$.

8.3.3 Proof of Lower Bounds

To prove Theorem 8.2 we can assume $u_n = O(1)$, $c_n = O(1)$, or $\lambda_n = O(1)$ which imply $z_n^2 = O(\log n)$ since the lower bounds are of the form $\beta_n(\alpha, V_n) \geq o(1)$ when $u_n \to \infty$, $c_n \to \infty$, or $\lambda_n \to \infty$.

Moreover, we can assume $u_n \asymp 1$, $c_n \asymp 1$, $\lambda_n \asymp 1$, respectively. In fact, assume one of these sequences is $o(1)$. Put

$$V_n(t) = V_n^{p,q}(t\rho_n, tR_n), \quad t > 1,$$

and note that by Lemma 2.4 we have the inequality

$$\beta_n(\alpha, V_n) \geq \beta_n(\alpha, V_n(t)).$$

This inequality and (8.71) imply that we can increase z_n without a change of h_n and without a decrease of $\beta_n(\alpha, V_n)$. One can easily check that if z_n increases and h_n is fixed, then the values λ_n, c_n, u_n increase, τ_n decreases, and, by making z_n larger, we can get the values λ_n, c_n, u_n arbitrarily small but bounded away from 0, $\tau_n \to \tau \geq 1$ (we omit these calculations; see Figure 9 in Section 8.1.6).

Let $k_n = nh_n$ be an integer. Consider the sets $\tilde{V}_n = V_n(z_n, k_n)$ from Section 8.2:

$$\tilde{V}_n = \{v \in R^n : v = z_n(t_1, \ldots, t_n),\ t_i \in \{-1, 0, 1\},\ \sum_{i=1}^n |t_i| = k_n\}.$$

It is clear that $\tilde{V}_n \subset V_n$, which implies the inequality

$$\beta_n(\alpha, V_n) \geq \beta_n(\alpha, \tilde{V}_n). \tag{8.83}$$

It was shown in Theorem 8.1 that if $k_n \to \infty$, then

$$\beta_n(\alpha, \tilde{V}_n) = \beta(\alpha, \pi^n) + o(1), \tag{8.84}$$

and if $k_n = O(1)$, then

$$\beta_n(\alpha, \tilde{V}_n) = (1-\alpha)(\Phi(\sqrt{2\log n} - z_n))^{k_n} + o(1),$$

which imply the lower bounds for an integer $k_n = nh_n$.

Let nh_n be no integer, $nh_n/\sqrt{\log n \log\log n} \to \infty$ (this implies $\tau \geq 1$). By making R_n larger $R_n^* = R_n(1+\delta_{n,0})$ (then $\beta_n(\alpha, \tilde{V}_n)$ decreases) we can get an integer $k_n = nh_n^*$; $nh_n + 1 > k_n > nh_n$ with

$$h_n^* = h_n(1+\delta_{n,1}), \quad z_n^* = z_n(1-\delta_{n,2}), \quad \delta_{n,0} \asymp \delta_{n,1} \asymp \delta_{n,2} = O((nh_n)^{-1}).$$

We would like to verify that the change $R_n \to R_n^*$ yields

$$\beta(\alpha, \pi_*^n) = \beta(\alpha, \pi^n) + o(1),$$

where $\pi_*^n = \pi^n(h_n^*, z_n^*)$, $\pi^n = \pi^n(h_n, z_n)$. It suffices to show that

$$u_n(h_n^*, z_n^*) \sim u_n(h_n, z_n),\ c_n(h_n^*, z_n^*) \sim c_n(h_n, z_n),\ \lambda_n(h_n^*, z_n^*) \sim \lambda_n(h_n, z_n).$$

Using the asymptotic continuity property which has been noted in the proof of Theorem 8.1 we see that it suffices to verify the relations $z_n^2 = o(nh_n)$ for $\tau > 1$ and $z_n d_n = o(nh_n)$ for $d_n = T_n - z_n \to \infty$; $\tau = 1$.

Let $\tau > 1$. Since $u_n \asymp 1$ for $\tau \geq 2$ and $c_n \asymp 1$ for $\tau \in (1, 2)$, it follows from (8.48), (8.50) that

$$z_n^2 \asymp \log n, \quad nh_n = n^{y_n}, \quad \liminf y_n = y > 0,$$

which yield the necessary relations.

Let $\tau = 1$. Since $\lambda_n \asymp 1$, $nh_n \to \infty$, using (8.19), (8.48), (8.50), we have

$$d_n \asymp \sqrt{\log nh_n}, \quad z_n \asymp \sqrt{\log n}.$$

Therefore the relation $z_n d_n = o(nh_n)$ follows from

$$nh_n / \sqrt{\log n \log\log n} \to \infty.$$

Theorem 8.2 is proved. □

8.4 Proof of Upper Bounds

The proofs of Theorems 8.3, 8.4 consist of three main steps.

First, we reject some sets of alternatives using the supreme tests. Next, we study the asymptotic distributions of test statistics (or their truncated versions) under the remaining alternatives. To evaluate the maximal type II errors we obtain extreme problems on the set of remaining alternatives. These problems are reduced to extreme problems on collections $\bar{r} = (r_1, \dots, r_n)$ of probability measures. Finally, we show that the collections $\bar{r}^*$ of the three-point measures $r_i^* = \pi(z_n, h_n)$, $i = 1, \dots, n$, are the solutions of the extreme problems. For the case $\tau \leq 2$ the extreme problems are somewhat different from those studied in Sections 5.2, 5.4. The proofs are based on Kuhn–Tucker Theorem (see [94], for example).

This is the principal point of the proof. If $\tau \geq 2$, then analogously to Section 5.2, the test statistics t_n are asymptotically $N(m_n(v_n), u_n^2)$-Gaussian under the null hypothesis ($v_n = 0$) as well as under the nonrejected alternatives v_n. Here $m_n(v_n) > 0$ and it depends on the alternative v_n. This implies the extreme problem: minimize $m_n(v_n)$ over nonrejected alternatives.

However, if $\tau \in (1, 2)$, then the log-likelihood ratios l_n are not asymptotically Gaussian and it is not clear how to compare these distributions. In this case we will show that under nonrejected alternatives the log-likelihood statistics l_n are asymptotically distributed as $\zeta = \zeta^1 + \eta$, where ζ^1 corresponds to the limit distribution of l_n under the Bayesian alternative and $\eta \geq 0$ a.s. This implies the required upper bounds for the maximal type II errors.

To prove the upper bounds we can assume $u_n \asymp 1$ for $\tau \geq 2$ or $c_n \asymp 1$ for $\tau \in (1, 2)$. This follows from the argument analogous to that which was presented at the beginning of Section 8.3.3.

Using the limit distributions of the statistics t_n, l_n under H_0 from the Remark to Propositions 8.1 and from Proposition 8.2 (3), and by the continuity property stated in Section 8.2.2, we get the relations for the tests $\psi_{n,\alpha}$ defined by (8.72), (8.73),

$$\alpha + o(1) \le \alpha(\psi_{n,\alpha}) \le \begin{cases} \alpha + o(1), & \text{if } n\Phi(-Q_n) \to 0, \\ \alpha + 2\delta + o(1), & \text{if } n\Phi(-Q_n) \le \delta. \end{cases}$$

Fix a small $\delta > 0$. By these relations, to prove the upper bounds of Theorems 8.3, 8.4 it suffices to evaluate the type II error probabilities $\beta_n(\psi_{n,\alpha}, V_n)$ for the tests $\psi_{n,\alpha} = \psi_{n,\alpha,\delta}$ such that $n\Phi(-Q_n) \le \delta$. Recall the inequality

$$\beta_n(\psi_{n,\alpha}, v) \le \min\{P_{v,1}, P_{v,2}\}, \tag{8.85}$$

where

$$P_{v,1} = P_v\left(\max_i |x_i| \le Q_n\right),$$

and

$$P_{v,2} = P_v(t_n \le u_n T_\alpha) \quad \text{or} \quad P_{v,2} = P_v(l_n \le T_{n,\alpha}).$$

Consider the probabilities $P_{v,1}$. Note that for any $\delta > 0$ there exists $B > 0$ such that $P_{v,1} < \delta$ if one of the two following conditions hold:

$$\max_i |v_i| > Q_n + B, \quad \sum_i \Phi(|v_i| - Q_n) > B. \tag{8.86}$$

This fact follows from the evaluations

$$P_v\left(\max_i |x_i| \le Q_n\right) = \prod_{i=1}^n (\Phi(Q_n - |v_i|) - \Phi(-Q_n - |v_i|))$$

$$\le \prod_{i=1}^n \Phi(Q_n - |v_i|) \le \Phi\left(Q_n - \max_i |v_i|\right) \le \Phi(-B); \tag{8.87}$$

and, by $1 - x \le \exp(-x)$, we have

$$P_v\left(\max_i |x_i| \le Q_n\right) = \prod_{i=1}^n (1 - \Phi(|v_i| - Q_n) - \Phi(-Q_n - |v_i|)) \tag{8.88}$$

$$\le \prod_{i=1}^n (1 - \Phi(|v_i| - Q_n)) \le \exp\left(-\sum_{i=1}^n \Phi(|v_i| - Q_n)\right) \le \exp(-B). \tag{8.89}$$

Set

$$V_B^n = \left\{ v \in R^n : \max_i |v_i| \le B + Q_n, \sum_i \Phi(|v_i| - Q_n) \le B \right\}$$

and we can consider the truncated versions of the test statistics

$$\hat{t}_n = t_n \mathbf{1}_{\{\max_i |x_i| \le Q_n\}}, \quad \hat{l}_n = l_n \mathbf{1}_{\{\max_i |x_i| \le Q_n\}}.$$

Taking into account relations (8.85), (8.86) we see that it suffices to evaluate the probabilities $P_v(\hat{t}_n \le T_\alpha)$ or $P_v(\hat{l}_n \le T_{n,\alpha})$ for $v \in V_n \cap V_B^n$ with large enough $B > 0$.

8.4.1 Upper Bounds for $\tau \ge 2$

Let us prove Theorem 8.3. It suffices to suppose that the u_n are bounded away from 0 since the upper bounds are trivial when $u_n \to 0$. Moreover, by the arguments analogous to those that were given at the beginning of Section 8.3.3 we can suppose $u_n \asymp 1$.

The scheme of the proof is the following. We specify Q_n below and we establish the inequality analogous to (5.22):

$$\beta_n(\psi_{n,\alpha}, v) \le \Phi(T_\alpha - h_{n,Q}(v)) + o(1), \quad v \in V_B^n, \tag{8.90}$$

where

$$h_{n,Q}(v) = u_n^{-1} h_n \sum_{i=1}^{n} 2\sinh^2(z_n v_i/2)\Phi(Q_n - z_n - |v_i|), \tag{8.91}$$

$$\tilde{u}_n^2 = 2nh_n^2 \sinh^2(z_n^2/2)\Phi(Q_n - 2z_n) \sim u_n^2, \tag{8.92}$$

and the parameters $z_n \to \infty$, $h_n \to 0$ are defined by (8.71), and the quantities u_n^2 are defined by (8.22). Next, we would like to show that the minimum of $h_{n,Q}(v)$ over $v \in V_n$ is $u_n + o(1)$. These yield the statement of Theorem 8.3.

We start with the case τ large enough (namely, let $\tau > 6$). Put $Q_n = \sqrt{2\log n}$. In this case, using (8.46), (8.48), one can take

$$\eta > 0, \quad \tau_0 > 0, \quad 3\eta/2 + \tau_0 < 1, \quad \zeta \in (0, 1 - (3\eta/2 + \tau_0)),$$

such that

$$Q_n = \sqrt{2\log n} \le \tau_0 T_n, \quad z_n \le \eta T_n, \quad Z_n = 3z_n/2 + Q_n + B - T_n < -\zeta T_n,$$

for any $B > 0$ and large enough n. In this case we can repeat the evaluations given in the proof of Proposition 5.3 (see Section A.2.2) for the case $i \in I_1$ (which corresponds to nontruncated versions of the statistics t_n).

Thus below we can consider the case that τ is small enough ($\tau \le 6$). We are bounded by a small extension of consideration in the proof of Proposition 5.3 in this case. By (8.23) we have $c_n = 2n\Phi(-T_n) \to 0$ and one can choose $Q_n = T_n - C_n/z_n$, where $C_n \to \infty$ is a sequence such that

$$C_n = o(z_n), \quad n\Phi(-Q_n) \asymp c_n e^{T_n C_n} \to 0. \tag{8.93}$$

Moreover, if $\tau = 2$, $d_n = 2z_n - T_n = T_n - 2z_n \to \infty$, then we assume $C_n = o(z_n/d_n)$ (recall that $d_n = o(z_n)$ for $\tau = 2$).

Consider the Q_n-truncated statistics

$$\hat{t}_n = \sum_i \hat{g}_n(X_i), \quad \hat{g}_n(x) = g_n(x)\mathbf{1}_{\{|x| \le Q_n\}};$$

here, analogously to the above, we set

$$g_n(x) = h_n \xi(x, z_n), \quad \xi(x, z_n) = \exp(-z^2/2)\cosh(xz) - 1.$$

By the definition of T_n and since $h_n = o(\exp(-C_n))$ (see (8.14)) we have

$$\hat{g}_n = \sup_x |\hat{g}_n(x)| = g_n(Q_n) = g_n(T_n - C_n/z_n) \asymp \exp(-C_n) \to 0. \quad (8.94)$$

Therefore all items in the sum $\hat{t}_n$ are uniformly small and we can apply the Central Limit Theorem. Denote

$$m_n(v) = E_{n,v}(\hat{t}_n), \quad D_n(v) = \mathrm{Var}_{n,v}(\hat{t}_n).$$

Let us show that, to prove Theorem 8.3, it suffices to verify the relations, for any sequence $v^{(n)} = v \in V_B^n$,

$$m_n(0) = o(1), \quad D_n(0) = u_n^2 + o(1), \quad \Delta D_n = o(\Delta m_n), \quad (8.95)$$
$$\Delta m_n \geq 2nh_n^2 \sinh^2(z_n^2/2)\Phi(Q_n - 2z_n) \sim u_n^2, \quad (8.96)$$

where

$$\Delta m_n = m_n(v) - m_n(0), \quad \Delta D_n = D_n(v) - D_n(0).$$

First, relations (8.95), (8.96) imply that

$$P_{n,v}(\hat{t}_n \leq u_n T_\alpha) \to 0 \quad \text{as} \quad \Delta m_n \to \infty.$$

In fact, it follows from (8.95), (8.96) that $m_n(v) \to \infty$ and $D_n(v) = o(m_n(v))$ as $\Delta m_n \to \infty$. Using the Chebyshev inequality and since $u_n T_\alpha \asymp 1$ we get

$$\begin{aligned} P_{n,v}(\hat{t}_n \leq u_n T_\alpha) &\leq P_{n,v}(|\hat{t}_n - m_n(v)| \geq m_n(v) - u_n T_\alpha) \\ &\leq D_n(v)/(m_n(v) - u_n T_\alpha)^2 \to 0. \end{aligned}$$

Therefore it suffices to assume that $\Delta m_n = O(1)$. Then the Central Limit Theorem yields $\hat{t}_n \sim N(\Delta m_n, u_n^2)$ in $P_{n,v}$-probability. Moreover, one can easily see that

$$\Phi(T_n - 2z_n) \sim \Phi(Q_n - 2z_n)$$

by the choice of Q_n. Therefore, we have

$$P_{n,v}(\hat{t}_n \leq u_n T_\alpha) = \Phi\left(T_\alpha - \frac{\Delta m_n}{u_n}\right) + o(1) \leq \Phi(T_\alpha - u_n) + o(1).$$

To prove (8.95), (8.96) note that it follows from (8.27), (8.28), (8.19), (8.24), (8.93) that

$$\begin{aligned} -m_n(0) &\sim nh_n\Phi(z_n - Q_n) \sim nh_n\Phi(z_n - T_n)\exp((\tau_n - 1)C_n) \\ &= \lambda_n \exp((\tau_n - 1)C_n) \asymp c_n \exp((\tau_n - 1)C_n) = o(1), \\ D_n(0) &\sim 2^{-1}nh_n^2 e^{z_n^2}\Phi(Q_n - 2z_n) \sim 2^{-1}nh_n^2 e^{z_n^2}\Phi(T_n - 2z_n) = u_n^2 \asymp 1, \end{aligned}$$

which yield the first relations in (8.95). Denote

$$\mu_n(t) = E\hat{g}_n(\eta + t) - E\hat{g}_n(\eta); \quad \sigma_n^2(t) = E\hat{g}_n^2(\eta + t) - E\hat{g}_n^2(\eta)$$

where $\eta \sim N(0,1)$. By (8.94) one has

$$\Delta m_n = \sum_i \mu_n(v_i), \quad \sup_t |\mu_n(t)| = o(1); \tag{8.97}$$

$$\begin{aligned} \Delta D_n &= \sum_i (\sigma_n^2(v_i) - (E_{v_i}\hat{g}_n)^2 + (E_0\hat{g}_n)^2) \\ &= \sum_i \sigma_n^2(v_i) + o\Big(\sum_i |\mu_n(v_i)|\Big). \end{aligned} \tag{8.98}$$

Below we use Proposition A.1 given in the Appendix, Section A.2.1. It follows from Proposition A.1 that for $|t| \le Q_n + B$ one has

$$\mu_n(t) \sim 2h_n \sinh^2(z_n t/2)\Phi(Q_n - z_n - |t|), \tag{8.99}$$

$$\sigma_n^2(t) = O(R_n(t) + h_n\mu_n(t)), \tag{8.100}$$

$$R_n(t) = h_n^2 e^{z_n^2} \sinh^2(z_n t)\Phi(Q_n - 2z_n - |t|). \tag{8.101}$$

Let us check the relation

$$R_n(t) = o(\mu_n(t)). \tag{8.102}$$

By (8.14), (8.99), (8.101), relation (8.102) is equivalent to

$$\exp(3z_n^2/2 - T_n z_n + z_n|t|)\Phi(Q_n - 2z_n - |t|) = o\left(\Phi(Q_n - z_n - |t|)\right). \tag{8.103}$$

To check (8.103) consider the different cases

$$\begin{aligned} &(a) : |t| < T_n - 3z_n/2 - \delta, \\ &(b) : T_n - 3z_n/2 - \delta \le |t| \le T_n - z_n + B, \\ &(c) : T_n - z_n + B < |t|. \end{aligned}$$

In case (a) relation (8.103) holds because $Q_n = T_n + o(1)$ and the argument of the function $\Phi(\cdot)$ in the right-hand side tends to ∞ and the argument of the exponent is bounded by $-z_n\delta \to -\infty$.

In case (b) the argument of the function $\Phi(\cdot)$ in the right-hand side is bounded away from $-\infty$ whenever the argument of the function $\Phi(\cdot)$ in the left-hand side tends to $-\infty$. By $Q_n < T_n$, using (8.19), we get that the left-hand side is bounded by $B\exp(-(T_n - z_n - |t|)^2/2)/(B + z_n) = o(1)$.

In case (c) by $Q_n = T_n - C_n/z_n$, using (8.19), we get that the left-hand side is of the rate

$$\delta_1 \frac{\exp(-(T_n - z_n - |t|)^2/2)}{B + z_n + x}, \quad \delta_1 = \exp(-C_n(|t| + 2z_n - T_n)/z_n),$$

where $x = |t| - Q_n + z_n - B = C_n/z_n > 0$, and the right-hand side is of the rate

$$\delta_2 \frac{\exp(-(T_n - z_n - |t|)^2/2)}{B + x}, \quad \delta_2 = \delta_1 \exp(C_n).$$

By $C_n \to \infty$, this yields (8.103).

It follows from (8.99)–(8.102) that

$$\sum_i \sigma_n^2(v_i) = o(\Delta m_n). \tag{8.104}$$

By (8.97), (8.98) this yields the last relation in (8.95).

Moreover, (8.97), (8.99) imply that (8.96) follows from the inequality

$$\inf_{v \in V_B^n} \sum_i 2\sinh^2(z_n v_i/2)\Phi(Q_n - z_n - |v_i|) \geq 2nh_n \sinh^2(z_n^2/2)\Phi(Q_n - 2z_n). \tag{8.105}$$

Inequality (8.105) could be presented in the following form. Let $p > q$, $\rho_n \leq R_n$. Denote by

$$\Xi_n = \Xi_n^{p,q}(\rho_n, R_n, H_n)$$

the set of collections $\bar{r} = (r_1, \dots, r_n)$ of probability measures on $(R^1, \mathcal{B})$ subject to the constraints

$$\sum_i \int |u|^p r_i(du) \geq \rho_n^p, \quad \sum_i \int |u|^q r_i(du) \leq R_n^q; \quad r_i([-H_n, H_n]) = 1.$$

Let the sequences $h_n \to 0$, $z_n \to \infty$ be defined by (8.71) and let the functions $\phi(t) = \phi_n(t)$ be given. Introduce the functionals

$$\Phi_n(r) = \int \phi_n(v) r(dv), \quad F_n(\bar{r}) = \sum_i \Phi_n(r_i). \tag{8.106}$$

Then for large enough n one often has the equality

$$\inf_{\bar{r} \in \Xi_n} F_n(\bar{r}) = F_n(\bar{r}_n^*) = n\Phi_n(\pi_n) = nh_n\phi(z_n), \tag{8.107}$$

where $\bar{r}_n^* = (\pi_n, \dots, \pi_n)$ and $\pi_n = \pi(h_n, z_n)$ is the symmetric three-point measure with the parameters (h_n, z_n) determined by (8.71). In particular, set $H_n = Q_n + B$ and

$$V_{n,1} = \{v = (v_1, \dots, v_n) \in V^n : \max_i |v_i| \leq H_n\},$$

(clearly, $V_B^n \subset V_{n,1}$). For $v \in V_{n,1}$ we set $\bar{r}_v = (\delta_{v_1}, \dots, \delta_{v_n})$. Then $\bar{r}_v \in \Xi_n$ and for large enough n one has

$$\inf_{v \in V_{n,1}} \sum_{i=1}^n \phi(v_i) \geq \inf_{\bar{r} \in \Xi_n} F_n(\bar{r}) = nh_n\phi(z_n). \tag{8.108}$$

We will specify below the functions $\phi(t) = \phi_n(t)$ and some additional constraint for inequalities (8.107), (8.108) to be fulfilled. In particular, inequality (8.105) follows from

Lemma 8.2 *Setting*

$$\phi(t) = \phi_n(t) = 2\sinh^2(z_n t/2)\Phi(Q_n - z_n - |t|), \tag{8.109}$$

let us consider the extreme problems in the left-hand sides of (8.107), (8.108) *which correspond to* (8.106). *Let a quantity* $\delta \in (0,1)$ *and a sequence* $d_n \to \infty$ *be given. Assume, for some* $B > 0$,

$$z_n < H_n \le (1-\delta)Q_n + \sqrt{2Q_n z_n - 3z_n^2} \le Bz_n, \;\; Q_n > 2z_n - d_n; \;\; d_n = o(z_n). \tag{8.110}$$

Then for a large enough n *equality* (8.107) *and inequality* (8.108) *hold.*

Proof of Lemma 8.2. This is given in the Appendix, Section A.6.

Note that the equality analogous to (8.107) follows from Theorem 5.3 for $\tau > 2$ and $\phi(t) = 2\sinh^2(z_n t/2)$ (formally it corresponds to (8.110), (8.109) with $H_n = Q_n = \infty$).

Thus Theorem 8.3 is proved. □

8.4.2 Upper Bounds for $\tau \in (1,2)$

Let us prove Theorem 8.4. It suffices to suppose that the c_n are bounded away from 0 since the upper bounds are trivial when $c_n \to 0$. Moreover, analogously to Section 8.4.1, we suppose $c_n \asymp 1$.

Rejection of Alternatives

Put $H_n = \eta z_n$, $\eta \in (1,\tau)$, and consider the sets

$$V_{n,1} = \left\{ v \in V_n \cap V_B^n : \max_i |v_i| \le H_n \right\},$$
$$V_{n,2} = \left\{ v \in V_n \cap V_B^n : \max_i |v_i| > H_n \right\}.$$

Take $Q_n = T_n + t(\delta)/z_n$ where $t(\delta)$ is such that $n\Phi(-Q_n) < \delta$.

Let $N_n(v) = \#\{i : |v_i| > H_n\}$ be the number of coordinates of $v \in R^n$ such that $|v_i| > H_n$. Let us verify the relation

$$\sup_{v \in V_{n,2}} N_n(v) = o(nh_n/z_n^3). \tag{8.111}$$

In fact, suppose there exist a sequence $v = v^{(n)} \in V_{n,2}$ and a quantity $A > 0$ such that $N_n(v^{(n)}) > Anh_n z_n^{-3}$. Note that for any $\eta \in (1,\tau)$ one has

$$h_n^{-1}\exp(-T_n^2/2) \asymp \exp(-(T_n - z_n)^2/2) = o(z_n^{-4}\exp(-(T_n - \eta z_n)^2/2)). \tag{8.112}$$

Since $c_n = 2n\Phi(-T_n) \asymp 1$, for some $b > 0$, $D > 0$ and for the sequence $v = v^{(n)}$ we get

$$\begin{aligned}
\sum_i \Phi(|v_i| - Q_n) &\ge bN_n(v)\Phi(\eta z_n - T_n) \\
&\ge Dz_n^{-4} nh_n \exp(-(T_n - \eta z_n)^2/2) \\
&\gg nT_n^{-1}e^{-T_n^2/2} \asymp n\Phi(-T_n) \asymp 1,
\end{aligned}$$

where the third rate inequality follows from (8.112). These relations contradict $v = v^{(n)} \in V_B^n$.

Set

$$\tilde{v}_i = v_i \mathbf{1}_{\{|v_i| \le H_n\}}, \quad \tilde{v} = (\tilde{v}_1, \dots, \tilde{v}_n) \in V_B^n.$$

Let $v \in V_{n,2}$. Then

$$\sum_i |\tilde{v}_i|^p \ge \sum_i |v_i|^p - N_n(v)(Q_n + B)^p \ge \rho_n^p - o(nh_n z_n^{p-3}) \ge \tilde{\rho}_n^p.$$

Using (8.71), (8.111) for large enough n one has

$$\tilde{v} \in \tilde{V}_n = V_n^{p,q}(\tilde{\rho}_n, R_n), \quad \tilde{\rho}_n = \rho_n(1 - z_n^{-3}).$$

Note that the set $X_{n,\alpha} = \{x \in R^n : l_n(x) \le T_{n,\alpha} = T_{c,\tau}^0(\alpha)\}$ is convex and symmetric in all coordinates x_i. Applying Lemma 2.4 we get the inequality

$$P_{n,v}(X_{n,\alpha}) \le P_{n,\tilde{v}}(X_{n,\alpha}) = P_{\tilde{v},2}.$$

Setting

$$\tilde{V}_{n,H_n} = \left\{ v \in \tilde{V}_n \cap V_B^n : \max_i |v_i| \le H_n \right\},$$

we have

$$\beta_n(\psi_{n,\alpha}, V_B^n) \le \sup_{v \in \tilde{V}_{n,H_n}} P_{v,2}. \tag{8.113}$$

Taking into account inequality (8.113), we evaluate below the probabilities $P_{n,v}(l_n \le T_{n,\alpha})$ for $v \in \tilde{V}_{n,H_n}$. We omit the tilde to simplify the notations of $\tilde{V}_n$.

An Outline of the Study

By (8.113), to obtain the upper bounds for $\tau \in (1,2)$, it suffices to verify that for any sequence $v^{(n)} = v \in V_{n,H_n}$ one has

$$\limsup \left(P_{n,v}\left(l_n \le T_{c,\tau}^0(\alpha)\right) - \beta(\alpha, \pi^n) \right) \le 0 \tag{8.114}$$

(we omit index n in $v^{(n)}$ to simplify the notation). Denote, as above,

$$L_n^0(t) = -2n\Phi(-T_n(t)), \quad L_n(t) = L_n^0(t) + \Delta L_n(t)$$

where $T_n(t)$ is determined by (8.18) and

$$\begin{aligned} \Delta L_n(t) &= -\sum_{i=1}^n (P_{1,v_i}(x : w_n(x) > t) - P_{1,0}(x : w_n(x) > t)) \qquad (8.115) \\ &= -\sum_{i=1}^n (\Phi(-T_n(t) + |v_i|) + \Phi(-T_n(t) - |v_i|) - 2\Phi(-T_n(t))), \end{aligned}$$

where $P_{1,v} = N(v,1)$ is the Gaussian measure on R_1.

Clearly, $\Delta L_n(t) \le 0$. Note that $\Delta L_n(t) \to 0$ as $t \to \infty$. Moreover, for any $t_0 > 0$, the functions $\Delta L_n(t)$ are bounded from below on $t \ge t_0$ and their derivatives $\Delta L_n'(t)$ are positive and bounded over $t \ge t_0$. In fact, the functions $|\Delta L_n(t)|$ and $\Delta L_n'(t)$ decrease in t and it suffices to check the case $t = t_0$. Note that $T_n(t_0) - |v_i| \ge z_n(\tau - \eta + o(1)) \ge cz_n$, $c > 0$. It easily follows from (8.18), (8.19) and from the choice Q_n, that the items in the sums (8.115) for $|\Delta L_n(t_0)|$ and in the analogous sums for their derivatives are of the rate $\Phi(-T_n(t) + |v_i|) \asymp \Phi(-Q_n + |v_i|)$ and it follows from the definition of the sets V_B^n that the sums of the last items are bounded by B. Thus the set of functions $\{\Delta L_n(t)\}$ is equicontinuous in the space $C\,[t_0, \infty)$ for every $t_0 > 0$ and, by passing to subsequences, we can assume that, for every $t > 0$,

$$\Delta L_n(t) \to \Delta L(t), \tag{8.116}$$

where $\Delta L(t)$ is continuous in $t \in (0, \infty)$ and fulfills the properties of the Levy spectrum supported on $(0, \infty)$, i.e., $\Delta L(t) \le 0$, $\Delta L(t) \to 0$ as $t \to \infty$, $\Delta L(t)$ is nondecreasing in $t > 0$, and (which will be shown later) for any $\delta > 0$,

$$\int_{+0}^{\delta} t^2 \, d\Delta L(t) < \infty. \tag{8.117}$$

Put

$$m_n = \sum_i m_{n,i} = 2h_n \sum_i \sinh^2(z_n v_i/2)\Phi(T_n - z_n - |v_i|). \tag{8.118}$$

The outline of the proof is as follows. We show that if $m_n \to \infty$, then $l_n \to \infty$ in $P_{n,v}$-probability. Let $m_n = O(1)$. Then, passing to a subsequence, we can assume l_n converges in $P_{n,v}$-probability to the sum $\zeta = \zeta^0 + \Delta\zeta$. Here ζ^0 and $\Delta\zeta$ are independent infinite divisible random variables. The variable ζ^0 corresponds to the P_0-limiting distribution of l_n and is described in Section 8.1.4. The variable $\Delta\zeta$ also has no Gaussian component and is supported on the half-line (b, ∞), $b \ge 0$, i.e., the characteristic function $\phi_{\Delta\zeta}(z)$ of $\Delta\zeta$ is of the form (compare with (8.5)):

$$\log \phi_{\Delta\zeta}(z) = i\Delta\gamma z + \int_0^\infty \left(\exp(izt) - 1 - \frac{izt}{1+t^2}\right) d\Delta L(t), \tag{8.119}$$

$$\Delta\gamma - \int_0^\infty \frac{t}{1+t^2}\, d\Delta L(t) = b \ge 0. \tag{8.120}$$

Then we obtain the relation

$$\Delta L(t) = L^{\Delta}(t) + \Lambda(t), \quad t > 0, \tag{8.121}$$

where $L^{\Delta}(t)$ is the Levy spectrum in (8.37), and $\Lambda(t)$ is also a Levy spectrum, i.e.,

$$\Lambda(t) \le 0, \quad d\Lambda(t)/dt \ge 0, \quad \Lambda(t) \to 0 \quad \text{as } t \to \infty, \quad \Lambda(t) = 0 \quad \text{for } t < 0,$$

and (8.117) implies the analogous relation for $\Lambda(t)$, for any $\delta > 0$,

$$\int_{+0}^{\delta} t^2\, d\Lambda(t) < \infty.$$

These relations imply the equality

$$\zeta = \zeta_0 + \zeta_\Delta + \zeta_\Lambda = \zeta^1 + \zeta_\Lambda, \tag{8.122}$$

where the random variables $\zeta_0,\ \zeta_\Delta,\ \zeta_\Lambda$ are independent, $\zeta^1 = \zeta_0 + \zeta_\Delta$ corresponds to the limiting distribution of l_n under the Bayesian alternatives P_{π^n}, and $\zeta_\Lambda \geq b \geq 0$ a.s. By (8.122) and Proposition 8.2 one has

$$P_{n,v^{(n)}}(l_n \leq T^0_{c,\tau}(\alpha)) \leq P(\zeta^1 \leq T^0_{c,\tau}(\alpha)) + o(1) = \beta(\alpha, \pi^n) + o(1)$$

which imply (8.114) and the upper bounds of Theorem 8.4.

The Study of the Limiting Distribution of l_n

To realize the above outline, introduce the functions

$$m_n(t) = \sum_i m_{n,i}(t) \;=\; 2h_n \sum_i \int_{|x|<T_n(t)} \xi(x, z_n)\xi(x, v_i)\, d\Phi(x),$$

$$s_n^2(t) = \sum_i s^2_{n,i}(t) \;=\; 2h^2_n \sum_i \int_{|x|<T_n(t)} \xi^2(x, z_n)\xi(x, v_i)\, d\Phi(x),$$

and put

$$s_n^2 = \sum_i s^2_{n,i} = 2h_n^2 e^{z_n^2} \sum_i \sinh^2(z_n v_i)\Phi(T_n - 2z_n - |v_i|).$$

Note that

$$\xi(x, v_i) = \frac{1}{2}\left(\frac{dP_{-v}}{dP_0}(x) + \frac{dP_v}{dP_0}(x)\right) - 1, \quad P_0(dx) = d\Phi(x),$$

and using the symmetry of $\xi(x, z_n)$ in x we have

$$\int_{|x|<T_n(t)} \xi(x, z_n)\xi(x, v_i)\, d\Phi(x) = \int_{|x|<T_n(t)} \xi(x, z_n)(P_v(dx) - P_0(dx))$$
$$= E_v\hat{\xi}(x, z_n) - E_0\hat{\xi}(x, z_n); \quad \hat{\xi}(x, z_n) = \xi(x, z_n)\mathbf{1}_{\{|x|<T_n(t)\}}.$$

Using (8.99), (8.100), (8.19) and relation (A.18) given in the proof of Proposition A.1 one can check that

$$m_{n,i}(t) \sim \begin{cases} m_{n,i}, & \text{if } |v_i| \leq T_n - z_n, \\ m_{n,i}(e^t - 1)^{|v_i|/z_n + 1 - \tau_n}, & \text{if } |v_i| > T_n - z_n, \end{cases} \tag{8.123}$$

and for large enough n one has

$$s^2_{n,i}(t) \leq C s^2_{n,i}(e^t - 1)^{|v_i|/z_n + 2 - \tau_n}$$

(we denote by C positive constants, possibly different, which do not depend on n). One can also check that

$$s_{n,i}^2 = \begin{cases} o(m_{n,i}), & \text{if } |v_i| \le T_n - z_n, \\ O(m_{n,i}), & \text{if } |v_i| > T_n - z_n, \end{cases}$$

which imply, for large enough n and small enough $t > 0$, the relation

$$s_n^2(t) \le Ctm_n(t), \quad C > 0. \tag{8.124}$$

Let us evaluate the differences between the means and variances

$$\Delta E_n = E_{n,v}(l_n) - E_{n,0}(l_n), \quad \Delta D_n = \mathrm{Var}_{n,v}(l_n) - \mathrm{Var}_{n,0}(l_n).$$

Note the equality

$$2\xi(x,v)\, d\Phi(x) = d\Phi(x + |v|) + d\Phi(x - |v|) - 2\, d\Phi(x). \tag{8.125}$$

It follows from the symmetry of the functions $w_n(x)$ in x that

$$\Delta E_n = \sum_i \int w_n(x)\xi(x, v_i)\, d\Phi(x). \tag{8.126}$$

Denote

$$\begin{aligned} \tilde{D}_{n,v} &= E_{n,v} l_n^2 = \frac{1}{2} \sum_i \int w_n^2(x) \left(d\Phi(x + |v_i|) + d\Phi(x - |v_i|) \right), \\ \tilde{D}_{n,0} &= E_{n,0} l_n^2 = \sum_i \int w_n^2(x)\, d\Phi(x). \end{aligned}$$

It follows from the evaluations in Section 8.1.4 (see (8.35)) that

$$E_{n,0} l_n = nE_{1,0} w_n(x) = O(1).$$

Clearly, $\Delta D_n \le \Delta \tilde{D}_n + o(1)$, where

$$\Delta \tilde{D}_n = \tilde{D}_{n,v} - \tilde{D}_{n,0} = \sum_i \int w_n^2(x)\xi(x, v_i)\, d\Phi(x). \tag{8.127}$$

For every $t > 0$ put

$$\Delta E_n(t) = \Delta E_n^+(t) + \Delta E_n^-(t), \quad \Delta \tilde{D}_n(t) = \Delta \tilde{D}_n^+(t) + \Delta \tilde{D}_n^-(t),$$

where the terms with the indexes "+" and "–" correspond to the sums of the integrals in (8.126) and (8.127) over the sets $\{|x| > T_n(t)\}$ and $\{|x| < T_n(t)\}$.

Let us obtain the relations, for every $t > 0$,

$$\begin{aligned} \Delta E_n^+(t) &= O(1), \quad \Delta \tilde{D}_n^+(t) = O(1), & (8.128) \\ \Delta E_n^-(t) &= m_n(t)(1 + O(t)), & (8.129) \\ \Delta \tilde{D}_n^-(t) &= O(m_n(t)). & (8.130) \end{aligned}$$

To obtain (8.128), we use equality (8.125) which implies

$$\Delta E_n^+(t) = A_n(t) - B_n(t); \tag{8.131}$$

the item $A_n(t)$ corresponds to the integration over $d\Phi(x+|v|)+d\Phi(x-|v|)$ and $B_n(t)$ corresponds to the integration over $d\Phi(x)$. It follows from (8.17), (8.18), (8.19) that the items are of the form

$$\begin{aligned} A_n(t) &\sim \sum_i (\tau_n - |v_i|/z_n)\Phi(|v_i| - T_n)I_{1,i} \\ &\quad + \sum_i (\tau_n + |v_i|/z_n)\Phi(-|v_i| - T_n)I_{2,i}, \qquad (8.132) \\ B_n(t) &\sim c_n\tau_n I_3, \qquad (8.133) \end{aligned}$$

where

$$\begin{aligned} I_{1,i} &= \int_{u>e^t-1} \log(1+u)u^{-(\tau_n+1-|v_i|/z_n)}\zeta_n(u)\,du; \qquad (8.134) \\ I_{2,i} &= \int_{u>e^t-1} \log(1+u)u^{-(\tau_n+1+|v_i|/z_n)}\zeta_n(u)\,du; \qquad (8.135) \\ I_3 &= \int_{u>e^t-1} \log(1+u)u^{-(\tau_n+1)}\zeta_n(u)\,du; \\ \zeta_n(u) &= e^{-(\log u/z_n)^2/2}. \end{aligned}$$

The boundness of $A_n(t),\ B_n(t)$ follows from the uniform boundness of the integrals above (for $|v_i|/z_n \le \eta < \tau$ by the definition of the set V_{n,H_n}) and from the boundness of the sums $\sum_i \Phi(-T_n \pm |v_i|)$ (these follow from (8.115), (8.116)). The boundness of the quantities $\Delta\tilde{D}_n^+(t)$ follows analogously.

Relations (8.129), (8.130) follow from (8.124) and from the inequalities: for any $C > 1$ there exist $C_1 > 0,\ C_2 > 0$ such that

$$|\log(1+z) - z| \le C_1 z^2, \quad \log^2(1+z) \le C_2 z^2 \quad \text{for } z > -1 + 1/C.$$

Assume $m_n \to \infty$. Then, by the boundness of $E_{n,0}(l_n),\ \mathrm{Var}_{n,0}(l_n)$, and from (8.123), (8.126)–(8.130), one has

$$E_{n,v}(l_n) \asymp m_n \to \infty, \quad \mathrm{Var}_{n,v}(l_n) \le BE_{n,v}(l_n),$$

and by the Chebyshev inequality $l_n \to \infty$ in $P_{n,v}$-probability.

Assume $m_n = O(1)$. Let us show that the following relations are fulfilled:

$$\lim_{n\to\infty} \sum_i P_{1,v_i}\{x : w_n(x) > t\} = -(L_0(t) + \Delta L(t)) \quad \forall\, t > 0; \qquad (8.136)$$

$$\begin{aligned} \lim_{t\to 0}\lim_{n\to\infty} \Delta\tilde{D}_n^-(t) &= 0, \qquad (8.137) \\ \liminf_{t\to 0}\liminf_{n\to\infty} \Delta E_n^-(t) &\ge 0, \qquad (8.138) \\ \lim_{t_2\to 0}\lim_{t_1\to 0}\lim_{n\to\infty} (\Delta E_n^-(t_2) - \Delta E_n^-(t_1)) &= 0. \qquad (8.139) \end{aligned}$$

Relation (8.136) follows from (8.116) and from the analogous relation (8.30) for the limiting $P_{n,0}$-spectrum L_0. Relations (8.123), (8.124) imply the evaluations

$$\Delta\tilde{D}^-(t) \le C_1 s_n^2(t) \le C_2 t m_n$$

which imply (8.137). Relation (8.138) follows from (8.129),(8.123).

To prove (8.139), consider the equality

$$\Delta E_n^-(t_2) - \Delta E_n^-(t_1) = \Sigma_{n,1}(t_1,t_2) + \Sigma_{n,2}(t_1,t_2).$$

Here $\Sigma_{n,l}(t_1,t_2)$, $l = 1,2$, correspond to the sums of the terms in (8.126) with i such that $|v_i| \le A_n$ (for $l = 1$) and $|v_i| \in (A_n, H_n]$ (for $l = 2$), where $A_n = \delta_n/T_n$, $\delta_n \to 0$. Since

$$|\xi(x,v)| \le \delta_n^2(1+o(1))/2 \quad \text{for } |v| \le A_n, \quad |x| \le T_n(t_2),$$

and using (8.16), (8.18) one has, for any $t_2 > t_1 > 0$,

$$\begin{aligned}\Sigma_{n,1}(t_1,t_2) \;&\le\; n\delta_n^2 \int_{T_n(t_1)}^{T_n(t_2)} w_n(x)\, d\Phi(x)(1+o(1)) \\ &\sim\; \delta_n^2 c\tau \int_{e^{t_1}-1}^{e^{t_2}-1} \log(1+u)u^{-(\tau+1)}\, du \to 0, \quad \text{as } n \to \infty. \end{aligned} \tag{8.140}$$

To evaluate $\Sigma_{n,2}(t_1,t_2)$, denote by $b_n(A_n)$ the number of terms in this sum. By (8.71) one has

$$b_n(A_n) \le A_n^{-q} \sum_i |v_i|^q \le (z_n/A_n)^q n h_n \asymp z_n^{2q}\delta_n^{-q} n h_n$$

and by (8.46), (8.50), for any $\varepsilon_1 > 0$, one can choose $\delta_n = n^{-\varepsilon_0}$ and $\varepsilon_0 < \varepsilon_1/q$ such that

$$m_n(A_n) \le B n^{1+\varepsilon_1} h_n = o(n). \tag{8.141}$$

By analogy with (8.115), (8.131) put

$$\Sigma_{n,2}(t_1,t_2) = A_n(t_1,t_2) - B_n(t_1,t_2),$$

where the item $A_n(t_1,t_2)$ corresponds to the integration over $d\Phi(x+|v|) + d\Phi(x-|v|)$ and $B_n(t_1,t_2)$ corresponds to the integration over $d\Phi(x)$ (compare with (8.131)). Analogously to the evaluation of $\Delta E_n^+(t)$ one can obtain the evaluations, for any $t_2 > t_1 > 0$,

$$B_n(t_1,t_2) \le (2+o(1))c\tau b_n(A_n)n^{-1}\phi_1(t_1,t_2) \to 0 \quad \text{as} n \to \infty,$$

where

$$\phi_1(t_1,t_2) = \int_{e^{t_1}-1}^{e^{t_2}-1} |\log(1+u)|u^{-(\tau+1)}\, du.$$

To evaluate $A_n(t_1,t_2)$ consider the presentation

$$A_n(t_1,t_2) = A_n^+(t_1,t_2) + A_n^-(t_1,t_2),$$

where (compare with (8.132), (8.133))

$$A_n^+(t_1,t_2)) \sim \sum_i (\tau_n - |v_i|/z_n)\Phi(|v_i| - T_n)I_{1,i},$$

$$A_n^-(t_1,t_2)) \sim \sum_i (\tau_n + |v_i|/z_n)\Phi(-|v_i| - T_n)I_{2,i},$$

and (compare with (8.134), (8.135))

$$I_{1,i} = \int_{e^{t_1}-1}^{e^{t_2}-1} \log(1+u)u^{-(\tau_n+1-|v_i|/z_n)}\zeta_n(u)\,du;$$

$$I_{2,i} = \int_{e^{t_1}-1}^{e^{t_2}-1} \log(1+u)u^{-(\tau_n+1+|v_i|/z_n)}\zeta_n(u)\,du,$$

and $\zeta_n(u) = e^{-(\log u/z_n)^2/2}$. It easily seen that

$$A_n^-(t_1,t_2)) = O(B_n(t_1,t_2)) \to 0.$$

To evaluate $A_n^+(t_1,t_2)$, set

$$A_n^+(t_1,t_2) = A_{n,1}^+(t_1,t_2) + A_{n,2}^+(t_1,t_2).$$

Here $A_{n,l}^+(t_1,t_2)$, $l = 1,2$, correspond to the sums of the terms with i such that $|v_i| \in (A_n, z_n(1-\varepsilon)]$ (for $l = 1$) and $|v_i| \in (z_n(1-\varepsilon), H_n]$ (for $l = 2$) where $0 < \varepsilon < 1$. For small enough $\varepsilon_1 > 0$, using (8.141), we get

$$\frac{b_n(A_n)}{z_n}e^{-(T_n - z_n(1-\varepsilon))^2/2} \le \frac{b_n(A_n)}{nh_n z_n}e^{-\varepsilon z_n^2(\tau_n - 1+\varepsilon/2)} \to 0, \quad \text{as } n \to \infty.$$

Setting

$$\phi_1(t_1,t_2) = \int_{e^{t_1}-1}^{e^{t_2}-1} |\log(1+u)|u^{-(\tau+\varepsilon)}\,du,$$

for small enough $t_2 > t_1 > 0$ one has

$$A_{n,1}^+(t_1,t_2) \le Bb_n(A_n)z_n^{-1}\exp(-(T_n - z_n(1-\varepsilon))^2/2)\phi_1(t_1,t_2) \to 0,$$

as $n \to \infty$.

Moreover, since $H_n = \eta z_n$, for small enough $t_2 > t_1 > 0$, one has

$$A_{n,2}^+(t_1,t_2) \le B_n\phi_\varepsilon(t_1,t_2); \quad B_n \le B\sum_i \Phi(T_n - |v_i|) = O(1) \quad \text{as } n \to \infty,$$

where

$$\phi_\varepsilon(t_1,t_2) = \int_{e^{t_1}-1}^{e^{t_2}-1} |\log(1+u)|u^{-(\tau+1-\eta)}du \asymp \int_{t_1}^{t_2} u^{\eta-\tau}\,du$$

and, by the choice of $\eta \in (1,\tau)$,

$$\lim_{t_2\to 0}\lim_{t_1\to 0} \phi_\varepsilon(t_1,t_2) \to 0.$$

The evaluations above imply (8.139).

By passing to a subsequence in (8.138), one can assume that there exists a limit $b \geq 0$ of the left-hand side of (8.138). Then (8.136)–(8.139) imply that l_n converges in $P_{n,v}$-distribution to the sum $\zeta = \zeta^0 + \Delta\zeta$.

In fact (see [86] and Section 8.1.1), relation (8.136) implies the necessary form of the Levy spectrums. Relations (8.137), (8.31), and (8.31) imply that the $P_{n,v}$-limiting distribution of l_n has no Gaussian component. Relation (8.117) follows from (8.137) as well.

Let us verify that relation (8.121) holds and that the function $\Lambda(t)$, $t > 0$, in (8.121) satisfies the properties of the Levi spectrum, i.e.,

$$\Lambda(t) \leq 0, \quad \frac{d}{dt}\Lambda(t) \geq 0; \quad \lim_{t \to \infty} \Lambda(t) = 0; \quad \int_{+0}^{\delta} t^2 \, d\Lambda(t) < \infty, \tag{8.142}$$

for some $\delta > 0$.

By passing to limits, as $n \to \infty$, we can assume $\Delta L_n(t) \to \Delta L(t)$ where ΔL_n is defined by (8.115). It was shown in Section 8.1.4 that

$$L^{\Delta}_{\pi_n}(t) = n\,(P_{1,0}(w_n(x) > t) - P_{1,\pi_n}(w_n(x) > t)) \to L^{\Delta}(t), \quad t > 0.$$

Therefore the first and second relations (8.142) follow from the inequalities

$$\Delta L_n \leq L^{\Delta}_{\pi_n}, \quad \frac{d}{dt}\Delta L_n \geq \frac{d}{dt}L^{\Delta}_{\pi_n}; \tag{8.143}$$

these yield the third relation in (8.142) as well. The fourth relation in (8.142) follows from (8.117) which has been verified previously.

One can rewrite inequalities (8.143) in the form

$$\begin{aligned}
&\inf_{v \in V_{n,1}} \sum_{i=1}^{n} (\Phi(-T_n(t) + v_i) + \Phi(-T_n(t) - v_i) - 2\Phi(-T_n(t))) \\
&\geq nh_n\,(\Phi(-T_n(t) + z_n) + \Phi(-T_n(t) - z_n) - 2\Phi(-T_n(t)))\,; \\
&\inf_{v \in V_{n,1}} \sum_{i=1}^{n} \left(e^{-(T_n(t)-v_i)^2/2} + e^{-(T_n(t)+v_i)^2/2} - 2e^{-T_n^2(t)/2}\right) \\
&\geq nh_n\left(e^{-(T_n(t)+z_n)^2/2} + e^{-(T_n(t)+z_n)^2/2} - 2e^{-T_n^2(t)/2}\right).
\end{aligned}$$

These inequalities follow from the following lemmas (we set $Q_n = T_n(t)$ in these lemmas).

Lemma 8.3 *Setting*

$$\phi(t) = \Phi(-Q_n + t) + \Phi(-Q_n - t) - 2\Phi(-Q_n), \tag{8.144}$$

let us consider the extreme problems in the left-hand sides of (8.107), (8.108) *which correspond to* (8.106). *Assume, for some* $B_1 > 0$, $B_2 > 0$, *that*

$$Q_n - z_n \geq d_n \to \infty, \qquad B_1 Q_n \geq H_n > z_n, \quad B_2 z_n > Q_n. \tag{8.145}$$

Then equality (8.107) *and inequality* (8.108) *hold for a large enough* n.

Lemma 8.4 *Setting*

$$\phi(t) = \exp(-(Q_n - t)^2/2) + \exp(-(Q_n + t)^2/2) - 2\exp(-Q_n^2/2), \quad (8.146)$$

let us consider extreme problems in the left-hand sides (8.107), (8.108) *which correspond to* (8.106). *Assume, for some* $B > 0$, *that*

$$Bz_n \geq Q_n - z_n \geq d_n \to \infty, \quad (2-\delta)Q_n - z_n \geq H_n > z_n. \quad (8.147)$$

Then for a large enough n *equality* (8.107) *and inequality* (8.108) *hold.*

Proofs of Lemmas 8.3 and 8.4. These are given in the Appendix, Section A.6.

Consider the component γ. Setting $\gamma_\zeta = \gamma^0 + \Delta\gamma$, for every $t > 0$ using (8.10), one has

$$\Delta\gamma = \lim_{n\to\infty} \sum_i \int_{|w_n|<t} w_n(x)\xi(x, v_i^{(n)})\, d\Phi(x) - \int_{+0}^{t} \frac{x^3}{1+x^2} d\Delta L(x) + \int_t^\infty \frac{x}{1+x^2} d\Delta L(x). \quad (8.148)$$

Since the quantity $\Delta\gamma$ does not depend on $t > 0$ in (8.148), suppose $t \to 0$ in (8.148). Relation (8.138) implies that the limit of the first term in the right-hand side is $b \geq 0$ (note that the quantity under the limit is $\Delta E_n^-(t)$). It also follows from (8.117) that the second term tends to 0. Moreover, we can write

$$\int_t^\infty \frac{x}{1+x^2} d\Delta L(x) = \int_{+0}^\infty \frac{x}{1+x^2} d\Delta L(x) - \int_{+0}^t \frac{x}{1+x^2} d\Delta L(x),$$

and using (8.139) one can pass to the limit in the second term

$$\begin{aligned} \lim_{t\to 0} \int_0^t \frac{x}{1+x^2} d\Delta L(x) &= \lim_{t_2\to 0} \lim_{t_1 \to 0} \int_{t_1}^{t_2} x d\Delta L(x) \\ &= \lim_{t_2\to 0} \lim_{t_1\to 0} \lim_{n\to\infty} (\Delta E_n^-(t_2) - \Delta E_n^-(t_1)) = 0. \end{aligned}$$

These imply equality (8.120), i.e.,

$$\Delta\gamma = b + \int_{+0}^\infty \frac{x}{1+x^2} d\Delta L(x).$$

Thus Theorem 8.4 is proved. □

8.4.3 Study of the Tests (8.41)

Put $\lambda_{n,\alpha} = \lambda_{n,\alpha}(z_n, h_n) = nh_n\Phi(z_n - H_{n,\alpha})$. Note that the quantity $\lambda_{n,\alpha}$ is of the form λ_n^* defined in Lemma 8.1.

Lemma 8.5 (1) *Assume either:* (i) $z_n - H_{n,\alpha} \to \infty$, *or* (ii) $nh_n \to \infty$ *and* $\lambda_{n,\alpha} \to \infty$. *Then* $\beta_n(\psi_{n,\alpha}, V_n) \to 0$.

(2) *Assume* $nh_n \to \infty$ *and* $\lambda_{n,\alpha} = O(1)$. *Then*

$$\beta_n(\psi_{n,\alpha}, V_n) \le (1-\alpha)\exp(-\lambda_{n,\alpha}) + o(1). \tag{8.149}$$

(3) *Assume* $nh_n = O(1)$. *Then*

$$\beta_n(\psi_{n,\alpha}, V_n) \le (1-\alpha)\Phi(\sqrt{2\log n} - z_n)^{nh_n} + o(1). \tag{8.150}$$

Proof of Lemma 8.5. We consider the case $\liminf \lambda_{n,\alpha} > 0$ below since the upper bounds are trivial for $\lambda_{n,\alpha} \to 0$. One has, for all $v \in V_n$,

$$\rho_n^p \le \sum_i |v_i|^p \le \max_i |v_i|^{p-q} \sum_i |v_i|^q \le \max_i |v_i|^{p-q} R_n^q.$$

Since $p > q$ this relation and (8.71) imply

$$\inf_{v \in V_n} \max_i |v_i| \ge z_n = (\rho_n^p / R_n^q)^{1/(p-q)}. \tag{8.151}$$

By analogy with (8.88) using (8.151) one has

$$\beta_n(\psi_{n,\alpha}, V_n) \le \Phi\left(H_{n,\alpha} - \inf_{v \in V_n} \max_i |v_i|\right) \le \Phi(H_{n,\alpha} - z_n) \to 0,$$

as $z_n - H_{n,\alpha} \to \infty$.

Let $nh_n \to \infty$. One can assume

$$H_{n,\alpha} \ge z_n + d_n \quad \text{for some sequence} \quad d_n \to \infty,\ d_n = o(z_n). \tag{8.152}$$

In fact, (8.152) holds when $\lambda_{n,\alpha} \asymp 1$. Let $\lambda_{n,\alpha} \to \infty$. If

$$\liminf (H_{n,\alpha} - z_n) < B < \infty,$$

then we can pass to the set $\tilde{V}_n = V_n(\tilde{\rho}_n, \tilde{R}_n)$ with

$$\tilde{\rho}_n = \rho_n(1-\delta_n), \quad \tilde{R}_n = R_n(1-\delta_n), \quad \delta_n = o(1), \quad \delta_n > 0.$$

By (8.1) this does not change h_n and leads to smaller $\tilde{z}_n = z_n(1-\delta_n)$ and we can take δ_n such that $\lambda_{n,\alpha} \to \infty$ and (8.152) is fulfilled. On the other hand, each point $\tilde{v} \in \tilde{V}_n$ is of the form

$$\tilde{v} = t_n v, \quad v \in V_n, \quad t_n = (1-\delta_n) < 1.$$

The admissible set of the tests (8.41) are convex and symmetric and it follows from Lemma 2.4 that $\beta_n(\psi_{n,\alpha}, \tilde{v}) \ge \beta_n(\psi_{n,\alpha}, v)$ which yields $\beta_n(\psi_{n,\alpha}, \tilde{V}_n) \ge \beta_n(\psi_{n,\alpha}, V_n)$.

Put $H_n = H_{n\alpha} + d_n/2$. Assume

$$v \in V_n, \quad \max_i |v_i| \ge H_n.$$

Then, analogously to (8.87), we have $\beta_n(\psi_{n,\alpha}, V_n) \to 0$. Assume

$$v \in V_n, \quad \max_i |v_i| \le H_n.$$

Introduce the functions (compare with Lemma 8.3)

$$F(v,Q)=\sum_{i=1}^{n}\phi(v_i,Q),\quad \phi(t,Q)=\Phi(-Q+t)+\Phi(-Q-t)-2\Phi(-Q).$$

One has (compare with (8.88))

$$\begin{aligned}\beta_n(\psi_{n,\alpha},v) &= \prod_{i=1}^{n}(1-\Phi(H_{n,\alpha}-|v_i|)-\Phi(-H_{n,\alpha}-|v_i|))\\ &= (1-\alpha)\prod_{i=1}^{n}\left(1-\frac{\phi(|v_i|,H_{n,\alpha})}{1-2\Phi(-H_{n,\alpha})}\right)\\ &\le (1-\alpha)\exp\left(-\frac{F(v,H_{n,\alpha})}{1-2\Phi(-H_{n,\alpha})}\right).\end{aligned} \tag{8.153}$$

Using Lemma 8.3 for $Q_n=H_{n,\alpha}$ one has

$$F(v,H_{n,\alpha})\ge\lambda_{n,\alpha}(1+o(1)),$$

which implies

$$\beta_n(\psi_{n,\alpha},v)\le(1-\alpha)\exp(-\lambda_{n,\alpha}(1+o(1)).$$

Thus statements (1) and (2) are proved.

Let $nh_n=O(1)$. Analogously to above, one can assume that $H_{n,\alpha}-z_n=O(1)$. Introduce the functions

$$F(v,Q_n)=\sum_{i=1}^{n}\phi(v_i,Q_n),$$

where

$$\phi(v,Q_n)=\log(\Phi(Q_n)-\Phi(-Q_n))-\log(\Phi(|v|-Q_n)-\Phi(-Q_n-|v|)). \tag{8.154}$$

Similarly to (8.89), (8.153) one has

$$\beta_n(\psi_{n,\alpha},v)=(1-\alpha)\exp(-F(v,H_{n,\alpha})).$$

Then inequality (8.150) follows from the relation $H_{n,\alpha}=\sqrt{2\log n}+o(1)$ and from the following lemma (we set $Q_n=H_{n,\alpha}$ in the lemma).

Lemma 8.6 *Let the functions* $\phi_n(t)=\phi_n(t,Q_n)$ *be defined by* (8.154). *Consider the extreme problems in the left-hand sides of* (8.107), (8.108) *which correspond to* (8.106). *Assume*

$$Q_n-z_n=O(1),\quad z_n+O(1)\ge H_n>z_n.$$

Then for a large enough n *equality* (8.107) *and inequality* (8.108) *hold true.*

Proof of Lemma 8.6. This is given in the Appendix, Section A.6.

Lemma 8.5 is proved. □

We obtain the upper bounds and consistency properties of Theorem 8.5 from Lemmas 8.5, 8.1. Recall only that $nh_n \geq 1$ (see the beginning of Section 8.3) and it follows from (8.21) that if $\tau < 1$, then either $nh_n \asymp 1$ and $z_n - H_{n,\alpha} \to \infty$ or $nh_n \to \infty$ and $\lambda_{n,\alpha} \to \infty$. Moreover, if $\tau \in (1, \infty)$, then, by (8.25), one has

$$c_n \sim (\tau - 1)\lambda_n/\tau \asymp \lambda_n.$$

Thus Theorem 8.5 is proved. □

8.5 Testing a Hypothesis which Is Close to a Simple Hypothesis

In Section 3.3.6 we have studied the problem of testing a simple null-hypothesis $H_0 : v = 0$ against an alternative which corresponds to the complement of an l^p-ball in R^n, i.e.,

$$H_1 : v \in R^n, \quad |v|_p \geq \rho_n, \quad 0 < p < \infty.$$

It was shown that distinguishability and nondistinguishability conditions in the problem are of the form $\rho_n/\rho_n^* \to \infty$ or $\rho_n/\rho_n^* \to 0$, and are determined by the rates (critical radii) ρ_n^* of the form (3.90). Now we would like to consider a composite null-hypothesis

$$H_0 : v \in R^n, \quad |v|_p \leq r_n,$$

which corresponds to an l^p-ball as well. The problem we would like to study is: *How small should radii r_n be to obtain distinguishability and nondistinguishability conditions with the same rates ρ_n^*?*

For example, let $p = 2$. Then one can easily see that χ^2-tests are minimax for this composite null-hypothesis as well, and the answer is $r_n = o(\rho_n^*)$. However, *what is it that holds for $p \neq 2$?*

The answer given below seems to be unexpected: *it holds for even p only.*

This result is extended to hypothesis testing problems under the functional Gaussian model and under smooth constraints on a signal. [3]

8.5.1 Statement and Main Results

Let us consider a hypothesis testing problem which corresponds to the null-hypothesis $H_0 : v \in V_{n,0}$ against the alternative $H_1 : v \in V_{n,1}$ where

$$V_{n,0} = \{v \in R^n : |v|_p \leq r_n\}, \quad V_{n,1} = \{v \in R^n : |v|_p \geq \rho_n\},$$

[3] Analogous phenomena have been established by Lepski, Nemirovski, and Spokoiny in [78] for the problem of the estimation of the functional $F(s) = \|s\|_p$.

where $0 < p < \infty$, $0 \le r_n < \rho_n$. We are interested in the distinguishability condition, i.e., in conditions either for $\gamma_n = \gamma(V_{n,0}, V_{n,1}) \to 0$ or for $\gamma_n = \gamma(V_{n,0}, V_{n,1}) \to 1$ (remember that the quantities $\gamma(V_{n,0}, V_{n,1})$ are defined in Section 2.4).

It was shown in Section 3.3.6 that for $r_n = 0$ we have the distinguishability condition:

$$\gamma_n \to 0 \quad \Leftrightarrow \quad \rho_n/\rho_n^* \to \infty; \quad \gamma_n \to 1 \quad \Leftrightarrow \quad \rho_n/\rho_n^* \to 0, \tag{8.155}$$

where the rates ρ_n^* are of the form

$$\rho_n^* = \rho_{n,p}^* = \begin{cases} n^{1/p-1/4}, & \text{if } p \le 2, \\ n^{1/2p}, & \text{if } p > 2. \end{cases} \tag{8.156}$$

We are interested in the question: *Which $r_n > 0$ preserve the conditions (8.155), (8.156)?*

The answer to this question is formulated in terms of the critical radii (rates) r_n^*, the form of which depends on the evenness of p. Namely, set

$$r_n^* = r_{n,p}^* = \begin{cases} n^{1/2p}, & \text{if } p \le 2, \\ n^{1/p-1/4k}, & \text{if } 2k \le p < 2(k+1), \ k = 1, 2, \dots. \end{cases} \tag{8.157}$$

The following theorem shows that the condition $r_n/r_n^* \to 0$ is necessary and sufficient for the conditions of distinguishability and nondistinguishability of form (8.156) in the problem.

Theorem 8.6

(1) *Lower bounds. Let $r_n/r_n^* \to \infty$. Then there exists a sequence ρ_n such that $\rho_n/\rho_n^* \to \infty$ and $\gamma_n \to 1$.*

(2) *Upper bounds. Let $r_n/r_n^* \to 0$. Then $\gamma_n \to 0$ for any sequence ρ_n such that $\rho_n/\rho_n^* \to \infty$.*

Observe that $r_n^* \asymp \rho_n^*$ only for even p; otherwise, $r_n^* = o(\rho_n^*)$. Moreover, the exponents of n in the definition of the rates $r_{n,p}^*$ are, as a function of p, discontinuous to the left for even $p > 2$.

8.5.2 Proof of Lower Bounds

It follows from Section 2.4.2 (see Proposition 2.9) that to obtain the relation $\gamma_n \to 1$ it suffices to construct sequences of priors $\pi_n^{(0)}, \pi_n^{(1)}$ on $(R^n, \mathcal{B}^n)$ meeting the conditions

$$\text{(i) } \pi_n^{(0)}(V_{n,0}) \to 1; \quad \text{(ii) } \pi_n^{(1)}(V_{n,1}) \to 1; \quad \text{(iii) } |P_{\pi_n^{(0)}} - P_{\pi_n^{(1)}}|_1 \to 0, \tag{8.158}$$

where $P_{\pi_n^{(l)}}$ are mixtures of the Gaussian measures $P_{n,v}$, $v \in R^n$, over the priors $\pi_n^{(l)}$, $l = 0, 1$, on $(R^n, \mathcal{B}^n)$ and

$$|P_{\pi_n^{(0)}} - P_{\pi_n^{(1)}}|_1 = E_{P_{n,0}} |L_{\pi_n^{(1)}}(x) - L_{\pi_n^{(0)}}(x)|$$

is the L_1-distance between the mixtures. We will search for the priors $\pi_n^{(l)}$ of product type with identical factors $\pi_{n,l}$ on $(R^1, \mathcal{B}^1)$, i.e.,

$$\pi_n^{(l)}(dv_1 \dots dv_n) = \pi_{n,l}(dv_1) \times \dots \times \pi_{n,l}(dv_n).$$

For these priors we have

$$\begin{aligned} L_{\pi_n^{(l)}}(x) &= \frac{dP_{\pi_n^{(l)}}}{dP_{n,0}} = \prod_{i=1}^n f_{\pi_{n,l}}(x_i), \\ f_{\pi_{n,l}}(t) &= \frac{dP_{\pi_{n,l}}}{dP_{1,0}}(t) = \int_{R^1} \exp(-u^2/2 + tu)\pi_{n,l}(du), \end{aligned}$$

where $P_{\pi_{n,l}}$ is the mixture of one-dimensional Gaussian measures $P_{1,t}$ over the one-dimensional prior $\pi_{n,l}$, and $f_{\pi_{n,l}}$ is the density of the mixture with respect to the standard Gaussian measure $P_{1,0}$, $l = 0, 1$.

It is easy to see, by means of the Chebyshev inequality, that to check the validity of conditions (i) and (ii) in (8.158) it suffices to establish, that for some constants $A > 1 > a > 0$, one has

$$nE_{\pi_{n,0}}|t|^p \le (ar_n)^p; \quad n\mathrm{Var}_{\pi_{n,0}}|t|^p = o(r_n^{2p}), \tag{8.159}$$

$$nE_{\pi_{n,1}}|t|^p \ge (A\rho_n)^p; \quad n\mathrm{Var}_{\pi_{n,1}}|t|^p = o(\rho_n^{2p}). \tag{8.160}$$

Remark. If conditions (8.159) and (8.160) hold and we have equality in the first inequalities, then, for any $B_0 > a > b_0 > 0$, $B_1 > A > b_1 > 0$, the priors $\pi_n^{(l)}$, $l = 0, 1$, are concentrated within the rings

$$R_{n,0} = \{b_0 \le |v|_p/r_n \le B_0\}, \quad R_{n,1} = \{b_1 \le |v|_p/\rho_n \le B_1\},$$

with the probability tending to 1, i.e., $\pi_n^{(l)}(R_{n,l}) \to 1$, $l = 0, 1$.

The following statement is convenient to verify condition (iii) in (8.158).

Lemma 8.7 *Let, for sufficiently large n and some $\delta > 0$,*

$$\inf_{t \in R^1} f_{\pi_{n,l}}(t) \ge \delta, \quad l = 0 \text{ or } 1. \tag{8.161}$$

Then in order for condition (iii) in (8.158) *to be true it suffices that*

$$n\|\pi_{n,1} - \pi_{n,0}\|^2 = nE_{P_{1,0}}(f_{\pi_{n,1}}(t) - f_{\pi_{n,0}}(t))^2 = o(1). \tag{8.162}$$

Recall that the norm $\|r\|$ is determined by the relation (see Section 3.3.3)

$$\|r\|^2 = \int_{R^1}\int_{R^1} (e^{uv} - 1)\, r(du)\, r(dv). \tag{8.163}$$

Proof of Lemma 8.7. We can assume $l = 0$ by symmetry. By inequality the $1 + x \le e^x$ we get

$$|P_{\pi_n^{(0)}} - P_{\pi_n^{(1)}}|_1^2 \le E_{P_{\pi_n^{(0)}}}\left(\frac{L_{\pi_n^{(1)}}(x)}{L_{\pi_n^{(0)}}(x)} - 1\right)^2 = E_{P_{\pi_n^{(0)}}}\left(\frac{L_{\pi_n^{(1)}}(x)}{L_{\pi_n^{(0)}}(x)}\right)^2 - 1$$

$$= \left(E_{P_{\pi_{n,0}}} \left(\frac{f_{\pi_{n,1}}(t)}{f_{\pi_{n,0}}(t)} - 1 \right)^2 + 1 \right)^n - 1$$

$$\leq \exp \left(n E_{P_{\pi_{n,0}}} \left(\frac{f_{\pi_{n,1}}(t)}{f_{\pi_{n,0}}(t)} - 1 \right)^2 \right) - 1,$$

and, therefore, in order for condition (iii) in (8.158) to be valid it suffices that

$$n E_{P_{\pi_{n,0}}} \left(\frac{f_{\pi_{n,1}}(t)}{f_{\pi_{n,0}}(t)} - 1 \right)^2 = n E_{P_{1,0}} \frac{(f_{\pi_{n,1}}(t) - f_{\pi_{n,0}}(t))^2}{f_{\pi_{n,0}}(t)} = o(1). \quad (8.164)$$

However under (8.161) relation (8.164) follows from (8.162). □

Recall the relation (see (3.72))

$$\|r\|^2 = \sum_{j=1}^{\infty} \frac{(\mu_j(r))^2}{j!}; \quad \mu_j(r) = \int_{R^1} u^j r(du),$$

where $\mu_j(r)$ are the moments of signed measure r and, if $\pi_{n,l}$, $l = 0, 1$, are symmetric (below we use symmetric measures only), then (8.162) may be rewritten as

$$n \sum_{j=1}^{\infty} \frac{(\mu_{2j}(\pi_{n,1}) - \mu_{2j}(\pi_{n,0}))^2}{(2j)!} = o(1). \quad (8.165)$$

The scheme we use below to construct the desired measures $\pi_{n,l}$, $l = 0, 1$, is based on the following idea. Assuming conditions (8.159), (8.160), we try to make equal to 0 as many as possible of the first terms of series (8.165).

For the measure π on $(R^1, \mathcal{B}^1)$ denote, as above, by

$$|\pi| = \inf\{T : \ \pi[-T, T] = 1\}/2,$$

one-half of the length of the symmetric support of the measure. It is convenient to use the following simple observation:

Lemma 8.8 *Let* $|\pi_{n,l}| = O(1)$, $l = 0, 1$.

(1) *Let* $\pi_{n,l}$, $l = 0, 1$, *be symmetric measures. Then* (8.161) *is valid.*

(2) *Let* $\rho_n \to \infty$, $r_n \to \infty$, *the first conditions* (8.159) *and* (8.160), *be valid, and in addition, the left- and right-hand sides in these relations have equal orders. Then the second conditions in* (8.159) *and* (8.160) *hold.*

(3) *Let* $\pi_{n,l}$, $l = 0, 1$, *be symmetric measures such that*

$$\mu_{2j}(\pi_{n,1}) = \mu_{2j}(\pi_{n,0}), \quad j \leq k, \quad k \geq 1,$$

and let $n(\mu_{2(k+1)}(\pi_{n,l}))^2 = o(1)$, $l = 0, 1$. *Then* (8.165) *holds.*

Proof. Statement (1) is a consequence of the following evaluation: for the symmetric probability measure π,

$$f_\pi(t) = \int_{R^1} e^{-u^2/2+ut} \pi(du)$$

$$= \frac{1}{2}\left(\int_{R^1} e^{-u^2/2+ut}\pi(du) + \int_{R^1} e^{-u^2/2-ut}\pi(du)\right)$$
$$= \int_{R^1} e^{-u^2/2}\cosh(tu)\pi(du) \geq \int_{R^1} e^{-u^2/2}\pi(du) \geq e^{-|\pi|^2/2}.$$

Statements (2) and (3) follow from the inequalities

$$E_\pi(|t|^{a+b}) \leq |\pi|^a E_\pi(|t|^b) \quad \text{for } a>0, \quad b>0; \quad \mathrm{Var}_\pi(|t|^p) \leq E_\pi(|t|^{2p}),$$

and from obvious estimates for the sum of series (8.165). □

Case $p \leq 2$

We illustrate this scheme for the case $p \leq 2$.

Let $r_n = U_n^{1/p} r_n^*$, $U_n \to \infty$, where we may assume that $U_n = o(n^a)$ for small enough $a > 0$. Put

$$\pi_{n,1} = (\delta_{-b_1} + \delta_{b_1})/2; \quad \pi_{n,0} = (1-h)\delta_0 + h(\delta_{-b_0} + \delta_{b_0})/2;$$

therefore the measures are symmetric and $|\pi_{n,l}| = b_l$, $l = 0,1$. Let $b_l = b_{l,n} = o(1)$, $h = h_n = o(1)$ be positive sequences subject to the conditions

$$\begin{gathered} b_1^2 = hb_0^2; \quad h = n^{-1/2}A_n, \quad 2A_n b_0^p = U_n, \\ A_n b_0^2 \to \infty, \quad A_n b_0^2 = o(n^{1/4}), \quad A_n b_0^4 \to 0. \end{gathered} \tag{8.166}$$

For example, if $p < 2$, then one can take

$$2A_n = U_n^{1+p/2}, \quad b_0 = U_n^{-1/2}, \quad a = 1/4.$$

If $p = 2$, then one can take

$$2A_n = U_n^{17/8}, \quad b_0 = U_n^{-7/16}, \quad a = 1/8.$$

Let us take

$$\rho_n = (n/2)^{1/p} b_1 = 2^{-1/p} A_n^{1/2} b_0 \rho_n^* \gg \rho_n^*.$$

Under this choice, conditions (8.159) and (8.160) are valid. In fact, one has

$$nE_{\pi_{n,0}}|t|^p = nhb_0^p = r_n^p/2; \quad nE_{\pi_{n,1}}|t|^p = nb_1^p = 2\rho_n^p.$$

The first condition in (8.166) allows us to cancel the difference of the second moments in (8.165) whereas for the fourth moments we have

$$\begin{aligned} n(\mu_4(\pi_{n,0}))^2 &= nh^2 b_0^8 = A_n^2 b_0^8 = o(1); \\ n(\mu_4(\pi_{n,1}))^2 &= nb_1^8 = n^{-1}(A_n b_0^2)^4 = o(1). \end{aligned}$$

Using Lemma 8.8 we have lower bounds from Theorem 8.6 for $p \leq 2$. □

Case $p \geq 2$

Let $2k \leq p < 2(k+1)$, $k = 1, 2, \ldots$. Take sequences of the real-valued tuples

$$z_1 > z_2 > \ldots > z_{k+1} > 0, \quad h_1,\ h_2, \ldots, h_{k+1}.$$

Here $z_i = z_{i,n}$, $h_i = h_{i,n} > 0$, $1 \leq i \leq k+1$ are quantities such that $H = \sum_i h_i \leq 1$ (below we refine their form). We set $\pi(z,h) = h(\delta_{-z} + \delta_z)/2$,

$$H_1 = \sum_{i:1\leq 2i-1\leq k+1} h_{2i-1}, \quad H_0 = \sum_{i:\ 2\leq 2i\leq k+1} h_{2i};$$

$$\pi_{n,1} = (1 - H_1)\delta_0 + \sum_{i:\ 1\leq 2i-1\leq k+1} \pi(z_{2i-1}, h_{2i-1}),$$

$$\pi_{n,0} = (1 - H_0)\delta_0 + \sum_{i:\ 2\leq 2i\leq k+1} \pi(z_{2i}, h_{2i}),$$

i.e., $\pi_{n,1}$ is a measure concentrated at points 0, $\pm z_1$, $\pm z_3, \ldots$, and $\pi_{n,0}$ is a measure concentrated at points 0, $\pm z_2$, $\pm z_4, \ldots$ and all measures are symmetric. Take $h = h_{k+1} = \text{const} \in (0,1)$ (e.g., $h = 1/2$).

Fix for a moment the tuples $\{z_i\}$. We select the tuples $\bar{h} = \{h_i,\ i = 1, \ldots, k\}$ in such a way as to annihilate the first k items is series (8.165), i.e., in such a way that

$$\mu_{2j}(\pi_{n,1}) = \mu_{2j}(\pi_{n,0}), \quad j = 1, \ldots, k.$$

This leads to the following system of equations, with respect to $\bar{h}$,

$$\sum_{i=1}^{k} (-1)^{(i-1)} h_i z_i^{2j} = (-1)^{k+1} h z_{k+1}^{2j}; \quad j = 1, \ldots, k. \tag{8.167}$$

Set

$$Q_i(z) = \frac{z^2}{z_i^2} \prod_{1\leq s\leq k,\ s\neq i} \frac{z^2 - z_s^2}{z_i^2 - z_s^2}.$$

Lemma 8.9 *If* $z_1 > z_2 > \ldots > z_{k+1} > 0$, *then the tuples*

$$h_i = (-1)^{k-i} h Q_i(z_{k+1}), \quad i = 1, \ldots, k,$$

solve system (8.167) *and* $h_i > 0$, $i = 1, \ldots, k$.

Proof. It is easy to check that if $0 < z < z_k$, then

$$(-1)^{k-i} h Q_i(z) > 0, \quad i = 1, \ldots, k.$$

To verify that h_i solve system (8.167), consider the functions

$$P_j(z^2) = \sum_{i=1}^{k} (-1)^{k-1} h Q_i(z) z_i^{2j} + (-1)^k h z^{2j}, \quad j = 1, \ldots, k,$$

as a polynomial of power k in $x = z^2$. Setting $z_0 = 0$ it is easy to see that $Q_j(z_i)$ equals 1 for $i = j$ and $Q_j(z_i)$ equals 0 for $i \neq j$; moreover, $Q_j(z_0) = 0$; $i, j = 1, \dots, k$. Hence it follows that $P_j(z_i^2) = 0$ for $i = 0, \dots, k$. Therefore $P_j(x) \equiv 0$ identically for $j = 1, \dots, k$. In particular, $P_j(z_{k+1}^2) = 0$, $j = 1, \dots, k$. This yields the statement of the lemma. □

We return to the choice of tuples $\{z_i\}$. We take the positive sequence $x = x_n = o(1)$, $y = y_n = o(1)$, $b = b_n = o(1)$, and put

$$z_1 = b; \quad z_i^2 = b^2 x y^{i-2}, \quad 2 \le i \le k+1.$$

It follows from Lemma 8.9 that in this case we have

$$h_1 = hxy^{k-1}\frac{(x - xy^{k-1})(xy - xy^{k-1})...(xy^{k-2} - xy^{k-1})}{(1-x)(1-xy)...(1-xy^{k-2})} \sim hx^k y^{\phi_1(k)}, \tag{8.168}$$

and, for $2 \le i \le k$,

$$h_i = h\frac{xy^{k-1}(1 - xy^{k-1})(x - xy^{k-1})...(xy^{i-3} - xy^{k-1})(xy^{i-1} - xy^{k-1})...}{xy^{i-2}(1 - xy^{i-2})(x - xy^{i-2})...(xy^{i-3} - xy^{i-2})(xy^{i-2} - xy^{i-1})...} \sim hy^{\phi_i(k)}, \tag{8.169}$$

where

$$\phi_i(k) = \begin{cases} k(k-1)/2 & \text{for } i = 1, \\ (k+1-i)(k+2-i)/2 & \text{for } 2 \le i \le k. \end{cases}$$

Note that $\phi_i(k) > 0$ for $1 \le i \le k$ and set $\phi_1(k) = \phi(k)$, $s = s_n = y^{\phi(k)}$. Then we get

$$\phi_2(k) = \phi(k), \quad \phi_i(k) = \phi(k) - (i-2)(2k-i+1)/2, \quad i = 3, \dots, k.$$

By (8.168), (8.169) the condition $H_1 + H_0 \sim h < 1$ holds for large n. Therefore $\pi_{n,l}$ are probability measures satisfying $|\pi_{n,1}| = b = o(1)$ and $|\pi_{n,0}| = bx^{1/2} = o(1)$.

Set

$$\psi_i(k,p) = \phi_i(k) + p(i-2)/2 \text{ for } 2 \le i \le k, \quad \psi_{k+1}(k,p) = (k-1)p/2.$$

Note that for $p \ge 2k$ we have $\psi_{k+1}(k,p) > \phi(k)$ and

$$\psi_2(k,p) = \phi(k), \quad \psi_i(k,p) = \phi(k) + (p - 2k + i - 1)(i-2)/2 > \phi(k),$$

for $i = 3, \dots, k$. Using relations (8.168), (8.169) once again we have

$$E_{\pi_{n,1}}|t|^p = \sum_{j:1 \le 2j+1 \le k+1} h_{2j+1} z_{2j+1}^p \sim hb^p(x^k y^{\phi_1(k)} + x^{p/2} y^{\psi_3(k,p)} + \dots + x^{p/2} y^{\psi_{2j+1}(k,p)} + \dots) \sim hb^p x^k s; \tag{8.170}$$

$$E_{\pi_{n,0}}|t|^p = \sum_{j:2 \le 2j \le k+1} h_{2j} z_{2j}^p \sim hb^p(x^{p/2} y^{\psi_2(k,p)} + \dots + x^{p/2} y^{\psi_{2j}(k,p)} + \dots) \sim hb^p x^{p/2} s. \tag{8.171}$$

Using analogous evaluations for $p = 2(k+1)$ we get

$$E_{\pi_{n,1}}|t|^{2(k+1)} = O(b^{2(k+1)}x^k s), \quad E_{\pi_{n,0}}|t|^{2(k+1)} = O(b^{2(k+1)}x^{(k+1)}s). \tag{8.172}$$

Set $r_n = U_n^{1/p}r_n^*$, $U_n \to \infty$, and $U_n = o(n^a)$ for small enough $a > 0$. Choose $x = x_n = o(1)$, $b = b_n = o(1)$, $y = y_n = o(1)$ in such a way that

$$2hnx^{p/2}b^p z = r_n^p, \quad nx^k b^p z \gg (\rho_n^*)^p = n^{1/2}, \quad x^k b^{2(k+1)} z = o(n^{-1/2}). \tag{8.173}$$

Setting $\rho_n^p = nhx^k b^p z/2$ and using (8.170)–(8.172) we see that

$$\begin{aligned} nE_{\pi_{n,1}}|t|^p &\sim 2\rho_n^p, \quad 2nE_{\pi_{n,0}}|t|^p \sim r_n^p, \\ n(E_{\pi_{n,1}}|t|^{2(k+1)})^2 &= O(nx^{2k}b^{4(k+1)}z^2) = o(1), \\ n(E_{\pi_{n,0}}|t|^{2(k+1)})^2 &= O(nx^{2(k+1)}b^{4(k+1)}z^2) = o(x^2) = o(1), \end{aligned}$$

the sequence satisfies $\rho_n \gg \rho_n^*$ and by Lemma 8.8 it meets the conditions of point (1) of Theorem 8.6.

It remains to show that it is possible to select the quantities satisfying (8.173). Take $b = U_n^{-\alpha}$, $x = n^{-1/2k}U_n^{\beta}$, and $2hz = U_n^{-\gamma}$. It suffices to find $\alpha > 0$, $\beta > 0$, $\gamma > 0$ such that

$$\beta = 2\alpha + 2(1+\gamma)/p, \quad \gamma + p\alpha < k\beta < \gamma + 2(k+1)\alpha.$$

If $p = 2k$, then to meet these relations it suffices to choose $\alpha > 1/2$ and any $\gamma > 0$. If $2k < p < 2(k+1)$, then it suffices to choose γ such that

$$\frac{2k}{p-2k} - \frac{2p\alpha}{p-2k} < \gamma < \frac{2k}{p-2k} - p\alpha.$$

To meet $\gamma > 0$ in this case it suffices to choose α such that $0 < \alpha \le k/p$. It is possible for $2k < p < 2(k+1)$. Thus point (1) of Theorem 8.6 is proved.

Remark. We have in fact proved the following. Let $r_n = U_n r_{n,p}^*$ where $U_n \to \infty$, $U_n = o(n^a)$ for some $0 < a \le a(p)$. Then there exists $\delta(p) > 0$ such that $\gamma_n \to 1$, if $\rho_n = O(U_n^{\delta}\rho_{n,p}^*)$ for $0 < \delta \le \delta(p)$. In addition, on account of the Remark on page 339 one may replace the sets $V_{n,l}$ in Theorem 8.6 with the rings $R_{n,l}$, $l = 0, 1$, defined in this Remark.

8.5.3 Upper Bounds

Let $r_n = o(r_n^*)$ and $\rho_n = U_n\rho_n^*$ where $U_n \to \infty$. We consider a sequence of tests

$$\psi_{n,p}(x) = \mathbf{1}_{\{L_{n,p}(x) > T_n\}}$$

based on the statistics

$$L_n = L_{n,p}(x) = n^{-1/2}\sum_{i=1}^n u_p(x_i)$$

and on the thresholds $T_n = U_n^{p/2} \to \infty$; the form of the functions $u_p(x)$ will be specified later. We would like to show that these tests are minimax consistent under the assumptions of point (2) of Theorem 8.6.

Using Proposition 3.1 we get that it suffices to show that

$$\sup_{v \in V_{n,0}} \mathrm{Var}_{n,v}(L_n) = O(1), \quad \sup_{v \in V_{n,0}} E_{n,v}(L_n) = o(1), \tag{8.174}$$

$$\inf_{v \in V_{n,1}} E_{n,v}(L_n)/T_n \to \infty, \quad \sup_{v \in V_{n,1}} \mathrm{Var}_{n,v}(L_n)/E^2_{n,v}(L_n) = o(1). \tag{8.175}$$

Below we make use of the inequality, for $0 \le a \le b$,

$$n^{-1}\sum_{i=1}^n |x_i|^a \le \left(n^{-1}\sum_{i=1}^n |x_i|^b\right)^{a/b}, \quad \left(\sum_{i=1}^n |x_i|^b\right)^{a/b} \le \sum_{i=1}^n |x_i|^a. \tag{8.176}$$

Case $p \le 2$

Set $u_p(x) = |x|^p - c(p)$ where $c(p) = E(|\xi|^p)$ and ξ is a standard Gaussian random variable. Note the following statement:

Lemma 8.10 *Let*

$$p \le 2, \quad h_p(t) = E(u_p(\xi + t)), \quad H_p(t) = \mathrm{Var}(u_p(\xi + t)), \quad x_+ = \max(x, 0).$$

There exist constants $C_l = C_l(p) > 0$, $l = 1, 2, 3$, such that, for any $t \in R^1$,

$$C_1 \min(t^2, |t|^p) \le h_p(t) \le C_2 |t|^p, \quad H_p(t) \le C_3(1 + |t|^{2(p-1)_+}).$$

Proof. It is not difficult to see that $h_p(t)$ is an even function which is strictly increasing on $|t|$, $h_p(0) = 0$, $h_p''(0) > 0$. In fact, we can rewrite the relation for $h_p(t)$ in the form

$$h_p(t) = (2\pi)^{-1/2} \int_R (|y|^p \exp(-(y-t)^2/2)\, dy - c(p)$$

which leads to $h_p(t)$ being infinitely differentiable and we have the relations

$$(2\pi)^{1/2}(h_p(t) - h_p(-t)) = \int_R |y|^p (\exp(-(y-t)^2/2)$$
$$- \exp(-(y+t)^2/2))\, dy = 2\int_R |y| \exp(-(y^2 + t^2)/2) \sinh(yt)\, dy = 0;$$

$$\begin{aligned}(2\pi)^{1/2} h_p'(t) &= p \int_R |x + t|^{p-1} \exp(-(x^2/2)\, dx \\ &= p \int_0^\infty y^{p-1}(\exp(-(y-t)^2/2) - \exp(-(y+t)^2/2))\, dy \\ &= 2p \int_0^\infty y^{p-1} \exp(-(y^2 + t^2)/2) \sinh(yt)\, dy > 0 \quad \text{for } t > 0.\end{aligned}$$

Setting

$$m(p) = \int_0^\infty y^p \exp(-y^2/2)\, dy, \quad p > 0,$$

we have

$$m'(p) = p \int_0^\infty y^{p-1} \exp(-y^2/2)\, dy > 0$$

and $m(p)$ is strictly increasing in $p > 0$. Therefore

$$(2\pi)^{1/2} h_p''(0) = \int_R |y|^p \exp(-y^2/2)(y^2 - 1)\, dy = 2(m(p+2) - m(p)) > 0.$$

Thus for any $b > 0$ one can find constants $B_i = B_i(p, b) > 0$ such that

$$B_1 t^2 \le h_p(t) \le B_2 t^2, \quad |t| \le b. \tag{8.177}$$

On the other hand, note the inequality

$$|a + b|^p \le A(p)(|a|^p + |b|^p), \quad A(p) = 2^{(p-1)_+}. \tag{8.178}$$

If $p \ge 1$, then inequality (8.178) follows from the convexity of the function x^p in $x > 0$; if $0 < p < 1$, then, setting

$$x = |a|/(|a| + |b|),\ y = |b|/(|a| + |b|), \quad x, y \in [0, 1], \quad x + y = 1,$$

and since $x^p \ge x,\ y^p \ge y$, (8.178) follows from the inequalities $1 \le x^p + y^p$.

Using (8.178) we obtain the evaluations

$$-2c(p) + |t|^p / A(p) \le h_p(t) \le (A(p) - 1)c(p) + A(p)|t|^p. \tag{8.179}$$

Relations (8.177), (8.179) yield the needed evaluations for $h_p(t)$. Further, applying Taylor's formula for $p > 1$,

$$||t + \xi|^p - |t|^p| = p|\theta\xi + t|^{p-1}|\xi| \le p(|\xi|^{p-1} + |t|^{p-1})|\xi|, \tag{8.180}$$

and the inequality

$$||t + \xi|^p - |t|^p| \le |\xi|^p \quad \text{for } 0 < p \le 1,$$

following from (8.178), and recalling the inequality

$$D(|t + \xi|^p) \le E(|t + \xi|^p - |t|^p)^2,$$

we deduce the needed evaluations for $H_p(t)$. $\square$

To evaluate the type I errors we check relations (8.174). By Lemma 8.10 we have, for $r_n \ll r_n^* = n^{1/2p}$,

$$\begin{aligned} E_{n,v}(L_n) &= n^{-1/2} \sum_{i=1}^n h_p(v_i) \le C_2 n^{-1/2} \sum_{i=1}^n |v_i|^p \le C_2 n^{-1/2} r_n^p = o(1), \\ \mathrm{Var}_{n,v}(L_n) &= n^{-1} \sum_{i=1}^n H_p(v_i) \le C_3 \left(1 + n^{-1} \sum_{i=1}^n |v_i|^{2(p-1)_+}\right) \end{aligned}$$

$$\le C_3\left(1+D_n^{2(p-1)_+/p}\right)=O(1), \quad D_n=n^{-1}\sum_{i=1}^n |v_i|^p, \tag{8.181}$$

(we have also used the first inequality in (8.176), since $2(p-1)_+ \le p$ for $0<p\le 2$, and the fact that $D_n \le n^{-1}r_n^p = o(1)$).

To evaluate the type II errors we check relations (8.175). Using Lemma 8.10 as before, we have

$$E_{n,v}(L_n) \ge C_1 n^{-1/2}\left(\sum_{i\in I} v_i^2 + \sum_{i\notin I}|v_i|^p\right), \quad I=I_n(v)=\{i:|v_i|\le 1\}.$$

Since

$$S_{n,1}+S_{n,2}=\sum_{i\in I}|v_i|^p+\sum_{i\notin I}|v_i|^p \ge \rho_n^p, \quad v\in V_{n,1},$$

one of the sums $S_{n,1}$, $S_{n,2}$ is not less than $\rho_n^p/2$. If this is the first sum, we have, by the first inequality (8.176),

$$n^{-1/2}\sum_{i\in I} v_i^2 \ge n^{1/2-2/p}S_{n,1}^{2/p} \ge n^{1/2-2/p}(\rho_n^p/2)^{2/p} \asymp U_n^2;$$

if this is the second sum, then

$$n^{-1/2}S_{n,2} \ge n^{-1/2}\rho_n^p/2 \asymp U_n^p n^{(2-p)/4}.$$

In any case, we get the first of the relationships in (8.175).

Further, by using the evaluations of variances given in (8.181), we have

$$\mathrm{Var}_{n,v}(L_n) = O(1) \quad \text{for } p\le 1 \quad \text{or } D_n=O(1).$$

Let $1<p\le 2$ and $D_n = D_{n,1}+D_{n,2}\to\infty$. Then, applying inequality (8.178), we get

$$\mathrm{Var}_{n,v}(L_n) = O\left(D_{n,1}^{2(p-1)/p}+D_{n,2}^{2(p-1)/p}\right)$$

where

$$D_{n,1}=n^{-1}\sum_{i\in I}|v_i|^p, \quad D_{n,2}=n^{-1}\sum_{i\notin I}|v_i|^p.$$

Note that

$$D_{n,1}^{2(p-1)/p} \le \left(n^{-1}\sum_{i\in I} v_i^2\right)^{p-1} \le 1 = o\left(E_{n,v}(L_n)\right),$$

since $p\le 2$. Therefore $D_{n,2}\to\infty$ and

$$D_{n,2}^{2(p-1)/p} \le D_{n,2} \ll n^{-1/2}\sum_{i\notin I}|v_i|^p \le E_{n,v}(L_n)/C_1,$$

since $2(p-1)/p\le 1$. In any case, the second relationship in (8.175) holds. Thus point (2) of Theorem 8.6 is proved for $p\le 2$.

Case $p \geq 2$

Let $2k \leq p < 2(k+1)$, $k \geq 1$, and k is an integer. Let ξ be a standard Gaussian random variable and let $f_p(t) = E(|\xi+t|^p)$. It is clear that $f_p(t)$ is an even infinitely differentiable function. Let $c_j(p) = f_p^{(2j)}(0)$ be derivatives of this function at point 0. Consider the functions $\phi_j(x)$, such that $E\phi_j(\xi+t) = t^{2j}$, $j = 0, 1, \dots, k-1$ (they are polynomials of degree $2j$; their explicit form is $\phi_j(x) = E(x+i\xi)^{2j}$, $i = \sqrt{-1}$, see [78], for example). Set

$$u_p(x) = |x|^p - \sum_{j=0}^{k-1} c_j(p)\phi_j(x)/(2j)! = |x|^p - Q_{k-1}(x^2),$$

where Q_{k-1} are polynomials of degree $k-1$. Note the following statement (compare with Lemma 3.2).

Lemma 8.11 *Let*

$$p \geq 2, \quad h_p(t) = Eu_p(\xi+t) = f_p(t) - P_{k-1}(t^2),$$

where P_{k-1} are polynomials of degree $k-1$ and $H_p(t) = Du_p(\xi+t)$. There exist constants $C_i = C_i(p) > 0$ such that, for any $t \in R^1$,

$$C_1|t|^p \leq h_p(t) \leq C_2(|t|^p + t^{2k}); \quad H_p(t) \leq C_3(1+|t|^{2(p-1)}).$$

Proof. One can easily see that $h_p(t)$ is an even infinitely differentiable function and that

$$h_p^{(j)}(0) = 0, \quad j = 0, 1, \dots, 2k-1; \quad h_p^{(2k)}(0) > 0.$$

Expanding $h_p(t)$ by Taylor's formula at point 0 and averaging we get

$$h_p(t) = d_k E(|\xi+\theta t|^{p-2k})t^{2k}, \quad \theta = \theta(t,\xi) \in [0,1], \tag{8.182}$$

where $d_k = p(p-1)\dots(p-2k+1)/(2k)! > 0$. Since the mean in (8.182) is bounded and separated from point 0 over any finite interval in t, there exist constants $B_i = B_i(p,b) > 0$ such that

$$B_0|t|^p \leq B_1 t^{2k} \leq h_p(t) \leq B_2 t^{2k}, \quad |t| \leq b. \tag{8.183}$$

Using inequality (8.178) we obtain, analogously to (8.179),

$$E(|\xi+\theta t|^{p-2k}) \leq A(p-2k)(E|\xi|^{p-2k} + |t|^{p-2k}). \tag{8.184}$$

Now the upper bounds for $h_p(t)$ follow from (8.182)–(8.184) .

Making use of inequality (8.178) once again we get

$$|\xi+t|^p \geq |t|^p/A(p) - |\xi|^p.$$

Therefore one has, for sufficiently large $b > 0$ and some $B > 0$,

$$h_p(t) \geq |t|^p/A(p) - P_{k-1}(t^2) - E|\xi|^p \geq B|t|^p \quad \text{for } |t| > b.$$

This gives the needed lower bounds for $h_p(t)$.

Using the inequality

$$||t+\xi|^p - |t|^p| \le p(|\xi|^{p-1} + |t|^{p-1})$$

(which is analogous to (8.180)) and taking into account the inequality

$$\mathrm{Var}(u_p(\xi+t)) = \mathrm{Var}(u_p(\xi+t)-|t|^p) \le 2E(|t+\xi|^p-|t|^p)^2+2EQ_{k-1}^2((\xi+t)^2)$$

and the relation

$$EQ_{k-1}^2((\xi+t)^2) = O(t^{4(k-1)}), \quad \text{as } t \to \infty,$$

we get the needed bounds for $H_p(t)$. □

To evaluate the type I errors we check the validity of (8.174). Applying Lemma 8.11 we have

$$E_{n,v}(L_n) = n^{-1/2}\sum_{i=1}^n h_p(v_i) \le C_2 n^{-1/2}(S_1+S_2),$$

where, by the first inequality in (8.176),

$$S_1 = \sum_{i=1}^n |v_i|^p \le r_n^p, \quad S_2 = \sum_{i=1}^n v_i^{2k} \le n^{1-2k/p} S_1^{2k/p}.$$

Therefore, if $r_n \ll r_n^* = n^{1/p-1/4k}$, then

$$\sup_{v\in V_{n,0}} E_{n,v}L_n \le n^{-1/2}(r_n^p + n^{1-2k/p}r_n^{2k}) = o(n^{(2k-p)/4k}+1) = o(1).$$

Further, using the the inequality $2(p-1) \ge p$ and second inequality in (8.176), we have

$$\begin{aligned}\mathrm{Var}_{n,v}(L_n) &= n^{-1}\sum_{i=1}^n H_p(v_i) < C_3\left(1+n^{-1}\sum_{i=1}^n |v_i|^{2(p-1)}\right)\\ &\le C_3\left(1+n^{-1}S_1^{2(p-1)/p}\right) = O(1+n^r) = O(1), \quad (8.185)\end{aligned}$$

since it is easy to check that

$$r = 1-2/p-(p-1)/2k < 0 \quad \text{for } p \ge 2k \ge 2.$$

To evaluate the type II errors we check the validity of (8.175). Using Lemma 8.11 once again we have

$$E_{n,v}(L_n) \ge C_1 n^{-1/2} S_1 \ge C_1 n^{-1/2}\rho_n^p \asymp U_n^p. \qquad (8.186)$$

Therefore the first of the relationships in (8.175) holds. By (8.185), (8.186) we have

$$\mathrm{Var}_{n,v}(L_n) \le Bn^{-1/p}(E_{n,v}L_n)^{2(p-1)/p} = o((E_{n,v}L_n)^2).$$

This yields the second of the relationships in (8.175). This completes the proof of Theorem 8.6. □

8.5.4 Extension to the Functional Gaussian Model

The results above could be extended to the functional Gaussian model (2.5) which corresponds to observation of a signal $f \in L_2(0,1)$ in white Gaussian noise,

$$dX_\varepsilon(t) = f(t)\,dt + \varepsilon\,dW(t), \quad \theta \in [0,1].$$

Let the null-hypothesis and the alternative be of the form

$$H_0:\ f \in F_{\varepsilon,0}, \quad H_1:\ f \in F_{\varepsilon,1},$$

where

$$F_{\varepsilon,0} = \{f \in L_2(0,1)\ :\ \|f\|_p \le \rho_{\varepsilon,0}\},$$
$$F_{\varepsilon,1} = \{f \in L_2(0,1):\ \|f\|_{(\eta,p)} \le C,\ \ \|f\|_p \ge \rho_{\varepsilon,1}\}.$$

Here $\|f\|_p$ is an L_p-norm and $\|\cdot\|_{(\eta,p)}$ is a Sobolev or Besov norms, $1 \le p < \infty$, $\eta > 0$ (see Section 2.7.2); $C > 0$ is a constant and $0 \le \rho_{\varepsilon,0} < \rho_{\varepsilon,1}$ are the given real-valued families.

We consider the case of a Besov norm, $\|\cdot\|_{(\eta,p)} = \|\cdot\|_{\eta,p,h}$ with some $h \in [1,\infty]$. Since the results we formulate below do not depend on parameter h, the same results for a Sobolev norm follow from inequalities (2.80).

We are interested, as above, in the conditions for distinguishability ($\gamma_\varepsilon \to 0$) and nondistinguishability ($\gamma_\varepsilon \to 1$) as $\varepsilon \to 0$ in the problem. Remember (see Section 3.5) that for the case of a simple hull-hypothesis (i.e., for $\rho_{\varepsilon,0} = 0$) the conditions are of the form

$$\gamma_\varepsilon \to 0 \quad \Leftrightarrow \quad \rho_{\varepsilon,1}/\rho^*_{\varepsilon,1} \to \infty; \quad \gamma_\varepsilon \to 1 \quad \Leftrightarrow \quad \rho_{\varepsilon,1}/\rho^*_{\varepsilon,1} \to 0, \tag{8.187}$$

with the rates (critical radii)

$$\rho^*_{\varepsilon,1} = \rho^*_{\varepsilon,1,p} = \begin{cases} \varepsilon^{4\eta/(4\eta+1)} & \text{for } p \le 2, \\ \varepsilon^{2\eta/(2\eta+1-1/p)} & \text{for } p > 2. \end{cases} \tag{8.188}$$

The question is: *Which $\rho_{\varepsilon,0}$ preserve these condition in the problem?* The answer is formulated in terms of the rates (critical radii)

$$\rho^*_{\varepsilon,0} = \rho^*_{\varepsilon,0,p} = \begin{cases} \varepsilon^{(4\eta-1+2/p)/(4\eta+1)} & \text{for } p \le 2, \\ \varepsilon^{(2\eta+1/2k-1/p)/(2\eta+1-1/p)} & \text{for } 2k \le p < 2(k+1). \end{cases} \tag{8.189}$$

Namely, we have

Theorem 8.7
(1) *Lower bounds. Let $\rho_{\varepsilon,0}/\rho^*_{\varepsilon,0} \to \infty$. Then there exists a family $\rho_{\varepsilon,1}$, such that $\rho_{\varepsilon,1}/\rho^*_{\varepsilon,1} \to \infty$ and $\gamma_\varepsilon \to 1$.*
(2) *Upper bounds. Let $\rho_{\varepsilon,0}/\rho^*_{\varepsilon,0} \to 0$. Then $\gamma_\varepsilon \to 0$ for any family $\rho_{\varepsilon,1}$, such that $\rho_{\varepsilon,1}/\rho^*_{\varepsilon,1} \to \infty$.*

Proof of Theorem 8.7. This is given in the following Section.

Lower Bounds

It suffices to consider the case $h = 1$ in the Besov norms; by (2.79) this corresponds to the "smallest" alternative. Take a function $\varphi(t)$, which is $[\eta+2]$-continuously differentiable on R^1, with the support on the interval $(0,1)$ and such that $\|\varphi\|_2 = 1$. For an integer $n > 0$ let us consider the tuple of functions

$$\varphi_{i,n}(t) = n^{1/2}\varphi(nt-i+1), \quad i = 1,\dots,n,$$

which is an orthonormal system in $L_2(0,1)$ with disjoint supports. Set

$$g_{\theta,n} = \sum_{i=1}^{n} \theta_i\varphi_{i,n}, \quad |\theta|_p = \left(\sum_{i=1}^{n}|\theta_i|^p\right)^{1/p}.$$

Using Lemma 3.8 we get the inequalities

$$\|g_{\theta,n}\|_{\eta,p,1} \le C_0|\theta|_p\, n^{\eta+1/2-1/p}, \quad \|g_{\theta,n}\|_p = C_1|\theta|_p\, n^{1/2-1/p}; \quad \theta \in R^n, \tag{8.190}$$

where $C_1 = \|\varphi\|_p$ and $C_0 > 0$ is a constant independent of n.

Denote by $\phi_0(p)$ the exponent of n in the quantity $r^*_{n,p}$ and by $\phi_1(p)$ the exponent of n in the quantity $\rho^*_{n,p}$, i.e.,

$$r^*_{n,p} = n^{\phi_0(p)}, \quad \rho^*_{n,p} = n^{\phi_1(p)},$$

where the quantities $\rho^*_{n,p}, r^*_{n,p}$ are defined by (8.156), (8.157). Set

$$\psi_l(p) = 1/2 - 1/p + \phi_l(p), \quad l = 0,1.$$

Observe that if either $p > 1$ or $l = 1$, then $\psi_l(p) > 0$, $l = 0,1$, and $\psi_0(1) = 0$. Put

$$n^*_\varepsilon = n_* = \varepsilon^{-1/(\eta+\psi_1(p))},$$

i.e., $\varepsilon n_*^{\eta+1/2-1/p}\rho^*_{n_*,p} = 1$. In addition,

$$\rho^*_{\varepsilon,1,p} = \varepsilon n_*^{1/2-1/p}\rho^*_{n_*,p} = \varepsilon n_*^{\psi_1(p)}, \quad \rho^*_{\varepsilon,0,p} = \varepsilon n_*^{1/2-1/p} r^*_{n_*,p} = \varepsilon n_*^{\psi_0(p)}. \tag{8.191}$$

Put $\rho_{\varepsilon,0} = W_\varepsilon\rho^*_{\varepsilon,0,p}$, where $W_\varepsilon \to \infty, W_\varepsilon = o(n_*^b)$ for small enough $b > 0$. Also put

$$n = n_\varepsilon \asymp W_\varepsilon^{-\alpha} n_* \to \infty, \quad \alpha > 0, \quad U_\varepsilon = W_\varepsilon^\beta, \quad \beta = 1 + \alpha\psi_0(p) > 0.$$

We have

$$\rho_{\varepsilon,0} \asymp U_\varepsilon \varepsilon n^{1/2-1/p} r^*_{n,p}. \tag{8.192}$$

Selecting γ such that

$$0 < \gamma < \min(\beta\delta(p) - \alpha\psi_1(p), \eta\alpha),$$

where $\delta(p)$ is the same as in Remark in the page 344 (this is possible for small enough $\alpha > 0$), we put

$$\rho_{\varepsilon,1} = W_\varepsilon^\gamma \rho^*_{\varepsilon,1,p} \asymp U_\varepsilon^\delta \varepsilon n^{1/2-1/p} \rho^*_{n,p}, \quad 0 < \delta = (\gamma + \alpha\psi_1(p))/\beta < \delta(p). \tag{8.193}$$

In addition,

$$\varepsilon U_\varepsilon^\delta \rho^*_{n,p} n^{\eta+1/2-1/p} \asymp W_\varepsilon^\gamma (n/n_*)^\eta \asymp W_\varepsilon^{\gamma-\eta\alpha} = o(1). \tag{8.194}$$

Let us show that the family $\rho_{\varepsilon,1}$ meets the conditions of point (1) of Theorem 8.7. Denote by G_n the linear closure of the system

$$\{\varphi_{i,n},\ i = 1,\dots,n\}.$$

Set $x = (x_1,\dots,x_n)$, $v = v(f) = (v_1,\dots,v_n)$, where

$$x_i = \varepsilon^{-1} \int \varphi_{i,n}(t)\, dX_\varepsilon(t), \quad v_i = \varepsilon^{-1}(f, \varphi_{i,n}),$$

and

$$V_{n,l} = \{v(f) : f \in G_n \cap F_{\varepsilon,l}\}, \quad l = 0, 1.$$

Clearly $x = v + \xi$, where $\xi \sim N(0, I_n)$ is the n-dimensional standard Gaussian vector. Recall that the $P_{\varepsilon,f}$-distributions of the random vectors x are the $P_{n,v}$-distributions on $(R^n, \mathcal{B}^n)$. In view of inequalities (8.190) and relations (8.191)–(8.194) the set $V_{n,0}$ contains the ring $\{b_0 \le |v|_p/r_n \le B_0\}$ for $r_n = U_\varepsilon r^*_{n,p}$ and the set $V_{n,1}$ contains the ring $\{b_1 \le |v|_p/\rho_n \le B_1\}$ for $\rho_n = U_\varepsilon^\delta \rho^*_{n,p}$; here $0 < b_l < B_l$, $l = 0, 1$, are some constants depending on C_0, C_1. In particular, it follows from (8.194) that if $b_1 \le |v|_p/\rho_n \le B_1$, then

$$\|f\|_{\eta,p,1} \le \varepsilon |v|_p n^{\eta+1/2-1/p} \le C_1 \varepsilon U_\varepsilon^\delta \rho^*_{n,p} n^{\eta+1/2-1/p} = o(1).$$

Therefore point (1) of Theorem 8.7 follows directly from the Remark on page 344.

Upper Bounds

Taking into account (2.79) it suffices to consider "widest" case which corresponds to $h = \infty$. Let us take an integer-valued family $n = n_\varepsilon \asymp n_*$, where n_* are the quantities defined in the proof of lower bounds. Let

$$\rho_{\varepsilon,0} = \varepsilon n^{1/2-1/p} r_n \ll \rho^*_{\varepsilon,0} \asymp \varepsilon n^{1/2-1/p} r^*_{n,p}, \tag{8.195}$$
$$\rho_{\varepsilon,1} = \varepsilon n^{1/2-1/p} \rho_n \gg \rho^*_{\varepsilon,1} \asymp \varepsilon n^{1/2-1/p} \rho^*_{n,p}; \tag{8.196}$$

note that $\rho^*_{\varepsilon,1} \asymp n^{-\eta}$ in view of the first relationship in (8.191) and the definition of n_*. Let Δ_n be a uniform partition of the interval $[0,1]$ into n intervals $\delta_{1,n},\dots,\delta_{n,n}$. Let $\varphi_{i,n} = n^{1/2} \mathbb{1}_{\delta_{i,n}}$ be an orthonormal system in $L_2(0,1)$ and Pr_n be the operator of the orthonormal projection of $L_2(0,1)$

to the space of step functions for the partition Δ_n:

$$Pr_n(f) = \sum_{i=1}^{n} (f, \varphi_{i,n})\varphi_{i,n}.$$

Denote $x = x(X_\varepsilon) = (x_1, \dots, x_n)$, $v = v(f) = (v_1, \dots, v_n)$,

$$x_i = \varepsilon^{-1} \int \varphi_{i,n}(t)\, dX_\varepsilon(t), \quad v_i = \varepsilon^{-1}(f, \varphi_{i,n}),$$

and set

$$V_{n,l} = \{v(f) : \ f \in F_{\varepsilon,l}\}, \quad l = 0, 1.$$

Clearly, the $P_{\varepsilon,f}$-distributions of x are the same as the $P_{n,v}$-distributions on $(R^n, \mathcal{B}^n)$ and

$$\varepsilon n^{1/2-1/p}|v(f)|_p = \|Pr_n(f)\|_p \le \|f\|_p. \tag{8.197}$$

In view of (8.197) the set $V_{n,0}$ lies in an l_p^n-ball of radius $r_n \ll r_{n,p}^*$.

Recall the inequality (see Proposition 2.16): if $\|f\|_{\eta,p,\infty} \le C$, then there exist positive constants C_1, C_2 such that

$$\|Pr_n f\|_p \ge C_1 \|f\|_p - C_2 C n^{-\eta}. \tag{8.198}$$

Since

$$\|f\|_p \ge \rho_{\varepsilon,1} \gg \rho_{\varepsilon,1}^* \asymp n^{-\eta} \quad \text{for } f \in F_{\varepsilon,1},$$

inequality (8.198) shows that the set $V_{n,1}$ lies outside an l_p^n-ball of radius of the order $\rho_n \gg \rho_{n,p}^*$. For this reason the needed statement follows directly from point (2) of Theorem 8.6. Moreover, let $\psi_{n,p}(x)$ be the tests constructed in the proof of point (2) of Theorem 8.6. Then the tests $\psi_{\varepsilon,p}(X_\varepsilon) = \psi_{n,p}(x(X_\varepsilon))$ are minimax consistent. □

A
Appendix

A.1 Proof of Proposition 2.16

In view of inequalities (2.79), (2.80) it suffices to consider widest case $h = \infty$ which corresponds to the Nikol'ski norm. Assume that we simplify $n = kr$ where an integer $r > \sigma + 2$ is fixed. Take a partition of $[0, 1]$ into intervals $\Delta_1, \ldots, \Delta_k$ of length $l = r/n$, with Δ_i being a union of r successive intervals $\delta_{i,n}$. Note (see [15]) that there exists a piecewise-polynomial approximation $P_{r,k}$ (on each Δ_j this is a polynomial $p_{j,r}$ of degree $\le r - 1$) such that, for some $C = C(p, \sigma) > 0$, one has

$$\|f - P_{r,k}\|_p \le C \|\Delta_l^{(r)}(f, \cdot)\|_{L_p(0,1-2l)} \le CRl^\sigma = Cr^\sigma Rn^{-\sigma}. \tag{A.1}$$

We give the proof of (A.1) for $\sigma = m + \delta \notin \mathbf{N}$. Let $1 < p < \infty$, $m \ge 1$; the case $m = 0$, $p = 1$, or $p = \infty$ is more simple. Set

$$p_{j,m}(t, x) = \sum_{s=0}^{m-1} f^{(s)}(x) \frac{(t-x)^s}{s!}, \quad p_{j,m}(t) = l^{-1} \int_{\Delta_j} p_{j,m}(t, x)\, dx.$$

Clearly, $p_{j,m}$ are polynomials in t of degree $\le m$ and by the Taylor formula we have, for $x, t \in \Delta_j$,

$$\begin{aligned} f(t) - p_{j,m}(t, x) &= \frac{1}{(m-1)!} \int_x^t (f^{(m)}(u) - f^{(m)}(x))(u - x)^{m-1}\, du, \\ f(t) - p_{j,m}(t) &= \frac{1}{l(m-1)!} \int_{\Delta_j} \int_x^t (f^{(m)}(u) - f^{(m)}(x))(u - x)^{m-1} du dx, \end{aligned}$$

and

$$|f(t) - p_{j,m}(t)| \frac{1}{l(m-1)!} \int_{\Delta_j} \int_{\Delta_j} |f^{(m)}(u) - f^{(m)}(x)||u - x|^{m-1}\, du\, dx.$$

Applying the Hölder inequality for $t \in \Delta_j$ we get

$$\begin{aligned} |f(t) - p_{j,m}(t)|^p &\le \int_{\Delta_j} |f^{(m)}(u) - f^{(m)}(x)|^p |u - x|^{-p\delta}\, du\, dx \\ &\quad \times \frac{1}{(l(m-1)!)^p} \int_{\Delta_j} \left(\int_{\Delta_j} \int_{\Delta_j} |u - x|^{p(\sigma-1)/(p-1)} du dx \right)^{p-1} \\ &\le Cl^{p\sigma-2} \int_{\Delta_j} \int_{\Delta_j} |f^{(m)}(u) - f^{(m)}(x)|^p |u - x|^{-p\delta}\, du\, dx, \end{aligned}$$

where

$$C = ((m-1)!)^{-p} \left(2(p-1)^2/(p\sigma - 1)(p\sigma + p - 2)\right)^{p-1}.$$

Therefore

$$\begin{aligned} \|f - P_{m,k}\|_p^p &= \sum_{j=1}^{k} \int_{\Delta_j} |f(t) - p_{j,m}(t)|^p\, dt \\ &\le Cl^{p\sigma-1} \sum_{j=1}^{k} \int_{\Delta_j} \int_{\Delta_j} |f^{(m)}(u) - f^{(m)}(x)|^p |u - x|^{-p\delta}\, du\, dx \\ &\le Cl^{p\sigma-1} \int_0^l \left(\int_0^{1-l} |f^{(m)}(u + h) - f^{(m)}(u)|^p y^{-p\delta} du \right) dy \\ &\le C \int_0^l (\omega_p^{(1)}(f^{(m)}, y)/y^\delta)^p\, dy \le Cl^{p\sigma-1} R^p \int_0^l dy = Cl^{p\sigma} R^p \end{aligned}$$

which yields (A.1).

By the Hölder's inequality one has

$$\|Pr_n(f - P_{r,k})\|_p \le \|f - P_{r,k}\|_p.$$

This and (A.1) together imply

$$\|Pr_n f\|_p \ge \|Pr_n P_{r,k}\|_p - Cr^\sigma Rn^{-\sigma}, \quad \|P_{r,k}\|_p \ge \|f\|_p - Cr^\sigma Rn^{-\sigma}.$$

It remains to show that

$$\|Pr_n P_{r,k}\|_p \ge B\|P_{r,k}\|_p.$$

Because $P_{r,k}$ is a polynomial $p_{j,r}$ of degree $\le r - 1$ on each Δ_j, by making the linear transform $\Delta_j \to [0, 1]$, we see that this follows from inequality

$$\|Pr_r p_r\|_p \ge B\|p_r\|_p, \quad B = B(r, p) > 0,$$

for any polynomial p_r on $[0, 1]$ of degree $\le r - 1$.

For the proof of the latter inequality, consider Pr_r as a linear operator $H_{r,p}$ mapping an r-dimensional space of polynomials of degree $\le r-1$ with a norm $\|\cdot\|_p$ into an r-dimensional space L_r of piecewise-constant functions for an equispaced partition of $[0,1]$ into r intervals with the same norm. This operator is injective (because $Pr_r p_r \equiv 0$ implies that p_r has at least r roots); hence the inverse operator $H_{r,p}^{-1}$ is bounded on its range, and one can set $B^{-1} = \|H_{r,p}^{-1}\| > 0$. □

A.2 Proof of Proposition 5.3

A.2.1 Properties of Statistics under Alternatives.

First we consider the asymptotic properties of statistics under alternatives. Let E_v be the expectation with respect to $(v,1)$-Gaussian probability on the real line. Recall the definitions

$$\xi(x,z) = (\nu(x,z) + \nu(x,-z))/2; \quad \nu(x,z) = e^{-z^2/2+xz} - 1; \quad z > 0. \tag{A.2}$$

Denote

$$\begin{aligned} m_\nu(z,v) &= E_v(\nu(x,z)), \quad \sigma_\nu^2(z,v) = \mathrm{Var}_v(\nu(x,z)), \\ m_\xi(z,v) &= E_v(\xi(x,z)), \quad \sigma_\xi^2(z,v) = \mathrm{Var}_v(\xi(x,z)). \end{aligned}$$

Using relations (3.76), (3.77) and we easily get

$$m_\nu(z,v) = e^{zv} - 1, \quad m_\xi(z,v) = 2\sinh^2(zv/2); \tag{A.3}$$

$$\sigma_\nu^2(z,v) = e^{2zv}(e^{z^2} - 1),$$

$$\sigma_\xi^2(z,v) = 2\sinh^2(z^2/2) + (e^{z^2} - 1)\sinh^2 zv. \tag{A.4}$$

Let

$$\hat\nu(x,z) = \nu(x,z) - m_\nu(v,z), \quad \hat\xi(x,z) = \xi(x,z) - m_\xi(v,z)$$

be P_v-centered statistics. Using (A.3) for any $v \in R^1$ we get, if $x \sim N(v,1)$, then

$$\hat\nu(x,z) = e^{zv}\nu(\eta,z), \quad \eta = x - v \sim N(0,1); \tag{A.5}$$

$$\hat\xi(x,z) = \frac{1}{2}(e^{zv}\nu(\eta,z) + e^{-zv}\nu(\eta,-z)). \tag{A.6}$$

Let us consider a presentation analogous to (A.5), (A.6):

$$\nu(x,z) = m_\nu(z,v) + \nu(\eta,z) + r_\nu(\eta;v,z), \tag{A.7}$$

$$\xi(x,z) = m_\xi(z,v) + \xi(\eta,z) + r_\xi(\eta;v,z), \quad \eta = x - v, \tag{A.8}$$

where

$$r_\nu(\eta;v,z) = (e^{zv} - 1)\nu(\eta,z),$$

$$r_\xi(\eta;v,z) = 2\sinh(vz/2)\left(\exp(-z^2/2)\sinh(z(\eta+v/2)) - \sinh(vz/2)\right).$$

The first and second moments of the difference $r(\eta; v, z)$ under $\eta \sim N(0,1)$ (which corresponds to $x \sim N(v,1)$) are

$$\begin{aligned} E(r_\nu(\eta; v, z)) &= E(r_\xi(\eta; v, z)) = 0, \\ E(r_\nu^2(\eta; v, z)) &= (e^{zv} - 1)^2 (e^{z^2} - 1), \end{aligned} \quad (A.9)$$

$$\begin{aligned} E(r_\xi^2(\eta; v, z)) &= 4\sinh^2(vz/2)\Big(\sinh z^2 + \sinh^2(vz/2)(e^{z^2} - 1)\Big) \\ &\le 4\sinh^2(vz/2)(e^{z^2} - 1)\exp(z|v|). \end{aligned} \quad (A.10)$$

Let us consider the asymptotics of moments of truncated statistics

$$\nu_H(x,z) = \nu(x,z)\mathbf{1}_{\{x<H\}}, \quad \xi_H(x,z) = \xi(x,z)\mathbf{1}_{\{|x|<H\}}.$$

First, evaluate the moments under the null-hypothesis. We have

$$\begin{aligned} E_0(\nu_H(x,z)) &= -(\Phi(z-H) - \Phi(-H)), \\ E_0(\xi_H(x,z)) &= -(\Phi(z-H) + \Phi(-z-H) - 2\Phi(-H)); \\ E_0(\nu_H^2(x,z)) &= e^{z^2}\Phi(H-2z) - 2\Phi(H-z) + \Phi(H); \\ E_0(\xi_H^2(x,z)) &= \frac{1}{2}\Big(e^{z^2}(\Phi(H-2z) - \Phi(-H-2z)) \\ &\quad + e^{-z^2}(\Phi(H) - \Phi(-H))\Big) + \Phi(H) - \Phi(-H) \\ &\quad + 2\Phi(-H+z) - 2\Phi(H+z). \end{aligned}$$

Assume $z \to \infty$ and $H > cz$ where $c > 2 - \sqrt{2} \approx 0.59$. Using the asymptotics (4.140) we have:

$$E_0(\nu_H(x,z)) \sim E_0(\xi_H(x,z)) \sim -\Phi(z-H), \quad (A.11)$$

$$\begin{aligned} E_0(\nu_H^2(x,z)) &\sim (e^{z^2} - 1)\Phi(H-2z), \\ E_0(\xi_H^2(x,z)) &\sim 2\sinh^2(z^2/2)\Phi(H-2z). \end{aligned} \quad (A.12)$$

Denote

$$\begin{aligned} \Delta m_{H,\nu}(v,z) &= E_v(\nu_H(x,z)) - E_0(\nu_H(x,z)), \\ \Delta\sigma_{H,\nu}^2(v,z) &= \mathrm{Var}_v(\nu_H(x,z)) - \mathrm{Var}_0(\nu_H(x,z)) + (\Delta m_{H,\nu}(v,z))^2; \\ \Delta m_{H,\xi}(v,z) &= E_v(\xi_H(x,z)) - E_0(\xi_H(x,z)), \\ \Delta\sigma_{H,\xi}^2(v,z) &= \mathrm{Var}_v(\xi_H(x,z)) - \mathrm{Var}_0(\xi_H(x,z)) + (\Delta m_{H,\xi}(v,z))^2. \end{aligned}$$

Proposition A.1 *Fix $a \in (0, 2^{-1/2})$ and let*

$$z \to \infty, \quad dH < z < H(1-a), \quad |v| \le H(1+a),$$

where $1 - a > d > 1 - (1-a^2)^{1/2}$. Then there exists $c = c(a) > 0$ such that for $\delta_H = \exp(-cH^2)$ we have

$$\begin{aligned} \Delta m_{H,\nu}(v,z) &= (e^{zv} - 1)\Phi(H - z - |v|)(1 + o(\delta_H)); \\ \Delta m_{H,\xi}(v,z) &= 2\sinh^2(zv/2)\Phi(H - z - |v|)(1 + o(\delta_H)), \end{aligned} \quad (A.13)$$

$$\Delta\sigma^2_{H,\nu}(v,z) = O(R_{H,\nu}(v,z) + \delta_H \Delta m_{H,\nu}(v,z));$$
$$\Delta\sigma^2_{H,\xi}(v,z) = O(R_{H,\xi}(v,z) + \delta_H \Delta m_{H,\xi}(v,z)), \quad (A.14)$$

where

$$R_{H,\nu}(v,z) = \exp(z^2 + z|v|))(\exp(zv) - 1)\Phi(H - 2z - |v|),$$
$$R_{H,\xi}(v,z) = \exp(z^2 + z|v|)\sinh^2(zv/2)\Phi(H - 2z - |v|). \quad (A.15)$$

Proof of Proposition A.1. We give the proof for the statistic $\xi(x,z)$ only; considerations for $\nu(x,z)$ are analogous and more simple. Note the equality: for any symmetric function $f(x) = f(-x)$, $x \in R^1$, one has

$$E_v f(x) - E_0 f(x) = E_0 f(x)\xi(x,v), \quad (A.16)$$

because of (remember)

$$\xi(x,v) = \frac{1}{2}\left(\frac{dP_v}{dP_0}(x) + \frac{dP_{-v}}{dP_0}(x)\right) - 1, \quad v \in R^1.$$

Using direct calculation and (A.16) we get

$$\begin{aligned}\Delta m_{H,\xi}(v,z) &= \frac{1}{2}e^{zv}(\Phi(H - z - v) - \Phi(-H - z - v))\\ &+ \frac{1}{2}e^{-zv}(\Phi(H - z + v) - \Phi(-H - z + v))\\ &- \Phi(H - z) + \Phi(-H - z)\\ &+ \Phi(-H - v) + \Phi(-H + v) - 2\Phi(-H). \quad (A.17)\end{aligned}$$

If $z|v| = o(1)$ or $z|v| = O(1)$, then, using the Taylor expansion and (4.140) we get

$$\Delta m_{H,\xi}(v,z) = 2\sinh^2(zv/2)\Phi(H - z - |v|)(1 + o(\delta_H)).$$

If $z|v| \to \infty$ and $|v| < \delta(H - z)$, where $0 < \delta < 1$, then using (4.140) we get

$$|\Delta m_{H,\xi}(v,z) - 2\sinh^2(zv/2)\Phi(H - z - |v|)| = o(\delta_H);$$

and if $\delta(H - z) \le |v| \le H(1 + a)$, then

$$|\Delta m_{H,\xi}(v,z) - 2\sinh^2(zv/2)\Phi(H - z - |v|)| = O(1),$$
$$2\sinh^2(zv/2)\Phi(H - z - |v|) > \exp(cH^2),$$

which implies (A.13). Analogously, one has

$$\begin{aligned}\mathrm{Var}_v(\xi_H(x,z)) - \mathrm{Var}_0(\xi_H(x,z)) &= E_v(\xi_H^2(x,z)) - E_0(\xi_H^2(x,z))\\ &- (\Delta m_{H,\xi}(v,z))^2\\ &- 2\Delta m_{H,\xi}(v,z)\, E_0(\xi_H(x,z));\\ E_v(\xi_H^2(x,z)) - E_0(\xi_H^2(x,z)) &= E_0(\xi_H^2(x,z)\xi(x,v));\\ E_0(\xi_H(x,z)) &\sim -\Phi(z - H) = o(\delta_H),\end{aligned}$$

(the last relation holds because of (A.11)). Using direct calculation and the Taylor expansion for $z|v| = o(1)$, analogously to the above, one can check

$$E_0(\xi_H^2(x,z)\xi(x,v)) = E_0(\xi^2(x,z)\xi(x,v)\mathbf{1}_{\{|x|\le H\}}) \le BR_{H,\xi}(v,z), \quad \text{(A.18)}$$

which yield (A.14) with (A.15). □

A.2.2 Evaluations of Type II Errors

We prove Proposition 5.3 in this section. We consider three-point factors only; considerations for the two-point case are analogous and based on the results of Section A.2.1. Since

$$\beta(\psi_{\varepsilon,S},v) = P_v\left(\overline{X}_{\varepsilon,S}\right) = P_v\left(\{t_\varepsilon \le S\} \cap \left\{\sup_i |x_i|/Q_{\varepsilon,i} \le 1\right\}\right),$$

for any $v \in l^2$, we get two inequalities

$$\beta(\psi_{\varepsilon,S},v) \le P_v(t_\varepsilon \le S); \quad \text{(A.19)}$$

$$\begin{aligned} \beta(\psi_{\varepsilon,S},v) &\le P_v\left(\sup_i |x_i|/Q_{\varepsilon,i} \le 1\right) \\ &= \prod_i (\Phi(Q_{\varepsilon,i} - |v_i|) - \Phi(-Q_{\varepsilon,i} - |v_i|)) \\ &\le \begin{cases} \inf_i \Phi(Q_{\varepsilon,i} - |v_i|) = \Phi(-\sup_i(|v_i| - Q_{\varepsilon,i})), \\ \exp(-\sum_i \Phi(|v_i| - Q_{\varepsilon,i})). \end{cases} \end{aligned} \quad \text{(A.20)}$$

Using (A.20) uniformly over S we get

$$\sup_{v\in V_{\varepsilon,1}} \beta(\psi_{\varepsilon,S},v) \to 0,$$

and in view of (A.19) it suffices to evaluate $P_v(t_\varepsilon \le S)$ for $v \in V_{\varepsilon,2}$.

The scheme of the evaluation is the following. We present the statistics (5.3) in the form $t_\varepsilon = l_{\varepsilon,v} + h_{\varepsilon,T}(v) + \delta_\varepsilon$ and we control that, under P_v-probability, $\delta_\varepsilon \to 0$ uniformly over $v \in V_{\varepsilon,2}$ and the statistics $l_{\varepsilon,v}$ are asymptotically Gaussian $N(0,1)$ uniformly over $v \in V_{\varepsilon,2}$, such that $h_{\varepsilon,T}(v)$ are not too large.

Denote (we omit the index ε to simplify notations below)

$$Y_i = z_i + |v_i| - T_i, \quad Z_i = 3z_i/2 + |v_i| - T_i = Y_i + z_i/2,$$
$$I_2 = I_{\varepsilon,\eta}, \quad I_1 = I \backslash I_{\varepsilon,\eta},$$

(this means that $I_1 = \{i : z_i \le \eta T_i\}$). By the choice of η and τ (recall that we take $\eta \in (0,\eta_0)$, $\eta_0 \in (0,1/4)$, $3\eta/2 + \tau < 1$) we have, for $i \in I_1$,

$$Q_i \le \tau T_i, \quad Z_i < -\zeta T_i, \quad \zeta \in (0, 1 - (3\eta/2 + \tau)). \quad \text{(A.21)}$$

The test procedure (5.4) can be presented in a slightly different form. Let

$$g_{\varepsilon,i,T}(x) = g_{\varepsilon,i}(x)\mathbf{1}_{|x|<T_{\varepsilon,i}}$$

be the $T_{\varepsilon,i}$-truncated statistics. Set

$$l_{\varepsilon,1} = u_\varepsilon^{-1} \sum_{i \in I_1} g_{\varepsilon,i}(x_i); \quad l_{\varepsilon,2} = u_\varepsilon^{-1} \sum_{i \in I_2} g_{\varepsilon,i,T}(x_i); \quad l_{\varepsilon,T} = l_{\varepsilon,1} + l_{\varepsilon,2}.$$

Since $T_{\varepsilon,i} > Q_{\varepsilon,i}$, we can change, in (5.4), the statistics t_ε of the form (5.3) to the statistics $l_{\varepsilon,T}$:

$$X_{\varepsilon,S} = \{l_{\varepsilon,T} > S\} \cup \left\{ \sup_i |x_i|/Q_{\varepsilon,i} > 1 \right\}. \tag{A.22}$$

Let us evaluate P_v-moments of the statistics $l_{\varepsilon,1}$, $v \in V_{\varepsilon,2}$ (which correspond to nontruncated items). Using equalities (A.3)–(A.4) and the presentations (5.14), (5.15), (5.18) we get

$$E_v l_{\varepsilon,1} = h_{1,\varepsilon}(v),$$

$$\begin{aligned} 0 \le \operatorname{Var}_v(l_{\varepsilon,1}) - u_{1,\varepsilon}^2/u_\varepsilon^2 &= u_\varepsilon^{-2} \sum_{i \in I_1} h_i^2(e^{z_i^2} - 1)\sinh^2(z_i v_i) \\ &\le 2u_\varepsilon^{-1} h_{1,\varepsilon}(v) \sup_{i \in I_1}(h_i e^{z_i|v_i|}(e^{z_i^2} - 1)) = 2u_\varepsilon^{-1} h_{1,\varepsilon}(v) \sup_{i \in I_1} d_i(v), \end{aligned}$$

where

$$d_i(v) = h_i \exp(z_i|v_i|)(\exp(z_i^2) - 1). \tag{A.23}$$

Let us show that there exists $a > 0$ such that

$$\delta_\varepsilon = \sup_{v \in V_{\varepsilon,2}} \sup_{i \in I_1} d_i(v) = o(\tau_\varepsilon), \quad \tau_\varepsilon = w_\varepsilon^a \to 0. \tag{A.24}$$

In fact, let $z_i \ge \theta T_i$ where $\theta > 0$ is small enough. Since $Z_i \le -\zeta T_i$ (see (A.21)), using (5.8), (5.10) we get, for $2\theta\zeta > a > 0$,

$$\begin{aligned} d_i(v) &\asymp \exp(3z_i^2/2 + z_i|v_i| - T_i z_i) = \exp(z_i Z_i) \\ &\le \exp(-\zeta\theta T_i^2) \le \exp(-\zeta\theta T_\varepsilon^2) = o(\tau_\varepsilon). \end{aligned} \tag{A.25}$$

Let $1 < z_i < \theta T_i$. By (5.8) we get

$$w_\varepsilon \ge u_{\varepsilon,i} \asymp h_i \exp(z_i^2/2) \asymp \exp(z_i(z_i - T_i)).$$

By (A.21) for $i \in I_1$ we have

$$T_i - |v_i| > T_i - Q_i - C_\varepsilon > \zeta_1 T_i, \quad \zeta_1 \in (0, 1 - \tau).$$

For small enough $\theta > 0$, $a_1 > a > 0$, such that $(\zeta_1 - a_1) > \theta(3/2 - a_1)$, these yield the relations

$$\begin{aligned} d_i(v) &\asymp \exp(3z_i^2/2 + z_i|v_i| - T_i z_i) \\ &\le \exp(z_i(3z_i/2 - \zeta_1 T_i)) < \exp(a_1 z_i(z_i - T_i)) = o(\tau_\varepsilon). \end{aligned}$$

Finally, let $z_i \le 1$. Then $w_\varepsilon \ge u_{\varepsilon,i} \asymp z_i^2 h_i \asymp z_i^2 \exp(-z_i T_i)$, and we have, for $0 < a < a_1 = \zeta_1 < 1$,

$$d_i(v) \asymp z_i^2 \exp(z_i(|v_i| - T_i)) \le z_i^2 \exp(-\zeta_1 z_i T_i)$$

$$\leq z_i^{2a_1} \exp(-a_1 z_i T_i)) = O(w_\varepsilon^{a_1}) = o(\tau_\varepsilon).$$

Recall that $u_{1,\varepsilon}^2/u_\varepsilon^2 = 1 + o(\tau_\varepsilon)$, $u_{2,\varepsilon}^2 = o(\tau_\varepsilon)$ because of A.1 or A.1a and (5.20). Therefore we get, uniformly over $v \in V_{\varepsilon,2}$,

$$E_v(l_{\varepsilon,1}) = h_{1,\varepsilon}(v), \quad \mathrm{Var}_v(l_{\varepsilon,1}) = 1 + o(\tau_\varepsilon h_{1,\varepsilon}(v)/u_\varepsilon) + o(\tau_\varepsilon). \qquad \text{(A.26)}$$

Divide the set I_2 into the sets

$$I_3 = \{i \in I_2 : \ Y_i \leq 1\}, \quad I_4 = I_2 \backslash I_3.$$

Put

$$h_{k,\varepsilon}(v) = u_\varepsilon^{-1} \sum_{i \in I_k} h_{\varepsilon,i,T}(v_i), \quad k = 3, 4; \quad h_{2,\varepsilon}(v) = h_{3,\varepsilon}(v) + h_{4,\varepsilon}(v);$$

recall that the quantities $h_{\varepsilon,i,T}(v_i)$ are refined by (5.19). Let us show that, uniformly over $v \in V_{\varepsilon,2}$,

$$E_v(l_{\varepsilon,2}) = h_{2,\varepsilon}(v)(1 + o(\tau_\varepsilon)) + o(\tau_\varepsilon), \quad h_{4,\varepsilon}(v) = o(\tau_\varepsilon); \qquad \text{(A.27)}$$

$$\mathrm{Var}_v(l_{\varepsilon,2}) = o(\tau_\varepsilon(1 + h_{2,\varepsilon}(v)/u_\varepsilon)), \qquad \text{(A.28)}$$

where, as previously, $\tau_\varepsilon = w_\varepsilon^a$ for some $a > 0$. In fact, using Proposition A.1 and (5.19) we get

$$E_v(l_{\varepsilon,2}) = u_\varepsilon^{-1} \sum_{i \in I_2} h_i(E_0 \xi_{T_i}(x_i, z_i) + \Delta m_{T_i,\xi}(z_i, v_i));$$

$$u_\varepsilon^{-1} \sum_{i \in I_2} h_i \Delta m_{T_i,\xi}(z_i, v_i)) = h_{2,\varepsilon}(v_i)(1 + o(\tau_\varepsilon));$$

$$0 \leq -\sum_{i \in I_2} h_i E_0 \xi_{T_i}(x_i, z_i) \sim \sum_{i \in I_2} h_i \Phi(z_i - T_i)$$

$$\asymp \sum_{i \in I_2} \Phi(-T_i) \leq \tau_\varepsilon \sum_{i \in I_2} \Phi(-Q_i) = o(\tau_\varepsilon)$$

(we used the relation $h_i \Phi(z_i - T_i) \asymp \Phi(-T_i)$ which follows from (4.140), (5.8), (5.10), the last relations follow from (5.11), (5.2)). Thus we get the first relation (A.27).

Using (4.140), (5.8) again, we get, for $i \in I_4$,

$$h_{\varepsilon,i,T}(v_i) \asymp u_\varepsilon^{-1} h_i e^{z_i |v_i|} \Phi(-Y_i) \asymp u_\varepsilon^{-1} \exp(-(|v_i| - T_i)^2/2)/Y_i.$$

Because of (5.11), taking δ_1 such that $0 < \delta_1 < \delta_0$, we get

$$T_i - |v_i| > Q_i - |v_i| + \delta_1 z_i > 0;$$

the second inequality follows from the definitions of the sets $V_{\varepsilon,2}$ and I_2.

Therefore, by $z_i > \eta T_\varepsilon$ and (5.10), we have

$$\begin{aligned} \exp(-(|v_i| - T_i)^2/2) &\leq \exp(-\delta_1^2 z_i^2/4 - (|v_i| - Q_i)^2/2) \\ &= o(\tau_\varepsilon) \Phi(|v_i| - Q_i)). \end{aligned}$$

By definition of the sets $V_{\varepsilon,2}$, these lead to the second relation in (A.27).

To evaluate the variances, using Proposition A.1 again, we get

$$|\mathrm{Var}_v(l_{\varepsilon,2}) - \mathrm{Var}_0(l_{\varepsilon,2})| = u_\varepsilon^{-2} \sum_{i\in I_2} h_i^2 |\Delta^2 \sigma^2_{T_i,\xi}(v_i,z_i) - (\Delta m_{T_i,\xi}(v_i,z_i))^2|$$

$$\le B u_\varepsilon^{-2} \sum_{i\in I_2} h_i^2 R_{T_i,\xi}(v_i,z_i) \tag{A.29}$$

$$+ B u_\varepsilon^{-1} \left(\sup_{i\in I_2} h_i |\Delta m_{T_i,\xi}(v_i,z_i)| + o(\tau_\varepsilon) \right) h_{2,\varepsilon}(v),$$

where

$$R_{T_i,\xi}(v_i,z_i) = \exp(z_i^2 + |v_i|z_i)\sinh^2(z_i v_i/2)\Phi(T_i - |v_i| - 2z_i).$$

Let us show that, uniformly over $i \in I_2$,

$$u_\varepsilon^{-1} h_i^2 R_{T_i,\xi}(v_i,z_i) = \begin{cases} o(\tau_\varepsilon h_{\varepsilon,i,T}(v_i)), & \text{if } i \in I_3, \\ O(h_{\varepsilon,i,T}(v_i)), & \text{if } i \in I_4. \end{cases} \tag{A.30}$$

In fact, if $Y_i \le 1$, then

$$Z_i = Y_i - z_i/2 \le -\eta_1 T_\varepsilon, \quad z_i Z_i \le -\eta_2 T_\varepsilon^2; \quad \eta_1 \in (0,\eta/2), \quad \eta_2 = \eta_1\eta,$$

and using (5.8) we get

$$\begin{aligned} u_\varepsilon^{-1} h_i^2 R_{T_i,\xi}(v_i,z_i) &\le u_\varepsilon^{-1} h_i^2 \exp(z_i^2 + |v_i|z_i)\sinh^2(z_i v_i/2) \\ &\asymp u_\varepsilon^{-1} h_i \sinh^2(z_i v_i/2)\exp(z_i Z_i) \\ &\le B h_{\varepsilon,i,T}(v_i))\exp(-\eta_2 T_\varepsilon^2), \end{aligned}$$

which yields the first relation in (A.30). If $Y_i > 1$, then

$$T_i - |v_i| - 2z_i = -(Y_i + z_i) \to -\infty,$$

and using (4.140), (5.8) we get

$$\begin{aligned} h_i^2 R_{T_i}(v_i,z_i) &= h_i^2 \exp(z_i^2 + |v_i|z_i)\sinh^2(z_i v_i/2)\Phi(T_i - |v_i| - 2z_i) \\ &\asymp h_i \exp(|v_i|z_i)\Phi(-Y_i)Y_i/(Y_i + z_i) \\ &= O(h_i \exp(|v_i|z_i)\Phi(-Y_i)). \end{aligned}$$

These yield the second relation in (A.30).

In view of (A.30), (A.27), the sum in (A.29) is $o(\tau_\varepsilon h_{2,\varepsilon}(v))$.

Let us evaluate the supremum in (A.29). Because of (A.13) we get

$$h_i \Delta m_{T_i}(v_i,z_i) \asymp h_i \exp(|v_i|z_i)\Phi(-Y_i);$$

if $Y_i < z_i/3$, then

$$\begin{aligned} h_i \exp(|v_i|z_i)\Phi(-Y_i) &\le h_i \exp(|v_i|z_i) \\ &\asymp \exp(z_i(Y_i - z_i/2)) \le \exp(-\eta^2 T_\varepsilon^2/6) = o(\tau_\varepsilon), \end{aligned}$$

and if $Y_i \ge z_i/3$, then, analogously to the above evaluations,

$$h_i \exp(|v_i|z_i)\Phi(-Y_i) \asymp \exp(-(|v_i| - T_i)^2/2)/Y_i = o(\tau_\varepsilon),$$

because of

$$T_i - |v_i| > T_i - Q_i + o(T_\varepsilon) > \delta_1 T_\varepsilon, \quad \delta_1 \in (0, \delta_0).$$

Thus the supremum in (A.29) is $o(\tau_\varepsilon)$.

Finally, using (A.12), (5.15), and (5.20) we get

$$u_\varepsilon^2 \operatorname{Var}_0(l_{\varepsilon,2}) \le \sum_{i \in I_2} E_0 g^2_{\varepsilon,i,T}(x_i) = \sum_{i \in I_2} h_i^2 E_0 \xi^2_{T_i}(x_i, z_i) \le u^2_{2,\varepsilon} = o(\tau_\varepsilon).$$

These yield (A.28).

In view of relations (A.26), (A.27)–(A.28) we get, for $\tau_\varepsilon = w_\varepsilon^a \to 0$,

$$E_v(l_{\varepsilon,T}) = h_\varepsilon(v,T)(1 + o(\tau_\varepsilon)), \quad \operatorname{Var}_v(l_{\varepsilon,T}) = 1 + o(\tau_\varepsilon(1 + h_{\varepsilon,T}(v)/u_\varepsilon)). \tag{A.31}$$

Take a family $D_\varepsilon = w_\varepsilon^{-a/2} \to \infty$ such that $D_\varepsilon \tau_\varepsilon = w_\varepsilon^{a/2} \to 0$ and denote

$$V_{\varepsilon,3} = \{v \in V_{\varepsilon,2} : h_{\varepsilon,T}(v) > D_\varepsilon\}; \quad V_{\varepsilon,4} = V_{\varepsilon,2} \backslash V_{\varepsilon,3}. \tag{A.32}$$

Since $S = o(w_\varepsilon^{-a/2}) = o(D_\varepsilon)$, using the Chebyshev inequality and the evaluations (A.31) we get, uniformly over $v \in V_{\varepsilon,3}$,

$$\begin{aligned} P_v(l_{\varepsilon,T} \le S) &= P_v(E_v(l_{\varepsilon,T}) - l_{\varepsilon,T} \ge E_v(l_{\varepsilon,T}) - S) \\ &\le P_v(|E_v(l_{\varepsilon,T}) - l_{\varepsilon,T}| \ge E_v(l_{\varepsilon,T}) - S) \\ &\le \operatorname{Var}_v(l_{\varepsilon,T})/(E_v(l_{\varepsilon,T}) - S))^2 \\ &\le B(1 + o(\tau_\varepsilon h_{\varepsilon,T}(v)/u_\varepsilon))/(h_{\varepsilon,T}(v) - S))^2 \to 0. \end{aligned}$$

To complete the proof it suffices to show that $l_{\varepsilon,2} - h_{2,\varepsilon}(v) \to 0$ and the statistics $l_{\varepsilon,1} - h_{1,\varepsilon}(v)$ are asymptotically $(0,1)$-Gaussian in P_v-probability uniformly over $v \in V_{\varepsilon,4}$. The first relation follows from (A.28) because of $\operatorname{Var}_v(l_{\varepsilon,2}) \to 0$ and $l_{\varepsilon,2} - h_{2,\varepsilon}(v) \to 0$ under P_v-probability uniformly over $v \in V_{\varepsilon,4}$.

Consider the statistics $l_{\varepsilon,1}$. Using the presentation (A.8) with $x_i = v_i + \eta_i$, we write

$$l_{\varepsilon,1} = h_{1,\varepsilon}(v) + l^v_{\varepsilon,1} + \Delta l_{\varepsilon,1}(v),$$

$$l^v_{\varepsilon,1} = u_\varepsilon^{-1} \sum_{i \in I_1} h_i \xi(\eta_i, z_i), \quad \Delta l_{\varepsilon,1}(v) = u_\varepsilon^{-1} \sum_{i \in I_1} h_i r_\xi(\eta_i; z_i, v_i).$$

Note that the P_v-probability of the random sequence $\{x_i\}$ corresponds to the P_0-probability of the standard Gaussian sequence $\{\eta_i\}$. Since assumptions A.2 or A.2a hold for the set $I_1 = I_{\varepsilon,\eta}$, it follows from Corollary 4.2 that $l^v_{\varepsilon,1} \to \eta \sim N(0,1)$ in P_0-probability. Thus we need to show that $\Delta l_{\varepsilon,1}(v) \to 0$ under P_0-probability. This follows from the relation

$$E_0(\Delta l_{\varepsilon,1}(v))^2 \to 0.$$

In view of (A.9) we have $E_0(\Delta l_{\varepsilon,1}(v)) = 0$. Thus, to control the required relation, it suffices to check that $\operatorname{Var}_0(\Delta l_{\varepsilon,1}(v)) \to 0$ uniformly over $v \in$

$V_{\varepsilon,4}$. Using (A.10) we get

$$\mathrm{Var}_0(\Delta l_{\varepsilon,1}(v)) \le 4u_\varepsilon^{-2} \sum_{i\in I_1} h_i^2 \sinh^2(v_i z_i/2)(e^{z_i^2}-1)\exp(z_i|v_i|).$$

Denote, analogously to (A.23),

$$d_i(v) = h_i \exp(z_i|v_i|)(\exp(z_i^2)-1).$$

Then we have

$$\mathrm{Var}_0(\Delta l_{\varepsilon,1}(v)) \le 2u_\varepsilon^{-2} h_{1,\varepsilon}(v) \sup_{i\in I_1} d_i(v),$$

and it follows from the evaluations (A.24) and from the definition (A.32) of the set $V_{\varepsilon,4}$ that

$$\mathrm{Var}_0(\Delta l_{\varepsilon,1}(v)) \le \delta_\varepsilon h_{1,\varepsilon}(v)) = o(w_\varepsilon^{a/2}) \quad \forall\, v \in V_{\varepsilon,4}. \qquad \square$$

A.3 Study of the Extreme Problem for Power Norms

A.3.1 Solution of the System (6.86), (6.87)

The system of equations (6.86), (6.87) could be presented in the form

$$4h_i \sinh^2 \frac{z_i^2}{2}\left(\frac{z_i^2 - p\tanh(z_i^2/2)}{z_i^q \tanh(z_i^2/2)}\right) = (p-q)Bi^{sq}, \tag{A.33}$$

$$4h_i \sinh^2 \frac{z_i^2}{2}\left(\frac{z_i^2 - q\tanh(z_i^2/2)}{z_i^p \tanh(z_i^2/2)}\right) = (p-q)Ai^{rp}. \tag{A.34}$$

Equations (A.33), (A.34) imply the solutions of (6.86), (6.87):

$$z_i^{p-q}\frac{z_i^2 - p\tanh(z_i^2/2)}{z_i^2 - q\tanh(z_i^2/2)} = \frac{B}{A} i^{sq-rp}, \tag{A.35}$$

$$h_i = Ai^{rp}\frac{z_i^p}{4\sinh^2(z_i^2/2)}\frac{(p-q)\tanh(z_i^2/2)}{(z_i^2 - q\tanh(z_i^2/2))}, \tag{A.36}$$

with the constraints

$$z_i^2 > p\tanh\frac{z_i^2}{2}, \quad \text{if } p>q, \quad z_i^2 < p\tanh\frac{z_i^2}{2}, \quad \text{if } p<q. \tag{A.37}$$

The constraints (A.37) are nontrivial for $p > 2$ corresponding to $z_i > z_p$, if $p > q$ and $z_i < z_p$, and if $p < q$ (see (5.119)). Recall that the quantities z_p are defined by equation (5.95) for $p > 2$ and we set $z_p = 0$ for $p \le 2$.

Introduce the functions

$$\phi_{p,q}(z) = z^{p-q}\frac{z^2 - p\tanh(z^2/2)}{z^2 - q\tanh(z^2/2)}; \tag{A.38}$$

$$\psi_{p,q}(z) = \frac{z^p}{4\sinh^2(z^2/2)}\frac{(p-q)\tanh(z^2/2)}{(z^2 - q\tanh(z^2/2)}. \tag{A.39}$$

It is possible to check directly that if $p > q$, then $\phi_{p,q}(z)$, $z > z_p$, is monotone increasing from 0 to ∞, and if $p < q$, then it is monotone decreasing on $0 < z < z_p$ from ∞ to 0 (recall that we formally assume $z_p = 0$ for $p \leq 2$). Note that we can deduce this from Lemma 5.3 (compare with Proposition A.4 below). Therefore it is possible to define the inverse function $\phi_{p,q}^{-1}(x)$, $x > 0$.

Recall that

$$\lambda = s/p - r/q, \quad \Delta = sq(4-p) - rp(4-q),$$

and set

$$x_i = i/n, \quad y_i = i/m, \quad i \geq 1.$$

It is convenient to replace the unknown parameters $A > 0$, $B > 0$ by other unknown parameters $n > 0$, $h_0 > 0$ for $\lambda \neq 0$, or by $m > 0$, $z_0 > 0$ for $\Delta \neq 0$:

$$n = n_\varepsilon(\kappa) = (A/B)^{1/pq\lambda}, \quad h_0 = h_{0,\varepsilon}(\kappa) = An^{rp}; \tag{A.40}$$

or

$$m = m_\varepsilon(\kappa) = \left(\frac{A^{4-q}}{B^{4-p}}\right)^{1/\Delta}, \quad z_0 = z_{0,\varepsilon}(\kappa) = \left(\frac{A^{sq}}{B^{rp}}\right)^{1/\Delta}. \tag{A.41}$$

It is clear that if $\lambda \neq 0$ and $\Delta \neq 0$, then one has

$$m = nh_0^{(p-q)/\Delta}, \quad z_0 = h_0^{pq\lambda/\Delta}, \quad y = h_0^{-(p-q)/\Delta}x. \tag{A.42}$$

The parameters n, h_0 and variables x are convenient to use for the solutions of (A.33), (A.34). However, the parameters m, z_0 and variables y are more convenient to use for solutions of (6.89) which are considered below. Usually we have solutions of both types. Therefore we need to consider both types of parameters.

We can rewrite (A.35), (A.36) for $\lambda \neq 0$:

$$\phi_{p,q}(z_i) = x^{pq\lambda}, \quad h_i = h_0 x^{rp}\psi_{p,q}(z_i), \quad i \geq 1,$$

or for $\Delta \neq 0$:

$$\phi_{p,q}(z_i) = y^{pq\lambda} z_0^{p-q}, \quad h_i = z_0^{4-p} y^{rp}\psi_{p,q}(z_i), \quad i \geq 1.$$

Thus the solutions of (A.33), (A.34) are of the form: if $\lambda \neq 0$, then

$$z_i = \phi_{p,q}^{-1}(x^{pq\lambda}) = z(x,\kappa), \quad h_i = h_0 x^{rp}\psi_{p,q}(z(x,\kappa)) = h_0\delta(x,\kappa), \tag{A.43}$$

where

$$z(x,\kappa) = \phi_{p,q}^{-1}(x^{pq\lambda}), \quad \delta(x,\kappa) = x^{rp}\psi_{p,q}(z(x,\kappa)), \tag{A.44}$$

and if $\Delta \neq 0$, then

$$z_i = \zeta(y,\kappa,z_0), \quad h_i = \eta(y,\kappa,z_0). \tag{A.45}$$

The functions ζ, η correspond to the change of variables $x \to y$:

$$\zeta(y,\kappa,z_0) = \phi_{p,q}^{-1}(z_0^{p-q}y^{\lambda}), \quad \eta(y,\kappa,z_0) = y^{rp}z_0^{4-p}\psi_{p,q}(\zeta(y,\kappa,z_0)). \quad (A.46)$$

Consider the subsets Ξ_1–Ξ_4 of the set Ξ_G,

$$\begin{aligned} \Xi_1 &= \{\kappa \in \Xi_G : p > 2,\ p > q\}, \\ \Xi_2 &= \{\kappa \in \Xi_G : p > 2,\ p < q\}, \\ \Xi_3 &= \{\kappa \in \Xi_G : p < 2,\ p > q\}, \\ \Xi_4 &= \{\kappa \in \Xi_G : p = 2,\ p > q\}. \end{aligned} \quad (A.47)$$

Recall that we do not consider the set $\kappa \in \Xi_G : p \le 2, p < q$, or $p = q$ because these cases were studied previously. Denote

$$\tilde{\Delta} = \tilde{\Delta}(\kappa) = 2(sq - (8-q)r), \quad C(p) = z_p^p/4\sinh^2(z_p^2/2).$$

Set

$$C_{p,q} = \left(\frac{|q-2|}{|p-2|}\right)^{(p-4)/(p-q)} \frac{|p-q|}{|2-q|}, \quad C(p,q) = \left(\frac{|q-2|}{|p-2|}\right)^{1/(p-q)},$$

and $C_q = (6(2-q))^{1/(6-q)}$. Note that if $\kappa \in \Xi_k$, $k \neq 3$, then $\lambda = \lambda(\kappa) > 0$ and if $\kappa \in \Xi_4$, then $I \ge 0$ (this follows from the definitions of the sets Ξ_{G_i}).

The following proposition describes the properties of the functions $z(x,\kappa)$ and $\delta(x,\kappa)$, $x > 0$, $\kappa \in \Xi_k$, $k = 1,\dots,4$ (if $\kappa \in \Xi_3$, then we assume $\lambda \neq 0$), and of the functions $\zeta(y,\kappa,z_0)$, $\eta(y,\kappa,z_0)$ for $\lambda \le 0$.

Remark. Note that $r < 0$, $s < 0$, $\Delta > 0$, $\kappa \in \Xi_3$ in the case $\lambda \le 0$. One can establish these inequalities by using simple calculations. However, they simply follow from consideration of Figures 10–13. Compare with the Remarks on pages 122 and 142.

Proposition A.2

(A). *The functions $z(x,\kappa)$ and $\delta(x,\kappa)$ are continuous positive smooth functions uniformly over a compact $K \subset R_+^1 \times \Xi_k$ which is disjoint with the set $R_+^1 \times \{\lambda = 0\}$. Uniformly over any such compact K the following rate relations* (1)–(4) *hold:*

1) *Let $p > 2,\ p > q$ ($\kappa \in \Xi_1$). Then $z(x,\kappa)$ is increasing on x and*

$$z(x,\kappa) \sim \begin{cases} z_p, & \text{if } x \to 0, \\ x^{pq\lambda/(p-q)}, & \text{if } x \to \infty, \end{cases}$$

$$\delta(x,\kappa) \sim \begin{cases} C(p)x^{rp}, & \text{if } x \to 0, \\ (p-q)x^{rp}z^{p-2}(x,\kappa)\exp(-z^2(x,\kappa)), & \text{if } x \to \infty. \end{cases}$$

(2) *Let $q > p > 2$ ($\kappa \in \Xi_2$). Then $z(x,\kappa)$ is decreasing in x and*

$$z(x,\kappa) \sim \begin{cases} z_p, & \text{if } x \to 0\ , \\ C(p,q)x^{pq\lambda/(p-q)}, & \text{if } x \to \infty, \end{cases}$$

$$\delta(x,\kappa) \sim \begin{cases} C(p)x^{rp}, & \text{if } x \to 0, \\ C_{p,q}x^{-\Delta/(p-q)}, & \text{if } x \to \infty. \end{cases}$$

(3) *Let $q < p < 2$ ($\kappa \in \Xi_3$). If $\lambda < 0$, then $z(x,\kappa)$ is increasing in x, and if $\lambda > 0$, then it is decreasing in x,*

$$z(x,\kappa) \sim \begin{cases} C(p,q)x^{pq\lambda/(p-q)}, & \text{if } x \to 0, \\ x^{pq\lambda/(p-q)}, & \text{if } x \to \infty\ , \end{cases}$$

$$\delta(x,\kappa) \sim \begin{cases} C_{p,q}x^{-\Delta/(p-q)}, & \text{if } x \to 0,\ \lambda > 0 \text{ or } x \to \infty,\ \lambda < 0, \\ (p-q)x^{rp}z^{p-2}\exp(-z^2), & \text{if } x \to \infty,\ \lambda > 0 \text{ or } x \to 0,\ \lambda < 0, \end{cases}$$

where $z = z(x,\kappa)$.

(4) *Let $2 = p > q$ ($\kappa \in \Xi_4$). Then $z(x,\kappa)$ is increasing in x and*

$$z(x,\kappa) \sim \begin{cases} C_q x^{2q\lambda/(6-q)}, & \text{if } x \to 0\ , \\ x^{2q\lambda/(2-q)}, & \text{if } x \to \infty, \end{cases}$$

$$\delta(x,\kappa) \sim \begin{cases} C_q^{-2}x^{-\tilde{\Delta}/(6-q)}, & \text{if } x \to 0\ , \\ (2-q)x^{2r}\exp(-z^2(x,\kappa)), & \text{if } x \to \infty. \end{cases}$$

(B) *Let $\lambda \le 0$ (recall that $\kappa \in \Xi_3$, $\Delta > 0$ in this case). If $\lambda = 0$, then $\zeta(y,\kappa,z_0)$ is constant. If $\lambda < 0$, then it is decreasing continuous in y. If $z_0 \to 0$, $z_0 m^{-rp/(1-p)} \to 0$, $y \ge m^{-1}$, $m \to \infty$, then $\zeta(y,\kappa,z_0) = z_0\tau_0(y,\kappa,z_0)$ where*

$$\tau_0(y,\kappa,z_0) \sim \tau_0(y,\kappa) = C(p,q)y^{pq\lambda/(p-q)}, \quad \eta(y,\kappa,z_0) \sim C_{p,q}y^{-\Delta/(p-q)}.$$

Proof of Proposition A.2. This is based on the standard properties of the inverse functions and on the standard asymptotic relations

$$\sinh x \sim \tanh x \sim x; \quad x^2 - 2\tanh(x^2/2) \sim x^6/12, \quad \text{as } x \to 0.$$

We omit these elementary calculations. □

Note that if $\kappa \in \Xi_1$, then $\Delta \le 0$. On the other hand, if $\kappa \in \Xi_l$, $l = 2,3,4$, then the quantity $\Delta = \Delta(\kappa)$ could be positive, negative, or zero. Denote

$$\begin{aligned} \Xi_\Delta^- &= \Xi_1 \cup \{\kappa \in \Xi_2 \cup \Xi_3 : \Delta \le 0\} \cup \left\{\kappa \in \Xi_4 : \tilde{\Delta} \le 0\right\}; \\ \Xi_\Delta^+ &= (\Xi_2 \cup \Xi_3 \cup \Xi_4) \setminus \Xi_\Delta^-. \end{aligned}$$

The partitions of the sets Ξ_l to sets $\Xi_l^- = \Xi_l \cap \Xi_\Delta^-$ and $\Xi_l^+ = \Xi_l \cap \Xi_\Delta^+$, $l = 2,3,4$, are presented in Figures 10–13 (compare with Figures 5–8 in Sections 6.4). Recall that the functions λ, μ, I and the points x^*, y^* are defined in Sections 6.4. Note that the straight line $\Delta = 0$ passes through the points 0 and x^*. If $2 < p < 4$, $p < q$, then the set Ξ_2^- is the triangle bounded by the lines $I = 0$, $\Delta = 0$, and $r = -1/2p$ (see Figure 10). If $\kappa \in \Xi_3$ (see Figure 12), then the inequality $\Delta \le 0$ yields $r \ge 0$, $\lambda > 0$. Therefore we have $r \ge 0$, $\lambda > 0$ for $\kappa \in \Xi_3^-$. Analogous relations hold for $\kappa \in \Xi_4$ (see Figure 13).

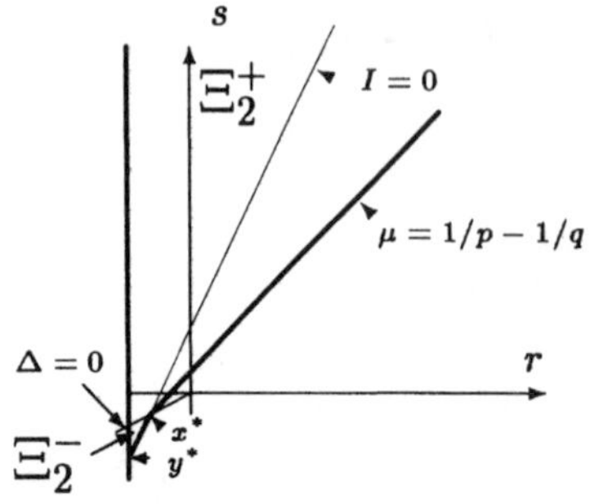

Fig. 10 $2 < p < q \leq \infty;\ p \leq 4$

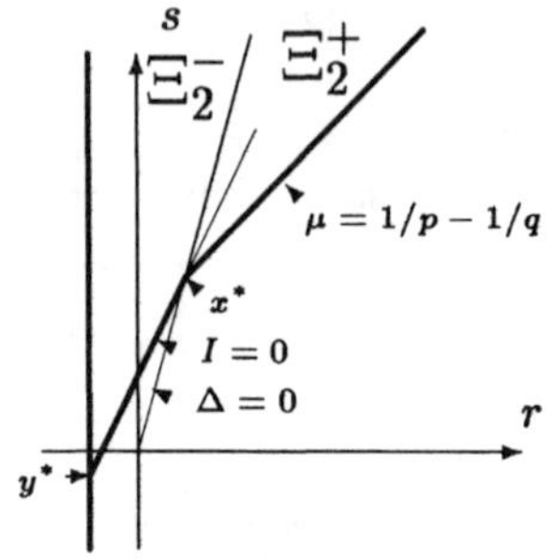

Fig. 11 $4 < p < q \leq \infty$

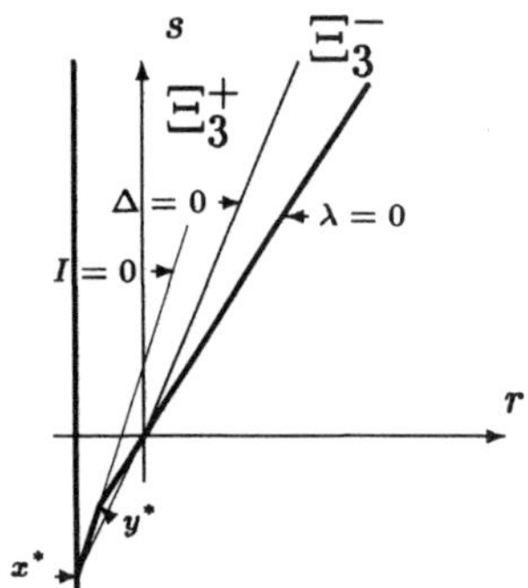

Fig. 12 $2 > p > q$

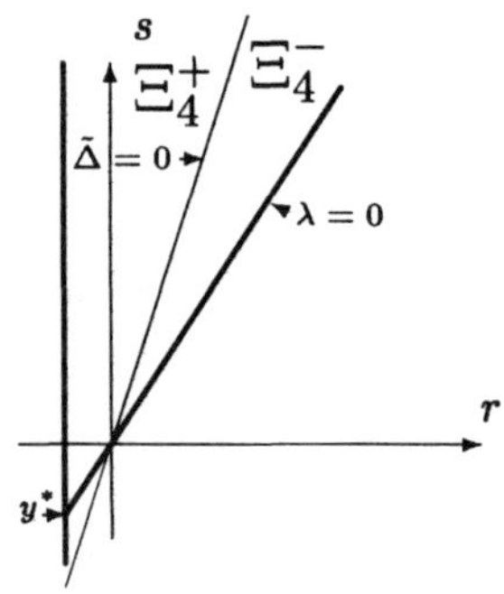

Fig. 13 $q < p = 2$

Proposition A.2 yields the relations for the solutions z_i, h_i of equations (A.33), (A.34). These are the solutions of (6.86), (6.87) for the case $h_i \leq 1$.

Corollary A.1

(1) *Assume* $\kappa \in \Xi_\Delta^-$ *and*

$$n \to \infty, \quad h_0 \to 0; \quad n^{-rp} h_0 \to 0 \ \text{ if } \ p > 2. \tag{A.48}$$

Then, for $\varepsilon > 0$ *small enough, the relations*

$$z_i = z(i/n, \kappa), \quad h_i = h_0 \delta(i/n, \kappa), \tag{A.49}$$

define the solutions of the system (6.86), (6.87) *and of the extreme problem* (6.83) *for all* $i \in \mathbf{N}$.

(2) *Assume* $\kappa \in \Xi_\Delta^+$ *and* $\lambda > 0$. *Then relations* (A.49) *define the solutions of the system* (6.86), (6.87) *and of the extreme problem* (6.83) *for* $i \in I_0$, *where the integer set* $I_0 = I_0(\kappa)$ *is defined by the relations*

$$I_0 = \begin{cases} i \geq 1 : i \leq n x_\varepsilon = m y_\varepsilon, & \text{if } \kappa \in \Xi_2^+, \\ i \geq 1 : i \geq n x_\varepsilon = m y_\varepsilon, & \text{if } \kappa \in \Xi_3^+,\ \lambda > 0 \ \text{ or } \ \kappa \in \Xi_4^+. \end{cases}$$

Here the quantity $x_\varepsilon = (m/n)y_\varepsilon$ *is determined by the equality* $h_0\delta(x_\varepsilon, \kappa) = 1$. *If* $p \neq 2$, *then this corresponds to*

$$y_\varepsilon = y_\varepsilon(\kappa) \sim y_1(\kappa) = C_{p,q}^{(p-q)/\Delta}, \quad as\ z_0 \to 0.$$

If $p = 2$, *then one has (by* $(6-q)/\tilde{\Delta} > (2-q)/\Delta$ *for* $\kappa \in \Xi_4$*):*

$$y_\varepsilon = y_\varepsilon(\kappa) \sim y_1(\kappa) = (C_q)^{(2q-12)/\tilde{\Delta}} h_0^{(6-q)/\tilde{\Delta}-(2-q)/\Delta} \to 0,$$

as $h_0 \to 0$, $\kappa \in \Xi_4$.

(3) *Assume*

$$\kappa \in \Xi_\Delta^+, \quad z_0 \to 0,\ m \to \infty, \quad z_0 m^{-pq\lambda/(p-q)} \to 0,$$

and $\lambda \leq 0$ *(this yields* $\kappa \in \Xi_3^+$*). Then the relations*

$$z_i = z_0\tau_0(i/m, \kappa, z_0), \quad h_i = \eta(i/m, \kappa, z_0),$$

define the solutions of the system (6.86), (6.87) *for*

$$i \in I_0 = \{i \geq my_\varepsilon\}, \quad y_\varepsilon \sim C_{p,q}^{(p-q)/\Delta}.$$

Remark. Recall that in view of Remark 3 from Section 6.5 the assumptions

$$n \to \infty, \quad h_0 \to 0; \qquad n^{-rp}h_0 \to 0, \ \text{if } p > 2;$$
$$z_0 \to 0, \quad m \to \infty, \quad z_0 m^{-pq\lambda/(p-q)} \to 0, \ \text{if } \kappa \in \Xi_3, \ \lambda \leq 0,$$

follow from the assumptions (6.58).

A.3.2 Solutions of the Extreme Problem for $\kappa \in \Xi_\Delta^-$

We use below that $I > 0$ for $\kappa \subset \Xi_\Delta^-$ (see (6.17) for a definition of the function $I = I(\kappa)$).

By Corollary A.1 it is enough, first, to find the values n, h_0 from equations (6.84) and to obtain u_ε from relation (6.90). Next, we need to control the assumptions (A.48).

We give outlines of the proofs and omit simple calculations which one can easily restore.

First, assume $r > r_p^*$. In this case $\kappa \in \Xi_{G_2}$ and, using Proposition A.2 and Corollary A.1 (1), we can rewrite equations (6.84) and relation (6.90) in the form

$$\rho_\varepsilon^p = h_0 n^{rp+1} C_{1,\varepsilon}(\kappa), \quad R_\varepsilon^q = h_0 n^{sq+1} C_{2,\varepsilon}(\kappa), \quad u_\varepsilon^2 = h_0^2 n C_{0,\varepsilon}(\kappa), \quad \text{(A.50)}$$

where

$$C_{1,\varepsilon}(\kappa) = C_{1,\varepsilon}(\kappa, n) = n^{-1}\sum_i \delta(x_i, \kappa) z^p(x_i, \kappa) x_i^{rp},$$

$$C_{2,\varepsilon}(\kappa) = C_{2,\varepsilon}(\kappa, n) = n^{-1}\sum_i \delta(x_i, \kappa) z^q(x_i, \kappa) x_i^{sq},$$

$$C_{0,\varepsilon}(\kappa) = C_{0,\varepsilon}(\kappa, n) = n^{-1}\sum_i \delta^2(x_i, \kappa) \sinh^2(z^2(x_i, \kappa)/2).$$

Here the functions $z(x,\kappa)$, $\delta(x,\kappa)$ are defined by (A.38), (A.39), (A.44). The functions $C_{l,\varepsilon}(\kappa,n)$, $l=0,1,2$, are continuous functions of κ,n which are bounded away from 0 and ∞ for n large enough.

Let us show that

$$\rho_\varepsilon^p \sim h_0 n^{rp+1} c_1^p(\kappa), \quad R_\varepsilon^q \sim h_0 n^{sq+1} c_2^q(\kappa), \quad u_\varepsilon^2 \sim h_0^2 n c_0(\kappa), \tag{A.51}$$

where the functions $c_l(\kappa)$, $l=0,1,2$, are of the forms

$$\begin{aligned} c_1^p(\kappa) &= \int_0^\infty \delta(x,\kappa) z^p(x,\kappa) x^{rp}\, dx, \\ c_2^q(\kappa) &= \int_0^\infty \delta(x,\kappa) z^q(x,\kappa) x^{sq}\, dx, \\ c_0(\kappa) &= 2\int_0^\infty \delta^2(x,\kappa)\sinh^2(z^2(x,\kappa)/2)\, dx. \end{aligned} \tag{A.52}$$

These correspond to substitution of the normalized sums by integrals in (A.50).

First, let us determine that the integrals in (A.52) are finite. In fact, since $z_i = z(x_i,\kappa) \to \infty$, as $x_i = i/n \to \infty$, $p>q$ (the last means $\kappa \notin \Xi_2^-$), using Proposition A.2 and the definition of the set Ξ_Δ^-, we get for all integrals in $c_l(\kappa)$, $l=0,1,2$,

$$\int_1^\infty \{\ldots\}\, dx \asymp \begin{cases} \int_1^\infty x^a \exp(-bx^c)\, dx, \ b>0,\ c>0, & \text{if } \kappa \notin \Xi_2^- \\ \int_1^\infty x^{(pqI/(p-q))-1}\, dx, \ I>0, & \text{if } \kappa \in \Xi_2^- \end{cases} = O(1), \tag{A.53}$$

where a, $b>0$, $c>0$ are some quantities depending on κ. If $\kappa \in \Xi_1^- \cup \Xi_2^- \cup \Xi_4^-$, then for integrals in $c_l(\kappa)$, $l=0,1$, one has

$$\int_0^1 \{\ldots\}\, dx \asymp \int_0^1 x^{2rp}\, dx = O(1) \tag{A.54}$$

and, for $c_2(\kappa)$,

$$\int_0^1 \{\ldots\}\, dx \asymp \begin{cases} \int_0^1 x^{2rp+pq\lambda}\, dx \asymp 1, & \text{if } \kappa \in \Xi_1 \cup \Xi_2, \\ \int_0^1 x^{4r+4pq\lambda/(6-q)}\, dx \asymp 1, & \text{if } \kappa \in \Xi_4. \end{cases} \tag{A.55}$$

If $\kappa \in \Xi_3^-$, then for all integrals $c_l(\kappa)$, $l=0,1,2$, one has, by $I>0$,

$$\int_0^1 \{\ldots\}\, dx \asymp \int_0^1 x^{(pqI/(p-q))-1}\, dx = O(1). \tag{A.56}$$

Next, the sums under consideration tend to the integrals; this follows from the continuity and piecewise monotonicity of the functions under the integrals. Note that relations (A.51) are uniform over all compacts $K \subset \{\kappa \in \Xi_\Delta^- : r > r_p^*\}$.

Let us show that relations (A.50), (A.51) imply that the solution of the first two equations in (A.50) exists and is of the form $n_\varepsilon \sim \tilde n_\varepsilon$, $h_{0,\varepsilon} \sim \tilde h_{0,\varepsilon}$, we also have the relation $u_\varepsilon^2 \sim \tilde{h}_{0,\varepsilon}^2 \tilde{n}_\varepsilon c_0(\kappa)$. Here $\tilde{n}_\varepsilon$ and $\tilde{h}_{0,\varepsilon}$ are defined by the equalities

$$\rho_\varepsilon^p = \tilde{h}_{0,\varepsilon} \tilde{n}_\varepsilon^{rp+1} c_1^p(\kappa), \quad R_\varepsilon^q = \tilde{h}_{0,\varepsilon} \tilde{n}_\varepsilon^{sq+1} c_2^q(\kappa).$$

In fact, introduce the variables

$$z_1 = (n/\tilde{n}_\varepsilon)^{1+rp} h_0/\tilde{h}_{0,\varepsilon} - 1, \quad z_2 = (n/\tilde{n}_\varepsilon)^{1+sq} h_0/\tilde{h}_{0,\varepsilon} - 1,$$

and consider the continuous function

$$f_n(z) = f(z) = (f_1(z), f_2(z)) \; : \; z = (z_1, z_2) \to R^2,$$

defined by (we omit the index n)

$$f_1(z_1, z_2) = \rho_\varepsilon^{-p} h_0 n^{rp+1} C_{1,\varepsilon}(\kappa, n) - 1 = z_1(1 + \delta_1(z_1, z_2)) + \delta_1(z_1, z_2),$$

$$f_2(z_1, z_2) = R_\varepsilon^{-q} h_0 n^{sq+1} C_{2,\varepsilon}(\kappa, n) - 1 = z_2(1 + \delta_2(z_1, z_2)) + \delta_2(z_1, z_2).$$

It follows from (A.50), (A.51) that $\delta_l(z_1, z_2) \to 0$, $l = 1, 2$, as $n \to \infty$, uniformly over any ball $D^2(a) \subset R^2$ with $a = O(1)$. Thus we have the relation $\|f_n(z) - z\| \to 0$, $n \to \infty$, uniformly over $z \in D^2(a)$ for any ball $D^2(a) = \{|z| \le a\} \subset R^2$. We can rewrite the equations in (A.50) in the form

$$\begin{cases} f_1(z_1, z_2) = 0, \\ f_2(z_1, z_2) = 0. \end{cases} \tag{A.57}$$

The existence of the solution (A.57) follows from the following topological statement:

Lemma A.1 *Let $f : D^k(a) \to R^k$, $k \ge 1$, be a continuous map such that $\|f(z) - z\| < b < a$ over the boundary sphere $z \in S^{k-1}(a)$. Then there exists $z_0 \in D^k(a)$ such that $f(z_0) = 0$.*

Proof. Set $\breve{R}^k = R^k \setminus \{0\}$. Suppose that there does not exist a point $z_0 \in D^k(a)$ such that $f(z_0) = 0$. This means that the image $f(D^k(a))$ does not contain the point 0. Let us consider the homotopy of the restriction $f = f_0$ on the sphere $S^{k-1}(a)$ to the unit map $f_1(z) = z$ of the form

$$f_t(z) = tz + (1-t)f(z) \; : \; z \to R^k, \quad t \in [0, 1].$$

Let us determine that the image $f_t(S^{k-1}(a))$, $t \in [0, 1]$, does not contain the point 0. In fact, let $f_{t_0}(z) = t_0 z + (1 - t_0) f(z) = 0$ for some $z \in S^{k-1}(a)$, $t_0 \in [0, 1]$. By $f(z) \ne 0$, $z \ne 0$ we get $t_0 \ne 0, 1$. Therefore

$$f(z) = -\frac{z t_0}{1 - t_0}; \quad \|f(z) - z\| = \frac{\|z\|}{1 - t_0} = \frac{a}{1 - t_0} > a.$$

Thus the families of maps

$$f_t(z) = tz + (1-t)f(z) \; : \; z \to \breve{R}^k, \quad t \in [0, 1]$$

provide a continuous homotopy of the restriction $f = f_0 : S^{k-1}(a) \to \breve{R}^k$ to the unit map $f_1(z) = z$, $z \in S^{k-1}(a)$. This leads to the contradiction because the map $f_1 : S^{k-1}(a) \to S^{k-1}(a)$ generates a nontrivial homotopy group of the set $\breve{R}^k$, i.e., it is impossible to extend this map to continuous map $f : D^k(a) \to \breve{R}^k$. □

Thus we have the existence of the solutions $h_0 = h_{0,\varepsilon}$, $n = n_\varepsilon$ of equations (A.50) with the asymptotics (6.21), (6.20) for $r > r_p^*$.

Next, let $r = r_p^*$. In this case for $\kappa \in \Xi_\Delta^-$ and $\kappa \in \Xi_{G_5}$, we have

$$p > 2, \quad r_p^* = -1/2p, \quad s > -1/2q.$$

The integral $c_2^q(\kappa)$ is of the same form, however the integrals $c_1^p(\kappa)$, $c_0(\kappa)$ diverge in (A.52). The relations for ρ_ε^p, u_ε^2 in (A.50) can be rewritten in the form

$$\begin{aligned} \rho_\varepsilon^p &= h_0 n^{1/2}\left(\int_{1/n}^1 \delta(x,\kappa) z^p(x,\kappa) x^{-1/2}\,dx + O(1)\right) \\ &\sim z_p^{2p} h_0 n^{1/2} \log n \left(4\sinh^2\frac{z_p^2}{2}\right)^{-1}, \end{aligned} \tag{A.58}$$

$$\begin{aligned} u_\varepsilon^2 &= 2nh_0^2\delta^2(x,\kappa)\left(\sinh^2\frac{z^2(x,\kappa)}{2}dx + O(1)\right) \\ &\sim z_p^{2p} h_0^2 n \log n \left(8\sinh^2\frac{z_p^2}{2}\right)^{-1}, \end{aligned} \tag{A.59}$$

which provide relations (6.45), (6.46). By the Remarks from Section 6.5, the assumption (6.58) implies that $h_0 = o(1)$, $n^{-1} = o(1)$.

A.3.3 Proofs of Theorems 6.1, 6.3 for $\kappa \in \Xi_\Delta^-$

Observe that $\Xi_\Delta^- \subset \Xi_{C_2} \cup \Xi_{C_5}$. In what follows we assume $u_\varepsilon^2 \asymp 1$, which is equivalent to $u_\varepsilon^2 \asymp nh_0^2 \asymp 1$ for $r > r_p^*$ and $u_\varepsilon^2 \asymp nh_0^2 \log n \asymp 1$ for $r = r_p^*$. To prove Theorems 6.1, 6.3 let us check the assumptions of Proposition 5.5.

Using Proposition A.2, the asymptotics (6.20), (6.21) and (6.45), (6.46) we can check that assumptions A.1, A.3 with any $\delta \in (0,1)$ for $p > q$, or A.4 for $p < q$, are fulfilled. Remember that assumptions A.1, A.2, A.3, A.4 are presented in Section 4.3.1.

In fact, let $p < q$ (this means $\kappa \in \Xi_2^-$). Then $z_i = z(x_i,\kappa) = O(1)$, which yields A.4. Let $p > q$ (this means $\kappa \notin \Xi_2^-$). It follows from Proposition A.2 that (compare with (A.53)–(A.56))

$$\begin{aligned} u_{\varepsilon,i}^2 &\sim 2h_0^2\delta^2(x_i,\kappa)\sinh^2(z^2(x_i,\kappa)/2) \\ &\asymp h_0^2 \begin{cases} x_i^{2rp}, & \text{if } x_i \le 1,\ \kappa \notin \Xi_3^-, \\ x_i^{(pqI/(p-q))-1}, & \text{if } x_i \le 1,\ \kappa \in \Xi_3^-, \\ x_i^{(pqI/(p-q))-1}\exp(-z(x_i,\kappa)^2), & \text{if } x_i > 1. \end{cases} \end{aligned} \tag{A.60}$$

Since $z_i = z(x_i, \kappa) \sim x^{pq\lambda/(p-q)} \to \infty$, as $x_i = i/n \to \infty$, we get

$$\exp(-z(x_i,\kappa)^2) \asymp \exp(-bx_i^c), \quad b > 0, \quad c > 0,$$

uniformly over $x_i \geq 1$. Therefore for any $\delta \in (0,1)$, $z_{\varepsilon,i} > 1$,

$$u_{\varepsilon,i}^2 \exp(\delta z_{\varepsilon,i}^2) \asymp h_0^2 x_i^a \exp(-(1-\delta)z^2(x_i,\kappa)) \asymp h_0^2 x_i^a \exp(-bx_i^c)$$

with $a = (pqI/(p-q)) - 1$, some $b = b(\delta) > 0$, $c > 0$. Analogously to (A.53) we get

$$\sum_{i:\, z_{\varepsilon,i}>1} u_{\varepsilon,i}^2 \exp(\delta z_{\varepsilon,i}^2) \asymp nh_0^2 \int_1^\infty x^a \exp(-bx^c)\, dx \asymp 1.$$

because the integral converges. This yields A.3 for $p > q$.

Note that $pqI/(p-q) \geq 1$ for $\kappa \in \Xi_3^-$ and $h_0 n^{-rp} \to 0$ for $\kappa \notin \Xi_3^-$. Therefore (A.60) yields assumption A.1:

$$\sup_{1 \leq i < \infty} u_{\varepsilon,i} \asymp h_0 \begin{cases} \max(1, n^{-rp}), & \text{if } \kappa \notin \Xi_3^- \\ 1, & \text{if } \kappa \in \Xi_3^- \end{cases} \to 0.$$

We need to control the assumptions (5.11), (5.29) for the thresholding (6.22). Consider the case $r > r_p^*$. Assuming $u_\varepsilon \asymp 1$, let us show that (5.11) holds for any $\eta_0 > 0, \tau \in (\sqrt{2/3}, 1)$ and inequalities (5.29) hold for any $\zeta_\varepsilon \to 0$ such that $(\log R_\varepsilon)^{1/2}\zeta_\varepsilon \to \infty$.

In fact, using (5.6), (5.7), and Proposition A.2, we have

$$T_{\varepsilon,i} \asymp \frac{\log(1+h_{\varepsilon,i}^{-1})}{z_{\varepsilon,i}} = \frac{\log \delta(x_i,\kappa)^{-1} + \log h_0^{-1}}{z_{\varepsilon,i}}, \quad \text{as } z_{\varepsilon,i} = O(1), \quad \text{(A.61)}$$

$$\begin{aligned} T_{\varepsilon,i} &= \frac{z_{\varepsilon,i}}{2} + \frac{\log h_{\varepsilon,i}^{-1}}{z_{\varepsilon,i}} + o(z_{\varepsilon,i}^{-1}) \\ &= \frac{3}{2} z_{\varepsilon,i} + \frac{\log h_0^{-1}}{z_{\varepsilon,i}} + O(\log z_{\varepsilon,i}/z_{\varepsilon,i}), \quad \text{as } z_{\varepsilon,i} \to \infty. \end{aligned} \quad \text{(A.62)}$$

Using (A.61)–(A.62), Proposition A.2, and Corollary A.1, since

$$\log R_\varepsilon \asymp \log n, \quad nh_0^2 \asymp u_\varepsilon^2 \asymp 1,$$

we easily get: if $x_i = O(1)$, then $z_{\varepsilon,i} = O(1)$ and for some $b = b(\kappa) > 0$,

$$T_{\varepsilon,i} \geq b \log n, \quad Q_{\varepsilon,i} = \sqrt{2\log i + c\log\log(i+1) + 4\log\log R_\varepsilon} \asymp \sqrt{\log n};$$

therefore $Q_{\varepsilon,i} = o(T_{\varepsilon,i})$ which yields (5.11). Let $x_i \to \infty$, $p < q$. Then for some $c = c(\kappa) > 0$ we have

$$z_{\varepsilon,i} \asymp x_i^{-c} \to 0, \quad T_{\varepsilon,i} \asymp x_i^c \log n, \quad Q_{\varepsilon,i} \sim \sqrt{2(\log n + \log x_i)} = o(T_{\varepsilon,i}).$$

Let $x_i \to \infty$, $p > q$. Then $z_{\varepsilon,i} \to \infty$ and for some $c = c(\kappa) > 0$ we have

$$Q_{\varepsilon,i} \sim \sqrt{2(\log n + \log x_i)}, \quad T_{\varepsilon,i} = \frac{3}{2}x_i^c + \frac{\log n}{2x_i^c} + O\left(\frac{\log x_i}{x_i^c}\right).$$

Therefore, if $x_i \to \infty$, $\log x_i = o(\log n)$, then by $a^2 + b^2 \ge 2ab$ we get

$$Q_{\varepsilon,i} \sim \sqrt{2\log n}, \quad \frac{3}{2}x_i^c + \frac{\log n}{2x_i^c} \ge \sqrt{3\log n},$$

which yields (5.11) with $\tau \in (\sqrt{2/3}, 1)$. If $x_i \to \infty$, $\log x_i > d\log n$ for any $d > 0$, then

$$T_{\varepsilon,i} \asymp x_i^c, \quad Q_{\varepsilon,i} \asymp \sqrt{\log x_i} = o(T_{\varepsilon,i}),$$

which yields (5.11) as well.

We need to control (5.29) for $p > q$, $x_i \to \infty$ only. It suffices to show that for $(\log R_\varepsilon)^{1/2}\zeta_\varepsilon \asymp (\log n)^{1/2}\zeta_\varepsilon \to \infty$ the following inequalities hold:

$$Q_{\varepsilon,i} \le T_{\varepsilon,i} - \frac{z_{\varepsilon,i}}{2} + \zeta_\varepsilon z_{\varepsilon,i}, \quad T_{\varepsilon,i} \ge \frac{3z_{\varepsilon,i}}{2} - \zeta_\varepsilon z_{\varepsilon,i}, \quad \text{if} \;\; z_{\varepsilon,i} > \eta T_{\varepsilon,i}. \tag{A.63}$$

The inequality $z_{\varepsilon,i} > \eta T_{\varepsilon,i}$ for small $\eta > 0$ yields $z_{\varepsilon,i} \asymp x_i^c > \eta_1\sqrt{\log n}$ for small $\eta_1 > 0$ and $\log x_i \ge b\log\log n$ for some $b > 0$. Using the evaluations above we get: if $\log x_i \asymp \log\log n$, then

$$\begin{aligned}
T_{\varepsilon,i} &\ge 3z_{\varepsilon,i}/2 + O(\log z_{\varepsilon,i}/z_{\varepsilon,i});\\
T_{\varepsilon,i} - z_{\varepsilon,i}/2 &= z_{\varepsilon,i} + \log n/2z_{\varepsilon,i} + O(\log z_{\varepsilon,i}/z_{\varepsilon,i})\\
&\ge \sqrt{2\log n} + O(\log z_{\varepsilon,i}/z_{\varepsilon,i})\\
&= \sqrt{2\log n} + O(\log\log n/\sqrt{\log n});\\
Q_{\varepsilon,i} &\le \sqrt{2\log n + 3\log x_i + 4\log\log R_\varepsilon}\\
&= \sqrt{2\log n} + O(\log\log n/\sqrt{\log n}),
\end{aligned}$$

which yield (A.63). If $\log x_i/\log\log n \to \infty$, then we get

$$T_{\varepsilon,i} \ge z_{\varepsilon,i}/2 \asymp x_i^c \gg (\log n)^B$$

for any $B > 0$, and $Q \asymp \sqrt{\log n + \log x_i}$. This yields $Q_{\varepsilon,i} = o(T_{\varepsilon,i})$, and the evaluations above yield (A.63) as well.

If $r = r_p^*$, then the study is analogous and is based on the relation $h_0 \asymp (n\log n)^{-1/2}$ for $u_\varepsilon \asymp 1$.

To obtain the lower bounds let us construct the families $\bar\pi_{\varepsilon,1} = \{\pi_{\varepsilon,i,1}\}$ which consist of symmetric three-point measures, satisfying A.1, A.2, and

$$\|\bar\pi_{\varepsilon,1}\| = u_\varepsilon + o(1), \quad \bar\pi_1^\varepsilon(V_\varepsilon) \to 1.$$

Analogously to Section 6.6.2, take values $\delta_\varepsilon = (\log\log n)^{-1} \to 0$ and the small change threshold constraints $\rho_\varepsilon, R_\varepsilon$:

$$\tilde\pi_\varepsilon = \bar\pi_\varepsilon(\kappa, (1+\delta_\varepsilon)\rho_\varepsilon, (1-\delta_\varepsilon)R_\varepsilon).$$

If $r > r_p^*$, then consider the "two-sided T_ε-truncated" sequences $\bar\pi_{\varepsilon,1} = \{\bar\pi_{\varepsilon,i,1}\}$ for $T_\varepsilon = n^a$ with $a > 0$ small enough,

$$\pi_{\varepsilon,i,1} = \begin{cases} \tilde\pi_{\varepsilon,i}, & \text{if } T_\varepsilon^- \le i/n \le T_\varepsilon,\\ \delta_0, & \text{otherwise,}\end{cases} \qquad \text{where } T_\varepsilon^- = T_\varepsilon^{-1}$$

and if $r = r_p^*$, then consider the "one-sided T_ε-truncated" sequences

$$\pi_{\varepsilon,i,1} = \begin{cases} \tilde{\pi}_{\varepsilon,i}, & \text{if } T_\varepsilon^- \le i/n \le T_\varepsilon, \\ \delta_0, & \text{if } i/n > T_\varepsilon, \end{cases} \quad \text{where } T_\varepsilon^- = 1/n.$$

Clearly, the family $\bar{\pi}_{\varepsilon,1}$ satisfies assumptions A.1, A.2 (recall that A.2. follows from A.1 and either A.3 or A.4). The relation $\|\bar{\pi}_{\varepsilon,1}\| = u_\varepsilon + o(1)$ follows from the asymptotics (6.20), (6.21) and (6.45), (6.46) and (A.53)–(A.56) (these correspond to the rejection of tails in the integrals). The relation $\bar{\pi}_1^\varepsilon(V_\varepsilon) \to 1$ follows from the Chebyshev inequality and relations (A.52)–(A.58) analogously to Section 6.6.2.

In fact, one can check that

$$E_{\bar{\pi}_1^\varepsilon}(G_1(v)) = (1 + \delta_\varepsilon - o(\delta_\varepsilon))^p H_{\varepsilon,1}, \quad E_{\bar{\pi}_1^\varepsilon}(G_2(v)) \le (1 - \delta_\varepsilon)^q H_{\varepsilon,2}.$$

Let us determine the following relations which are needed to use the Chebyshev inequality:

$$\begin{aligned} \mathrm{Var}_{\bar{\pi}_1^\varepsilon}(G_1(v)) &= \sum_i \tilde{z}_i^{2p} \tilde{h}_i (1 - \tilde{h}_i) i^{2rp} \\ &\asymp \tilde{n}^{1+2rp} \tilde{h}_0 \int_{T_\varepsilon^-}^{T_\varepsilon} \delta(\tilde{x}, \kappa) z^{2p}(\tilde{x}, \kappa) \tilde{x}^{2rp}\, d\tilde{x} = o(H_{\varepsilon,1}^2 \delta_\varepsilon^2), \end{aligned} \tag{A.64}$$

$$\begin{aligned} \mathrm{Var}_{\bar{\pi}_1^\varepsilon}(G_2(v)) &= \sum_i \tilde{z}_i^{2q} \tilde{h}_i (1 - \tilde{h}_i) i^{2sq} \\ &\asymp \tilde{n}^{1+2sq} \tilde{h}_0 \int_{T_\varepsilon^-}^{T_\varepsilon} \delta(\tilde{x}, \kappa) z^{2q}(\tilde{x}, \kappa) \tilde{x}^{2sq}\, d\tilde{x} = o(H_{\varepsilon,2}^2 \delta_\varepsilon^2). \end{aligned} \tag{A.65}$$

We need to verify only the last relations in (A.64), (A.65). Note that

$$\begin{aligned} H_{\varepsilon,1} &\asymp \begin{cases} n^{1+rp} h_0, & \text{if } r > r_p^*, \\ n^{1/2} h_0 \log n, \ \kappa \in \Xi_1 \cup \Xi_2, & \text{if } r = r_p^* = -1/2p, \end{cases} \\ H_{\varepsilon,2} &\asymp n^{1+sq} h_0. \end{aligned} \tag{A.66}$$

We omit the tilde below. It suffices to show that for both integrals $I_{1,2}$ in (A.64), (A.65) one has: if $r > r_p^*$, then $I_{1,2} = o(nh_0\delta_\varepsilon^2)$ and if $r = r_p^*$, then

$$I_1 = o(nh_0 \log^2 n \delta_\varepsilon^2), \quad I_2 = o(nh_0 \delta_\varepsilon^2).$$

Let $r > r_p^*$. Then $nh_0 \asymp n^{1/2} u_\varepsilon$ and the needed relations follow from the evaluations

$$\int_{T_\varepsilon^{-1}}^{T_\varepsilon} \{\dots\}\, dx = O(T_\varepsilon^A) = o(n^{\delta_0})$$

with some $A = A(\kappa) > 0$ and one can make $\delta_0 > 0$ arbitrary small by choosing $a > 0$ small enough. Let $r = r_p^*$. Then

$$p > 2, \quad r_p^* = -1/2p, \quad \kappa \in \Xi_1 \cup \Xi_2$$

and we also have $H_{\varepsilon,1} \asymp (\log n)^{1/2}$ for $u_\varepsilon \asymp nh_0^2 \log n \asymp 1$. Therefore

$$\int_{n^{-1}}^{T_\varepsilon} \{\ldots\}\, dx \le O(T_\varepsilon^{A_2}) + \int_{n^{-1}}^{1} \{\ldots\}\, dx$$

and

$$\int_{1/n}^{1} \delta(x,\kappa) z^{2p}(x,\kappa) x^{2rp}\, dx \asymp \int_{1/n}^{1} x^{3rp}\, dx \asymp n^{1/2}. \tag{A.67}$$

Thus we obtain

$$\mathrm{Var}_{\tilde\pi^\varepsilon}(G_1(v)) \le h_0(O(T_\varepsilon^{A}) + Bn^{1/2}) \asymp (\log n)^{-1/2} = o(H_{\varepsilon,1}^2 \delta_\varepsilon^2).$$

Analogously, relation (A.65) holds by $nh_0\delta_\varepsilon^2 \asymp n^{1/2}\delta_\varepsilon^2/\log^{1/2} n$ and for some $0 < \delta_0 < 1/2$ we have

$$\int_{1/n}^{1} \delta(x,\kappa) z^{2q}(x,\kappa) x^{2sq}\, dx \asymp \int_{1/n}^{1} x^{3rp+2pq\lambda}\, dx = O(n^{1/2-\delta_0}). \tag{A.68}$$

Thus Theorems 6.1, 6.3 are proved for the case $\kappa \in \Xi_\Delta^-$.

A.3.4 Solution of the Extreme Problem (6.88)

Using the notations (A.41) we can rewrite equation (6.89):

$$2z_0^{-4} z_i^2 \sinh(z_i^2) + qy^{sq}(z/z_0)^q = py^{rp}(z/z_0)^p. \tag{A.69}$$

We need to select the root of (A.69) which provides a minimum in (6.88). By Corollary A.1 we need to study (A.69) for $\kappa \in \Xi_\Delta^+$:

$$\Xi_\Delta^+ = \{\kappa \in \Xi_2 \cup \Xi_3 : \Delta > 0\} \cup \{\kappa \in \Xi_4 : \tilde\Delta > 0\} = \Xi_2^+ \cup \Xi_3^+ \cup \Xi_4^+.$$

One can easily check that if $\kappa \in \Xi_4^+$ (i.e., $\kappa \in \Xi_4$ and $\tilde\Delta > 0$), then $\Delta > 0$. Therefore $\Delta > 0$ for any $\kappa \in \Xi_\Delta^+$.

We use the parameters z_0, m and the variables $y = y_i = i/m$ jointly with h_0, n and $x = x_i = i/n$ here (see (A.40)–(A.42)). We need to consider $i \in I_1(\kappa)$, $y \in Y_\varepsilon(\kappa)$, where the integer set $I_1 = I_1(\kappa)$ is the complement of I_0 defined in Corollary A.1:

$$I_1 = I_1(\kappa) = \begin{cases} \{i \ge my_\varepsilon\}, & \text{if } \kappa \in \Xi_2^+, \\ \{i \le my_\varepsilon\}, & \text{if } \kappa \in \Xi_3^+ \text{ or } \kappa \in \Xi_4^+, \end{cases}$$

and

$$Y_{1,\varepsilon}(\kappa) = Y_1 = \begin{cases} [y_\varepsilon, \infty), & \text{if } \kappa \in \Xi_2^+, \\ (0, y_\varepsilon], & \text{if } \kappa \in \Xi_3^+ \text{ or } \kappa \in \Xi_4^+, \end{cases}$$

the values $y_\varepsilon = y_\varepsilon(\kappa) \sim y_1(\kappa)$ are defined in Corollary A.1. Denote by Y_0 the complement of the set Y_1, i.e., $Y_0 = R_+^1 \backslash Y_1$ and consider the set of parameters x corresponding to Y_0,

$$X_\varepsilon(\kappa) = \begin{cases} (0, x_\varepsilon], & \text{if } \kappa \in \Xi_2^+ \quad, \\ [x_\varepsilon, \infty), & \text{if } \kappa \in \Xi_3^+ \text{ or } \kappa \in \Xi_4^+. \end{cases}$$

Let $h_0 \to 0$. If $\kappa \in \Xi_2^+$, then $x_\varepsilon \to \infty$, and if $\kappa \in \Xi_3^+$, $\lambda > 0$ or $\kappa \in \Xi_4^+$, then $x_\varepsilon \to 0$ (note that $z_0 \to 0$ in these cases). Let $z_0 \to 0$. If $\kappa \in \Xi_3^+$, $\lambda < 0$, then $x_\varepsilon \to \infty$.

Proposition A.3

(1) *Let $\kappa \in \Xi_\Delta^+$, $i \in I_1(\kappa)$. There exists the unique solution of* (A.69) *which provides a minimum in* (6.88) *and is of the form*

$$z_i = z_1(y_i, \kappa, z_0) = z_0\tau(y_i, z_0, \kappa) > 0,$$

where $\tau(y, z_0, \kappa)$ is the continuous positive smooth function in z_0, $y \in Y_{1,\varepsilon}$, $\kappa \in \Xi_\Delta^+$ and the boundary continuity condition holds

$$z_0\tau(y_\varepsilon, z_0, \kappa) = \zeta(y_\varepsilon, \kappa, z_0) = z(x_\varepsilon, \kappa).$$

Here the function $\zeta = \zeta(y, \kappa, z_0)$ is defined by (A.46).

(2) *Assume $z_0 = o(1)$ and $z_0 m^{-rp/(4-p)} = o(1)$ for $p \leq 2$, $r < 0$. Then, uniformly over z_0, $y \in Y_{1,\varepsilon}$, $\kappa \in \Xi_\Delta^+$, the function $\tau(y, z_0, \kappa) \sim \tau(y, \kappa)$ asymptotically does not depend on z_0:*

$$z_i \sim z_0\tau_i = z_0\tau(y_i, \kappa), \quad y_i = i/m,$$

where $\tau = \tau(y, \kappa)$ is the maximal solution of the equation (compare with (4.95))

$$f(\tau, y) = py^{rp}, \quad \text{where } f(\tau, y) = 2\tau^{4-p} + qy^{sq}\tau^{q-p}. \tag{A.70}$$

The function $\tau(y, \kappa)$ is a continuous smooth function on the sets $Y_{1,\varepsilon}(\kappa)$ with the asymptotical properties

$$\tau(y, \kappa) \sim \begin{cases} (p/2)^{1/(4-p)} y^{rp/(4-p)}, & z_i \to 0, \quad \text{if } y \to 0, \quad \kappa \in \Xi_3 \cup \Xi_4, \\ (p/q)^{1/(q-p)} y^{-pq\lambda/(q-p)}, & z_i \to 0, \quad \text{if } y \to \infty, \quad \kappa \in \Xi_2 \, . \end{cases} \tag{A.71}$$

Proof of Proposition A.3. We give the outline of the proof only. Remember that the existence and uniqueness of the solution follows from Lemma 5.3. We can rewrite equation (A.69) in the form: $f(z) = py^{rp}z_0^{4-p}$ where

$$f(z) = f(z; y, z_0, \kappa) = 2z^{2-p}\sinh(z^2) + qy^{sq}z_0^{4-q}z^{q-p}.$$

Note that $f(z) \to \infty$ as $z \to \infty$. If $p \leq 4$, $q > p$, then the function $f(z)$ increases in $z > 0$ and $f(z) \to 0$ as $z \to 0$ for $p < 4$, $q > p$. If $p = 4$, $q > p$, then $f(z) \to 2$ as $z \to 0$ and, by $r > 0$ in this case, the solution exists for $y > (1/2)^{4r}$ (note that $y > 1$ for $y \in Y_{1,\varepsilon}$). Therefore, if $p \leq 4$, $q > p$, then there exists the unique solution of (A.69), $z_i = z(y, \kappa, z_0) > 0$.

If $q > p > 4$ or $q < p \leq 2$, then $f(z) \to \infty$ as $z \to 0$ and equation (A.69) has two positive roots $0 < z_i^- < z_i^+$; the first one corresponds to the local maximum of the function in (6.88) and we need to select the second root $z_i^+ = z(y, \kappa, z_0) > 0$ which corresponds to the minimum in (6.88) and provides a global minimum by Lemma 5.3.

Using the asymptotics $\sinh x \sim x$, $\tanh x \sim x$, as $x \to 0$, one can easily obtain the asymptotics of Proposition A.3 (2) and (A.70) from (A.69). If $p < 4$, $q > p$, then $f(\tau, y)$ is increasing from 0 to ∞ and the derivative $f'_\tau(\tau, y)$ in τ is strictly positive. This yields the smoothness of $\tau(y) = \tau(y, \kappa)$. If $q > p = 4$, then these hold for $y > (1/2)^{4r} + \delta$ with any $\delta > 0$. If $q > p > 4$ or $q < p \le 2$, then it is possible to check that the quantity

$$\min_\tau f(\tau, y) = f(\tau_{\min}, y)$$

is bounded away from the right-hand side of equation (A.70) (i.e., from $py^{\tau p}$) and for any $\delta > 0$ the derivative $f'_\tau(\tau, y)$ is bounded away from 0 uniformly over $\tau > \tau_{\min} + \delta$ and on $y \in Y_{1,\varepsilon}(\kappa)$. This yields the smoothness of $\tau(y, \kappa)$ over $y \in Y_{1,\varepsilon}(\kappa)$ as well. □

Using relations (A.42) between the variables x, n and y, m, Propositions A.2, A.3, and Remark 3 from Section 6.5 we obtain the following

Corollary A.2 *Assume $\kappa \in \Xi_\Delta^+$, $u_\varepsilon \asymp 1$ and the assumptions on z_0, m of the Proposition A.3, (2) hold. Then the minimum points in the extreme problem (6.83) are defined by the relations:*

(1) If $\lambda > 0$, then we can express the values z_i, h_i in terms of the variables $x = i/n$,

$$z_i = \begin{cases} z(i/n, \kappa), & \text{if } i \in I_0, \\ z(i/n, \kappa, h_0) = z_1(i/m, \kappa, z_0), & \text{if } i \in I_1, \end{cases}$$

where the function $z(x, \kappa, h_0) = z_1(y, \kappa, z_0)$ is of the rate

$$z(x, \kappa, h_0) \sim \begin{cases} (p/2)^{1/(4-p)} h_0^{1/(4-p)} x^{rp/(4-p)}, & \text{if } x \to 0, \ \kappa \in \Xi_3, \\ (p/q)^{1/(q-p)} x^{-pq\lambda/(q-p)}, & \text{if } x \to \infty, \ \kappa \in \Xi_2, \end{cases}$$

and for $\kappa \in \Xi_4$ one has

$$z(x, \kappa, h_0) \sim h_0^{1/2} x^r, \quad \text{if } x \le x_\varepsilon \sim (C_q^{-2} h_0)^{(6-q)/\tilde{\Delta}},$$

where

$$h_i = \begin{cases} h_0 \delta(i/n, \kappa), & \text{if } i \in I_0, \\ 1, & \text{if } i \in I_1. \end{cases}$$

The functions $\delta(x, \kappa)$, $z(x, \kappa)$ are described by Proposition A.2.

(2) Let $\lambda \le 0$ (this is possible for $\kappa \in \Xi_3$, $I < 0$ only. [1]*) Then we can express the values z_i, h_i in terms of the variables $y = i/m$,*

$$z_i = \begin{cases} \zeta(i/m, \kappa, z_0), & \text{if } i \in I_0, \\ z_1(i/m, \kappa, z_0), & \text{if } i \in I_1, \end{cases}$$

and

$$h_i = \begin{cases} \eta(i/m, \kappa, z_0), & \text{if } i \in I_0, \\ 1, & \text{if } i \in I_1, \end{cases}$$

[1] See the Remark on page 366.

where the functions $\zeta(y,\kappa,z_0),\eta(y,\kappa,z_0)$ are defined by Proposition A.2. The properties of the function $z_1(y,\kappa,z_0)$ are described by Proposition A.3.

Note that the relations for z_i, $i \in I_1$, correspond to the results for $p \le q$, $p \le 2$ from Section 4.3.3. Also the assumption $u_\varepsilon \asymp 1$ could be extended to (6.58).

A.3.5 Solution of the Extreme Problem for $\kappa \in \Xi_\Delta^+$

Propositions A.2, A.3 and Corollaries A.1, A.2 have determined the minimum point in the extreme problem (6.83) for $\kappa \in \Xi_\Delta^+$ as functions of the unknown parameters z_0, m or h_0, n, if $\lambda \neq 0$. These parameters must be determined from relations (6.84), (6.90). Let us consider the different cases $I>0$, $I=0$, $I<0$.

Note that the relation $I>0$ corresponds to the case when the solutions with $h_i<1$ have the main "mass" in the sum for u_ε^2 (this relation holds for $\kappa \in \Xi_\Delta^-$); the relation $I<0$ correspond to case when the solutions with $h_i=1$ have the main "mass." This defines different types of asymptotics in following cases.

Case $I>0$

Note that $\lambda>0$ and $\kappa \in \Xi_2 \cup \Xi_3 \cup \Xi_4$ in this case. Assume $n\to\infty$, $h_0 \to 0$. Then $z_0 \to 0$. We can rewrite relations (6.84), (6.90) in the form

$$\begin{aligned}
\rho_\varepsilon^p &= h_0 n^{rp+1}(C_{1,\varepsilon}(\kappa) + h_0^{pqI/\Delta} D_{1,\varepsilon}(\kappa)),\\
R_\varepsilon^q &= h_0 n^{sq+1}(C_{2,\varepsilon}(\kappa) + h_0^{pqI/\Delta} D_{2,\varepsilon}(\kappa)), \qquad (A.72)\\
u_\varepsilon^2 &= h_0^2 n(C_{0,\varepsilon}(\kappa) + h_0^{pqI/\Delta} D_{0,\varepsilon}(\kappa)),
\end{aligned}$$

where

$$\begin{aligned}
C_{1,\varepsilon}(\kappa) &= C_1(\kappa,n) = n^{-1}\sum_{i\in I_0}(i/n)^{rp} z^p(i/n,\kappa)\delta(i/n,\kappa),\\
C_{2,\varepsilon}(\kappa) &= C_2(\kappa,n) = n^{-1}\sum_{i\in I_0}(i/n)^{sq} z^q(i/n,\kappa)\delta(i/n,\kappa), \qquad (A.73)\\
C_{0,\varepsilon}(\kappa) &= C_0(\kappa,n) = (2n)^{-1}\sum_{i\in I_0}\delta^2(i/n,\kappa)\sinh^2(z^2(i/n,\kappa)/2);
\end{aligned}$$

and, by $z_i = z_0\tau^p(i/m,z_0,\kappa) \sim z_0\tau^p(i/m,\kappa)$,

$$\begin{aligned}
D_{1,\varepsilon}(\kappa) &\sim D_1(\kappa,m) = m^{-1}\sum_{i\in I_1}(i/m)^{rp}\tau^p(i/m,\kappa),\\
D_{2,\varepsilon}(\kappa) &\sim D_2(\kappa,m) = m^{-1}\sum_{i\in I_1}(i/m)^{sq}\tau^q(i/m,\kappa), \qquad (A.74)\\
D_{0,\varepsilon}(\kappa) &\sim D_0(\kappa,m) = (2m)^{-1}\sum_{i\in I_1}\tau^4(i/m,\kappa),
\end{aligned}$$

(the last asymptotics correspond to $2\sinh^2(z^2/2) \sim z^4/2$ as $z \to 0$.)

Let $r > r_p^*$ (this means that $\kappa \in \Xi_{G_2}$). Using the asymptotics from Propositions A.2, A.3, from Corollary A.1, and replacing the sums by integrals, we can check that

$$C_{1,\varepsilon}(\kappa) \sim c_1^p(\kappa), \quad C_{2,\varepsilon}(\kappa) \sim c_1^q(\kappa), \quad C_{0,\varepsilon}(\kappa) \sim c_0(\kappa).$$

Here the values $c_l(\kappa)$ are defined by (A.52) and the integrals are finite by the constraints on κ. In fact, we can replace the sums by integrals over the sets X_ε with vanishing errors, as $n \to \infty$. The difference δ_ε with $c_l(\kappa)$ is of the rate

$$\delta_\varepsilon \asymp \begin{cases} \int_{x_\varepsilon}^{\infty} x^{(pqI/(p-q))-1}\,dx, & \text{if } \kappa \in \Xi_2 \\ \int_0^{x_\varepsilon} x^{(pqI/(p-q))-1}\,dx, & \text{if } \kappa \in \Xi_3 \\ \int_0^{x_\varepsilon} x^{4r}\,dx, & \text{if } \kappa \in \Xi_4, \quad l=0,1, \\ \int_0^{x_\varepsilon} x^{4r+4pq\lambda/(6-q)}\,dx, & \text{if } \kappa \in \Xi_4, \quad l=2. \end{cases}$$

Recall that $r > -1/4$, $\lambda > 0$ if $\kappa \in \Xi_4$, $I > 0$. Using the properties of the quantities x_ε noted before in Proposition A.3 one can see that $\delta_\varepsilon = o(1)$.

Analogously, one can check that $D_{l,\varepsilon} = O(1)$. In fact, if $\kappa \in \Xi_2$, then, by (see the Remark on page 366)

$$\mu = s - r > 1/p - 1/q, \quad \Delta > 0, \quad y_\varepsilon \asymp 1, \quad m \asymp n h_0^{(p-q)/\Delta} \to \infty,$$

we get

$$D_{0,\varepsilon} \asymp \int_{y_\varepsilon}^{\infty} y^{-(\mu+\Delta)/(q-p)}\,dy \asymp 1, \quad D_{l,\varepsilon} \asymp \int_{y_\varepsilon}^{\infty} y^{-\mu/(q-p)}\,dy \asymp 1, \quad l = 1,2.$$

Let $\kappa \in \Xi_3 \cup \Xi_4$. If $m = o(1)$, then $I_1 = \emptyset$ and $D_{l,\varepsilon} = 0$; if $m = O(1)$, then $D_{l,\varepsilon} = O(1)$ by $y_\varepsilon \asymp 1$. If $m \to \infty$, then, by $r > r_p^* = 1/4 - 1/p$, $\Delta > 0$, one has

$$D_{2,\varepsilon} \asymp \int_0^{y_\varepsilon} y^{(4rp+\Delta)/(4-p)}\,dy \asymp 1, \quad D_{l,\varepsilon} \asymp \int_0^{y_\varepsilon} y^{4rp/(4-p)}\,dy \asymp 1, \; l = 0,1.$$

These imply the relations analogous to (A.50). Analogously to Section A.3.2 we show that these provide the existence of the solutions $h_0 = h_{0,\varepsilon}$, $n = n_\varepsilon$ with the asymptotics (6.21), (6.20).

Let $r = r_p^*$. By $I > 0$, this means $\kappa \in \Xi_2$, $p < 4$, $\kappa \in \Xi_{G_5}$, and $r_p^* = -1/2p$. By $s > -1/2q$, the sum $C_{2,\varepsilon}(\kappa)$ is equivalent to the finite integral $c_2(\kappa)$ which is of the form (A.52) in this case. The evaluations for $D_{l,\varepsilon}$ are analogous to above. The integrals for $c_0(\kappa)$, $c_1(\kappa)$ diverge, however, relations (A.58), (A.59) hold true.

Case $I < 0$

Note that $\kappa \in \Xi_2 \cup \Xi_3$ in this case. Assume $r > r_p^*$ (this corresponds to $\kappa \in \Xi_{G_1}$), $m \to \infty$, $z_0 \to 0$, and if $p < 2$, then $m^{-rp/(4-p)} z_0 \to 0$. Note that if $u_\varepsilon \asymp 1$, then these assumptions hold and $n = mz_0^{-(p-q)/pq\lambda} \to \infty$ for $\lambda > 0$ (note that if $\kappa \in \Xi_3$, $\lambda < 0$, then $h_0 \to \infty$, $x_\varepsilon \to \infty$, $n \to 0$).

Let $\kappa \in \Xi_2 \cup \Xi_3$ and $\lambda > 0$. We get

$$\begin{aligned} \rho_\varepsilon^p &= z_0^p m^{rp+1}(D_{1,\varepsilon}(\kappa) + z_0^{-I/\lambda}C_{1,\varepsilon}(\kappa)), \\ R_\varepsilon^q &= z_0^q m^{sq+1}(D_{2,\varepsilon}(\kappa) + z_0^{-I/\lambda}C_{2,\varepsilon}(\kappa)), \\ u_\varepsilon^2 &= z_0^4 m(D_{0,\varepsilon}(\kappa) + z_0^{-I/\lambda}C_{0,\varepsilon}(\kappa)), \end{aligned} \tag{A.75}$$

where $C_{l,\varepsilon}(\kappa)$, $D_{l,\varepsilon}(\kappa)$ are defined in (A.73), (A.74). Let us show that

$$\begin{aligned} D_{1,\varepsilon}(\kappa) &= \int_{Y_1} \tau^p(y,\kappa)y^{rp}\,dy + o(1), \\ D_{2,\varepsilon}(\kappa) &= \int_{Y_1} \tau^q(y,\kappa)y^{sq}\,dy + o(1), \\ D_{0,\varepsilon}(\kappa) &= \frac{1}{2}\int_{Y_1} \tau^4(y,\kappa)\,dy + o(1). \end{aligned} \tag{A.76}$$

In fact, if $\kappa \in \Xi_3$, then $4rp/(4-p) > -1$; if $\kappa \in \Xi_2$, then $\mu > 1/p - 1/q$. Therefore

$$\int_{Y_1} \tau^p(y,\kappa)y^{rp}\,dy \asymp \begin{cases} \int_{y_\varepsilon}^\infty y^{-\mu pq/(q-p)}\,dy, & \text{if } \kappa \in \Xi_2, \\ \int_0^{y_\varepsilon} y^{4rp/(4-p)}\,dy, & \text{if } \kappa \in \Xi_3, \end{cases} = O(1),$$

$$\int_{Y_1} \tau^q(y,\kappa)y^{sq}dy \asymp \begin{cases} \int_{y_\varepsilon}^\infty y^{-\mu pq/(q-p)}\,dy, & \text{if } \kappa \in \Xi_2, \\ \int_0^{y_\varepsilon} y^{(4rp+\Delta)/(4-p)}\,dy, & \text{if } \kappa \in \Xi_3, \end{cases} = O(1),$$

$$\int_{Y_1} \tau^4(y,\kappa)\,dy \asymp \begin{cases} \int_{y_\varepsilon}^\infty y^{-(\mu pq+\Delta)/(q-p)}\,dy, & \text{if } \kappa \in \Xi_2, \\ \int_0^{y_\varepsilon} y^{4rp/(4-p)}\,dy, & \text{if } \kappa \in \Xi_3, \end{cases} = O(1).$$

Let us show that

$$\begin{aligned} \tilde{C}_{1,\varepsilon}(\kappa) = z_0^{-I/\lambda}C_{1,\varepsilon}(\kappa) &= \frac{|p-q|}{|pqI|}C^p(p,q)C_{p,q}^{1+pqI/\Delta} + o(1), \\ \tilde{C}_{2,\varepsilon}(\kappa) = z_0^{-I/\lambda}C_{2,\varepsilon}(\kappa) &= \frac{|p-q|}{|pqI|}C^q(p,q)C_{p,q}^{1+pqI/\Delta} + o(1), \\ \tilde{C}_{0,\varepsilon}(\kappa) = z_0^{-I/\lambda}C_{0,\varepsilon}(\kappa) &= \frac{|p-q|}{|pqI|}C^4(p,q)C_{p,q}^{2+pqI/\Delta} + o(1). \end{aligned} \tag{A.77}$$

Let $\kappa \in \Xi_2$. Then

$$I_0 = \{i \le nx_\varepsilon = my_\varepsilon\}, \quad x_\varepsilon \to \infty.$$

Using Proposition A.2 and Corollary A.1 one can check

$$\begin{aligned} C_{1,\varepsilon} &= n^{-1}\sum_{i\in I_0}(i/n)^{rp}z^p(i/n,\kappa)\delta(i/n,\kappa) \\ &= \int_{1/n}^1 x^{rp}z^p(x,\kappa)\delta(x,\kappa)\,dx + \int_1^{x_\varepsilon} x^{rp}z^p(x,\kappa)\delta(x,\kappa)\,dx + o(1). \end{aligned}$$

Observe that

$$\int_{1/n}^{1} x^{rp} z^p(x,\kappa)\delta(x,\kappa)\,dx \asymp \int_0^1 x^{2rp}\,dx = O(1);$$

$$\int_1^{x_\varepsilon} x^{rp} z^p(x,\kappa)\delta(x,\kappa)\,dx \asymp \int_1^{x_\varepsilon} x^{pqI/(p-q)-1}\,dx \asymp z_0^{I/\lambda} \to \infty.$$

Taking small enough $d > 0$, consider a family $x_{\varepsilon,1} \to \infty$, such that

$$x_{\varepsilon,1} = o(x_\varepsilon), \quad \int_1^{x_{\varepsilon,1}} x^{pqI/(p-q)-1}\,dx \asymp z_0^{I/(\lambda(1+d))}. \tag{A.78}$$

Using the asymptotics of the functions $z(x,\kappa)$, $\delta(x,\kappa)$ as $x \to \infty$ for the evaluation of the integral over the interval $[x_{\varepsilon,1}, x_\varepsilon]$, we get

$$\int_1^{x_\varepsilon} x^{rp} z^p(x,\kappa)\delta(x,\kappa)\,dx = z_0^{I/\lambda}\frac{|p-q|}{|pqI|}C^p(p,q)C_{p,q}^{1+pqI/\Delta} + O(z_0^{I/\lambda(1+d)}). \tag{A.79}$$

These imply the first relation in (A.77). The second and third relations in (A.77) are proved in a similar way.

Let $\kappa \in \Xi_3$, $\lambda > 0$. Then

$$I_0 = \{i \geq nx_\varepsilon = my_\varepsilon\}, \quad x_\varepsilon \to 0.$$

Using Proposition A.2 and Corollary A.1 analogously to the above, take a family $x_{\varepsilon,1} \to 0$ such that

$$x_{\varepsilon,1}/x_\varepsilon \to \infty, \quad \int_{x_{\varepsilon,1}}^1 x^{pqI/(p-q)-1}\,dx \asymp z_0^{I/\lambda(1+d)}. \tag{A.80}$$

Then we use the relations

$$\begin{aligned} C_{1,\varepsilon} &= n^{-1}\sum_{i\in I_0}(i/n)^{rp}z^p(i/n,\kappa)\delta(i/n,\kappa) \\ &= \int_{x_\varepsilon}^1 x^{rp}z^p(x,\kappa)\delta(x,\kappa)\,dx + \int_1^\infty x^{rp}z^p(x,\kappa)\delta(x,\kappa)\,dx + o(1), \end{aligned}$$

where

$$\int_1^\infty x^{rp}z^p(x,\kappa)\delta(x,\kappa)\,dx = O(1)$$

and

$$\begin{aligned} \int_{x_\varepsilon}^1 x^{rp}z^p(x,\kappa)\delta(x,\kappa)\,dx &\sim C^p(p,q)C_{p,q}\int_{x_\varepsilon}^{x_{\varepsilon,1}} x^{pqI/(p-q)-1}\,dx \\ &\asymp z_0^{I/\lambda} \to \infty. \end{aligned}$$

The evaluations analogous to the above imply the first relation in (A.77). The second and third relations are proved in a similar way.

Let $\kappa \in \Xi_3$, $\lambda \leq 0$. Then

$$I_0 = \{i \geq my_\varepsilon\}, \quad y_\varepsilon \sim C_{p,q}^{(p-q)/\Delta}.$$

Using Proposition A.2 (B) and Corollary A.1 we can rewrite relations (6.84), (6.90) in the form

$$\begin{aligned} \rho_\varepsilon^p &= z_0^p m^{rp+1}(\tilde{C}_{1,\varepsilon}(\kappa) + D_{1,\varepsilon}(\kappa)), \\ R_\varepsilon^q &= z_0^p m^{sq+1}(\tilde{C}_{2,\varepsilon}(\kappa) + D_{2,\varepsilon}(\kappa)), \\ u_\varepsilon^2 &= z_0^4 m(\tilde{C}_{0,\varepsilon}(\kappa) + D_{0,\varepsilon}(\kappa)), \end{aligned} \tag{A.81}$$

where

$$\begin{aligned} \tilde{C}_{1,\varepsilon}(\kappa) &= m^{-1}\sum_{i\in I_0}(i/m)^{rp}\tau_0^p(i/m,\kappa,z_0)\eta(i/m,\kappa,z_0), \\ \tilde{C}_{2,\varepsilon}(\kappa) &= m^{-1}\sum_{i\in I_0}(i/m)^{sq}\tau_0^q(i/m,\kappa,z_0)\eta(i/m,\kappa,z_0), \\ \tilde{C}_{0,\varepsilon}(\kappa) &\sim (2m)^{-1}\sum_{i\in I_0} h_0^2(i/m,\kappa,z_0)\tau_0^4(i/m,\kappa,z_0); \end{aligned}$$

(the last asymptotics correspond to $2\sinh^2(z^2/2) \sim z^4/2,\ z \to 0$).

The values $D_{l,\varepsilon}(\kappa)$ are defined in (A.74). The asymptotics of these values are given in (A.76).

The values $\tilde{C}_{l,\varepsilon}$, $l = 0, 1, 2$, satisfy (A.77). In fact,

$$\begin{aligned} \tilde{C}_{1,\varepsilon} &= m^{-1}\sum_{i\in I_0}(i/m)^{rp}\tau^p(i/m,\kappa,z_0)\eta(i/m,\kappa,z_0) \\ &\sim C^p(p,q)C_{p,q}\int_{y_\varepsilon}^\infty y^{pqI/(p-q)-1}dy = \frac{|p-q|}{|pqI|}C^p(p,q)C_{p,q}^{1+pqI/\Delta}. \end{aligned}$$

The second and third relations in (A.77) are proved in a similar way.

Relations (A.75) or (A.81), jointly with (A.76), (A.77), imply the asymptotics (6.18), (6.19) with

$$\begin{aligned} c_1^p(\kappa) &= \frac{|p-q|}{pq|I|}C^p(p,q)C_{p,q}^{1+pqI/\Delta} + \int_{Y_1}\tau^p(y,\kappa)y^{rp}\,dy, \\ c_2^q(\kappa) &= \frac{|p-q|}{pq|I|}C^q(p,q)C_{p,q}^{1+pqI/\Delta} + \int_{Y_1}\tau^q(y,\kappa)y^{sq}\,dy, \\ c_0(\kappa) &= \frac{|p-q|}{2pq|I|}C^4(p,q)C_{p,q}^{2+pqI/\Delta} + \frac{1}{2}\int_{Y_1}\tau^4(y,\kappa)\,dy, \end{aligned} \tag{A.82}$$

where Y_1 is either $(0, y_1]$ or $[y_1, \infty)$, $y_1 = C_{p,q}^{(p-q)/\Delta}$.

Let $r = r_p^*$. This means $\kappa \in \Xi_3$ (i.e., $q < p < 2$), $\kappa \in \Xi_{G_4}$, $r_p^* = 1/4 - 1/p$, $s > 1/4 - 1/q$. The quantity $c_2(\kappa)$ is defined by (A.82). However, the integrals diverge in relation (A.82) for $c_1(\kappa)$, $c_0(\kappa)$. These integrals correspond to the items $D_{1,\varepsilon}(\kappa), D_{0,\varepsilon}(\kappa)$. The relations for these quantities should be changed by the following:

$$D_{1,\varepsilon}(\kappa) \sim (p/2)^{p/(4-p)}\int_{1/m}^1 y^{-1}\,dy = (p/2)^{p/(4-p)}\log m,$$

$$2D_{0,\varepsilon}(\kappa) \sim (p/2)^{4/(4-p)} \int_{1/m}^{1} y^{-1}\, dy = (p/2)^{4/(4-p)} \log m.$$

This implies the asymptotics (6.43), (6.44) with

$$c_1(\kappa) = (p/2)^{1/(4-p)}, \quad c_0(\kappa) = (p/2)^{4/(4-p)}/2.$$

Case $I = 0$

We have: either

$$\kappa \in \Xi_{G_3}, \quad \kappa \in \Xi_2 \cup \Xi_3, \quad r > r_p^*,$$

or

$$\kappa \in \Xi_{G_5}, \quad \kappa \in \Xi_4, \quad r = r_p^* = -1/4.$$

Note that $\lambda > 0$ (see the Remark on page 366) and $m \to \infty$, $n \to \infty$ in these cases.

Let $r > r_p^*$. Then relations (A.72)–(A.74) hold with $D_{\varepsilon,l} = O(1)$ and

$$\begin{aligned} C_{1,\varepsilon}(\kappa) &\sim C_{p,q} C^p(p,q) \int_{X_\varepsilon^*} x^{-1}\, dx \sim \\ &\sim C_{p,q} C^p(p,q) \frac{|p-q|}{\Delta} \log h_0^{-1} = c_1^p(\kappa) \log h_0^{-1}, \quad \text{(A.83)} \\ C_{2,\varepsilon}(\kappa) &\sim C_{p,q} C^q(p,q) \int_{X_\varepsilon^*} x^{-1}\, dx \\ &\sim C_{p,q} C^q(p,q) \frac{|p-q|}{\Delta} \log h_0^{-1} = c_2^q(\kappa) \log h_0^{-1}, \quad \text{(A.84)} \\ C_{0,\varepsilon}(\kappa) &\sim \frac{1}{2} C_{p,q}^2 C^4(p,q) \int_{X_\varepsilon^*} x^{-1}\, dx \\ &\sim \frac{1}{2} C_{p,q}^2 C^4(p,q) \frac{|p-q|}{\Delta} \log h_0^{-1} = c_0(\kappa) \log h_0^{-1}, \quad \text{(A.85)} \end{aligned}$$

where

$$X_\varepsilon^* = \begin{cases} [x_\varepsilon(\kappa), 1], & \text{if } \kappa \in \Xi_3, \\ [1, x_\varepsilon(\kappa)), & \text{if } \kappa \in \Xi_2. \end{cases}$$

Let $r = r_p^* = -1/4$. Then $x_\varepsilon = m y_\varepsilon / n \to 0$, $y_\varepsilon \to 0$. We have

$$\rho_\varepsilon^2 \sim h_0 n^{1/2} \left(\int_{x_\varepsilon}^{1} x^{-1}\, dx + \int_{1/m}^{y_\varepsilon} y^{-1}\, dy + O(1) \right) \sim h_0 n^{1/2} \log n,$$

$$u_\varepsilon^2 \sim \frac{1}{2} h_0^2 n \left(\int_{x_\varepsilon}^{1} x^{-1}\, dx + \int_{1/m}^{y_\varepsilon} y^{-1}\, dy + O(1) \right) \sim \frac{1}{2} h_0^2 n \log n.$$

By $h_0 n^{sq+1} = z_0^q m^{sq+1}$, we get

$$z_0^q m^{sq+1} D_{2,\varepsilon}(\kappa) \asymp z_0^q m^{sq+1} \int_{1/m}^{y_\varepsilon} y^{-1+\Delta/2}\, dy = o(h_0 n^{sq+1})$$

and

$$R_\varepsilon^q \sim h_0 n^{sq+1} c_2^q(\kappa),$$

where $c_2(\kappa)$ is defined by (A.52).

These relations imply the asymptotics (6.45), (6.46).

A.3.6 Proofs of Theorems 6.1, 6.3 for $\kappa \in \Xi_\Delta^+$

To prove Theorems 6.1, 6.3 it suffices to verify the assumptions of Proposition 5.5, assumptions (5.11), (5.29) for the thresholding (6.22), assuming $u_\varepsilon \asymp 1$.

Using Propositions A.2, A.3 and Corollaries A.1, A.2 one can easily check that assumptions A.1, A.4 for $p < q$ and A.3 with any $\delta \in (0,1)$ for $p > q$ are fulfilled. These considerations are analogous to those that were given in Section A.3.3.

To verify (5.11) and (5.29) with $\zeta_\varepsilon \to 0$ such that $\zeta_\varepsilon \gg \log\log R_\varepsilon/\log R_\varepsilon$ for the thresholding (6.22) we use relations (A.61), (A.62) for $i \in I_0$. If $i \in I_1$, then $z_{\varepsilon,i} = z_0\tau(y_i,\kappa) = o(1)$ and, using (5.7), (5.8), we have

$$T_{\varepsilon,i} \asymp z_{\varepsilon,i}{}^{-1}. \tag{A.86}$$

We omit the evaluations which are analogous to Section A.3.3. Note only the case $\kappa \in \Xi_3^+$ and $\lambda > 0$ where $z_{\varepsilon,i} \to \infty$, as $x = i/n \to \infty$. If $I \geq 0$ in this case, the assumption $u_\varepsilon \asymp 1$ leads to $2\log h_0^{-1} \sim \log n$ and we repeat the evaluations from Section A.3.3 which were based on the relation $2\log h_0^{-1} \geq \log n + O(1))$ for $r > r_p^*$ mainly. If $I < 0$, then the assumption $u_\varepsilon \asymp 1$ leads to $4\log z_0^{-1} \sim \log m$. Using relations (A.12) we can write this in the form

$$\log h_0^{-1} \sim C\log n, \quad C = \frac{\Delta}{4pq\lambda + p - q},$$

and the inequality $I < 0$ is equivalent to $C > 1/2$. Therefore we can repeat the evaluations from Section A.3.3 as well.

Let us construct the families $\bar\pi_{\varepsilon,1} = \{\pi_{\varepsilon,i,1}\}$ such that

$$\|\bar\pi_{\varepsilon,1}\| = u_\varepsilon + o(1), \quad \pi_1^\varepsilon(V_\varepsilon) \to 1.$$

It suffices to assume that $u_\varepsilon \asymp 1$. Analogously to Sections 6.6.2, A.3.3, introduce the quantities $\delta_\varepsilon = (\log\log R_\varepsilon)^{-1}$ and put

$$\tilde\pi_\varepsilon = \bar\pi_\varepsilon(\kappa, (1+\delta_\varepsilon)\rho_\varepsilon, (1-\delta_\varepsilon)R_\varepsilon).$$

If $\lambda \leq 0$, $\kappa \in \Xi_3$ (recall that $I < 0$ in this case) or $I = 0$, $\kappa \in \Xi_4$, $r = r_p^* = -1/4$, then we set $\bar\pi_{\varepsilon,1} = \tilde\pi_\varepsilon$.

For $I > 0$ we take a family $T_\varepsilon = R_\varepsilon^a$ with $a = a(\kappa) > 0$ small enough. For $I \leq 0$, $\lambda > 0$, $\kappa \notin \Xi_4$, we take the families T_ε and T_ε^{-1} in the following way:

(1) Let $\kappa \in \Xi_2$. Then

$$T_\varepsilon^{-1} = y_\varepsilon x_{\varepsilon,1}/x_\varepsilon; \quad T_\varepsilon^{-pq(I+\mu)/(p-q)} = o(m\delta_\varepsilon^2), \quad \text{if } I+\mu \neq 0;$$
$$\log(T_\varepsilon) = o(m\delta_\varepsilon^2), \quad \text{if } I+\mu = 0.$$

(2) Let $\kappa \in \Xi_3$. Then

$$T_\varepsilon = y_\varepsilon x_{\varepsilon,1}/x_\varepsilon; \quad T_\varepsilon^{pq(I+\mu)/(p-q)} = o(m\delta_\varepsilon^2), \quad \text{if } I+\mu \neq 0;$$
$$\log(T_\varepsilon) = o(m\delta_\varepsilon^2), \quad \text{if } I+\mu = 0;$$

the family $x_{\varepsilon,1}$ satisfies (A.78), (A.80) for $I < 0$ and $x_{\varepsilon,1} = 1$ for $I = 0$.

Let $r > r_p^*$, $I > 0$. Then we consider the "two-sided T_ε-truncated" sequences $\bar{\pi}_{\varepsilon,1}$:

$$\pi_{\varepsilon,i,1} = \begin{cases} \tilde{\pi}_{\varepsilon,i}, & \text{if } T_\varepsilon^{-1} \le i/n \le T_\varepsilon, \\ \delta_0, & \text{otherwise.} \end{cases}$$

Let either $r = r_p^*$ or $I \le 0$ ($\kappa \notin \Xi_4$). Then we consider a "one-sided T_ε-truncated" family of the following type. If $I > 0$, $r = r_p^*$, then

$$\pi_{\varepsilon,i,1} = \begin{cases} \tilde{\pi}_{\varepsilon,i}, & \text{if } 1/n \le i/n \le T_\varepsilon, \\ \delta_0, & \text{otherwise.} \end{cases}$$

If $I \le 0$, $\lambda > 0$, $\kappa \in \Xi_3$, then

$$\pi_{\varepsilon,i,1} = \begin{cases} \tilde{\pi}_{\varepsilon,i}, & \text{if } 1/m \le i/m \le T_\varepsilon, \\ \delta_0, & \text{otherwise.} \end{cases}$$

If $I \le 0$, $\kappa \in \Xi_2$, then

$$\pi_{\varepsilon,i,1} = \begin{cases} \tilde{\pi}_{\varepsilon,i}, & \text{if } i/m \ge T_\varepsilon^{-1}, \\ \delta_0, & \text{otherwise.} \end{cases}$$

The relation $\|\bar{\pi}_{\varepsilon,1}\| = u_\varepsilon + o(1)$ follows from the evaluations which were given in Section A.3.5. Analogously to Sections 6.6.2, A.3.3, the relation $\pi_1^\varepsilon(V_\varepsilon) \to 1$ follows from the Chebyshev inequality and from the relations

$$E_{\pi_1^\varepsilon} G_1(v) = (1+\delta_\varepsilon - o(\delta_\varepsilon))^p H_{\varepsilon,1}, \quad E_{\pi_1^\varepsilon} G_2(v) \le (1-\delta_\varepsilon)^q H_{\varepsilon,2},$$

where $H_{\varepsilon,1} = \rho_\varepsilon^p$, $H_{\varepsilon,2} = R_\varepsilon^q$. Omitting the tilde, we use the following evaluations of variances (the last equalities in the following we will determine below):

$$\mathrm{Var}_{\pi_1^\varepsilon}(G_1(v)) = \sum_{i \in I_0} z_i^{2p} i^{2rp} h_i(1-h_i) \asymp \begin{cases} n^{1+2rp} h_0 I_1, & \text{if } \lambda > 0, \\ m^{1+2rp} z_0^{2p} I_2, & \text{if } \lambda \le 0, \end{cases}$$
$$= o(H_{\varepsilon,1}^2 \delta_\varepsilon^2), \tag{A.87}$$

and

$$\mathrm{Var}_{\pi_1^\varepsilon}(G_2(v)) = \sum_{i \in I_0} z_i^{2q} i^{2sq} h_i(1-h_i) \asymp \begin{cases} n^{1+2sq} h_0 I_3, & \text{if } \lambda > 0, \\ m^{1+2sq} z_0^{2q} I_4, & \text{if } \lambda \le 0, \end{cases}$$
$$= o(H_{\varepsilon,2}^2 \delta_\varepsilon^2). \tag{A.88}$$

We denote in (A.87), (A.88)

$$I_1 = \int_{x\in X_0^*} \delta(x,\kappa) z^{2p}(x,\kappa) x^{2rp}\,dx, \tag{A.89}$$

$$I_2 = \int_{y\in Y_0^*} \eta(y,\kappa,z_0) \tau_0^{2p}(y,\kappa) y^{2rp}\,dy, \tag{A.90}$$

$$I_3 = \int_{x\in X_0^*} \delta(x,\kappa) z^{2q}(x,\kappa) x^{2sq}\,dx, \tag{A.91}$$

$$I_4 = \int_{y\in Y_0^*} \eta(y,\kappa,z_0) \tau_0^{2q}(y,\kappa) y^{2sq}\,dy. \tag{A.92}$$

Here

$$X_0^* = \begin{cases} X_\varepsilon \cap [T_\varepsilon^{-1}, T_\varepsilon], & \text{if } I > 0,\ r > r_p^*, \\ X_\varepsilon \cap [n^{-1}, T_\varepsilon], & \text{if } I > 0,\ r = r_p^*, \\ X_\varepsilon, & \text{if } I = 0,\ \kappa \in \Xi_4, \\ X_\varepsilon \cap [x_{\varepsilon,1}, \infty), & \text{if } I \le 0,\ \kappa \in \Xi_2, \\ X_\varepsilon \cap [n^{-1}, x_{\varepsilon,1}], & \text{if } I \le 0,\ \kappa \in \Xi_3,\ \lambda > 0; \end{cases}$$

$$Y_0^* = Y_0 \cap [m^{-1}, T_\varepsilon], \quad \text{if } \kappa \in \Xi_3, \quad \lambda \le 0.$$

To check the last equalities in (A.87), (A.88), first assume $I \ge 0$. Then $\log h_0^{-1} \asymp \log n$, $\lambda > 0$ (see the Remark on page 366) and $\log h_0^{-1} \asymp \log n$ and

$$H_{\varepsilon,1} \asymp \begin{cases} n^{1+rp} h_0, & \text{if } r > r_p^* \text{ and } I > 0, \\ n^{1+rp} h_0 \log h_0^{-1}, & \text{if } r = r_p^* \text{ or } I = 0, \end{cases}$$

$$H_{\varepsilon,2} \asymp \begin{cases} n^{1+sq} h_0, & \text{if } I > 0 \text{ or } I = 0,\ r = r_p^*,\ \kappa \in \Xi_4, \\ n^{1+sq} h_0 \log h_0^{-1}, & \text{if } I = 0,\ \kappa \notin \Xi_4. \end{cases}$$

It suffices to obtain the following relations for the integrals $I_{1,3}$ in (A.89), (A.91): if $r > r_p^*$, $I > 0$, then $I_{1,3} = o(nh_0\delta_\varepsilon^2)$; if $r = r_p^*$ or $I = 0$, then $I_1 = o(nh_0 \log^2 h_0^{-1} \delta_\varepsilon^2)$; if $I = 0$, $\kappa \notin \Xi_4$, then $I_3 = o(nh_0 \log^2 h_0^{-1} \delta_\varepsilon^2)$; if $r = r_p^*$, then $I_3 = o(nh_0\delta_\varepsilon^2)$.

Let $r > r_p^*$, $I > 0$. Then $nh_0 \asymp n^{1/2} u_\varepsilon \asymp R_\varepsilon^{\delta_0}$ for some $\delta_0 > 0$ and the needed relations follow from the evaluations

$$\int_{X_\varepsilon \cap [T_\varepsilon^{-1}, T_\varepsilon]} \{\dots\}\,dx = O(T_\varepsilon^{A_2}) = o(R_\varepsilon^{\delta});$$

here $A_2 = A_2(\kappa) > 0$ and one can make $\delta > 0$ arbitrarily small by choosing $a > 0$ small enough.

Let $r = r_p^*$, $I > 0$. Then $p > 2$, $r_p^* = -1/2p$, $\kappa \in \Xi_2$ and, for some $\delta_0 > 0$,

$$nh_0 \log^2 h_0^{-1} \asymp (n \log^3 h_0^{-1})^{1/2} u_\varepsilon \asymp R_\varepsilon^{\delta_0} (\log R_\varepsilon)^{3/2},$$
$$nh_0 \asymp (n/\log h_0^{-1})^{1/2} u_\varepsilon \asymp R_\varepsilon^{\delta_0} (\log R_\varepsilon)^{-1/2}.$$

We have

$$\int_{X_\varepsilon\cap[n^{-1},T_\varepsilon]}\{\dots\}\,dx \le O(T_\varepsilon^{A_2}) + \int_{n^{-1}}^{1}\{\dots\}\,dx$$

and, analogously to (A.67)–(A.68),

$$\int_{1/n}^{1}\delta(x,\kappa)z^{2p}(x,\kappa)x^{2rp}\,dx \asymp \int_{1/n}^{1} x^{3rp}\,dx \asymp n^{1/2} = o(nh_0\log^2 h_0^{-1}\delta_\varepsilon^2), \tag{A.93}$$

$$\int_{1/n}^{1}\delta(x,\kappa)z^{2q}(x,\kappa)x^{2sq}\,dx \asymp \int_{1/n}^{1} x^{3rp+2pq\lambda}\,dx = o(n^{1/2}\log^{-1/2} h_0^{-1}\delta_\varepsilon^2). \tag{A.94}$$

Let $I=0,\ r>r_p^*$ (this means that $\kappa\in\Xi_2\cup\Xi_3,\ \lambda>0$). If $\kappa\in\Xi_2$, then $X_0^*=[1,x_\varepsilon],\ \mu>0$ and we calculate the variances directly:

$$\begin{aligned}
\mathrm{Var}_{\pi_1^\varepsilon}(G_1(v)) &\asymp h_0 n^{2rp+1}\int_{X_0^*} x^{2rp}z^{2p}(x,\kappa)\delta(x,\kappa)\,dx\\
&\asymp h_0 n^{2rp+1}\int_1^{x_\varepsilon} x^{pq\mu/(p-q)-1}dx \asymp h_0 n^{2rp+1}O(1)\\
&= o(n^{2rp+1}\log h_0^{-1}\delta_\varepsilon^2); \qquad (A.95)\\
\mathrm{Var}_{\pi_1^\varepsilon}(G_2(v)) &\asymp h_0 n^{2sq+1}\int_{X_0^*} x^{2sq}z^{2q}(x,\kappa)\delta(x,\kappa)dx\\
&\asymp h_0 n^{2sq+1}\int_1^{x_\varepsilon} x^{pq\mu/(p-q)-1}\,dx \asymp h_0 n^{2sq+1}O(1)\\
&= o(n^{2sq+1}\log h_0^{-1}\delta_\varepsilon^2). \qquad (A.96)
\end{aligned}$$

If $\kappa\in\Xi_3$, then $X_0^*=[x_\varepsilon,1]$ and it is possible that $\mu\le 0$. By repeating the evaluations (A.95), (A.96) we obtain the same results for $\mu>0$; if $\mu=0$, then we have a small difference at the point $s=r=-1/4$ which leads to an inessential additional log-factor. If $\mu<0$, then, analogously, by

$$h_0\int_{x_\varepsilon}^{1} x^{\mu/(p-q)-1}\,dx \asymp h_0^{4pq\lambda/\Delta},\quad 4pq\lambda/\Delta>0,$$

we get relations (A.87), (A.88). Recall that in these cases one has

$$H_{\varepsilon,1}^2 \asymp n^{2rp+1}\log h_0^{-1},\quad H_{\varepsilon,2}^2\asymp n^{2sq+1}\log h_0^{-1}.$$

Let $I=0$, $r=r_p^*=-1/4$ (this means that $\kappa\in\Xi_4$, $p=2$). Then we get

$$\begin{aligned}
\mathrm{Var}_{\pi_1^\varepsilon}(G_1(v)) &= \sum_{i\in I_0} z_i^4 i^{4r}h_i(1-h_i) \asymp h_0\int_{x_\varepsilon}^{1} x^{(8q\lambda-\tilde\Delta)/(6-q)-1}\,dx\\
&\asymp \begin{cases} h_0, & \text{if } 8q\lambda-\tilde\Delta>0,\\ h_0\log h_0^{-1}, & \text{if } 8q\lambda-\tilde\Delta=0,\\ h_0^{8q\lambda/\tilde\Delta}, & \text{if } 8q\lambda-\tilde\Delta<0,\end{cases} = o(\log n\,\delta_\varepsilon^2),
\end{aligned}$$

$$\begin{aligned}
\mathrm{Var}_{\pi_1^\varepsilon}(G_2(v)) &= \sum_{i\in I_0} z_i^{2q} i^{2sq} h_i(1-h_i) \asymp h_0 n^{2sq+1} \int_{x_\varepsilon}^1 x^{(16q\lambda+\tilde\Delta)/(6-q)-2}\,dx \\
&\asymp \begin{cases} h_0 n^{2sq+1}, & \text{if } \tilde\Delta + 16q\lambda - 6 + q > 0, \\ h_0 n^{2sq+1}\log h_0^{-1}, & \text{if } \tilde\Delta + 16q\lambda - 6 + q = 0, \\ h_0^{24q\lambda/\tilde\Delta} n^{2sq+1}, & \text{if } \tilde\Delta + 16q\lambda - 6 + q < 0, \end{cases} \\
&= o(n^{2sq+1}\delta_\varepsilon^2/\log n),
\end{aligned}$$

(recall that $H_{\varepsilon,1}^2 \asymp \log n$, $H_{\varepsilon,2}^2 \asymp n^{2sq+1}/\log n$ in this case).

Let $I < 0$ (recall that $\kappa \in \Xi_2 \cup \Xi_3$ in this case). Then we get

$$\begin{aligned}
H_{\varepsilon,1} &\asymp \begin{cases} m^{1+rp} z_0^p, & \text{if } r > r_p^*, \\ m^{1+rp} z_0^p \log m, & \text{if } r = r_p^*, \end{cases} \\
H_{\varepsilon,2} &\asymp m^{1+sq} z_0^q,
\end{aligned} \tag{A.97}$$

and relations (A.88), (A.87) follow from evaluations: if $\lambda > 0$, then

$$\begin{aligned}
\mathrm{Var}_{\pi_1^\varepsilon}(G_1(v)) &\asymp h_0 n^{2rp+1} \int_{X_0^*} x^{2rp} z^{2p}(x,\kappa)\delta(x,\kappa)\,dx \\
&\asymp m^{2rp+1} z_0^{2p} \int_{Y_0^*} y^{(pq(I+\mu)/(p-q))-1}\,dy \\
&= o(m^{2rp+2} z_0^{2p}\delta_\varepsilon^2),
\end{aligned} \tag{A.98}$$

$$\begin{aligned}
\mathrm{Var}_{\pi_1^\varepsilon}(G_2(v)) &\asymp h_0 n^{2sq+1} \int_{X_0^*} x^{2sq} z^{2q}(x,\kappa)\delta(x,\kappa)\,dx \\
&\asymp m^{2sq+1} z_0^{2q} \int_{Y_0^*} y^{(pq(I+\mu)/(p-q))-1}\,dy \\
&= o(m^{2sq+2} z_0^{2q}\delta_\varepsilon^2),
\end{aligned} \tag{A.99}$$

where $Y_0^* = (T_\varepsilon^{-1}, y_\varepsilon)$ for $\kappa \in \Xi_2$ and $Y_0^* = (y_\varepsilon, T_\varepsilon)$ for $\kappa \in \Xi_3$.

If $\lambda \le 0$, then $\mu < 0$ and using Proposition A.2 we obtain the relations analogous to (A.98)–(A.99) directly without any truncations.

These evaluations hold for $I < 0$, $r = r_p^*$ as well (recall that $\kappa \in \Xi_3$ in this case). These yield relations (A.87), (A.88).

Thus Theorems 6.1, 6.3 are proved for the case $\kappa \in \Xi_\Delta^+$. □

A.3.7 Extreme Problem for $q = \infty$

The outline of the study given Section 6.7.2 leads to the following results.

Proof of Theorems 6.1, 6.3 for $p \le 2$

In this case we have $r \ge 1/4 - 1/p, s > r + 1/p \ge 1/4$. Introduce the variables $m = m_\varepsilon(\kappa)$, $z_0 = z_{0,\varepsilon}(\kappa)$ from the relations

$$z_0 m^s = R_\varepsilon, \quad z_0 m^{-rp/(4-p)} = (Ap/2)^{1/(4-p)}, \tag{A.100}$$

and assume $z_0 \to 0$, $m \to \infty$, and $z_0 m^{-rp/(4-p)} \to 0$. Then we have $h_i = 1$,

$$z_i \sim \begin{cases} z_0(i/m)^{rp/(4-p)}, & \text{if } i \le m, \\ z_0(i/m)^{-s}, & \text{if } i \ge m. \end{cases}$$

The constraints (6.93) imply the relations

$$\rho_\varepsilon^p = \sum_i i^{rp} h_i z_i^p \sim z_0^p m^{rp+1} \begin{cases} c_1^p(\kappa), & \text{if } r > r_p^* = 1/4 - 1/p, \\ \log m, & \text{if } r = r_p^* = 1/4 - 1/p, \end{cases}$$

and we have

$$u_\varepsilon^2 = 2 \sum_i h_i^2 \sinh(z_i^2/2) \sim m z_0^4 \begin{cases} c_0(\kappa), & \text{if } r > r_p^* = 1/4 - 1/p, \\ (\log m)/2, & \text{if } r = r_p^* = 1/4 - 1/p, \end{cases}$$

where

$$\begin{aligned} c_1^p(\kappa) &= \int_0^1 y^{4rp/(4-p)}\, dy + \int_1^\infty y^{-p(s-r)}\, dy \\ &= \frac{4-p}{4rp+4-p} + \frac{1}{p(s-r)-1}, \\ 2c_0(\kappa) &= \int_0^1 y^{4rp/(4-p)}\, dy + \int_1^\infty y^{-4s}\, dy = \frac{4-p}{4rp+4-p} + \frac{1}{4s-1}, \end{aligned}$$

which yields the asymptotics (6.18), (6.19) for $r > r_p^*$ and (6.43), (6.44) for $r = r_p^*$. Recall that the relations $z_0 \to 0$, $m \to \infty$ follow from the assumptions $u_\varepsilon \asymp 1$.

It is not difficult to check the assumptions of Proposition 5.5 (note that $\pi^\varepsilon(V_\varepsilon) = 1$ by $h_i \equiv 1$ for all i).

Theorems 6.1, 6.3 are proved for $q = \infty$, $p \le 2$. □

Proof of Theorems 6.1, 6.3 for $p > 2$

Note that $r \ge -1/2p$, $s > 0$ in this case. Put

$$\Delta = \Delta(\kappa) = s(4-p) + rp, \quad n = (R_\varepsilon/z_p)^{1/s}, \quad h_0 = An^{rp},$$

and assume $h_0 \to 0$, $n \to \infty$. Recall that $I = 2s(p-2) - 2rp - 1$ for $q = \infty$.

First, consider the case $\Delta \le 0$ and note that $I > 0$ in this case. For $x = i/n$ we have

$$z_i = \begin{cases} z_p, & \text{if } x \le 1, \\ z_p x^{-s}, & \text{if } x \ge 1, \end{cases} \tag{A.101}$$

$$h_i = h_0 x^{rp} \psi_p(z_i) = h_0 \delta(x, \kappa), \tag{A.102}$$

where $\delta(x, \kappa) = x^{rp}\psi_p(z_i)$, the function $\psi_p(z)$ is defined by (6.95),

$$\delta(x, \kappa) \sim z_p^{p-4} x^\Delta, \quad \text{as } x \to \infty.$$

The constraints (6.93) imply the relations

$$\rho_\varepsilon^p \sim h_0 n^{rp+1} \begin{cases} c_1^p(\kappa), & \text{if } r > r_p^* = -1/2p, \\ z_p^p \psi_p(z_p) \log n, & \text{if } r = r_p^* = -1/2p, \end{cases}$$

$$u_\varepsilon^2 \sim nh_0^2 \begin{cases} c_0(\kappa), & \text{if } r > r_p^* = -1/2p, \\ z_p^p \psi_p(z_p) \log n/2, & \text{if } r = r_p^* = -1/2p, \end{cases}$$

where

$$\begin{aligned} c_1^p(\kappa) &= z_p^p \left(\psi_p(z_p) \int_0^1 x^{2rp}\,dx + \int_1^\infty x^{(r-s)p}\delta(x,\kappa)\,dx \right) \\ &= z_p^p(\psi_p(z_p)/(2rp+1) + I_1), \\ c_0(\kappa) &= \psi_p(z_p)\frac{z_p^p}{2}\int_0^1 x^{2rp}\,dx + 2\int_1^\infty \delta^2(x,\kappa)\sinh^2\left(\frac{z_p^2}{2x^{2s}}\right)dx \\ &= \psi_p(z_p) z_p^p/2(2rp+1) + I_0 \end{aligned}$$

and the integrals are finite

$$I_l \le B\int_1^\infty x^{-pI-1}\,dx = B/pI, \quad l = 0, 1.$$

These imply the asymptotics (6.20), (6.21) for $r > r_p^*$ and (6.45), (6.46) for $r = r_p^*$; by Remark 3 from Section 6.5, the relations $h_i \le 1,\ h_0 \to 0,\ n \to \infty$ follow from the assumptions (6.58).

Let $\Delta > 0$. Then we put $m = nh_0^{-1/\Delta}$, $z_0 = z_p h_0^{s/\Delta}$. We have, for $x = i/n,\ y = i/m$,

$$z_i = \begin{cases} z_p, & \text{if } x \le 1, \\ z_p x^{-s} = z_0 y^{-s}, & \text{if } x \ge 1, \end{cases} \tag{A.103}$$

$$h_i = \begin{cases} h_0 C(p) x^{rp}, & \text{if } x \le 1, \\ h_0\delta(x,\kappa), & \text{if } 1 \le x \le x_\varepsilon, \\ 1, & \text{if } x \ge x_\varepsilon. \end{cases} \tag{A.104}$$

where x_ε is defined by the relation $h_0\delta(x_\varepsilon,\kappa) = 1$. Using the asymptotics of $\delta(x,\kappa)$ as $x \to \infty$ we get

$$x_\varepsilon = (m/n)y_\varepsilon \sim h_0^{-1/\Delta} z_p^{(4-p)/\Delta} \to \infty, \quad y_\varepsilon \sim z_p^{(4-p)/\Delta}.$$

As in Section A.3.5, we study the cases $I > 0,\ I = 0,\ I < 0$. Using the considerations analogous to those that were used in Section A.3.5 for $\kappa \in \Xi_2$ and relations (A.103) , (A.104), we obtain the required asymptotics (6.18), (6.19) and (6.45), (6.46). Verification of the assumptions of Proposition 5.5 is carried out analogously to Section A.3.6 as well (note that we need to evaluate the π_1^ε-variation of the functional $G_1(v)$ only).

Theorems 6.1 6.3 are proved for $q = \infty$, $p > 2$ as well. □

A.3.8 Proofs of Propositions 6.1, 6.2

Proof of Proposition 6.1. One can easily check that the inequality $z_i(\kappa) \le BQ_{\varepsilon,i}$ is equivalent to $z_i(\kappa) \le B(1+o(1))\sqrt{\log \tilde{n}}$ (remember that $\log n \asymp \log R_\varepsilon$). Therefore we get for ε small enough that if

$z_i^2(\kappa_0) < B_1 \log n$, $B_1 < B$, then $z_i^* \le B\sqrt{\log n}$, and if $z_i^2(\kappa_0) > B_2 \log n$, $B_2 > B$, then $z_i^* = 0$.

Recall that the rate properties of these sequences are defined by Propositions A.2, A.3 and by Corollaries A.1, A.2. Using Propositions A.2, A.3 and Corollaries A.1, A.2 we get that if the $z_i(\kappa_0)$ are bounded over i (this holds for $p < q$), then $z_i^* = \hat{z}_i$ and these values are bounded over i. If $z_i(\kappa_0) \to \infty$ as $i \to \infty$ (this is possible for $p > q$, $\lambda = s/p - r/q > 0$), then $z_i(\kappa) \asymp (i/\tilde{n})^a$, where $a = a(\kappa) = (sq - rp)/(p - q)$ is the Lipschitz function in $\kappa \in K$ and if $a(\kappa_0) > 0$, then $a(\kappa)$ is positive and bounded away from 0 under constraints (6.99), (6.100). These yield the relations for $i \in I_\varepsilon^*$:

$$\begin{aligned}\log h_i^* &= \log h_i(\kappa_0)(1 + O(L_2 + L_1 \log\log n)) = \log h_i(\kappa_0)(1 + o(1)),\\ z_i(\kappa_0) &\le z_i^* \le z_i(\kappa_0)(1 + O(L_2 + L_1 \log\log n)) = z_i(\kappa_0)(1 + o(1)).\end{aligned}$$

Using the relations above analogously to the evaluations in Sections A.3.3, A.3.6 we easily determine assumptions A.1, A.4 for $p < q$ or $2 > p > q$, $\lambda < 0$, and if $p > q$, $sq > rp$, then we get A.3 and relations (5.11). To verify (5.29) we need to consider an i such that $z_{i,0} \asymp \sqrt{\log n} \asymp \sqrt{\log R_e}$ only. Using evaluations analogous to (A.62) we can check that

$$T_i^* = T_{\varepsilon,i}(z_i^*, h_i^*) = T_{\varepsilon,i}(z_{i,0}, h_{i,0}) + z_{i,0} O(L_1 \log\log n).$$

Note that, using (5.10), we have $T_\varepsilon \ge \sqrt{\log R_e}$. These yield (5.29) with

$$\zeta_\varepsilon^* = \zeta_\varepsilon(k_0) + BL_1 \log\log R_\varepsilon, \quad B > 0,$$

satisfying the assumptions of Proposition 5.4. Statement (1) is proved.

The scheme of the proof of statement (2) is the following. Let $K \subset \Xi_{G_{02}}$ (the considerations are analogous for $K \subset \Xi_{G_{01}}$). It is clear that

$$u_L \ge \hat{u}_0, \text{ where } \hat{u}_0^2 = \sum_{i \in I_\varepsilon^*} u_{0,i}^2, \quad u_{0,i}^2 = 2h_{i,0}^2 \sinh^2(z_{i,0}^2/2).$$

Moreover, for any $B_1 \in (0, B)$, $b_1 \in (0, B_1)$ and some $c > 0, b_{1,2} > 0$, one can easily see that

$$0 \le u_0^2 - \hat{u}_0^2 \le Bh_0^2 \sum_{i: z_{i,0} > B_1\sqrt{\log n}} z_{i,0}^c \exp(-z_{i,0}^2) = o(n^{-b_1}) = o(R_\varepsilon^{-b_2}).$$

To compare u_L and $\hat{u}_0$, let us divide the index set I_ε^* into three subsets:

$$I_{\varepsilon,1}^* = \{i \in I_\varepsilon^* : z_i \ge b\sqrt{\log\log n}\}, \quad I_{\varepsilon,2}^* = \{i \in I_\varepsilon^* : i \le n/\log n\},$$

(i. e., if $i \in I_{\varepsilon,2}^*$, then $x_i(\kappa_0) = i/n \le 1/\log n$, $n = n(\kappa_0)$), and

$$I_{\varepsilon,3}^* = I_\varepsilon^* \backslash (I_{\varepsilon,1}^* \cup I_{\varepsilon,2}^*.$$

This division corresponds to the case of either $\kappa \in \Xi_1$ or $\kappa \in \Xi_3$, $\lambda > 0$ when $z \to \infty$ as $x \to \infty$. If $\kappa \in \Xi_2$, then $z \to 0$ as $x \to \infty$ and we set

$$I_{\varepsilon,1}^* = \{i \in I_\varepsilon^* : z_i(\kappa_0) \le 1/\log n\}.$$

We give consideration to the first case; for the second case the consideration is analogous (a small difference arises for $\kappa \in \Xi_2$ when $h_i = 1$ for large enough i).

Divide the sum for $\hat{u}_0^2$ into three parts $\hat{u}_0^2 = \hat{u}_{0,1}^2 + \hat{u}_{0,2}^2 + \hat{u}_{0,3}^2$ where

$$\hat{u}_{0,1}^2 = \sum_{i \in I_{\varepsilon,1}^*} u_{0,i}^2, \quad \hat{u}_{0,2}^2 = \sum_{i \in I_{\varepsilon,2}^*} u_{0,i}^2, \quad \hat{u}_{0,3}^2 = \sum_{i \in I_{\varepsilon,3}^*} u_{0,i}^2.$$

Analogously, we set $u_L^2 = u_{L,1}^2 + u_{L,2}^2 + u_{L,3}^2$.

Let us show that the quantities $u_{0,1}^2$, $u_{0,2}^2$ and $u_{L,1}^2$, $u_{L,2}^2$ are small, $u_{0,3}^2$ is close to $u_{L,3}^2$. We have

$$\hat{u}_{0,1}^2 = B_1 h_0^2 \sum_{i \in I_{\varepsilon,1}^*} z_{i,0}^c \exp(-z_{i,0}^2) = u_0^2 o(\log^{-b_3} n) = u_0^2 o(\log^{-b_3} R_\varepsilon),$$

and

$$\hat{u}_{0,2}^2 = \sum_{i \in I_{\varepsilon,2}^*} u_{0,i}^2 = B_2 h_0^2 n \int_0^{1/\log n} x^{2rp}\, dx = o\left(\frac{u_0^2}{\log^{b_4} n}\right) = o\left(\frac{u_0^2}{\log^{b_4} R_\varepsilon}\right).$$

To evaluate $u_{L,1}^2$, $u_{L,2}^2$, let us consider the corresponding sums for any $\tilde{\kappa} \in K$. Note that

$$u_L^2 \le 2h_0^{*2}\delta^{*2}(x,\kappa)\sinh^2(z_i^{*2}/2),$$

where the asterisk corresponds to the maximum over the set $\Delta_\varepsilon = \Delta_\varepsilon(\kappa_0, L)$. Recall that if $(\kappa, \tilde{\rho}_\varepsilon) \in \Delta_\varepsilon$, then $\tilde{h}_0 = h_0(1 + O(L_2))$. If $i \in I_{\varepsilon,1}^*$ or $i \in I_{\varepsilon,2}^*$, then for $x^a = (i/n)^a$ one has

$$x^{a - B_1 L_1}(1 + O(L_2 n^{-1})) < \tilde{x}^{\tilde{a}} < x^{a + B_1 L_1}(1 + O(L_2 n^{-1})). \qquad \text{(A.105)}$$

Here $B_1 > 0$ for $a > 0$ or $B_1 < 0$ for $a < 0$. Therefore, if $\tilde{x} \to 0$, $\tilde{h}_i < 1$, then for $\delta^2(\tilde{x}, \tilde{\kappa}) \sim C\tilde{x}^{\tilde{a}_1}$ (A.105) holds, and if $z \to 0$ for $x \to 0$, then $\sinh^2(\tilde{z}_i^2/2) \sim \tilde{z}_i^4/4$, and if $z \to z_p$ for $x \to 0$, then $\sinh^2(\tilde{z}_i^2/2) = \sinh^2(z_{\tilde{p}}^2/2) + o(1)$. Using these relations and evaluations for the calculation of $u_\varepsilon^2(\kappa)$ for $i \in I_{\varepsilon,2}^*$ we have

$$u_{L,2}^2 = u_0^2\, o(\log^{-b_5} R_\varepsilon). \qquad \text{(A.106)}$$

If $\tilde{h}_i = 1$ for $\tilde{x} \to 0$, then $y = i/m \le y_\varepsilon$ and $y_\varepsilon \to 0$ or $y_\varepsilon = \text{const}$. We have $z_i = z_0 y_1^a \to 0$ in this case and the relation analogous to (A.105) holds for y^{a_1}. Analogously to the above we have (A.106).

Let $i \in I_{\varepsilon,1}^*$. Then $\tilde{z} = \tilde{B}\tilde{x}^{\tilde{a}} = B(i/\tilde{n})^{\tilde{a}}$ and relation (A.105) holds. Also we have

$$\exp(\tilde{z}^2) < \exp(B^2 x^{2a + 2B_1 L_1}(1 + O(l_2 n^{-1})).$$

Using these relations and the asymptotics

$$\sinh^2(\tilde{z}_i^2/2) \sim \exp(\tilde{z}_i^2)/4, \quad \delta^2(\tilde{x}, \tilde{\kappa}) \asymp \tilde{x}^{\tilde{d}} \exp(-\tilde{z}^2),$$

we have $u_{L,1}^2 = u_0^2 o(\log^{-b_6} R_\varepsilon)$.

Here and above, $b_k > 0,\ k = 1,2,3,4,5,6$, are some constants.

To evaluate the difference $u_{L,3}^2 - \hat{u}_{0,3}^2$ we consider the inequality

$$u_{L,3}^2 - \hat{u}_{0,3}^2 \le \sum_{i \in I_{\varepsilon,3}^*} (u_i^{*,2} - u_i^2),$$

where the quantities

$$u_i^{*2} = 2h_0^{*2}\delta^{*2}(x,\kappa)\sinh(z^{*2}/2), \quad u_i^2 = 2h_i^2 \sinh(z^2/2),$$

are close to one another for all $i \in I_{\varepsilon,3}^*$. Namely,

$$\tilde{x}^{\tilde{a}} = x^a(1+O(L_2)+O(L_1 \log\log n)); \quad \tilde{z}^{\tilde{a}} = z^a(1+O(L_2)+O(L_1 \log\log n)).$$

If $z = O(1)$ we have

$$\begin{aligned} \delta(\tilde{x},\tilde{\kappa}) &= \tilde{B}\tilde{x}^{\tilde{a}} = \delta(x,\kappa)(1+O(L_2)+O(L_1 \log\log n)), \\ \exp(\tilde{z}^2) &= \exp(z^2)(1+O(L_2)+O(L_1 \log\log n)). \end{aligned}$$

If $\log\log n > z \to \infty$, we have, for any $\delta > 0$,

$$\begin{aligned} \delta(\tilde{x},\tilde{\kappa}) &= \tilde{B}\tilde{z}^{\tilde{a}} \exp(-\tilde{z}^2) = \delta(x,\kappa)(1+O(L_2)+O(L_1 \log^{2+\delta} \log n)); \\ \exp(\tilde{z}^2) &= \exp(z^2)(1+O(L_2)+O(L_1 \log^{2+\delta} \log n)). \end{aligned}$$

Therefore

$$u_i^{*2} = u_i^2(1+O(L_2)+O(L_1 \log\log n)).$$

There relations yield that, for some $c > 0,\ d > 0$ and any $\delta > 0$, one has

$$u_L^2 = u_0^2(1+o(R_\varepsilon^{-c})+o(\log^{-d} R_\varepsilon)+O(L_2)+O(L_1 \log^{2+\delta} \log n)). \quad \square$$

Proof of Proposition 6.2. We give the proof for the correlation coefficients $\rho_\varepsilon(\kappa_1,\kappa_2)$; considerations for $\tilde{\rho}_\varepsilon(\kappa_1,\kappa_2)$ are analogous.

Let $\kappa_l \in K_2 = K$. Note that

$$I = I(\kappa_l) > 0, \quad \|\bar{\pi}_\varepsilon(\kappa_l)\| = u_\varepsilon(\kappa_l) \asymp h_{0,l} n_l^{1/2},$$

and using the evaluations of Section A.3.5, for some $\delta_2 > 0,\ C_1 > 0$ uniformly over K for ε small enough, one has

$$\sum_{i \in I_1(\kappa_l)} h_{i,l}^2 \sinh^2(z_{i,l}^2/2) \le C_1 u_\varepsilon^2(\kappa_l) R_\varepsilon^{-\delta_2}, \tag{A.107}$$

where $h_{i,l} = 1$, and the set $I_1 = I_1(\kappa)$ is the complement of the set $I_0 = I_0(\kappa)$ which is defined in Corollary A.1. Note the inequality

$$\begin{aligned} &\left(\sum_{i \in I_1} h_{i,1} h_{i,2} \sinh^2(z_{i,1} z_{i,2}/2)\right)^2 \\ &\qquad \le \sum_{i \in I_1} h_{i,1}^2 \sinh^2(z_{i,1}^2/2) \sum_{i \in I_1} h_{i,2}^2 \sinh^2(z_{i,2}^2/2) \end{aligned} \tag{A.108}$$

which corresponds to the inequality $(\bar{\pi}_1, \bar{\pi}_2) \le \|\bar{\pi}_1\|\|\bar{\pi}_2\|$. Taking into account relation (A.107) and inequality (A.108) we can consider the items in the numerator with $i \in I_0(\kappa_1) \cap I_0(\kappa_2)$ only. Also, one can choose $C_2 > 0$ such that, for every $\kappa_l \in K$, $l = 1, 2$ and $i \in I_0(\kappa_l)$,

$$h_{i,l} \sinh(z_{i,l}^2/2) \le C_2 h_{0,l} \begin{cases} (i/n_l)^{a_l - 1/2}, & \text{if } i < n_l, \\ (i/n_l)^{-b_l - 1/2}, & \text{if } i \ge n_l, \end{cases}$$

where $a_l > 0$, $b_l > 0$ are bounded away from 0 uniformly over $\kappa_l \in K$.

In fact, by Proposition A.2 and Corollary A.1 we can put

$$a_l = \begin{cases} r_l p_l + 1/2, & \text{if } \kappa_l \in \Xi_1 \cup \Xi_2 \cup \Xi_4, \\ pqI(\kappa_l)/2(p_l - q_l), & \text{if } \kappa_l \in \Xi_3, \end{cases}$$

$$b_l = \begin{cases} pqI(\kappa_l)/2(q_l - p_l), & \text{if } \kappa_l \in \Xi_2, \\ D, & \text{if } \kappa_l \in \Xi_1 \cup \Xi_3 \cup \Xi_4, \end{cases}$$

with any $D > 0$ by the exponential decrease of the items for $\kappa_l \in \Xi_1 \cup \Xi_3 \cup \Xi_4$.

It suffices to assume that $n_1/n_2 < c$, $n_1 > C$ for c small enough and C large enough. Let A_1, A_2, A_3 be the sums of the type

$$\sum_{i \in I_0(\kappa_1) \cap I_0(\kappa_2)} h_{i,1} h_{i,2} \sinh^2(z_{i,1} z_{i,2}/2)$$

with, respectively, $i < n_1$, $n_1 \le i \le n_2$, $i > n_2$. Using the evaluations above we have, uniformly over $\kappa_l \in K$,

$$\begin{aligned} A_1 &\le C_2^2 h_{0,1} h_{0,2} (n_1/n_2)^{a_2 - 1/2} \sum_{i < n_1} (i/n_1)^{a_1 + a_2 - 1} \\ &\asymp h_{0,1} h_{0,2} (n_1 n_2)^{1/2} (n_1/n_2)^{a_2}, \\ A_3 &< C_2^2 h_{0,1} h_{0,2} (n_2/n_1)^{-b_1 - 1/2} \sum_{i > n_2} (i/n_2)^{-b_1 - b_2 - 1} \\ &\asymp h_{0,1} h_{0,2} (n_1 n_2)^{1/2} (n_1/n_2)^{b_2}. \end{aligned}$$

Analogously, if $a_2 < b_1$, then

$$\begin{aligned} A_2 &\le C_2^2 h_{0,1} h_{0,2} (n_1/n_2)^{a_2 - 1/2} \sum_{i \ge n_1} (i/n_1)^{a_2 - b_1 - 1} \\ &\asymp h_{0,1} h_{0,2} (n_1 n_2)^{1/2} (n_1/n_2)^{a_2}, \end{aligned}$$

if $a_2 > b_1$, then

$$\begin{aligned} A_2 &\le C_2^2 h_{0,1} h_{0,2} (n_2/n_1)^{-b_1 - 1/2} \sum_{i \le n_2} (i/n_2)^{a_2 - b_1 - 1} \\ &\asymp h_{0,1} h_{0,2} (n_1 n_2)^{1/2} (n_1/n_2)^{b_1}, \end{aligned}$$

and if $a_2 = b_1$, then

$$A_2 \le C_2^2 h_{0,1} h_{0,2} (n_1/n_2)^{a_2 - 1/2} \sum_{n_2 \ge i \ge n_1} (i/n_1)^{-1}$$

$$\asymp h_{0,1}h_{0,2}(n_1n_2)^{1/2}(n_1/n_2)^{a_2}\log(n_2/n_1).$$

These relations imply the statement of Proposition 6.2 for $\kappa_l \in K_2$.

Let $\kappa_l \in K_1 = K$. In this case we have

$$I = I(\kappa_l) < 0, \quad \|\bar{\pi}_{\kappa_l,\varepsilon}(\kappa_l)\| = u_\varepsilon(\kappa_l) \asymp z_{0,l}^2 m_l^{1/2}.$$

Set $I_\varepsilon(\delta, l) = \{i : \ z_{i,l} > \delta\}$. Using the evaluations analogous to Sections A.3.3, A.3.6, Propositions A.2, A.3, and Corollaries A.1, A.2 one can choose positive $\delta_2, \ C_1, \ \delta_\varepsilon \to 0$ such that, uniformly over K,

$$\sum_{i\in I_\varepsilon(\delta_\varepsilon,l)} h_{i,l}^2 \sinh^2(z_{i,l}^2/2) \le C_1 u_\varepsilon^2(\kappa_l) R_\varepsilon^{-\delta_2}.$$

Taking into account this relation and the inequality (A.108), we can consider the items in the numerator with $i \in I^\varepsilon(\delta_\varepsilon, 1) \cap I^\varepsilon(\delta_\varepsilon, 2)$ only, where $I^\varepsilon(\delta_\varepsilon, l)$ is the complement of the set $I_\varepsilon(\delta_\varepsilon, l)$.

Then one can choose $C_2 > 0$ such that, for every $\kappa_l \in K, \ l = 1, 2$, and $i \in I_\varepsilon(\delta_\varepsilon, l)$,

$$h_{i,l}\sinh(z_{i,l}^2/2) \le C_2 z_{0,l}^2 \begin{cases} (i/m_l)^{a_l - 1/2}, & \text{if } i < m_l, \\ (i/m_l)^{-b_l - 1/2}, & \text{if } i \ge m_l, \end{cases}$$

where $a_l > 0, \ b_l > 0$ are bounded away from 0 uniformly over $\kappa_l \in K$. In fact, by Propositions A.2, A.3 and by Corollaries A.1, A.2 we can put

$$a_l = \begin{cases} (2r_l p_l/(4 - p_l)) + 1/2, & \text{if } \kappa_l \in \Xi_0 \cup \Xi_3, \\ pqI(\kappa_l)/2(p_l - q_l), & \text{if } \kappa_l \in \Xi_2, \end{cases}$$

$$b_l = \begin{cases} -pqI_l/2(p_l - q_l), & \text{if } \kappa_l \in \Xi_3, \\ (2pq\lambda(\kappa_l)/(q_l - p_l)) + 1/2, & \text{if } \kappa_l \in \Xi_0 \cup \Xi_2, \end{cases}$$

where we denote $\Xi_0 = \{\kappa \in \Xi_G : p \le 2, \ q \ge p\}$. Then the evaluations are analogous to the above. □

A.4 Study of the Extreme Problem for Besov Norms

A.4.1 Solution of the System (6.110), (6.111)

The system (6.110), (6.111) could be presented in the form

$$2^{j+2}h_j^{2-t/q}\sinh^2\frac{z_j^2}{2}\left(\frac{z_j^2 - p\tanh(z_j^2/2)}{z_j^t\tanh(z_j^2/2)}\right) = \frac{Bt(p-q)}{q}2^{j(st+t/q)}, \quad (A.109)$$

$$2^{j+2}h_j^{2-h/p}\sinh^2\frac{z_j^2}{2}\left(\frac{z_j^2 - q\tanh(z_j^2/2)}{z_j^h\tanh(z_j^2/2)}\right) = \frac{Ah(p-q)}{p}2^{j(rh+h/p)}. \quad (A.110)$$

Equations (A.109), (A.110) imply the solutions of (6.110), (6.111):

$$\sinh^{2l}\frac{z_j^2}{2}z_j^{\gamma-4l}\left(\frac{z_j^2-p\tanh(z_j^2/2)}{z_j^2-q\tanh(z_j^2/2)}\right)^{2-h/p}\left|\frac{z_j^2}{\tanh(z_j^2/2)-q}\right|^l=B_0 2^{j\zeta}\frac{B^{2-h/p}}{A^{2-t/q}}, \tag{A.111}$$

$$h_j^l=\frac{Aqh}{Btp}2^{-jR}z_j^{h-t}\left(\frac{z_j^2-p\tanh(z_j^2/2)}{z_j^2-q\tanh(z_j^2/2)}\right), \tag{A.112}$$

here we recall that

$$\zeta=2(st-rh)+t/q-h/p-th(s/p-r/q)$$

(see (6.47)) and we denote

$$l=t/q-h/p,\quad R=l+st-rh,\quad \tilde\sigma=(p-q)ht/pq,$$
$$\gamma=4l-\tilde\sigma+2(h-t),\quad B_0=(|p-q|/4)^l\,(h/p)^{t/q-2}\,(q/t)^{h/p-2}. \tag{A.113}$$

We assume $z_j\in Z_{p,q}$, where $Z_{p,q}\subset R^1_+$ are the sets of the form

$$Z_{p,q}=\begin{cases}\{z>z_p\}, & \text{if } p>2,\quad p>q,\\ \{0<z<z_p\}, & \text{if } p>2,\quad p<q,\\ \{z>0\}, & \text{if } p\le 2,\quad p>q.\end{cases}$$

Recall that z_p is the positive root of the equation $z^2-p\tanh(z^2/2)=0$, $p>2$ (see (5.95)). Introduce the functions

$$\phi_{p,q,h,t}(z)=z^{\gamma-4l}\sinh^{2l}\frac{z^2}{2}\left(\frac{z^2-p\tanh(z^2/2)}{z^2-q\tanh(z^2/2)}\right)^{2-h/p}\left|\frac{z^2}{\tanh(z^2/2)}-q\right|^l, \tag{A.114}$$

$$\psi_{p,q,h,t}(z)=z^{h-t}\left(\frac{z^2-p\tanh(z^2/2)}{z^2-q\tanh(z^2/2)}\right). \tag{A.115}$$

Proposition A.4 *The function $\phi_{p,q,h,t}(z)$ is strictly monotone over the set $Z_{p,q}$ and there exists an inverse function $\phi^{-1}_{p,q,h,t}(x)$ defined for all $x>0$ and taking values in $Z_{p,q}$.*

Proof. First, note that the function $\phi_{p,q.h,t}(z)$ is continuous and that it takes all positive values when the variable z takes values in $Z_{p,q}$ This fact follows from the study of this function near the ends of the intervals $Z_{p,q}$ (i.e., near the points 0, z_p, or ∞). These provide the existence of the solutions $(h_j,z_j)\in\bar\Theta$ of equations (A.111), (A.112) (recall that the set $\bar\Theta$ is defined by (5.120)). If $h_j<1$ in (A.112), then these correspond to the minimum point $(x_j,y_j)\in\Lambda$ (see (5.118)) of the minimized function in (6.107). This point is unique by Lemma 5.3 for any $A>0$, $B>0$.

Suppose that the function $\phi_{p,q,h,t}(z)=\phi(z)$ is not strictly monotone, i.e., there exist $z_1\in Z_{p,q}$, $z_2\in Z_{p,q}$ such that $\phi(z_1)=\phi(z_2)=D>0$. Note that for any fixed $j\ge 1$ we can find $A>0$, $B>0$ such that the right-hand side of (A.111) equals D and the right-hand side of (A.112), with $z_j = z_1$ and $z_j = z_2$, is smaller than 1. These provide two minimum points in the extreme problem (6.107) and $h_j \in (0,1)$ is defined by (A.112). This contradicts Lemma 5.3. □

Put

$$\tilde{r} = rh + h/p - 1, \quad \tilde{s} = st + t/q - 1, \quad \Delta_1 = \tilde{s}(4-h) - \tilde{r}(4-t). \quad \text{(A.116)}$$

It is convenient to change the unknown parameters $A > 0$, $B > 0$ by other unknown parameters $n = 2^{j_0}$, $j_0 \geq 1$, $h_0 > 0$ for $\zeta \neq 0$, or by $m = 2^{j_1}$, $j_1 \geq 1$, $z_0 > 0$, for $\Delta_1 \neq 0$,

$$2^{j_0} = n_\varepsilon(\tau) = \left(B_0^{-1}\frac{A^{2-t/q}}{B^{2-h/p}}\right)^{1/\zeta}, \quad h_0^l = h_{0,\varepsilon}^l(\tau) = \frac{hq}{tp}\frac{A}{B}n^{-R}, \quad \text{(A.117)}$$

and

$$2^{j_1} = m_\varepsilon(\tau) = \left(\frac{A^{4-t}}{B^{4-h}}\right)^{1/\Delta_1}, \quad z_0 = z_{0,\varepsilon}(\tau) = \left(\frac{A^{\tilde{s}}}{B^{\tilde{r}}}\right)^{1/\Delta_1}. \quad \text{(A.118)}$$

We denote below

$$x = x_j = 2^{j-j_0}; \quad y = y_j = 2^{j-j_1}.$$

It is clear that if $\zeta \neq 0$ and $\Delta_1 \neq 0$, then one has

$$2^{j_1-j_0} = h_0^{\gamma/\Delta_1}\left(\frac{tp}{hq}\right)^{\gamma/l\Delta_1} B_0^{-(h-t)/l\Delta_1} = z_0^{\gamma/\zeta}B_0^{1/\zeta},$$

$$z_0 = h_0^{\zeta/\Delta_1}\left(\frac{tp}{hq}B_0^{-R/\zeta}\right)^{\zeta/l\Delta_1}, \quad y = 2^{j_0-j_1}x. \quad \text{(A.119)}$$

It is convenient to use the parameters $2^{j_0} = n$, h_0, and variables x for the study of the solutions of system (6.110), (6.111). However, it is more convenient to use the parameters $2^{j_1} = m$, z_0, and variables y in the case when we operate with solutions of equation (6.113) (this case corresponds to $h_j = 1$ and is considered in the next subsection). We often have solutions of both types. Therefore, we need to consider at the same time both types of parameters.

We can rewrite (A.111), (A.112) for $\zeta \neq 0$:

$$\phi_{p,q,h,t}(z_j) = x_j^\zeta, \quad h_j = h_0 x_j^{-R/l}\psi_{p,q,h,t}^{1/l}(z_j), \quad \text{(A.120)}$$

or, for $\Delta_1 \neq 0$, one has

$$\phi_{p,q,h,t}(z_j) = B_0 y_j^\zeta z_0^\gamma, \quad h_j = \left(\frac{hq}{tp}z_0^{t-h}y_j^{-R}\psi_{p,q,h,t}(z_j)\right)^{1/l}. \quad \text{(A.121)}$$

Using Proposition A.4 we can write the solutions of (A.120) and (A.121) in the form: if $\zeta \neq 0$, then

$$z_j = z(x_j, \tau), \quad h_j = h_0\,\delta(x_j, \tau), \quad \text{(A.122)}$$

where

$$z(x,\tau) = \phi^{-1}_{p,q,h,t}(x^\zeta), \quad \delta(x,\tau) = (x^{-R}\psi_{p,q,h,t}(z(x,\tau)))^{1/l}, \tag{A.123}$$

and, if $\Delta_1 \neq 0$, then

$$z_j = \varpi(y_j,\tau,z_0), \quad h_j = \eta(y_j,\tau,z_0), \tag{A.124}$$

where

$$\varpi(y,\tau,z_0) = \phi^{-1}_{p,q,h,t}(B_0 z_0^\gamma y^\zeta), \quad \eta^l(y,\tau,z_0) = \frac{hq}{tp} y^{-R} z_0^{t-h} \psi_{p,q,h,t}(\varpi(y,\tau,z_0)).$$

Let $\kappa \in \Xi_l$, $l = 1,2,3,4$, where the sets Ξ_l are determined by (A.47). Denote

$$\tilde{\Delta}_1 = \tilde{\Delta}_1(\tau) = \Delta_1 - 4\tilde{r}, \quad \gamma_1 = \gamma + 2(4-h), \quad a = \tilde{r}/(2-h/p),$$
$$C(p,q,h) = |(p^2 - z_p^4)/((p-q)z_p^{4-h}|^{1/(2-h/p)},$$

and set

$$\begin{aligned} B_2 &= 2^{2l}|p-2|^{h/p-2}|q-2|^{2-t/q}, \\ B_3 &= |p-2|^{(4-h)/\gamma}\,|q-2|^{(t-4)/\gamma}\,4^{(t-h)/\gamma}, \\ B_4 &= 2^{2l}6^{2-h/2}(2-q)^{2-t/q}, \\ B_5 &= (4^{4+h-t}6^{h-4}(2-q)^{t-8})^{1/\gamma_1}. \end{aligned}$$

Note that $a \geq -1/2$ and if $\kappa \in \Xi_k$, $k \neq 3$, then $\zeta > 0$, if $\kappa \in \Xi_4$, then $I \geq 0$, if $\kappa \in \Xi_k$, $k \neq 2$, then $\gamma > 0$, if $\kappa \in \Xi_2$, then $\gamma < 0$.

Remark. One can study the relations between the functions $\lambda, I, \Delta_1, \tilde{\Delta}_1, R$ by using geometrical consideration (compare with the Remarks on pages 122, 142, and 366). One can draw illustrations analogous to Figures 10–13 for the power norms. One needs to add the points $u^* = (1/h - 1/p,\ 1/t - 1/q)$ and $z^* = (-1/p, -1/q)$ to the illustrations. As above, the line $I = 0$ contains the points x^*, y^*, the line $\lambda = 0$ contains the points y^*, z^* and the origin. The new line $\Delta_1 = 0$ contains the points x^*, u^*, the line $\zeta = 0$ contains the points y^*, u^*, the line $R = 0$ contains the points u^*, z^*, and the line $\tilde{\Delta}_1 = 0$ contains the point u^*. The sign of the function $\gamma = \gamma(p,q,h,t)$ is studied analytically by taking the derivative in t and by taking the sign for $t = q$. We omit these illustrations and calculations.

The following proposition describes the properties of the functions $z(x,\tau)$ and $\delta(x,\tau)$, $x > 0$, $\kappa \in \Xi_k$, $k = 1,2,3,4$ (if $\kappa \in \Xi_3$, then we assume $\zeta \neq 0$), and of the functions $\varpi(y,\tau,z_0)$, $\eta(y,\tau,z_0)$ for $\zeta \leq 0$. Note that in the cases under consideration we have

$$r < 0, \quad s < 0, \quad \gamma > 0, \quad I > 0, \quad \Delta_1 > 0, \quad \kappa \in \Xi_3,$$

(see the Remark above). Recall that $x = x_j = 2^{j-j_0}$, $y = y_j = 2^{j-j_1}$.

Define the quantities $x_\varepsilon = x_\varepsilon(\tau)$, $y_\varepsilon = y_\varepsilon(\tau)$ and $\tilde{j}_1 = \tilde{j}_1(\tau)$ by the equalities

$$h_0\delta(x_\varepsilon, \tau) = 1, \quad y_\varepsilon = 2^{j_0 - j_1} x_\varepsilon, \quad 2^{\tilde{j}_1} = x_\varepsilon\, 2^{j_0} = y_\varepsilon\, 2^{j_1}. \tag{A.125}$$

Proposition A.5 (A) *The functions $z(x,\tau)$ and $\delta(x,\tau)$ are uniformly continuous positive smooth functions over $x \in R^1_+, \tau = (\kappa, h, t) \in \Gamma$, $\kappa \in K$, for any compact $\Gamma \subset \Xi_k \times R^3_+$ such that the constraints (6.8) hold, and K is disjoint with the set $\{\lambda = 0\} \times R^1_+$. Uniformly over any compact Γ of such type the following rate relations (1) – (4) hold:*

(1) *Let $p > 2$, $p > q$ (i.e., $\kappa \in \Xi_1$). Then $z(x,\tau)$ is increasing in x and*

$$z(x,\tau) \sim \begin{cases} z_p, & \text{if } x \to 0, \\ \sqrt{(\zeta \log x)/l}, & \text{if } x \to \infty, \end{cases}$$

$$\delta(x,\tau) \sim \begin{cases} C(p,q,h)x^a, & \text{if } x \to 0, \\ (\zeta \log x/l)^{(h-t)/2l} x^{-R/l}, & \text{if } x \to \infty, \end{cases}$$

and

$$\sinh^2 \frac{z^2}{2} \sim x^{\zeta/l} \left(\frac{\zeta}{l} \log x \right)^{1-\gamma/2l}, \quad \text{if } x \to \infty.$$

(2) *Let $p > 2$, $p < q$ (i.e., $\kappa \in \Xi_2$). Then $z(x,\tau)$ is decreasing in x and*

$$z(x,\tau) \sim \begin{cases} z_p, & \text{if } x \to 0\,, \\ B_2^{1/\gamma} x^{\zeta/\gamma}, & \text{if } x \to \infty\,, \end{cases}$$

$$\delta(x,\tau) \sim \begin{cases} C(p,q,h)x^a, & \text{if } x \to 0, \\ B_3 x^{-\Delta_1/\gamma}, & \text{if } x \to \infty. \end{cases}$$

(3) *Let $p < 2$, $p > q$ (i.e., $\kappa \in \Xi_3$). If $\zeta < 0$, then $z(x,\tau)$ is increasing on x , if $\zeta > 0$, then it is decreasing on x and if $\zeta = 0$, then $z(x,\tau) \sim z_0\,(B_0 B_2)^{1/\gamma} = \text{const}$,*

$$z(x,\tau) \sim \begin{cases} B_2^{1/\gamma} x^{\zeta/\gamma}, & \text{if } x \to 0 \text{ for } \zeta > 0 \text{ or } x \to \infty \text{ for } \zeta < 0, \\ \sqrt{(\zeta \log x)/l}, & \text{if } x \to \infty \text{ for } \zeta > 0 \text{ or } x \to 0 \text{ for } \zeta < 0, \end{cases}$$

$$\delta(x,\tau) \sim \begin{cases} B_3 x^{-\Delta_1/\gamma}, & \text{if } x \to 0 \text{ for } \zeta > 0 \text{ or } x \to \infty \text{ for } \zeta < 0, \\ (\zeta \log x/l)^{(h-t)/2l} x^{-R/l}, & \text{if } x \to \infty \text{ for } \zeta > 0 \text{ or } x \to 0 \text{ for } \zeta < 0, \end{cases}$$

$$\sinh^2 \frac{z^2}{2} \sim x^{\zeta/l} \left(\frac{\zeta}{l} \log x \right)^{1-\gamma/2l}, \quad \text{if } x \to \infty \text{ for } \zeta > 0 \text{ or } x \to 0 \text{ for } \zeta < 0.$$

(4) *Let $2 = p > q$ (i.e., $\kappa \in \Xi_4$). Then $z(x,\tau)$ is increasing in x and*

$$z(x,\tau) \sim \begin{cases} B_4^{1/\gamma_1} x^{\zeta/\gamma_1}, & \text{if } x \to 0, \\ \sqrt{(\zeta \log x)/l}, & \text{if } x \to \infty; \end{cases}$$

$$\delta(x,\tau) \sim \begin{cases} B_5 x^{-\tilde{\Delta}_1/\gamma_1}, & \text{if } x \to 0\,, \\ (\zeta \log x/l)^{(h-t)/2l} x^{-R/l}, & \text{if } x \to \infty, \end{cases}$$

$$\sinh^2 \frac{z^2}{2} \sim x^{\zeta/l} \left(\frac{\zeta}{l} \log x\right)^{1-\gamma/2l}, \quad \text{if } x \to \infty.$$

(B) *Let $\zeta \le 0$ (recall that $\kappa \in \Xi_3$, $\Delta_1 > 0$, $I < 0$ in this case). If $\zeta = 0$, then $\varpi(y,\tau,z_0)$ is constant. If $\zeta < 0$, then it is a decreasing continuous function in y.*

Let us take $\delta > 0$ such that $-\gamma/\zeta > \delta$. Set $y_{0,\delta} \asymp z_0^{-\gamma/\zeta-\delta} \to 0$. Introduce the sets $Y_{\varepsilon,\delta} = \{y > y_{0,\delta}\}$. Put

$$c_\tau = (B_0 B_2)^{1/\gamma}, \quad c_h = B_0^{(h-t)/l\gamma} B_3 (hq/tp)^{1/l}.$$

In this case, if $z_0 \to 0$, then $h_0 \to \infty$, and we have

$$\varpi(y,\tau,z_0) = z_0 \vartheta_0(y,\tau,z_0).$$

If either $\zeta = 0$, $y > 0$, or $y \in Y_{\varepsilon,\delta}$, then, uniformly over y, we have

$$\vartheta_0(y,\tau,z_0) \sim \vartheta_0(y,\tau) = c_\tau y^{\zeta/\gamma}, \quad \eta(y,\tau,z_0) \sim \eta_0(y,\tau) = c_h y^{-\Delta_1/\gamma}.$$

In particular, for any $c > 0$, we have $c \in Y_{\varepsilon,\delta}$ for small enough $\varepsilon > 0$, and $\{y : y \ge c\} \subset Y_{\varepsilon,\delta}$.

Proof. This is based on the standard properties of inverse functions and on the standard asymptotic relations

$$\sinh x \sim \tanh x \sim x; \quad x^2 - 2\tanh(x^2/2) \sim x^6/12, \quad \text{as } x \to 0. \quad \square$$

Denote by Γ_G the set which consists of $\tau = (\kappa, h, t)$ such that the constraints (6.8) are fulfilled (i.e., $t > q$, $p > h > 0$) and $\kappa \in \Xi_G$.

Let $\Gamma_{\Delta_1}^- \subset \Gamma_G$ be the set consisting of τ such that either $\kappa \in \Xi_1$ or $\kappa \in \Xi_2 \cup \Xi_3$, $\Delta_1 = \Delta_1(\tau) \le 0$ or $\kappa \in \Xi_4$, $\tilde{\Delta}_1 = \tilde{\Delta}_1(\tau) \le 0$.

Let the set $\Gamma_{\Delta_1}^+$ be the complement of $\Gamma_{\Delta_1}^-$ in Γ_G,

$$\Gamma_{\Delta_1}^+ = \Gamma_G \backslash \Gamma_{\Delta_1}^-.$$

Note that if $\kappa \in \Xi_3$, then the inequality $\Delta_1(\tau) \le 0$ yields that $r \ge 0$, $\zeta > 0$. Recall that the quantities x_ε, y_ε, $\tilde{j}_1$ are defined by (A.125).

From Proposition A.5 we obtain

Corollary A.3

(1) *Assume $\tau \in \Gamma_{\Delta_1}^-$ and*

$$j_0 \to \infty, \quad h_0 \to 0; \quad 2^{-j_0 a} h_0 \to 0, \quad \text{if } p > 2. \tag{A.126}$$

Then for $\varepsilon > 0$ small enough the relations

$$z_j = z(2^{j-j_0}, \tau) = z(x,\tau), \quad h_i = h_0 \delta(2^{j-j_0}, \tau) = h_0 \delta(x,\tau), \tag{A.127}$$

define the solutions of the system (6.110), (6.111) for all j.

(2) *Assume* $\tau \in \Gamma^+_{\Delta_1}$ *and* $\zeta > 0$. *Then relations* (A.127) *define the solutions of the system* (6.110), (6.111) *for* $j \in \mathcal{J}_0$ *where the integer set* $\mathcal{J}_0 = \mathcal{J}_0(\tau)$ *is defined for any* $\tau \in \Gamma^+_{\Delta_1}$ *by the relations*

$$\mathcal{J}_0 = \begin{cases} j : 1 < 2^j \le 2^{j_0} x_\varepsilon = 2^{j_1} y_\varepsilon, & \text{if } \kappa \in \Xi_2, \\ j : 2^j \ge 2^{j_0} x_\varepsilon = 2^{j_1} y_\varepsilon, & \text{if } \kappa \in \Xi_3,\ \zeta > 0 \text{ or } \kappa \in \Xi_4, \\ j : 2^j \ge 2^{j_1} y_\varepsilon, & \text{if } \kappa \in \Xi_3,\ \zeta \le 0. \end{cases}$$

If $p \neq 2$, *then we have, as* $z_0 \to 0$,

$$y_\varepsilon = y_\varepsilon(\tau) \sim y_1(\tau) = c(p,q,h,t) = (B_0^{(h-t)/l\gamma} B_3 \, (hq/tp)^{1/l})^{\gamma/\Delta_1}. \quad (A.128)$$

Let $p = 2$, *i.e.,* $\kappa \in \Xi_4$. *Note that if* $\tilde{\Delta}_1 > 0$, *then* $\Delta_1 > 0$. *One has, as* $h_0 \to 0$,

$$y_\varepsilon = y_\varepsilon(\tau) \sim y_1(\tau) = (B_0^{(4+h-t)/l} B_5^{\gamma_1})^{1/\tilde{\Delta}_1} h_0^{4\zeta(4-h)/\tilde{\Delta}_1\Delta_1} \to 0. \quad (A.129)$$

Moreover, in this case for $j \in \mathcal{J}_0$, *we have*

$$z_j = z(2^{j-j_0}, \tau) = \varpi(2^{j-j_1}, \tau, z_0), \quad h_j = h_0 \delta(2^{j-j_0}, \tau) = \eta(2^{j-j_1}, \tau, z_0). \quad (A.130)$$

(3) *Assume* $\tau \in \Gamma^+_{\Delta_1}$, $z_0 \to 0$, $j_1 \to \infty$ *and* $\zeta \le 0$, *i.e.,* $\kappa \in \Xi_3$. *Then the solutions of the system* (6.110), (6.111) *for* $j \in \mathcal{J}_0$ *are determined by the relations*

$$z_j = z_0 \vartheta_0(2^{j-j_1}, \tau, z_0), \quad h_j = \eta(2^{j-j_1}, \tau, z_0).$$

Remark. In view of Remark 3 from Section 6.5, assumptions (6.58) yield the relations which were used above:

$$j_0 \to \infty, \quad h_0 \to 0; \quad 2^{-j_0 a} h_0 \to 0, \quad \text{if } p > 2; \quad z_0 \to 0,$$

and $j_1 \to \infty$ for $\kappa \in \Xi_3$, $\zeta \le 0$.

A.4.2 Solution of the Extreme Problem (6.112)

We need to study equations (6.113) and to select the roots which correspond to a minimum in the extreme problems (6.112). The considerations are analogous to the case of the power norms. We rewrite (6.113) in the form

$$2z_0^{-4} z^2 \sinh(z^2) + t y^{\tilde{s}} (z/z_0)^t = h y^{\tilde{r}} (z/z_0)^h \quad (A.131)$$

which is the same as (A.69) replacing p, q, rp, sq by h, t, $\tilde{r}$, $\tilde{s}$. By Corollary A.3 we need to study (A.131) for $\tau \in \Gamma^+_{\Delta_1}$. Note that if $\tau \in \Gamma^+_{\Delta_1}$, then

$$\kappa \in \Xi_2 \cup \Xi_3, \quad \Delta_1(\tau) > 0, \quad \text{or} \quad \kappa \in \Xi_4, \quad \tilde{\Delta}_1(\tau) > 0.$$

Observe that if $\kappa \in \Xi_4$, $\tilde{\Delta}_1 > 0$, then $\Delta_1 > 0$. Therefore $\Delta_1 > 0$ for any $\tau \in \Gamma^+_{\Delta_1}$.

We need to consider $j \in \mathcal{J}_1(\tau)$, $y \in Y_{1,\varepsilon}(\tau)$, where the integer set $\mathcal{J}_1 = \mathcal{J}_1(\tau)$ is the complement of $\mathcal{J}_0$ defined in Corollary A.3:

$$\mathcal{J}_1 = \mathcal{J}_1(\tau) = \begin{cases} \{j : j > \tilde{j}_1\}, & \text{if } \kappa \in \Xi_2,\ \Delta_1 > 0 \\ \{j : 1 \le j < \tilde{j}_1\}, & \text{if } \kappa \in \Xi_3,\ \Delta_1 > 0 \text{ or } \kappa \in \Xi_4,\ \tilde{\Delta}_1 > 0, \end{cases}$$

and

$$Y_{1,\varepsilon}(\tau) = Y_1 = \begin{cases} (y_\varepsilon, \infty), & \text{if } \kappa \in \Xi_2,\ \Delta_1 > 0, \\ (0, y_\varepsilon), & \text{if } \kappa \in \Xi_3,\ \Delta_1 > 0 \ \text{ or } \kappa \in \Xi_4,\ \tilde{\Delta}_1 > 0, \end{cases}$$

the values $y_\varepsilon = y_\varepsilon(\tau) = 2^{\tilde{j}_1 - j_1} \sim y_1(\tau)$ and the values $y_1(\tau)$ are defined in Corollary A.3.

It is also convenient to consider the sets

$$\mathcal{J}_0^* = \begin{cases} \{j : -\infty < j \le \tilde{j}_1\}, & \text{if } \kappa \in \Xi_2, \\ \{j : \tilde{j}_1 \le j < \infty\}, & \text{if } \kappa \in \Xi_3 \cup \Xi_4, \end{cases}$$

$$\mathcal{J}_1^* = \begin{cases} \{j : \tilde{j}_1 < j < \infty\}, & \text{if } \kappa \in \Xi_2, \\ \{j : -\infty < j < \tilde{j}_1\}, & \text{if } \kappa \in \Xi_3 \cup \Xi_4. \end{cases}$$

Let $h_0 \to 0$. If $\kappa \in \Xi_2$, $\Delta_1 > 0$, then $x_\varepsilon = 2^{\tilde{j}_1 - j_0} \to \infty$, and if $\kappa \in \Xi_3$, $\Delta_1 > 0$, $\zeta > 0$ or $\kappa \in \Xi_4$, $\tilde{\Delta}_1 > 0$, then $x_\varepsilon \to 0$ (note that $z_0 \to 0$ in these cases). Let $z_0 \to 0$. If $\kappa \in \Xi_3$, $\Delta_1 > 0$, $\zeta < 0$, then $x_\varepsilon \to \infty$, $h_0 \to \infty$.

Proposition A.6

(1) *Let $\tau \in \Gamma^+_{\Delta_1}$, $j \in \mathcal{J}_1(\tau)$. There exists a unique solution of* (A.131) *which provides a minimum in* (6.112). *This is of the form*

$$z_j = \varpi_1(y, \tau, z_0) = z_0 \vartheta(y, z_0, \tau) > 0,$$

where $\vartheta(y, z_0, \tau)$ is a continuous positive smooth function in z_0, $y \in Y_{1,\varepsilon}$, $\tau \in \Gamma^+_{\Delta_1}$, and the boundary continuity condition holds:

$$\varpi_1(y_\varepsilon, \tau, z_0) = z_0 \vartheta(y_\varepsilon, z_0, \tau) = \varpi(y_\varepsilon, \tau, z_0) = z(x_\varepsilon, \tau).$$

(2) *Assume $z_0 = o(1)$ and if $p \le 2$, $\tilde{r} < 0$, then $z_0 2^{-j_1 \tilde{r}/(4-h)} = o(1)$. Then, uniformly over z_0, $y \in Y_{1,\varepsilon}$, $\tau \in \Gamma^+_{\Delta_1}$, one has $z_j \sim z_0 \vartheta$; here $\vartheta = \vartheta(y, \tau)$ is the solution of the equation*

$$2\vartheta^{4-h} + t y^{\tilde{s}} \vartheta^{t-h} = h y^{\tilde{r}}, \quad y = 2^{j - j_1}. \tag{A.132}$$

The function $\vartheta(y, \tau)$ is a continuous smooth function on the sets $Y_{1,\varepsilon}(\tau)$ with the asymptotic properties: $z_j \to 0$ and

$$\vartheta(y, \tau) \sim \begin{cases} (h/2)^{1/(4-h)} y^{\tilde{r}/(4-h)}, & \text{if } y \to 0,\ \kappa \in \Xi_3 \cup \Xi_4, \\ (h/t)^{1/(t-h)} y^{-R/(t-h)},\ R > 0, & \text{if } y \to \infty,\ \kappa \in \Xi_2. \end{cases} \tag{A.133}$$

Proof. We only give the outline of this proof. Remember that $\Delta_1 > 0$. Assuming $t \ne 4$, let us set $u = (z/z_0) y^{-\tilde{s}/(4-t)}$. Then, by $\sinh z^2 \sim z^2$, one can rewrite (A.131) in the form

$$c(z, \tau) \frac{2}{h} u^{4-h} + \frac{t}{h} u^{t-h} = y^{-\Delta_1/(4-t)}, \tag{A.134}$$

where $c(z,\tau) \sim 1$ as $z \to 0$. We consider the equation

$$\frac{2}{h}u^{4-h} + \frac{t}{h}u^{t-h} = y^{-\Delta_1/(4-t)}. \qquad (A.135)$$

By Lemma 5.3 there exists a unique point which provides the minimum in the extreme problem of interest. Therefore there exists the unique solution $z = z(\tau)$ of equation (A.131) corresponding to the minimum in the extreme problem of interest. Also there exists the unique solution $u = u(\tau)$ of equation (A.134) and, for small enough $z > 0$, there exists unique solution $\tilde{u}(\tau) \sim u(\tau)$ of equation (A.135) which corresponds to $z = z(\tau)$.

If $t > h$, $h \leq 4$, then, by monotonicity of the left-hand side of (A.135), there exists a unique solution of (A.135). In other cases (i.e., for $t < h$ or $h > 4$) the left-hand side of (A.135) has two intervals of monotonicity. Therefore (A.135) has two roots corresponding to $z_{j,1}$, $z_{j,2}$ of the form

$$z_{j,1} \sim z_0(hy^{-R}/t)^{1/(t-h)} \quad \text{or} \quad z_{j,2} \sim z_0(hy^{\tilde{r}}/2)^{1/(4-h)}. \qquad (A.136)$$

Since the left-hand side of (A.135) is positive as $u \to \infty$, we need to select the larger solution. To obtain its asymptotics, one can use (A.136) and the asymptotics of the ratio

$$z_{j,1}/z_{j,2} \asymp y^{\Delta_1/(t-h)(h-4)}, \quad \Delta_1 > 0.$$

Let $t = 4$. Then $\tilde{s} > 0$ and, analogously, the solution of equation (A.132) is in the form of (A.133). □

Using Propositions A.5, A.6, the relationships (A.119) between variables x, n and y, m, we obtain the following:

Corollary A.4 *Assume $\tau = (\kappa, h, t) \in \Gamma^+_{\Delta_1}$, the assumptions on z_0, $m = 2^{j_1}$ of Proposition A.6, n.2 hold, and the family u_ε satisfies assumptions* (6.58). *Then the solutions of* (6.110), (6.111) *are defined by the relations:*

(1) If $\zeta > 0$, then we can express the values z_j, h_j in terms of the variables $x_j = 2^{j-j_0}$,

$$z_j = \begin{cases} z(x_j,\tau), & \text{if } j \in \mathcal{J}_0, \\ z(x_j,\tau,h_0) = \varpi_1(y_j,\tau,z_0), & \text{if } j \in \mathcal{J}_1, \end{cases}$$

where $z(x,\tau,h_0) = \varpi_1(y,\tau,z_0)$ (recall that $y_j = 2^{j-j_1}$, $x_j = 2^{j_1-j_0}y_j$). If $x \to 0$, $\kappa \in \Xi_3$, then

$$z(x,\tau,h_0) \sim (B_0\,(h/2)^l\,(tp/hq)^{(2-h/p)})^{1/(4-h)l}\,h_0^{(2-h/p)/(4-h)}\,x^{\tilde{r}/(4-h)}.$$

If $x \to \infty$, $\kappa \in \Xi_2$, then

$$z(x,\tau,h_0) \sim (p/q)^{1/(t-h)}\,h_0^{l/(t-h)}\,x^{-R/(t-h)}.$$

For $\kappa \in \Xi_4$, if $x \leq x_\varepsilon \sim B_0^{(t-h)/\Delta_1 l}\,(tp/hq)^{\gamma/\Delta_1 l}\,h_0^{\gamma/\Delta_1}\,y_\varepsilon$, then one has

$$z(x,\tau,h_0) \sim (h/2)^{1/(4-h)}\,(2t/hq)^{1/2l}\,h_0^{1/2}\,x^{\tilde{r}/(4-h)}.$$

Here y_ε is defined in (A.129) *and*

$$h_j = \begin{cases} h_0 \delta(x_j, \tau), & \text{if } j \in \mathcal{J}_0 \ , \\ 1, & \text{if } j \in \mathcal{J}_1 \ . \end{cases}$$

The properties of the functions $\delta(x,\tau)$, $z(x,\tau)$ are determined by Proposition A.5.

(2) *Let $\zeta \le 0$, this corresponds to $\kappa \in \Xi_3$, $I < 0$, see the Remark on page 399. Then we can express the values z_j, h_j in terms of variables $y_j = 2^{j-j_1}$,*

$$z_j = \begin{cases} \varpi(y_j, \tau, z_0) = z_0 \vartheta_0(y, \tau, z_0), & \text{if } j \in \mathcal{J}_0, \\ \varpi_1(y_j, \tau, z_0), & \text{if } j \in \mathcal{J}_1, \end{cases}$$

and

$$h_j = \begin{cases} \eta(y_j, \tau, z_0), & \text{if } \ j \in \mathcal{J}_0, \\ 1, & \text{if } \ j \in \mathcal{J}_1, \end{cases}$$

where the functions $\vartheta_0(y,\tau,z_0)$, $\eta(y,\tau,z_0)$ are determined by Proposition A.5 (B). The properties of the function $\varpi_1(y,\tau,z_0)$ are determined by Proposition A.6.

Observe that, if $\zeta = 0$, $z_0 \to 0$ (recall that $\kappa \in \Xi_3$ and $I < 0$ in this case), $j \in \mathcal{J}_0$, then

$$z_j = c_\tau z_0, \qquad h_j \sim (y_j / y_\varepsilon)^{-R/l}.$$

To study equations (6.108), relation (6.109), and the properties of the solution of the extreme problem it is convenient to formulate the following proposition which describes rate relations for the quantities z_j, h_j and follows from Propositions A.5, A.6 and Corollaries A.3, A.4.

Proposition A.7

I. *Let $\kappa \in \Xi_1$, $j_0 \to \infty$, $h_0 \to 0$. Then we have*

$$z_j \asymp \begin{cases} z_p, & \text{if } j \le j_0, \\ \sqrt{1 + j - j_0}, & \text{if } j > j_0, \end{cases} \tag{A.137}$$

$$h_j \asymp h_0 \begin{cases} 2^{(j-j_0)a}, & \text{if } j \le j_0, \\ 2^{-(j-j_0)R/l} (j - j_0)^{(h-t)/2l}, & \text{if } j > j_0. \end{cases}$$

II. *Let $\kappa \in \Xi_2$. Then we have $\gamma < 0$.*

(a) *If $\Delta_1 \le 0$, $j_0 \to \infty$, $h_0 \to 0$, then*

$$z_j \asymp \begin{cases} z_p, & j \le j_0, \\ 2^{(j-j_0)\zeta/\gamma}, & j > j_0, \end{cases} \quad h_j \asymp h_0 \begin{cases} 2^{(j-j_0)a}, & j \le j_0, \\ 2^{-(j-j_0)\Delta_1/\gamma}, & j > j_0. \end{cases} \tag{A.138}$$

(b) *If $\Delta_1 > 0$, $j_0 \to \infty$, $h_0 \to 0$, $z_0 \to 0$, then*

$$z_j \asymp \begin{cases} z_p, & \text{if } j \le j_0, \\ 2^{(j-j_0)\zeta/\gamma} \asymp z_0 2^{(j-\tilde{j}_1)\zeta/\gamma}, & \text{if } j_0 < j \le \tilde{j}_1, \\ z_0 2^{-(j-\tilde{j}_1)R/(t-h)}, & \text{if } j > \tilde{j}_1, \end{cases} \tag{A.139}$$

$$h_j \asymp \begin{cases} h_0 2^{(j-j_0)a}, & \text{if } j \le j_0, \\ h_0 2^{-(j-j_0)\Delta_1/\gamma} \asymp 2^{-(j-j_1)\Delta_1/\gamma}, & \text{if } j_0 < j \le \tilde{j}_1, \end{cases}$$

$$h_j = 1, \quad \text{if } j > \tilde{j}_1.$$

III. *Let* $\kappa \in \Xi_3$. *Then* $\gamma > 0$ *and we have*

(a) *If* $\zeta > 0,\ \Delta_1 \le 0,\ j_0 \to \infty,\ h_0 \to 0$, *then*

$$z_j \asymp \begin{cases} 2^{(j-j_0)\zeta/\gamma}, & \text{if } j \le j_0, \\ \sqrt{1+j-j_0}, & \text{if } j > j_0, \end{cases} \tag{A.140}$$

$$h_j \asymp h_0 \begin{cases} 2^{-(j-j_0)\Delta_1/\gamma}, & \text{if } j \le j_0, \\ 2^{-(j-j_0)R/l}(j-j_0)^{(h-t)/2l}, & \text{if } j > j_0. \end{cases}$$

(b) *If* $\zeta > 0,\ \Delta_1 > 0,\ z_0 \to 0,\ j_0 \to \infty,\ h_0 \to 0$, *then*

$$z_j \asymp \begin{cases} z_0 2^{(j-\tilde{j}_1)\tilde{r}/(4-h)}, & \text{if } j \le \tilde{j}_1, \\ z_0 2^{(j-\tilde{j}_1)\zeta/\gamma} \asymp 2^{(j-j_0)\zeta/\gamma}, & \text{if } \tilde{j}_1 < j \le j_0, \\ \sqrt{1+j-j_0}, & \text{if } j > j_0, \end{cases} \tag{A.141}$$

$$h_j = 1, \quad \text{if } j \le \tilde{j}_1,$$

$$h_j \asymp \begin{cases} 2^{-(j-\tilde{j}_1)\Delta_1/\gamma} \asymp h_0 2^{-(j-j_0)\Delta_1/\gamma}, & \text{if } \tilde{j}_1 < j \le j_0, \\ h_0 2^{-(j-j_0)R/l}(j-j_0)^{(h-t)/2l}, & \text{if } j > j_0. \end{cases}$$

(c) *If* $\zeta \le 0,\ j_1 \to \infty,\ z_0 \to 0$, *then* $\Delta_1 > 0,\ \tilde{j}_1 > j_0,\ h_0 \to \infty,\ I < 0$, *and*

$$z_j \asymp \begin{cases} z_0 2^{(j-\tilde{j}_1)\tilde{r}/(4-h)}, & \text{if } j \le \tilde{j}_1, \\ z_0 2^{(j-\tilde{j}_1)\zeta/\gamma} & \text{if } j > \tilde{j}_1, \end{cases} \tag{A.142}$$

$$h_j = 1, \quad \text{if } j \le \tilde{j}_1,$$

$$h_j \asymp 2^{-(j-\tilde{j}_1)\Delta_1/\gamma}, \text{if } j > \tilde{j}_1.$$

IV. *Let* $\kappa \in \Xi_4$. *Then* $\gamma_1 > 0$ *and we have*

(a) *If* $\tilde{\Delta}_1 \le 0,\ j_0 \to \infty,\ h_0 \to 0$, *then*

$$z_j \asymp \begin{cases} 2^{(j-j_0)\zeta/\gamma_1}, & \text{if } j \le j_0, \\ \sqrt{1+j-j_0}, & \text{if } j > j_0, \end{cases} \tag{A.143}$$

$$h_j \asymp h_0 \begin{cases} 2^{-(j-j_0)\tilde{\Delta}_1/\gamma_1}, & \text{if } j \le j_0, \\ 2^{-(j-j_0)R/l}(j-j_0)^{(h-t)/2l}, & \text{if } j > j_0. \end{cases}$$

(b) *If* $\tilde{\Delta}_1 > 0$, *then* $\Delta_1 > 0$ *and if, moreover,* $z_0 \to 0,\ j_0 \to \infty,\ h_0 \to 0$, *then*

$$z_j \asymp \begin{cases} z_0 2^{(j-j_1)\tilde{r}/(4-h)}, & \text{if } j \le \tilde{j}_1, \\ 2^{(j-j_0)\zeta/\gamma_1}, & \text{if } \tilde{j}_1 < j \le j_0, \\ \sqrt{1+j-j_0}, & \text{if } j > j_0, \end{cases} \tag{A.144}$$

$$h_j = 1, \quad \text{if } j \le \tilde{j}_1,$$

$$h_j \asymp \begin{cases} h_0 2^{-(j-j_0)\tilde{\Delta}_1/\gamma_1}, & \text{if } \tilde{j}_1 < j \le j_0, \\ h_0 2^{-(j-j_0)R/l}(j-j_0)^{(h-t)/2l}, & \text{if } j > j_0. \end{cases}$$

All the relations presented above are uniform over j *under given constraints on* j.

Note that $\zeta \geq 0$ implies $\lambda > 0$ and $\lambda \geq 0$ implies $R = R(\tau) > 0$ (see the Remark on page 399). Moreover, $\tilde{r}/(4-h) \geq -1/4$ and if $p \leq 2$, $\tau \in \Gamma^+_{\Delta_1}$, then

$$z_0 2^{-j_1\tilde{r}/(4-h)} \asymp 2^{-j_1 4h(r+1/p-1/4)/(4-h)} = o(1) \text{ for } \kappa \in \Xi_{G_1};$$
$$z_0 2^{-j_1\tilde{r}/(4-h)} \asymp 2^{-j_0 h(r+1/2p)} = o(1) \text{ for } \kappa \in \Xi_{G_2}.$$

A.4.3 Solution of the Extreme Problem for $\tau \in \Gamma^-_{\Delta_1}$

By Corollary A.3, it is enough to find the values j_0, h_0 from equations (6.108), to obtain u_ε from relations (6.109) and to check assumptions (A.126). We give the outlines of proofs and omit simple calculations which one can easily work out.

First, assume $r > r_p^*$. In this case $\kappa \in \Xi_{G_2}$ and we can rewrite relations (6.108) and (6.109) in the form

$$\begin{aligned} \rho_\varepsilon^h &= h_0^{h/p} 2^{j_0(rh+h/p)} C_{1,\varepsilon}(\tau), \\ R_\varepsilon^t &= h_0^{t/q} 2^{j_0(st+t/q)} C_{2,\varepsilon}(\tau), \\ u_\varepsilon^2 &= h_0^2 2^{j_0} C_{0,\varepsilon}(\tau), \end{aligned} \tag{A.145}$$

where $C_{k,\varepsilon}(\tau) = C_{k,\varepsilon}(\tau, j_0)$, $k = 0, 1, 2$, are continuous functions of the variables τ, j_0, the functions are bounded away from 0 and ∞ for 2^{-j_0} small enough, and these are determined by the relations

$$\begin{aligned} C_{1,\varepsilon}(\tau) &= \sum_{j=1}^{\infty} \delta^{h/p}(2^{j-j_0},\tau) z^h(2^{j-j_0},\tau) 2^{(j-j_0)(rh+h/p)} \\ C_{2,\varepsilon}(\tau) &= \sum_{j=1}^{\infty} \delta^{t/q}(2^{j-j_0},\tau) z^t(2^{j-j_0},\tau) 2^{(j-j_0)(st+t/q)} \\ C_{0,\varepsilon}(\tau) &= 2\sum_{j=1}^{\infty} 2^{j-j_0} \delta^2(2^{j-j_0},\tau) \sinh^2 \frac{z^2(2^{j-j_0},\tau)}{2}. \end{aligned} \tag{A.146}$$

Note that

$$C_{k,\varepsilon}(\tau) \to c_k^{a_k}(\tau, j_0), \quad k = 0,1,2, \quad a_0 = 1, \quad a_1 = h, \quad a_2 = t, \tag{A.147}$$

where $c_k(\tau, j_0)$, $k = 0, 1, 2$, are continuous 1-periodical functions on j_0 determined by the relations

$$\begin{aligned} c_1^h(\tau,j_0) &= c_1^+(\tau,j_0) + c_1^-(\tau,j_0) = \sum_{-\infty}^{\infty} x_j^{rh+h/p} \delta^{h/p}(x_j,\tau) z^h(x_j,\tau), \\ c_2^t(\tau,j_0) &= c_2^+(\tau,j_0) + c_2^-(\tau,j_0) = \sum_{-\infty}^{\infty} x_j^{st+t/q} \delta^{t/q}(x_j,\tau) z^t(x_j,\tau), \\ c_0(\tau,j_0) &= c_0^+(\tau,j_0) + c_0^-(\tau,j_0) = 2\sum_{-\infty}^{\infty} x_j \delta^2(x_j,\tau) \sinh^2 \frac{z^2(x_j,\tau)}{2}, \end{aligned} \tag{A.148}$$

where $x_j = 2^{j-j_0}$, the functions $z(x,\tau)$, $\delta(x,\tau)$ are determined by (A.114), (A.115), (A.123) and are continuous in x. For the quantities c_k^+, $k = 0,1,2$ the summation is taken over $j > j_0$; for c_k^-, $k = 0,1,2$, one is taken over $j \le j_0$. Relations (A.147) are uniform over all compacts

$$\Gamma \subset \{\tau \in \Gamma_{\Delta_1}^- : r > r_p^*\}.$$

Using Proposition A.5 and the definition of the set Ξ_{G_2}, one can check that the sums in (A.148) are finite. In fact, for all sums $c_k^+(\tau)$, $k = 0,1,2$ one has

$$c_k^+ = \sum_{j>j_0}\{\ldots\} \asymp \begin{cases} \sum_{j>j_0} 2^{-(j-j_0)th\lambda/l}(j-j_0)^{n_k}, \ l\lambda > 0, & \text{if } \kappa \notin \Xi_2 \\ \sum_{j>j_0} 2^{(j-j_0)htI/\gamma}, \ I\gamma < 0, & \text{if } \kappa \in \Xi_2 \end{cases} \asymp 1, \tag{A.149}$$

where $n_1 = n_2 = \tilde{\sigma}/2l$, $n_0 = n_1 - 1$.

If $\kappa \in \Xi_1 \cup \Xi_2$, then for sums $c_k^-(\tau)$, $k = 0,1,2$ one has:

$$c_k^- = \sum_{j\in(-\infty,j_0]}\{\ldots\} \asymp \begin{cases} \sum\limits_{j\ge -j_0} 2^{-(j+j_0)h(a+rp+1)/p}, \ a+rp+1>0, & \text{if } k=1, \\ \sum\limits_{j\ge -j_0} 2^{-(j+j_0)t(a+sq+1)/q}, \ a+sq+1>0, & \text{if } k=2, \\ \sum\limits_{j\ge -j_0} 2^{-(j+j_0)(2a+1)}, \ 2a+1>0, & \text{if } k=0, \end{cases} \tag{A.150}$$

it is clear that $c_k^- \asymp 1$, $k = 0,1,2$, in these cases.

If $\kappa \in \Xi_4$, then $I > 0$, $\gamma_1 > 0$, $r > -1/4$, $s > -1/2q$, and for sums $c_k^-(\tau)$, $k = 0, 1, 2$, one has, for $k = 0, 1$,

$$d_k = 2h(4r+1)/\gamma_1 > 0, \quad d_2 = t(d_0/2+(4-h)(2sq+1)/\gamma_1)/q > 0, \tag{A.151}$$

$$c_k^- = \sum_{j\in(-\infty,j_0]}\{\ldots\} \asymp \begin{cases} \sum\limits_{j\ge -j_0} 2^{-(j+j_0)(htI/\gamma_1+d_1)}, & \text{if } k=1 \\ \sum\limits_{j\ge -j_0} 2^{-(j+j_0)(htI/\gamma_1+d_2)}, & \text{if } k=2 \\ \sum\limits_{j\ge -j_0} 2^{-(j+j_0)(htI/\gamma_1+d_0)}, & \text{if } k=0 \end{cases} \asymp 1. \tag{A.152}$$

If $\kappa \in \Xi_3$, then for all sums $c_k^-(\tau)$, $k = 0,1,2$, one has, by $I\gamma > 0$,

$$c_k^-(\tau) = \sum_{j\in(-\infty,j_0]}\{\ldots\} \asymp \sum_{j\ge -j_0} 2^{-(j+j_0)htI/\gamma} \asymp 1. \tag{A.153}$$

It is clear that we have in all cases

$$c_k^-(\tau) \sim \sum_{j\in[1,j_0]}\{\ldots\}, \quad k = 0,1,2. \tag{A.154}$$

These relations imply the existence of the solutions of (A.145)

$$n_\varepsilon = \tilde{n}_\varepsilon(1 + O(\tilde{n}_\varepsilon^{-\delta})), \quad h_{0,\varepsilon} = \tilde{h}_{0,\varepsilon}(1 + O(\tilde{n}_\varepsilon^{-\delta})),$$

and the relation

$$u_\varepsilon^2 = \tilde h_{0,\varepsilon}^2 \tilde n_\varepsilon (c_0(\tau, j_0) + O(\tilde n_\varepsilon^{-\delta}))$$

where $\tilde n_\varepsilon$ and $\tilde h_{0,\varepsilon}$ are defined by the relations

$$\rho_\varepsilon^h = \tilde h_{0,\varepsilon} \tilde n_\varepsilon^{rh+h/p} c_1^h(\tau, j_0), \quad R_\varepsilon^t = \tilde h_{0,\varepsilon} \tilde n_\varepsilon^{st+t/q} c_2^t(\tau, j_0).$$

Proofs are analogous to proofs for the power norms (see Lemma A.1).

Next, let $r = r_p^*$. In this case we have $p > 2$, $r_p^* = -1/2p$, $s > -1/2q$ for $\tau \in \Gamma_{\Delta_1}^-$ and $\kappa \in \Xi_{G_5}$. The second relation in (A.148) is of the same form. However, the sums $c_1(\tau, j_0)$, $c_0(\tau, j_0)$ diverge in (A.148), because the sums c_k^-, $k = 0, 1$, diverge in (A.150). The relation for ρ_ε^h, u_ε^2 in (A.145) could be rewritten in the form

$$\rho_\varepsilon^h = h_0^{h/p} 2^{j_0 h/2p} \left(\sum_{1 \le j \le j_0} 2^{(j-j_0)(h/2p)} \delta^{h/p}(2^{j-j_0}, \tau) z^h(2^{j-j_0}, \tau) + O(1) \right)$$

$$\sim h_0^{h/p} 2^{j_0 h/2p} j_0 \, z_p^h C^{h/p}(p, q, h), \tag{A.155}$$

$$u_\varepsilon^2 = h_0^2 2^{j_0} \left(\sum_{1 \le j \le j_0} 2^{j-j_0} \delta^2(2^{j-j_0}, \tau) \sinh^2 \frac{z^2(2^{j-j_0}, \tau)}{2} + O(1) \right)$$

$$\sim 2h_0^2 2^{j_0} j_0 \, C^2(p, q, h) \sinh^2(z_p^2/2), \tag{A.156}$$

which provides relations (6.53), (6.54), and (6.48).

A.4.4 *Solution of the Extreme Problem for* $\tau \in \Gamma_{\Delta_1}^+$

Propositions A.5, A.6 and Corollaries A.3, A.4 determine the solutions of equations (6.110), (6.111) for $\tau \in \Gamma_{\Delta_1}^+$ as functions of the unknown parameters z_0, $m = 2^{j_1}$ or h_0, $n = 2^{j_0}$, if $\zeta \ne 0$. These parameters should be determined from relations (6.108). Let us consider differently the cases $I > 0$, $I = 0$, $I < 0$.

Note that the relation $I > 0$ corresponds to the case when solutions with $h_j < 1$ have a main part in the sum for u_ε^2 (this holds for $\tau \in \Gamma_{\Delta_1}^-$); the relation $I < 0$ corresponds to the opposite case: solutions with $h_j = 1$ have an essential or main "mass" in the sum. This defines different types of asymptotics in these cases.

As above, we give a scheme of investigations and omit elementary calculations which one can easily work out using Propositions A.5, A.6 and Corollaries A.3, A.4.

Case $I > 0$

Note that $\zeta > 0$ and $\kappa \in \Xi_2 \cup \Xi_3 \cup \Xi_4$ in this case (see the Remark on page 399).

First, consider the case $r > r_p^*$ (note that this holds for $\kappa \in \Xi_3 \cup \Xi_4$ and for $\kappa \in \Xi_2$, $p \geq 4$). Assume that $j_0 \to \infty$, $h_0 \to 0$. Then $z_0 \to 0$. We can rewrite relations (6.108), (6.109) in the form

$$\begin{aligned}(\rho_\varepsilon)^h &= h_0^{h/p} n^{rh+h/p}(C_{1,\varepsilon}(\tau) + \tilde{d}_1(\tau)\, h_0^{htI/\Delta_1} D_{1,\varepsilon}(\tau)),\\ (R_\varepsilon)^t &= h_0^{t/q} n^{st+t/q}(C_{2,\varepsilon}(\tau) + \tilde{d}_2(\tau) h_0^{htI/\Delta_1} D_{2,\varepsilon}(\tau)), \qquad (A.157)\\ u_\varepsilon^2 &= h_0^2 n (C_{0,\varepsilon}(\tau) + \tilde{d}_0(\tau) h_0^{htI/\Delta_1} D_{0,\varepsilon}(\tau)).\end{aligned}$$

Denote

$$b_1 = Iht/l\Delta_1, \quad b_2 = -ht(\mu + 1/q - 1/p)/l\Delta_1.$$

For $\tilde{d}_l = \tilde{d}_l(\tau)$ we have

$$\tilde{d}_1 = (tp/hq)^{h/pl+b_1} B_0^{b_2}, \quad \tilde{d}_2 = tp\tilde{d}_1/hq, \quad \tilde{d}_0 = (tp/hq)^{2/l+b_1} B_0^{-1/l+b_2}.$$

The functions $C_{l,\varepsilon}(\tau)$, $l = 0, 1, 2$, are defined by the relations analogous to (A.146) where the sums are taken over $j \in \mathcal{J}_0$. The functions $D_{l,\varepsilon}(\tau)$, $l = 0, 1, 3$, are defined by the relations, for some $\delta > 0$,

$$\begin{aligned}D_{1,\varepsilon}(\tau) &= \sum_{j \in \mathcal{J}_1} y_j^{rh+h/p} \vartheta^h(y_j, \tau, z_0) \sim \sum_{j \in \mathcal{J}_1^*} y_j^{rh+h/p} \vartheta^h(y_j, \tau),\\ D_{2,\varepsilon}(\tau) &= \sum_{j \in \mathcal{J}_1} y_j^{st+t/q} \vartheta^t(y_j, \tau, z_0) \sim \sum_{j \in \mathcal{J}_1^*} y_j^{st+t/q} \vartheta^t(y_j, \tau), \qquad (A.158)\\ D_{0,\varepsilon}(\tau) &= 2^{-1} \sum_{j \in \mathcal{J}_1} y_j \vartheta^4(y_j, \tau, z_0)(1 + O(z_0^\delta)) \sim 2^{-1} \sum_{j \in \mathcal{J}_1^*} y_j \vartheta^4(y_j, \tau).\end{aligned}$$

The relation for $D_{0,\varepsilon}(\tau)$ corresponds to the asymptotics $2\sinh^2(z^2/2) = z^4(1 + O(z^2))/2$, as $z \to 0$. One can easily check the relations

$$\sum_{j \in \mathcal{J}_1^* \setminus \mathcal{J}_1} y_j^{(\dots)} \vartheta^{(\dots)}(y_j, \tau) = o(1)$$

which yield the equivalences in (A.158). Recall that the sets $\mathcal{J}_0^*$, $\mathcal{J}_1^*$ were defined on page 403.

Using the asymptotics of Propositions A.5, A.6 and of Corollaries A.3, A.4, the evaluations from Section 6.10 one can check

$$C_{l,\varepsilon}(\tau) \sim c_l^{a_l}(\tau, j_0), \quad l = 0, 1, 2,$$

where $a_0 = 1$, $a_1 = h$, $a_2 = t$, the values $c_l(\tau, j_0)$ are defined by (A.148) and the sums are finite by the constraints on τ.

In fact, the difference $\delta_{l,\varepsilon} = \delta_{l,1,\varepsilon} + \delta_{l,2,\varepsilon}$, $l = 0, 1, 2$, between $c_l(\tau)$ and $C_l(\tau)$ is of the rate

$$\delta_{l,1,\varepsilon} \sim \sum_{j \in \mathcal{J}_0^* \setminus \mathcal{J}_0} (\dots) = o(1),$$

if $\mathcal{J}_0^* \neq \mathcal{J}_0$, then the last relation follows from (A.154).

Let us evaluate $\delta_{l,2,\varepsilon}$. We get

$$\delta_{l,2,\varepsilon} \sim \sum_{j\in\mathcal{J}_1^*}(\ldots)$$

$$\asymp \begin{cases} x_\varepsilon^{\varsigma_1}\sum_{j\in\mathcal{J}_1^*}2^{-|(j-\bar{j}_1)\varsigma_1|}, & \text{if } \kappa\in\Xi_2 \text{ or } \kappa\in\Xi_3,\ \zeta>0,\\ x_\varepsilon^{\varsigma_2+d_l}\sum_{j<\bar{j}_1}2^{(j-\bar{j}_1)(\varsigma_2+d_l)}, & \text{if } \kappa\in\Xi_4,\end{cases}$$

where $\varsigma_1 = htI/\gamma$, $\varsigma_2 = htI/\gamma_1$ and the values $d_l > 0$, $l = 0,1,2$, are defined by (A.151). Using the following properties of the values x_ε, γ, or γ_1:

$$\begin{array}{ll} x_\varepsilon\to\infty,\ \gamma<0, & \text{if } \kappa\in\Xi_2,\\ x_\varepsilon\to 0,\ \gamma>0,\ \zeta>0, & \text{if } I>0,\ \kappa\in\Xi_3,\\ x_\varepsilon\to 0,\ \gamma_1>0, & \text{if } \kappa\in\Xi_4,\end{array}$$

one can see that $\delta_\varepsilon = o(1)$.

One can also check that $D_{l,\varepsilon} = O(1)$. In fact, if $\kappa\in\Xi_2$, then $\mathcal{J}_1^* = \mathcal{J}_1$, $t>h$ and by

$$\mu > 1/p - 1/q,\quad \Delta_1>0,\quad y_\varepsilon \sim 2^{\bar{j}_1-j_1}\asymp 1,\quad m = 2^{j_1}\asymp nh_0^{\gamma/\Delta_1}\to\infty,$$

then for $y_j = 2^{j-j_1}$ and $\varsigma_3 = (1/p-1/q-\mu)/(1/h-1/t) < 0$ one has:

$$D_{0,\varepsilon}\asymp\sum_{j>\bar{j}_1} y_j^{\varsigma_3-\Delta_1/(t-h)}\asymp 1,\quad D_{l,\varepsilon}\asymp\sum_{j>\bar{j}_1} y_j^{\varsigma_3}\asymp 1,\ l=1,2. \qquad \text{(A.159)}$$

Let $\kappa\in\Xi_3\cup\Xi_4$. If $m = 2^{j_1} = O(1)$, then $D_{l,\varepsilon} = O(1)$ by $y_\varepsilon = 2^{\bar{j}_1-j_1} = O(1)$. Recall that $\Delta_1>0$ for $\kappa\in\Xi_4$ and $\tilde{\Delta}_1>0$. Therefore, in the cases under consideration, we have

$$r > r_p = 1/4 - 1/p,\quad \Delta_1 > 0,\quad h<2.$$

Denoting

$$c_r = (4rh + 4h/p - h)/(4-h)$$

and noting that $c_r > 0$, we have, as $m\to\infty$,

$$D_{2,\varepsilon}\asymp\sum_{1\le j<\bar{j}_1} y_j^{c_r+\Delta_1/(4-h)},\quad D_{l,\varepsilon}\asymp\sum_{1\le j<\bar{j}_1} y_j^{c_r},\quad l=0,1. \qquad \text{(A.160)}$$

Moreover, for any $m = 2^{j_1}$, we have

$$\sum_{j\in\mathcal{J}_1^*\setminus\mathcal{J}_1} y_j^{(\ldots)}\vartheta^{(\ldots)}(y_j,\tau)\asymp\sum_{-\infty}^{0}2^{(j-j_1)A_l} = o(1),$$

where $A_l>0$ is the exponent which corresponds to the lth sum, $l=0,1,2$. These yield the second asymptotic equality in (A.158).

Recall that $y_j = 2^{j-j_1}$ and if $\kappa \notin \Xi_4$, then $y_\varepsilon = 2^{\tilde{j}_1 - j_1} \sim c(p,q,h,t)$, if $\kappa \in \Xi_4$, then $y_\varepsilon = 2^{\tilde{j}_1 - j_1} \to 0$. Therefore, we get, for $l = 0,1,2$,

$$D_{l,\varepsilon} \asymp 1, \text{ if } \kappa \in \Xi_3, \quad D_{l,\varepsilon} = o(1), \quad \text{if } \kappa \in \Xi_4. \tag{A.161}$$

These imply the relations analogous to (A.145)–(A.148).

Let $r = r_p^*$. Then $\kappa \in \Xi_2, p < 4, r = -1/2p$, and $\kappa \in \Xi_{G_5}$. Relations (A.157) hold. The quantities $D_{l,\varepsilon}(\tau)$, $l = 0,1,2$ satisfy (A.159). The quantities

$$C_{2,\varepsilon}(\tau) \sim c_2^t(\tau, j_0) = c_2^+(\tau, j_0) + c_2^-(\tau, j_0)$$

and $c_2^t(\tau, j_0)$, $c_2^+(\tau, j_0)$, $c_2^-(\tau, j_0)$ satisfy (A.148), (A.149), and (A.150). The relation for ρ_ε^h, u_ε^2 in (A.157) could be rewritten in the forms (A.155) and (A.156). These provide relation (6.53), (6.54), and (6.48).

Case $I < 0$

Note that $\kappa \in \Xi_2 \cup \Xi_3$ in this case (since $I \geq 0$ for $\kappa \in \Xi_4$) and if $\kappa \in \Xi_2$, then $r > r_p^*$. Let $r > r_p^*$ (this corresponds to $\kappa \in \Xi_{G_1}$), $j_1 \to \infty, z_0 \to 0$ and if $p \leq 2$, then $2^{-j_1 \tilde{r}/(4-h)} z_0 = o(1)$. Note that these assumptions hold under (6.58) and $2^{j_0} = B_0^{-1/\zeta} 2^{j_1} z_0^{-\gamma/\zeta} \to \infty$ for $\zeta > 0$. If $\kappa \in \Xi_3$, $\zeta < 0$, then $h_0 \to \infty$, $x_\varepsilon \to \infty$.

For $\zeta > 0$ one has

$$\begin{aligned} (\rho_\varepsilon)^h &= z_0^h 2^{j_1(rh+h/p)} (D_{1,\varepsilon}(\tau) + \tilde{c}_1 z_0^{-htI/\zeta} C_{1,\varepsilon}(\tau)), \\ (R_\varepsilon)^t &= z_0^t 2^{j_1(st+t/q)} (D_{2,\varepsilon}(\tau) + \tilde{c}_2 z_0^{-htI/\zeta} C_{2,\varepsilon}(\tau)), \\ u_\varepsilon^2 &= z_0^4 2^{j_1} (D_{0,\varepsilon}(\tau)(1 + O(z_0^\delta)) + \tilde{c}_0 z_0^{-htI/\zeta} C_{0,\varepsilon}(\tau)). \end{aligned} \tag{A.162}$$

Here, if $S = ht\lambda/l\zeta$, then

$$\tilde{c}_1 = (tp/hq)^{-h/pl} B_0^S, \quad \tilde{c}_2 = (tp/hq)^{-t/ql} B_0^S, \quad \tilde{c}_0 = (tp/hq)^{-2/l} B_0^{1/l+S}.$$

The functions $C_{l,\varepsilon}(\tau)$, $l = 0,1,3$, are defined by the relations analogous to (A.146), where the sums are considered over $j \in \mathcal{J}_0$; the functions $D_{l,\varepsilon}(\tau)$, $l = 0,1,3$, are defined in (A.158) and $D_{l,\varepsilon}(\tau) \asymp 1$, $l = 0,1,3$, according (A.159) or (A.161).

Denote $H = B_0^{\tilde{\sigma}/l\gamma}$ and $H_1 = tp/hq$. Remember that $\gamma < 0$ for $\kappa \in \Xi_2$ and $\gamma > 0$ for $\kappa \in \Xi_3$. One can show that

$$\begin{aligned} \tilde{C}_{1,\varepsilon}(\tau) &= \tilde{c}_1 z_0^{-htI/\zeta} C_{1,\varepsilon}(\tau) \sim H H_1^{-h/pl} B_2^{h/\gamma} B_3^{h/p} \sum_{j \in \tilde{\mathcal{J}}_0} 2^{(j-j_1)htI/\gamma} \\ &\sim H H_1^{-h/pl} B_2^{h/\gamma} B_3^{h/p} \sum_{j \in \mathcal{J}_0^*} 2^{(j-j_1)htI/\gamma}; \\ \tilde{C}_{2,\varepsilon}(\tau) &= \tilde{c}_2 z_0^{-htI/\zeta} C_{2,\varepsilon}(\tau) \sim H H_1^{-t/ql} B_2^{t/\gamma} B_3^{t/q} \sum_{j \in \tilde{\mathcal{J}}_0} 2^{(j-j_1)htI/\gamma}; \end{aligned} \tag{A.163}$$

$$\sim HH_1^{-t/ql}B_2^{t/\gamma}B_3^{t/q}\sum_{j\in\mathcal{J}_0^*}2^{(j-j_1)htI/\gamma}; \tag{A.164}$$

$$\begin{aligned}\tilde{C}_{0,\varepsilon}(\tau) &= \tilde{c}_0 z_0^{-htI/\zeta}C_{0,\varepsilon}(\tau) \sim HH_1^{-2/l}B_0^{1/l}B_2^{4/\gamma}B_3^2\sum_{j\in\tilde{\mathcal{J}}_0}2^{(j-j_1)(htI/\gamma)-1}\\ &\sim HH_1^{-2/l}B_0^{1/l}B_2^{4/\gamma}B_3^2\sum_{j\in\mathcal{J}_0^*}2^{(j-j_1)(htI/\gamma)-1},\end{aligned} \tag{A.165}$$

where

$$\tilde{\mathcal{J}}_0 = \begin{cases}\{j : (\tilde{j}_1+j_0)/2 < j \le \tilde{j}_1\}, & \text{if } \kappa\in\Xi_2,\\ \{j : \tilde{j}_1 \le j < (\tilde{j}_1+j_0)/2\}, & \text{if } \kappa\in\Xi_3.\end{cases}$$

The second relation is proved analogously to the proof for the case of the power norms. The third relation follows from the properties of geometric progressions.

In this case we have

$$\begin{aligned}\tilde{C}_{1,\varepsilon}(\tau) &= z_0^{-h}\sum_{j\in\mathcal{J}_0}2^{(j-j_1)(rh+h/p)}\varpi^h(y_j,\tau,z_0)\eta^{h/t}(y_j,\tau,z_0),\\ \tilde{C}_{2,\varepsilon}(\tau) &= z_0^{-t}\sum_{j\in\mathcal{J}_0}2^{(j-j_1)(st+t/q)}\varpi^t(y_j,\tau,z_0)\eta^{t/q}(y_j,\tau,z_0),\\ \tilde{C}_{0,\varepsilon}(\tau) &= \frac{z_0^{-4}}{2}\sum_{j\in\mathcal{J}_0}2^{j-j_1}\varpi^4(y_j,\tau,z_0)\eta^2(y_j,\tau,z_0)(1+O(z_0^\delta)).\end{aligned}$$

If $j\in\tilde{\mathcal{J}}_0$, we have (remember that $y_j = 2^{j-j_1}$, $y_\varepsilon = 2^{\tilde{j}_1-j_1}$)

$$\begin{aligned}z_0^{-h}y_j^{rh+h/p}\varpi^h(y_j,\tau,z_0)\eta^{h/t}(y_j,\tau,z_0) &\sim HH_1^{-h/pl}B_2^{h/\gamma}B_3^{h/p}y_j^{htI/\gamma},\\ z_0^{-t}y_j^{st+t/q}\varpi^t(y_j,\tau,z_0)\eta^{t/q}(y_j,\tau,z_0) &\sim HH_1^{-t/ql}B_2^{t/\gamma}B_3^{t/q}y_j^{htI/\gamma},\\ z_0^{-4}y_j^{j-j_1}\varpi^4(y_j,\tau,z_0)\eta^2(y_j,\tau,z_0) &\sim HH_1^{-2/l}B_0^{1/l}B_2^{4/\gamma}B_3^2y_j^{htI/\gamma},\end{aligned}$$

and

$$\varpi(y_\varepsilon,\tau,z_0) = \varpi_1(y_\varepsilon,\tau,z_0) = z_0\vartheta(y_\varepsilon,\tau,z_0), \quad \eta(y_\varepsilon,\tau,z_0) = 1. \tag{A.166}$$

Let $\kappa\in\Xi_3$, $\zeta\le 0$. Then

$$J_0 = \{j : 2^j \ge 2^{j_1}y_\varepsilon\}, \quad y_\varepsilon \sim c(p,q,h,t).$$

Using Proposition A.5 (B) and Corollary A.3 we can rewrite relations (6.108), (6.109) in the form

$$\begin{aligned}(\rho_\varepsilon)^h &= z_0^h 2^{j_1(rh+h/p)}(\tilde{C}_{1,\varepsilon}(\tau)+D_{1,\varepsilon}(\tau)),\\ (R_\varepsilon)^t &= z_0^t 2^{j_1(st+t/q)}(\tilde{C}_{2,\varepsilon}(\tau)+D_{2,\varepsilon}(\tau)),\\ u_\varepsilon^2 &= z_0^4 2^{j_1}(\tilde{C}_{0,\varepsilon}(\tau)+D_{0,\varepsilon}(\tau)),\end{aligned} \tag{A.167}$$

where

$$\tilde{C}_{1,\varepsilon}(\tau) = \sum_{j\in\mathcal{J}_0} 2^{(j-j_1)(rh+h/p)}\vartheta_0^h(y_j,\tau,z_0)\eta^{h/t}(y_j,\tau,z_0),$$

$$\tilde{C}_{2,\varepsilon}(\tau) = \sum_{j\in\mathcal{J}_0} 2^{(j-j_1)(st+t/q)}\vartheta_0^t(y_j,\tau,z_0)\eta^{t/q}(y_j,\tau,z_0),$$

$$\tilde{C}_{0,\varepsilon}(\tau) = 2^{-1}\sum_{j\in\mathcal{J}_0} 2^{j-j_1}\vartheta_0^4(y_j,\tau,z_0)\eta^2(y_j,\tau,z_0)(1+O(z_0^\delta));$$

the last relation corresponds to the asymptotics

$$2\sinh^2(z^2/2) = z^4(1+O(z^2))/2, \quad \text{as } z\to 0.$$

The values $D_{l,\varepsilon}(\tau)$ are defined in (A.158). The sharp and rate asymptotics of these values are presented in (A.160), (A.161).

The values $\tilde{C}_{l,\varepsilon}$, $l=0,1,2$, satisfy (A.163)–(A.165). In fact,

$$\tilde{C}_{1,\varepsilon} = \sum_{j\in\mathcal{J}_0} 2^{(j-j_1)(rh+h/p)}\vartheta_0^h(y_j,\tau,z_0)\eta^{h/p}(y_j,\tau,z_0)$$

$$\sim \sum_{j\in\mathcal{J}_0} y_j^{(rh+h/p)} y_j^{\zeta h/\gamma} c_\tau^h y_j^{-h\Delta_1/p\gamma} c_h^{h/p}$$

$$\sim c_\tau^h c_h^{h/p}\sum_{j\geq\tilde{j}_1} 2^{(j-j_1)htI/\gamma} \sim HH_1^{-h/pl}B_2^{h/\gamma}B_3^{h/p}\sum_{j\in\mathcal{J}_0^*} 2^{(j-j_1)htI/\gamma}.$$

Relations (A.164), (A.165) can be proved in a similar way.

The relations (A.162) or (A.167) jointly with (A.159), (A.160), (A.163)–(A.165) imply the asymptotics (6.18), (6.19) and the functions $c_l(\tau,j_1)$, $l=0,1,2$, are determined by relation (6.33):

$$c_l^{a_l}(\tau,j_1) = \sum_{-\infty}^{\infty}\omega_{\tau,l}(j-j_1), \quad l=0,1,2; \quad a_0=1, \quad a_1=h, \quad a_2=t, \tag{A.168}$$

where

$$\omega_{\tau,1}(v) = \begin{cases} HH_1^{-h/pl}B_2^{h/\gamma}B_3^{h/p}2^{vhtI/\gamma}, & \text{if } v+j_1\in\mathcal{J}_0^*, \\ 2^{v(rh-h/p)}\vartheta^h(2^v,\tau), & \text{if } v+j_1\in\mathcal{J}_1^*, \end{cases}$$

$$\omega_{\tau,2}(v) = \begin{cases} HH_1^{-t/ql}B_2^{t/\gamma}B_3^{t/q}2^{vhtI/\gamma}, & \text{if } v+j_1\in\mathcal{J}_0^*, \\ 2^{v(st+t/q)}\vartheta^t(2^v,\tau), & \text{if } v+j_1\in\mathcal{J}_1^*, \end{cases}$$

$$\omega_{\tau,0}(v) = \begin{cases} HH_1^{-2/l}B_0^{1/l}B_2^{4/\gamma}B_3^2 2^{vhtI/\gamma}, & \text{if } v+j_1\in\mathcal{J}_0^*, \\ 2^v\vartheta^4(2^v,\tau))/2, & \text{if } v+j_1\in\mathcal{J}_1^*. \end{cases}$$

The functions $c_l^{a_l}(\tau,j_1)$, $l=0,1,2$, are 1-periodical and are continuous in j_1 by (A.166).

Let $r=r_p^*$. Then $\kappa\in\Xi_3$, $r_p^*=1/4-1/p$, and $\kappa\in\Xi_{G_4}$. In this case the quantities $\tilde{C}_{l,\varepsilon}(\tau)$, $l=0,1,2$, satisfy (A.163)–(A.165). The values $D_{l,\varepsilon}(\tau)$ are defined in (A.158). The sharp and rate asymptotics of $D_{2,\varepsilon}(\tau)$ are presented in (A.160), (A.161). For $D_{l,\varepsilon}(\tau)$, $l=0,1$, by $j_1 \sim \tilde{j}_1$, we have

$$D_{1,\varepsilon}(\tau) = \sum_{j=1}^{j<\tilde{j}_1} 2^{(j-j_1)(rh+hp)}\vartheta^h(y_j,\tau)(1+O(z_0^\delta)) \sim j_1(h/2)^{h/(4-h)},$$

$$D_{0,\varepsilon}(\tau) = 2^{-1}\sum_{j=1}^{j<\tilde{j}_1} 2^{(j-j_1)}\vartheta^4(y_j,\tau)(1+O(z_0^\delta)) \sim 2^{-1}j_1(h/2)^{4/(4-h)}.$$

Therefore

$$\begin{aligned}
\rho_\varepsilon^h &\sim z_0^h 2^{j_1h/4} j_1(h/2)^{h/(4-h)},\\
R_\varepsilon^t &\sim c_2^t(\tau,j_1) z_0^t 2^{j_1(st+t/q)},\\
u_\varepsilon^2 &\sim 2^{-1}z_0^4 2^{j_1} j_1(h/2)^{4/(4-h)},
\end{aligned}$$

where the quantity $c_2^t(\tau,j_1)$ is defined by (A.168).

Case $I=0$

In this case we have either

$$\kappa\in\Xi_{G_3}\cap(\Xi_2\cup\Xi_3),\quad \lambda>0,\quad \zeta>0,\quad r>r_p,\quad m\to\infty,\quad n\to\infty,$$

or

$$\kappa\in\Xi_{G_5}\cap\Xi_4,\ \lambda>0,\ \zeta>0,\ r=r_p=-1/4,\ s>-1/2q,\ m\to\infty,\ n\to\infty,$$

(see the Remark on page 399).

Let $r>r_p$. Then relations (A.157), (A.158) hold with $D_{l,\varepsilon}=O(1)$ for $l=0,1,2$; note that we have evaluated the quantities $D_{l,\varepsilon}$ in relations (A.159), (A.160). We also have

$$\begin{aligned}
C_{1,\varepsilon}(\tau) &\sim B_2^{h/\gamma}B_3^{h/p}\#(\mathcal{J}^*)\\
&\sim B_2^{h/\gamma}B_3^{h/p}|\gamma|\log_2 h_0^{-1}/\Delta_1 = c_1^h(\tau), && \text{(A.169)}\\
C_{2,\varepsilon}(\tau) &\sim B_2^{t/\gamma}B_3^{t/q}\#(\mathcal{J}^*)\\
&\sim B_2^{t/\gamma}B_3^{t/q}|\gamma|\log_2 h_0^{-1}/\Delta_1 = c_2^t(\tau), && \text{(A.170)}\\
C_{0,\varepsilon}(\tau) &\sim 2^{-1}B_2^{4/\gamma}B_3^2\#(\mathcal{J}^*)\\
&\sim 2^{-1}B_2^{4/\gamma}B_3^2|\gamma|\log_2 h_0^{-1}/\Delta_1 = c_0(\tau), && \text{(A.171)}
\end{aligned}$$

where

$$\mathcal{J}^* = \begin{cases}[\tilde{j}_1(\tau), j_0(\tau)], & \text{if } \kappa\in\Xi_3,\\ [j_0(\tau),\tilde{j}_1(\tau)], & \text{if } \kappa\in\Xi_2.\end{cases}$$

The proof is analogous to that for the case of the power norm with $\Delta>0$, $I<0$; we take $x_{\varepsilon,1}=2^{\hat{j}_{\varepsilon,1}-j_0}$, where $\hat{j}_{\varepsilon,1}=j_0+\log(\tilde{j}_1-j_0)$ for $\kappa\in\Xi_2$ and $\hat{j}_{\varepsilon,1}=j_0-\log(j_0-\tilde{j}_1)$ for $\kappa\in\Xi_3$.

These relations imply the asymptotics (6.49), (6.50).

Let $r = r_p = -1/4$. Then $\kappa \in \Xi_4$, $x_\varepsilon \to 0$, $y_\varepsilon \to 0$, $\tilde{j}_1 \to \infty$, and

$$\begin{aligned}(\rho_\varepsilon)^h &\sim h_0^{h/2} 2^{j_0 h/4} (B_4^{h/\gamma_1} B_5^{h/2} (j_0 - \tilde{j}_1) + \tilde{d}_1 (h/2)^{h/(4-h)} \tilde{j}_1) \\ &\sim h_0^{h/2} 2^{j_0 h/4} \log_2 h_0^{-1} \tilde{\Delta}_1^{-1} (\gamma_1 B_4^{h/\gamma_1} B_5^{h/2} + \tilde{d}_1 4\zeta (h/2)^{h/(4-h)}), \\ u_\varepsilon^2 &\sim 2^{-1} h_0^2 2^{j_0} (B_4^{4/\gamma_1} B_5^2 (j_0 - \tilde{j}_1) + \tilde{d}_0 (h/2)^{4/(4-h)} \tilde{j}_1) \\ &\sim h_0^2 2^{j_0} \log_2 h_0^{-1} (2\tilde{\Delta}_1)^{-1} (\gamma_1 B_4^{4/\gamma_1} B_5^2 + \tilde{d}_0 4\zeta (h/2)^{4/(4-h)}).\end{aligned}$$

Finally, observe that, by (A.161), one has

$$\tilde{d}_2 h_0^{t/q} 2^{j_0(st+t/q)} D_{2,\varepsilon}(\tau) = z_0^t 2^{j_1(st+t/q)} D_{2,\varepsilon}(\tau) = o(h_0^{t/q} 2^{j_0(st+t/q)}).$$

We get

$$(R_\varepsilon)^t \sim h_0^{t/q} 2^{j_0(st+t/q)} c_2^t(\tau, j_0),$$

where $c_2^t(\tau, j_0)$ is defined by (A.148). These relations imply the asymptotics (6.53), (6.54).

Auxiliary Proposition

To verify the assumptions on the solution of the extreme problem, which are needed to apply the theory from previous chapters, it is convenient to formulate the following proposition, which follows from the results presented above.

Denote

$$\begin{aligned}&\varrho_0 = 1 + 2a, \quad \varrho_1 = (rp + 1 + a)h/p, \quad \varrho_2 = (sq + 1 + a)t/q, \\ &v_1 = c_r, \quad v_2 = c_r + \Delta_1/(4-h), \quad n_0 = \tilde{\sigma}/2l - 1, \quad n_1 = \tilde{\sigma}/2l, \\ &\varsigma_1 = htI/\gamma, \quad \varsigma_2 = htI/\gamma_1, \quad \varsigma_3 = (1/p - 1/q - \mu)/(1/h - 1/t),\end{aligned}$$

recall that $c_r = (4rh + 4h/p - h)/(4-h)$ and

$$d_0 = d_1 = 2h(4r+1)/\gamma_1, \quad d_2 = t(d_1/2 + (4-h)(2sq+1)/\gamma_1)/q.$$

Below we consider sums of the form

$$S_{1,k} = \sum_{1 \le j \le j_0} 2^{(j-j_0)\varrho_k}, \ k = 0,1,2; \quad S_{2,k} = \sum_{1 \le j \le \tilde{j}_1} 2^{(j-j_1)v_k}, \ k = 1,2;$$

$$S_{3,k} = \sum_{j > j_0} 2^{-(j-j_0)ht\lambda/l} (1 + j - j_0)^{n_k}, \ k = 0,1; \quad S_4 = \sum_{j > \tilde{j}_1} 2^{(j-j_1)\varsigma_3};$$

$$S_{5,k} = \sum_{j_0 < j \le \tilde{j}_1} 2^{(j-j_k)\varsigma_1}, \ k = 0,1; \qquad S_6 = \sum_{j > j_0} 2^{(j-j_0)\varsigma_1};$$

$$S_{7,k} = \sum_{\tilde{j}_1 < j \le j_0} 2^{(j-j_k)\varsigma_1}, \ k = 0,1; \qquad S_8 = \sum_{1 \le j \le j_0} 2^{(j-j_0)\varsigma_1};$$

$$S_9 = \sum_{j > \tilde{j}_1} 2^{(j-j_1)\varsigma_1}; \qquad S_{10,k} = \sum_{1 \le j \le j_0} 2^{(j-j_0)(\varsigma_2 + d_k)}, \ k = 1,2;$$

$$S_{11,k} = \sum_{\tilde{j}_1 < j \le j_0} 2^{(j-j_0)(\varsigma_2+d_k)},\ k=1,2;\ S_{12} = \sum_{j>\tilde{j}_1} 2^{(j-j_1)(\varsigma_3-\Delta_1/(t-h))}.$$

Proposition A.8
I. *Let* $\kappa \in \Xi_1$. *Then we have*

$$u_\varepsilon^2 \asymp h_0^2\, 2^{j_0}(S_{1,0}+S_{3,0}),$$
$$\rho_\varepsilon^h \asymp h_0^{h/p} 2^{j_0(rh+h/p)}(S_{1,1}+S_{3,1}),$$
$$R_\varepsilon^t \asymp h_0^{t/q} 2^{j_0(st+t/q)}(S_{1,2}+S_{3,1}).$$

If $r > r_p^*$, *then* $S \asymp 1$ *for all sums here. If* $r = r_p^*$, *then these hold for all sums, except the sums* $S_{1,0} \asymp S_{1,1} \asymp j_0$.
II. *Let* $\kappa \in \Xi_2$.
(a) *If* $\Delta_1 \le 0$, *then*

$$u_\varepsilon^2 \asymp h_0^2\, 2^{j_0}(S_{1,0}+S_6),$$
$$\rho_\varepsilon^h \asymp h_0^{h/p} 2^{j_0(rh+h/p)}(S_{1,1}+S_6),$$
$$R_\varepsilon^t \asymp h_0^{t/q} 2^{j_0(st+t/q)}(S_{1,2}+S_6).$$

If $r > r_p^*$, *then* $S \asymp 1$ *for all sums here. If* $r = r_p^*$, *then these hold for all sums, except the sums* $S_{1,0} \asymp S_{1,1} \asymp j_0$.
(b) *If* $\Delta_1 > 0,\ I > 0$, *then*

$$u_\varepsilon^2 \asymp h_0^2\, 2^{j_0}(S_{1,0}+S_{5,0}+h_0^{htI/\Delta_1} S_{12}),$$
$$\rho_\varepsilon^h \asymp h_0^{h/p} 2^{j_0(rh+h/p)}(S_{1,1}+S_{5,0}+h_0^{htI/\Delta_1} S_4),$$
$$R_\varepsilon^t \asymp h_0^{t/q} 2^{j_0(st+t/q)}(S_{1,2}+S_{5,0}+h_0^{htI/\Delta_1} S_4).$$

If $I > 0,\ r > r_p^*$, *then* $S \asymp 1$ *for all sums here. If* $I = 0$, *then these hold for all sums, except the sum* $S_{5,0} \asymp \tilde{j}_1 - j_0$. *If* $p < 4,\ r = r_p^* = -1/2p$, *then* $S \asymp 1$ *for all sums, except the sums* $S_{1,0} \asymp S_{1,1} \asymp j_0$.
(c) *If* $\Delta_1 > 0,\ I < 0$, *then*

$$u_\varepsilon^2 \asymp z_0^4\, 2^{j_1}(z_0^{-htI/\zeta} S_{1,0}+S_{5,1}+S_{12}),$$
$$\rho_\varepsilon^h \asymp z_0^h 2^{j_1(rh+h/p)}(z_0^{-htI/\zeta} S_{1,1}+S_{5,1}+S_4),$$
$$R_\varepsilon^t \asymp z_0^t 2^{j_1(st+t/q)}(z_0^{-htI/\zeta} S_{1,2}+S_{5,1}+S_4).$$

All sums are of the rate $S \asymp 1$ *here.*
III. *Let* $\kappa \in \Xi_3$.
(a) *If* $\Delta_1 \le 0$, *then*

$$u_\varepsilon^2 \asymp h_0^2\, 2^{j_0}(S_8+S_{3,0}),$$
$$\rho_\varepsilon^h \asymp h_0^{h/p} 2^{j_0(rh+h/p)}(S_8+S_{3,1}),$$
$$R_\varepsilon^t \asymp h_0^{t/q} 2^{j_0(st+t/q)}(S_8+S_{3,1}).$$

All sums are of the rate $S \asymp 1$ *here.*

(b) *If* $\Delta_1 > 0,\ I \geq 0$, *then*

$$\begin{aligned} u_\varepsilon^2 &\asymp h_0^2\, 2^{j_0}(h_0^{htI/\Delta_1} S_{2,1} + S_{7,0} + S_{3,0}), \\ \rho_\varepsilon^h &\asymp h_0^{h/p} 2^{j_0(rh+h/p)}(h_0^{htI/\Delta_1} S_{2,1} + S_{7,0} + S_{3,1}), \\ R_\varepsilon^t &\asymp h_0^{t/q} 2^{j_0(st+t/q)}(h_0^{htI/\Delta_1} S_{2,2} + S_{7,0} + S_{3,1}). \end{aligned}$$

If $I > 0$, *then* $S \asymp 1$ *for all sums here. If* $I = 0$, *then these hold for all sums, except the sum* $S_{7,0} \asymp j_0 - \tilde{j}_1$.

(c) *If* $\Delta_1 > 0,\ I < 0,\ \zeta > 0$, *then*

$$\begin{aligned} u_\varepsilon^2 &\asymp z_0^4\, 2^{j_1}(S_{2,1} + S_{7,1} + z_0^{-htI/\zeta} S_{3,0}), \\ \rho_\varepsilon^h &\asymp z_0^h 2^{j_1(rh+h/p)}(S_{2,1} + S_{7,1} + z_0^{-htI/\zeta} S_{3,1}), \\ R_\varepsilon^t &\asymp z_0^t 2^{j_1(st+t/q)}(S_{2,2} + S_{7,1} + z_0^{-htI/\zeta} S_{3,1}). \end{aligned}$$

If $r > r_p^*$, *then* $S \asymp 1$ *for all sums here. If* $r = r_p^*$, *then these hold for all sums, except the sum* $S_{2,1} \asymp \tilde{j}_1$.

(d) *If* $\zeta \leq 0$, *then* $\Delta_1 > 0,\ I < 0$, *and*

$$\begin{aligned} u_\varepsilon^2 &\asymp z_0^4\, 2^{j_1}(S_{2,1} + S_9), \\ \rho_\varepsilon^h &\asymp z_0^h 2^{j_1(rh+h/p)}(S_{2,1} + S_9), \\ R_\varepsilon^t &\asymp z_0^t 2^{j_1(st+t/q)}(S_{2,2} + S_9). \end{aligned}$$

If $r > r_p^*$, *then* $S \asymp 1$ *for all sums here. If* $r = r_p^*$, *then these hold for all sums, except the sum* $S_{2,1} \asymp \tilde{j}_1$.

IV. *Let* $\kappa \in \Xi_4$.

(a) *If* $\Delta_1 \leq 0$, *then*

$$\begin{aligned} u_\varepsilon^2 &\asymp h_0^2\, 2^{j_0}(S_{10,1} + S_{3,0}), \\ \rho_\varepsilon^h &\asymp h_0^{h/p} 2^{j_0(rh+h/p)}(S_{10,1} + S_{3,1}), \\ R_\varepsilon^t &\asymp h_0^{t/q} 2^{j_0(st+t/q)}(S_{10,2} + S_{3,1}). \end{aligned}$$

All sums are of the rate $S \asymp 1$ *here.*

(b) *If* $\Delta_1 > 0$, *then* $I \geq 0$ *and*

$$\begin{aligned} u_\varepsilon^2 &\asymp h_0^2\, 2^{j_0}(h_0^{htI/\Delta_1} S_{2,1} + S_{11,1} + S_{3,0}), \\ \rho_\varepsilon^h &\asymp h_0^{h/p} 2^{j_0(rh+h/p)}(h_0^{htI/\Delta_1} S_{2,1} + S_{11,1} + S_{3,1}), \\ R_\varepsilon^t &\asymp h_0^{t/q} 2^{j_0(st+t/q)}(h_0^{htI/\Delta_1} S_{2,2} + S_{11,2} + S_{3,1}). \end{aligned}$$

If $I > 0$, *then* $S \asymp 1$ *for all sums here, except the sums* $S_{2,1} = o(1),\ S_{2,2} = o(1)$. *If* $I = 0$, *then* $r = r_p^*$ *and* $S \asymp 1$ *for all sums here, except the sums* $S_{2,1} \asymp \tilde{j}_1,\ S_{11,1} \asymp j_0 - \tilde{j}_1$.

A.4.5 Upper Bounds

We need to check the relations (5.11), i.e., to verify that there exist $\delta_0 > 0$, $\eta_0 > 0$, $\tau \in (0,1)$, such that

$$Q_{\varepsilon,j} \le T_{\varepsilon,j} - \delta_0 z_{\varepsilon,j} + o(1), \quad Q_{\varepsilon,j} \le \tau T_{\varepsilon,j}, \quad \text{if } z_{\varepsilon,j} \le \eta_0 T_{\varepsilon,j}. \tag{A.172}$$

for the thresholds $Q_{\varepsilon,j}$ determined by (6.32),

$$Q_{\varepsilon,j} = \sqrt{2(j\log 2 + c\log(j+) + 2\log\log R_\varepsilon)}, \quad c > 1/2.$$

It is clear that (5.2) holds

$$\sum_{i,j} \Phi(-Q_{\varepsilon,j}) \asymp \sum_j 2^j \exp(-Q_{\varepsilon,j}^2/2)/Q_{\varepsilon,j} \le (\log\log R_\varepsilon)^{1/2}/\log R_\varepsilon \to 0.$$

Recall (see (5.6)) that, as $z_{\varepsilon,j} \to \infty$,

$$T_{\varepsilon,j} \sim \frac{z_{\varepsilon,j}}{2} + \frac{\log(h_{\varepsilon,j}^{-1}+1)}{z_{\varepsilon,j}} + o(z_{\varepsilon,j}^{-1}).$$

We also need to check assumptions A.1, A.3 or A.4, A.5.

For the Besov norm case assumption A.3 is of the form: there exists $\delta > 0$ such that

$$\sum_j 2^j u_{\varepsilon,j}^2 \exp(\delta z_{\varepsilon,j}^2) = O(u_\varepsilon^2), \quad u_{\varepsilon,j}^2 \asymp h_{\varepsilon,j}^2 \sinh^2(z_{\varepsilon,j}^2/2).$$

If $p < q$ or $p > q$, $\zeta \le 0$, then assumption A.3 follows from A.4 by $\sup_j z_{\varepsilon,j} = O(1)$. In other cases, one has $\lambda > 0$ and, if $j \le j_0$, then $z_j = O(1)$. However, if $x_j = 2^{j-j_0} \to \infty$, then $z_j \to \infty$. Let $\kappa \in \Xi_1 \cup \Xi_4$ or $\kappa \in \Xi_3$, $\zeta > 0$. Then assumption A.3 follows from Proposition A.8. In fact, using the relations for u_ε^2 from Proposition A.8, we see that the sum $S_{3,0}$ corresponding to "large" $z_j \to \infty$ arises for $\lambda > 0$ only. It follows from the form of the items in $S_{3,0}$ that we can add the factor $2^{\delta z_{\varepsilon,j}^2}$ of the form

$$\exp(\delta z^2) \sim 2^{(j-j_0)\zeta\delta/l}, \quad \zeta l > 0, \quad j > j_0,$$

with $\delta > 0$ small enough in each item without a loss of convergence. All other sums $S_{i,k}$ with $i \ne 3$ or S_i corresponding to the index sets with $z_j = O(1)$ uniformly over these index sets, and for $0 < \delta_0 = \zeta\delta/l < ht\lambda/l$ we have

$$\sum_j 2^j u_{\varepsilon,j}^2 \exp(\delta z_{\varepsilon,j}^2) \asymp \sum_{j\le j_0} 2^j u_{\varepsilon,j}^2 + 2^{j_0} h_0^2 \sum_{j>j_0} 2^{-(j-j_0)(ht\lambda/l-\delta_0)} = O(u_\varepsilon^2).$$

This yields A.3.

Thus for any $\tau \in \Gamma_G$ we have verified either assumption A.3 or A.4.

Below we assume $u_\varepsilon \asymp 1$ to simplify calculations; we can extend the results to the case when assumptions (6.58) are fulfilled.

Verification of assumption A.1

Assumption A.1 is of the form

$$\sup_j \|\pi_{\varepsilon,j}\|^2 = 2h_j^2 \sinh^2 \frac{z_j^2}{2} = o(1). \tag{A.173}$$

To control this relation recall that

$$u_\varepsilon^2 = \sum_{j=1}^{\infty} 2^j \|\pi_{\varepsilon,j}\|^2 \asymp 1$$

and we need to verify relation (A.173) for $j = O(1)$ only.

All the rate relations that we use below are presented in Proposition A.7. The required relations between the functions I, Δ_1, ζ, and so on, are verified according to the Remark on page 399.

(a) Let $\kappa \in \Xi_1$ or $\kappa \in \Xi_2$, $I \geq 0$. In this case, one has $j_0 \to \infty$, $h_0 \to 0$. Let $r > r_p^*$, $I > 0$. Then $u_\varepsilon^2 \asymp h_0^2 2^{j_0} \asymp 1$. If $j < j_0$, then

$$h_j \asymp h_0 2^{(j-j_0)a}, \quad z_j \asymp 1, \tag{A.174}$$

where $a = \tilde{r}/(2 - h/p)$, $a > -1/2$. Therefore,

$$\|\pi_{\varepsilon,j}\|^2 \asymp h_0^2\, 2^{2(j-j_0)a} \asymp 2^{(j-j_0)(2a+1)} 2^{-j} \leq \max(2^{-j_0(2a+1)}, 2^{-j_0}) = o(1).$$

Let $I = 0$. Then $u_\varepsilon^2 \asymp h_0^2 2^{j_0} \log h_0^{-1}$ and, by $r > r_p^*$, we have $a > -1/2$. Therefore, analogously to the above, we get $\|\pi_{\varepsilon,j}\|^2 = o(1)$.

Let $r = r_p^*$. We have

$$u_\varepsilon^2 \asymp h_0^2 2^{j_0} j_0, \quad a = -1/2.$$

If $j < j_0$, then relations (A.174) hold and we have

$$\|\pi_{\varepsilon,j}\|^2 \asymp h_0^2\, 2^{-(j-j_0)} \asymp 2^{-j}/j_0 = o(1).$$

(b) Let $\kappa \in \Xi_2$, $I < 0$. In this case we have $j_1 \to \infty$, $z_0 \to 0$, $\Delta_1 > 0$, $\zeta > 0$, $s - r > 1/p - 1/q$, $r > 1/4 - 1/p$, $t > q > p > h$, and

$$u_\varepsilon^2 \asymp z_0^4\, 2^{j_1}, \quad h_0 \asymp z_0^{\Delta_1/\zeta}, \quad 2^{j_0} \asymp z_0^{-(4+\gamma/\zeta)},$$

where

$$4 + \gamma/\zeta = (htI + 2\Delta_1)/\zeta.$$

Hence, we have

$$h_0 \to 0, \quad 2^{-j_0} \asymp h_0^{(4\zeta+\gamma)/\Delta_1}.$$

Thus, by $s > r + 1/p - 1/q$, we get

$$4\zeta + \gamma > (4r + 4/p - 1)(t - h + h(t/q - 1) + t(1 - h/p)) > 0.$$

Therefore $j_0 \to \infty$ and

$$h_0^2 2^{j_0} \asymp z_0^{-htI/\zeta} = o(1).$$

If $j < j_0$, then the quantities z_j, h_j satisfy (A.174) and we have

$$\|\pi_{\varepsilon,j}\|^2 \asymp h_0^2\, 2^{2(j-j_0)a} \asymp 2^{(j-j_0)(2a+1)} 2^{-j} o(1) = o(1).$$

Therefore we get that relation (A.173) is fulfilled in this case.

(c) Let $\kappa \in \Xi_3$, $\zeta > 0$. In this case, we have $\gamma > 0$, $h < p < 2$. If $I > 0$, then

$$r > 1/4 - 1/p, \quad h_0 \to 0, \quad j_0 \to \infty, \quad u_\varepsilon^2 \asymp h_0^2\, 2^{j_0}.$$

Let $\Delta_1 \le 0$. Then, for $j < j_0$, we have

$$z_j \asymp 2^{(j-j_0)\zeta/\gamma}, \quad h_j \asymp h_0 2^{-(j-j_0)\Delta_1/\gamma}. \tag{A.175}$$

Therefore

$$\|\pi_{\varepsilon,j}\|^2 \asymp 2^{(j-j_0)htI/\gamma} 2^{-j} \le \max(2^{-j_0 htI/\gamma}, 2^{-j_0}) = o(1).$$

Thus, in this case, we have

$$\sup_j \|\pi_{\varepsilon,j}\|^2 = o(1).$$

If $I > 0$, $\Delta_1 > 0$, then

$$u_\varepsilon^2 \asymp h_0^2\, 2^{j_0}, \quad z_0 \to 0.$$

For $\tilde{j}_1 \le j < j_0$ the quantities z_j, h_j satisfy (A.175) and we get, analogously to the case $\Delta_1 < 0$, that

$$\sup_{\tilde{j}_1 \le j} \|\pi_{\varepsilon,j}\|^2 = o(1).$$

Let $j < \tilde{j}_1$. Then the quantities z_j, h_j satisfy

$$z_j \asymp z_0 2^{(j-\tilde{j}_1)\tilde{r}/(4-h)}, \quad h_j = 1. \tag{A.176}$$

Let $\tilde{j}_1 = O(1)$. Then

$$\sup_{j \le \tilde{j}_1} \|\pi_{\varepsilon,j}\|^2 \asymp z_0^4 = o(1).$$

Let $\tilde{j}_1 \to \infty$. Note the relations $z_0^4\, 2^{\tilde{j}_1} = o(1)$ and $y_\varepsilon = 2^{\tilde{j}_1 - j_1} \sim c(p,q,h,t)$ (see (A.128)). Using these relations we have, for $j \le \tilde{j}_1$,

$$\|\pi_{\varepsilon,j}\|^2 \asymp z_0^4 2^{(j-j_1)4\tilde{r}/(4-h)} \asymp 2^{(j-j_1)4h(r+1/p-1/4)/(4-h)} 2^{-j} o(1) = o(1).$$

Therefore, in this case, we have

$$\sup_j \|\pi_{\varepsilon,j}\|^2 = o(1).$$

If $I \le 0$, then $\Delta_1 > 0$, $\tilde{j}_1 \to \infty$, $z_0 \to 0$ and for $I < 0$, $r > r_p^*$ we have $u_\varepsilon^2 \asymp z_0^4 2^{j_1}$. If either $r = r_p^*$ or $I = 0$, then $u_\varepsilon^2 \asymp z_0^4 2^{j_1} \log z_0^{-1}$. If $j \le \tilde{j}_1$, then the values z_j, h_j satisfy (A.176) and we have

$$\|\pi_{\varepsilon,j}\|^2 \asymp z_0^4 2^{(j-j_1)4\tilde{r}/(4-h)} = O(2^{(j-j_1)4h(r+1/p-1/4)/(4-h)} 2^{-j}) = o(1).$$

Therefore we obtain relation (A.173) for point (c).

(d) Let $\kappa \in \Xi_3$, $\zeta \le 0$. In this case, we have $I < 0$, $\tilde{j}_1 \to \infty$, $z_0 \to 0$. Also if $r > r_p^*$, then $u_\varepsilon^2 \asymp z_0^4 2^{j_1}$ and if $r = r_p^*$, then $u_\varepsilon^2 \asymp z_0^4 2^{j_1} j_1$.

If $j \le \tilde{j}_1$, then the values z_j, h_j satisfy (A.176) and we have

$$\|\pi_{\varepsilon,j}\|^2 \asymp z_0^4 2^{(j-j_1)4\tilde{r}/(4-h)} = O(2^{(j-j_1)4h(r+1/p-1/4)/(4-h)}2^{-j}) = o(1).$$

These yield relation (A.173) in this case.

(e) Let $\kappa \in \Xi_4$. In this case, we have $I \ge 0$, $r \ge -1/4$, and if $I = 0$, then $r = -1/4$; moreover, $\zeta > 0$, $\gamma_1 > 0$, and

$$j_0 \to \infty, \quad h_0 \to 0, \quad x_\varepsilon = 2^{\tilde{j}_1 - j_0} \to 0, \quad y_\varepsilon = 2^{\tilde{j}_1 - j_1} \to 0,$$

(see (A.129)). If $I > 0$, then $u_\varepsilon^2 \asymp h_0^2 2^{j_0}$. If $I = 0$, then $u_\varepsilon^2 \asymp h_0^2 2^{j_0} \log h_0^{-1}$.

If $\tilde{\Delta}_1 \le 0$, then $I > 0$ (this is equivalent to $r > r_p^*$ in the case of $\kappa \in \Xi_4$). If $j \le j_0$, then one has

$$h_j \asymp h_0 2^{-(j-j_0)\tilde{\Delta}_1/\gamma_1}, \quad z_j \asymp 2^{(j-j_0)\zeta/\gamma_1},$$

and

$$\|\pi_{\varepsilon,j}\|^2 \asymp h_0^2 2^{(j-j_0)(4\zeta-2\tilde{\Delta}_1)/\gamma_1} \asymp 2^{-j}2^{(j-j_0)h(tI+8r+2)/\gamma_1} = o(1). \quad \text{(A.177)}$$

If $\tilde{\Delta}_1 > 0$, then $z_0 \to 0$, $j_1 \to \infty$, $j_1 - \tilde{j}_1 \to \infty$.

Let $j \le \tilde{j}_1$. In this case, the quantities z_j, h_j satisfy (A.176) and

$$\|\pi_{\varepsilon,j}\|^2 \asymp z_0^4 2^{(j-j_1)4\tilde{r}/(4-h)} \asymp 2^{(j-j_1)4h(r+1/p-1/4)/(4-h)}2^{-j}o(1) = o(1).$$

Let $\tilde{j}_1 < j \le j_0$. If $r > -1/4$, then relations (A.177) hold. If $r = -1/4$, then

$$\|\pi_{\varepsilon,j}\|^2 \asymp h_0^2 2^{(j-j_0)(4\zeta-2\tilde{\Delta}_1)/\gamma_1} \asymp 2^{-j}/\log(h_0^{-1}) = o(1).$$

Relation (A.173) is proved.

Remark. One can easily see that $h_j^2 \sinh^2(z_j^2/2) = O(\log^{-1} R_\varepsilon)$.

Verification of Relation (A.172)

To simplify the notation we omit the index ε. Remember that if $z_j = O(1)$, then

$$T_j = z_j^{-1} \log(1 + h_j^{-1}) + O(1), \quad z_j T_j \ge \log(1 + h_j^{-1}) \ge \log 2.$$

(a) Let $j = O(\log\log R_\varepsilon)$. Then it follows from Proposition A.7 that $z_j = O(1)$. In fact, if $z_j \to \infty$, then $j > j_0 > a \log R_\varepsilon$, $a = q/(sq+1)$. Moreover, assumption A.1 holds and $h_j^2 \sinh^2(z_j^2/2) = O(\log^{-1} R_\varepsilon)$ (see the Remark above). If $0 < B^{-1} < z_j < B$, then for small enough ε we have

$$\log(1 + h_j^{-1}) \ge \log\log R_\varepsilon/2 + O(1).$$

Let $z_j = o(1)$. If $z_j \le \log^{-1} \log R_\varepsilon$, then

$$z_j^{-1} \log(1 + h_j^{-1}) > 2^{-1} \log\log R_\varepsilon.$$

If $1 > z_j > \log^{-1}\log R_\varepsilon$, then

$$z_j^{-1}\log(1+h_j^{-1}) > \log\log R_\varepsilon(1/2+o(1)).$$

Therefore we have, for some $B > 0$ and small enough ε,

$$T_j = z_j^{-1}\log(1+h_j^{-1}) + O(1) > B\log\log R_\varepsilon \gg Q_j \asymp \sqrt{\log\log R_\varepsilon}.$$

(b) If $z_j = o(1)$, $j \to \infty$, $\log\log R_\varepsilon = o(j)$, then $2^j h_j^2 z_j^4 = O(1)$. Let $z_j \le 1/j$. Therefore,

$$T_j \ge z_j^{-1}\log 2 + O(1) \ge j\log 2 + O(1) \gg Q_j \asymp \sqrt{j}.$$

Let $B > z_j > 1/j$. Then

$$h_j = O(1)2^{-j/2}z_j^{-2} < O(1)2^{-j/2}j^2$$

and

$$T_j > \log(1+h_j^{-1})/B + O(1) \ge (j\log 2/2 - 2\log j)/B + O(1) \gg Q_j \asymp \sqrt{j}.$$

This yields relation (A.172) in this case.

(c) Let $j \to \infty$, $\log\log R_\varepsilon = o(j)$, and $z_j \asymp 1$ (i.e., $0 < B^{-1} < z_j < B$). Then $h_j = O(1)2^{-j/2} > 0$ and

$$T_j > j\log 2/2B + O(1) \gg Q_j \asymp \sqrt{j}.$$

Therefore relation (A.172) holds.

(d) Let $z_j \to \infty$. Then $Q_j \sim \sqrt{2j\log 2}$. Denote

$$b = \sqrt{\zeta\log 2/l}, \quad B = th\lambda\log 2/l > 0, \quad d = B/2b\sqrt{2\log 2} > 0.$$

In this case we have (see Proposition A.5), $j \gg j_0$, $\lambda > 0$, and $z_j \sim \sqrt{b(j-j_0)}$,

$$T_j = \frac{z_j}{2} + \frac{\log(1+h_j^{-1}) + O(1)}{z_j}, \qquad h_j \sim h_0(b(j-j_0))^{h-t)/2l}2^{-(j-j_0)R/l},$$

$$\frac{\log(1+h_j^{-1})}{z_j} \sim \frac{j_0\sqrt{\log 2}}{(b(j-j_0))^{1/2}} + \frac{R\sqrt{\log 2(j-j_0)}}{\sqrt{\zeta l}} + A(\tau),$$

where $A(\tau) \ge o(1)$ (recall that $h_0 = 2^{-j_0/2}O(1)$).

Since $2R = \zeta + l + th\lambda$, we get

$$T_j \sim \sqrt{j-j_0}\left(\frac{b}{2} + \frac{R\log 2}{lb}\right) + \frac{j_0\log 2}{2b\sqrt{j-j_0}} + A(\tau),$$

$$\frac{R\log 2}{l} = \frac{b^2 + \log 2 + B}{2}.$$

Here we assume $A(\tau) \to \infty$, otherwise the quantity $A(\tau)$ is absent in the relation above.

Let $j = j_0 + o(j_0)$. Then

$$T_j \asymp \sqrt{j_0}/o(1) + A(\tau) \gg Q_j \asymp \sqrt{j_0}.$$

Let $j_0 = o(j)$. By $x^2 + y^2 \geq 2xy$, we have

$$\begin{aligned} T_j &\sim \sqrt{j - j_0}\left(b + \frac{\log 2}{2b} + \frac{B}{2b}\right) + A(\tau) \\ &\geq \sqrt{j - j_0}\left(2\sqrt{\frac{\log 2}{2}} + \frac{B}{2b}\right) \sim \sqrt{2j \log 2}(1 + d). \end{aligned}$$

Let $0 < c_1 j_0 < j - j_0 < c_2 j_0$. Set $j = j_0(1 + c)$, $c_2 > c > c_1 > 0$. Then, analogously, we have

$$\begin{aligned} T_j &\sim \sqrt{j_0}\left(b\sqrt{c} + \frac{\log 2(c+1)}{2b\sqrt{c}} + \frac{B\sqrt{c}}{2b}\right) + A(\tau) \\ &\geq \sqrt{2 \log 2(c+1) j_0} + \frac{B\sqrt{j_0 c}}{2b}. \end{aligned}$$

These yield relation (A.172) in this case.

Verification of Assumption A.5

Assumption A.5 is of the form

$$\sup_{v \in V_\varepsilon^H(H_{1,\varepsilon}, H_{2,\varepsilon})} \sum_{j \in \mathcal{J}_{\varepsilon,\eta}} F_{1,j}(\bar{f}_{1,j}(v_j)) = o(H_{1,\varepsilon}),$$

where $\mathcal{J}_{\varepsilon,\eta} = \{j : z_{\varepsilon,j} > \eta T_{\varepsilon,j}\}$, and we can assume

$$V_\varepsilon^H = \{v \in V_\varepsilon : |v_{ij}| \leq H_{\varepsilon,j}\}, \quad H_{\varepsilon,j} = Q_{\varepsilon,j} + C_\varepsilon,$$

and C_ε is a family such that $C_\varepsilon \to \infty$, $C_\varepsilon = O(Q_{\varepsilon,j})$.

The sets $\mathcal{J}_{\varepsilon,\eta}$ are nonempty for $p > q$, $\zeta > 0$ only, because $z_{\varepsilon,j} = O(1)$ in other cases. Also the inequality $\zeta > 0$ yields $\lambda > 0$.

It follows from Proposition A.7 and from the evaluations above, that $z_{\varepsilon,j} = O(1)$ for $j \leq j_0$, and if $j > j_0$, then

$$T_{\varepsilon,j} \sim z_{\varepsilon,j}/2 + \log h_{\varepsilon,j}^{-1}/z_{\varepsilon,j}, \quad \mathcal{J}_{\varepsilon,\eta} = \{j : z_{\varepsilon,j} > \eta T_{\varepsilon,j}\} \subset \{j \geq (1 + \gamma_0) j_0\},$$

where $\gamma_0 > 0$, $z_{\varepsilon,j}^2 \asymp j - j_0$.

Let $v \in V_\varepsilon^H$. Using the inequality

$$2^{jrp} \sum_l |v_{lj}|^p \leq \max_l |v_{lj}|^{p-q} 2^{-jpq\lambda} 2^{jsq} \sum_l |v_{lj}|^q \leq \max_l |v_{lj}|^{p-q} 2^{-jpq\lambda} R_\varepsilon^q,$$

and either the relations, for $\kappa \in G_2$,

$$\rho_\varepsilon^p \asymp h_0 2^{j_0(rp+1)} = 2^{j_0(rp-sq)} h_0 2^{j_0(sq+1)} \asymp 2^{j_0(rp-sq)} R_\varepsilon^q,$$

or the relations, for $\kappa \in G_1$ (recall that $I < 0$ in this case),

$$\rho_\varepsilon^p \asymp z_0^p 2^{j_1(rp+1)} = 2^{j_1(rp-sq)} z_0^p 2^{j_1(sq+1)} \asymp z_0^{p-q} 2^{j_1(rp-sq)} R_\varepsilon^q,$$

where $2^{(j_1-j_0)pq\lambda}z_0^{q-p} \asymp z_0^{-Ipql/\zeta} \to 0$, we get

$$F_{1,j}(\bar{f}_{1,j}(v_j)) = \left(\sum_{l=1}^{2^j} 2^{jpr}|v_{lj}|^p\right)^{h/p} \leq B_1 Q_{\varepsilon,j}^{h(p-q)/p} 2^{(j-j_0)(rp-sq)h/p}\rho_\varepsilon^h.$$

By $Q_{\varepsilon,j} \asymp j^{1/2}$ for $j \in \mathcal{J}_{\varepsilon,\eta}$ we have (recall that $H_{\varepsilon,1} = \rho_\varepsilon^h$)

$$\sup_{v \in V_\varepsilon^H} \sum_{j \in \mathcal{J}_{\varepsilon,\eta}} F_{1,j}(\bar{f}_{1,j}(v_j)) \leq B_2\rho_\varepsilon^h \sum_{j \geq (1+\gamma_0)j_0} j^{h(p-q)/2p} 2^{-(j-j_0)qh\lambda} = o(H_{\varepsilon,1}).$$

For $\kappa \in \Xi_{G_3}$, $\kappa \in \Xi_{G_4}$, or $\kappa \in \Xi_{G_5}$ the additional log-factor is inessential for the proof.

The upper bounds of Theorem 6.2 for Besov norms are proved.

A.4.6 Lower Bounds

First, let us obtain the rate lower bounds from Theorem 6.2 (2). It is enough to construct sequences of three-point measures $\bar{\pi}_\varepsilon = \{\pi_{\varepsilon,i,j}\} = \bar{\pi}_\varepsilon(\tau)$, $\tau = (\kappa, t, h)$ such that $\|\bar{\pi}_\varepsilon\| \asymp u_\varepsilon$, $\pi^\varepsilon(V_\varepsilon(\tau)) \to 1$, and assumptions A1, A2 hold. We assume (6.58). Set

$$\pi_{\varepsilon,i,j} = \begin{cases} \delta_0, & \text{if } j \neq j^*, \\ (1-h_{j^*})\delta_0 + (h_{j^*}/2)(\delta_{z_{j^*}} + \delta_{-z_{j^*}}), & \text{if } j = j^*, \end{cases} \quad 1 \leq i \leq 2^j,$$

where δ_z is the Dirac mass at point $z \in R^1$,

$$j^* = j_0,\ h_{j^*} = h_0,\ z_{j^*} = 1, \text{ if } I > 0; \quad j^* = j_1,\ h_{j^*} = 1,\ z_{j^*} = z_0, \text{ if } I < 0,$$

and the values $n = 2^{j_0}$, h_0, $m = 2^{j_1}$, z_0 are determined by the relations analogous to (6.19), (6.21) with changed $\rho_\varepsilon' = B\rho_\varepsilon$, $R_\varepsilon' = R_\varepsilon/B$, for any $B > 1$,

$$n^{rp+1}h_0 = (\rho_\varepsilon')^p, \quad n^{sq+1}h_0 = (R_\varepsilon')^q,$$

or

$$m^{rp+1}z_0^p = (\rho_\varepsilon')^p, \quad m^{sq+1}z_0^q = (R_\varepsilon')^q.$$

It is clear that $\|\bar{\pi}_\varepsilon\| \asymp u_\varepsilon$ where u_ε is determined by (6.18), (6.20). By the product measures π^ε are supported on the resolution j^* only, the relation $\pi^\varepsilon(V_\varepsilon(\tau)) \to 1$ follows from the relations

$$\pi^\varepsilon\left\{2^{rpj^*}\sum_{i=1}^{2^{j^*}}|v_{ij^*}|^p > \rho_\varepsilon^p\right\} \to 1, \quad \pi^\varepsilon\left\{2^{sqj^*}\sum_{i=1}^{2^{j^*}}|v_{ij^*}|^q < R_\varepsilon^q\right\} \to 1. \tag{A.178}$$

If $I < 0$, then one can easily check that these relations hold with π^ε-probability equal to 1. If $I > 0$, then one can check these relations using the Chebyshev inequality by

$$E_{\pi^\varepsilon}\left(2^{rpj^*}\sum_{i=1}^{2^{j^*}}|v_{ij^*}|^p\right) = n^{rp+1}h_0 = (B\rho_\varepsilon)^p,$$

$$E_{\pi^\varepsilon}\left(2^{sqj^*}\sum_{i=1}^{2^{j^*}}|v_{ij^*}|^q\right) = n^{sq+1}h_0 = (R_\varepsilon/B)^q,$$

$$\mathrm{Var}_{\pi^\varepsilon}\left(2^{rpj^*}\sum_{i=1}^{2^{j^*}}|v_{ij^*}|^p\right) < n^{2rp+1}h_0 = o(\rho_\varepsilon^{2p}),$$

$$\mathrm{Var}_{\pi^\varepsilon}\left(2^{sqj^*}\sum_{i=1}^{2^{j^*}}|v_{ij^*}|^q\right) < n^{2sq+1}h_0 = o(R_\varepsilon^{2q}),$$

if $nh_0 \to \infty$; the last relation holds under (6.58).

To obtain the sharp lower bounds, analogously to the case of power norms, let us take a small positive value δ (this will tend to 0 later) and consider an extreme problem with the change

$$\rho_{\varepsilon,\delta} = \rho_\varepsilon(1+2\delta), \quad R_{\varepsilon,\delta} = R_\varepsilon(1-2\delta).$$

First, let $\bar\pi_{\varepsilon,\delta,1} = \{\pi(h_{j,\varepsilon,\delta}, z_{j,\varepsilon,\delta})\}$ be the solutions of the changed extreme problem (remember that each factor $\pi(h_{j,\varepsilon,\delta}, z_{j,\varepsilon,\delta})$ is repeated 2^j times). Next, let us reject these sequences putting $z_{j,\varepsilon,\delta} = 0$ for $j \notin \mathcal{J}_{\varepsilon,1}$ where $\mathcal{J}_{\varepsilon,1}$ are index sets where the sums in the extreme problem are asymptotically concentrated:

$$\begin{aligned} \sum_{j\notin \mathcal{J}_{\varepsilon,1}} 2^j h_{j,\varepsilon,\delta}^2 \sinh^2(z_{j,\varepsilon,\delta}^2/2) &= o(1), \\ \sum_{j\notin \mathcal{J}_{\varepsilon,1}} 2^{j(rh+h/p)} h_{j,\varepsilon,\delta}^{h/p} z_{j,\varepsilon,\delta}^h &< (\tau_\varepsilon\rho_\varepsilon)^h, \\ \sum_{j\notin \mathcal{J}_{\varepsilon,1}} 2^{j(st+t/q)} h_{j,\varepsilon,\delta}^{t/q} z_{j,\varepsilon,\delta}^t &< (\tau_\varepsilon R_\varepsilon)^t, \end{aligned} \tag{A.179}$$

for some $\tau_\varepsilon \to 0$ (one can find such sets by the properties of the solutions of the extreme problems which were studied above). Let $\tilde\pi^\varepsilon = \pi_\delta^\varepsilon$ be the product measures which correspond to $j \in \mathcal{J}_{\varepsilon,1}$. We need to control the assumptions of Proposition 5.5.

It follows from the convexity and asymptotical continuity (see Section 5.2.3) that $\|\bar\pi_{\varepsilon,\delta}\| \sim \|\bar\pi_\varepsilon\|$, assumptions A.1, A.2 are fulfilled by evaluations from the previous section (recall that A.2 follows from A.1 and A.3). Thus we need to verify that $\pi_\delta^\varepsilon(V_\varepsilon) \to 1$. This relation follows from

$$S_\varepsilon = \sum_{j \in \tilde{\mathcal{J}}_{\varepsilon,1}} (2^j h_{j,\varepsilon,\delta})^{-1} = o(1), \quad \tilde{\mathcal{J}}_{\varepsilon,1} = \mathcal{J}_{\varepsilon,1} \cap \mathcal{J}_{\varepsilon,0}, \quad \mathcal{J}_{\varepsilon,0} = \{j : h_j \neq 1\}, \tag{A.180}$$

and under (A.180) we can take $\delta = \delta_\varepsilon \to 0$ such that $\delta_\varepsilon^{-2} S_\varepsilon \to 0$, $\tau_\varepsilon = o(\delta_\varepsilon)$.

In fact, it suffices to verify the relations

$$\pi_\delta^\varepsilon\left(\sum_j F_{1,j}(v_j) < \rho_\varepsilon^h\right) \to 0, \quad \pi_\delta^\varepsilon\left(\sum_j F_{2,j}(v_j) > R_\varepsilon^t\right) \to 0, \tag{A.181}$$

where the sums could be taken over the index sets $\mathcal{J}_{\varepsilon,1}$,

$$F_{1,j}(v_j) = 2^{jrh}\left(\sum_{i=1}^{2^j} |v_{ij}|^p\right)^{h/p}, \quad F_{2,j}(v_j) = 2^{jst}\left(\sum_{i=1}^{2^j} |v_{ij}|^q\right)^{t/q}.$$

Consider the first relation in (A.181). Denote

$$\rho_{j,\varepsilon,\delta}^p = 2^{j(rp+1)} h_{j,\varepsilon,\delta} z_{j,\varepsilon,\delta}^p.$$

It follows from the construction and from the study extreme problem above, that for small ε,

$$\sum_j \rho_{j,\varepsilon,\delta}^h > (\rho_\varepsilon(1+\delta))^h.$$

Therefore, for a constant $c = c(h)$, $b = b(h,p)$, we get

$$\begin{aligned}
\pi_\delta^\varepsilon\left(\sum_j F_{1,j}(v_j) < \rho_\varepsilon^h\right) &\le \pi_\delta^\varepsilon\left(\sum_j F_{1,j}(v_j) < \sum_j \rho_{j,\varepsilon,\delta}^h(1-c\delta)\right) \\
&\le \sum_j \pi_\delta^\varepsilon\left(F_{1,j}(v_j) < \rho_{j,\varepsilon,\delta}^h(1-c\delta)\right) \\
&= \sum_j \pi_\delta^\varepsilon\left(F_{1,j}^{p/h}(v_j) < \rho_{j,\varepsilon,\delta}^p(1-b\delta)\right) \\
&\le \sum_j \pi_\delta^\varepsilon\left(|F_{1,j}^{p/h}(v_j) - \rho_{j,\varepsilon,\delta}^p| > b\delta\rho_{j,\varepsilon,\delta}^p\right).
\end{aligned}$$

Since $E_{\pi_\delta^\varepsilon} F_{1,j}^{p/h}(v_j) = \rho_{j,\varepsilon,\delta}^p$, it follows from the Chebyshev inequality that the last sum is smaller than

$$\begin{aligned}
\sum_j \frac{\mathrm{Var}_{\pi_\delta^\varepsilon}(F_{1,j}^{p/h}(v_j))}{(b\delta)^2 \rho_{j,\varepsilon,\delta}^{2p}} &= (b\delta)^{-2} \sum_j \frac{2^{j(2pr+1)} h_{j,\varepsilon,\delta}(1-h_{j,\varepsilon,\delta}) z_{j,\varepsilon,\delta}^{2p}}{2^{2j(pr+1)} h_{j,\varepsilon,\delta}^2 z_{j,\varepsilon,\delta}^{2p}} \\
&= (b\delta)^{-2} \sum_j (1-h_{j,\varepsilon,\delta})(2^j h_{j,\varepsilon,\delta})^{-1} \le (b\delta)^{-2} S_\varepsilon.
\end{aligned}$$

The second relation in (A.181) is considered in an analogous way.

Verification of Relation (A.180)

We would like to define the index sets $\tilde{\mathcal{J}}_{\varepsilon,1}$ such that (A.180) holds and that relations (A.179) are fulfilled for the set $\mathcal{J}_{\varepsilon,1} = \tilde{\mathcal{J}}_{\varepsilon,1} \cup \{j : h_j = 1\}$. Below we omit the indexes ε, δ in $h_{j,\varepsilon,\delta}$ to simplify notations; also for real $a \le b$ we set $\sum_a^b$ as the sum over the integer $a \le j \le b$.

Let $\kappa \in \Xi_1$. By Proposition A.8 we have $u_\varepsilon^2 \asymp h_0^2 2^{j_0}(S_{1,0} + S_{3,0})$ and

$$\rho_\varepsilon^h \asymp h_0^{h/p} 2^{j_0(rh+h/p)}(S_{1,1} + S_{3,1}), \quad R_\varepsilon^t \asymp h_0^{t/q} 2^{j_0(st+t/q)}(S_{1,2} + S_{3,1}).$$

where the sums are defined on page 416. If $r > r_p^*$, then all sums $S \asymp 1$ here. If $r = r_p^*$, then these hold for all sums except $S_{1,1} \asymp S_{1,0} \asymp j_0$.

Let $r > r_p^* = -1/2p$. Then we can evaluate the sums for "small" $j \le j_0(1-d)$, $0 < d < 1$,

$$S_{1,1} = \sum_{j=1}^{j_0}(\dots) : \sum_{j=1}^{j_0(1-d)} 2^{(j-j_0)h(rp+1+a)/p} \asymp 2^{-dj_0h(rp+1+a)/p} = o(1),$$

$$S_{1,2} = \sum_{j=1}^{j_0}(\dots) : \sum_{j=1}^{j_0(1-d)} 2^{(j-j_0)t(sq+1+a)/p} \asymp 2^{-dj_0t(st+1+a)/q} = o(1),$$

$$S_{1,0} = \sum_{j=1}^{j_0}(\dots) : \sum_{j=1}^{j_0(1-d)} 2^{(j-j_0)(1+2a)} \asymp 2^{-dj_0(1+2a)} = o(1).$$

as well as for "large" $j \ge j_0(1+d)$; $k = 1, 2$,

$$S_{3,k} = \sum_{j_0}^{\infty}(\dots) : \sum_{j_0(1+d)}^{\infty} 2^{(j-j_0)ht\lambda/l}(j-j_0)^{n_k} \asymp 2^{-dj_0ht\lambda/l}(dj_0)^{n_k} = o(1).$$

Thus $\tilde{\mathcal{J}}_{\varepsilon,1} = \{j : j_0(1-d) \le j \le j_0(1+d)\}$ in this case.

Let $\kappa \in \Xi_{G_2}$. Recall that $j_0 \to 0$, $2^{j_1} \asymp 2^{j_0(1-\gamma/1\Delta_1)}$, and for some $d_0 > 0$, for all $\tau \in \Gamma_{\Delta_1}^+$, we have, if $\gamma < 0$, then $\tilde{j}_1 > (1+d_0)j_0$, if $\gamma > 0$, then $\tilde{j}_1 < (1-d_0)j_0$. Then, analogously to the above, for some $d \in (0,1)$, we get

$$\tilde{\mathcal{J}}_{\varepsilon,1} = \{j : j_0(1-d) \le j \le j_0(1+d)\}.$$

Using analogous evaluations, we get, if $\kappa \in \Xi_{G_1}$ and $\kappa \in \Xi_2$, then

$$\tilde{\mathcal{J}}_{\varepsilon,1} = \{j : \tilde{j}_1(1-d) \le j \le \tilde{j}_1).$$

If $\kappa \in \Xi_{G_1}$ and $\kappa \in \Xi_3 \cup \Xi_4$, then

$$\tilde{\mathcal{J}}_{\varepsilon,1} = \{j : \tilde{j}_1 \le j \le \tilde{j}_1(1+d)\}.$$

If $\kappa \in \Xi_{G_3}$, then the sets $\tilde{\mathcal{J}}_{\varepsilon,1}$ are of the form

$$\tilde{\mathcal{J}}_{\varepsilon,1} = \begin{cases} (j_0, \tilde{j}_1), & \text{if } \tilde{j}_1 > j_0, \quad \kappa \in \Xi_2, \\ (\tilde{j}_1, j_0), & \text{if } j_0 > \tilde{j}_1, \quad \kappa \in \Xi_3. \end{cases}$$

If $\kappa \in \Xi_{G_4}$, then

$$\tilde{\mathcal{J}}_{\varepsilon,1} = \{j : \tilde{j}_1 \le j \le \tilde{j}_1(1+d)\}.$$

If $\kappa \in \Xi_{G_5}$ and $\kappa \in \Xi_1 \cup \Xi_2$, then

$$\tilde{\mathcal{J}}_{\varepsilon,1} = \{j : \sqrt{j_0} \le j \le j_0(1+d)\}.$$

If $\kappa \in \Xi_{G_5}$ and $\kappa \in \Xi_4$ then

$$\tilde{\mathcal{J}}_{\varepsilon,1} = \{j : \tilde{j}_1 \le j \le j_0(1+d)\}.$$

Let us consider the sum S_ε defined by (A.180). To simplify, we study the case $u_\varepsilon \asymp 1$; one can obtain the results under assumption (6.58) without difficulty.

(a) Let $\kappa \in \Xi_1$.

If $r > r_p^*$, then $\kappa \in \Xi_{G_2}$ and

$$S_\varepsilon = \sum_{j_0(1-d)}^{j_0(1+d)} \frac{1}{2^j h_j} \asymp \sum_{j_0(1-d)}^{j_0} \frac{1}{h_0 2^j 2^{(j-j_0)a}} + \sum_{j_0+1}^{j_0(1+d)} \frac{j^{(t-h)/2l}}{h_0 2^j 2^{-(j-j_0)R/l}},$$

where $R > 0$, $l > 0$, $a > -1/2$. Here we have $h_0 \asymp 2^{-j_0/2}$. Therefore, there exists $A > 0$, such that for $0 < d < \min((2(1+a))^{-1}, l/2R)$,

$$S_\varepsilon < A(j_0 2^{-j_0(1/2-d(1+a))} + j_0^{1+(t-h)/2l} 2^{-j_0(1/2-dR/l)}) = o(1).$$

Let $r = r_p^* = -1/2p$. Then $a = -1/2$, $\kappa \in \Xi_{G_5}$, and

$$S_\varepsilon = \sum_{\sqrt{j_0}}^{j_0(1+d)} \frac{1}{2^j h_j} \asymp \sum_{\sqrt{j_0}}^{j_0} \frac{1}{h_0 2^j 2^{-(j-j_0)/2}} + \sum_{j_0+1}^{j_0(1+d)} \frac{j^{(t-h)/2l}}{h_0 2^j 2^{-(j-j_0)R/l}},$$

where $R > 0$, $l > 0$. Here we have $h_0 \asymp 2^{-j_0/2} j_0^{-1/2}$ and there exists $A > 0$ such that for the first sum in this relation we have

$$\sum_{\sqrt{j_0}}^{j_0} \frac{1}{h_0 2^j 2^{-(j-j_0)/2}} < A j_0^{3/2} 2^{-\sqrt{j_0}/2} = o(1).$$

The second sum is evaluated in an analogous way.

(b) Let $\kappa \in \Xi_2$. Then $\gamma < 0$.

Let $\Delta_1 \le 0$ or $\Delta_1 > 0$, $I > 0$, and $r > r_p^*$. In this case, $\kappa \in \Xi_{G_2}$ and we have

$$S_\varepsilon = \sum_{j_0(1-d)}^{j_0(1+d)} \frac{1}{2^j h_j} \asymp \sum_{j_0(1-d)}^{j_0} \frac{1}{h_0 2^j 2^{(j-j_0)a}} + \sum_{j_0+1}^{j_0(1+d)} \frac{1}{h_0 2^j 2^{-(j-j_0)\Delta_1/\gamma}}.$$

If $\Delta_1 < 0$, then the proof $S_\varepsilon = o(1)$ is analogous to the cases $\kappa \in \Xi_1$, $r > r_p^*$: we replace $-R/l$ by $-\Delta_1/\gamma$. If $\Delta_1 \ge 0$, $I > 0$, then for $0 < d < 2(1+a))^{-1}$ we have that there exists $A > 0$ such that,

$$S_\varepsilon < A(j_0 2^{-j_0(1/2-d(1+a))} + j_0 2^{-j_0/2}) = o(1).$$

Let $\Delta_1 > 0,\ I < 0$. Then

$\kappa \in \Xi_{G_1},\ j_0 < \tilde{j}_1,\ \tilde{j}_1 \to \infty,\ j_1 \sim \tilde{j}_1,\ \tilde{\mathcal{J}}_{\varepsilon,1} = ((1-d)\tilde{j}_1,\ \tilde{j}_1),\ u_\varepsilon^2 \asymp z_0^4\, 2^{j_1},$

and we have

$$S_\varepsilon = \sum_{\tilde{j}_1(1-d)}^{\tilde{j}_1} \frac{1}{2^j h_j} \asymp \sum_{\tilde{j}_1(1-d)}^{\tilde{j}_1} \frac{1}{2^j 2^{-(j-j_1)\Delta_1/\gamma}} \le A\tilde{j}_1 2^{-\tilde{j}_1(1-d(1-\Delta_1/\gamma))},$$

where $A > 0$ is large enough. Then $S_\varepsilon = o(1)$ for any $0 < d < 1/(1-\Delta_1/\gamma))$.

Let $I = 0$. Then $\kappa \in \Xi_{G_3},\ \tilde{j}_1 > j_0 \to \infty,\ \tilde{\mathcal{J}}_{\varepsilon,1} = (j_0,\ \tilde{j}_1)$. We consider the sets $\tilde{\mathcal{J}}_{\varepsilon,1}$ defined above because the "majors parts" of the sums for the quantities $u_\varepsilon^2,\ \rho^h,\ R^t$ are concentrated over the set $j \in (j_0,\ \tilde{j}_1)$. In this case we have

$$(\tilde{j}_1 - j_0) \asymp \log h_0^{-1}, \quad h_0 \asymp 2^{-j_0/2}(\tilde{j}_1 - j_0)^{-1/2} \to 0, \quad j_0 \to \infty, \quad \Delta_1 > 0,$$

and

$$S_\varepsilon = \sum_{j_0}^{\tilde{j}_1} \frac{1}{2^j h_j} \asymp \sum_{j_0}^{\tilde{j}_1} \frac{1}{h_0 2^j 2^{-(j-j_0)\Delta_1/\gamma}} < B(\tilde{j}_1 - j_0)^{3/2} 2^{-j_0/2} = o(1).$$

Let $r = r_p^* = -1/2p$. Then

$$\kappa \in \Xi_{G_5}, \quad a = -1/2, \quad \Delta_1 < 0, \quad h_0 \asymp 2^{-j_0/2} j_0^{-1/2}.$$

For $A > 0$ large enough and $0 < d < \gamma/2\Delta_1$, we have

$$\begin{aligned} S_\varepsilon &< \sum_{\sqrt{j_0}}^{j_0(1+d)} \frac{1}{2^j h_j} \asymp \sum_{\sqrt{j_0}}^{j_0} \frac{1}{h_0 2^j 2^{-(j-j_0)/2}} + \sum_{j_0+1}^{j_0(1+d)} \frac{1}{h_0 2^j 2^{-(j-j_0)\Delta_1/\gamma}} \\ &< A j_0^{3/2}(2^{-\sqrt{j_0}/2} + 2^{-j_0(1/2 - d\Delta_1/\gamma)}) = o(1). \end{aligned}$$

(c) Let $\kappa \in \Xi_3$. Then $\gamma > 0$.

If $\Delta_1 \le 0$ or $\Delta_1 > 0,\ I > 0$, then $\kappa \in \Xi_{G_2},\ j_0 > \tilde{j}_1,\ j_0 \to \infty$, and

$$S_\varepsilon = \sum_{j_0(1-d)}^{j_0(1+d)} \frac{1}{2^j h_j} \asymp \sum_{j_0(1-d)}^{j_0} \frac{1}{h_0 2^j 2^{-(j-j_0)\Delta_1/\gamma}} + \sum_{j_0+1}^{j_0(1+d)} \frac{j^{(t-h)/2l}}{h_0 2^j 2^{-(j-j_0)R/l}}.$$

Here we have $h_0 \asymp 2^{-j_0/2}$. The second sum has been evaluated in point (a); the first sum is evaluated analogously to the above. Taking $d > 0$ small enough, we get $S_\varepsilon = o(1)$.

Let $I < 0,\ r > r_p^*$. Then $\kappa \in \Xi_{G_1},\ j_1 \sim \tilde{j}_1 \to \infty,\ \Delta_1/\gamma > 0$. In this case, we have

$$S_\varepsilon = \sum_{\tilde{j}_1}^{\tilde{j}_1(1+d)} \frac{1}{2^j h_j} \asymp \sum_{\tilde{j}_1}^{\tilde{j}_1(1+d)} \frac{1}{2^j 2^{-(j-j_1)\Delta_1/\gamma}} < A\tilde{j}_1 2^{-\tilde{j}_1(1-d\Delta_1/\gamma)} = o(1),$$

where $A > 0$ is large enough and $0 < d < \gamma/\Delta_1$.

Let $I = 0$. Then $\kappa \in \Xi_{G_3}$, $\tilde{j}_1 < j_0$, $\tilde{j}_1 \to \infty$,

$$(j_0 - \tilde{j}_1) \asymp \log h_0^{-1}, \quad h_0 \asymp ((j_0 - \tilde{j}_1)2^{j_0})^{-1/2}, \quad \tilde{\mathcal{J}}_{\varepsilon,1} = (\tilde{j}_1, j_0)$$

and $\Delta_1/\gamma = 2\zeta/\gamma + 1/2 > 1/2$. In this case, we have

$$S_\varepsilon = \sum_{\tilde{j}_1}^{j_0} \frac{1}{2^j h_j} \asymp \sum_{\tilde{j}_1}^{j_0} \frac{1}{h_0 2^j 2^{-(j-j_0)\Delta_1/\gamma}}.$$

If $1 - \Delta_1/\gamma \le 0$, then, for $A > 0$ large enough, we have

$$S_\varepsilon < A(j_0 - \tilde{j}_1)^{3/2} 2^{-j_0/2} = o(1).$$

If $1 - \Delta_1/\gamma > 0$, then, for $A > 0$ large enough, we have

$$S_\varepsilon < A(j_0 - \tilde{j}_1)^{3/2} 2^{-(j_0 - \tilde{j}_1)(\Delta_1/\gamma - 1/2)} 2^{-\tilde{j}_1/2} = o(1).$$

Let $r = r_p^* = 1/4 - 1/p$. Then

$$\Delta_1 > 0, \quad I < 0, \quad j_1 \sim \tilde{j}_1 \to \infty, \quad \kappa \in \Xi_{G_4}.$$

For $A > 0$ large enough and $0 < d < \gamma/\Delta_1$, we have

$$S_\varepsilon - \sum_{\tilde{j}_1}^{\tilde{j}_1(1+d)} \frac{1}{2^j h_j} \asymp \sum_{\tilde{j}_1}^{\tilde{j}_1(1+d)} \frac{1}{2^j 2^{-(j-\tilde{j}_1)\Delta_1/\gamma}} < A\tilde{j}_1 2^{-\tilde{j}_1(1-d\Delta_1/\gamma)} = o(1).$$

(d) Let $\kappa \in \Xi_4$. If $I > 0$. Then

$$\kappa \in \Xi_{G_2}, \quad j_0 > j_1 > \tilde{j}_1, \quad j_0 \to \infty, \quad \tilde{\mathcal{J}}_{\varepsilon,1} = ((1-d)j_0, (1+d)j_0).$$

In this case, we have

$$S_\varepsilon = \sum_{j_0(1-d)}^{j_0(1+d)} \frac{1}{2^j h_j} \asymp \sum_{j_0(1-d)}^{j_0} \frac{1}{h_0 2^j 2^{-(j-j_0)\tilde{\Delta}_1/\gamma_1}} + \sum_{j_0+1}^{j_0(1+d)} \frac{j^{(t-h)/2l}}{h_0 2^j 2^{-(j-j_0)R/l}}.$$

Here we have $h_0 \asymp 2^{-j_0/2}$. The second sum has been evaluated in point (a). Let us evaluate the first sum. If $\tilde{\Delta}_1/\gamma_1 < 0$, then, for $0 < d < 1/2(1-\tilde{\Delta}_1/\gamma_1)$ and $A > 0$ large enough, one gets

$$S_{\varepsilon,1} = \sum_{j_0(1-d)}^{j_0} \frac{1}{h_0 2^j 2^{-(j-j_0)\tilde{\Delta}_1/\gamma_1}} < Aj_0 2^{-j_0(1/2 - d(1-\tilde{\Delta}_1/\gamma_1)} = o(1).$$

If $\tilde{\Delta}_1/\gamma_1 \ge 0$, then

$$S_{\varepsilon,1} < Aj_0 2^{-j_0(1/2-d)} = o(1)$$

for $0 < d < 1/2$ and $A > 0$ large enough. Thus $S_\varepsilon = o(1)$ in this case.

Let $I = 0$. Then $r = -1/4$, $\kappa \in \Xi_{G_5}$, $j_0 > \tilde{j}_1 \to \infty$,

$$\tilde{\mathcal{J}}_{\varepsilon,1} = (\tilde{j}_1, (1+d)j_0),$$

and

$$\tilde{\Delta}_1/\gamma_1 = 2\zeta/\gamma_1 + 1/2 > 1/2, \quad h_0 \asymp ((j_0 - \tilde{j}_1)2^{j_0})^{-1/2}.$$

We have

$$S_\varepsilon = \sum_{\tilde{j}_1}^{j_0(1+d)} \frac{1}{2^j h_j} \asymp \sum_{\tilde{j}_1}^{j_0} \frac{1}{h_0 2^j 2^{-(j-j_0)\tilde{\Delta}_1/\gamma_1}} + \sum_{j_0+1}^{j_0(1+d)} \frac{j^{(t-h)/2l}}{h_0 2^j 2^{-(j-j_0)R/l}}.$$

The first item of this sum is evaluated analogously to the case $\kappa \in \Xi_3$, $I = 0$: we replace Δ_1/γ by $\tilde{\Delta}_1/\gamma_1$. The second item of this sum is evaluated as in case (a). Therefore we have $S_\varepsilon = o(1)$. This completes the proof. □

A.4.7 Proof of Proposition 6.3

We use the results given in Propositions A.8, A.7.

I. Note that $\tilde{j}_1 = j + 1 + O(1)$ in this case. If $\kappa \in \Xi_{G_1}$, then either $\kappa \in \Xi_2$ or $\kappa \in \Xi_3$ and $z_0 \asymp 2^{-j_1/4}$. Moreover, if $p < 2$, then $r > 1/4 - 1/p$. Recall that if $\kappa \in \Xi_2$, then $\gamma < 0$ and if $\kappa \in \Xi_3$, then $\gamma > 0$. It follows from (A.139), (A.141), (A.142) that there exists $\delta > 0$ such that the required relation holds.

II. Let $\kappa \in \Xi_{G_2}$ and $\kappa \in \Xi_2$. Then $z_0 = B_0^{-1/\gamma} 2^{(\tilde{j}_1 - j_0)\zeta/\gamma} y_\varepsilon^{-\zeta/\gamma}$, where $y_\varepsilon \sim c(p,q,h,t)$ (see (A.128)). It follows from (A.138), (A.139) that, for any $\eta > 0$, there exist $A > 0$ and $\delta > 0$ such that the required relation holds.

III. Let $\kappa \in \Xi_{G_2}$ and $\kappa \in \Xi_1 \cup \Xi_3 \cup \Xi_4$. Then it follows from (A.137), (A.140), (A.141), (A.143), (A.144) that there exists $A > 0$ such that the required relation holds.

IV.

(a) Let (see Proposition A.8)

$$u_\varepsilon^2(\tau) = \sum_{j=1}^{\infty} u_{\varepsilon,j}^2(\tau) = A u_\varepsilon^2(\tau)(S_1 + S_2),$$

where S_l, $l = 1, 2$, are $S_{i,k}$ or S_i (these quantities are defined on page 416). Then

$$u_{\varepsilon,j}^2(\tau) \le A_j u_\varepsilon^2(\tau) 2^{-B(\mathcal{J}_l)|j - j_\varepsilon^*|}, \tag{A.182}$$

where $\mathcal{J}_l$, $l = 1, 2$, are the index sets corresponding to S_l, $0 < A_0 < A_j < A_1 < \infty$ for all integers $j > 0$ and ε small enough, $B(\mathcal{J}_l) > 0$, $l = 1, 2$. It is clear that (6.114) holds.

(b) Let (see Proposition A.8)

$$u_\varepsilon^2(\tau) = \sum_{j=1}^{\infty} u_{\varepsilon,j}^2(\tau) = A u_\varepsilon^2(\tau)(S_1 + S_2 + \delta_0 S_3),$$

where S_l, $l = 1,2,3$, are $S_{i,k}$ or S_i (see the definition on page 416) and

$$\delta_0 = \begin{cases} z_0^{-htI/\zeta}, & z_0 \to 0, \quad I < 0, \quad \text{if } \kappa \in \Xi_{G_1} \\ h_0^{htI/\Delta_1}, & h_0 \to 0, \quad I > 0, \quad \text{if } \kappa \in \Xi_{G_2} \end{cases} \asymp 2^{-ht|(j_1-j_0)I/\gamma|}.$$

Then (A.182) holds for $j \in \mathcal{J}_1 \cup \mathcal{J}_2$. If $j \in \mathcal{J}_3$, then we have

$$u_{\varepsilon,j}^2(\tau) \le A_j u_\varepsilon^2(\tau)\delta_0 2^{-B(\mathcal{J}_3)|j-\tilde{j}_\varepsilon|},$$

where $B(\mathcal{J}_3) > 0$, $\tilde{j}_\varepsilon \ne j_\varepsilon^*$, and

$$\tilde{j}_\varepsilon = \begin{cases} j_0, & \text{if } \kappa \in \Xi_{G_1}, \\ j_1, & \text{if } \kappa \in \Xi_{G_2},. \end{cases}$$

Let us take $\delta > 0$ such that $\delta < \min(ht|I/\gamma|, B(\mathcal{J}_3))$. Set

$$N = |j - \tilde{j}_\varepsilon| + |j_1 - j_0| - |j - j_\varepsilon^*|.$$

Clearly, $N \ge 0$. We have

$$\frac{u_{\varepsilon,j}^2}{u_\varepsilon^2} < A 2^{-|j-j_\varepsilon^*|\delta} 2^{-|j-\tilde{j}_\varepsilon|(B(\mathcal{J}_3)-\delta)} 2^{-|j_1-j_0|(ht|I/\gamma|-\delta)} 2^{-\delta N} < A 2^{-|j-j_\varepsilon^*|\delta},$$

where $A > 0$ and it is clear that (6.114) holds. □

A.5 Proof of Lemma 7.4

Let $\tilde{u}_\varepsilon$ be the values $u_\varepsilon(\tau_\varepsilon, \tilde{r}_\varepsilon(\tau_\varepsilon))$ corresponding to $\tilde{r}_\varepsilon(\tau) = (1-\delta_\varepsilon)r_\varepsilon(\tau)$, the values $\delta_\varepsilon = o(1)$ will be defined concretely below. It follows from the study of the extreme problem in Chapter 6 that under assumptions (7.69) one has

$$u_\varepsilon(\tau, \tilde{r}_\varepsilon) \sim u_\varepsilon(\tau, r_\varepsilon)$$

uniformly over $\tau \in \Gamma$ which yields

$$\tilde{u}_\varepsilon \sim u_\varepsilon = u_\varepsilon(\tau_\varepsilon, r_\varepsilon(\tau_\varepsilon)).$$

Remember that

$$u_\varepsilon(\tau, r_\varepsilon) = \|\bar{\pi}_\varepsilon(\tau, r_\varepsilon)\| = \inf_{\bar{\pi} \in \Pi_\varepsilon(\tau, r_\varepsilon)} \|\bar{\pi}\|, \tag{A.183}$$

where

$$\Pi_\varepsilon(\tau, r_\varepsilon) = \left\{ \bar{\pi} : \sum_j \left(2^{rhj} \sum_\iota E_{\pi_{\iota j}} |v|^p \right)^{h/p} \ge (r_\varepsilon/\varepsilon)^h, \right.$$
$$\left. \sum_j \left(2^{sqj} \sum_\iota E_{\pi_{\iota j}} |v|^q \right)^{t/q} \le (R/\varepsilon)^t \right\}.$$

The extreme sequences $\bar{\pi}_\varepsilon$ in (A.183) consist of the three-point measures $\pi_{\iota j;\varepsilon}(\tau, r_\varepsilon) = \pi(h_j(\tau, r_\varepsilon), z_j(\tau, r_\varepsilon))$ and there exist the quantities

$$z_0 = z_{0,\varepsilon}(\tau, r_\varepsilon), \quad j_1 = j_{1,\varepsilon}(\tau, r_\varepsilon), \text{ if } \kappa \in \Xi_{G_1},$$

or

$$h_0 = h_{0,\varepsilon}(\tau, r_\varepsilon), \quad j_0 = j_{0,\varepsilon}(\tau, r_\varepsilon), \text{ if } \kappa \in \Xi_{G_2},$$

and the functions $d_l(\tau)$, $l = 0, 1, 2$, which are bounded away from 0 and ∞ such that, for $\kappa \in \Xi_{G_1}$,

$$u_\varepsilon^2(\tau, r_\varepsilon) \sim d_0(\tau) z_0^4 2^{j_1},$$

$$\left(\frac{r_\varepsilon}{\varepsilon}\right)^p \sim d_1(\tau) z_0^p 2^{(rp+1)j_1}, \quad \left(\frac{R}{\varepsilon}\right)^q \sim d_2(\tau) z_0^q 2^{(sq+1)j_1} \tag{A.184}$$

or, for $\kappa \in \Xi_{G_2}$,

$$u_\varepsilon^2(\tau, r_\varepsilon) \sim d_0(\tau) h_0^2 2^{j_0},$$

$$\left(\frac{r_\varepsilon}{\varepsilon}\right)^p \sim d_1(\tau) h_0 2^{(rp+1)j_0}, \quad \left(\frac{R}{\varepsilon}\right)^q \sim d_2(\tau) h_0 2^{(sq+1)j_0}. \tag{A.185}$$

Note that relations (A.184), (A.185) and assumption (7.69) imply that, uniformly over $\tau \in \Gamma$,

$$j_\varepsilon^* \asymp \log \varepsilon^{-1}, \quad j_\varepsilon^* = \begin{cases} j_1, & \text{if } \kappa_\varepsilon \in \Xi_{G_1}, \\ j_0, & \text{if } \kappa_\varepsilon \in \Xi_{G_2}. \end{cases} \tag{A.186}$$

Also if $\kappa_\varepsilon \in \Xi_{G_1}$, then $r + 1/p - 1/4 > 0$, $s + 1/q - 1/4 > 0$, and

$$r_\varepsilon/\varepsilon \asymp u_\varepsilon^{1/2} 2^{j_1(r+1/p-1/4)}, \quad R/\varepsilon \asymp u_\varepsilon^{1/2} 2^{j_1(s+1/q-1/4)}; \tag{A.187}$$

if $\kappa_\varepsilon \in \Xi_{G_2}$, then $r + 1/2p > 0$, $s + 1/2q > 0$ and

$$r_\varepsilon/\varepsilon \asymp u_\varepsilon^{1/p} 2^{j_0(r+1/2p)}, \quad R/\varepsilon \asymp u_\varepsilon^{1/q} 2^{j_0(s+1/2q)}. \tag{A.188}$$

Let $\tilde{\pi}_\varepsilon = \{\tilde{\pi}_{\varepsilon,j}\} = \bar{\pi}_\varepsilon(\tau_\varepsilon, \tilde{r}_\varepsilon)$ be the family of sequences which corresponds to $\tilde{r}_\varepsilon$:

$$\|\tilde{\pi}_\varepsilon\| \sim u_\varepsilon(\tau_\varepsilon, r_\varepsilon) \sim C\sqrt{\log\log \varepsilon^{-1}}. \tag{A.189}$$

Recall the inequality (see (5.41))

$$\|\bar{\pi}_\varepsilon(\tau, r_\varepsilon)\|^{-1} \inf_{v \in V_\varepsilon(\tau, r_\varepsilon)} (\bar{\pi}_\varepsilon(\tau, r_\varepsilon), \bar{\delta}_v) \geq \|\bar{\pi}_\varepsilon(\tau, r_\varepsilon)\|. \tag{A.190}$$

Using this inequality for the set $\tilde{V}_\varepsilon$ we get that for all $\tilde{v} = \tilde{v}_\varepsilon \in \tilde{V}_\varepsilon$ one has

$$\begin{aligned} \|\tilde{\pi}_\varepsilon\|^{-1} \sum_j 2h_{\varepsilon,j} \sum_\iota \sinh^2(z_{\varepsilon,j}\tilde{v}_{\iota j}/2) &= \sum_j r_{\varepsilon,j}(\bar{\pi}^*_{\varepsilon,j}, \bar{\delta}_{\tilde{v}_\varepsilon}) \\ &\geq C\sqrt{\log\log \varepsilon^{-1}}(1 + o(1)), \end{aligned} \tag{A.191}$$

where

$$r_{\varepsilon,j}^2 = \frac{2^j \|\bar{\pi}_{\varepsilon,j}\|^2}{\|\bar{\pi}_\varepsilon\|^2} = \frac{2^j h_{\varepsilon,j}^2 \sinh^2(z_{\varepsilon,j}^2/2)}{\sum_j 2^j h_{\varepsilon,j}^2 \sinh^2(z_{\varepsilon,j}^2/2)}; \quad \sum_j r_{\varepsilon,j}^2 = 1,$$

and $\bar{\pi}_{\varepsilon,j}^* = \bar{\pi}_j^*(z_{\varepsilon,j})$, $\|\bar{\pi}_{\varepsilon,j}^*\| = 1$ are normalized sequences defined by (7.75).

It follows from the study of the extreme problem in Section 6.9 (see Proposition 6.3) that for all $\tau_\varepsilon \in \Gamma$ there exist values $a = a(\Gamma) > 0$ and $b = b(\Gamma) > 0$ such that for small enough ε one has

$$r_{\varepsilon,j} \le a 2^{-b|j-j_\varepsilon^*|}, \tag{A.192}$$

where j_ε^* is determined by (A.186). Relation (A.192) yields

$$\sum_j r_{\varepsilon,j}(1 + |j - j_\varepsilon^*|) \le B = B(\Gamma)$$

and using (A.191) we get

$$\sup_j (\bar{\pi}_{\varepsilon,j}^*, \delta_{\bar{v}_{\varepsilon,j}})/(1 + |j - j_\varepsilon^*|) \ge C_1 \sqrt{\log\log \varepsilon^{-1}}(1 + o(1)), \quad C_1 = C/B. \tag{A.193}$$

We can assume later without loss of generality that $\kappa = \kappa(\tau_\varepsilon) \in \Xi_{G_1}$ or $\kappa = \kappa(\tau_\varepsilon) \in \Xi_{G_2}$, and consider these cases differently. Also without loss of generality we can assume that later various necessary relations between parameters $\kappa = (p, q, r, s)$ hold true for all small enough $\varepsilon > 0$.

The following simple proposition will be used later:

Proposition A.9 *For a set $\mathcal{J}_\varepsilon \subset \mathcal{J}$ denote*

$$a_j^p = 2^{rpj} \sum_\iota |v_{\iota j}|^p, \quad D_\varepsilon(\mathcal{J}_\varepsilon) = \sum_{j \in \mathcal{J}_\varepsilon} a_j^h/(r_\varepsilon/\varepsilon)^h.$$

Assume that (7.69) *holds and $v = v_\varepsilon$ satisfy* (7.86). *Set $\mathcal{J}_\varepsilon^- = \{j < J_{\varepsilon,0}\}$. Then $D_\varepsilon(\mathcal{J}_\varepsilon^-) = o(1)$.*

Proof. Note that under assumption (7.86) we have

$$a_j^p = O(J_{\varepsilon,0}^{p/2} 2^{(rp+1)j}), \quad j < J_{\varepsilon,0},$$

which implies

$$\sum_{j < J_{\varepsilon,0}} a_j^h = O(J_{\varepsilon,0}^{h/2} 2^{(rh+h/p)J_{\varepsilon,0}}) = O((\log \varepsilon^{-1})^B)$$

for some $B > 0$. On the other hand, under assumptions (7.69), relations (A.187), (A.188), (A.186) imply $r_\varepsilon/\varepsilon \ge B\varepsilon^{-b}$ for some $b > 0$. These relations imply $D_\varepsilon(\mathcal{J}_\varepsilon^-) = o(1)$. □

Remark. Assume

$$p > q, \quad \delta = s - r > 0 \quad \text{or } p \le q, \quad \delta = s - r + 1/q - 1/p > 0,$$

and put $\mathcal{J}_\varepsilon^+ = \{j > J_{\varepsilon,1}\}$. Then analogous evaluations show that $D_\varepsilon(\mathcal{J}_\varepsilon^+) = o(1)$. In fact, note the inequalities

$$2^{sqj} \sum_\iota |v_{\iota j}|^q \le (R/\varepsilon)^q$$

and

$$\left(\sum_{i=1}^n |x_i|^p\right)^{1/p} \le \left(\sum_{i=1}^n |x_i|^q\right)^{1/q}, \quad \text{if } p > q,$$
$$\left(n^{-1}\sum_{i=1}^n |x_i|^p\right)^{1/p} \le \left(n^{-1}\sum_{i=1}^n |x_i|^q\right)^{1/q}, \quad \text{if } p \le q.$$

Using these inequalities, we get $a_j \le 2^{-\delta j}(R/\varepsilon)$. It follows from (A.187), (A.188), (A.186) that $(R/r_\varepsilon) = O(\varepsilon^{-B})$ for some $B = B(\Gamma) > 0$ which imply

$$D_\varepsilon(\mathcal{J}_\varepsilon^+) = O((R/r_\varepsilon)^h 2^{-\delta h J_{\varepsilon,1}}) = o(1).$$

A.5.1 *Case* $\kappa \in \Xi_{G_1}$

Note that this case corresponds to all $p \le q$, $p \le 2$ and is possible for $q < p \le 2$ and for $2 < p < q$ (see Figures 1–8 in Section 6.4).

First, let $p \le q$, $p \le 2$ (we omit index ε). It follows from the study extreme problem in Section 4.3.4 that $h_{\varepsilon,j} \equiv 1$ and that there exists $\delta = \delta(\Gamma) > 0$ such that if $\tau = (\kappa, h, t) \in \Gamma$ and $\kappa \in \Xi_{G_1}$, then

$$\sup_j z_{\varepsilon,j}(\tau_\varepsilon) 2^{\delta \max(j,j_1)} = o(1). \tag{A.194}$$

Set

$$\mathcal{J}_\varepsilon = \{j \notin \mathcal{J}_\varepsilon^-\}, \quad \delta_\varepsilon^- = D_\varepsilon(\mathcal{J}_\varepsilon^-).$$

Using Proposition A.9 we have $\tilde{v}_\varepsilon \in \tilde{V}_\varepsilon$ with $\delta_\varepsilon \sim \delta_\varepsilon^-/h = o(1)$. Using (A.194) and (7.86) one has

$$\sup_{\iota j} |v_{\iota j}| z_{\varepsilon,j} = o(1),$$

and by $\sinh x \sim x$ as $x = o(1)$, we get

$$(\bar{\pi}_{\varepsilon,j}^*, \bar{\delta}_{v_j}) \sim H_j(v), \tag{A.195}$$

where the quantities $H_j(v)$ are determined by (7.85). Using (A.193), we get

$$\sup_j H_j(\tilde{v}_\varepsilon)/(1 + |j - j_\varepsilon^*|) \ge C_1 \sqrt{\log\log \varepsilon^{-1}}(1 + o(1)), \quad C_1 = C/B(\Gamma). \tag{A.196}$$

If $C_1 > 2$, then (A.196) and (A.186) imply (7.87). In fact, by (A.186) we have $\sqrt{\log j_\varepsilon^*} \sim \sqrt{\log\log \varepsilon^{-1}}$, and it suffices to verify that

$$\sqrt{\log j} \le c_\varepsilon(1 + |j - j_\varepsilon^*|)\sqrt{\log j_\varepsilon^*}, \quad c_\varepsilon \sim 1 \quad \forall\, j.$$

Clearly, this holds for $j \le Bj_\varepsilon^*$, because $\sqrt{\log Bj_\varepsilon^*} \sim \sqrt{\log j_\varepsilon^*}$ for any $B > 1$. If $j > Bj_\varepsilon^*$, then $|j - j_\varepsilon^*| > bj$, $b = (B-1)/B > 0$, and $\sqrt{\log j} = o(j)$.

Next, let $q < p \le 2$ or $2 < p < q$. Recall that, by $\kappa \in \Xi_{G_1}$, this yields the inequality $I = I(\kappa) < 0$. Denote

$$b_{\varepsilon,j} = 2^{rpj} \sum_\iota |v_{\iota,j}|^p/(r_\varepsilon/\varepsilon)^p. \tag{A.197}$$

Fix $\eta > 0$ and let $\mathcal{J}^* = \mathcal{J}_\varepsilon^*(v_\varepsilon) = \mathcal{J}_\varepsilon^1(v_\varepsilon) \cap \mathcal{J}_\varepsilon^0$ where

$$\begin{aligned} \mathcal{J}_\varepsilon^1(v_\varepsilon) &= \{j \ge J_{\varepsilon,0} : H_j(v_\varepsilon)/T_j < 2\}, \\ \mathcal{J}_\varepsilon^0 &= \begin{cases} \{j > j_1(1+\eta)\}, & \text{if } q < p \le 2,, \\ \{J_{\varepsilon,0} \le j < j_1(1-\eta)\}, & \text{if } q > p > 2. \end{cases} \end{aligned}$$

Proposition A.10 *Under the above assumptions there exists $\tau > 0$, $\tau = \tau(\eta, \Gamma)$, such that*

$$\sup_{j \in \mathcal{J}^*} b_{\varepsilon,j} 2^{\tau|j-j_1|} = o(1).$$

Proof. Using the Hölder inequality, for $q < p \le 2$ or $2 < p < q$, one can obtain that

$$\sum_\iota |v_{\iota,j}|^p \le \left(\sum_\iota |v_{\iota,j}|^2\right)^{(q-p)/(q-2)} \left(\sum_\iota |v_{\iota,j}|^q\right)^{(p-2)/(q-2)}.$$

Because

$$2^{sqj} \sum_\iota |v_{\iota,j}|^q \le (R/\varepsilon)^q, \quad \sum_\iota |v_{\iota,j}|^2 \le 4\sqrt{2}(\log j)^{1/2} 2^{j/2},$$

and using relations (A.187), (7.69), after some simple arithmetical calculations, we get

$$b_{\varepsilon,j} \le Bu_\varepsilon^{(q-p)/2}(\log j)^{(q-p)/2(q-2)} 2^{-I(j-j_1)/2(q-2)}$$

which yields the required relation. □

Put

$$\mathcal{J}_\varepsilon = \{j \notin (\mathcal{J}_\varepsilon^- \cup \mathcal{J}^*)\}, \quad \delta_\varepsilon^* = D_\varepsilon(\mathcal{J}_\varepsilon^- \cup \mathcal{J}^*).$$

Using Propositions A.9, A.10 we easily get $\tilde{v}_\varepsilon \in \tilde{V}_\varepsilon$ with $\delta_\varepsilon \sim \delta_\varepsilon^*/h = o(1)$. Note that if $H_j(v_\varepsilon)/T_j \ge 2$ for $j \in \mathcal{J}_\varepsilon^0$, then (7.87) holds.

Let $\mathcal{J}_\varepsilon^0 = \mathcal{J}^*$. It follows from Corollary 6.3 that there exists $\delta = \delta(\Gamma) > 0$ such that if $\tau = (\kappa, h, t) \in \Gamma, \kappa \in \Xi_{G_1},\ q < p \le 2$, or $2 < p < q$, then

$$\sup_{j \notin \mathcal{J}_\varepsilon^0,\ j \ge J_{\varepsilon,0}} z_{\varepsilon,j}(\tau_\varepsilon) 2^{\delta \max(j,j_1)} = o(1). \tag{A.198}$$

These imply that $(\bar\pi^*_{\varepsilon,j}, \bar\delta_{v,j}) \sim H_j(v)$ uniformly over $j \notin \mathcal{J}^0_\varepsilon$, $j \ge J_{\varepsilon,0}$. Using (A.193), we obtain (A.196) which implies (7.87).

A.5.2 Case $\kappa \in \Xi_{G_2}$

This case corresponds to $I > 0$, $\lambda > 0$; this holds for all $p > q$, $p > 2$; this is possible for $2 < p \le q$ and for $2 \ge p > q$ (see Figures 1–8 in Section 6.4).

First, suppose that $2 < p \le q$. Put

$$\mathcal{J}_\varepsilon = \{j \notin \mathcal{J}^-_\varepsilon\}, \quad \delta^-_\varepsilon = D_\varepsilon(\mathcal{J}_\varepsilon).$$

Using Proposition A.9 we have $\tilde v_\varepsilon \in \tilde V_\varepsilon$ with $\delta_\varepsilon \sim \delta^-_\varepsilon / h = o(1)$.

Fix a small enough $\eta > 0$. It follows from Proposition 6.3 that there exist $A = A(\Gamma) > 0$, $\delta = \delta(\Gamma) > 0$ such that if $\tau = (\kappa, h, t) \in \Gamma$, $\kappa \in \Xi_{G_2}$, $2 < p \le q$, then

$$z_{\varepsilon,j} \le \begin{cases} A, & \text{if } j \le j_0(1+\eta), \\ 2^{-\delta(j-j_0)}, & \text{if } j > j_0(1+\eta), \end{cases} \tag{A.199}$$

which imply $(\bar\pi^*_{\varepsilon,j}, \bar\delta_v) \sim H_j(v)$ uniformly over $j > j_0(1+\eta)$.

Let $J_{\varepsilon,0} \le j \le j_0(1+\eta)$. If $z_{\varepsilon,j} < z_{j,1} = 1/\sqrt{j}$, then we put $k_j = 1$. In the other case we set

$$k_j = k_{\varepsilon,j} = \max\{k \,:\, z_{j,k} \le z_{\varepsilon,j}\}.$$

Note that $1 \le k_j \le O(K(c,j))$ because of (A.199). If $z_{\varepsilon,j} < z_{j,1} = 1/\sqrt{j}$, then $z_{\varepsilon,j}|v_{\iota j}| < 2$, and by the choice of z_{j,k_j}, we have

$$\sinh^2(z_{\varepsilon,j} v_{\iota j}/2)/\sinh(z^2_{\varepsilon,j}/2) \ge c_1 \sinh^2(z_{j,k_j} v_{\iota j}/2)/\sinh(z^2_{j,k_j}/2)$$

for some (absolute) constant c_1. This relation implies that

$$H_{j,k_{\varepsilon,j}}(\tilde v_\varepsilon) = (\bar\pi^*_{j,k_{\varepsilon,j}}, \bar\delta_{\tilde v_{\varepsilon,j}}) \ge c_1 (\bar\pi^*_{\varepsilon,j}, \bar\delta_{\tilde v_{\varepsilon,j}}). \tag{A.200}$$

Thus, using (A.193), we get

$$\begin{aligned} \max&\left\{ \sup_{j \le j_0(1+\eta)} H_{j,k_{\varepsilon,j}}(\tilde v_\varepsilon)/(1+|j-j^*_\varepsilon|),\ \sup_{j \ge j_0(1+\eta)} H_j/(1+|j-j^*_\varepsilon|) \right\} \\ &\ge C_1\sqrt{\log\log\varepsilon^{-1}}(1+o(1)), \\ &C_1 = \min\left(c_1 C/B(\Gamma),\ C/B(\Gamma)\right) > 2, \end{aligned}$$

for large enough $C > 0$. This relation and (A.186) imply (7.87) or (7.88).

Next, assume $2 < p$, $q < p$, or $2 \ge p > q$. It follows from Corollary 6.3 that there exists $A = A(\Gamma) > 0$ such that if $\tau = (\kappa, h, t) \in \Gamma$, $\kappa \in \Xi_{G_2}$, $q < p \le 2$ or $p > 2$, $p > q$, then

$$z_{\varepsilon,j} \le \begin{cases} A, & \text{if } j \le j_0, \\ A\sqrt{1+j-j_0}, & \text{if } j > j_0. \end{cases} \tag{A.201}$$

Fix a small enough $\eta > 0$, such that

$$z_{\varepsilon,j} < \sqrt{K(c,j) - K} = \sqrt{cj} \quad \text{for } J_{\varepsilon,0} \le j \le j_0(1+\eta).$$

Denote $\mathcal{J}^1 = \{j > j_0(1+\eta)\}$.

Proposition A.11 *Under the above assumptions there exists $\tau > 0$, $\tau = \tau(\eta, \Gamma)$, such that*

$$\sup_{j \in \mathcal{J}^1} b_{\varepsilon,j} 2^{\tau(j-j_0)} = o(1).$$

Proof. By $p \geq q$, using the inequality

$$\sum_{\iota} |v_{\iota,j}|^q \leq \max_{\iota} |v_{\iota,j}|^{p-q} \sum_{\iota} |v_{\iota,j}|^q,$$

and because

$$2^{sqj} \sum_{\iota} |v_{\iota,j}|^q \leq (R/\varepsilon)^q, \quad \max_{\iota} |v_{\iota,j}| \leq \sqrt{2(\log 2)j}(1+o(1)),$$

using also relations (A.187), (7.69), after some simple arithmetical calculations, we get

$$b_{\varepsilon,j} \leq B(j)^{(p-q)/2} 2^{-\lambda(j-j_0)}, \quad \lambda = \lambda(\kappa) > 0,$$

which implies the necessary relation. □

Put

$$\mathcal{J}_\varepsilon = \{j : \ j \notin \mathcal{J}^- \cup \mathcal{J}^1\}, \quad \delta_\varepsilon^1 = D_\varepsilon(\mathcal{J}^- \cup \mathcal{J}^1).$$

Using Propositions A.9, A.11 we easily get $\tilde{v}_\varepsilon \in \tilde{V}_\varepsilon$ with $\delta_\varepsilon \sim \delta_\varepsilon^1/h = o(1)$.

Let $J_{\varepsilon,0} \leq j \leq j_0(1+\eta)$. If $z_{\varepsilon,j} < z_{j,1} = 1/\sqrt{j}$, then we set $k_j = 1$. In the other case we put

$$k_j = k_{\varepsilon,j} = \max\{k : z_{j,k} \leq z_{\varepsilon,j}\}.$$

Note that $1 \leq k_j \leq K(c,j)$ because of (A.201). Then, analogously to (A.200), we can find an (absolute) constant c_1 such that

$$H_{j,k_\varepsilon}(\tilde{v}_\varepsilon) = (\bar{\pi}^*_{j,k_\varepsilon}, \bar{\delta}_{\tilde{v}_{\varepsilon,j}}) \geq c_1 (\bar{\pi}^*_{\varepsilon,j}, \bar{\delta}_{\tilde{v}_{\varepsilon,j}}).$$

Thus, using (A.193), we get

$$\sup_j H_{j,k_{\varepsilon,j}}(\tilde{v}_\varepsilon)/(1+|j-j_\varepsilon^*|) \geq C_2 \sqrt{\log\log\varepsilon^{-1}}(1+o(1)), \ C_2 = c_1 C/B(\Gamma).$$

This relation and (A.186) imply (7.88). □

A.6 Proofs of Lemmas 8.2, 8.3, 8.4, 8.6

We give this outline of the proofs and omit some calculations which are rather simple.

The lemmas deal with linear convex minimization problems: to minimize the linear functional

$$\inf_{\bar{r}\in\Xi_n} F(\bar{r}), \quad F(\bar{r}) = \sum_{i=1}^{n} \int \phi(v) r_i(dv),$$

where $\Xi_n = \{\bar{r} = (r_1, \dots, r_n)\}$ is a convex set of collections of probability measures r_i on the real line, subject to the constraints

$$\sum_{i=1}^{n} \int \phi_1(v) r_i(dv) \geq W_1, \quad \sum_{i=1}^{n} \int \phi_2(v) r_i(dv) \leq W_2, \quad r_i([-H_n, H_n]) = 1, \tag{A.202}$$

$i = 1, \dots, n$. The functions $\phi(v)$ are defined by (8.109), (8.144), (8.146), (8.154) and satisfy

$$\phi(t) = \phi(-t) \geq 0, \quad \phi(0) = 0,$$

the functions $\phi_l(v)$ and quantities W_l, $l = 1, 2$, are of the form

$$\phi_1(v) = |v|^p, \quad \phi_2(v) = |v|^q, \quad W_1 = W_{1,n} = \rho_n^p, \quad W_2 = W_{2,n} = R_n^q.$$

By symmetry of the problems the infimum is attained on the collection $\bar{r}^* = (r^*, \dots, r^*) \in \Xi_n$ of identical symmetric measures r^*. Furthermore, using the method of subdifferentials and the Kuhn–Tucker Theorem (see, e.g., [63], pp. 76–77) one gets the sufficient conditions for an infimum: there exist $\lambda = \lambda_n > 0$, $\mu = \mu_n > 0$, $\zeta = \zeta_n$, such that

$$\phi(v) - \lambda\phi_1(v) + \mu\phi_2(v) \geq \zeta \quad \forall\, v \in [-H_n, H_n], \tag{A.203}$$

$$r^*\left(\{v : \phi(v) - \lambda\phi_1(v) + \mu\phi_2(v) = \zeta\}\right) = 1, \tag{A.204}$$

and

$$n \int \phi_1(v) r^*(dv) = W_1, \quad n \int \phi_2(v) r^*(dv) = W_2. \tag{A.205}$$

Certainly, this fact is easily seen directly. In fact, let $\bar{r}(r_1, \dots, r_n) \in \Xi_n$. Then, by integrating inequality (A.203) over r_i and summing, we get

$$\sum_{i=1}^{n} \int (\phi(v) - \lambda\phi_1(v) + \mu\phi_2(v) - \zeta) r_i(dv) \geq 0$$

which yields

$$F(\bar{r}) \geq \sum_{i=1}^{n} \left(\lambda \int \phi_1(v) r_i(dv) - \mu \int \phi_2(v) r_i(dv) + \zeta \right) \geq \lambda W_1 - \mu W_2 + n\zeta$$

and we have a strict inequality here if

$$\exists\, i \in \{1, \dots, n\} : r_i\left(\phi(v) - \lambda\phi_1(v) + \mu\phi_2(v) > \zeta\right) > 0.$$

On the other hand, equalities (A.204), (A.205) yield the equality

$$F(\bar{r}^*) = \lambda W_1 - \mu W_2 + n\zeta.$$

It suffices to check that the measure $r^* = \pi(z_n, h_n)$ satisfies (A.204), (A.205), and that (A.203) holds for some $\lambda > 0$, $\mu > 0$, and $\eta \in R^1$. Relation (A.205) follows from the definition of the parameters z_n, h_n in (8.71). Since $0 < z_n < H_n$ in the lemmas, relation (A.204) implies the equality in (A.203) for $v = 0$, $v = z_n$, and $v = -z_n$. Since the functions $\phi(v)$, $\phi_1(v) = |v|^p$, $\phi_2(v) = |v|^q$ are symmetric, it suffices to verify inequality (A.203) for $v \in [0, H_n]$. We put $\zeta = 0$ and take λ, μ from the equations

$$\phi(z_n) = \lambda z_n^p - \mu z_n^q, \quad \phi'(z_n) = p\lambda z_n^{p-1} - q\mu z_n^{q-1},$$

which corresponds to

$$\lambda = \frac{z_n\phi'(z_n) - q\phi(z_n)}{z_n^p(p-q)}, \quad \mu = \frac{z\phi'(z_n) - p\phi(z_n)}{z_n^q(p-q)}. \tag{A.206}$$

Since $\phi(0) = 0$, under this choice we get equality (A.204). Moreover, these imply that the line $w = 0$ at the half-plane $\{(v, w),\ v > 0\}$ is tangent to the line $w = \phi(v) - \lambda v^p + \mu v^q$ at the point $(z_n, 0)$. We need to check the inequalities $\lambda > 0$, $\mu > 0$ and, for $0 \le v \le H_n$,

$$\psi(v) = \phi(v) + t_n^q \frac{z_n\phi'(z_n) - p\phi(z_n)}{p-q} - t_n^p \frac{z_n\phi'(z_n) - q\phi(z_n)}{p-q} \ge 0; \tag{A.207}$$

here and below we set $t_n = t_n(v) = v/z_n$. For large enough z_n the inequalities $\lambda > 0$, $\mu > 0$ follow from the relations

$$\phi'(z_n) > 0, \quad \eta(z_n) = \phi(z_n)/z_n\phi'(z_n) \to 0, \quad \text{as } z_n \to \infty. \tag{A.208}$$

We leave for the reader to verify relations (A.208) for the functions defined by (8.109), (8.144), (8.146), (8.154) under the assumptions of the lemmas (the verification is based on simple calculations and on relation (8.19)).

To prove relation (A.207), let us divide the interval $(0, H_n]$ into three (or two) subintervals

$$\Delta_n^- = (0, z_n^-), \quad \Delta_n^0 = [z_n^-, z_n^+]; \quad \Delta_n^+ = (z_n^+, H_n]$$

(the interval Δ_n^+ is empty under the assumptions of Lemma 8.6).

Let $v \in \Delta_n^-$, where $z_n^- = z_n(1 - B\eta(z_n))$ and $B = B(p,q)$ is large enough. Note that $\mu z_n^q = \lambda z_n^p + \phi(z_n)$ and $\phi(v) \ge 0$ for $v \in \Delta_n^-$. Then relation (A.207) follows from

$$t_n^q \frac{z_n\phi'(z_n) - p\phi(z_n)}{p-q} - t_n^p \frac{z_n\phi'(z_n) - q\phi(z_n)}{p-q} = \phi(z_n)t_n^p(A(z_n) - 1),$$

where $A(z) = (t_n^{p-q} - 1)(\eta^{-1}(z) - p)/(p-q)$ and we have, as $z \to \infty$,

$$A(z) > ((1 - B\eta(z))^{p-q} - 1)\frac{\eta^{-1}(z) - p}{p-q} \sim B(1 - p\eta(z)) \to B.$$

Fix $\delta > 0$ small enough. Let

$$Q_n \ge z_n^+ = z_n + \delta d_n, \quad H_n > z_n^+,$$

(this holds under the assumptions of Lemmas 8.2, 8.3, 8.4) and let $v \in \Delta_n^+ = [z_n^+, H_n]$. To satisfy (A.207) for $v \in \Delta_n^+$ it is sufficient that, for some $B > 0$,

$$Bz_n \geq H_n, \quad \phi(v) \geq \phi(z_n^+) > 0, \quad v \in \Delta_n^+, \quad z_n\phi'(z_n)/\phi(z_n^+) \to 0, \tag{A.209}$$

as $z_n \to \infty$, In fact, under (A.209) for $v \in \Delta_n^+$ we have $1 < t_n < B$. Setting $g(v) = t_n^p - t_n^q$ we get,

$$0 < g(v) < (H_n/z_n)^p \leq B^p,$$

and

$$\begin{aligned} \psi(v) &\geq \phi(z_n^+) + \phi(z_n)g(v)\left(1 - \frac{z_n\phi'(z_n)}{\phi(z_n)(p-q)}\right) \\ &> \phi(z_n^+)\left(1 - \frac{z_n\phi'(z_n)}{\phi(z_n^+)}\frac{g(v)}{(p-q)}\right) = \phi(z_n^+)(1 + o(1)). \end{aligned}$$

The first relation (A.209) is given by the assumptions of Lemmas 8.2, 8.3, 8.4, 8.6. The third relation (A.209) is verified by using simple calculations and (8.19) under the assumptions of Lemmas 8.2, 8.3, 8.4. By the function $\phi(v)$, increases in v in Lemma 8.3, and we have the relation

$$\phi(v) \geq \phi(z_n^+), \quad v \in \Delta_n^+, \tag{A.210}$$

which yields the second relation in (A.209). Under the assumptions of Lemmas 8.2, 8.4 the function $\phi(v)$ increases in $v \leq v^* \sim Q_n$ and decreases in $v \geq v^*$; $v^* > z_n^+$. If $v^* \geq H_n$, then this yields the second relation (A.210). If $v^* < H_n$, then $\phi(H_n) \geq \phi(z_n^+)$ under the constraints on H_n in Lemmas 8.2, 8.3, 8.4, 8.6. This yields the second relation (A.210) as well.

Let $v \in \Delta_n^0 = (z_n^-, z_n^*)$ where $z_n^* = \min(H_n, z_n^+)$ (note that $z_n^* = H_n$ under the assumptions of Lemma 8.6). Recall that the line $w = 0$ is tangent to the line $w = \psi(v)$ at the point $v = z_n$. Inequality (A.207) follows from the convexity of $\psi(v)$, $v \in \Delta_n^0$.

Setting $\phi_n'' = \inf_{v \in \Delta_n^0} \phi''(v)$ note that

$$\psi''(v) \geq \phi_n''\left(1 + \frac{\phi'(z_n)}{z_n\phi_n''}\left(A(v, z_n) + \frac{\phi(z_n)}{z_n\phi'(z_n)}B(v, z_n)\right)\right),$$

where

$$\begin{aligned} A(v, z_n) &= \frac{1}{p-q}\left(q(q-1)t_n^{q-2} - p(p-1)t_n^{p-2}\right); \\ B(v, z_n) &= \frac{pq}{p-q}\left((p-1)t_n^{p-2} - (q-1)t_n^{q-2}\right). \end{aligned}$$

The functions $A(v, z_n)$, $B(v, z_n)$ are bounded over $v \in \Delta_n^0$ for large enough z_n. Therefore the convexity of $\psi(v)$, $v \in \Delta_n^0$, follows from condition (A.208) and from the following conditions:

$$\phi_n'' = \inf_{v \in \Delta_n^0} \phi''(v) > 0, \quad \phi'(z_n)/z_n\phi_n'' \to 0 \quad \text{as } z_n \to \infty. \tag{A.211}$$

One can easily verify conditions (A.211) under the assumptions of Lemmas 8.2, 8.3, 8.4, 8.6. In fact, $\phi_n'' = \phi''(z_n^-) > 0$ under the assumptions of Lemmas 8.2, 8.3, 8.4; $\phi_n'' \asymp \phi'(z_n) \asymp 1$ under the assumptions of Lemma 8.6. The second relations (A.211) are verified by using simple calculations and (8.19) under the assumptions of Lemmas 8.2, 8.3, 8.4, 8.6. We leave these calculations to the reader.

Lemmas 8.2, 8.3, 8.4, 8.6 are proved. □

References

[1] Barra, J.-R. *Notions Fondamentales de Statistique Mathématique.* Dunod, Paris, 1971.

[2] Bickel, P.J., Chibisov, D.M., and van Zwet W.R. On efficiency of first and second-order. *Internat. Statist. Rev.*, **49** (1981), 169–175.

[3] Borovkov, A.A. On the power of the χ^2 test with increasing number of class intervals. *Theory Probab. Appl.*, **22** (1977), 366–370.

[4] Borovkov, A.A. *Mathematical Statistics.* Nauka, Moscow, 1984. (In Russian.)

[5] Borovkov, A.A. *Mathematical Statistics: Supplementary Chapters.* Nauka, Moscow, 1984. (In Russian.)

[6] Brown, L.D. and Low, M.G. Asymptotic equivalence of nonparametric regression and white noise. *Ann. Statist.*, **24** (1996), 2384–2398.

[7] Bretagnolle, J. and Huber, C. Estimation de densités: Risque minimax. *Z. Wahrsch. Verw. Gebiete*, **47** (1979), 119–137.

[8] Burnashev, M.V. On the minimax detection of an inaccurately known signal in a Gaussian noise background. *Theory Probab. Appl.*, **24** (1979), 107–119.

[9] Cartan, H. *Théorie Élémentaire des Fonctions Analytiques d'une ou Plusieurs Variables Complexes.* Hermann, Paris, 1961.

[10] Chentsov N.N. *Statistical Decision Rules and Optimal Inference.* Nauka, Moscow, 1972. (In Russian.)

[11] Cohen, A., Daubechies, I., Jewerth, B., and Vial, P. Multiresolution analysis, wavelets and fast algorithms on an interval. *C. R. Acad. Sci. Paris (A)*, **316** (1993), 417–421.

[12] Cohen, A., Daubechies, I., and Vial, P. Wavelets on an interval and fast wavelet transforms. *Appl. Comput. Harmonic Anal. (A)*, **1** (1993), 54–81.

[13] Darling, D.A. The Kolmogorov–Smirnov, Cramér–von Mises tests. *Ann. Math. Statist.*, **28** (1957), 823–838.

[14] Delyone, B. and Juditsky, A. Wavelet estimators, global error measures: revisited. Technical Report 782 (1993), IRISA, Rennes.

[15] Devore, R.A. and Popov, V.A. Interpolation of Besov spaces. *Trans. Amer. Math. Soc.*, **305** (1988), No. 1, 397–413.

[16] Donoho, D.L. Asymptotic minimax risk for sup-norm loss: solution via optimal recovery. *Probab. Theory Related Fields*, **99** (1993), 145–170.

[17] Donoho, D.L. and Johnstone, I.M. Minimax estimation via wavelet shrinkage. *Ann. Statist.*, **26** (1998), No 3, 879–921.

[18] Donoho, D.L, Johnstone, I.M., Kerkyacharian, G., and Picard, D. Wavelet shrinkage: asymptopia ? (with discussion). *J. Royal Statist. Soc., Ser. B*, **57** (1995), No 2, 301–369.

[19] Donoho, D.L., Liu, R.S., and Mac Gibbon, B. Minimax risk of hyperrectangles and implications. *Ann. Statist.*, **18** (1990), 1416–1437.

[20] Efroimovich, S.Yu. and Pinsker, M.S. Estimation of square-integrable probability density of a random variable. *Problems Inform. Transmission*, **18** (1982), 175–189.

[21] Ermakov, M.S. Asymptotically minimax tests for testing nonparametric hypotheses on a distribution density. *Zap. Nauchn. Sem. (LOMI)*, **166** (1988), 44–53. (In Russian.)

[22] Ermakov, M.S. Minimax detection of a signal in a Gaussian white noise. *Theory Probab. Appl.*, **35** (1990), 667–679.

[23] Ermakov, M.S. Minimax hypotheses testing on a spectral density. *Zap. Nauchn. Sem. (LOMI)*, **184** (1990), 115–125.

[24] Ermakov, M.S. Asymptotic minimaxity of usual goodness of fit tests. In: *Proc. 5th Vilnius Conference on Probab. Theory and Math. Stat. VSP/Mokslas*, **1** (1990), 322–331.

[25] Ermakov, M.S. On asymptotic minimaxity of rank tests. *Statist. Probab. Lett.*, **15** (1992), 191–196.

[26] Ermakov, M.S. Minimax nonparametric testing of a hypotheses on a distribution density. *Theory Probab. Appl.*, **39** (1994), No 3, 396–416.

[27] Ermakov, M.S. Asymptotic minimaxity of tests of Kolmogorov and omega-squared types. *Theory Probab. Appl.*, **40** (1995), 54–67.

[28] Ermakov, M.S. Asymptotic minimaxity of chi-squared tests. *Theory Probab. Appl.*, **42** (1997), 668–695.

[29] Fan, Ky. Minimax theorems. *Proc. Nat. Acad. USA*, **39** (1953), 42–47.

[30] Hájek, J. and Sidak, Z. *Theory of Rank Tests*. Academia, Prague, 1967.

[31] Härdle, W., Kerkyacharian, G., Picard, D., and Tsybakov, A.B. *Wavelets: Properties and Statistical Applications*. Lectures Notes in Statistics. Springer-Verlag, New York, 1997.

[32] Horowitz, J.L. and Spokoiny, V.G.. An adaptive, rate-optimal test for a parametric model against a non-parametric alternative. *Econometrica*, **69** (2001), No 3, 599–631.

[33] Huber, P. J. *Robust Statistics*. Wiley, New York, 1981.

[34] Huber, P. J. and Strassen, V. Minimax tests and the Neyman-Pearson lemma for capacities, *Ann. Statist.*, **1** (1973), 251–263; **2** (1974), 223–224.

[35] Ibragimov, I.A. and Khasminskii, R.Z. One problem of statistical estimation in a white Gaussian noise. *Soviet Math. Dokl.*, **236** (1977), No 4, 333–337.

[36] Ibragimov, I.A. and Khasminskii, R.Z. Asymptotic properties of some non-parametric estimaters in a Gaussian white noise. In: *Proc 3rd Summer School on Probab. Theory and Math. Stat. Varna 1978, Sofia* (1980), 31–64.

[37] Ibragimov, I.A. and Khasminskii, R.Z. On estimation of a probability density. *Zap. Nauchn. Sem. (LOMI)*, **98** (1980), 66–85.

[38] Ibragimov, I.A. and Khasminskii, R.Z. *Statistical Estimation: Asymptotic Theory*. Springer-Verlag, Berlin, 1981.

[39] Inglot, T. and Ledwina, T. Asymptotic optimality of data-driven Neyman's tests for uniformity. *Ann. Statist.*, **24** (1996), No 5, 1996–2019.

[40] Inglot, T., Kallenberg, W.C.M. and Ledwina, T. Vanishing shortcoming and asymptotic relative efficiency. *Ann. Statist.*, **28** (2000), No 1, 215–238.

[41] Ingster, Yu.I. Minimax nonparametric detection of signals in white Gaussian noise. *Problems Inform. Transmission*, **18** (1982), 130–140.

[42] Ingster, Yu.I. Asymptotically optimal Bayes tests for composite hypotheses. *Theory Probab. Appl.*, **28** (1983), 775–794.

[43] Ingster, Yu.I. Asymptotic minimax testing of nonparametric hypotheses on the distribution density of an independent sample. *Zap. Nauchn. Sem. (LOMI)*, **136** (1984), 74–96. (In Russian.)

[44] Ingster, Yu.I. An asymptotic minimax test of nonparametric hypotheses about spectral density. *Theory Probab. Appl.*, **29** (1984), 846–847.

[45] Ingster, Yu.I. Asymptotically optimal tests for testing composite finite-parametrical hypotheses. *Theory Probab. Appl.*, **30** (1985), 289–308.

[46] Ingster, Yu.I. Minimax testing of nonparametric hypotheses on a distribution density in L_p-metrics. *Theory Probab. Appl.*, **31** (1986), 333–337.

[47] Ingster, Yu.I. Asymptotically minimax testing of the hypothesis of independence, *Zap. Nauchn. Sem. (LOMI)*, **153** (1986), 60–72.

[48] Ingster, Yu.I. On comparison of the minimax properties of Kolmogorov's, ω^2 test and χ^2 test. *Theory Probab. Appl.*, **32** (1987), 346–350.

[49] Ingster, Yu.I. Asymptotically minimax testing of nonparametric hypotheses, In: *Proc. 4th Vilnius Conference on Probab. Theory and Math. Stat.*, VNU Science Press, **1** (1987), 553–573.

[50] Ingster, Yu.I. Minimax testing of a simple hypothesis against a weakly nonparametric set of alternatives, *Theory Probab. Appl.*, **33** (1988), 552–556.

[51] Ingster, Yu.I. Minimax detection of a signals in l_p-metrics. *Zap. Nauchn. Sem. (LOMI)*, **184** (1990), 152–168 (In Russian, Transl: *J. Soviet Math.*, **68** (1994), No 4).

[52] Ingster, Yu.I. Asymptotically minimax hypothesis testing for nonparametric alternatives. I, II, III. *Math. Methods Statist.*, **2** (1993), 85–114, 171–189, 249–268.

[53] Ingster, Yu.I. Minimax testing of the hypothesis of independence for ellipsoids in l_p, *Zap. Nauchn. Sem. (POMI)*, **207** (1993), 77–97. (In Russian.)

[54] Ingster, Yu.I. Minimax hypotheses testing on a probability density for ellipsoids in l_p. *Theory Probab. Appl.*, **39** (1994), No 3, 417–430.

[55] Ingster, Yu.I. Minimax hypotheses testing for non-degenerate loss functions and extreme convex problems. *Zap. Nauchn. Sem. (POMI)*, **228** (1996), 162–188. (In Russian, Transl: *J. Math. Sci. (New York)*, **93** (1999), 354–371).

[56] Ingster, Yu.I. Some problems of hypothesis testing leading to infinitely divisible distributions. *Math. Methods of Statist.*, **6** (1997), No 1, 47–69.

[57] Ingster, Yu.I. Adaptive chi-square tests. *Zap. Nauchn. Sem. (POMI), Probab. and Statist. 2.*, **244** (1997), 150–166.

[58] Ingster, Yu.I. Minimax detection of a signal for l^n-balls. *Math. Methods of Statist.*, **7** (1998), No 4, 401–428.

[59] Ingster, Yu.I. Adaptation in minimax non-parametric hypothesis testing. WIAS, Preprint No. 419. Berlin, 1998.

[60] Ingster, Yu.I. On testing of hypothesis which is close to simple hypothesis. *Theory Probab. Appl.*, **45** (2000), No 2, 356–368.

[61] Ingster, Yu.I. and Suslina, I.A. Minimax signal detection for Besov balls and bodies. *Problems Inform. Transmission*, **34** (1998), No 1, 56–68.

[62] Ingster, Yu.I. and Suslina, I.A. Minimax nonparametric hypothesis testing for ellipsoids and Besov bodies. *ESAIM: Probab. and Statist.*, **4** (2000), 53–135.

[63] Ioffe, A.D. and Tikhomirov, V.M. *The Theory of Extreme Problems.* Nauka, Moscow, 1974 (In Russian)

[64] Juditsky, A. Wavelet estimators: adapting to unknown smoothness. *Math. Methods of Statist.*, **6** (1997), No 1, 1–25.

[65] Kac, M., Kiefer, J. and Wolfowitz, J. On tests of normality and other tests of goodness of fit based on distance methods, *Ann. Math. Statist.*, **26** (1955), 189–211.

[66] Kallenberg, W.C.M. Intermediate efficiency, theory and examples. *Ann. Statist.*, **11** (1983), 170–182.

[67] Kendall, M. G. and Stuart, A. *The Advanced Theory of Statistics, Vol. 2: Inferences and Relationships.* Hafner, New York, 1974.

[68] Korostelev, A.P. Asymptotically minimax regression estimator in uniform norm up to exact constant. *Theory Probab. Appl.*, **38** (1993), 737–743.

[69] Korostelev, A.P., Tsybakov, A.B. *Minimax Theory of Image Reconstruction.* Lectures Notes in Statistics, Vol. 93, Springer-Verlag, New York, 1993.

[70] Le Cam, L. Locally asymptotically normal families of distributions. *Univ. Calif. Publ. Statist.*, **3** (1960), 37–98.

[71] Le Cam, L. *Asymptotic methods in statistical decision theory.* Springer-Verlag, Berlin, 1986.

[72] Lehmann, E.L. *Testing Statistical Hypotheses.* Wiley, New York, 1959.

[73] Lepski, O.V. One problem of adaptive estimation in Gaussian white noise. *Theory Probab. Appl.*, **35** (1990), 459–470.

[74] Lepski, O.V. Asymptotic minimax adaptive estimation. 1. Upper bounds. *Theory Probab. Appl.*, **36** (1991), 654–659.

[75] Lepski, O.V. Asymptotic minimax adaptive estimation. 2. Statistical model without optimal adaptation. Adaptive estimators. *Theory Probab. Appl.*, **37** (1992), 468–481.

[76] Lepski, O.V. On asymptotical exact testing of nonparametric hypotheses. CORE D.P. 9329 (1993), Universitè Catholigue de Louvan.

[77] Lepski, O.V., Mammen, E., and Spokoiny, V.G. Optimal spatial adaptation to inhomogeneous smoothness: an approach based on kernel estimates with variable bandwidth selectors. *Ann. Statist.*, **25** (1997), No 3, 929–947.

[78] Lepski, O.V., Nemirovski,A. and Spokoiny, V.G. On estimation of L_p norm of a regression function. *Probab. Theory Related Fields*, **113** (1999), 221–253.

[79] Lepski, O.V. and Spokoiny, V.G. Minimax nonparametric hypothesis testing: the case of an inhomogeneous alternative. *Bernoulli*, **5** (1999), 333–358.

[80] Lepski, O.V. and Tsybakov, A.B. Asymptotically exact nonparametric hypothesis testing in sup-norm and at a fixed point. *Probab. Theory Related Fields*, **117** (2000), No 1, 17–48.

[81] Lepski, O.V. and Spokoiny, V.G. Optimal pointwize adaptive methods in nonparametric estimation. *Ann. Statist.*, **25** (1997), No 6, 2512–2546.

[82] Mann, H. B. and Wald, A. On the choice of the number of intervals in the application of the chi-square test. *Ann. Math. Statist.*, **13** (1942), 306–317.

[83] Meyer, Y. *Ondelettes*. Hermann. Paris, 1990.

[84] Nikitin, Ya.Yu. *Asymptotic Efficiency of Nonparametric Tests.* Cambridge University Press, Cambridge, 1995.

[85] Nussbaum, M. Asymptotic equivalence of density estimation and Gaussian white noise. *Ann. Statist.*, **24** (1996), 2399–2430.

[86] Petrov, V.V. *Sums of Independent Random Variables.* Springer-Verlag, New York, 1975.

[87] Pinsker, M.S. Optimal filtration of square-integrable signals in Gaussian noise. *Problems Inform. Transmission*, **16** (1980), 120–133.

[88] Roussas, G. *Contiguity of Probability Measures: Some Applications in Statistics.* Cambridge University Press, Cambridge, 1972.

[89] Sion, M. On general minimax theorems. *Pacific J. Math.*, **58** (1958), 171–176.

[90] Spokoiny, V.G. Adaptive hypothesis testing using wavelets. *Ann. Stat.*, **24** (1996), No 6, 2477–2498.

[91] Spokoiny, V.G. Adaptive and spatially adaptive testing of nonparametric hypothesis. *Math. Methods of Statist.*, **7** (1998), No 3, 245–273.

[92] Suslina, I.A. Minimax detection of a signal for l_q-ellipsoids with a removed l_p-ball. *Zap. Nauchn. Sem. (POMI)*, **207** (1993), 127–137. (In Russian, Tranl: *J. Math. Sci. (New York)* **81** (1996), 2442–2449).

[93] Suslina, I.A. Extreme problems arising in minimax detection of a signal for l_q-ellipsoids with a removed l_p-ball. *Zap. Nauchn. Sem. (POMI)*, **228** (1996), 312–332. (In Russian, Tranl: *J. Math. Sci. (New York)*, **93** (1999), 454–469).

[94] Tikhomirov, V.M. *Some Problems of Approximation Theory.* Moscow State University Press, Moscow, 1976. (In Russian.)

[95] Triebel, H. *Theory of Functional Space. Vol. 2.* Birkhauser, Basel, 1992.

[96] Tsybakov, A.B. Asymptotically efficient signal estimation in L_2 with general loss functions. *Problems Inform. Transmission*, **33** (1997), No 1, 78–88.

[97] Tsybakov, A.B. Pointwise and sup-norm sharp adaptive of function on the Sobolev classes. *Ann. Statist.*, **26** (1998), No 6, 2420–2469.

[98] Wald, A. Tests of statistical hypotheses concerning several parameters when the number of observations is large. *Trans. Amer. Math. Soc.*, **54** (1943), 426–482.

[99] Wald, A. *Statistical Decision Functions.* Wiley, New York, 1950.

Subject Index

Adaptive problems, 36, 263
Approach
  asymptotic, 2, 8, 23, 63, 137, 145
  Bayesian, 47, 48, 55, 64, 65, 268
  minimax, 12, 22, 47, 53, 55, 67
    asymptotic variant, 63
  Neyman and Pearson, 5, 6, 8, 16, 42, 63
Assumption
  A.1, 156, 159
  A.1a, 158
  A.2, 156, 159
  A.2a, 158
  A.3, 156, 159
  A.4, 156, 159
  A.5, 199
  A.6, 201
Asymptotics
  of classical type, 30, 36, 80, 81, 111, 115–117, 244, 245
  of degenerate type, 30, 31, 36, 80, 129, 136, 173, 176, 178, 180, 181, 184, 244, 245, 265, 268
  of Gaussian type, 30, 31, 36, 129, 161, 162, 164, 166, 167, 171, 196, 226, 228
    G_1, 235, 237, 244, 246, 263, 267, 268, 278, 279, 287, 380, 407, 412, 428, 430, 432, 433
    G_2, 235, 237, 244, 246, 263, 267, 268, 278, 279, 369, 380, 407, 408, 428–433
    G_3, 241, 244, 384, 415, 425, 428, 430–432
    G_4, 241, 244, 383, 414, 425, 429, 431
    G_5, 241, 244, 245, 372, 380, 384, 409, 412, 415, 425, 429–431
  of infinite-divisible type, 36, 292, 303, 307

Basis
  Fourier, 27, 73, 94, 166, 168
  wavelet, 76, 134, 172, 239
Bonferroni method, 32, 280

Constraints
  one-sided, 7, 36, 81, 104, 224
  two-sided, 7, 23, 28, 34, 36, 79, 81, 106, 117, 123, 128, 136, 140, 141, 161, 162, 168, 226
Criterion
  $\beta(\alpha)$, 42, 58, 60
  γ, 42, 58–60
  Bayesian, 10
  minimax, 9, 11, 22

Distance
  Fisher, 66
  for measures
    L_1, 45, 48, 58, 64, 65, 67, 97, 99
    L_2, 64, 65, 97
Distinguishability, 17, 18, 20, 21, 29, 31, 64, 67–71, 79, 80, 85, 89, 93, 96, 97, 104, 108, 128, 129, 145, 177, 180, 226, 228, 229, 234, 238, 286, 292, 299, 304, 306–309, 337, 338, 350
Distribution
  χ^2, 12, 51
    noncentral, 12, 51

infinite divisible, 293, 294, 301–303, 307, 327
Kolmogorov, 16, 18, 20
Smirnov, 16

Extreme problem, 11, 34, 138, 139, 141, 160, 162, 166, 169, 195, 200, 204, 205, 208, 209, 224, 226
Lagrange, 205–207, 219, 249, 253, 254, 259
one-dimensional, 35, 36, 204, 205, 217–219

Hilbert
norm
of measures, 33, 99
of sequence of measures, 98, 99, 102, 155, 185
structure, 33, 35, 100, 101, 138, 185, 194

Inner radii, 35, 93

Kolmogorov diameter, 35, 93, 95

Levi spectral function, 293, 294, 299, 301–303, 333

Measure
one-point, nonsymmetric, 116, 123, 135, 214
three-point, 30, 36, 103, 112, 116, 120, 124, 133, 155, 185, 186, 194, 196, 214, 217, 275, 276
two-point, 36, 102–104, 106, 111, 116, 132, 154, 155, 159, 185, 186, 194, 196, 214, 217, 275
two-point, nonsymmetric, 116, 123, 135, 214, 218

Nondistinguishability, 22, 29, 64, 65, 70, 71, 93, 94, 104, 108, 129, 177, 270, 280, 307, 337, 338, 350
Nonempty conditions, 118, 119
Norm
L_2, 16, 40, 78
L_p, 16, 27, 70, 71, 73, 88, 128
l_2, 22, 38
l_p, 71, 72
Besov
in functional space, 27, 31, 32, 36, 74–76, 78, 79, 127, 128, 130, 132, 134, 171, 184, 239, 263
in sequence space, 32, 35, 36, 72, 76, 116, 123, 134, 136, 154, 160, 168, 171, 173, 180, 184, 185, 191, 197, 199, 206, 263
Hölder, 75, 78, 184
Nikol'skii, 75, 130
of Kolmogorov type, 68, 71
power, 28, 36, 72, 116, 136, 154, 160, 164, 173, 185, 191, 197, 206, 262
Sobolev, 27, 31, 32, 36, 73, 75, 76, 78, 79, 127, 128, 130, 171, 184, 263
supreme, 78

Positive alternative, 60, 88, 89, 104, 116, 117, 123, 135, 136, 140, 141, 192, 196, 198, 201–204, 208, 218, 223, 224, 228
Prior
orthogonal, 36, 106, 107, 113, 133, 135, 136, 173, 264
asymptotically, 269, 270
product, 30, 34, 36, 98, 102–104, 111, 120, 124, 132, 136, 154, 159, 185, 186, 196, 263, 269, 275, 276
semiorthogonal, 106, 107, 115
asymptotically, 106, 107, 175

Rates
- adaptive, 263, 267, 268
- for l_p^n-balls, 104
- for positive smooth signal
  - r_ε^{*+}, 135
- for power and Besov norms in sequence space
  - ρ_ε^*, 177, 180, 239, 243
- for smooth signal
  - r_ε^*, 25, 31, 94, 128, 172, 184, 240
- of estimation
  - $r_\varepsilon^{\text{est}}$, 26, 31, 77, 130, 184

Set
- Ξ_C, 118, 231
- Ξ_C^+, 118
- Ξ_D, 176
- Ξ_G, 231, 249
- Ξ_T, 119
- Ξ_T^+, 123
- Ξ_i, $i = 1, 2, 3, 4$, 366
- Ξ_{G_0}, 256, 265
- Ξ_{G_1}, 232, 235
- Ξ_{G_2}, 232, 235
- Ξ_{G_k}, $k = 3, 4, 5$, 232, 240
- $\Xi_{G_{0k}}$, $k = 1, 2$, 256, 265

Shortcoming, 12

Signal
- multichannel, 36, 307, 314
- positive, 36

Test
- χ^+, 88, 89, 135
- χ^+-type, 116
- χ^2, 10, 12, 13, 24, 35, 51, 58, 81, 85, 96, 106, 134, 286, 290, 337
- χ^2 Pearson, 15, 18–21, 32, 130, 287
- χ^2-type, 84, 88, 113, 116, 129, 136, 154, 160, 161, 164, 172
- χ^p, 81, 105, 134
- χ^p-type, 86, 87, 114, 116, 129
- adaptive, 31, 32, 280, 285, 287
- admissible, 42, 47, 48
- asymptotically Bayesian, 13
- asymptotically minimax, 13, 25, 29–31, 34, 36, 63, 66, 154, 161, 164, 173, 177, 178, 262, 265, 280, 282
- based on t_ε-statistic, 33, 82, 83, 186, 286, 290
- based on linear statistic, 59, 81, 82, 135–138
- Bayesian, 10, 11, 47, 48, 50, 51, 58, 65
- consistent, 36, 69, 75, 87–92, 172, 180, 181, 184, 262, 263, 285, 345, 353
- Cramér–von Mises–Smirnov, 15–18, 20, 69, 130
- Kolmogorov, 15–18, 69, 80, 130
- maximum likelihood, 8, 68
- minimax, 11, 53, 56, 58–60, 138
- Neyman–Pearson, 21
- nonrandomized, 4, 41
- optimal, 42–48, 56, 59, 65, 66
- plug-in, 79, 80
- randomized, 4, 41, 63, 91, 177, 178, 180, 184, 280
- Student, 7
- supreme, 81, 89–92, 108, 114, 116, 129, 135, 136, 173, 286, 290

Trivial problem, 30, 36, 70, 71, 81, 106, 109–111, 115–117, 119, 123, 128, 164, 166, 169, 171, 244

Wavelet transform, 28, 36, 76, 78, 171, 172, 184, 239, 263

Parameter and Function Index

a, in Section A.4, 399

Δ, 120, 162, 365
Δ_+, 143
Δ_1, in Section A.4, 398
$\tilde{\Delta}$, 366
$\tilde{\Delta}_1$, in Section A.4, 399
$\delta(x,\kappa)$, in Sections A.3, 365
$\delta(x,\tau)$, in Sections A.4, 400

η, 29, 127, 172, 239

$G_n(x,t^2)$, 10
$g(\bar{a},h)$, in Section 3.4, 111
$g_p(\bar{a})$, in Section 3.4, 110
$g_p^+(\bar{a})$, in Section 3.4, 116
γ_1, in Section A.4, 399
γ, in Section A.4, 397

h_0, 235, 237, 365, 398

I, 119, 235, 243
$I_{\varepsilon,\eta}$, 190, 199
1_A, 5
I_+, 123

j_0, 150, 166, 237, 398
j_1, 237, 398
$\tilde{j}_1$, in Section A.4, 400

κ, in Chapters 6, 7, Sections A.3, A.4, 231, 262, 365, 399

l, in Section A.4, 397
λ, 119, 243, 365

m, 143, 163, 235, 365
μ, 119, 243

n, 235, 365
$\nu(x,z)$, 102, 356

$\Phi(t)$, 2, 5

$Q_{\varepsilon,i}$, 236
$Q_{\varepsilon,ij} = Q_{\varepsilon,j}$, 238

R, in Section A.4, 397
$\tilde{r}$, in Section A.4, 398
$r^*_{p,+}$, 61, 116
r^*_p, 111, 244

$\tilde{s}$, in Section A.4, 398
σ, 29, 127, 172, 239
$\tilde{\sigma}$, in Section A.4, 397

T_n, in Chapter 8, 295
$T_{\varepsilon,i}$, 187
T_α, 5
$T_{n,\alpha}$, 10
τ, in Chapter 8, 296
τ, in Chapters 6, 7, Section A.4, 231, 266, 398
$\tau(y,\kappa)$, in Section A.3, 377
$\vartheta(y,\tau)$, in Sections A.3, 403

$u(y)$, 143, 147
u^*, in Section A.4, 399

$V_\varepsilon^H(H_{1,\varepsilon},H_{2,\varepsilon})$, 200
V_ε^H, 192
v_0, 143, 150

x^*, 244
x_ε, in Sections A.3, A.4, 368, 400
$\xi(x,z)$, 31, 103, 356

y^*, 244
y_ε, in Sections A.3, A.4, 368, 400

z^*, in Section A.4, 399
z_0, 163, 166, 235, 237, 365, 398
z_p, 209, 237, 397
ζ, 241, 397
$z(x,\kappa)$, in Sections A.3, 365
$z(x,\tau)$, in Sections A.4, 400
$z(y)$, in Section 4, 163

Lecture Notes in Statistics

For information about Volumes 1 to 117, please contact Springer-Verlag

118: Radford M. Neal, Bayesian Learning for Neural Networks. xv, 183 pp., 1996.

119: Masanao Aoki and Arthur M. Havenner, Applications of Computer Aided Time Series Modeling. ix, 329 pp., 1997.

120: Maia Berkane, Latent Variable Modeling and Applications to Causality. vi, 288 pp., 1997.

121: Constantine Gatsonis, James S. Hodges, Robert E. Kass, Robert McCulloch, Peter Rossi, and Nozer D. Singpurwalla (Editors), Case Studies in Bayesian Statistics, Volume III. xvi, 487 pp., 1997.

122: Timothy G. Gregoire, David R. Brillinger, Peter J. Diggle, Estelle Russek-Cohen, William G. Warren, and Russell D. Wolfinger (Editors), Modeling Longitudinal and Spatially Correlated Data. x, 402 pp., 1997.

123: D. Y. Lin and T. R. Fleming (Editors), Proceedings of the First Seattle Symposium in Biostatistics: Survival Analysis. xiii, 308 pp., 1997.

124: Christine H. Müller, Robust Planning and Analysis of Experiments. x, 234 pp., 1997.

125: Valerii V. Fedorov and Peter Hackl, Model-Oriented Design of Experiments. viii, 117 pp., 1997.

126: Geert Verbeke and Geert Molenberghs, Linear Mixed Models in Practice: A SAS-Oriented Approach. xiii, 306 pp., 1997.

127: Harald Niederreiter, Peter Hellekalek, Gerhard Larcher, and Peter Zinterhof (Editors), Monte Carlo and Quasi-Monte Carlo Methods 1996. xii, 448 pp., 1997.

128: L. Accardi and C.C. Heyde (Editors), Probability Towards 2000. x, 356 pp., 1998.

129: Wolfgang Härdle, Gerard Kerkyacharian, Dominique Picard, and Alexander Tsybakov, Wavelets, Approximation, and Statistical Applications. xvi, 265 pp., 1998.

130: Bo-Cheng Wei, Exponential Family Nonlinear Models. ix, 240 pp., 1998.

131: Joel L. Horowitz, Semiparametric Methods in Econometrics. ix, 204 pp., 1998.

132: Douglas Nychka, Walter W. Piegorsch, and Lawrence H. Cox (Editors), Case Studies in Environmental Statistics. viii, 200 pp., 1998.

133: Dipak Dey, Peter Müller, and Debajyoti Sinha (Editors), Practical Nonparametric and Semiparametric Bayesian Statistics. xv, 408 pp., 1998.

134: Yu. A. Kutoyants, Statistical Inference For Spatial Poisson Processes. vii, 284 pp., 1998.

135: Christian P. Robert, Discretization and MCMC Convergence Assessment. x, 192 pp., 1998.

136: Gregory C. Reinsel, Raja P. Velu, Multivariate Reduced-Rank Regression. xiii, 272 pp., 1998.

137: V. Seshadri, The Inverse Gaussian Distribution: Statistical Theory and Applications. xii, 360 pp., 1998.

138: Peter Hellekalek and Gerhard Larcher (Editors), Random and Quasi-Random Point Sets. xi, 352 pp., 1998.

139: Roger B. Nelsen, An Introduction to Copulas. xi, 232 pp., 1999.

140: Constantine Gatsonis, Robert E. Kass, Bradley Carlin, Alicia Carriquiry, Andrew Gelman, Isabella Verdinelli, and Mike West (Editors), Case Studies in Bayesian Statistics, Volume IV. xvi, 456 pp., 1999.

141: Peter Müller and Brani Vidakovic (Editors), Bayesian Inference in Wavelet Based Models. xiii, 394 pp., 1999.

142: György Terdik, Bilinear Stochastic Models and Related Problems of Nonlinear Time Series Analysis: A Frequency Domain Approach. xi, 258 pp., 1999.

143: Russell Barton, Graphical Methods for the Design of Experiments. x, 208 pp., 1999.

144: L. Mark Berliner, Douglas Nychka, and Timothy Hoar (Editors), Case Studies in Statistics and the Atmospheric Sciences. x, 208 pp., 2000.

145: James H. Matis and Thomas R. Kiffe, Stochastic Population Models. viii, 220 pp., 2000.

146: Wim Schoutens, Stochastic Processes and Orthogonal Polynomials. xiv, 163 pp., 2000.

147: Jürgen Franke, Wolfgang Härdle, and Gerhard Stahl, Measuring Risk in Complex Stochastic Systems. xvi, 272 pp., 2000.

148: S.E. Ahmed and Nancy Reid, Empirical Bayes and Likelihood Inference. x, 200 pp., 2000.

149: D. Bosq, Linear Processes in Function Spaces: Theory and Applications. xv, 296 pp., 2000.

150: Tadeusz Caliński and Sanpei Kageyama, Block Designs: A Randomization Approach, Volume I: Analysis. ix, 313 pp., 2000.

151: Håkan Andersson and Tom Britton, Stochastic Epidemic Models and Their Statistical Analysis. ix, 152 pp., 2000.

152: David Ríos Insua and Fabrizio Ruggeri, Robust Bayesian Analysis. xiii, 435 pp., 2000.

153: Parimal Mukhopadhyay, Topics in Survey Sampling. x, 303 pp., 2000.

154: Regina Kaiser and Agustín Maravall, Measuring Business Cycles in Economic Time Series. vi, 190 pp., 2000.

155: Leon Willenborg and Ton de Waal, Elements of Statistical Disclosure Control. xvii, 289 pp., 2000.

156: Gordon Willmot and X. Sheldon Lin, Lundberg Approximations for Compound Distributions with Insurance Applications. xi, 272 pp., 2000.

157: Anne Boomsma, Marijtje A.J. van Duijn, and Tom A.B. Snijders (Editors), Essays on Item Response Theory. xv, 448 pp., 2000.

158: Dominique Ladiray and Benoît Quenneville, Seasonal Adjustment with the X-11 Method. xxii, 220 pp., 2001.

159: Marc Moore (Editor), Spatial Statistics: Methodological Aspects and Some Applications. xvi, 282 pp., 2001.

160: Tomasz Rychlik, Projecting Statistical Functionals. viii, 184 pp., 2001.

161: Maarten Jansen, Noise Reduction by Wavelet Thresholding. xxii, 224 pp., 2001.

162: Constantine Gatsonis, Bradley Carlin, Alicia Carriquiry, Andrew Gelman, Robert E. Kass Isabella Verdinelli, and Mike West (Editors), Case Studies in Bayesian Statistics, Volume V. xiv, 448 pp., 2001.

163: Erkki P. Liski, Nripes K. Mandal, Kirti R. Shah, and Bikas K. Sinha, Topics in Optimal Design. xii, 164 pp., 2002.

164: Peter Goos, The Optimal Design of Blocked and Split-Plot Experiments. xiv, 244 pp., 2002.

165: Karl Mosler, Multivariate Dispersion, Central Regions and Depth: The Lift Zonoid Approach. xii, 280 pp., 2002.

166: Hira L. Koul, Weighted Empirical Processes in Dynamic Nonlinear Models, Second Edition. xiii, 425 pp., 2002.

167: Constantine Gatsonis, Alicia Carriquiry, Andrew Gelman, David Higdon, Robert E. Kass, Donna Pauler, and Isabella Verdinelli (Editors), Case Studies in Bayesian Statistics, Volume VI. xiv, 376 pp., 2002.

168: Susanne Rässler, Statistical Matching: A Frequentist Theory, Practical Applications and Alternative Bayesian Approaches. xvii, 238 pp., 2002.

169: Yu. I. Ingster and I.A. Suslina, Nonparametric Goodness-of-Fit Testing Under Gaussian Models. xiv, 453 pp., 2003.

170: Tadeusz Caliński and Sanpei Kageyama, Block Designs: A Randomization Approach, Volume II: Design. xii, 351 pp., 2003.